Lecture Notes in Computer Scien

T0238797

Commenced Publication in 1973
Founding and Former Series Editors:
Gerhard Goos, Juris Hartmanis, and Jan van Leeuwen

Beniamino Murgante   Sanjay Misra
Maurizio Carlini   Carmelo M. Torre
Hong-Quang Nguyen   David Taniar
Bernady O. Apduhan   Osvaldo Gervasi (Eds.)

Computational Science and Its Applications – ICCSA 2013

13th International Conference
Ho Chi Minh City, Vietnam, June 24-27, 2013
Proceedings, Part II

 Springer

Volume Editors

Beniamino Murgante, Università degli Studi della Basilicata, Potenza, Italy
E-mail: beniamino.murgante@unibas.it

Sanjay Misra, Covenant University, Canaanland OTA, Nigeria
E-mail: sanjay.misra@covenantuniversity.edu.ng

Maurizio Carlin, Università degli Studi della Tuscia, Viterbo, Italy
E-mail: maurizio.carlini@unitus.it

Carmelo M. Torre, Politecnico di Bari, Italy
E-mail: torre@poliba.it

Hong-Quang Nguyen, Int. University VNU-HCM, Ho Chi Minh City, Vietnam
E-mail: htphong@hcmiu.edu.vn

David Taniar, Monash University, Clayton, VIC, Australia
E-mail: david.taniar@infotech.monash.edu.au

Bernady O. Apduhan, Kyushu Sangyo University, Fukuoka, Japan
E-mail: bob@is.kyusan-u.ac.jp

Osvaldo Gervasi, University of Perugia, Italy
E-mail: osvaldo@unipg.it

ISSN 0302-9743   e-ISSN 1611-3349
ISBN 978-3-642-39642-7   e-ISBN 978-3-642-39643-4
DOI 10.1007/978-3-642-39643-4
Springer Heidelberg Dordrecht London New York

Library of Congress Control Number: 2013942720

CR Subject Classification (1998): C.2.4, C.2, H.4, F.2, H.3, D.2, F.1, H.5, H.2.8, K.6.5, I.3

LNCS Sublibrary: SL 1 – Theoretical Computer Science and General Issues

Typesetting: Camera-ready by author, data conversion by Scientific Publishing Services, Chennai, India

Printed on acid-free paper

Springer is part of Springer Science+Business Media (www.springer.com)

Preface

These multiple volumes (LNCS volumes 7971, 7972, 7973, 7974, and 7975) consist of the peer-reviewed papers from the 2013 International Conference on Computational Science and Its Applications (ICCSA2013) held in Ho Chi Minh City, Vietnam, during June 24–27, 2013.

ICCSA 2013 was a successful event in the International Conferences on Computational Science and Its Applications (ICCSA) conference series, previously held in Salvador, Brazil (2012), Santander, Spain (2011), Fukuoka, Japan (2010), Suwon, South Korea (2009), Perugia, Italy (2008), Kuala Lumpur, Malaysia (2007), Glasgow, UK (2006), Singapore (2005), Assisi, Italy (2004), Montreal, Canada (2003), (as ICCS) Amsterdam, The Netherlands (2002), and San Francisco, USA (2001).

Computational science is a main pillar of most of the present research, industrial, and commercial activities and plays a unique role in exploiting ICT innovative technologies; the ICCSA conference series have been providing a venue to researchers and industry practitioners to discuss new ideas, to share complex problems and their solutions, and to shape new trends in computational science.

Apart from the general track, ICCSA 2013 also included 33 special sessions and workshops, in various areas of computational sciences, ranging from computational science technologies, to specific areas of computational sciences, such as computer graphics and virtual reality. We accepted 46 papers for the general track, and 202 in special sessions and workshops, with an acceptance rate of 29.8%. We would like to express our appreciation to the Workshops and Special Sessions Chairs and Co-chairs.

The success of the ICCSA conference series, in general, and ICCSA 2013, in particular, is due to the support of many people: authors, presenters, participants, keynote speakers, Workshop Chairs, Organizing Committee members, student volunteers, Program Committee members, International Liaison Chairs, and people in other various roles. We would like to thank them all. We would also like to thank Springer for their continuous support in publishing ICCSA conference proceedings.

May 2013

David Taniar
Beniamino Murgante
Hong-Quang Nguyen

Message from the General Chairs

On behalf of the ICCSA Organizing Committee it is our great pleasure to welcome you to the proceedings of the 13th International Conference on Computational Science and Its Applications (ICCSA 2013), held June 24–27, 2013, in Ho Chi Minh City, Vietnam.

ICCSA is one of the most successful international conferences in the field of computational sciences, and ICCSA 2013 was the 13th conference of this series previously held in Salvador da Bahia, Brazil (2012), in Santander, Spain (2011), Fukuoka, Japan (2010), Suwon, Korea (2009), Perugia, Italy (2008), Kuala Lumpur, Malaysia (2007), Glasgow, UK (2006), Singapore (2005), Assisi, Italy (2004), Montreal, Canada (2003), (as ICCS) Amsterdam, The Netherlands (2002), and San Francisco, USA (2001).

The computational science community has enthusiastically embraced the successive editions of ICCSA, thus contributing to making ICCSA a focal meeting point for those interested in innovative, cutting-edge research about the latest and most exciting developments in the field. It provides a major forum for researchers and scientists from academia, industry and government to share their views on many challenging research problems, and to present and discuss their novel ideas, research results, new applications and experience on all aspects of computational science and its applications. We are grateful to all those who have contributed to the ICCSA conference series.

For the successful organization of ICCSA 2013, an international conference of this size and diversity, we counted on the great support of many people and organizations.

We would like to thank all the workshop organizers for their diligent work, which further enhanced the conference level and all reviewers for their expertise and generous effort, which led to a very high quality event with excellent papers and presentations.

We especially recognize the contribution of the Program Committee and local Organizing Committee members for their tremendous support, the faculty members of the School of Computer Science and Engineering and authorities of the International University (HCM-VNU), Vietnam, for allowing us to use the venue and facilities to realize this highly successful event. Further, we would like to express our gratitude to the Office of the Naval Research, US Navy, and other institutions/organizations that supported our efforts to bring the conference to fruition.

We would like to sincerely thank our keynote speakers who willingly accepted our invitation and shared their expertise.

We also thank our publisher, Springer-Verlag, for accepting to publish the proceedings and for their kind assistance and cooperation during the editing process.

Finally, we thank all authors for their submissions and all conference attendees for making ICCSA 2013 truly an excellent forum on computational science, facilitating an exchange of ideas, fostering new collaborations and shaping the future of this exciting field.

We thank you all for participating in ICCSA 2013, and hope that you find the proceedings stimulating and interesting for your research and professional activities.

Osvaldo Gervasi
Bernady O. Apduhan
Duc Cuong Nguyen

Organization

ICCSA 2013 was organized by The Ho Chi Minh City International University (Vietnam), University of Perugia (Italy), University of Basilicata (Italy), Monash University (Australia), and Kyushu Sangyo University (Japan).

Honorary General Chairs

Phong Thanh Ho	International University (VNU-HCM), Vietnam
Antonio Laganà	University of Perugia, Italy
Norio Shiratori	Tohoku University, Japan
Kenneth C.J. Tan	Qontix, UK

General Chairs

Osvaldo Gervasi	University of Perugia, Italy
Bernady O. Apduhan	Kyushu Sangyo University, Japan
Duc Cuong Nguyen	International University (VNU-HCM), Vietnam

Program Committee Chairs

David Taniar	Monash University, Australia
Beniamino Murgante	University of Basilicata, Italy
Hong-Quang Nguyen	International University (VNU-HCM), Vietnam

Workshop and Session Organizing Chair

Beniamino Murgante	University of Basilicata, Italy

Local Organizing Committee

Hong Quang Nguyen	International University (VNU-HCM), Vietnam (Chair)
Bao Ngoc Phan	International University (VNU-HCM), Vietnam

Van Hoang   International University (VNU-HCM),
                              Vietnam
Ly Le   International University (VNU-HCM),
                              Vietnam

International Liaison Chairs

Jemal Abawajy   Deakin University, Australia
Ana Carla P. Bitencourt   Universidade Federal do Reconcavo da Bahia,
                              Brazil
Claudia Bauzer Medeiros   University of Campinas, Brazil
Alfredo Cuzzocrea   ICAR-CNR and University of Calabria, Italy
Marina L. Gavrilova   University of Calgary, Canada
Robert C.H. Hsu   Chung Hua University, Taiwan
Andrés Iglesias   University of Cantabria, Spain
Tai-Hoon Kim   Hannam University, Korea
Sanjay Misra   University of Minna, Nigeria
Takashi Naka   Kyushu Sangyo University, Japan
Ana Maria A.C. Rocha   University of Minho, Portugal
Rafael D.C. Santos   National Institute for Space Research, Brazil

Workshop Organizers

Advances in Web-Based Learning (AWBL 2013)

Mustafa Murat Inceoglu   Ege University, Turkey

Big Data: Management, Analysis, and Applications (Big-Data 2013)

Wenny Rahayu   La Trobe University, Australia

Bio-inspired Computing and Applications (BIOCA 2013)

Nadia Nedjah   State University of Rio de Janeiro, Brazil
Luiza de Macedo Mourell   State University of Rio de Janeiro, Brazil

Computational and Applied Mathematics (CAM 2013)

Ana Maria Rocha   University of Minho, Portugal
Maria Irene Falcao   University of Minho, Portugal

Computer-Aided Modeling, Simulation, and Analysis (CAMSA 2013)

Jie Shen   University of Michigan, USA
Yanhui Wang   Beijing Jiaotong University, China
Hao Chen   Shanghai University of Engineering Science,
                              China

Computer Algebra Systems and Their Applications (CASA 2013)

Andres Iglesias   University of Cantabria, Spain
Akemi Galvez   University of Cantabria, Spain

Computational Geometry and Applications (CGA 2013)

Marina L. Gavrilova   University of Calgary, Canada
Han Ming Huang   Guangxi Normal University, China

Chemistry and Materials Sciences and Technologies (CMST 2013)

Antonio Laganà   University of Perugia, Italy

Cities, Technologies and Planning (CTP 2013)

Giuseppe Borruso   University of Trieste, Italy
Beniamino Murgante   University of Basilicata, Italy

Computational Tools and Techniques for Citizen Science and Scientific Outreach (CTTCS 2013)

Rafael Santos   National Institute for Space Research, Brazil
Jordan Raddickand   Johns Hopkins University, USA
Ani Thakar   Johns Hopkins University, USA

Econometrics and Multidimensional Evaluation in the Urban Environment (EMEUE 2013)

Carmelo M. Torre   Polytechnic of Bari, Italy
Maria Cerreta   Università Federico II of Naples, Italy
Paola Perchinunno   University of Bari, Italy

Energy and Environment - Scientific, Engineering and Computational Aspects of Renewable Energy Sources, Energy Saving and Recycling of Waste Materials (ENEENV 2013)

Maurizio Carlini   University of Viterbo, Italy
Carlo Cattani   University of Salerno, Italy

Future Computing Systems, Technologies, and Applications (FISTA 2013)

Bernady O. Apduhan   Kyushu Sangyo University, Japan
Rafael Santos   National Institute for Space Research, Brazil
Jianhua Ma   Hosei University, Japan
Qun Jin   Waseda University, Japan

Geographical Analysis, Urban Modeling, Spatial Statistics (GEOG-AN-MOD 2013)

Giuseppe Borruso   University of Trieste, Italy
Beniamino Murgante   University of Basilicata, Italy
Hartmut Asche   University of Potsdam, Germany

International Workshop on Biomathematics, Bioinformatics and Biostatistics (IBBB 2013)

Unal Ufuktepe   Izmir University of Economics, Turkey
Andres Iglesias   University of Cantabria, Spain

International Workshop on Agricultural and Environmental Information and Decision Support Systems (IAEIDSS 2013)

Sandro Bimonte   IRSTEA, France
Andr Miralles   IRSTEA, France
Franois Pinet   IRSTEA, France
Frederic Flouvat   University of New Caledonia, New Caledonia

International Workshop on Collective Evolutionary Systems (IWCES 2013)

Alfredo Milani   University of Perugia, Italy
Clement Leung   Hong Kong Baptist University, Hong Kong

Mobile Communications (MC 2013)

Hyunseung Choo   Sungkyunkwan University, Korea

Mobile Computing, Sensing, and Actuation for Cyber Physical Systems (MSA4CPS 2013)

Moonseong Kim   Korean Intellectual Property Office, Korea
Saad Qaisar   NUST School of Electrical Engineering and
                          Computer Science, Pakistan

Mining Social Media (MSM 2013)

Robert M. Patton   Oak Ridge National Laboratory, USA
Chad A. Steed   Oak Ridge National Laboratory, USA
David R. Resseguie   Oak Ridge National Laboratory, USA
Robert M. Patton   Oak Ridge National Laboratory, USA

**Parallel and Mobile Computing in Future Networks
(PMCFUN 2013)**

Al-Sakib Khan Pathan   International Islamic University Malaysia,
                                    Malaysia

**Quantum Mechanics: Computational Strategies and Applications
(QMCSA 2013)**

Mirco Ragni   Universidad Federal de Bahia, Brazil
Vincenzo Aquilanti   University of Perugia, Italy
Ana Carla Peixoto Bitencourt   Universidade Federal do Reconcavo da Bahia,
                                    Brazil
Roger Anderson   University of California, USA
Frederico Vasconcellos
    Prudente   Universidad Federal de Bahia, Brazil

**Remote Sensing Data Analysis, Modeling, Interpretation and
Applications: From a Global View to a Local Analysis (RS 2013)**

Rosa Lasaponara   Institute of Methodologies for Environmental
                                    Analysis - National Research Council, Italy
Nicola Masini   Archaeological and Monumental Heritage
                                    Institute - National Research Council, Italy

**Soft Computing for Knowledge Discovery in Databases
(SCKDD 2013)**

Tutut Herawan   Universitas Ahmad Dahlan, Indonesia

Software Engineering Processes and Applications (SEPA 2013)

Sanjay Misra   Covenant University, Nigeria

**Spatial Data Structures and Algorithms for Geoinformatics
(SDSAG 2013)**

Farid Karimipour   University of Tehran, Iran and
                                    Vienna University of Technology, Austria

Software Quality (SQ 2013)

Sanjay Misra   Covenant University, Nigeria

Security and Privacy in Computational Sciences (SPCS 2013)

Arijit Ukil   Tata Consultancy Services, India

Technical Session on Computer Graphics and Geometric Modeling (TSCG 2013)

Andres Iglesias   University of Cantabria, Spain

Tools and Techniques in Software Development Processes (TTSDP 2013)

Sanjay Misra   Covenant University, Nigeria

Virtual Reality and Its Applications (VRA 2013)

Osvaldo Gervasi   University of Perugia, Italy
Lucio Depaolis   University of Salento, Italy

Wireless and Ad-Hoc Networking (WADNet 2013)

Jongchan Lee   Kunsan National University, Korea
Sangjoon Park   Kunsan National University, Korea

Warehousing and OLAPing Complex, Spatial and Spatio-Temporal Data (WOCD 2013)

Alfredo Cuzzocrea   Istituto di Calcolo e Reti ad Alte Prestazioni -
                                  National Research Council, Italy and
                                  University of Calabria, Italy

Program Committee

Jemal Abawajy   Deakin University, Australia
Kenny Adamson   University of Ulster, UK
Filipe Alvelos   University of Minho, Portugal
Hartmut Asche   University of Potsdam, Germany
Md. Abul Kalam Azad   University of Minho, Portugal
Assis Azevedo   University of Minho, Portugal
Michela Bertolotto   University College Dublin, Ireland
Sandro Bimonte   CEMAGREF, TSCF, France
Rod Blais   University of Calgary, Canada
Ivan Blecic   University of Sassari, Italy
Giuseppe Borruso   University of Trieste, Italy
Yves Caniou   Lyon University, France
José A. Cardoso e Cunha   Universidade Nova de Lisboa, Portugal
Carlo Cattani   University of Salerno, Italy
Mete Celik   Erciyes University, Turkey
Alexander Chemeris   National Technical University of Ukraine
                                  "KPI", Ukraine
Min Young Chung   Sungkyunkwan University, Korea
Gilberto Corso Pereira   Federal University of Bahia, Brazil
M. Fernanda Costa   University of Minho, Portugal

Wenny Rahayu	La Trobe University, Australia
Jerzy Respondek	Silesian University of Technology, Poland
Ana Maria A.C. Rocha	University of Minho, Portugal
Humberto Rocha	INESC-Coimbra, Portugal
Alexey Rodionov	Institute of Computational Mathematics and Mathematical Geophysics, Russia
Cristina S. Rodrigues	University of Minho, Portugal
Haiduke Sarafian	The Pennsylvania State University, USA
Ricardo Severino	University of Minho, Portugal
Jie Shen	University of Michigan, USA
Qi Shi	Liverpool John Moores University, UK
Dale Shires	U.S. Army Research Laboratory, USA
Ana Paula Teixeira	University of Tras-os-Montes and Alto Douro, Portugal
Senhorinha Teixeira	University of Minho, Portugal
Graça Tomaz	University of Aveiro, Portugal
Carmelo Torre	Polytechnic of Bari, Italy
Javier Martinez Torres	Centro Universitario de la Defensa Zaragoza, Spain
Giuseppe A. Trunfio	University of Sassari, Italy
Unal Ufuktepe	Izmir University of Economics, Turkey
Mario Valle	Swiss National Supercomputing Centre, Switzerland
Pablo Vanegas	University of Cuenca, Equador
Paulo Vasconcelos	University of Porto, Portugal
Piero Giorgio Verdini	INFN Pisa and CERN, Italy
Marco Vizzari	University of Perugia, Italy
Krzysztof Walkowiak	Wroclaw University of Technology, Poland
Robert Weibel	University of Zurich, Switzerland
Roland Wismüller	Universität Siegen, Germany
Xin-She Yang	National Physical Laboratory, UK
Haifeng Zhao	University of California, Davis, USA
Kewen Zhao	University of Qiongzhou, China

Additional Reviewers

Antonio Aguilar	Universitat de Barcelona, Spain
José Alfonso Aguilar Caldern	Universidad Autnoma de Sinaloa, Mexico
Vladimir Alarcon	Geosystems Research Institute, Mississippi State University, USA
Margarita Alberti	Universitat de Barcelona, Spain
Vincenzo Aquilanti	University of Perugia, Italy
Takefusa Atsuko	National Institute of Advanced Industrial Science and Technology, Japan
Raffaele Attardi	University of Napoli Federico II, Italy

Sansanee Auephanwiriyakul	Chiang Mai University, Thailand
Assis Azevedo	University of Minho, Portugal
Thierry Badard	Université Laval, Canada
Marco Baioletti	University of Perugia, Italy
Daniele Bartoli	University of Perugia, Italy
Paola Belanzoni	University of Perugia, Italy
Massimiliano Bencardino	University of Salerno, Italy
Priyadarshi Bhattacharya	University of Calgari, Canada
Massimo Bilancia	University of Bari, Italy
Gabriele Bitelli	University of Bologna, Italy
Letizia Bollini	University of Milano Bicocca, Italy
Alessandro Bonifazi	University of Bari, Italy
Atila Bostam	Atilim University, Turkey
Maria Bostenaru Dan	University of Bucharest, Romania
Thang H. Bui	Ho Chi Minh City University of Technology, Vietnam
Michele Campagna	University of Cagliari, Italy
Francesco Campobasso	University of Bari, Italy
Maurizio Carlini	University of Tuscia, Italy
Simone Caschili	University College of London, UK
Sonia Castellucci	University of Tuscia, Italy
Filippo Celata	University of Rome La Sapienza, Italy
Claudia Ceppi	Polytechnic of Bari, Italy
Ivan Cernusak	Comenius University of Bratislava, Slovakia
Maria Cerreta	University of Naples Federico II, Italy
Aline Chiabai	Basque Centre for Climate Change, Spain
Andrea Chiancone	University of Perugia, Italy
Eliseo Clementini	University of L'Aquila, Italy
Anibal Zaldivar Colado	Universidad Autonoma de Sinaloa, Mexico
Marco Crasso	Universidad Nacional del Centro de la provincia de Buenos Aires, Argentina
Ezio Crestaz	Saipem, Italy
Maria Danese	IBAM National Research Council, Italy
Olawande Daramola	Covenant University, Nigeria
Marcelo de Alemida Maia	Universidade Federal de Uberlândia, Brazil
Roberto De Lotto	University of Pavia, Italy
Lucio T. De Paolis	University of Salento, Italy
Pasquale De Toro	University of Naples Federico II, Italy
Hendrik Decker	Universidad Politécnica de Valencia, Spain
Margherita Di Leo	Joint Research Centre, Belgium
Andrea Di Carlo	University of Rome La Sapienza, Italy
Arta Dilo	University of Twente, The Netherlands
Alberto Dimeglio	CERN, Switzerland
Young Ik Eom	Sungkyunkwan University, South Korea
Rogelio Estrada	Universidad Autonoma de Sinaloa, Mexico
Stavros C. Farantos	University of Crete, Greece

Yong-Wan Roh	Korean Intellectual Property Office, South Korea
Luiz Roncaratti	Universidade de Brasilia, Brazil
Marzio Rosi	University of Perugia, Italy
Francesco Rotondo	Polytechnic of Bari, Italy
Catherine Roussey	National Research Institute of Science and Technology for Environment and Agriculture, France
Rafael Oliva Santos	Universidad de La Habana, Cuba
Valentino Santucci	University of Perugia, Italy
Dario Schirone	University of Bari, Italy
Michel Schneider	Institut Supérieur d'Informatique de Modélisation et de leurs Applications, France
Gabriella Schoier	University of Trieste, Italy
Francesco Scorza	University of Basilicata, Italy
Nazha Selmaoui	Université de la Nouvelle-Calédonie, New Caledonia
Ricardo Severino	University of Minho, Portugal
Vladimir V. Shakhov	Institute of Computational Mathematics and Mathematical Geophysics SB RAS, Russia
Sungyun Shin	National University Kunsan, South Korea
Minhan Shon	Sungkyunkwan University, South Korea
Ruchi Shukla	University of Johannesburg, South Africa
Luneque Silva Jr.	State University of Rio de Janeiro, Brazil
V.B. Singh	University of Delhi, India
Michel Soares	Federal University of Uberlândia, Brazil
Changhwan Son	Sungkyunkwan University, South Korea
Henning Sten Hansen	Aalborg University, Denmark
Emanuele Strano	University of the West of England, UK
Madeena Sultana	Jahangirnagar University, Bangladesh
Setsuo Takato	Toho University, Japan
Kazuaki Tanaka	Kyushu Institute of Technology, Japan
Xueyan Tang	Nanyang Technological University, Singapore
Sergio Tasso	University of Perugia, Italy
Luciano Telesca	IMAA National Research Council, Italy
Lucia Tilio	University of Basilicata, Italy
Graça Tomaz	Instituto Politécnico da Guarda, Portugal
Melanie Tomintz	Carinthia University of Applied Sciences, Austria
Javier Torres	Universidad de Zaragoza, Spain
Csaba Toth	University of Calgari, Canada
Hai Tran	U.S. Government Accountability Office, USA
Jim Treadwell	Oak Ridge National Laboratory, USA

Chih-Hsiao Tsai	Takming University of Science and Technology, Taiwan
Devis Tuia	Laboratory of Geographic Information Systems, Switzerland
Arijit Ukil	Tata Consultancy Services, India
Paulo Vasconcelos	University of Porto, Portugal
Flavio Vella	University of Perugia, Italy
Mauro Villarini	University of Tuscia, Italy
Christine Voiron-Canicio	Université Nice Sophia Antipolis, France
Kira Vyatkina	Saint Petersburg State University, Russia
Jian-Da Wu	National Changhua University of Education, Taiwan
Toshihiro Yamauchi	Okayama University, Japan
Iwan Tri Riyadi Yanto	Universitas Ahmad Dahlan, Indonesia
Syed Shan-e-Hyder Zaidi	Sungkyunkwan University, South Korea
Vyacheslav Zalyubouskiy	Sungkyunkwan University, South Korea
Alejandro Zunino	National University of the Center of the Buenos Aires Province, Argentina

Sponsoring Organizations

ICCSA 2013 would not have been possible without tremendous support of many organizations and institutions, for which all organizers and participants of ICCSA 2013 express their sincere gratitude:

 Ho CHi Minh City International University, Vietnam
(http://www.hcmiu.edu.vn/HomePage.aspx)

 University of Perugia, Italy
(http://www.unipg.it)

 Monash University, Australia
(http://monash.edu)

 Kyushu Sangyo University, Japan
(www.kyusan-u.ac.jp)

 University of Basilicata, Italy (http://www.unibas.it)

 The Office of Naval Research, USA
(http://www.onr.navy.mil/Science-technology/onr-global.aspx)

ICCSA 2013 Invited Speakers

Dharma Agrawal
University of Cincinnati, USA

Manfred M. Fisher
Vienna University of Economics and Business, Austria

Wenny Rahayu
La Trobe University, Australia

Selecting LTE and Wireless Mesh Networks for Indoor/Outdoor Applications

Dharma Agrawal*

School of Computing Sciences and Informatics, University of Cincinnati, USA
dharmaagrawal@gmail.com

Abstract. The smart phone usage and multimedia devices have been increasing yearly and predictions indicate drastic increase in the upcoming years. Recently, various wireless technologies have been introduced to add flexibility to these gadgets. As data plans offered by the network service providers are expensive, users are inclined to utilize freely accessible and commonly available Wi-Fi networks indoors.

LTE (Long Term Evolution) has been a topic of discussion in providing high data rates outdoors and various service providers are planning to roll out LTE networks all over the world. The objective of this presentation is to compare usefulness of these two leading wireless schemes based on LTE and Wireless Mesh Networks (WMN) and bring forward their advantages for indoor and outdoor environments. We also investigate to see if a hybrid LTE-WMN network may be feasible. Both these networks are heterogeneous in nature, employ cognitive approach and support multi hop communication. The main motivation behind this work is to utilize similarities in these networks, explore their capability of offering high data rates and generally have large coverage areas.

In this work, we compare both these networks in terms of their data rates, range, cost, throughput, and power consumption. We also compare 802.11n based WMN with Femto cell in an indoor coverage scenario, while for outdoors; 802.16 based WMN is compared with LTE. The main objective is to help users select a network that could provide enhanced performance in a cost effective manner.

---

* More information can be found at http://www.iccsa.org/invited-speakers

Neoclassical Growth Theory, Regions and Spatial Externalities

Manfred M. Fisher[*]

Vienna University of Economics and Business, Austria
manfred.fischer@wu.ac.at

Abstract. The presentation considers the standard neoclassical growth model in a Mankiw-Romer-Weil world with externalities across regions.

The reduced form of this theoretical model and its associated empirical model lead to a spatial Durbin model, and this model provides very rich own- and cross-partial derivatives that quantify the magnitude of direct and indirect (spillover or externalities) effects that arise from changes in regions characteristics (human and physical capital investment or population growth rates) at the outset in the theoretical model.

A logical consequence of the simple dependence on a small number of nearby regions in the initial theoretical specification leads to a final-form model outcome where changes in a single region can potentially impact all other regions. This is perhaps surprising, but of course we must temper this result by noting that there is a decay of influence as we move to more distant or less connected regions.

Using the scalar summary impact measures introduced by LeSage and Pace (2009) we can quantify and summarize the complicated set of non-linear impacts that fall on all regions as a result of changes in the physical and human capital in any region. We can decompose these impacts into direct and indirect (or externality) effects. Data for a system of 198 regions across 22 European countries over the period 1995 to 2004 are used to test the predictions of the model and to draw inferences regarding the magnitude of regional output responses to changes in physical and human capital endowments.

The results reveal that technological interdependence among regions works through physical capital externalities crossing regional borders.

---

[*] More information can be found at http://www.iccsa.org/invited-speakers

Global Spatial-Temporal Data Integration to Support Collaborative Decision Making

Wenny Rahayu*

La Trobe University, Australia
W.Rahayu@latrobe.edu.au

Abstract. There has been a huge effort in the recent years to establish a standard vocabulary and data representation for the areas where a collaborative decision support is required. The development of global standards for data interchange in time critical domains such as air traffic control, transportation systems, and medical informatics, have enabled the general industry in these areas to move into a more data-centric operations and services. The main aim of the standards is to support integration and collaborative decision support systems that are operationally driven by the underlying data.

The problem that impedes rapid and correct decision-making is that information is often segregated in many different formats and domains, and integrating them has been recognised as one of the major problems. For example, in the aviation industry, weather data given to flight en-route has different formats and standards from those of the airport notification messages. The fact that messages are exchanged using different standards has been an inherent problem in data integration in many spatial-temporal domains. The solution is to provide seamless data integration so that a sequence of information can be analysed on the fly.

Our aim is to develop an integration method for data that comes from different domains that operationally need to interact together. We especially focus on those domains that have temporal and spatial characteristics as their main properties. For example, in a flight plan from Melbourne to Ho Chi Minh City which comprises of multiple international airspace segments, a pilot can get an integrated view of the flight route with the weather forecast and airport notifications at each segment. This is only achievable if flight route, airport notifications, and weather forecast at each segment are integrated in a spatial temporal system.

In this talk, our recent efforts in large data integration, filtering, and visualisation will be presented. These integration efforts are often required to support real-time decision making processes in emergency situations, flight delays, and severe weather conditions.

---

* More information can be found at http://www.iccsa.org/invited-speakers

Table of Contents – Part II

Roto-torsional Levels
for Symmetric and Asymmetric Systems:
Application to HOOH and HOOD Systems

Ana Carla Peixoto Bitencourt[1],
Frederico Vasconcellos Prudente[2], and Mirco Ragni[1]

[1] Department of Physics
Universidade de Feira de Santana, UEFS
Feira de Santana, Bahia (BR)
ana.bitencourt@gmail.com
[2] Institute of Physics
Universidade Federal da Bahia, UFBA
Salvador, Bahia (BR)

Abstract. Two pictures of separation of torsional mode in intramolecular dynamics are given for the treatment of hindered rotations of molecular systems like ABCD, which present a large amplitude motion associated with the torsional mode. The energy profile (torsional potential) is described by a dihedral angle and the chosen coordinates are based on orthogonal local vectors. Our model consists of two linear rigid rotors AB and CD that rotate around the Jacobi vector connecting the centers of mass of the diatoms AB and CD. We have used two procedures to calculate the roto-torsional energy levels. The first, referred to bi-rotor, uses the Hamiltonian as function of the azimuth angles of the AB and CD rotors. In the second one, referred to roto-torsion, we separate the internal rotation (torsional mode) from the overall rotation around the Jacobi vector. For the cases where the two moments of inertia are equal, e.g. HOOH, conservation of both energy and angular momentum for a system viewed as involving either torsion plus external rotation or interaction of two rotors requires correlation of levels with symmetries $\tau = 1$ and 4 with zero or even values of the external rotation angular momentum quantum number k in units of $\hbar$. Conversely, torsional energy levels that belong to the $\tau = 2$ and 3 symmetries, correlate with odd values of k. In HOOD the two rotors have different moments of inertia, and this causes further level splitting for $\tau = 2$ and 3 only. Here we apply the two procedures to understanding the roto-torsional levels for HOOH and HOOD molecules.

Keywords: Orthogonal coordinates, roto-torsional levels.

1 Introduction

Recently there has been a renewed interest in the studies of internal rotation (torsional mode) in small molecules of the type ABCD, such as HOOH and

B. Murgante et al. (Eds.): ICCSA 2013, Part II, LNCS 7972, pp. 1–16, 2013.
© Springer-Verlag Berlin Heidelberg 2013

HOSH [2,11,16,19,15]. For hydrogen peroxide and its isotopomers the torsional dynamics are best described in diatom-diatom vectors, a particular local orthogonal coordinate set, [10], leading to the interpretation of the OH and OD groups as semi-rigid rotors executing torsion motions around a (Jacobi) vector joining their centers of mass. In previous works [10,11] we have described the torsional mode of the HOOH using the Jacobi dihedral angle as variable, while the other five degrees of freedom were frozen at the equilibrium configuration. Formulas relating geometrical and local vector parameters are given in Ref. [10]. The particular choice of the parametrization rigorously eliminates couplings terms in the Hamiltonian [14]. Some recent results for hydrogen peroxide, based on the diatom-diatom vectors and using a full-dimensional quantum calculation of the vibrational energy levels, can be found in [3], for a $J = 0$ Hamiltonian, for deuterated isotopomers [4,5], and in the case of $J \neq 0$ [9,8,17]. As discussed in Ref. [14], the diatom-diatom vectors not describe the torsional mode in general ABCD molecules [12]. Here we use the HOOH and HOOD as prototypes molecules to study the symmetric and not-symmetric systems [18,13]. Both are near-prolate symmetric tops with the principal axis corresponding to the smallest moment of inertia being coincident with the second vector of Jacobi, see Fig. 1. Since HOOD does not possess the symmetry of H_2O_2 it is necessary to derive

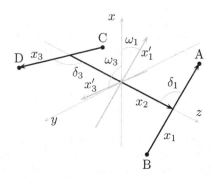

Fig. 1. *Jacobi H* scheme. $\delta_2 = 0$.

the periodicity of the torsional eigenfunctions. The purpose of this paper is to characterize the torsional levels around the O-O bond in the HOOH and HOOD molecules. The principal difference between these two systems is that, in the second case, the inertia moments of the two dimers are different. For symmetric systems, like the HOOH molecule, the torsional levels are well characterized and it is well know that these are subdivided in four symmetries, generally indicated with $\tau = 1, 2, 3$ and 4. For not symmetric system, like HOOD, further symmetries appear. This aspect can be described through the so called H scheme of orthogonal local vectors for four bodies [10,14]. This scheme consists of two vectors, each one joining an oxygen atom to the respective hydrogen (or deuterium in the case of the OD dimer) plus a vector that joins the two centers of mass of the diatoms, see Fig. 1. The torsional potential of the HOOH and HOOD is well described by the dihedral angle ω_1' when the others coordinates are fixed to their values at the equilibrium. The inertia moments of the two diatoms permit to obtain the periodicity of ω_1' and to introduce all the torsional symmetries around the O-O bonds. These symmetries are associated to the projection of the total angular momentum on the $\mathbf{x}_2$ vector, which prolongation is taken as z axis. Two methodologies are used to calculate the torsional levels. The first, referred by "bi-rotors" (BR), consists in solving the problem of two planar and rigid rotors coupled by a torsional potential. A proper combination of the two angles ω_1 and ω_3 that describe the rotation of $\mathbf{x}_1$ and $\mathbf{x}_3$ around the z axis, permits to separate the external rotation and to define a second approach, referred by as "roto-torsion" (RT), that includes the correct periodicity of the dihedral angle and permits to define all the symmetries of the torsion.

The paper is structured as follows. In the next section the Jacobi H scheme is discussed and constrains are applied to the coordinates to reduce the problem in a useful form to describe the torsional problem. In section 3 the BR scheme is presented while section 4 describes the RT one. The torsional levels of the HOOH and HOOD molecules are given in section 5. Section 6 reports remarks and conclusions.

2 Methodology

In this section we give the kinetic energy operator for two linear rigid rotors AB and CD that rotate around the vector $\mathbf{x}_2$, see Fig. 1. The reduced problem is treated in two ways leading to the BR and RT schemes. In the orthogonal local vectors parametrization [14], also called diatom-diatom vectors, we have

$$\mathbf{x}_1 = \mathbf{r}_A - \mathbf{r}_B , \qquad \mathbf{x}_3 = \mathbf{r}_C - \mathbf{r}_D ,$$

$$\mathbf{x}_2 = \frac{1}{m_A + m_B}(m_A \mathbf{r}_A + m_B \mathbf{r}_B) - \frac{1}{m_C + m_D}(m_C \mathbf{r}_C + m_D \mathbf{r}_D) ,$$

$$\mathbf{x}_4 = \frac{1}{m}(m_A \mathbf{r}_A + m_B \mathbf{r}_B + m_C \mathbf{r}_C + m_D \mathbf{r}_D) , \tag{1}$$

where m_i and $\mathbf{r}_i$ are the masses and the position vectors of the particles ($i = A, B, C, D$), respectively; $\mathbf{x}_4$ gives the position of the center of mass and m is the total mass of the molecule. The kinetic energy operator is expressed as

$$\hat{T}(\mathbf{x}) = -\frac{\hbar^2}{2}\left[\frac{1}{\mu_1}\frac{\partial^2}{\partial \mathbf{x}_1^2} + \frac{1}{\mu_2}\frac{\partial^2}{\partial \mathbf{x}_2^2} + \frac{1}{\mu_3}\frac{\partial^2}{\partial \mathbf{x}_3^2} + \frac{1}{m}\frac{\partial^2}{\partial \mathbf{x}_4^2}\right] , \tag{2}$$

where

$$\frac{1}{\mu_1} = \frac{1}{m_A} + \frac{1}{m_B} , \quad \frac{1}{\mu_2} = \frac{1}{m_A + m_B} + \frac{1}{m_C + m_D} , \quad \frac{1}{\mu_3} = \frac{1}{m_C} + \frac{1}{m_D} . \tag{3}$$

Neglecting the center of mass and in spherical coordinates, we have

$$\hat{T} = -\frac{\hbar^2}{2}\sum_{l=1}^{3}\frac{1}{\mu_l}\left[\frac{1}{r_l^2}\frac{\partial}{\partial r_l}r_l^2\frac{\partial}{\partial r_l} + \frac{1}{r_l^2}\left(\frac{1}{\sin \delta_l}\frac{\partial}{\partial \delta_l}\sin \delta_l\frac{\partial}{\partial \delta_l} + \frac{1}{\sin^2 \delta_l}\frac{\partial^2}{\partial \omega_l^2}\right)\right] , \tag{4}$$

where $r_l \geq 0$, $0 \leq \delta_l \leq \pi$ and $0 \leq \omega_l < 2\pi$. Fixing r_l, δ_l and ω_2 in the previous equation we reduce the problem in a useful form to represent the BR scheme. In fact the kinetic energy operator is

$$\hat{T}^{br} = -\frac{\hbar^2}{2} \left(\frac{1}{I_1} \frac{\partial^2}{\partial \omega_1^2} + \frac{1}{I_3} \frac{\partial^2}{\partial \omega_3^2} \right) , \tag{5}$$

where I_1 and I_3 are the effective moments of inertia of the rotors. They are obtained from those of the rotors multiplying them by $\sin^2 \delta_1$ and $\sin^2 \delta_3$, respectively. Then

$$I_1 = \mu_1 r_1^2 \sin^2 \delta_1 , \qquad I_3 = \mu_3 r_3^2 \sin^2 \delta_3 . \tag{6}$$

For a system like ABCD (see Fig. 1) we have $I_1 = I_{AB}$, $I_3 = I_{CD}$ and the z axis coincides with the vector $\mathbf{x}_2$. The angles ω_1 and ω_3 are the rotation (azimuth) angles of AB and CD rotors, respectively.

In the RT scheme, we separate the torsional mode from the overall rotation defining two new coordinates as combination of ω_1 and ω_3 [19,7]:

$$\omega_1' = \omega_3 - \omega_1 , \tag{7}$$

$$\omega_3' = \frac{I_1 \omega_1 + I_3 \omega_3}{I_1 + I_3} . \tag{8}$$

The ranges of the variables ω_1 and ω_3 lead to those of the new variables:

$$-2\pi \leq \omega_1' < 2\pi , \qquad 0 \leq \omega_3' < 2\pi . \tag{9}$$

The angle ω_3' represents the external rotation of the system around the z axis and has a periodicity of 2π. The dihedral angle ω_1' has a periodicity of 4π but the torsional potential, that is a function of ω_1', has a periodicity 2π. However, as explained in section 4, to define the periodicity of the eigenfunctions it is necessary to consider the inertia moments of the two rigid rotors.

Using eqs. (7) and (8) the kinetic energy operator for the RT scheme can be written as

$$\hat{T}^{rt} = -\frac{\hbar^2}{2} \left[\left(\frac{1}{I_1} + \frac{1}{I_3} \right) \frac{\partial^2}{\partial \omega_1'^2} + \frac{1}{I_1 + I_3} \frac{\partial^2}{\partial \omega_3'^2} \right] . \tag{10}$$

Fixing the ω_3' value we impose that the total angular momentum is zero, so the second term of the eq. (10) vanishes. In particular, if the two effective moments of inertia are equal as in H-O-O-H, the torsional part of the eq. (10) can be written as

$$\hat{T}^t = -\frac{\hbar^2}{\mu_{OH} r_{OH}^2 \sin^2 \delta_1} \frac{\partial^2}{\partial \omega_1'^2} . \tag{11}$$

Identical results can be obtained starting from the mass scaled *Jacobi H* coupling scheme vectors.

In the following we treat the BR and RT models separately. In both cases we illustrate first the "free" situation and then we introduce the torsional potential that is a function of the dihedral angle ω_1', eq. (7).

3 Bi-rotot (BR) Model

We consider first the "free" situation, in order to find a basis set for the treatment of the hindered rotation when the torsional potential is introduced.

3.1 "Free" Bi-rotor

The Schröndinger equation for the "free" bi-rotor motion is obtained by eq. (5) and results in:

$$\left[-\frac{\hbar^2}{2I_1}\frac{\partial^2}{\partial\omega_1^2} - \frac{\hbar^2}{2I_3}\frac{\partial^2}{\partial\omega_3^2} \right] \psi_{k_1,k_3}^0(\omega_1,\omega_3) = E_{k_1,k_3}^{br}\psi_{k_1,k_3}^0(\omega_1,\omega_3) \ . \tag{12}$$

The superscript 0 of the eigenfunctions ψ indicates the "free" situation. The generic eigenfunction $\psi_{k_1,k_3}^0(\omega_1,\omega_3)$ can be written as

$$\psi_{k_1,k_3}^0(\omega_1,\omega_3) = \frac{1}{2\pi}\,e^{i(k_1\omega_1+k_3\omega_3)} \ , \tag{13}$$

where $k_1, k_3 = 0, \pm1, \pm2, \pm3, \ldots$ are the quantum numbers of the two rotors. The total energy is given by the sum of the two energies $E_{k_1}^{r1}$ and $E_{k_3}^{r2}$ of the separated rotors:

$$E_{k_1,k_3}^{br} = E_{k_1}^{r1} + E_{k_3}^{r2} = \frac{\hbar^2}{2}\left(\frac{k_1^{\,2}}{I_1} + \frac{k_3^{\,2}}{I_3}\right) \ , \tag{14}$$

while the angular momenta is given by

$$\hat{l}\psi_{k_1,k_3}^0(\omega_1,\omega_3) = -i\hbar\left(\frac{\partial}{\partial\omega_1} + \frac{\partial}{\partial\omega_3}\right)\psi_{k_1,k_3}^0 = \hbar(k_1+k_3)\psi_{k_1,k_3}^0 \ . \tag{15}$$

For symmetric systems we have $I_1 = I_3 = I$, so

$$E_{k_1,k_3}^{br} = \frac{\hbar^2}{2I}\left(k_1^2 + k_3^2\right) \ . \tag{16}$$

3.2 Hindered Bi-rotor Model

Introducing the torsional potential in the Schrödinger equation (12) we have

$$\left[-\frac{\hbar^2}{2}\left(\frac{1}{I_1}\frac{\partial^2}{\partial\omega_1^2} + \frac{1}{I_3}\frac{\partial^2}{\partial\omega_3^2}\right) + V(\omega_1,\omega_3) \right]\psi_j = E_j^{br}\psi_j \ , \tag{17}$$

Let's expand the potential in a cosine series:

$$V(\omega_1,\omega_3) = \sum_{l=0}^{v} V_l\cos(l(\omega_3-\omega_1)) \ , \qquad l = 0,1,2,\ldots \tag{18}$$

The eigenfunctions ψ_j is expanded using the basis set $\psi^0_{k_1,k_3}(\omega_1,\omega_3)$ of eq. (13):

$$\psi_j = \sum_{k_1,k_3} f^j_{k_1,k_3} \psi^0_{k_1,k_3}(\omega_1,\omega_3) \ . \tag{19}$$

where $f^j_{k_1,k_3}$ are the coefficients of the expansion. The kinetic energy matrix elements are given by

$$T_{k_1,k_3;k'_1,k'_3} = \frac{\hbar^2}{2}\left(\frac{k_1^2}{I_1}+\frac{k_3^2}{I_3}\right)\delta_{k_1,k'_1}\delta_{k_3,k'_3} \tag{20}$$

while the potential energy matrix elements are given by

$$V_{k_1,k_3;k'_1,k'_3} = \sum_l \frac{V_l}{2}\left(\delta_{0,k'_1-k_1-l}\,\delta_{0,k'_3-k_3+l} + \delta_{0,k'_1-k_1+l}\,\delta_{0,k'_3-k_3-l}\right) \ . \tag{21}$$

The angular momentum eigenvalues of each eigenfunction is obtained as follows.

$$\left\langle \psi_j^* \left| \hat{l} \right| \psi_j \right\rangle = \hbar \sum_{k_1,k_3}(k_1+k_3)(f_{k_1,k_3})^2 \ . \tag{22}$$

From eq. (21) it can be found that the V_l term gives a non-zero contribution to the V_{k_1,k_3,k'_1,k'_3} element only if $l = k'_1 - k_1 = k_3 - k'_3$ or $l = k_1 - k'_1 = k'_3 - k_3$. Consequently it must be $k = k_1 + k_3 = k'_1 + k'_3$. This result reflects that the potential couples only basis set functions with the same value of k, that means with the same value of the total angular momentum, see eq. (15). Therefor, the Hamiltonian matrix can be factorized in sub-matrices, one for each value of the total angular momentum number k, with consequently reduction of the calculation time.

An interesting result, presented in the next section, can be anticipated here observing what follows. A fixed value of k means that the eigenvalues of the corresponding matrix gives the torsional energies plus a fixed contribution of the overall rotation energy. Analogously, the eigenfunctions are product of a well defined overall rotation eigenfunction times appropriated torsional eigenfunctions. As described in the next sections, the overall rotation eigenfunction is given by $e^{ik\omega'_3}$, depending by k and by the coordinate ω'_3, eq. (8). This discussion permits to conclude that the torsional basis set for a particular value of k is

$$\frac{e^{i(k_1\omega_1+k_3\omega_3)}}{e^{ik\omega'_3}} = e^{i(k_1\omega_1+k_3\omega_3-k\omega'_3)} = e^{i[k_1\omega_1+k_3\omega_3-k(I_1\omega_1+I_3\omega_3)/(I_1+I_3)]}$$

$$= e^{i\omega'_1(k_3I_1-k_1I_3)/(I_1+I_3)} = e^{i\omega'_1(k_3-kI_3/(I_1+I_3))} \ . \tag{23}$$

4   Roto-torsion (RT) Model

As in the previous section, also the roto-torsion problem is initially tackled using a zero torsional potential. The eigenfunctions of the "free" situation are then used to expand the solution of the problem when the torsional potential is introduced.

4.1 "Free" Roto-torsion

To treat separately the torsional mode and the overall rotation around the Jacobi vector $\mathbf{x}_2$ we have to use the kinetic energy operator of eq. (10). The "free" Schrödinger equation is

$$\left[-\frac{\hbar^2}{2\mathcal{I}} \frac{\partial^2}{\partial \omega_1'^2} - \frac{\hbar^2}{2(I_1 + I_3)} \frac{\partial^2}{\partial \omega_3'^2} \right] \Psi_{n,k}^0(\omega_1', \omega_3') = E_{n,k}^{rt} \Psi_{n,k}^0(\omega_1', \omega_3') \, , \qquad (24)$$

where

$$\frac{1}{\mathcal{I}} = \frac{1}{I_1} + \frac{1}{I_3} \, . \qquad (25)$$

The eigenfunction $\Psi^0(\omega_1', \omega_3')$ can be written as

$$\Psi_{n,k}^0(\omega_1', \omega_3') = \xi(\omega_1')\eta(\omega_3') = \frac{1}{2\pi} \, e^{in\omega_1'} \, e^{ik\omega_3'} \, , \qquad (26)$$

where the correct values of n and k are obtained with appropriated considerations about the periodicity of ω_1' and ω_3', respectively. Imposing a null value of ω_3' in eq. (8) we have $\omega_1 I_1 + \omega_3 I_3 = 0$, that, by a classical point of view, corresponds to a null value of the total angular momentum. In other words, the null value of the total angular momentum is guaranteed if

$$\omega_1 = -\omega_3 \frac{I_3}{I_1}. \qquad (27)$$

If we consider $I_1 < I_3$, it easy to see that if the rotor with inertia I_3 spans a full rotation ($\omega_3 = 2\pi$) and back to an indistinguishable position, the other rotor have to do an angle of $\omega_1 = -2\pi I_3/I_1$ to guaranteed a null value of the total angular momentum. Substituting eq. (27) in eq. (7) it is one obtained

$$\omega_1' = \omega_3 - \omega_1 = \omega_3 \frac{I_1 + I_3}{I_1} \, . \qquad (28)$$

The exact periodicity is obtained when both ω_1 and ω_3 are multiples of 2π, so the system oscillates between two indistinguishable positions. Therefore, the period of ω_1' must be $2\pi p(I_1 + I_3)/I_1$, where p is an integer chosen so that $p(I_1+I_3)/I_1 = N$ is approximatively an integer. This boundary condition implies that

$$e^{in\omega_1'} = e^{in(\omega_1' + 2\pi p(I_1+I_3)/I_1)} \, ,$$

$$e^{in2\pi p(I_1+I_3)/I_1} = \cos(n\, 2\,\pi\, p(I_1 + I_3)/I_1) + i\sin(n\, 2\,\pi\, p(I_1 + I_3)/I_1) = 1 \, ,$$

$$n2\pi p(I_1 + I_3)/I_1 = \pm 2\pi j \quad ; \qquad j = 0, 1, 2, \ldots$$

$$n = \pm \frac{I_1}{p(I_1 + I_3)} j = \pm \frac{j}{N} \, . \qquad (29)$$

Starting from eqs. (24) and (26), it is found that the rotational energy is

$$E_k^r = \frac{\hbar^2}{2(I_1 + I_3)} k^2 \quad ; \qquad k = 0, \pm 1, \pm 2, \pm 3, \ldots \qquad (30)$$

and the angular momentum is

$$\hat{l}'\Psi^0_{n,k}(\omega'_1,\omega'_3) = -i\hbar\frac{\partial}{\partial\omega'_3}\Psi^0_{n,k}(\omega'_1,\omega'_3) = \hbar k\Psi^0_{n,k}(\omega'_1,\omega'_3) \ . \tag{31}$$

In fact ω'_1 is an internal coordinate and does not carry information regarding the total angular momentum. Obviously the eigenvalues of the two operators $\hat{l}$, eq. (15), and $\hat{l}'$, eq. (31), need to be equal,

$$k_1 + k_3 = k \ . \tag{32}$$

Concordantly to that discussed at the end of sec. 3, the torsional energy levels can be found observing that:

$$E^t_n = E^{br}_{k_1,k_3} - E^r_k \ , \tag{33}$$

where $E^{br}_{k_1,k_3}$ are the energy of the bi-rotor, eq. (14), and E^r_k are the rotational energy, eq. (30). Concordantly to eqs. (24) and (26), the torsional levels can be written as

$$E^t_n = \frac{\hbar^2}{2}\frac{1}{\mathcal{I}}n^2 \ , \tag{34}$$

and using eq. (33) we find the possible values for n:

$$E^t_n = \frac{\hbar^2}{2\mathcal{I}}\left(k_1 - \frac{I_1}{I_1 + I_3}k\right)^2 = \frac{\hbar^2}{2\mathcal{I}}\left(k_3 - \frac{I_3}{I_1 + I_3}k\right)^2 \ . \tag{35}$$

Consequently we identify

$$n = -k_1 + \frac{I_1}{I_1 + I_3}k = k_3 - \frac{I_3}{I_1 + I_3}k \ . \tag{36}$$

This last result was anticipated in eq. (23). For a given system, I_1 and I_3 are fixed while k must be fixed to an integer value concordantly to the considered total angular momentum. This implies that the possibles values of n are determined by k_1 or k_3, that are also integer. As an example, for a null value of the total angular momentum ($k = 0$), we have $n = -k_1 = k_3$,

$$E^t_n = \frac{\hbar^2}{2\mathcal{I}}k_1^2 \ , \tag{37}$$

and only integer values of n are possible.

Symmetric Case Considering $I_1 = I_3 = I$, the kinetic energy operator of eq. (24) is

$$\hat{T}^{rt} = -\frac{\hbar^2}{I}\frac{\partial^2}{\partial\omega'^2_1} - \frac{\hbar^2}{4I}\frac{\partial^2}{\partial\omega'^2_3} \ , \tag{38}$$

where I is given by eq. (6) and the eigenvalues of the first term are given by eq. (34):

$$E^t_n = \frac{\hbar^2}{I}\left(k_3 - \frac{1}{2}k\right)^2 = \frac{\hbar^2}{I}n^2 \ , \tag{39}$$

- For even k, $n = 0, \pm 1, \pm 2, \pm 3, \ldots$
- For odd k, $n = \pm \frac{1}{2}, \pm \frac{3}{2}, \pm \frac{5}{2}, \ldots$

Another way to write the torsional energy levels is by separation in four symmetries with the quantum number $\tau = 1, 2, 3, 4$ [6,11]:

$$E_{j,1}^t = \frac{\hbar^2}{I} j^2 \qquad j = 0, 1, 2, \ldots$$

$$E_{j,2}^t = \frac{\hbar^2}{I} \left(j + \frac{1}{2} \right)^2 \qquad j = 0, 1, 2, \ldots$$

$$E_{j,3}^t = \frac{\hbar^2}{I} \left(j + \frac{1}{2} \right)^2 \qquad j = 0, 1, 2, \ldots$$

$$E_{j,4}^t = \frac{\hbar^2}{I} j^2 \qquad j = 1, 2, 3 \ldots \tag{40}$$

The eigenvalues of the second term of eq. (38) (external rotation) are

$$E_k^r = \frac{\hbar^2}{4I} k^2 \; ; \qquad k = 0, \pm 1, \pm 2, \pm 3, \ldots \tag{41}$$

and the total energy is

$$E_{j,\tau,k}^{rt} = E_{j,\tau}^t + E_k^r . \tag{42}$$

Resuming $\tau = 2, 3$ symmetries are compatible only with $k = \pm 1, \pm 3, \pm 5, \ldots$ while $\tau = 1, 4$ ones are compatible only with $k = 0, \pm 2, \pm 4, \ldots$

4.2 Hindered Roto-torsional

Introducing the potential, the Schrödinger equation is written as

$$\left[-\frac{\hbar^2}{2\mathcal{I}} \frac{\partial^2}{\partial \omega_1'^2} - \frac{\hbar^2}{2(I_1 + I_3)} \frac{\partial^2}{\partial \omega_3'^2} + V(\omega_1') \right] \Psi_{j,k} = \left(E_j^t + E_k^r \right) \Psi_{j,k} . \tag{43}$$

The eigenfunctions $\Psi_{j,k}$ are expanded using the basis set $\Psi_{n,k}^0 (\omega_1', \omega_3')$, eq. (26). The kinetic energy matrix elements are obtained using eq. (24)

$$T_{nk;n'k'} = \left(\frac{n^2 \hbar^2}{2\mathcal{I}} + E_k^r \right) \delta_{n,n'} \delta_{k,k'} , \tag{44}$$

and the potential energy matrix elements are

$$V_{nk;n'k'} = \sum_l \frac{V_l}{2} (\delta_{0,l+n'-n} + \delta_{0,l-n'+n}) \delta_{k,k'} . \tag{45}$$

where N depends of the period of ω_1', see eqs. (28)-(29). The last equation shows that the matrix is factorized in sub-matrices, one for each value of k. This because, as anticipated at the end of section 3, the potential does not couple basis set functions with different angular momenta k. Each sub-matrices presents in the diagonal elements the contribution of the external rotation for that value of k. For each sub-matrix this contribution is constant and can be neglect to obtain only the torsional energy levels. The other important thing to be observed is that each block is build up only with k-compatible functions, as imposed by eq. (36).

5   Examples: HOOH and HOOD

As explained in previous papers [10,11], the Jacobi H scheme can be used to predict the torsional path of HOOH system. The strategy adopted is to fix all the Jacobi parameters to those of the equilibrium. The torsional path is obtained varying only the dihedral angle ω_1'. This angle depends on the masses of the system. So the torsional path is different for HOOH and HOOD systems. As can be see in Fig. 2, the differences between the two predicted torsional path and the optimized path are negligible, especially for our purpose. Tab. 1 reports

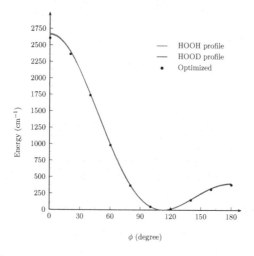

Fig. 2. Torsional profile predicted by the angle ω_1' for HOOH and HOOD systems, red and blue lines respectively. To compare with the optimized profile (Black dots), the potential is presented in function of the geometrical dihedral angle HOOH (ϕ).

the geometry of the minimum of the hydrogen peroxide expressed in internal parameters and calculated at UMP2=full/aug-cc-pvqz level of theory. Jacobi H parameters for the HOOH and HOOD are also reported. Tab. 2 reports the coefficient V_l of eq. (18) for the energy profile of HOOH and HOOD systems presented in Fig. 2. These coefficients are found by the Newton-Raphson algorithm fitting the ab-*initio* points (available on request from the authors). For the HOOH system, $I_1 = I_3$, $p = 1$ and the periodicity of ω_1' is 4π, see eqs. (28) and (29). For HOOD, the approximated inertia moments in Tab. 2 lead to a finite periodicity of ω_1'. This periodicity depends by the level of accuracy of the values of the two inertias. Expressing the inertia with a greater number significant figures, a greater periodicity of ω_1' is found. Consequently, the representation of the torsional energy levels improves. We give the values of the inertia with three-four figures because this level of accuracy is sufficient for us purpose and we found $p = 200$. This means that the basis set derived for $k = 0$ is approximatively

Table 1. Geometrical parameters of the minimum of the hydrogen peroxide calculated at ump2=full/aug-cc-pvqz level. Jacobi parameters for HOOH and HOOD are also presented. For the HOOD case, $|\mathbf{x}_3|$ join the atoms O and D. Angles are expressed in degrees and lengths in Å.

Geometrical parameters	r_{HO}	r_{OO}	r_{OH}	$\angle HOO$	$\angle OOH$	$\angle HOOH$
	0.9627	1.4433	0.9627	99.94	99.94	112.59

Jacobi H scheme	$|\mathbf{x}_1|$	$|\mathbf{x}_2|$	$|\mathbf{x}_3|$	φ_1	φ_3	ω_1'
HOOH	0.9627	1.4660	0.9627	102.98	102.98	111.76
HOOD	0.9627	1.4782	0.9627	103.68	104.89	111.34

Table 2. Values, in cm^{-1}, of the coefficients V_l of eq. (18) for HOOH and HOOD energy profiles of Fig. 2. Effective inertia moments I_1 and I_3 in u.m.a.Å^2 are also presented. The masses in u.m.a. of O, H and D atoms are 15.9994, 1.0079 and 2.01363 respectively.

	V_0	V_1	V_2	V_3	V_4	V_5	I_1	I_3
HOOH	837.551	1072.064	687.812	65.383	8.973	1.601	0.834	0.834
HOOD	834.975	1061.922	689.156	65.023	8.076	1.328	0.830	1.548

correct for $|k| = 200$ too. Analogously, $|k| = 1$ and $|k| = 201$ are near compatible with the same symmetry and so on for all the values of k. Eq. (36) permits to calculate the compatible values of n for every k. In section 5.2 we present how to tackle this type of problems.

The torsional symmetries found for each k can be further separated in even and odd functions as suggested by the second equality of eq. (45). In fact the torsional potential, due to the symmetry around π, can be expanded in a cosine series and l assumes only integer values. Three types of integrals are found: A first type is of the form

$$\int \cos(n\,\omega_1')\cos(l\,\omega_1')\sin(\,n'\omega_1')\,d\omega_1' \,, \tag{46}$$

and is always zero. In other words, cosines and sines are not coupled by a symmetric torsional potential. The other two types of integrals are

$$\int \cos(n\,\omega_1')\cos(l\,\omega_1')\cos(\,n'\omega_1')\,d\omega_1' \,, \tag{47}$$

$$\int \sin(n\,\omega_1')\cos(l\,\omega_1')\sin(\,n'\omega_1')\,d\omega_1' \,. \tag{48}$$

In summary, for a given value of k, the possible values of n are calculated with eq. (36). Furthermore the torsional matrix can be factorized in two sub-matrices, one of them representation of even eigenfunctions (expanded in cosine functions) while the other is the representation of the odd eigenfunctions (expanded in sine functions).

5.1 Symmetric Systems: HOOH

BR Model. The problem is tackled following the factorization described at the end of section 3. For a selected value of k only values of k_1 and k_3 that respect the condition $k_1 + k_3 = k$ are taken. These were introduced in eqs. (20)-(21) to found the energy levels of the bi-rotors E_k^{br}. The rotational contribution E_k^r is found with eq. (30), while the torsional contribution E_j^t is simply $E_j^{br} - E_k^r$. Results for the torsional energy levels E_j^t are reported in Tab. 3.

Table 3. Torsional energy contributions of the bi-rotors energy levels for the HOOH system, obtained with eq. (42), $E_j^t(cm^{-1})$. The bi-rotors levels were calculated with basis set of 120 eigenfunctions for each k.

$k = 0, \pm 2, \pm 4, \ldots$	$k = \pm 1, \pm 3, \ldots$	$k = 0, \pm 2, \pm 4, \ldots$	$k = \pm 1, \pm 3, \ldots$
172.860197	172.860204	2171.396396	2182.712934
184.505254	184.505245	2432.736930	2395.268584
432.933494	432.933550	2589.873299	2685.168639
551.304401	551.304165	2946.351412	2771.972723
754.236097	754.237293	2980.969544	3223.361043
965.131881	965.125746	3520.964623	3234.172685
1194.968428	1194.999239	3523.889768	3841.034956
1435.192291	1435.043177	4183.740018	3841.755708
1681.497440	1682.181378	4183.905322	4548.679360
1932.273870	1929.351857	-	4548.715075

RT Model. When $I_1 = I_3$ according with section 4.1 and eq. (39), in the torsional basis set $\xi(\omega_1')$, eq. (26), n assumes both integer and half integer values. The decomposition of ξ in sines and cosines leads to

$$\frac{1}{1 + (\sqrt{2} - 1)\delta_{j,0}} \frac{1}{\sqrt{2\pi}} \cos(j\,\omega_1') \qquad (49a)$$

$$\frac{1}{\sqrt{2\pi}} \cos[(j + 1/2)\,\omega_1'] \qquad (49b)$$

$$\frac{1}{\sqrt{2\pi}} \sin[(j + 1/2)\,\omega_1'] \qquad (49c)$$

$$\frac{1}{\sqrt{2\pi}} \sin(j\,\omega_1') \qquad (49d)$$

with $j = 0, \pm 1, \pm 2, \ldots$. Note that these equations have a period of 4π and this justifies the normalization factors. In the eq. (49d) $j = 0$ loses meant. As describe above in this section, the potential coupling only eigenfunction with the same parity, (cosine with cosine and sine with sine); moreover the expansion in a serie of cosine of the torsional potential does not couple the $\cos[j\varphi]$ functions with the $\cos[(j'+1/2)\varphi]$ ones, as so as it does not couple $\sin[j\varphi]$ and $\sin[(j'+1/2)\varphi]$ functions. So the eigenfunctions of the problem become:

$$\xi_{\nu,1} = \frac{1}{\sqrt{2\pi}} \sum_{j=0} a_j^\nu \cos(j\ \omega_1') \tag{50a}$$

$$\xi_{\nu,2} = \frac{1}{\sqrt{2\pi}} \sum_{j=0} b_j^\nu \cos[(j+1/2)\ \omega_1'] \tag{50b}$$

$$\xi_{\nu,3} = \frac{1}{\sqrt{2\pi}} \sum_{j=0} c_j^\nu \sin[(j+1/2)\ \omega_1'] \tag{50c}$$

$$\xi_{\nu,4} = \frac{1}{\sqrt{2\pi}} \sum_{j=1} d_j^\nu \sin(j\ \omega_1') \tag{50d}$$

These basis sets were used to calculated the torsional energy levels that are presented in Tab. 4. As expected, the values obtained with the two different but equivalent procedure are identical.

Table 4. Torsional energy levels of H_2O_2 system calculated with the RT procedure. Basis set of 200 eigenfunctions for each symmetry. Values in cm^{-1}.

$\nu\backslash\tau$	1	2	3	4
0	172.860197	172.860204	184.505245	-
1	432.933494	432.933550	551.304165	184.505254
2	754.236097	754.237293	965.125746	551.304401
3	1194.968428	1194.999239	1435.043177	965.131881
4	1681.497440	1682.181378	1929.351857	1435.192291
5	2171.396396	2182.712934	2395.268584	1932.273870
6	2589.873299	2685.168639	2771.972723	2432.736930
7	2980.969544	3223.361043	3234.172685	2946.351412
8	3523.889768	3841.034956	3841.755708	3520.964623
9	4183.905322	4548.679360	4548.715075	4183.740018

5.2 Non Symmetric System: HOOD

The mass of the deuterium atom is approximatively double respect that of the hydrogen. This affects the torsional energy levels despite the torsional potential (written in Jacobi coordinates) be approximatively the same for the two cases. More exactly, the higher mass of the deuterium thicken the levels, but the difference in mass between the two systems is not so relevant respect the characteristic of the torsional potentials. This means that the energies of the lower torsional levels of the HOOD are expected to be of the same order of those of the HOOH.

BR Model. As in the HOOH case, torsional energy contributions E_j^t of the bi-rotors energy levels for the HOOD system are obtained neglecting the rotational energy contribution E_k^r from E_j^{br}, see eqs. (30) and (17). The bi-rotors levels are obtained considering the factorization in sub-matrices described at the end of section 3. The torsional energy levels of HOOD calculated with the BR model, for $k = 0, \pm1, \pm2$, are showed in Tab. 5.

Table 5. Torsional energy contributions of the bi-rotors energy levels for the HOOD system, obtained with eq. (42). The bi-rotors levels were calculated with basis set of 120 eigenfunctions for each k.

torsional contribution (cm^{-1}).		
$k = 0$	$k = \pm1$	$k = \pm2$
157.414847	157.414847	157.414847
162.768676	162.768676	162.768676
398.858888	398.858892	398.858892
477.489056	477.489043	477.489045
649.776606	649.776673	649.776662
821.080170	821.079836	821.079891
1013.417658	1013.419336	1013.419060
1216.217830	1216.209519	1216.210885
1427.194799	1427.234678	1427.228122
1643.461487	1643.278270	1643.308352
1861.261273	1862.054670	1861.923699
2081.283525	2078.067422	2078.587141
2284.306288	2295.664206	2293.689023
2515.689746	2479.169064	2484.280015
2633.830757	2703.865455	2690.318948
2962.571619	2830.604867	2847.777383
2984.792085	3134.217450	3112.129466
3462.205628	3282.450595	3306.038508
3464.013067	3653.391059	3627.387613

RT Model For HOOD the possible values of n for the torsional problem are evaluated with eq. (36) and $I_3/(I_1 + I_3) = 0.6511$. The approximative values of I_1 and I_3 for HOOD are given in Tab. 2). Considering eqs. (47) and (48) we can derive the following "symmetries" (basis sets) for k equal to $0, \pm1, \pm2$:

- $\tau = 0c$: $\cos[j\,\omega_1']$ with $j = 0, 1, 2, \ldots$ and $k = 0$
- $\tau = 0s$: $\sin[j\,\omega_1']$ with $j = 1, 2, \ldots$ and $k = 0$
- $\tau = 1c$: $\cos[(j+\epsilon)\,\omega_1']$ with $j = 0, \pm1, \pm2, \ldots$; $\epsilon = 0.6511$ and $k = \pm1$
- $\tau = 1s$: $\sin[(j+\epsilon)\,\omega_1']$ with $j = 0, \pm1, \pm2, \ldots$; $\epsilon = 0.6511$ and $k = \pm1$
- $\tau = 2c$: $\cos[(j+\epsilon)\,\omega_1']$ with $j = 0, \pm1, \pm2, \ldots$; $\epsilon = 1.3022$ and $k = \pm2$
- $\tau = 2s$: $\sin[(j+\epsilon)\,\omega_1']$ with $j = 0, \pm1, \pm2, \ldots$; $\epsilon = 1.3022$ and $k = \pm2$

Another possible way to write these basis sets is the following:

- $\tau = 0c$: $\cos[j\,\omega_1']$ with $j = 0, 1, 2, \ldots$ and $k = 0$
- $\tau = 0s$: $\sin[j\,\omega_1']$ with $j = 1, 2, \ldots$ and $k = 0$

- $\tau = 1c$: $\cos[(j + \epsilon)\,\omega_1']$ with $j = 0, 1, 2, \ldots$; $\epsilon = 0.6511, 0.3489$ and $k = \pm 1$
- $\tau = 1s$: $\sin[(j + \epsilon)\,\omega_1']$ with $j = 0, 1, 2, \ldots$; $\epsilon = 0.6511, 0.3489$ and $k = \pm 1$
- $\tau = 2c$: $\cos[(j + \epsilon)\,\omega_1']$ with $j = 0, 1, 2, \ldots$; $\epsilon = 1.3022, 0.6978$ and $k = \pm 2$
- $\tau = 2s$: $\sin[(j + \epsilon)\,\omega_1']$ with $j = 0, 1, 2, \ldots$; $\epsilon = 1.3022, 0.6978$ and $k = \pm 2$

With these basis functions the torsional levels given in Tab. 6 are found.

Our results shows that, under the *trans* barrier, the torsional energy levels are degenerate. This is independent by the quantum number k and, consequently by the symmetry τ. Significant splitting between different symmetries τ for the same level are predicted starting from the sixth level, just and under the *cis* barrier. This means that the experimental observation of the separation in symmetries of the torsional problem could be not so easy.

Table 6. Torsional energy levels of HOOD system calculated with the RT procedure. Basis set of 400 eigenfunctions for each symmetry. Values in cm^{-1}.

k	0		±1		±2	
Levels τ	0c	0s	1c	1s	2c	2s
0	157.414847		157.414847	157.414847	157.414847	157.414847
0		162.768676	162.768676	162.768676	162.768676	162.768676
1	398.858888		398.858892	398.858892	398.858892	398.858892
1		477.489056	477.489043	477.489043	477.489045	477.489045
2	649.776606		649.776673	649.776673	649.776662	649.776662
2		821.080170	821.079836	821.079836	821.079891	821.079891
3	1013.417658		1013.419336	1013.419336	1013.419060	1013.419060
3		1216.217830	1216.209519	1216.209519	1216.210885	1216.210885
4	1427.194799		1427.234678	1427.234678	1427.228122	1427.228122
4		1643.461487	1643.278270	1643.278270	1643.308352	1643.308352
5	1861.261273		1862.054670	1862.054670	1861.923699	1861.923699
5		2081.283525	2078.067422	2078.067422	2078.587141	2078.587141
6	2284.306288		2295.664207	2295.664207	2293.689021	2293.689021
6		2515.689746	2479.169061	2479.169061	2484.280021	2484.280021
7	2633.830757		2703.865462	2703.865462	2690.318933	2690.318933
7		2962.571619	2830.604858	2830.604858	2847.777402	2847.777402
8	2984.792085		3134.217463	3134.217463	3112.129442	3112.129442
8		3462.205628	3282.450582	3282.450582	3306.038534	3306.038534
9	3464.013067		3653.391073	3653.391073	3627.387585	3627.387585

6   Conclusions and Perspective

In this work we have shown how torsional energies can be calculated with both bi-rotor and by the roto-torsion schemes. We remark that the two schemes are equivalent and related one to the other. The separation of the overall rotation and the consequent factorization in symmetries of the torsional problem is possible in an easy way due to the properties of the Jacobi coordinates (H scheme). This factorization greatly improves the calculation of the torsional levels, also describing spectral lines of non symmetric systems like HOOD. Obviously, a full calculation, including all the degree of freedom, for not symmetric systems could be of extreme interest, especially if we consider the origin of the life, see [1]. In fact, a variety of organic and inorganic molecules, indispensable for the development of the live, present one or more torsional degree of freedom. Of interest is that, frequently, the inertia moments of the two dimers involved in the torsional mode are different. This means that further level splitting can be expected with consequently modification of the partition functions and rate constants.

Acknowledgments. This work has been supported by Conselho Nacional de Desenvolvimento Científi- co e Tecnológico (CNPq - Brazil) and by Fundação de Amparo à Pesquisa do Estado da Bahia (FAPESB - Brazil).

References

1. Aquilanti, V., Maciel, G.S.: Orig. Life Evol. Biosph. 36, 435 (2006)
2. Bitencourt, A.C.P., Ragni, M., Maciel, G.S., Aquilanti, V., Prudente, F.V.: J. Chem. Phys. 129, 154316 (2008)
3. Chen, R., Ma, G., Guoa, H.: Chem. Phys. Lett. 320, 567–574 (2000)
4. Chen, R., Ma, G., Guoa, H.: J. Chem. Phys. 114, 4763–4774 (2001)
5. Fehrensen, B., Luckhaus, D., Quack, M.: Chem. Phys. Litt. 300, 312–320 (1999)
6. Hunt, R.H., Leacock, R.A., Peters, C.W., Hecht, K.T.: J. Chem. Phys. 42, 1931–1946 (1965)
7. Koehler, J.S., Dennison, D.M.: Phys. Rev. 57, 1006–1021 (1940)
8. Koput, J., Carter, S., Handy, N.C.: J. Chem. Phys. 115, 8345–8350 (2001)
9. Lin, S.Y., Guo, H.: J. Chem. Phys. 119, 5867–5873 (2003)
10. Maciel, G.S., Bitencourt, A.C.P., Ragni, M., Aquilanti, V.: Chem. Phys. Lett. 432, 383–390 (2006)
11. Maciel, G.S., Bitencourt, A.C.P., Ragni, M., Aquilanti, V.: Int. J. Quant. Chem. 107, 2697–2707 (2007)
12. Maciel, G.S., Bitencourt, A.C.P., Ragni, M., Aquilanti, V.: J. Phys. Chem. A 111, 12604–12610 (2007)
13. Pelz, G., Yamada, K.M.T., Winnewisser, G.: J. Mol. Spectrosc. 159, 507 (1993)
14. Ragni, M., Bitencourt, A.C.P., Aquilanti, V.: Int. J. Quant. Chem. 107, 2870–2888 (2007)
15. Roncaratti, L.F., Aquilanti, V.: Int. J. Quant. Chem. 110, 716 (2010)
16. Ross, S.C., Yamada, K.M.T.: Phys. Chem. Chem. Phys. 9, 5809–5813 (2007)
17. Senent, M.L., Fernández-Herrera, S., Smeyers, Y.G.: Spectrochimica Acta Part A 56, 1457–1468 (2000)
18. Winnewisser, G., Yamada, K.M.T.: Vib. Spectrosc. 1, 263 (1991)
19. Yamada, K.M.T., Winnewisser, G., Jensen, P.: J. Mol. Struct. 695, 323–337 (2004)

Carbon Oxides in Gas Flows and Earth and Planetary Atmospheres: State-to-State Simulations of Energy Transfer and Dissociation Reactions

Andrea Lombardi*, Antonio Laganà, Fernando Pirani,
Federico Palazzetti, and Noelia Faginas Lago

Dipartimento di Chimica,
Università di Perugia, Perugia, Italy
{ebiu2005,lagana05}@gmail.com, pirani@unipg.it,
fede_75it@yahoo.it, piovro@gmail.com
http://www.chm.unipg.it/gruppi?q=node/48

Abstract. In this paper we illustrate an approach to the study of the molecular collision dynamics, suited for massive calculations of vibrational state-specific collision cross sections and rate constants of elementary gas phase processes involving carbon oxides. These data are used in the theoretical modeling of the Earth and planetary atmospheres and of non-equilibrium reactive gas flows containing the CO_2 and CO molecules. The approach is based on classical trajectory simulations of the collision dynamics and on the bond-bond semi-empirical description of the intermolecular interaction potential, that allows the formulation of full dimension potential energy surfaces (the main input of simulations) for small and medium size systems. The bond-bond potential energy surfaces account for the dependence of the intermolecular interaction on some basic physical properties of the colliding partners, including modulations induced by the monomer deformation. The approach has been incorporated into a Grid empowered simulator able to handle the modeling of the $CO_2 + CO_2$ collisions, while extensions to other processes relevant for the modeling of gaseous flows and atmospheres, such as $CO + CO \rightarrow C + CO_2$ and $CO_2 + N_2$, are object of current work. Here the case of $CO_2 + CO_2$ collisions will be illustrated in detail to exemplify an application of the method.

Keywords: Intermolecular interactions, molecular dynamics, carbon oxides, gas flows, Earth and planetary atmospheres.

1 Introduction

The dynamics of molecules in gaseous systems is dominated by bimolecular collisions events which generate roto-vibrational energy exchange and are therefore responsible for the energy relaxation and the state population of molecules.

---

* Corresponding author.

B. Murgante et al. (Eds.): ICCSA 2013, Part II, LNCS 7972, pp. 17–31, 2013.
© Springer-Verlag Berlin Heidelberg 2013

In turn, such energy–relaxing and population–altering effects play a key role in determining the energy balance of several chemical processes. This is precisely the case of the processes involving carbon oxides, such as $CO_2 + CO_2$, $CO_2 + N_2$ and $CO + CO \rightarrow C + CO_2$, which are object of interest in connection to the study of Earth's and planetary atmospheres [1,2] (their presence on Venus and Mars is well documented). Carbon monoxide is also one of the most abundant molecules in interstellar space. Recently CO molecules have been found in the 10-million-degree gas associated with the young supernova remnant Cassiopeia A (Cas A, see [3]). Beside their role in astrochemistry and atmospheric chemistry, the above mentioned collisions involving CO and CO_2 deserve the interest of the gas dynamics community, since they give rise to a series of elementary processes relevant to the kinetic and fluid dynamic study of shock waves in connection to the spacecraft reentry problem (see the FP7 EU project [4]). Especially in the cases in which the speed of the vehicles exceeds the local speed of sound, the consequent formation of a shock wave leads to strong excitation of the molecular internal degrees of freedom (rotational, vibrational and electronic) and promotes a strong energy transfer that activates many chemical reactions [5]. The reacting gas or plasma, which interacts with the surface of the thermal protection system, is often characterized by thermal and chemical non equilibrium conditions and the only useful collision observable, that can be safely used in models, is the collision cross section, a quantity which is not averaged over an energy distribution.

Further motivations for work on such processes (in addition to their relevance to the goals of Ref. [4]) come from other innovative fields in which there is need for carrying out dynamical simulations, namely, supercritical fluids for extraction processes, the study of the environmental impact of green-house gas, the exploitation of plasma chemistry.

The quantum mechanical solutions to the problem of the energy transfer are mainly based on the so called close-coupling approach, where the wave function describing the scattering process is expanded in a suitable set of basis functions, most of the times obtained as direct product of the internal states of the colliding molecules and a set of angular functions. Then the resulting problem is solved numerically. A main difficulty with applying the close-coupling methods to problems other than those involving three atoms is that the vibro-rotational basis sets required become too large for the calculation to be feasible. Improvements of the feasibility are based on an appropriate choice of the coordinate system and of the related internal and rotational state basis set [6]. High dimensional scattering problems are quite common in kinetic modeling of gas phase systems and the processes involving the CO_2 molecules are a paradigmatic example of the difficulties encountered. Recent attempts in the gas dynamics community for refining kinetic models at a state-specific level [7], require the calculation of state-to-state cross sections and rate constant for a representative set of the many possible state-to-state energy exchange processes. The necessary massive calculation are very demanding and the use of trajectories, motivated by the fact that accurate quantum calculations of inelastic and reactive cross sections are unfeasible for any system made up by more than four atoms, as well as three-atom processes become unfeasible when massive calculations, required to form a data base of state-to-state vibrational exchange cross sections and thermal rate coefficients, have to be performed.

The bimolecular collisions are driven by weak non-covalent intermolecular forces [8,9]. As a consequence, the development of an intermolecular potential energy surface (PES) is a crucial issue in the theoretical study of molecular dynamics, since the description of the interaction is the main input of both quantum and classical scattering calculations [8,9]. Even for relatively simple systems, high level *ab initio* calculations are computationally so demanding that an adequate investigation of the full configuration space is still, in practice, out of reach. In order to give a realistic description of the intermolecular interaction for these cases, semi-empirical approaches are the only viable alternative. An effective formulation of the intermolecular interaction potential is that based on the so called bond-bond approach (see [10] and references therein), which expresses the intermolecular interaction in terms of bond properties and parameters characterizing the internal molecular structure, such as charge distributions and polarizabilities [11]. For the case of the CO_2 + CO_2 collisions treated in Ref. [11], of fundamental importance, due to the widespread presence of this molecule in Earth and planetary atmospheres, the effects of vibrations and rotations of the molecule have to be included in models, since modify polarizabilities and charge distributions, strongly affecting the interactions. In this respect, the work in Ref. [11] has improved the existing CO_2 dimer PESs (see e.g. [12]), by releasing the frozen stretching and bending constraints for the CO_2 monomers. The resulting intermolecular interaction part of the PES is given a flexible analytic formulation including its dependence on the internal degrees of freedom (stretching and/or bending) of the monomers. The PESs formulated according to the bond-bond and related approaches are parametrized using quantities having a well defined physical meaning, are usable in different environments and conditions, and are extensible to the description of systems of increasing complexity (see for instance Refs. [13,14,15]). For these reasons, they can easily be used in conjunction with computational technologies that nowadays make available impressive amounts of computing time, such as when codes are implemented on the Grid in an organized form (possibly a work-flow [16]). A work-flow for simulations CO_2 + CO_2 collisions has recently actually been developed as part of the activities of the virtual organization COMPCHEM [17].

In this paper, we will consider the case of vibrational energy transfer in CO_2 + CO_2 collisions, while the other above mentioned processes, CO + CO and CO_2 + N_2, are currently object of ongoing work devoted to build up the required potential energy surfaces and to perform QCT calculations for cross sections and rate constants.

The paper is organized as follows. In section 2 background information on the bond-bond approach to the intermolecular PES is given and the application to the case of CO_2 + CO_2 collisions is illustrated. In section 3 details of QCT collision dynamics calculations are given and an illustration of typical outcomes of molecular dynamics simulations for the case of $CO_2 + CO_2$ collisions, is proposed. Section 4 contains conclusions and future perspectives.

2   The Representation of the PES

The main input of any scattering calculation is the potential energy function which represents the interaction energy of the system as a function of the positions of the atoms. In collision problems there are two atomic or molecular fragments that, initially at large distance, collide exchanging energy among their translational, vibrational and rotational degrees of freedom, and eventually get far apart. At large distances (greater than typical bond distances) long range attractive interaction dominates, whereas at short distances repulsive forces overcome the attraction. The collisions are therefore strongly influenced by the combined effect of these attractive and repulsive non-covalent forces, and much of the effort in the theoretical scattering studies have to be spent for a realistic description of these interactions. An appropriate representation of a PES for the collision of two molecules is obtained expressing the interaction potential as a function of the distance R between the two fragments (centers of mass), of a set of angles Ω defining the mutual orientation of the fragments, and a set of internal coordinates, collectively denoted as ρ, that uniquely defines the configuration of each colliding partner. This representation permits the formulation of the PES as a sum of two main terms accounting for the internal interaction energy and for the intermolecular interaction energy, respectively:

$$V(R, \Omega, \rho) = V_{intra}(\rho) + V_{inter}(R, \Omega; \rho) \tag{1}$$

where $V_{intra}(\rho)$ is the internal interaction energy and coincides with the potential energy of the two isolated molecules and $V_{inter}(R, \Omega; \rho)$ represents the intermolecular part of the interaction, that is supposed to depend parametrically on the internal structure of the interacting partners.

The term $V_{inter}(r, \Omega; \rho)$ can be conveniently expressed as a sum of two effective interaction components (the coordinates are omitted for simplicity):

$$V_{inter} = V_{vdW} + V_{elect}. \tag{2}$$

where V_{vdW} and V_{elect} represent the van der Waals (size repulsion plus dispersion attraction) and the electrostatic interaction components, respectively. V_{elect} originates from the anisotropic molecular charge distributions of the two bodies, which asymptotically tend to the permanent quadrupole-permanent quadrupole interaction. Both V_{vdW} and V_{elect} strongly depend on the distance R between the centers of mass of the two monomers (say the molecules a and b), and also vary as a function of the additional coordinates (in general taken as angular coordinates, collectively denoted as Ω in Eq. 1) defining their mutual orientation. A careful definition of a suitable set of angular coordinates is important in many cases, for example when stereochemistry is concerned [18,19,20,21,22], or when an expansion in terms of special angular functions (i.e. hyperspherical harmonics [23,24,25,26,27,28,29,30,31]) is needed, similarly to what is usually done in quantum and classical mechanics to improve the separability of the internal degrees of freedom of molecules or clusters (see e. g. [32]).

2.1 The Bond-Bond Approach to the vdW Term

In the framework of the bond-bond approach (see [10] and references therein) the van der Waals term, V_{vdW} of Eq. 2, is expressed as a sum over the contributions of all the possible interaction center pairs between the two monomers V_{vdW}^i

$$V_{vdW}(R, \gamma) = \sum_i^N V_{vdW}^i (R_i, \gamma_i) . \tag{3}$$

where R_i is the distance between the reference points of the two interacting centers of the i-th pair, γ_i denotes collectively the angular coordinates defining the related mutual orientation and N is the number of interaction pairs. In most common applications the interaction centers are chemical bonds, but they can be atoms, group of atoms or bonds, depending on the structure of the monomers. Such representation is based on the additive character of the various bond polarizability components in contributing to the overall molecular polarizability (a fundamental feature of the vdW interactions), that can be extended to the related interaction contributions. Moreover, the interaction terms, so formulated, indirectly accounts for nonlinear three body effects [33], since bond polarizability components are not merely the sum of the contributions of the isolated atoms.

The V_{vdW}^i term is formulated as an extension and a generalization of the atom-bond pairwise additive property discussed in Refs [33,34]. It is important to note that each of the interacting centers considered here is assumed to be an independent sub-unit having a definite polarizability and a given electronic charge distribution (it can be, for example, of nearly cylindrical symmetry when the center is a bond).

The explicit formulation adopted for V_{vdW}^i is of the Improved Lennard-Jones (ILJ) type [35]:

$$\frac{V_{vdW}^i(R_i, \gamma_i)}{\varepsilon_i(\gamma_i)} = f(x_i) = \left[\frac{m}{n(x_i) - m} \left(\frac{1}{x_i} \right)^{n(x_i)} - \frac{n(x_i)}{n(x_i) - m} \left(\frac{1}{x_i} \right)^m \right] \tag{4}$$

where x_i is the reduced distance defined as

$$x_i = \frac{R_i}{R_{mi}(\gamma_i)}. \tag{5}$$

while γ_i again denotes collectively the angular coordinates (see Eq. 3) and ε_i and R_{mi} are the well depth of the interaction potential and the related equilibrium distance respectively. The parameter m takes pair specific values (e.g. the value is equal to 6 for neutral interacting centers). It is worth emphasizing here that the ILJ function [35] is definitely more realistic than the original Lennard-Jones(12,6) one, because it represents more accurately both the size repulsion (first term within the square brackets) and the long range dispersion attraction (second term within the square brackets) [27,36].

As a matter of fact, the n exponent is expressed as a function of both the distance R_i and the angles γ_i (see Eq. 3) using the following empirical equation [34]:

$$n(x_i) = \beta + 4.0 x_i^2. \tag{6}$$

in which β is a parameter depending on the nature and the hardness of the interacting centers that introduces the metric of a more ambient-like characteristic (that can be named as the hardness of the interacting partners [34,35]) by modulating the repulsion and controlling the strength of the attraction. The introduction of this modulation (absent in the classical Lennard-Jones potential function) provides the ILJ expression with the possibility of indirectly taking into account induction, charge transfer and atom clustering effects.

2.2   The Electrostatic Component

The V_{elect} term of Eq. 2 originates from the anisotropic molecular charge distributions of the two bodies, which asymptotically tend to the permanent quadrupole-permanent quadrupole interaction, and can be formulated as a sum of Coulomb potential terms for each pair of interacting molecules, say a and b. For a given pair one has:

$$V_{elect}(R, \gamma) = \sum_{jk} \frac{q_{ja} q_{kb}}{r_{jk}} \tag{7}$$

where q_{ja} and q_{kb} are point charges located on the interacting molecules a and b and r_{jk} is the distance between them. This improves considerably the usual expression depending on the product of the molecular quadrupoles Q (see Eq. 9 of Ref. [10]).

In the above given Eq. 7, instead, the charge distributions on each molecular monomer are taken to be compatible with the corresponding calculated molecular quadrupoles.

Such formulation of V_{elect} must be used for cases in which the molecular dimensions are not negligible with respect to the intermolecular distance R [9]. The choice of the spatial distribution of the charge is not complicated for relatively simple molecules (e.g. triatomics), especially when the dominant role of strongly charged atoms (e.g. oxygen in H_2O) is well evident [37,38,39]. The choice of the charge centers has instead a certain extent of arbitrariness for more complex systems (see, e.g., Ref [40,41]).

2.3   The Intermolecular Interaction of the $CO_2 + CO_2$ System

Both the V_{vdW} and V_{elect} terms of Eq. 2 depend, as above mentioned, on the distance R between the centers of mass of the two molecules (say a and b), and on a set of angular coordinates, that can be conveniently set up to be the Jacobi angular coordinates Θ_a, Θ_b and Φ describing the $a - b$ mutual orientation as well (for examples of the importance of the coordinate choice for the representation of the molecular potential energy see [42,43] and references therein). To construct the entire potential energy a set of leading configurations, which differ for the mutual orientation of the two monomers, can be individuated by the following values of the angular variables $(\Theta_a, \Theta_b, \Phi) = (90°, 90°, 0°), (90°, 90°, 90°)$, $(0°, 90°, 0°), (0°, 0°, 0°), (45°, 45°, 0°)$ and $(60°, 60°, 0°)$. The corresponding configurations are also referred to as H, X, T, L, S^{45} and S^{60} (see Fig. 1 of Ref. [11]). The van der Waals term, V_{vdW} of Eq. 2, is in turn expressed as a sum of the contributions V_{vdW}^i for all the different possible bond$_{ai}$-bond$_{bi}$ i-th pairs between the two monomers.

$$V_{vdW}(R, \Theta_a, \Theta_b, \Phi) = \sum_i^K V_{vdW}^i (R_i, \theta_{ai}, \theta_{bi}, \phi_i). \qquad (8)$$

in which R_i is in this case the distance between reference points conveniently placed on the two bonds of the i-th pair, θ_{ai}, θ_{bi}, the ϕ_i are the related mutual orientation angles, and $K = 4$ is the number of CO-CO pairs for the two CO_2 molecules. This formula enforces the additivity of the various bond polarizability components that contribute to form the overall molecular polarizability (a fundamental feature of the vdW interactions) and, as pointed out in section 2.1 indirectly accounts for three body effects [33].

The V_{vdW}^i terms are formulated as in Eq. 4 exploiting the pairwise additivity discussed in Refs [33,34]. It is important to note that each of the bonds considered here is assumed to be an independent diatomic sub-unit having a definite polarizability and a given electronic charge distribution of nearly cylindrical symmetry. In addition, it has been assumed that the reference point of each bond is set at about the geometric bond center (more precisely the reference point of the CO pair has been displaced of 0.1278 Å towards the O-end) because the dispersion and the bond centers do not usually coincide. One has [35]:

$$\frac{V_{vdW}^i(R_i, \gamma_i)}{\varepsilon_i(\gamma_i)} = f(x_i) = \left[\frac{6}{n(x_i) - 6} \left(\frac{1}{x_i} \right)^{n(x_i)} - \frac{n(x_i)}{n(x_i) - 6} \left(\frac{1}{x_i} \right)^6 \right] \qquad (9)$$

where x_i, the reduced distance and the exponent n are as in Eqs. 5 and 6 respectively, γ_i denotes collectively the triple of angles $(\theta_{ai}, \theta_{bi}, \phi_i)$ and ε_i and R_{mi} are, as usual, respectively the well depth of the interaction potential and the related equilibrium distance. Again we would like to emphasize here that the ILJ function [35] is much more realistic than the o Lennard-Jones(12,6) function because it better reproduces both the size repulsion (first term within the square brackets) and the long range dispersion attraction (second term within the square brackets) [36,27].

To obtain a more flexible formulation of the V_{vdW}^i it has been found convenient to expand the parameters ε_i and R_{mi} in terms of bipolar spherical harmonics $A^{L_1 L_2 L}(\gamma)$ [11]. In this way $f(x_i)$, the reduced form of the bond-bond potentials (see Ref. [34]) is taken to be the same for all the relative orientations (see Refs. [44,45,46,47]). For the CO_2–CO_2 system it was found sufficient to truncate the expansion to the fifth order:

$$\varepsilon_i(\gamma) = \varepsilon_i^{000} + \varepsilon_i^{202} A^{202}(\gamma) + \varepsilon_i^{022} A^{022}(\gamma) + \varepsilon_i^{220} A^{220}(\gamma) + \varepsilon_i^{222} A^{222}(\gamma) \quad (10)$$

$$R_{mi}(\gamma) = R_{mi}^{000} + R_{mi}^{202} A^{202}(\gamma) + R_{mi}^{022} A^{022}(\gamma) + R_{mi}^{220} A^{220}(\gamma) + R_{mi}^{222} A^{222}(\gamma) \quad (11)$$

In Appendix A of Ref. [10] it is illustrated a method to estimate the ε_i and R_{mi} expansion parameters from the values of bond (or diatomic molecule) polarizability. In the case we are showing here, the five limiting configurations (mutual orientations) of each i bond-bond pair that have been chosen, are: $H_i(\theta_{ai}=90°,\theta_{bi}=90°,\phi_i=0°)$, $X_i(\theta_{ai}=90°,\theta_{bi}=90°,\phi_i=90°)$, $T_{ai}(\theta_{ai}=90°,\theta_{bi}=0°,\phi_i=0°)$, $T_{bi}(\theta_{ai}=0°,\theta_{bi}=90°,\phi_i=0°)$, and $L_i(\theta_{ai}=0°,\theta_{bi}=0°,\phi_i=0°)$ [48], are considered for obtaining the ε_i and R_{mi} parameters (the method gives the same parameters for the X_i and H_i geometries and in the case of rigid molecules also for T_{ai} and T_{bi}). For the five selected geometries of each i-th bond-bond pair the coefficients $\varepsilon_i^{L_1L_2L}$ and $R_{mi}^{L_1L_2L}$ can be obtained by a simple inversion of Eqs. 10 and 11 [10]. The global function obtained in this way is a tentative full dimensional PES whose accuracy can be improved by a fine tuning of the ε_i and R_{mi} parameters by fitting experimental data and by comparing the model predictions with accurate ab initio electronic structure calculations (see Ref. [11] for details about comparisons and parameter optimization).

2.4   Dependence of the Intermolecular Energy on Monomer Deformations

Unlike the PESs available from the literature [49,50] reproducing the interaction of two rigid linear molecules, the parameters of the formulation illustrated in the above sections depend respectively on the C-O bond polarizability, say α, and on the monomer charge distribution (that determines the molecular multipole moments) components (see Appendix A of Ref. [10] and Eq. 7). Therefore the effect of the modification of the bond length r and of the monomer bending angle δ on the bond polarizability α, and on the positions and values of the point charges, localized along the CO on the O and C atom (see [11]), can be explicitly evaluated, and so the consequent changes in the value of the interaction parameters. For the C-O bond stretching and shrinking an empirical radial dependence of α on r (see the Appendix B of Ref.[10] for a similar problem in the case of H_2 and N_2) was worked out (see also detailed discussions in Appendix A of Ref. [11]). A multipole moment representation can be obtained for the charge distribution with the aid of ab initio calculations (see, e. g., Fig. 3 in Ref. [11]). To this end, the key point is to fit ab initio data using appropriate analytic functions to obtain the radial dependence of the point charges on the bond lengths r of each monomer. (This procedure is shown in detail in Appendix B of Ref. [11]).

The dependence of the intermolecular interactions on the bending of the monomer, has been incorporated into the PES modeling each monomer as an 'effective' linear molecule whose bond length r is shorter and whose total electronic charge, responsible for the molecular polarizability, is decreased along the molecular axis and increased perpendicularly to it with respect to the unbent molecule, averaging the interaction over oscillations and rotations around the main molecular axis, for each value of the monomer bending angle. This procedure has been detailed in Appendix C of Ref. [11], where a dependence on the bending angle has been included in both the bond polarizability α and the point charges.

3   Molecular Dynamics Simulations

The availability of a realistic intermolecular interaction potential energy surface allows the dynamical evaluation of the energy exchange occurring in molecular collisions between external (translation) and internal (rotation, vibration) degrees of freedom. The energy exchange is primarily due to the effects of the intermolecular interactions and is in general enhanced as the energy available to molecular collisions increases (like when warming up a system), although it also depends in different ways on the initial allocation of energy in the various modes, since, as seen in previous sections, physical properties such as polarizabilities and multipole moments can vary significantly with the vibrational and/or rotational excitation of the molecules. For the reasons mentioned in the above introductory sections, massive calculations on the systems CO_2 + CO_2 are performed using the quasi-classical trajectory (QCT) method. This allows us to determine state-to-state probabilities and cross sections for processes resulting in a energy exchange between vibrational degrees of freedom. For the CO_2 + CO_2 systems, this required to simulate the following non reactive collisions:

$$CO_2(v_{a1}, v_{a2}, v_{a3}) + CO_2(v_{b1}, v_{b2}, v_{b3}) \rightarrow CO_2(v'_{a1}, v'_{a2}, v'_{a3}) + CO_2(v'_{b1}, v'_{b2}, v'_{b3}) \tag{12}$$

where the $v_{a(b)i}$ ($i = 1, 2, 3$) are the quantum labels of a normal-mode model for symmetric stretching, bending and asymmetric stretching respectively, before (unprimed) and after (primed) the collision event involving the two (a and b) CO_2 molecules. CO_2 is linear in its equilibrium geometry, with degenerate (ground) bending states. The rotation around the O–C–O molecular axis may occur when the bending mode is excited and a quantum number for the total molecular angular momentum projection on the quantization axis z, usually denoted as l is needed. Nevertheless, the amount of energy associated with rotations around the molecular axis is neglected in our model treatment of CO_2 because it is in general small and, in any case, smaller than the statistical error of the QCT calculations (that amounts up to 5 %) [11].

In our model, separability between rotations and vibrations is allowed, and the CO_2 molecule is considered to be a linear rotor. However, the effects induced by the modification of the molecular shape occurring during the collision are properly taken into account by computing the true intermolecular potential V_{inter}, as described in Sec. 2.4.

The QCT calculations illustrated here have been performed using the VENUS96 program [51]. This program has been properly modified in order to incorporate the CO_2–CO_2 potential energy routine obtained according to the approach illustrated in section 2 and containing a potential energy function for the isolated molecules (V_{intra}) [52] and an intermolecular potential energy function V_{inter} (see Eq. 1). The code required additional modifications for the selection of the trajectory initial conditions.

The initial conditions of the batches of trajectories run for the study reported here were, in fact, selected as follows: the collision energy E was given a fixed value; the initial rotational angular momenta of the two molecules were selected through a random sampling of the Boltzmann distribution corresponding to the rotational temperature T_{rot} (that for our calculations was set equal to the translational one, as usually done for this type of massive computational campaigns); the initial vibrational states of the two molecules were defined by choosing two triples of integer numbers (one for each molecule) corresponding to the $v_{a(b)1}$, $v_{a(b)2}$, $v_{a(b)3}$ quantum numbers. Then initial coordinates and momenta for the relative motion were set by assigning a random value to the impact parameter b in the range $[0, b_{max}]$, where b_{max}, the maximum impact parameter,

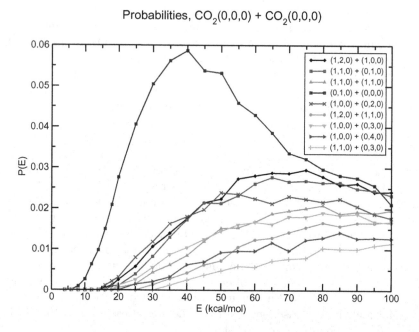

Fig. 1. Transition probabilities, of a series of low probability processes, as a function of the collision energy for CO_2+CO_2 collisions with the two molecules initially in their ground vibrational states, and with initial angular momenta randomly sampled from a Boltzmann distribution in the hypothesis of rotational temperature equal to the translational temperature

Probabilities, $CO_2(0,0,0)$ + $CO_2(0,0,0)$

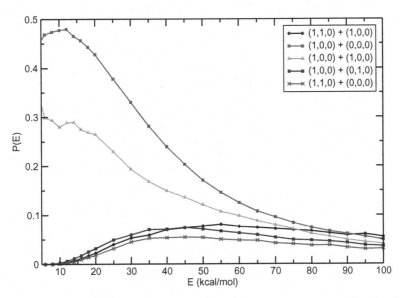

Fig. 2. Transition probabilities, of a series of high probability processes, as a function of the collision energy for CO_2+CO_2 collisions with the two molecules initially in their ground vibrational states, and with initial angular momenta randomly sampled from a Boltzmann distribution in the hypothesis of rotational temperature equal to the translational temperature

was taken as a truncation limit. The molecules were then randomly oriented, with the initial distances being large enough to make the interaction between them negligible and the rotation of each of them that of a linear rigid rotor, with no coupling between rotations and vibrations.

Figures 1 and 2 below, show vibrational transition probabilities as a function of the collision energy, for collisions of molecules initially in their ground vibrational state. Figure 1 shows the series of low probability transitions, whose collective behavior is that of a convergent trend, as the energy increases, to probability values ranging between 1 and 3 %. A somewhat anomalous behavior is that of the $(0,1,0)$ + $(0,0,0)$ transition probability, corresponding to the transfer of one energy quantum in the bending mode of one of the two molecules. The probability profile starts to increase quickly yet at a collision energy of 5 kcal/mol and and exhibits a well pronounced maximum around 40 kcal/mol. Figure 2 reports a series of similar plots for the largest probability transitions, showing a slightly different behavior. At low energies (say up to 15 kcal/mol) the symmetric stretching excitation is the only active process, and the transitions $(1,0,0)$ + $(0,0,0)$ and $(1,0,0)$ + $(1,0,0)$ dominates, with probabilities around 45 and 30 % respectively. At higher energies, probabilities for these processes decrease, while the probability of different excitation processes, involving also the

Table 1. Probabilities and cross sections for CO_2+CO_2 collisions at a collision energy $E= 65.0$ kcal/mol, initial angular momenta of the molecules sampled randomly from a Boltzmann distribution at a rotational temperature $T_{rot}=32500$ K. The initial vibrational state of CO_2 is the ground vibrational state (0,0,0) for both molecules

v'_{a1}	v'_{a2}	v'_{a3}	v'_{b1}	v'_{b2}	v'_{b3}	Prob.	Cross section (Å^2)
1	0	0	0	0	0	0.10755	147.68918
1	0	0	1	0	0	0.08809	121.89021
1	1	0	1	0	0	0.07424	105.09936
1	0	0	0	1	0	0.05932	81.49853
1	1	0	0	0	0	0.04761	65.18432
0	1	0	0	0	0	0.03854	52.54149
0	0	0	0	0	0	0.03310	45.82671
1	2	0	1	0	0	0.02873	39.98163
1	1	0	0	1	0	0.02766	37.31915
1	0	0	0	2	0	0.02123	28.89951
1	1	0	1	1	0	0.01876	26.79776
1	2	0	0	0	0	0.01682	23.52837
1	0	0	0	3	0	0.01612	22.80238
0	2	0	0	0	0	0.01418	20.15736
1	2	0	1	1	0	0.01278	18.16008
1	3	0	1	0	0	0.01307	17.54502
1	0	0	0	4	0	0.01163	16.71428
0	1	0	0	1	0	0.01204	16.32544

bending mode, start to increase. There is however a collective behavior at high energies, similar to that of lower probability transitions of Figure 1, with a trend convergent to a range of values between 4 and 6 %. A proper averaging of the probabilities over a sufficiently large set of impact parameters, leads to cross sections for the energy exchange processes occurring upon $CO_2 + CO_2$ collisions. As an example, the probabilities and the corresponding cross sections for the most relevant vibrational transitions upon collisions of two CO_2 molecules, initially at their ground vibrational states, for an energy of 65 kcal/mol, are shown in Table 1. The approach illustrated here is in principle applicable to even more complex systems (e.g. larger molecules or ions, see Refs [53,54]) to obtain cross sections, provided that a description of the interaction potential is given by means of the method illustrated in previous Section 2.

Cross sections, at a state-to-state level, are very basic quantities and can hardly ever be measured, having to be obtained by dynamics simulations of the collision processes. Rate constants can be obtained at any temperature from cross sections, provided they have been calculated over a wide range of collision energies, by averaging them over the appropriate energy distribution.

4 Conclusions

The aim of this paper was to illustrate the validity of a combined method for state-to-state cross section calculations, obtained using the well established bond-bond approach for the construction of intermolecular potential energy surfaces and the QCT method for the simulation of collisions. Due to the portability of the bond-bond method, based on the additivity of bond polarization components, and the versatility of the QCTs, this approach has been implemented as a Grid empowered simulator for massive calculations of cross sections and rate constants of use in modeling of atmospheres and gaseous flows. To exemplify the possible applications, the specific case of massive calculations of state-to-state vibrational energy transfer cross section for $CO_2 + CO_2$ collisions has been illustrated, a choice motivated by the importance of this process in Earth and planetary atmosphere modeling and in the kinetic studies of gaseous flows encountered in spacecraft reentry problem studies. Other processes involving carbon oxides are being currently studied with a similar approach and a particular care is being dedicated to the $CO + CO \rightarrow C + CO_2$ reaction. For the $CO_2 + CO_2$ collisions we have shown vibrational energy transfer probabilities over a wide range of collision energies, taking advantage of the dependence of the intermolecular interaction part on the internal geometry of the two molecules, explicitly included in the PES. Moreover, a list of all the most probable transitions for collisions at energy equal to 65 kca/mol has been shown as a sample of the huge amount of data that can be obtained by extensive use of the Grid simulator.

These results confirm that the possibility of obtaining realistic PESs and the computational efficiency of QCTs (compared to quantum calculations) makes it possible systematic studies of inelastic and reactive collision processes involving diatomic and triatomic molecules (the majority of those occurring in the atmospheres and in shock waves during reentry), properly implementing the approach as a Grid simulator. The resulting tools are then sources for the generation of cross section and rate constant databases for the kinetic modeling of atmospheres and gas flows.

Acknowledgments. The authors acknowledge financial support from MIUR PRIN 2008 (contract 2008KJX4SN_003), MIUR PRIN 2010-2011 (contract 2010ERFKXL_002), Phys4entry FP72007-2013 (contract 242311) and EGI Inspire. Thanks are also due to IGI and the COMPCHEM virtual organization and to CINECA for the allocated computing time.

References

1. Khalil, M.A., Rasmussen, R.A.: Nature 332, 242 (1988)
2. Palazzetti, F., Maciel, G.S., Lombardi, A., Grossi, G., Aquilanti, V.: J. Chin. Chem. Soc.-Taip. 59, 1045–1052 (2012)
3. http://www.astronomy.com/~/link.aspx?_id=9c5fef44-c7a0-4333-baeb-628add917d08
4. Project Phys4entry FP7242311, http://users.ba.cnr.it/imip/cscpal38phys4entry/activities.html
5. Capitelli, M., Ferreira, C.M., Gordiets, B.F., Osipov, R.: Plasma kinetics in atmospheric gases. Springer (2000)

6. Lombardi, A., Ragni, M., De Fernandes, I. F.: Proceedings - 12th International Conference on Computational Science and Its Applications, ICCSA 2012, art. no. 6257613 , pp. 77–82 (2012)
7. Kustova, E., Nagnibeda, E.: State-to-state theory of vibrational kinetics and dissociation in three-atomic gases. In: Bartel, T., Gallis, M. (eds.) Rarefied Gas
8. Hirschfelder, J.O.: Intermolecular Forces. Adv. Chem. Phys. 12 (1967)
9. Maitland, G.C., Rigby, M., Smith, E.B., Wakeham, W.A.: Intermolecular Forces. Clarendon Press, Oxford (1987); Dynamics, AIP Conference Proceedings, vol. 585, pp. 620–627 (2001)
10. Cappelletti, D., Pirani, F., Bussery-Honvault, B., Gomez, L., Bartolomei, M.: Phys. Chem. Chem. Phys. 10, 4281 (2008)
11. Bartolomei, M., Pirani, F., Laganà, A., Lombardi, A.: J. Comp. Chem. 33, 1806 (2012)
12. Barton, A.E., Chablo, A., Howard, B.J.: Chem. Phys. Lett. 60, 414 (1979); Phys. Chem. Chem. Phys. 100, 4281–4293 (2000)
13. Bruno, D., Catalfamo, C., Capitelli, M., Colonna, G., De Pascale, O., Diomede, P., Gorse, C., Laricchiuta, A., Longo, S., Giordano, D., Pirani, F.: Phys. Plasmas 17, 112315 (2010)
14. Albertí, M., Huarte-Larrañaga, F., Aguilar, A., Lucas, J.M., Pirani, F.: Phys. Chem. Chem. Phys. 13, 8251 (2011)
15. Lombardi, A., Lago, N.F., Laganà, A., Pirani, F., Falcinelli, S.: A bond-bond portable approach to intermolecular interactions: Simulations for N-methylacetamide and carbon dioxide dimers. In: Murgante, B., Gervasi, O., Misra, S., Nedjah, N., Rocha, A.M.A.C., Taniar, D., Apduhan, B.O. (eds.) ICCSA 2012, Part I. LNCS, vol. 7333, pp. 387–400. Springer, Heidelberg (2012)
16. Manuali, C., Rampino, S., Laganà, A.: Comp. Phys. Comm. 181, 1179 (2010); Manuali, C., Laganà, A.: Future Gen. Comp. Syst. 27, 315 (2011)
17. Laganá, A., Riganelli, A., Gervasi, O.: On the structuring of the computational chemistry virtual organization COMPCHEM. In: Gavrilova, M.L., Gervasi, O., Kumar, V., Tan, C.J.K., Taniar, D., Laganá, A., Mun, Y., Choo, H. (eds.) ICCSA 2006. LNCS, vol. 3980, pp. 665–674. Springer, Heidelberg (2006), http://www.compchem.unipg.it
18. Elango, M., Maciel, G.S., Lombardi, A., Cavalli, S., Aquilanti, V.: Int. J. Quantum Chem. 111, 1784–1791 (2011)
19. Elango, M., Maciel, G.S., Palazzetti, F., Lombardi, A., Aquilanti, V.: J. Phys. Chem. A. 114, 9864–9874 (2010)
20. Lombardi, A., Palazzetti, F., Maciel, G.S., Aquilanti, V., Sevryuk, M.B.: Int. J. Quantum Chem. 111, 1651 (2011)
21. Lombardi, A., Maciel, G.S., Palazzetti, F., Grossi, G., Aquilanti, V.: J. Vacuum Soc. Japan 53, 645 (2010)
22. Aquilanti, V., Grossi, G., Lombardi, A., Maciel, G.S., Palazzetti, F.: Phys. Scripta 78, 058119 (2008)
23. Barreto, P.R.P., Albernaz, A.F., Caspobianco, A., Palazzetti, F., Lombardi, A., Grossi, G., Aquilanti, V.: Comput. Theor. Chem. 990, 56–61 (2012)
24. Barreto, P.R.P., Albernaz, A.F., Palazzetti, F., Lombardi, A., Grossi, G., Aquilanti, V.: Phys. Scripta 84, 028111 (2011)
25. Aquilanti, V., Grossi, G., Lombardi, A., Maciel, G.S., Palazzetti, F.: Rendiconti Lincei 22, 125 (2011)
26. Palazzetti, F., Munusamy, E., Lombardi, A., Grossi, G., Aquilanti, V.: Int. J. Quantum Chem. 118, 318–332 (2011)

27. Barreto, P.R.P., Palazzetti, F., Grossi, G., Lombardi, A., Maciel, G.S., Vilela, A.F.A.: Int. J. Quantum Chem. 110, 777 (2010)
28. Aquilanti, V., Lombardi, A., Yurtsever, E.: Phys. Chem. Chem. Phys., 4, 5040–5051 (2002)
29. Sevryuk, M.B., Lombardi, A., Aquilanti, V.: Phys. Rev. A, 72, 033201 (2005)
30. Barreto, P.R.P., Vilela, A.F.A., Lombardi, A., Maciel, G.S., Palazzetti, F., Aquilanti, V.: J. Phys. Chem. A, 111, 12754 (2007)
31. Aquilanti, V., Lombardi, A., Sevryuk, M.B.: J. Chem. Phys. 121, 5579–5589 (2004)
32. Calvo, F., Gadéa, X., Lombardi, A., Aquilanti, V.: J. Chem. Phys. 125, 114307 (2006)
33. Pirani, F., Cappelletti, D., Liuti, G.: Chem. Phys. Lett. 350, 286 (2001)
34. Pirani, F., Albertí, M., Castro, A., Moix Teixidor, M., Cappelletti, D.: Chem. Phys. Lett. 37, 394 (2004)
35. Pirani, F., Brizi, S., Roncaratti, L., Casavecchia, P., Cappelletti, D., Vecchiocattivi, F.: Phys. Chem. Chem. Phys. 10, 5489 (2008)
36. Lombardi, A., Palazzetti, F.: Journal of Molecular Structure: THEOCHEM 852, 22 (2008)
37. Faginas Lago, N., Huarte-Larrañaga, F., Albertí, M.: Eur. Phys. J. D 55, 75 (2009)
38. Albertí, M., Faginas Lago, N., Pirani, F.: Chem. Phys. 399, 232 (2012)
39. Albertí, M., Faginas Lago, N.: J. Phys. Chem. A 116, 3094 (2012)
40. Albertí, M., Aguilar, A., Lucas, J.M., Pirani, F., Coletti, C., Re, N.: J. Phys. Chem. A 113, 14606 (2009)
41. Albertí, M., Faginas Lago, N.: European Phys. Journal D 67, 73 (2013)
42. Maciel, G.S., Barreto, P.R.P., Palazzetti, F., Lombardi, A., Aquilanti, V.: J. Chem. Phys. 129, 164302 (2008)
43. Ragni, M., Lombardi, A., Pereira Barreto, P.R., Peixoto Bitencourt, A.C.: J. Phys. Chem. A 113, 15355 (2009)
44. Pack, R.T.: Chem. Phys. Lett. 55, 197 (1978)
45. Candori, R., Pirani, F., Vecchiocattivi, F.: Chem. Phys. Lett. 102, 412 (1983)
46. Beneventi, L., Casavecchia, P., Volpi, G.G.: J. Chem. Phys. 85, 7011 (1986)
47. Beneventi, L., Casavecchia, P., Pirani, F., Vecchiocattivi, F., Volpi, G.G., Brocks, G., van der Avoird, A., Heijmen, B., Reuss, J.: J. Chem. Phys. 95, 195 (1991)
48. Gomez, L., Bussery-Honvault, B., Cauchy, T., Bartolomei, M., Cappelletti, D., Pirani, F.: Chem. Phys. Lett. 445, 99 (2007)
49. Oakley, M.T., Wheatley, R.J.: J. Chem. Phys. 130, 034110 (2009)
50. Bukowski, R., Sadlej, J., Jeziorski, B., Jankowski, P., Szalewicz, K., Kucharski, S.A., Williams, H.L., Rice, B.M.: J. Chem. Phys. 110, 3785 (1999)
51. Hase, W.L., Duchovic, R.J., Hu, X., Komornicki, A., Lim, K.F., Lu, D.-H., Peslherbe, G.H., Swamy, K.N., Vande Linde, S.R., Zhu, L., Varandas, A., Wang, H., Wolf, R.J.: J. Quantum Chemistry Program Exchange Bulletin 16, 671 (1996)
52. Carter, S., Murrell, J.N.: Croat. Chem. Acta 57, 355 (1984)
53. Faginas Lago, N., Albertí, M., Laganà, A., Lombardi, A.: Water $(H_2O)_m$ or Benzene $(C_6H_6)_n$ Aggregates to Solvate the K^+? In: Murgante, B., Misra, S., Carlini, M., Torre, C.M., Quang, N.H., Taniar, D., Apduhan, B.O., Gervasi, O. (eds.) ICCSA 2013, Part I. LNCS, vol. 7971, pp. 1–15. Springer, Heidelberg (2013)
54. Falcinelli, S., Rosi, M., Candori, P., Vecchiocattivi, F., Bartocci, A., Lombardi, A., Faginas Lago, N., Pirani, F.: Modeling the Intermolecular Interactionsand Characterization of the Dynamics of Collisional Autoionization Processes. In: Murgante, B., Misra, S., Carlini, M., Torre, C.M., Quang, N.H., Taniar, D., Apduhan, B.O., Gervasi, O. (eds.) ICCSA 2013, Part I. LNCS, vol. 7971, pp. 69–83. Springer, Heidelberg (2013)

Continuous and Discrete Algorithms in Quantum Chemistry: Polynomial Sets, Spin Networks and Sturmian Orbitals

Danilo Calderini[3], Cecilia Coletti[2], Gaia Grossi[1], and Vincenzo Aquilanti[1]

[1] Universitá di Perugia, via Elce di Sotto, 8 - 06183 Perugia, Italy
vincenzoaquilanti@yahoo.it, gaia.grossi@unipg.it
http://www.chm.unipg.it/chimgen/mb/theo2/home.html
[2] Dipartimento di Farmacia, Università G. d'Annunzio, Via dei Vestini,
66100 Chieti, Italy
ccoletti@unich.it
[3] Scuola Normale Superiore di Pisa, Piazza dei Cavalieri, 7 - 56126 Pisa, Italy
danilo.calderini@sns.it

Abstract. An effort is accounted for in the present paper to exhibit the recently actively investigated connection between the search and use of "orbitals" as basis sets in applied quantum mechanics and current advances in the mathematics of special functions and orthogonal polynomials, which are in turn motivated by the developments of the quantum theory of angular momentum. The latter theory in modern applications forms the basis for the class of "spin-network" algorithms. These "orbitals" enjoy important properties regarding orthogonality and completeness. In configuration space, they are often designated as Kepler-Coulomb Sturmian orbitals, in momentum space they are intimately connected with hyperspherical harmonics. The paper contains a brief presentation including also computational results and a discussion oriented towards the numerical use of these orbitals.

Keywords: Classical orthogonal polynomials, Kepler-Coulomb Sturmians, configuration space, hydrogenoid orbitals.

1 Introduction

Basic connections among the 6j symbols of angular momentum theory, both with the theory of superposition coefficients of hyperspherical harmonics and with the theory of discrete orthogonal polynomials, have been studied in [1, 2].

As we will see in the next section, expanding work in reference [3], we also sketched a connection between the Askey scheme of orthogonal polynomials and the tools of angular momentum theory (6j, 3j, rotation D-matrices,. . .); descending along the scheme corresponds in quantum mechanics to an itinerary towards the semiclassical limit, while ascending the ladder provides discretization algorithms for quantum mechanical calculations (for example, the hyperquantization algorithm). This work (Sec. 2) starts by updating the previous one

B. Murgante et al. (Eds.): ICCSA 2013, Part II, LNCS 7972, pp. 32–45, 2013.
© Springer-Verlag Berlin Heidelberg 2013

[3] by accounting on recent progress and then presents illustrations of exact and asymptotic formulae of relevance for applications. The main focus of the paper is on the role of orthogonal polynomials and special functions as orbitals in quantum chemistry and is tackled in Sec. 3. Concluding remarks follow in Sec. 4. This work extends recent work on polynomials [4], spin networks [5–7] and Sturmian orbitals [8].

2 Polynomials and Spin Networks: Beyond the Quantum Mechanical Jacobi Ladder

According to the book of Genesis, 28: 10, the biblical patriarch Jacob dreamed of "a ladder set up on earth, and the top of it reached to heaven". The standard presentation of orthogonal polynomial families (e. g. ref. [9], Chapter 22) ascends from the basic polynomials of Hermite, through those of Laguerre up to Jacobi's, the latter named after the great mathematician of the Nineteenth century. Together with their particular cases, these families form what we can call *Jacobi ladder*: as is well known, those polynomials where recognized at the birth of quantum mechanics as occurring in the solution of Schrödinger's equation, and are important tools as expansion basis sets for atomic, molecular and nuclear wavefunctions. Fig. 1 is a sketch of how the mathematicians of the late Twentieth Century extended the Jacobi ladder in different directions, remarkably by introducing polynomials of a discrete variable, or *discrete polynomials* for short. Interestingly, these mathematical developments were inspired by the quantum theory of angular momentum due to Wigner, Racah and others, and form the topics which include the powerful tools modernly known as "spin networks" [10, 11]: some time ago we elaborated such a parallelism [2], as illustrated in Fig. 2.

From a computational viewpoint, explicit expressions for the 6j coefficients can be written according to the series expressions of the Racah type, or as generalized hypergeometric series, or in connection with the Racah polynomials, and similarly for all functions appearing in Figs. 1 and 2. Orthogonal polynomials of a discrete variable are important tools of numerical analysis for the representation of functions on grids. As matrix elements of the overlap between spherical and hyperspherical harmonics [2, 12], they occur in quantum chemistry as momentum space (or Sturmian) orbitals, see following Section. We exploited both their connection with the coupling and recoupling coefficients of angular momentum theory [13] and their asymptotic relationships (semiclassical limit) [14, 15] to develop a discretization procedure, the hyperquantization algorithm, applied to the study of anisotropic interactions and of reactive scattering as a quantum mechanical n-body problem [16]. Use for fitting of potential energy surfaces has been also proposed [17] and further applications have been made to stereodirectional dynamics of chemical processes via an exact representation for scattering matrix [18], as well as to the characterization of atomic and molecular polarizations [19].

Of great relevance from the viewpoint of quantum chemistry and of applications in atomic and molecular science are the dual sets of Laguerre and Charlier

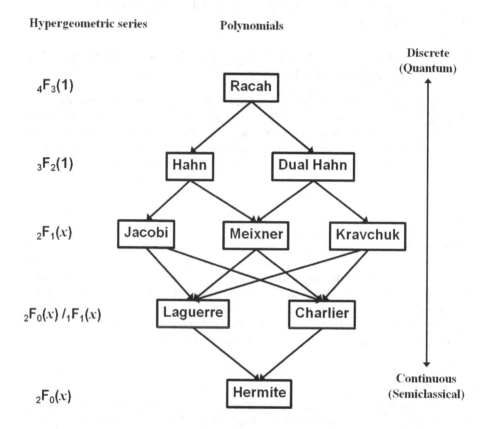

Fig. 1. Classification of orthogonal hypergeometric polynomials (a partial view of schemes in [1] and [20]). The first column lists the correspondence with the generalized hypergeometric series. Polynomials at the same level are identical, but the role of the three-term recurrence relation and of the finite difference equation varies - Hahn and dual Hahn differ in the role played by the "degree" and the "variable" parameters, which correspond to the dual role that angular momentum and projection quantum numbers play in vector coupling coefficients. Similarly for the Jacobi polynomials, where the Meixner and Kravchuk polynomials are the "discrete" counterparts, and the Charlier polynomials are the "discrete" dual of Laguerre polynomials. For the Hermite polynomials, a "duality" connects the continuous variable and the discrete polynomial order: in their occurrence as wavefunctions of the quantum harmonic oscillator (Fig. 2), the duality is between elongation and vibrational level of the oscillator [21].

Discrete and Continuous Wavefunctions and Spin Networks

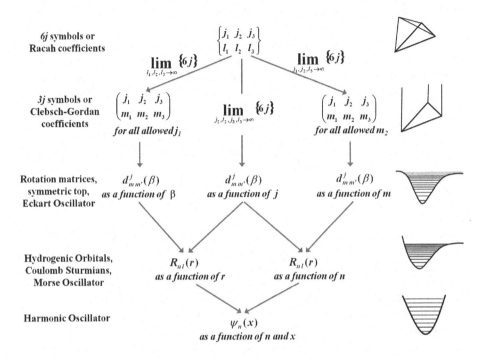

Fig. 2. Occurrence of continuous and discrete polynomials in quantum mechanics. The first two downward connections are limiting relationships which physically correspond to semiclassical limits, while the remaining ones are progressive confinements in space due to the form of the potential. Earlier versions of this scheme were given in [2, 3], which also provide explicit asymptotic relationships corresponding to the arrows. Recent advances are reported in refs. [5–7], where the main properties are illustrated in detail.

polynomials, which describe the configuration space radial wavefunctions of the Kepler-Coulomb problem. (For the classification of hydrogenoid wavefunctions in momentum space see ref. [22]: they involve D matrices as harmonics on the S_3 hypersphere). The Laguerre and Charlier polynomials occur in the Sturmian orbitals under focus in the next Section. Their name was derived from their being solutions of a Sturm-Liouville problem, involving a second order differential equation of the Schrödinger type. All *continuous* functions in Figs.1 and 2 enjoy the same property, and provide orthonormal complete basis sets.

It is remarkable that a modern viewpoint looks at the *discrete* functions of Figs. 1 and 2 as orthonormal complete basis sets, which are solutions of three-term difference equations, considered as discrete Sturm-Liouville problems, and obtained by diagonalization of Jacobi (again!) matrices [23] (a Jacobi matrix is real, symmetric and tridiagonal).

3 Kepler-Coulomb Sturmian Orbitals

As a theory of electronic structure of atoms and molecules, quantum chemistry uses basis sets for the solution of Schrödinger equation. Because the Schrödinger equation for the non-relativistic hydrogen atom is analytically soluble in a closed form, the use of hydrogen-like orbitals as expansion basis sets to calculate the electronic wavefunctions of atoms and molecules has been extensively considered. In order to tackle the obstacles encountered for the efficient computation of matrix elements involved in the calculations, preference has been given over the years to Gaussian orbitals, more convenient for such purposes.

However, recent progresses concerning the use of Slater orbitals [24–28] have revived the interest in the more *natural* [29] (from both the physical and mathematical viewpoint) exponential type orbitals. Indeed, the use of a procedure based on Gaussian orbitals suffers of several disadvantages, among which the calculation of long-range interactions and of the properties of the excited states, and the slow convergence in reproducing the cusps in the electronic densities at nuclei.

Among exponential type orbitals, Sturmian basis sets have been considered by several authors [30–32] both in their generalized form [30, 33–35] and as Coulomb Sturmians [36–38, 31] for applications to atomic and molecular structure.

There is a close relationship between Sturmians and Slater type orbitals (see for instance [39]) and the former can be expressed as a simple linear combination of the latter, so that progresses in the computation of integrals over Slater type orbitals can also be exploited for Sturmian basis sets.

The full strength of the approach based on Kepler-Coulomb Sturmians comes from the fact that their counterparts in momentum space are hyperspherical harmonics and their overlaps or superposition integrals can be easily reduced to sums over discrete analogues of classical orthogonal polynomials [1], which, in turn, can be related to generalizations of vector coupling and recoupling coefficients ($3j$, $6j$ or $9j$ coefficients) [2]. As depicted in [2], the radial part of Coulomb Sturmians [2, 3] appears as a step in the ladder of Askey scheme [20, 21] of orthogonal polynomials of Figure 2.

All other functions involving angles or hyperangles can be analytically handled within this scheme, including limits (downward arrows), explicit matrix elements for integrals, etc. The reader is invited to see [21].

The quantum mechanics of multielectron atoms and molecules can indeed be discussed in terms of the breaking of the hyperspherical symmetry of a d-dimensional hydrogenic atom with $d=3(N-1)$ for N body Coulomb problems, due to the introduction of further charged particles (electrons and/or nuclei) [31]. In this case, the generalization of Fock's treatment [40] to spaces of mathematical dimensions higher than the physical one allows to study atomic and molecular structure from the point of view of the broken symmetry of hyperspheres. In configuration space, Sturmian basis functions can thus be used as expansion basis sets to build up atomic and molecular orbitals.

This can be done by using the hyperspherical approach, where the motion of an N body system is reconducted to that of a single particle with reduced mass

$M = (\prod m_i / \sum m_i)^{1/(N-1)}$ in a d-dimensional space. In this space the position of the particle is given by an hyperradius ρ and a set of (d-1) hyperangles, collectively denoted as ω.

This method has been long known [41–44] and successfully applied for the treatment of interactions among few particles (in particular for problems involving nuclear dynamics [45, 46]). The formulation described in ref. [31], which will be sketched here, has the advantage of fully exploiting the benefits one can obtain from the use of Coulomb Sturmian basis sets. In particular, (i) The secular equation can be very compactly formulated. All integrals can be written in closed form as matrix elements corresponding to coupling, recoupling or transformation coefficients of hyperangular momenta algebra. Note that this is feasible also for the so called 'radial' integrals over Laguerre polynomials (see eq. (7) in the following). As a consequence, the structure of the relevant matrices, and in particular of their zeroes, can be foreseen and exploited, sparseness being crucial in applications; (ii) One can resort to alternative basis sets, with different symmetry properties, to accelerate convergence.

The method is general and thus can be applied to any N-body Coulomb problem, i.e. to whatsoever mass ratios: The central symmetry of the hydrogen atom can be broken either by introducing electrons, leading to a multielectron atomic system, or by introducing further nuclei, leading to a molecular multi-center problem, to be treated without the Born-Oppenheimer approximation. Applications so far have concerned three-body problems, considering in particular the bielectronic series and the H_2^+ molecular ion [31].

In the general case one can write the Schrödinger equation for a system of N particles interacting through Coulomb forces as (atomic units will be implied throughout):

$$\left[\frac{1}{r^{d-1}} \frac{\partial}{\partial r} r^{d-1} \frac{\partial}{\partial r} - \frac{\Delta(\Omega_{d-1})^2}{r^2} + 2V - 2E \right] \Psi(\mathbf{r}) = 0. \tag{1}$$

$\Delta(\Omega_{d-1})$ is the Beltrami-Laplace operator on the d-dimensional sphere, with eigenvalues $\lambda(\lambda + d - 2)$. Thus, λ here plays the role of the grand orbital angular momentum quantum number and σ collectively represents the set of d-2 projections of λ (i.e. the quantum numbers σ are eigenvalues of the rotation operators of the subgroups related to the chosen chain reduction). The Coulomb potential in hyperspherical coordinates takes the simple factorized form:

$$V = -\frac{Z(\Omega)}{r} \tag{2}$$

where $Z(\Omega)$ plays the role of an anisotropic charge. Owing to this term equation (1) is not separable: if $Z(\Omega)$ were a constant equation (1) would coincide with the Schrödinger equation of the multidimensional hydrogen atom, whose solutions can be obtained exactly as d-dimensional Sturmian basis functions [47]. Therefore, in configuration space the many-body Coulomb problem is isomorphic to that of a multidimensional hydrogen atom with an anisotropic charge.

For this reason the use of d-dimensional Sturmian basis sets to expand multi-electronic orbitals $\Psi(\mathbf{r})$ seems appropriate:

$$\Psi(\mathbf{r}) = \sum_{n\lambda\sigma} c_{n\lambda\sigma} u_{n\lambda\sigma}(\mathbf{r}). \tag{3}$$

where $u_{n\lambda\sigma}(\mathbf{r})$ are polar Sturmians [47].

If the expansion (3) is inserted into equation (1), after some manipulations, one gets the following secular equation:

$$|\mathbf{A} - p_o\mathbf{I}|\mathbf{c} = 0 \tag{4}$$

where the eigenvalue is the momentum p_0, as appropriate for Sturmian basis sets. The energy spectrum can be recovered by using the relation $E = -p_0^2/2$. In eq. (4) $\mathbf{I}$ is the unit matrix, $\mathbf{c}$ is the eigenvector of the expansion coefficients and $\mathbf{A}$, the matrix to be diagonalized, is written in terms of radial and angular integrals:

$$A_{nn'\lambda\lambda'mm'} = R_{nn'\lambda\lambda'}\Omega_{\lambda\lambda'mm'} \tag{5}$$

where $\Omega_{\lambda\lambda'mm'}$ is the integral over the ω angular variables

$$\Omega_{\lambda'\sigma'\lambda\sigma} = \int Y^*_{\lambda'\sigma'}(\omega)Z(\Omega)Y_{\lambda\sigma}(\omega)\mathrm{d}\omega \tag{6}$$

and $R_{nn'\lambda\lambda'}$ is the radial integral

$$R_{n\lambda n'\lambda'} = 2\sqrt{\frac{(n'-\lambda'-1)!(n-\lambda-1)!}{(2n+d-3)(2n'+d-3)[(n'+\lambda'+d-3)!(n+\lambda+d-3)!]^3}} \times \tag{7}$$

$$\times \int \exp(2p_0r)(2p_0r)^{\lambda+\lambda'+d-2} L^{2\lambda'+d-2}_{n'-\lambda'-1}(2p_0r) L^{2\lambda+d-2}_{n-\lambda-1}(2p_0r)\mathrm{d}(2p_0r).$$

The result of the integration (6) depends on the selected parametrization for the angular variables and on the particular Coulomb system under analysis. In any case, it can always be expressed in terms of vector coupling and recoupling coefficients of hyperangular momenta, because equation (6) can be handled to give an integral over three hyperspherical harmonics.

The explicit expression of the radial integral $R_{n\lambda n'\lambda'}$ depends only on the number of particles (i.e. on the dimension of configuration space) and not on the nature of the particles. Furthermore, $R_{n\lambda n'\lambda'}$ can be written as a linear combination of vector coupling coefficients [31]. In the case of the three-body problem, with d=6, its explicit expression is:

$$R_{nn'\lambda\lambda'} = \frac{1}{(n'+\lambda+3)^{\frac{1}{2}}}\left[(n+\lambda+3)^{\frac{1}{2}}F + (n-\lambda)^{\frac{1}{2}}B\right] \tag{8}$$

with

$$F = \langle \frac{n' + \lambda' + 2}{2}, \frac{-n + \lambda' + 2\lambda + 4}{2}, \frac{n' - n + \lambda' - \lambda}{2}, \frac{n' - n + \lambda' - \lambda}{2} | \frac{n' + \lambda + 2}{2}, \frac{-n' + 2\lambda' + \lambda + 4}{2} \rangle$$

$$B = \langle \frac{n' + \lambda' + 3}{2}, \frac{\lambda - n' + 2\lambda + 3}{2}, \frac{n' - n + \lambda' - \lambda - 1}{2}, \frac{n' - n + \lambda' - \lambda + 1}{2} | \frac{n' + \lambda + 2}{2}, \frac{2\lambda' - n' + \lambda + 4}{2} \rangle$$

where F and B are Clebsch-Gordan coefficients, so $R_{nn'\lambda\lambda'}$ for a three-body Coulomb problem depends on the involved quantum numbers only and not on the nature of the system. Even if this result might seem surprising from a configuration space point of view, if we look at the problem from the momentum space perspective the whole matrix $A_{nn'\lambda\lambda'mm'}$ can be in fact obtained from the integration over angular variables only, parametrizing the S_6 hypersphere embedded in a seven-dimensional Euclidean space. Thus the quantum numbers n and λ label the eigenvalues of the generators of the seven-dimensional hypersphere S_6 and of a six-dimensional subspace, respectively. Under this perspective, the possibility of exploiting different parametrizations of S_6, whose symmetry properties are closer to those of the system, represents an extremely attractive feature.

Being essentially a continuation of an extensive review presented a dozen years ago, we dedicate next Section to assess progress in straightforward computational schemes, indicating the need of "alternative", or "non canonical" strategies within the same mathematical framework.

4   Numerical Convergence

It is relevant to document the convergence of the direct procedure by updating the results [31] exploiting the presently available computational power. The system used as benchmark is the helium atom, studied at a non relativistic level of theory by using the symmetrical parametrization [48, 38], which proved to give better convergence upon truncation. Such a parametrization is labeled by three quantum numbers: n the hyperangular quantum number (analogue to the principal quantum number for the hydrogen atom); λ, whose limiting value is $\lambda \leq n - 1$ and which, in the case of zero total angular momentum, is restricted to even values; and μ, which has the following range $0 \leq \mu \leq \lambda$ and must have the same parity as $\lambda/2$, i.e. it changes as $\mu = \lambda, \lambda - 4, \ldots, 0$ (or 2 according to the parity of λ).

In this Section, we also present some new results concerning the energy of the first excited states of He for zero total angular momentum (singlet states). Indeed, it is worth pointing out that the method has a variational behavior and is effective not only for the ground state but also for the excited states, unlike most of the popular variational methods.

Here, we use as benchmark -2.903724 Hartree for the ground state [49], and -2.1458 Hartree and -2.0611 Hartree for the first and second singlet excited states, $1s2s$ and $1s3s$, respectively [50]. As mentioned before, the matrix is symmetrical and, in principle, infinite: using the proper truncation of the wavefunction expansion we obtain a finite dimensional matrix that can be diagonalized with standard algorithms. Different truncation schemes were tested: We started by using a closed shell truncation, that is by truncating on the quantum number n, considering all the allowed states in the shell (Table 1). Such a truncation scheme clearly leads to identical results for all the possible parametrizations that can be employed (e.g. the asymmetrical one or others [31]). The convergence is very slow for both ground and excited states, with better numerical results for the ground state (differences from the benchmark data are reported as Δ in the Tables).

Table 1. Closed Shell Truncation

electronic state	n_{max}	dim	E_{calc}(Hartree)	Δ(Hartree)
ground	80	11900	-2.8937	0.0100
ground	90	16744	-2.8952	0.0085
1s2s	80	11900	-2.0797	0.0661
1s2s	90	16744	-2.0882	0.0576
1s3s	80	11900	-1.9063	0.1548
1s3s	90	11900	-1.9233	0.1378

The second truncation scheme corresponds to a cut upon the λ quantum number and include all the allowed values of μ (Table 2). In [31] it was found that cutting upon appropriately chosen λ values could improve the convergence for the ground state energy. Indeed, also in the present calculations convergence gets much better for the ground state, however for the excited states it is worse than in the closed shell approach, even if the basis set dimension is quite larger. This suggests, not unsurprisingly, that larger λ (or μ) states are needed for a correct description of the excited states energy with respect to that of the ground state.

Table 2. Truncation on λ

electronic state	n_{max},λ_{max}	dim	E_{calc}(Hartree)	Δ(Hartree)
ground	280, 32	20976	-2.9001	0.0036
ground	300, 40	33110	-2.9015	0.0022
ground	350, 40	39160	-2.9016	0.0021
1s2s	280, 32	20976	-2.0692	0.0766
1s2s	300, 40	33110	-2.0934	0.0524
1s2s	350, 40	39160	-2.0934	0.0524
1s3s	280, 32	20976	-1.8262	0.2349
1s3s	300, 40	33110	-1.8820	0.1791
1s3s	350, 40	39160	-1.8820	0.1791

Although it is possible to cut upon both λ and μ, here we decided to truncate on the μ quantum number only, mostly to understand which quantum number is determinant to reach a faster convergence on the excited states (Table 3). The results of ref. [31] indicated that the poor convergence for the ground state energy obtained by cutting too drastically on λ was in fact due to the subsequent elimination of higher μ states.

Table 3. Truncation on μ

electronic state	$n_{max}, \lambda_{max}, \mu_{max}$	dim	E_{calc}(Hartree)	Δ(Hartree)
ground	100, 98, 44	19292	-2.8963	0.0074
ground	100, 98, 52	20592	-2.8963	0.0074
1s2s	100, 98, 44	19292	-2.0945	0.0513
1s2s	100, 98, 52	20592	-2.0949	0.0509
1s3s	100, 98, 44	20592	-1.9296	0.1315
1s3s	100, 98, 52	39160	-1.9356	0.1255

Results of Table 3 show that, although very large μ values are not needed to improve convergence for the ground state energy (results published on [31] indicated that the inclusion of states with μ_{max} larger than 24 were not necessary to reach convergence, at least with $n \sim 100$), high values of μ are mandatory to accelerate convergence for excited state energies. This fact also explains the poor performances obtained through the λ truncation scheme. Further calculations are currently being carried out to improve convergence on the excited states.

In any case, it is clear that alternative ways are to be found: the procedure is mathematically sound and physically motivated, but the progress towards the description of the simplest systems is so slow that the search for alternative ways is the most important direction motivating the further study.

Alternative Sturmian bases (mostly in configuration space) have been constructed, often as the product of one-electron Sturmians and applied for extensive calculations of the energy spectrum of atoms and ions [51], also in strong external fields [52, 53]. Convergence of these sets is rapid, although the spectrum of the generating eigenproblem is mixed, containing also, apart from discrete eigenvalues, a continuum part: therefore there is no guarantee that every well-behaved function of the same variables obeying the same boundary conditions can be expanded in terms of this basis [54].

Remarkable performances for atomic bound states have been reached through the introduction of *optimal* Sturmian basis functions, possessing the correct physical behavior at short and long distances from the nucleus [33].

5   Concluding Remarks

In general, the construction of families of continuous and discrete polynomials can be based, according to Favard theorem, on three-term recurrence relationships, which can be interpreted as finite difference equations. The actual construction is thus equivalent to the diagonalization of a Jacobi matrix (see Section 2). This is the standard procedure for the Wigner-Racah coefficients of quantum angular momentum and for the Racah polynomials considered in [2, 3, 5–7] and references therein, but also for their continuous extensions and their q-analogues in the Askey (or Nikiforov) schemes [1, 20]. The unique property of these families is that, belonging to the so-called hypergeometric or classical type, such diagonalization can be performed in closed form.

Parallel and future work involves cases where the diagonalization cannot be performed explicitly, yet mathematically interesting polynomial sets can be characterized, and the motivation for their investigation comes from physical problems. We refer the reader to cases arising from the separation of the hydrogen wavefunction in elliptic coordinates [55–57, 32, 58] and from the study of the spectrum of the quantum mechanical volume operator [4].

Regarding Sturmian approaches, the momentum space perspective, not considered here but amply cited in the previous referenced literature, is also used to shed light on the nature and symmetry properties of Coulomb Sturmians and to obtain important mathematical connections (in terms of angular momentum algebra) between these sets and their momentum space counterparts, which involve hyperspherical harmonics. Ref. [8] contains both an updated review and a sketch of future perspectives.

References

1. Nikiforov, A., Suslov, S., Uvarov, V.: Classical Orthogonal Polynomials of a Discrete Variable. Springer, Berlin (1991)
2. Aquilanti, V., Cavalli, S., Coletti, C.: Angular and hyperangular momentum recoupling, harmonic superposition and Racah polynomials: a recursive algorithm. Chem. Phys. Lett. 344, 587 (2001)
3. Ragni, M., Bitencourt, A.C.P., da S. Ferreira, C., Aquilanti, V., Anderson, R.W., Littlejohn, R.G.: Exact computation and asymptotic approximations of 6j symbols: Illustration of their semiclassical limits. Int. J. Quantum Chem. 110, 731 (2010)
4. Aquilanti, V., Marinelli, D., Marzuoli, A.: Hamiltonian dynamics of a quantum of space: hidden symmetries and spectrum of the volume operator, and discrete orthogonal polynomials. J. Phys. A: Math. Theor. 46, 175303 (2013)
5. Bitencourt, A.C.P., Marzuoli, A., Ragni, M., Anderson, R.W., Aquilanti, V.: Exact and Asymptotic Computations of Elementary Spin Networks: Classification of the Quantum–Classical Boundaries. In: Murgante, B., Gervasi, O., Misra, S., Nedjah, N., Rocha, A.M.A.C., Taniar, D., Apduhan, B.O. (eds.) ICCSA 2012, Part I. LNCS, vol. 7333, pp. 723–737. Springer, Heidelberg (2012)
6. Anderson, R.W., Aquilanti, V., Bitencourt, A.C.P., Marinelli, D., Ragni, M.: The screen representation of spin networks: 2D recurrence, eigenvalue equation for 6j symbols, geometric interpretation and hamiltonian dynamics. In: Murgante, B., Misra, S., Carlini, M., Torre, C.M., Quang, N.H., Taniar, D., Apduhan, B.O., Gervasi, O. (eds.) ICCSA 2013, Part II. LNCS, vol. 7972, pp. 46–59. Springer, Heidelberg (2013)

7. Ragni, M., Littlejohn, R.G., Bitencourt, A.C.P., Aquilanti, V., Anderson, R.W.: The Screen representation of spin networks. Images of 6j symbols and semiclassical features. In: Murgante, B., Misra, S., Carlini, M., Torre, C.M., Quang, N.H., Taniar, D., Apduhan, B.O., Gervasi, O. (eds.) ICCSA 2013, Part II. LNCS, vol. 7972, pp. 60–72. Springer, Heidelberg (2013)
8. Coletti, C., Calderini, D., Aquilanti, V.: d-dimensional Kepler-Coulomb Sturmians and Hyperspherical Harmonics as Complete Orthonormal Atomic and Molecular Orbitals. Adv. Quantum Chem. (2013) (in press)
9. Abramowitz, M., Stegun, I.A.: Handbook of Mathematical Function. Dover, New York (1964)
10. Aquilanti, V., Bitencourt, A.C.P., da S. Ferreira, C., Marzuoli, A., Ragni, M.: Quantum and semiclassical spin networks: from atomic and molecular physics to quantum computing and gravity. Physica Scripta 78, 058103 (2008)
11. Aquilanti, V., Bitencourt, A.C.P., da S. Ferreira, C., Marzuoli, A., Ragni, M.: Combinatorics of angular momentum recoupling theory: spin networks, their asymptotics and applications. Theor. Chem, Acc. 123, 237–247 (2009)
12. Aquilanti, V., Coletti, C.: 3nj-symbols and harmonic superposition coefficients: an icosahedral abacus. Chem. Phys. Lett. 344, 601 (2001)
13. De Fazio, D., Cavalli, S., Aquilanti, V.: Orthogonal Polynomials of a discrete variable as expansion basis sets in quantum mechanics. The hyperquantization algorithm. Int. J. Quantum Chem. 93, 91–111 (2003)
14. Aquilanti, V., Haggard, H.M., Littlejohn, R.G., Yu, L.: Semiclassical analysis of Wigner 3j-symbol. J. Phys. A 40, 5637–5674 (2007)
15. Aquilanti, V., Haggard, H.M., Hedeman, A., Jeevanjee, N., Littlejohn, R.G., Yu, L.: Semiclassical mechanics of the Wigner 6J-symbol. Journal of Physics A 45, 065209 (2012)
16. De Fazio, D., Aquilanti, V., Cavalli, S., Aguilar, A., Lucas, J.M.: Exact state-to-state quantum dynamics of the F + HD → HF + D reaction on model potential energy surfaces. J. Chem. Phys. 129, 064303 (2008)
17. Aquilanti, V., Capecchi, G.: Harmonic analysis and discrete polynomials. From semiclassical angular momentum theory to the hyperquantization algorithm. Theor. Chem. Acc. 104, 183 (2000)
18. Aquilanti, V., Cavalli, S., Grossi, G., Anderson, R.W.: Stereodirected States in Molecular Dynamics: A Discrete Basis Representation for the Quantum Mechanical Scattering Matrix. J. Phys. Chem. 95, 8184–8193 (1991)
19. Anderson, R.W., Aquilanti, V.: The discrete representation correspondence between quantum and classical spatial distributions of angular momentum vectors. J. Chem. Phys. 124, 214104 (2006)
20. Koekoek, R., Swarttouw, R.: The Askey-scheme of hypergeometric orthogonal polynomials and its q-analogue, TU Delft, The Netherlands (1998), Anonymous ftpsite:ftp.twi.tudelft.nl, directory:/pub/publications/tech-reports
21. Askey, R.: Duality for classical orthogonal polynomials. J. Comp. App. Math. 178, 37–43 (2005)
22. Aquilanti, V., Cavalli, S., Coletti, C.: Hyperspherical Symmetry of Hydrogenic Orbitals and Recoupling Coefficients among Alternative Bases. Phys. Rev. Lett. 80, 3209 (1998)
23. Garcia, A.G., Hernandez-Medina, M.A.: Discrete Sturm–Liouville problems, Jacobi matrices and Lagrange interpolation series. J. Math. Anal. Appl. 280, 221–231 (2003)
24. Safouhi, H., Hoggan, P.E.: New methods for accelerating the convergence of molecular electronic integrals over exponential type orbitals. Mol. Phys. 101, 19 (2003)

25. Hoggan, P.E.: General two-electron exponential type orbital integrals in poly-atomics without orbital translations. Int. J. Quantum Chem. 109, 2926 (2009)
26. Fernández Rico, J., López, R., Aguado, A., Ema, I., Ramírez, G.: Reference program for molecular calculations with Slater-type orbitals. J. Comp. Chem. 19, 1284 (1998)
27. Fernández Rico, J., López, R., Ema, I., Ramírez, G.: Electrostatic potentials and fields from density expansions of deformed atoms in molecules. J. Comp. Chem. 25, 1347 (2004)
28. Ruiz, M.B., Peuker, K.: Analytical expressions of exchange and three-center nuclear attraction integrals over 1s and 2s Slater orbitals with different exponents. Chem. Phys. Lett. 412, 244 (2005)
29. Shull, H., Löwdin, P.O.: Superposition of Configurations and Natural Spin Orbitals. Applications to the He Problem. J. Chem. Phys. 30, 617 (1959)
30. Avery, J.: Hyperspherical Harmonics and Generalized Sturmians. Kluwer Academic Publishers, Dordrecht (2000)
31. Aquilanti, V., Cavalli, S., Coletti, C., Di Domenico, D., Grossi, G.: Hyperspherical harmonics as Sturmian orbitals in momentum space: a systematic approach to the few-body Coulomb problem. Int. Rev. Phys. Chem. 20, 673 (2001)
32. Aquilanti, V., Caligiana, A., Cavalli, S., Coletti, C.: Hydrogenic Orbitals in Momentum Space and Hyperspherical Harmonics. Elliptic Sturmian Basis Sets. Int. J. Quantum Chem. 92, 212 (2003)
33. Randazzo, J.M., Ancarani, L.U., Gasaneo, G., Frappicini, A.L., Colavecchia, F.D.: Optimal Sturmian basis functions for atomic three-body systems. Phys. Rev. A 81, 042520 (2010)
34. Randazzo, J.M., Frappicini, A.L., Colavecchia, F.D., Gasaneo: Discrete sets of Sturmian functions applied to two-electron atoms. Phys. Rev. A 79, 022507 (2009)
35. Mitnik, D.M., Colavecchia, F.D., Gasaneo, G., Randazzo, J.M.: Computational methods for Generalized Sturmans basis. Comp. Phys. Comm. 182, 1145 (2011)
36. Avery, J., Avery, J.: Generalized Sturmian Solutions for Many-Particle Schrödinger Equations. J. Phys. Chem. A 108, 8848 (2004)
37. Avery, J., Avery, J.: Coulomb Sturmians as a basis for molecular calculations. Mol. Phys. 110, 1593 (2012)
38. Aquilanti, V., Cavalli, S., Coletti, C., Grossi, G.: Alternative Sturmian Bases and Momentum Space Orbitals: an Application to the Hydrogen Molecular Ion. Chem. Phys. 209, 405 (1996)
39. Calderini, D., Cavalli, S., Coletti, C., Grossi, G., Aquilanti, V.: Hydrogenoid orbitals revisited: From Slater orbitals to Coulomb Sturmians. J. Chem. Sci. 124, 187 (2012)
40. Fock, V.: Zur Theorie des Wasserstoffatoms. Z. Phys. 98, 145 (1935)
41. Macek, J.: Properties of autoionizing states of He. J. Phys. B: Atom. Molec. Phys. 1, 831 (1968)
42. Smith, F.T.: Generalized Angular Momentum in Many-Body Collisions. Phys. Rev. 120, 1058 (1960)
43. Smith, F.T.: A Symmetric Representation for Three Body Problems. I. Motion in a Plane. J. Math. Phys. 3, 735 (1962)
44. Ballot, L., Fabre de la Ripelle, M.: Application of the hyperspherical formalism to the trinucleon bound state problems. Ann. Phys. 127, 62 (1980)
45. Aquilanti, V., Cavalli, S., Coletti, C., De Fazio, D., Grossi, G.: In: Tsipis, C.A., Popov, V.S., Herschbach, D.R. (eds.) New Methods in Quantum Theory, pp. 233–250. Kluwer (1996)

46. Aquilanti, V., Cavalli, S., De Fazio, D.: Hyperquantization algorithm. II. Implementation for the F+H2 reaction dynamics including open-shell and spin-orbit interactions. J. Chem. Phys. 109, 3805 (1998)
47. Aquilanti, V., Cavalli, S., Coletti, C.: The d-dimensional hydrogen atom: hyperspherical harmonics as momentum space orbitals and alternative Sturmian basis sets. Chem. Phys. 214, 1 (1997)
48. Aquilanti, V., Cavalli, S., Grossi, G.: Hyperspherical coordinates for molecular dynamics by the method of trees and the mapping of potential energy surfaces for triatomic systems. J. Chem. Phys. 85, 1362 (1986)
49. Drake, G.W.F.: Atomic, Molecular and Optical Physics Handbook. AIP, New York (1996)
50. Morton, D.C., Wu, Q., Drake, G.W.F.: Energy Levels for the Stable Isotopes of Atomic Helium (^{4}He I and ^{3}He I). Can. J. Phys. 84, 83–105 (2006)
51. Avery, J., Avery, J.: The Generalized Sturmian Method for Calculating Spectra of Atoms and Ions. J. Math. Chem. 33, 145 (2003)
52. Avery, J., Coletti, C.: Generalized Sturmians applied to atoms in strong external fields. J. Math. Chem. 27, 43 (2000)
53. Avery, J., Coletti, C.: New Trends in Quantum Systems in Chemistry and Physics, vol. I, p. 77. Kluwer, Dordrecht (Maruani, J., et al., eds.)
54. Szmytkowski, R.: The Dirac-Coulomb Sturmian functions in the $Z = 0$ limit: properties and applications to series expansions of the Dirac Green function and the Dirac plane wave. J. Phys. A 33, 427 (2000)
55. Patera, J., Winternitz, P.: A new basis for the representations of the rotation group. Lamé and Heun polynomials. J. Math. Phys. 14, 1130 (1973)
56. Kalnins, E.G., Miller, W., Winternitz, P.: The Group $O(4)$, Separation of Variables and the Hydrogen Atom. S.I.A.M. J. Appl. Math. 30, 630 (1976)
57. Miller, W.: Symmetry and Separation of Variables. Addison-Wesley Publishing Company, Reading Mass (1977)
58. Aquilanti, V., Caligiana, A., Cavalli, S.: Hydrogenic elliptic orbitals, Coulomb Sturmian sets, and recoupling coefficients among alternative bases. Int. J. Quantum Chem. 92, 99 (2003)

The Screen Representation of Spin Networks: 2D Recurrence, Eigenvalue Equation for $6j$ Symbols, Geometric Interpretation and Hamiltonian Dynamics

Roger W. Anderson[1], Vincenzo Aquilanti[2,3], Ana Carla Peixoto Bitencourt[4], Dimitri Marinelli[5,6], and Mirco Ragni[4]

[1] Department of Chemistry, University of California, Santa Cruz, CA 95064, U.S.A.
anderso@ucsc.edu
[2] Dipartimento di Chimica, Università di Perugia, Italy
vincenzoaquilanti@yahoo.it
[3] Istituto Metodologie Inorganiche e Plasmi CNR, Roma, Italy
[4] Departamento de Física, Universidade Estadual de Feira de Santana, Brazil
[5] Dipartimento di Fisica, Università degli Studi di Pavia, Italy
[6] INFN, sezione di Pavia, 27100 Pavia, Italy

Abstract. This paper treats $6j$ symbols or their orthonormal forms as a function of two variables spanning a square manifold which we call the "screen". We show that this approach gives important and interesting insight. This two dimensional perspective provides the most natural extension to exhibit the role of these discrete functions as matrix elements that appear at the very foundation of the modern theory of classical discrete orthogonal polynomials. Here we present 2D and 1D recursion relations that are useful for the direct computation of the orthonormal $6j$, which we name U. We present a convention for the order of the arguments of the $6j$ that is based on their classical and Regge symmetries, and a detailed investigation of new geometrical aspects of the $6j$ symbols. Specifically we compare the geometric recursion analysis of Schulten and Gordon with the methods of this paper. The 1D recursion relation, written as a matrix diagonalization problem, permits an interpretation as a discrete Shrödinger-like equations and an asymptotic analysis illustrates semiclassical and classical limits in terms of Hamiltonian evolution.

1 Introduction

Continuing and extending previous work [1–4] on $6j$ symbols, (or on the equivalent Racah coefficients), of current use in quantum mechanics and recently also of interest as the elementary building blocks of spin networks [5–7], in this paper we *(i)* - adopt a representation (the "screen") accounting for exchange and Regge symmetries; *(ii)* - introduce a recurrence relationship in two variables, allowing not only a computational algorithm for the generation of the $6j$ symbols to be plotted on the screen, but also representing a partial difference equation allowing us to interpret the dynamics of the two dimensional system. *(iii)* - introduce

B. Murgante et al. (Eds.): ICCSA 2013, Part II, LNCS 7972, pp. 46–59, 2013.
© Springer-Verlag Berlin Heidelberg 2013

a recurrence relationship as an equation in one variable, extending the known ones which are also computationally interesting; *(iv)* - give a formulation of the difference equation as a matrix diagonalization problem, allowing its interpretation as a discrete Schrödinger equation; *(v)* - discuss geometrical and dynamical aspects from an asymptotic analysis. We do not provide here detailed proofs of these results, but give sufficient hints for the reader to work out the derivations. For some of the topics we refer to a recent problem recently tackled [8]; numerical and geometrical illustrations are presented on a companion paper [9]. A concluding section introduces aspects of relevance for the general spin networks by sketching some features of the $9j$ symbols.

2   The Screen: Classical and Regge Symmetries, Canonical Form

The Wigner $6j$ symbols $\begin{Bmatrix} j_1 & j_2 & j_{12} \\ j_3 & j & j_{23} \end{Bmatrix}$ are defined as a matrix element beetween alternative angular momentum coupling schemes [10] by the relation

$$\langle j_1 j_2 (j_{12}) j_3 j m \mid j_1 j_2 j_3 (j_{23}) j' m' \rangle = (-1)^{j_1 + j_2 + j_3 + j} \, \delta_{jj'} \delta_{mm'} U (j_1 j_2 j j_3; j_{12} j_{23}) ,$$

where the orthonormal transformation U is

$$U (j_1 j_2 j j_3; j_{12} j_{23}) = \sqrt{(2 j_{12} + 1)(2 j_{23} + 1)} \begin{Bmatrix} j_1 & j_2 & j_{12} \\ j_3 & j & j_{23} \end{Bmatrix} \tag{1}$$

For given values of j_1, j_2, j_3, and j the U will be defined over a range for both j_{12} and j_{23}. These ranges are given by

$$
\begin{aligned}
j_{12\ min} &= \max(\mid j_1 - j_2 \mid, \mid j - j_3 \mid), & j_{12\ max} &= \min(j_1 + j_2, j + j_3), \\
j_{23\ min} &= \max(\mid j_1 - j \mid, \mid j_2 - j_3 \mid), & j_{23\ max} &= \min(j_1 + j, j_2 + j_3), \\
\text{and} \quad & j_{12\ min} \le j_{12} \le j_{12\ max}, & j_{23\ min} &\le j_{23} \le j_{23\ max}.
\end{aligned}
\tag{2}
$$

The screen corresponds to the $6j$ or, as we specify below, the U values for all possible values of j_{12} and j_{23} .

The range for j_{12} and j_{23} is determined by the values of the independent variables: j_1, j_2, j_3, and j. In the remainder of this paper we make this clear by introducing new variables a, b, c, d, x and y to replace the j values. We specify the new variables by establishing a correspondence:

$$\begin{Bmatrix} a & b & x \\ c & d & y \end{Bmatrix} \Leftrightarrow \begin{Bmatrix} j_1 & j_2 & j_{12} \\ j_3 & j & j_{23} \end{Bmatrix} \tag{3}$$

Assuming that x and y remain respectively in the upper and lower right side of the $6j$ symbols, there are four classical and one Regge relevant symmetries:

$$\begin{Bmatrix} a & b & x \\ c & d & y \end{Bmatrix} = \begin{Bmatrix} b & a & x \\ d & c & y \end{Bmatrix} = \begin{Bmatrix} d & c & x \\ b & a & y \end{Bmatrix} = \begin{Bmatrix} c & d & x \\ a & b & y \end{Bmatrix} = \begin{Bmatrix} s-a & s-b & x \\ s-c & s-d & y \end{Bmatrix}, \tag{4}$$

where $s = (a+b+c+d)/2$. It can be shown [2, 1] that $x_{max} - x_{min} = y_{max} - y_{min} = 2\min(a, b, c, d, s-d, s-c, s-b, s-a) = 2\kappa$. The square screen will contain $(2\kappa+1)^2$ values. The canonical ordering for $6j$ screens can now be specified by considering the two sets of values: a, b, c, d and its Regge transform $a' = s-a$, $b' = s-b$, $c' = s-c$, and $d' = s-d$. Take the set with the smallest entry and use the classical $6j$ symmetries to place this smallest value in the upper left corner of the $6j$ symbol. The placement of the other $6j$ arguments are determined by the symmetry relations. The resulting symbol has the property that $x_{min} = b-a \leq x \leq b+a = x_{max}$ and $y_{min} = d-a \leq y \leq d+a = y_{max}$. Furthermore we require that $a \leq b \leq d$ for the Canonical form. This may require using Eq. 5 to "orient" the screen in this way.

$$\begin{Bmatrix} a & b & x \\ c & d & y \end{Bmatrix} = \begin{Bmatrix} a & d & y \\ c & b & x \end{Bmatrix} \tag{5}$$

It can be shown that any symbol to be studied as a function of two entries can be reduced to the canonical form of Eq. 5 where $a \leq b \leq d \leq b+c-a$ and $c_{min} = d-a+b \leq c \leq d+a-b = c_{max}$.

Regge transformation for the parameters of the screen is a linear $O(4)$ transformation:

$$\frac{1}{2}\begin{pmatrix} -1 & 1 & 1 & 1 \\ 1 & -1 & 1 & 1 \\ 1 & 1 & -1 & 1 \\ 1 & 1 & 1 & -1 \end{pmatrix}\begin{pmatrix} a \\ b \\ c \\ d \end{pmatrix} = \begin{pmatrix} s-a \\ s-b \\ s-c \\ s-d \end{pmatrix}. \tag{6}$$

It can be checked that several functions appearing below (caustics, ridges, etc.) are invariant under such symmetry and also when represented on the screen (See [9]).

3   Tetrahedra and 6j Symbols

In the following when we consider $6j$ properties as correlated to those of the tetrahedron of Figure 1a [3], we use the substitutions $A = a+1/2$, $B = b+1/2$, $C = c+1/2$, $D = d+1/2$, $X = x+1/2$, $Y = y+1/2$ which greatly improves all asymptotic formulas down to surprisingly low values of the entries. We show the argument ranges where the correspondence with the tetrahedron breaks down in section 5.2.

The area of each triangular face is given by the Heron formula:

$$F(A, B, C) = \frac{1}{4}\sqrt{(A+B+C)(-A+B+C)(A-B+C)(A+B-C)} \tag{7}$$

where A, B, C are the sides of the face. Upper case letters are used here to stress that geometric lengths are used in the equation. The square of the area can be also expressed as a Cayley-Menger determinant. Similarly, the square of the volume of an irregular tetrahedron, can also be written as a Cayley-Menger determinant (Eq. 8 or as a Gramian determinant [11]. The latter determinant embodies a clearer relationship with a vectorial picture but with partial spoiling of the symmetry.

$$V^2 = \frac{1}{288} \begin{vmatrix} 0 & C^2 & D^2 & Y^2 & 1 \\ C^2 & 0 & X^2 & B^2 & 1 \\ D^2 & X^2 & 0 & A^2 & 1 \\ Y^2 & B^2 & A^2 & 0 & 1 \\ 1 & 1 & 1 & 1 & 0 \end{vmatrix}. \tag{8}$$

An explicit formula, due to Piero della Francesca, will be used in the companion paper [9]. Additionally mirror symmetry [1], can be used to extend screens to cover a larger range of arguments. The appearance of squares of tetrahedron edges entails that the invariance with respect to the exchange $X \leftrightarrow -X$ implies formally $x \leftrightarrow -x - 1$ with respect to entries in the $6j$ symbol. Although this is physically irrelevant when the js are pseudo-vectors, such as physical spins or orbital angular momenta, it can be of interest for other (e.g. discrete algorithms) applications. Regarding the screen, it can be seen that actually by continuation of X and Y to negative values, one can have replicas that can be glued by cutting out regions shaded in Fig. 6 in [12], allowing mapping onto the S^2 manifold.

Figure 1b illustrates V^2 for values of $a, b, c,$ and d used later in this paper.

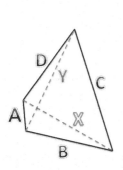

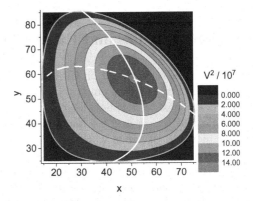

(a) Ponzano-Regge tetrahedron built with the six angular momenta in the $6j$ symbol.

(b) V^2 (contours for Eq. 8), caustics Eq. 12 (gray boundary), ridges (solid white Eq. 9, dashed Eq. 13) for $a = 30$, $b = 45$, $c = 60$, and $d = 55$.

Fig. 1.

The following equations were first introduced in Refs. [1] and [3], but they are rewritten here with changed notation. When the values of A, B, C, D and X are fixed, the maximum value for the volume as a function of Y is given by the "ridge" curve

$$Y^{Vmax} = \left(\frac{(A^2 - B^2)(C^2 - D^2) + (A^2 + B^2 + C^2 + D^2)X^2 - X^4}{2X^2} \right)^{1/2}, \quad (9)$$

the corresponding volume is

$$V^{max}(A, B, C, D, X) = \frac{\sqrt{\Lambda_{A,B,X} \Lambda_{C,D,X}}}{24X}, \quad (10)$$

where

$$\Lambda_{\alpha,\beta,\gamma} = \left(\alpha^2 - \beta^2 \right)^2 - 2\gamma^2 \left(\alpha^2 + \beta^2 \right) + \gamma^4. \quad (11)$$

Therefore the two values of Y for which the volume is zero are:

$$Y^z = \left(\left(Y^{Vmax} \right)^2 \pm \frac{\sqrt{\Lambda_{A,B,X} \Lambda_{C,D,X}}}{2X^2} \right)^{1/2}. \quad (12)$$

The values for Y^z mark the boundaries between classical and nonclassical regions, and therefore called "caustics".

Also when the values of A, B, C, D and Y are fixed, the maximum value for the volume as a function of X is given by the other "ridge" curve:

$$X^{Vmax} = \left(\frac{(A^2 - D^2)(C^2 - B^2) + (A^2 + B^2 + C^2 + D^2)Y^2 - Y^4}{2Y^2} \right)^{1/2}. \quad (13)$$

4   Recursion Formulas and Exact Calculations

The U values that are represented on the screen must be calculated by efficient and accurate algorithms, and we employed several methods that we have previously discussed and tested. Explicit formulas are available either as sums over a single variable and series, and we have used such calculations with multiple precision arithmetic in previous work [13],[3], [14], [15] . These high accuracy calculations are entirely reliable for all U that we have considered in the past, and the results provide a stringent test for other methods. However recourse to recursion formulas appears most convenient for fast accurate calculations and -as we will emphasize- also for semiclassical analysis, in order to understand high j limit and in reverse to interpret them as discrete wavefunctions obeying Schrödinger type of difference (rather than differential) equations.

The goal is to determine the elements of the ortho-normal transformation matrix:

$$U(x, y) = \sqrt{(2x + 1)(2y + 1)} \begin{Bmatrix} a & b & x \\ c & d & y \end{Bmatrix}. \quad (14)$$

Two approaches can be used to evaluate $U(x, y)$: evaluate the $6j$ from recursion formulas and then apply the normalization or to use direct calculation from explicit formulas.

4.1 2D (x,y) Recursion for U

In this work, we first derive and computationally implement a two variable recurrence that permits construction of the whole orthonormal matrix. The derivation follows our paper in [14] and is also of interest for other $3nj$ symbols.

By setting $h = 0$ in the formula 43 in section 6 , we obtain a five term recurrence relation for $U(x, y)$:

$$(-1)^{2x} \sqrt{\frac{2x-1}{2y+1}} \begin{Bmatrix} b & x-1 & a \\ 1 & a & x \end{Bmatrix} \begin{Bmatrix} d & x-1 & c \\ 1 & c & x \end{Bmatrix} U(x-1,y)$$

$$+(-1)^{2x} \sqrt{\frac{2x+1}{2y+1}} \begin{Bmatrix} b & x & a \\ 1 & a & x \end{Bmatrix} \begin{Bmatrix} d & x & c \\ 1 & c & x \end{Bmatrix} U(x,y)$$

$$+(-1)^{2x} \sqrt{\frac{2x+3}{2y+1}} \begin{Bmatrix} b & x+1 & a \\ 1 & a & x \end{Bmatrix} \begin{Bmatrix} d & x+1 & c \\ 1 & c & x \end{Bmatrix} U(x+1,y)$$

$$= (-1)^{2y} \sqrt{\frac{2y-1}{2x+1}} \begin{Bmatrix} b & y-1 & c \\ 1 & c & y \end{Bmatrix} \begin{Bmatrix} d & y-1 & a \\ 1 & a & y \end{Bmatrix} U(x,y-1)$$

$$+(-1)^{2y} \sqrt{\frac{2y+1}{2x+1}} \begin{Bmatrix} b & y & c \\ 1 & c & y \end{Bmatrix} \begin{Bmatrix} d & y & a \\ 1 & a & y \end{Bmatrix} U(x,y)$$

$$+(-1)^{2y} \sqrt{\frac{2y+3}{2x+1}} \begin{Bmatrix} b & y+1 & c \\ 1 & c & y \end{Bmatrix} \begin{Bmatrix} d & y+1 & a \\ 1 & a & y \end{Bmatrix} U(x,y+1) \qquad (15)$$

This recurrence relation Eq. 15 will yield the entire set of $U(x, y)$ that constitute the screen. Replacing the $6j$ symbols of unit argument with the algebraic expressions in Varshalovich [10], we obtain an effective method to calculate the screen.

5 1D (x) Symmetric Recursion for U

Starting with the recurrence relation in Neville [16] and Schulten and Gordon [17] for the $6j$ and carefully converting it into a recurrence relation for U, we can write a three term symmetric recursion relationship, which is here conveniently represented as an eigenvalue equation:

$$p_+(x) U(x+1,y) + w(x)U(x,y) + p_-(x) U(x-1,y) = \lambda(y) U(x,y) , \quad (16)$$

where

$$p_+(x) = \{(a+b+x+2)(a+b-x)(a-b+x+1)(-a+b+x+1)\}^{\frac{1}{2}}$$
$$\times \{(d+c+x+2)(d+c-x)(d-c+x+1)(-d+c+x+1)\}^{\frac{1}{2}} \quad (17)$$
$$\times (x+1)^{-1}[(2x+1)(2x+3)]^{-\frac{1}{2}}$$

$$p_- (x) = p_+ (x - 1) \tag{18}$$

$$
\begin{aligned}
w(x) = {}& [b(b+1) - a(a+1) + x(x+1)] \\
& \times [d(d+1) - c(c+1) - x(x+1)] / [x(x+1)]
\end{aligned}
\tag{19}
$$

$$\lambda (y) = 2 [y(y+1) - b(b+1) - c(c+1)] . \tag{20}$$

For convenience we can also define:

$$w_\lambda = w(x) - \lambda(y) \tag{21}$$

A row of the screen may be efficiently and accurately calculated from these equations. Diagonalization of the symmetric tridiagonal matrix given by the $p_+ (x)$, $w(x)$, $p_- (x)$ provides an accurate check: the eigenvalues of the tridiagonal matrix precisely match those expected from Eq. 20 and eigenvectors generate $U(x, y)$. Stable results are obtained with double precision arithmetic.

5.1   Potential Functions and Hamiltonian Dynamics

For the eigenvalue equation (Eq. 16), interpreted as discrete Schrödinger-like equation, two potentials $\mathbf{W}^+ (x)$ and $\mathbf{W}^- (x)$ can be defined:

$$\mathbf{W}^\pm (x) = w(x) \pm 2 \mid p(x) \mid, \tag{22}$$

where [18]

$$p(x) = \frac{1}{2} (p_+ (x) + p_- (x)) \tag{23}$$

or [19]

$$p(x) = \sqrt{(p_+ (x) \, p_- (x))}. \tag{24}$$

The two definitions agree well except for x near the limits x_{min} or x_{max}. With the second choice for $\bar{p}(x)$ the values for $\mathbf{W}^\pm$ are the same at the limits, but there are differences with the first choice. See the figures 2a and 2b. Compare with Ref. [8] where Hamiltonian dynamics is developed for a similar system. Braun's potential functions are closely related to the caustics illustrated in [1] and [9].

5.2   Geometric Interpretation

The geometrical interpretations of the $6j$ symbols provide fundamental understanding and important semiclassical limits. This approach originates from Ponzano and Regge [20] and elaborated by others, notably Schulten and Gordon [17].

The three-term recursion relationship Eq. 16, for U admits an illustration in terms of a geometric interpretation: with some approximations to be detailed below one has finite difference equations (see Ref.[16], Eq.(67) for relationships

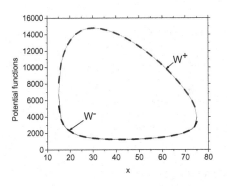
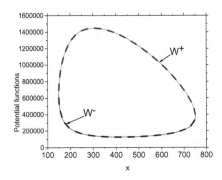

(a) Angular momenta corresponding to Figure 1b

(b) Angular momenta: $a = 300$, $b = 450$, $c = 600$, and $d = 550$.

Fig. 2. Potential functions corresponding to Eq. 23 (dashed blue and black lines) and Eq.24 (thin solid orange and red lines)

between recursions and finite difference). Consider the Schulten-Gordon relationships Eq.(66) and Eq.(67)(Ref. [17]). Here we show new geometric representations of the recursion relationships.

By setting $a = A - \frac{1}{2}$, $b = B - \frac{1}{2}$, $c = C - \frac{1}{2}$, $d = D - \frac{1}{2}$, $x = X - \frac{1}{2}$, and $y = Y - \frac{1}{2}$ one can write Eq. 16 in terms of triangle areas, a length X', and the cosine of a dihedral angle θ_3. The accuracy of this approximation is excellent, and depends slightly on the choice for X'.

$$\frac{F(X - \frac{1}{2}, A, B)F(X - \frac{1}{2}, C, D)}{\left(X - \frac{1}{2}\right)^2}U\left(x - 1, y\right)$$

$$+\frac{F(X + \frac{1}{2}, A, B)F(X + \frac{1}{2}, C, D)}{\left(X + \frac{1}{2}\right)^2}U\left(x + 1, y\right)$$

$$-2\cos\theta_3\frac{F(X', A, B)F(X', C, D)}{X'^2}U\left(x, y\right) \approx 0 \qquad (25)$$

and Eq.(69)[17]

$$\cos\theta_3 = \frac{2X'^2Y^2 - X'^2\left(-X'^2 + D^2 + C^2\right) - B^2\left(X'^2 + D^2 - C^2\right) - A^2\left(X'^2 - D^2 + C^2\right)}{16F\left(X', B, A\right)F\left(X', D, C\right)}, \qquad (26)$$

where $F(a, b, c)$ is "area" of abc triangle (Eq. 7). (This recursion relation Eq. 25 must be multiplied through by 8 to compare precisely with Eq. 16.)

Here we consider two choices for X' in Eq. 19:

$$X'^2 = \left(X - \frac{1}{2}\right)\left(X + \frac{1}{2}\right) = X^2 - \frac{1}{4}, \qquad (27)$$

$$X' = X \qquad (28)$$

The first choice (Eq. 27) provides an almost exact approximation to Eqns. 17, 18,19, and 20 coefficients in Eq. 16. The second Eq. 28 uses only integer or half integer arguments, and for most X works as well as the first. The figures 3a and 3b show the errors and their significance. In these figures $w_\lambda \, (approx)$ is specified as:

$$w_\lambda \, (approx) = -2 \cos \theta_3 \frac{F(X', A, B) F(X', C, D)}{X'^2} \tag{29}$$

For either choice of X', the recursion coefficients are connected to the geometry of tetrahedra [20]:

$$\frac{3}{2} V X' = F(X', A, B) F(X', C, D) \sin \theta_3 \, , \tag{30}$$

where V is the tetrahedral volume.

Equations 25 can be recast by the geometric mean approximation:

$$\frac{F(X \pm \frac{1}{2}, A, B)}{(X \pm \frac{1}{2})} \simeq \frac{\sqrt{F(X \pm 1, A, B) F(X, A, B)}}{\sqrt{X(X \pm 1)}}, \tag{31}$$

where A and B can be also replaced by C and D.

With Eq. 31 Eq. 25 becomes:

$$\frac{\sqrt{F(X - 1, A, B) F(X, A, B) F(X - 1, C, D) F(X, C, D)}}{X(X - 1)} U(x - 1, y)$$
$$+ \frac{\sqrt{F(X + 1, A, B) F(X, A, B) F(X + 1, C, D) F(X, C, D)}}{X(X + 1)} U(x + 1, y)$$
$$- 2 \cos \theta_3 \frac{F(X, A, B) F(X, C, D)}{X^2} U(x, y) \approx 0, \quad (32)$$

This equation is useful, but definitely less accurate than Eq. 25 (See Figures 4a and 4b).

With cancellation of terms in X, this Eq. 32 becomes:

$$\frac{\sqrt{F(X - 1, A, B) F(X - 1, C, D)}}{(X - 1)} U(x - 1, y)$$
$$+ \frac{\sqrt{F(X + 1, A, B) F(X + 1, C, D)}}{(X + 1)} U(x + 1, y)$$
$$- 2 \cos \theta_3 \frac{\sqrt{F(X, A, B) F(X, C, D)}}{X} U(x, y) \approx 0. \tag{33}$$

This is equivalent to the recursion relation of Schulten and Gordon [17], that they use to establish their semiclassical approximations for $6j$ symbols. Their equation is accurate enough for $x_{min} \ll x \ll x_{max}$, but not so accurate near the limits.

In terms of the finite difference operator, Eq. 32 becomes after using Eq. 30: $\Delta^2(x) f(x) = f(x + 1) - 2f(x) + f(x - 1)$:

$$[\Delta^2(X) + 2 - 2 \cos \theta_3] f(X) \simeq 0 \, , \tag{34}$$

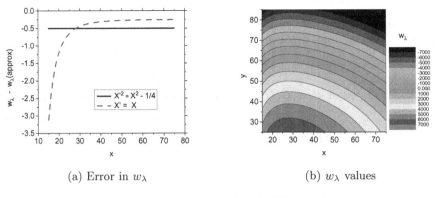

(a) Error in w_λ (b) w_λ values

Fig. 3. Parameters a, b, c, d of Figure 1b

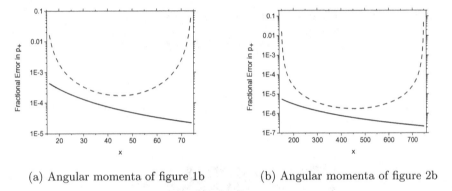

(a) Angular momenta of figure 1b   (b) Angular momenta of figure 2b

Fig. 4. Fractional errors in p_+ using Eq. 25, solid black line and Eq. 32, dashed blue line.

where

$$f(X) = \frac{\sqrt{F(X,A,B)F(X,C,D)}}{X} U(x,y) = \sqrt{\frac{V}{X \sin \theta_3}} U(x,y) \ . \tag{35}$$

We have, explicitly

$$\cos \theta_3 = \pm \sqrt{1 - \left(\frac{3VX}{2F(X,A,B)F(X,C,D)} \right)^2} \ . \tag{36}$$

Our Eq. 35 is only slightly different from that of Schulten and Gordon, because we have an extra X in the denominator of the definition of $f(X)$. This occurs because we use the recursion for U instead of that for $6j$.

5.3 Semiclassical Approximation

The following developments parallel those in [3]. From the above formulas, and from that of the volume, we have that

- $V = 0$ implies $\cos\theta_3 = \pm 1$ and establishes the classical domain between X_{min} and X_{max}
- $F(X, A, B) = 0$ or $F(X, C, D) = 0$ establish the definition limits x_{min} and x_{max}.

For a Schrödinger type equation

$$\frac{d^2\psi}{dx^2} + p^2\psi = 0 \ , \qquad \hbar^2/2m = 1 \ , \tag{37}$$

its discrete analog in a grid having one as a step,

$$\psi_{n+1} + (p^2 - 2)\psi_n + \psi_{n-1} = 0, \tag{38}$$

and we then have after comparing Eq. 38 with Eq. 34

$$f(X+1) - 2\cos\theta_3 f(X) + f(X-1) = 0. \tag{39}$$

The identification

$$p = \pm(2 - 2\cos\theta_3)^{1/2} \tag{40}$$

is then evident. Here we present a x,y plot Fig. 5 of $1 - \cos\theta_3$ that clearly shows this definition of the classical region.

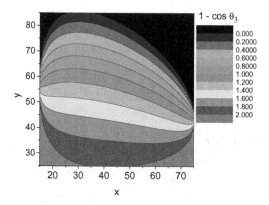

Fig. 5. x,y plot of $\cos\theta_3$ for angular momenta of Fig 1(b)

Evidentially, on the closed loop, we can enforce Bohr-Sommerfeld phase space quantization:

$$\oint p\,dx = (n + 1/2)\,\pi \ . \tag{41}$$

The eigenvalues n obtained in this way may be easily related to the allowed y. These formulas are illustrated in Fig. 6 and 7 of Ref [3].

The Ponzano-Regge formula for the $6j$ in the classical region is

$$\begin{Bmatrix} a & b & x \\ c & d & y \end{Bmatrix} \approx \frac{1}{\sqrt{12\pi|V|}} \cos(\varPhi), \qquad (42)$$

where the Ponzano-Regge phase is: $\varPhi = A\theta_1 + B\theta_2 + X\theta_3 + C\eta_1 + D\eta_2 + Y\eta_3 + \frac{\pi}{4}$. The angles are determined by rearranged equations: Eq. 30. The various dihedral angles are found from the equations in [20].

To the extent that the Ponzano-Regge approximation is valid we see that the $6j$ symbols have a magnitude envelop given by V and a phase that is a function of X and Y determined by $\varPhi$. Eq. 42 works quite well for X and Y well within the classical region. However its use near the caustics is limited because of two factors:

1. The approximate recursion relation given by Eq. 32 differs most from the exact recursion Eqs. 17,18,19,20 near the caustics.
2. The semiclassical approximation for the $6j$ also breaks down near the caustics.

For piece-wise extensions , see [20] and for uniformly valid formulas see [17].

6 $9j$ and Higher Spin Networks

In this work, we first have derived and computationally implemented a two variable recurrence that permits construction of the whole orthonormal matrix The derivation follows our paper in [14] and is also of interest for other $3nj$ symbols.

We find in [14]; see also [10], the following 2D recurrence relationship for $9j$ symbols:

$$\frac{A_{c+1}(ab,fj)}{(c+1)(2c+1)} \begin{Bmatrix} a & b & c+1 \\ d & e & f \\ g & h & j \end{Bmatrix} + \frac{A_c(ab,fj)}{c(2c+1)} \begin{Bmatrix} a & b & c-1 \\ d & e & f \\ g & h & j \end{Bmatrix} - \frac{A_{d+1}(ef,ag)}{(d+1)(2d+1)} \begin{Bmatrix} a & b & c \\ d+1 & e & f \\ g & h & j \end{Bmatrix}$$

$$- \frac{A_d(ef,ag)}{d(2d+1)} \begin{Bmatrix} a & b & c \\ d-1 & e & f \\ g & h & j \end{Bmatrix} = \left[\frac{B_d(ag,fe)}{d(d+1)} - \frac{B_c(ab,jf)}{c(c+1)} \right] \begin{Bmatrix} a & b & c \\ d & e & f \\ g & h & j \end{Bmatrix}$$

$$A_q(pr,st) = [(-p+r+q)(p-r+q)(p+r-q+1)(p+r+q+1)]^{\frac{1}{2}}$$
$$\times [(-s+t+q)(s-t+q)(s+t-q+1)(s+t+q+1)]^{\frac{1}{2}}$$
$$B_q(pr,st) = [q(q+1) - p(p+1) + r(r+1)][q(q+1) - s(s+1) + t(t+1)]$$

$$(43)$$

Geometrical interpretations of A's as proportional to products of areas of triangular faces and of B's as angular functions of associated structures, will serve for further work on the dynamical description of general spin networks. As noted in

[14], Eq. 15 can be derived by setting $h = 0$ in Eq. 43, and using the property that a $3nj$ symbol downgrades to a $(3n - 1)j$ symbol when one of its entries is zero. In conclusion, expanding the discussion of Eq. 43 in [14], we suggest that the "screen" for the above $9j$ symbols is three-dimensional, and generalization to higher spin networks should be straight forward.

Acknowledgement. We thank Professor Annalisa Marzuoli for many productive discussions during this research.

References

1. Bitencourt, A.C.P., Marzuoli, A., Ragni, M., Anderson, R.W., Aquilanti, V.: Exact and asymptotic computations of elementary spin networks: Classification of the quantum–classical boundaries. In: Murgante, B., Gervasi, O., Misra, S., Nedjah, N., Rocha, A.M.A.C., Taniar, D., Apduhan, B.O. (eds.) ICCSA 2012, Part I. LNCS, vol. 7333, pp. 723–737. Springer, Heidelberg (2012); See arXiv:1211.4993[math-ph]

2. Aquilanti, V., Haggard, H.M., Hedeman, A., Jeevanjee, N., Littlejohn, R.G., Yu, L.: Semiclassical mechanics of the Wigner 6j-symbol. J. Phys. A 45(6), 065209 (2012)

3. Ragni, M., Bitencourt, A.C., Aquilanti, V., Anderson, R.W., Littlejohn, R.G.: Exact computation and asymptotic approximations of 6j symbols: Illustration of their semiclassical limits. Int. J. Quantum Chem. 110(3), 731–742 (2010)

4. Littlejohn, R.G., Mitchell, K.A., Reinsch, M., Aquilanti, V., Cavalli, S.: Internal spaces, kinematic rotations, and body frames for four-atom systems. Phys. Rev. A 58, 3718–3738 (1998)

5. Aquilanti, V., Bitencourt, A.C.P., da S. Ferreira, C., Marzuoli, A., Ragni, M.: Quantum and semiclassical spin networks: from atomic and molecular physics to quantum computing and gravity. Phys. Scripta 78(5), 058103 (2008)

6. Aquilanti, V., Bitencourt, A., da S. Ferreira, C., Marzuoli, A., Ragni, M.: Combinatorics of angular momentum recoupling theory: spin networks, their asymptotics and applications. Theor. Chem. Acc. 123, 237–247 (2009)

7. Aquilanti, V., Capecchi, G.: Harmonic analysis and discrete polynomials. From semiclassical angular momentum theory to the hyperquantization algorithm. Theor. Chem. Accounts 104, 183–188 (2000)

8. Aquilanti, V., Marinelli, D., Marzuoli, A.: Hamiltonian dynamics of a quantum of space: hidden symmetries and spectrum of the volume operator, and discrete orthogonal polynomials. J. Phys. A: Math. Theor. 46, 175303 (2013), arXiv:1301.1949v2 [math-ph]

9. Ragni, M., Littlejohn, R.G., Bitencourt, A.C.P., Aquilanti, V., Anderson, R.W.: The screen representation of spin networks. Images of 6j symbols and semiclassical features. In: Murgante, B., Misra, S., Carlini, M., Torre, C.M., Quang, N.H., Taniar, D., Apduhan, B.O., Gervasi, O. (eds.) ICCSA 2013, Part II. LNCS, vol. 7972, pp. 60–72. Springer, Heidelberg (2013)

10. Varshalovich, D., Moskalev, A., Khersonskii, V.: Quantum Theory of Angular Momentum. World Scientific, Singapore (1988)

11. Freidel, L., Louapre, D.: Asymptotics of 6j and 10j symbols. Classical and Quantum Gravity 20, 1267–1294 (2003)

12. Aquilanti, V., Haggard, H.M., Hedeman, A., Jeevangee, N., Littlejohn, R., Yu, L.: Semiclassical mechanics of the Wigner $6j$-symbol. J. Phys. A 45(065209) (2012), arXiv:1009.2811v2 [math-ph]

13. Anderson, R.W., Aquilanti, V.: The discrete representation correspondence between quantum and classical spatial distributions of angular momentum vectors. J. Chem. Phys. 124, 214104 (9 pages) (2006)

14. Anderson, R.W., Aquilanti, V., Marzuoli, A.: 3nj morphogenesis and semiclassical disentangling. J. Phys. Chem. A 113, 15106–15117 (2009)

15. Anderson, R., Aquilanti, V., da S. Ferreira, C.: Exact computation and large angular momentum asymptotics of $3nj$ symbols: semiclassical disentangling of spinnetworks. J. Chem. Phys. 129, 161101 (5 pages) (2008)

16. Neville, D.: A technique for solving recurrence relations approximately and its application to the $3 - j$ and $6 - j$ symbols. J. Math. Phys. 12, 2438 (1971)

17. Schulten, K., Gordon, R.: Semiclassical approximations to 3j- and 6j-coefficients for quantum-mechanical coupling of angular momenta. J. Math. Phys. 16, 1971–1988 (1975)

18. Braun, P.A.: Discrete semiclassical methods in the theory of Rydberg atoms in external fields. Rev. Mod. Phys. 65, 115–161 (1993)

19. Braun, P.: WKB method for three-term recursion relations and quasienergies of an anharmonic oscillator. Sov. Phys. Theor. Math. Phys. 16, 1070–1081 (1978)

20. Ponzano, G., Regge, T.: Semiclassical limit of Racah coefficients. In: Bloch, F., et al. (eds.) Spectroscopic and Group Theoretical Methods in Physics, pp. 1–58. North-Holland, Amsterdam (1968)

The Screen Representation of Spin Networks: Images of $6j$ Symbols and Semiclassical Features

Mirco Ragni[1], Robert G. Littlejohn[2], Ana Carla Peixoto Bitencourt[1], Vincenzo Aquilanti[3,4], and Roger W. Anderson[5]

[1] Departamento de Física, Universidade Estadual de Feira de Santana, Brazil
[2] Department of Physics, University of California, Berkeley, California 94720, U.S.A
[3] Dipartimento di Chimica, Università di Perugia, Italy
vincenzoaquilanti@yahoo.it
[4] Istituto Metodologie Inorganiche e Plasmi CNR, Roma, Italy
[5] Department of Chemistry, University of California, Santa Cruz, CA 95064, U.S.A.
anderso@ucsc.edu

Abstract. This article presents and discusses in detail the results of extensive exact calculations of the most basic ingredients of spin networks, the Racah coefficients (or Wigner $6j$ symbols), exhibiting their salient features when considered as a function of two variables - a natural choice due to their origin as elements of a square orthogonal matrix - and illustrated by use of a projection on a square screen introduced recently. On these screens, shown are images which provide a systematic classification of features previously introduced to represent the caustic and ridge curves (which delimit the boundaries between oscillatory and evanescent behaviour according to the asymptotic analysis of semiclassical approaches). Particular relevance is given to the surprising role of the intriguing symmetries discovered long ago by Regge and recently revisited; from their use, together with other newly discovered properties and in conjunction with the traditional combinatorial ones, a picture emerges of the amplitudes and phases of these discrete wavefunctions, of interest in wide areas as building blocks of basic and applied quantum mechanics.

1 Introduction

In this paper, extensive computational results serve to illustrate the main features of the well known Wigner $6j$ symbols (or equivalently of the related Racah coefficients). Their importance has transcended the context of the quantum theory of angular momentum, where they were introduced originally: they appear as the building blocks of spin network structures, of widespread relevance in quantum science and its applications [1], [2].

In their introduction as matrix elements between alternative angular momentum coupling schemes, Wigner and Racah had the insight of associating the six entries of a 6-j symbol with the lengths of the edges of a (generally irregular) tetrahedron and established asymptotic (or semi-classical) relationships with the geometrical properties, such as volumes and dihedral angles, of

B. Murgante et al. (Eds.): ICCSA 2013, Part II, LNCS 7972, pp. 60–72, 2013.
© Springer-Verlag Berlin Heidelberg 2013

such a tetrahedron. In 1968, Ponzano and Regge [3] initiated the study of the functional dependence of the $6j$-symbol on one of its six entries, arguing for a role of the tetrahedral volume and dihedral angles in the amplitude and phase of a (discrete) WKB-type of approximation of a wave-function. Independently, Neville [4] and Schulten and Gordon [5] provided rigorous derivations: the latter also introducing efficient computational procedures [6] (for account of progress see [7]) and numerical illustrations from this one-dimensional perspective (see also [8], [9], [10]). The closely related Racah polynomials [11], [12], [13], [14] are at the foundation of modern approaches in the theory and applications of special functions and orthogonal polynomials. In this account, we present for the first time illustrations from the two-dimensional perspective, which is naturally based on the view [4], [9], [10] of the 6-j symbols as matrix elements enjoying a self dual property. The basic ideas of this approach are referred to in [9], [10] as the 4-j model : accordingly here plots as a function of two discrete variables are given in a square screen (see [15]), in a sense generalizing the traditional presentations in square numerical tables [16]. After a presentation of the general case in Section II, we illustrate symmetric and limiting cases in Section III (an important case being that of the Clebsch-Gordan coefficients, also known as Wigners 3-j symbols, see [17]). In the order of presentation, we are closely following the previous classification of the classical-quantum boundaries [15]. In section IV, we provide additional and concluding remarks.

2   Some Theory and Methods

We present screen images in the next section that include the values of the $6j$ symbols or more precisely the $U(x,y) = \sqrt{(2x+1)(2y+1)} \begin{Bmatrix} a & b & x \\ c & d & y \end{Bmatrix}$. The U values have been calculated with a variety of methods: direct summation with multi-precision arithmetic [18], [19], exact integer arithmetic, three- and five-term recursion relations [20], and checked by solving the eigenvalue equation [20]. All the calculations give precise agreement with each other.

2.1   Canonical Ordering

We choose the Canonical ordering of the a, b, c, d as proposed in [20]. In this ordering a is the smallest of the eight values $a, b, c, d, a', b', c', d'$ with the primed quantities the Regge conjugate values of the unprimed values. The ordering assures that the screen has dimension $(2a+1) \times (2a+1)$ and the ranges $b - a \leq x \leq b + a$ and $d - a \leq y \leq d + a$. For most of the cases considered in this paper, this Canonical form agrees with a slightly different ordering proposed in [21].

We have also found that the following expressions for s, r, u, and v are useful for describing the topology of the screens corresponding to different values for a, b, c and d. See Ref. [21] for more discussion.

$$s \equiv \left[(a+c) + (b+d)\right]/2 \tag{1}$$
$$r \equiv \left[(a+c) - (b+d)\right]/2 \tag{2}$$
$$u \equiv \left[(a+b) - (c+d)\right]/2 \tag{3}$$
$$v \equiv \left[(a+d) - (b+c)\right]/2 \tag{4}$$

For these definitions, s is the semiperimeter, r is the difference in the sums of column values, u is the difference in the sums of row values, and v is the difference between sums of diagonals. The definitions for r, u, v in equations 2, 3, and 4 constrain the values for c in the Canonical ordering such that either v or r must be equal or less than 0, and u must also be less than 0.

Ponzano-Regge Theory. The screen images show many features most of which are explained with the Ponzano-Regge theory and some symmetry considerations. The Ponzano-Regge estimate for $6j$ in the classical region ($V^2 > 0$) is

$$\begin{Bmatrix} a\ b\ x \\ c\ d\ y \end{Bmatrix} \approx \frac{1}{\sqrt{12\pi|V|}} \cos\left(\varPhi\right), \tag{5}$$

Hence the $6j$ symbols have a magnitude envelope determined by the tetrahedron volume, V, and oscillations given by the Cosine of the Ponzano-Regge phase $\varPhi$. Both the volume and the phase are given by the geometry of the tetrahedron with sides: $A, B, C, D, X,$ and Y (See [3] for details. Here $A = a + 1/2, ..., X = x + 1/2, Y = y + 1/2$).

The square volume of the tetrahedron can be calculated with a Cayley-Menger or a Gram determinant, but the results obtained with both determinants are equivalent to the famous formula known to Euler but first found five centuries ago by the Renaissance mathematician, architect and painter Piero della Francesca. We give his formula arranged as needed in the following.

$$\begin{aligned} 288V^2 = {} & 2A^2C^2(-A^2 + B^2 + X^2 + Y^2 + D^2 - C^2) \\ & +2B^2D^2(A^2 - B^2 + X^2 + Y^2 - D^2 + C^2) \\ & +2X^2Y^2(A^2 + B^2 - X^2 - Y^2 + D^2 + C^2) \\ & -(A^2 + C^2)(B^2 + D^2)(X^2 + Y^2) \\ & -(A^2 - C^2)(B^2 - D^2)(X^2 - Y^2) \end{aligned} \tag{6}$$

We write the Ponzano-Regge phase as $\varPhi = A\theta_1 + B\theta_2 + X\theta_3 + C\eta_1 + D\eta_2 + Y\eta_3 + \frac{\pi}{4}$ (See [3], [5]).

2.2 Piero Line Symmetries

An important screen symmetry is the possible presence of a Piero line where the $6j$ and U are symmetric with respect to interchange of x and y. The classical

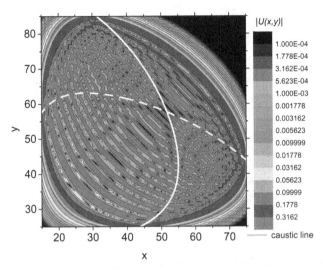

Fig. 1. Plot of $|U(x, y)|$ for $a = 30, b = 45, c = 60, d = 55$. Ranges are $15 \leq x \leq 75$ and $25 \leq y \leq 85$. There are $2a + 1$ values of x and y. This corresponds to fig. 1a in [15]. The caustic closed curve (light gray), corresponding to zero volume (Eq. 6) encircles the classical region of positive volume, showing oscillatory behaviour, while outside in the four nonclassical regions the values are exponentially decaying,. This is the canonical form with $a \leq b \leq d$ namely the screen is oriented, the caustic touching the sides at four points, denoted North, West, South and Eastern gates(see Sec Concluding remarks). The corresponding Regge conjugate can be shown to be $a = 50, b = 65, c = 40, d = 35$. None of the primed quantities are smaller than a.

symmetry relation: $\begin{Bmatrix} a & b & x \\ c & d & y \end{Bmatrix} = \begin{Bmatrix} a & d & y \\ c & b & x \end{Bmatrix}$ shows that the screen will be invariant to this interchange if $b = d$, if the $6j$ is written in the conventional order where a is equal to the smallest argument. The Piero line is the diagonal corresponding to $x = y$, and the $6j$ or U are symmetric with respect to this line. The Piero equation for V^2, Eq. (6) shows this symmetry very clearly. The first four lines in Eq. (6) are symmetric with respect to interchange of X and Y, but the symmetry is broken with the term in the last row unless $B = D$. (It seems that the other possible case where $A = C$, can not occur for conventional ordering unless B is also equal to D). There appear no exact symmetries for U or $6j$ with respect to the line $x + y = x_{min} + y_{max}$. There will be a Piero line symmetry whenever $u = v$ (See Eqns. 3, 4). Piero line symmetries are found in Figures 6a, 7, 8a, 8b, and 9b in Section 3.

2.3 Regge Symmetry

The Regge conjugate will be the same as the original if the sum of any two of a, b, c, d is the same as the sum of the other two. Hence the Regge conjugate will

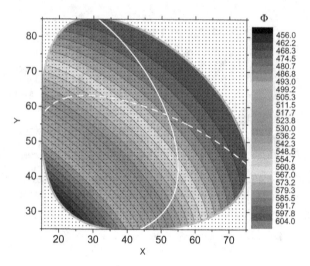

Fig. 2. The figure shows the region of the x, y plane relevant for the $6j$-symbol with the same values of a, b, c and d as in Fig. 1. The small spots are the quantized values of x and y, at which the $6j$-symbol is defined. The heavy light gray oval curve is the caustic line, which surrounds the classically allowed region. The lighter lines and color changes inside the classically allowed region are the contours of the Ponzano-Regge phase Φ. Contour lines are separated by a phase difference of 2π. As we move across a horizontal line, varying x while holding y fixed, we can see how many spots lie between two contour values of Φ. For example, near the upper right side of the caustic curve there are up to nine spots between contour values. This means that if the $6j$-symbol is plotted in a stick diagram, as in Fig. 1 of Ref. [9], then there will be several sticks under a single lobe of oscillation of the $6j$-symbol, as shown in the right side of that figure. But near the bottom of the caustic curve, there are approximately only two spots per 2π increment of phase, which means that the sticks alternate in sign, as shown on the left side of Fig. 1 of Ref. [9].

be the same if the product $ruv = 0$. This equivalence between the original and Regge conjugate is found for the cases corresponding to Figures 5a, 5b, 6a, 8a, 8b, 9a, and 9b.

2.4   Location of Gates

We define a gate as the place along each of the four sides of the screen where the caustic line, $V^2 = 0$ touches the side. The gate will generally be some where in the center of a side, but for specific choices for a, b, c, d the gates may coalesce at the corners of the screen or occupy an entire side. Eqns. 2, 3, and 4 also yield information about the location of the gates. We start with the approximate equation (7) for the value of y that gives the maximum volume of the tetrahedron

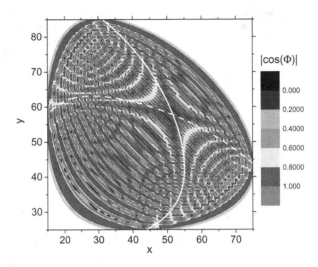

Fig. 3. $|\cos(\Phi)|$ for angular momenta of Figure 1. This figure is based on the Φ in figure 2.

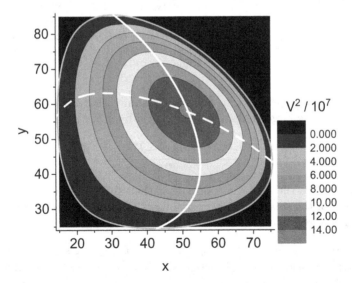

Fig. 4. Volume, caustics, ridges for angular momenta of Figure 1

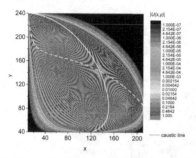

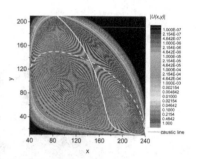

(a) Plot of $|U(x,y)|$ for $a = 100, b = 110, c = 130$ and $d = 140$. Ranges are $10 \leq x \leq 210$ and $40 \leq y \leq 240$. This corresponds to fig 1b in [15]. Caustic and ridge lines are shown. . In this case $v = 0$ and therefore the two Regge conjugates are identical. Note the coalescence of Northern and Western gates at the upper left corner, also because $v = 0$, Eq. 4.

(b) Plot of $|U(x,y)|$ for $a = 100, b = 140, c = 130$ and $d = 110$. $40 \leq x \leq 240$ and $10 \leq y \leq 210$. This corresponds to fig. 1c in [15]. As in the previous case Fig. 5a, the values spanned by x and y are both $2a + 1$, but they are interchanged, namely the convention for the orientation of the screen is not adopted. Again, since $u = 0$, the two Regge conjugates are identical. Actually the previous case and this one are connected by a classical exchange symmetry, and the two figures are related by reflection with respect to the diagonal of the screen connecting lower left and upper right corners. Now the coalescence is between the East and South gates, which are moved to the lower right corner of the screen, because $u = 0$, Eq. 3.

Fig. 5

for a given x. This approximate equation assumes that all of the quantities A, B, C, D, X, Y are large enough to be replaced by a, b, c, d, x, y.

$$y_{Vmax}^2 = \frac{(a^2 - b^2)(c^2 - d^2) + (a^2 + b^2 + c^2 + d^2)x^2 - x^4}{2x^2}. \tag{7}$$

This equation gives the analytic results that the values of y giving positive V^2 in the corners of the screen are given as follows (We are assuming conventional ordering). These results were first described in Ref. [15].

1. For $r = 0$ (Eq. 2), For $x = b - a$, $y_{Vmax} = d - a$. Positive V^2 is found at the south-west corner. As $|r|$ increases, the lower branch of the caustic line is found further from this corner. See figure 6a.
2. For $u = 0$ (Eq. 3), For $x = b + a$, $y_{Vmax} = d - a$. Positive V^2 is found at the south-east corner. See figure 5b.
3. For $v = 0$ (Eq. 4), For $x = b - a$, $y_{Vmax} = d + a$. Positive V^2 is found at the north-west corner. See figure 5a.

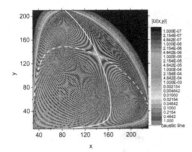

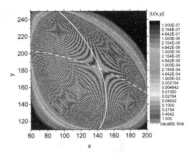

(a) Plot of $|U(x,y)|$ for $a = 100$, $b = 130$, $c = 140$ and $d = 110$. $30 \leq x \leq 230$ and $10 \leq y \leq 210$. This corresponds to fig.1d in [15], where only the caustic and ridge curves were given. The values spanned by x and y are both $2a + 1$. As in the previous two cases, a relationship holds: here we have $r = 0$ (Eq. 2), and the two Regge conjugates are again identical, but the coalescence is now between the West and South gates, in the lower left corner of the screen. A reflection symmetry is to be noted with respect to the diagonal of the screen connecting lower left and upper right corners.

(b) Plot of $|U(x,y)|$ for $a = 70$, $b = 130$, $c = 180$ and $d = 180$. $60 \leq x \leq 200$ and $110 \leq y \leq 250$. This plot of $|U(x,y)|$ corresponds to Fig. 2 in [15], where only the caustic and ridge curves were given for $a = 100, b = 100, c = 150$ and $d = 210$. Ranges are $60 = x = 200$ and $110 = y = 250$. This latter plot is not canonical: in fact, the Regge conjugate has $a = 70; b = 130; c = 180, d = 180$.

Fig. 6.

We can note that positive V^2 will never occur in the north-east corner. More than one of these conditions may be present. See figures 8a and 8b for $u = v = 0$, figure 9a for $r = v = 0$ (here the west side of the caustic line is the line $x = b - a$, and figure 9b for $r = u = v = 0$.

3   Images and Discussion

This section reports and discusses a series of graphs where the plots of caustics and ridges published in Ref. [15] are superimposed on $x - y$ color plots of true U. The following features are common to all of the plots except Figures 2, 3, and 4.

1. The angular momenta a, b, c, d are written in Canonical form, which means that the screens have dimension $(2a + 1) \times (2a + 1)$, and the ranges of x and y are $b - a \leq x \leq b + a$ and $d - a \leq y \leq d + a$.
2. All the figures show the caustic line (light gray oval)that encircles the central, classical regions where $(V^2 > 0)$.
3. All the figures also show the ridge lines solid and dashed white lines.
4. All of the figures except Figures 2, 3, and 4 display $|U(x,y)|$.

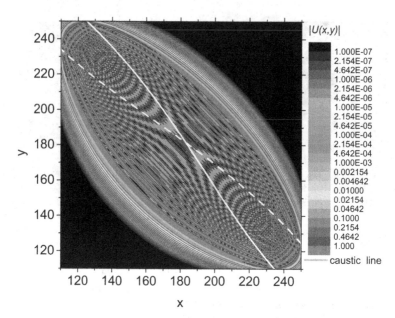

Fig. 7. Plot of $|U(x,y)|$ for $a = 70$, $b = 180$, $c = 130$ and $d = 180$. $110 \leq x \leq 250$ and $110 \leq y \leq 250$. Plot of $|U(x,y)|$ corresponding to Fig. 3 in [15], where only the caustic and ridge curves were given, for $a = 100; b = 150; c = 100; d = 210$. Since the Regge conjugate has $a = 70; b = 180; c = 130$ and $d = 180$, the plot can be taken as canonical, at variance with Fig 8a. Now since $u = v = -30$ there is a Piero line which is the diagonal from the lower left corner to the upper right corner, making the plot symmetrical by reflection.

3.1   General Case

The general case is found when there are no special symmetries, and Figure 1 shows an example. This plot of $|U(x,y)|$ shows several striking features that can be explained inside the caustics line (boundary between $V^2 \geq 0$ and $V^2 < 0$) with the Ponzano-Regge theory [3]. The figure shows that the U are small where V^2 is maximum. This occurs at the intersection of the two ridge lines, which show the This occurs where the V^2 is maximum as a function of y for given x and as a function of x for given y (see Figure 4). The magnitude of U tends to increase as the caustic line is approached the x, y point corresponding to the maximum. Most of the structure in Fig. 1 is a consequence of the Ponzano-Regge phase (Cosine term in equation 5). Figure 2 shows this phase for the same values of a, b, c, d. Figure 3 gives the $|\cos \Phi|$ values for the screen, and it is obvious that the structure in Fig. 1 agrees well with the expectations of the Ponzano-Regge theory. We have found that the same argument can also explain all of the images in Figures 5-9 in this paper.

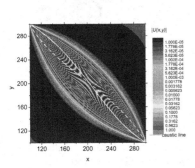

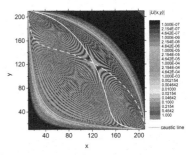

(a) Plot of $|U(x,y)|$ for $a = 100$, $b = 200$, $c = 100$ and $d = 200$. $100 \leq x \leq 300$ and $100 \leq y \leq 300$. Plot of $|U(x,y)|$ for the case of Fig. 4a in [15], where only the caustic and ridge curves were given, the canonical form being endorsed when parameters are rewritten exchanging columns as follows: $a = 100; b = 200; c = 100$ and $d = 200$. Here, since $u = v = 0$, Regge symmetry makes conjugates identical, and there is a Piero line symmetry. Note for comparison to the following case that $r = -100$.

(b) Plot of $|U(x,y)|$ for $a = 100$, $b = 110$, $c = 100$ and $d = 110$. $10 \leq x \leq 210$ and $10 \leq y \leq 210$. Plot of $|U(x,y)|$ for the case of Fig. 4b in [15], where only the caustic and ridge curves were given, the canonical form being endorsed when parameters are rewritten exchanging columns as follows: $a = 100; b = 110; c = 100$ and $d = 110$. As in Fig 9, since $u = v = 0$, Regge symmetry makes conjugates identical, and there is Piero symmetry. Note for comparison to the previous case that now $r = -10$:the lower difference between sums of columns shows qualitative shape changes for mismatch between columns.

Fig. 8

Figure 1 also shows the general fact that the magnitude of the U are oscillatory in the classical region, $V^2 \geq 0$, and exponentially decreasing as x, y is moved deeper into the nonclassical region. The values for U can be estimated with a suitable extension of the Ponzano-Regge theory (See Ref. [5]).

3.2 Symmetric Cases

Figures 5-9 show images of screens with different symmetries. They illustrate cases where the gates coalesce in the northwest, southwest, and southeast corners, and Piero line symmetries. They also show cases where the Regge conjugates are the same as the $6j$ with the original arguments, and where the $6j$ approximate $3j$ symbols.

4 Discussion, Additional and Concluding Remarks

The extensive images of the exactly calculated $6j$'s on the square screens illustrate how the caustic curves separate the classical and nonclassical regions,

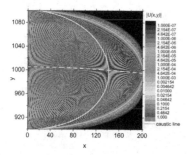

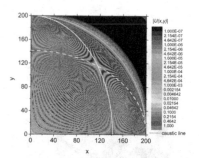

(a) Plot of $|U(x,y)|$ for $a = 100$, $b = 100$, $c = 1000$ and $d = 1000$. $0 \leq x \leq 200$ and $900 \leq y \leq 1100$. Case when both $r = 0$ and $v = 0$. This corresponds to fig.5 in [15], where only the caustic and ridge curves were given. Now $r = v = 0$, and the two Regge conjugates are again identical, but the coalescence is now of both the North and South gates with the West gate, on the full line from the lower left to the upper left corners of the screen, As noted in [15], since a, b and x are smaller than c, d and y, we can regard this plot as that of a $3j$ symbol, (: : :) where the entries in the upper row are the angular momenta $100, 100, x$ and the corresponding projections in the lower row are y - 1000, $1000 - y$, 0. Note that a reflection along the y line by mirror symmetry would lead to a replica of the image on the screen whereby the plane would consist of a classically allowed region limited by an ellipse as a caustic curve. In the view of the plot as that of the $3j$ symbol, described above, the operation corresponds to that of allowing one of the nonzero projection to change sign.

(b) Plot of $|U(x,y)|$ for the fully symmetric case: $a = 100$, $b = 100$, $c = 100$ and $d = 100$. $0 \leq x \leq 200$ and $0 \leq y \leq 200$. This corresponds to fig.6 in [15], where only the caustic and ridge curves were given. The three relationships as in figures 4, 5 and 6 occur here, since $r = u = v = 0$. The two Regge conjugates are again identical: one of the coalescences is again as in Fig 11 of both the North and South gates with the West gate, on the full line from the lower left to the upper left corners of the screen; :now another coalescence is of both the East and West gates with the South gate, on the full line from the lower left to the lower right corners of the screen. Also since $ruv = 0$ there is a Piero line as the diagonal from the lower left corner to the upper right corner, making the plot symmetrical by reflection along this line. As noted in [15], repeated reflections along the x and y lines by mirror symmetry would lead to replicas of the image on the screen, whereby the plane would consist of classically allowed regions limited by circles as caustics, tangent in four points.

Fig. 9

where they show wavelike and evanescent behaviour respectively. Limiting cases, and in particular those referring to $3j$ and Wigner's d matrix elements can be analogously depicted and discussed. Interesting also are the ridge lines, which separate the images in the screen tending to qualitatively different foldings of the quadrilateral, namely convex in the upper right region, concave in the upper left and lower right ones, and crossed in the lower left region.

Catastrophe theory classification. The pictures of 6-j on the screen in the previous Section exhibit most clearly features amenable to be classified in terms of catastrophe theory, with a panorama of valley bottoms, ridges and both elliptic and hyperbolic umbilics arising in the two-dimensional membrane-like modes. See [9], [22].

Chirality gates. This remark concerns the formal analogy between the present problem of four angular momenta arranged as vectors having a (not necessarily) planar quadrilater structure and those of the motion of tetra-atomic or four center structures where bonds can be treated as *rigid* while bending and torsion modes are allowed. If we consider A, B, C, D, as the lengths of the four bonds, and X and Y as the diagonals of the quadrilateral, i.e. the distances between atoms not connected by bonds, a *screen* representation can be set up, the *caustic* corresponding to allowed planar configurations. It is known that transition between chirality pairs in tetrahedral structure corresponds to *flattened* structures, and therefore the *caustic* curves shows configurations through which such a system would find its way to chirality exchange modes. In particular the four points where in the generic case (figures 1-4) the caustic touches the screen, are labeled accordingly North, West, South and East *gates*, since they mark where and how a planar structure should fold to perform such chirality interchange mode. In Ref. [21] a similarity is also pointed out with the celebrated problem of the kinematics of the four-bar linkage, the fundamental mobile mechanism of engines.

Alternative mappings. Motivated from the phase-space analysis of semiclassical dynamics in Ref. [10], alternatively to the x, y *screen* it is interesting to consider other conjugated variables such as e.g. x and the associated momentum p_x, corresponding to a dihedral torsional mode. The corresponding mapping, rather than on a square, is on a spherical triangle on the surface of the sphere S^2. We are also exploring a third type of mappings involving the modes of torsion angles corresponding e.g. to p_x and p_y, of interest for intramolecular dynamics.

Work on extensions to $3nj$ symbols, to q analogues, and to alternative coordinates of elliptic type is in progress.

References

1. Aquilanti, V., Bitencourt, A., da S. Ferreira, C., Marzuoli, A., Ragni, M.: Quantum and semiclassical spin networks: from atomic and molecular physics to quantum computing and gravity. Physica Scripta 78, 058103 (2008)
2. Aquilanti, V., Bitencourt, A., da S. Ferreira, C., Marzuoli, A., Ragni, M.: Combinatorics of angular momentum recoupling theory: spin networks, their asymptotics and applications. Theor. Chem. Accounts 123, 237 (2009)
3. Ponzano, G., Regge, T.: Semiclassical limit of Racah coefficients. In: Bloch, F., et al. (eds.) Spectroscopic and Group Theoretical Methods in Physics, pp. 1–58. North–Holland, Amsterdam (1968)
4. Neville, D.: A technique for solving recurrence relations approximately and its application to the $3 - j$ and $6 - j$ symbols. J. Math. Phys. 12, 2438 (1971)
5. Schulten, K., Gordon, R.: Semiclassical approximations to 3j- and 6j-coefficients for quantum-mechanical coupling of angular momenta. J. Math. Phys. 16, 1971–1988 (1975)

6. Schulten, K., Gordon, R.: Exact recursive evaluation of 3j- and 6j-coefficients for quantum mechanical coupling of angular momenta. J. Math. Phys. 16, 1961–1970 (1975)
7. Ragni, M., Bitencourt, A.C., Aquilanti, V., Anderson, R.W., Littlejohn, R.G.: Exact computation and asymptotic approximations of 6j symbols: Illustration of their semiclassical limits. Int. J. Quantum Chem. 110(3), 731–742 (2010)
8. Aquilanti, V., Cavalli, S., Coletti, C.: Angular and hyperangular momentum recoupling, harmonic superposition and Racah polynomials. a recursive algorithm. Chem Phys. Letters 344, 587–600 (2001)
9. Littlejohn, R.G., Yu, L.: Uniform semiclassical approximation for the Wigner 6j-symbol in terms of rotation matrices. J. Phys. Chem. A 113, 14904–14922 (2009)
10. Aquilanti, V., Haggard, H.M., Hedeman, A., Jeevangee, N., Littlejohn, R., Yu, L.: Semiclassical mechanics of the Wigner 6j-symbol. J. Phys. A 45(065209) (2012), arXiv:1009.2811v2 [math-ph]
11. Aquilanti, V., Capecchi, G.: Harmonic analysis and discrete polynomials. From semiclassical angular momentum theory to the hyperquantization algorithm. Theor. Chem. Accounts 104, 183–188 (2000)
12. De Fazio, D., Cavalli, S., Aquilanti, V.: Orthogonal polynomials of a discrete variable as expansion basis sets in quantum mechanics. The hyperquantization algorithm. Int. J. Quantum Chem. 93, 91–111 (2003)
13. Aquilanti, V., Cavalli, S., De Fazio, D.: Angular and Hyperangular Momentum Coupling Coefficients as Hahn Polynomials. J. Phys. Chem. 99(42), 15694–15698 (1995)
14. Koekoek, R., Lesky, P., Swarttouw, R.: Hypergeometric orthogonal polynomials and their q-analogues. Springer (2010)
15. Bitencourt, A.C.P., Marzuoli, A., Ragni, M., Anderson, R.W., Aquilanti, V.: Exact and asymptotic computations of elementary spin networks: Classification of the quantum–classical boundaries. In: Murgante, B., Gervasi, O., Misra, S., Nedjah, N., Rocha, A.M.A.C., Taniar, D., Apduhan, B.O. (eds.) ICCSA 2012, Part I. LNCS, vol. 7333, pp. 723–737. Springer, Heidelberg (2012), See arXiv:1211.4993[math-ph]
16. Varshalovich, D., Moskalev, A., Khersonskii, V.: Quantum Theory of Angular Momentum. World Scientific, Singapore (1988)
17. Aquilanti, V., Haggard, H.M., Littlejohn, R.G., Yu, L.: Semiclassical analysis of Wigner 3 j -symbol. J. Phys. A 40(21), 5637–5674 (2007)
18. Anderson, R.W., Aquilanti, V.: The discrete representation correspondence between quantum and classical spatial distributions of angular momentum vectors. J. Chem. Phys. 124, 214104 (9 pages) (2006)
19. Anderson, R.W., Aquilanti, V., da Silva Ferreira, C.: Exact computation and large angular momentum asymptotics of 3nj symbols: semiclassical disentangling of spin networks. J. Chem. Phys. 129, 161101–161105 (2008)
20. Anderson, R.W., Aquilanti, V., Bitencourt, A.C.P., Marinelli, D., Ragni, M.: The screen representation of spin networks: 2D recurrence, eigenvalue equation for 6j symbols, geometric interpretation and hamiltonian dynamics. In: Murgante, B., Misra, S., Carlini, M., Torre, C.M., Quang, N.H., Taniar, D., Apduhan, B.O., Gervasi, O. (eds.) ICCSA 2013, Part II. LNCS, vol. 7972, pp. 46–59. Springer, Heidelberg (2013)
21. Aquilanti, V., Marinelli, D., Marzuoli, A.: Hamiltonian dynamics of a quantum of space: hidden symmetries and spectrum of the volume operator, and discrete orthogonal polynomials. J. Phys. A: Math. Theor. 46, 175303 (2013), arXiv:1301.1949v2 [math-ph]
22. Gilmore, R.: Catastrophe Theory for Scientists and Engineers. Dover, New York (1993)

Unit Disk Cover Problem in 2D

Rashmisnata Acharyya[1], Basappa Manjanna[2], and Gautam K. Das[2]

[1] Department of Computer Science and Engineering,
Tezpur University, Assam, 786028, India
[2] Department of Mathematics, Indian Institute of Technology Guwahati
Guwahati - 781039, India

Abstract. In this paper we consider the *discrete unit disk cover* problem and the *rectangular region cover* problem as follows.

Given a set $\mathcal{P}$ of points and a set $\mathcal{D}$ of unit disks in the plane such that $\cup_{D_i \in \mathcal{D}} D_i$ covers all the points in $\mathcal{P}$, select minimum cardinality subset $\mathcal{D}^* \subseteq \mathcal{D}$ such that each point in $\mathcal{P}$ is covered by at least one disk in $\mathcal{D}^*$.

Given rectangular region $\mathcal{R}$ and a set $\mathcal{D}$ of unit disks in the plane such that $\mathcal{R} \subseteq \cup_{D_i \in \mathcal{D}} D_i$, select minimum cardinality subset $\mathcal{D}^{**} \subseteq \mathcal{D}$ such that each point of a given rectangular region $\mathcal{R}$ is covered by at least one disk in $\mathcal{D}^{**}$.

For the first problem, we propose an $(9 + \epsilon)$-factor $(0 < \epsilon \le 6)$ approximation algorithm. The previous best known approximation factor was 15 [Fraser, R., López-Ortiz, A.: The within-strip discrete unit disk cover problem, Can. Conf. on Comp. Geom. 61–66 (2012)]. For the second problem, we propose (i) an $(9 + \epsilon)$-factor $(0 < \epsilon \le 6)$ approximation algorithm, (ii) an 2.25-factor approximation algorithm in reduce radius setup, improving previous 4-factor approximation result in the same setup [Funke, S., Kesseelman, A., Kuhn, F., Lotker, Z., Segal, M.: Improved approximation algorithms for connected sensor cover. Wir. Net. 13, 153–164 (2007)].

The solution of the *discrete unit disk cover* problem is based on a *polynomial time approximation scheme* (PTAS) for the subproblem *line separable discrete unit disk cover*, where all the points in $\mathcal{P}$ are on one side of a line and covered by the disks centered on the other side of that line.

Keywords: Discrete Unit Disk Cover, Approximation Algorithm, Computational Geometry.

1 Introduction

In the *unit disk cover* (UDC) problem, we consider two problems, namely the *discrete unit disk cover* (DUDC) problem and the *rectangular region cover* (RRC) problem. In the DUDC problem, given a set $\mathcal{P} = \{p_1, p_2, \ldots, p_n\}$ of n points and a set $\mathcal{D} = \{d_1, d_2, \ldots, d_m\}$ of m unit disks in the plane, we wish to determine the minimum cardinality set $\mathcal{D}^* \subseteq \mathcal{D}$ such that $\mathcal{P} \cap \mathcal{D}^* = \mathcal{P}$. In the

B. Murgante et al. (Eds.): ICCSA 2013, Part II, LNCS 7972, pp. 73–85, 2013.

rectangular region cover (RRC) problem, given a rectangular region $\mathcal{R}$ and a set $\mathcal{D} = \{d_1, d_2, \ldots, d_m\}$ of m unit disks in the plane, the objective is to determine the minimum cardinality set $\mathcal{D}^{**} \subseteq \mathcal{D}$ such that $\mathcal{R} \cap \mathcal{D}^{**} = \mathcal{R}$. The DUDC and the RRC problems are geometric versions of the general set cover problem which is known to be NP-complete [14]. The general set cover problem is not approximable within a factor of $c \log n$, for some constant c, where n is the size of input. However, the DUDC and the RRC problems admit constant factor approximation results. These two problems have been studied extensively due to their wide applications in wireless networks [6, 11, 19].

1.1   Related Work

The DUDC problem has a long history in the literature. It is an NP-complete problem [14]. The first constant factor approximation algorithm has been proposed by Brönnimann and Goodrich [4] using the concept of epsilon net. After that many authors proposed constant factor approximation algorithm for the DUDC problem [3, 6–8, 17, 18]. A summary of such results are available in [10]. Using local search, Mustafa and Ray [17] proposed a PTAS for the DUDC problem. The time complexity of their PTAS is $O(m^{2 \cdot (\frac{8\sqrt{2}}{\epsilon})^2 + 1} n)$ for $0 < \epsilon \leq 2$. Thus for $\epsilon = 2$, we have a 3-factor approximation result in $O(m^{65} n)$ time, which is not practical. This leads to further research on the DUDC problem for finding constant factor approximation algorithm with reasonable running time. Das et al. [10] proposed a 18-factor approximation algorithm. The running time of their algorithm is $O(mn + n \log n + m \log m)$. Recently, Fraser and López-Ortiz [12] proposed a 15-factor approximation algorithm for the DUDC problem, which runs in $O(m^6 n)$ time. Das et al. [9] studied a restricted version of the DUDC problem, where the centers of all the disks in $\mathcal{D}$ are within a unit disk and all the points in $\mathcal{P}$ are outside of that unit disk. They proposed a 2-factor approximation algorithm for this restricted version of the DUDC problem, which runs in $O((m + n)^2)$ time.

In the way to solve the DUDC problem, some authors consider a restricted version of the DUDC problem, which is known as *line-separable discrete unit disk cover* (LSDUDC) problem in the literature [10]. In this problem, the plane being divided into two half-planes ℓ^+ and ℓ^- defined by a line ℓ, all the points in $\mathcal{P}$ are in ℓ^- and the centers of disks in $\mathcal{D}$ are in $\ell^+ \cup \ell^-$ such that each point in $\mathcal{P}$ is covered by at least one disk centered in ℓ^+. Carmi et al. [8] described a 4-factor approximation algorithm for the LSDUDC problem. Later, Claude et al. [6] proposed a 2-factor approximation algorithm for the LSDUDC problem. Another restricted version of the DUDC problem is *within strip discrete unit disk cover* (WSDUDC) problem, where all the points in $\mathcal{P}$ and the centers of all the disks in $\mathcal{D}$ lie inside a strip of width h. Das et al. [10] proposed a 6-factor approximation algorithm for $h = 1/\sqrt{2}$. Later, Fraser and López-Ortiz [12] proposed a $3\lceil 1/\sqrt{1 - h^2}\rceil$-factor approximation result for $0 \leq h < 1$. They also proposed a 3-factor (resp. 4-factor) algorithm for $h \leq 4/5$ (resp. $h \leq 2\sqrt{2}/3$).

Agarwal and Sharir [2] studied the Euclidean k-center problem. Here, a set $\mathcal{P}$ of n points, a set $\mathcal{O}$ of m points, and an integer k are given. The objective is to find k disks centered at the points in $\mathcal{O}$ such that each point in $\mathcal{P}$ is covered by at least one disk and the radius of largest disk is minimum. This problem is known to be NP-hard [2]. For fixed k, Hwang et al. [15] presented a $m^{O(\sqrt{k})}$-time algorithm. Latter, Agarwal and Procopiuc [1] presented $m^{O(k^{1-1/d})}$-time algorithm for the d-dimensional points. Fowler et al. [13] considered the minimum geometric disk cover problem, where input is a set $\mathcal{P}$ of points in the plane and the objective is to compute minimum cardinality set $\mathcal{X}$ of unit disks such that each point in $\mathcal{P}$ is covered by at least one disks in $\mathcal{X}$. They proved that the problem is NP-hard. Hochbaum and Maass [16] proposed a PTAS for the geometric disk cover problem.

A *sector* is a maximal region formed by the intersection of a set of disks i.e., all points within the sector are covered by the same set of disks. Funke et al. [11] proposed *greedy sector cover* algorithm for the RRC problem. The approximation factor of their algorithm is $O(\log w)$, where w is the maximum number of sectors covered by a single disk. They proved that the greedy sector cover algorithm has an approximation algorithm no better than $\Omega(\log w)$. In the same paper, they proposed grid placement algorithm (based on the algorithm proposed by Bose et al. [5]) and proved that their algorithm produce 18π-factor approximation result. Though the algorithm is not guaranteeing full coverage of the region of interest, the area that remains uncovered can be bounded by the number of chosen grids. In the same paper, they have also considered the RRC problem in different setup. We denote this setup as *reduced radius* setup. Here, we assume that the region of interest $\mathcal{R}$ is also covered by the disks in $\mathcal{D}$ after reducing their radius to $(1 - \delta)$. δ is said to be *reduce radius parameter*. The reduce radius setup has many applications in wireless sensor networks, where coverage remains stable under small perturbations of sensing ranges/positions. In this setup an algorithm $\mathcal{A}$ is said to be β-factor approximation if $\frac{|\mathcal{A}_{out}|}{|opt|} \leq \beta$, where $\mathcal{A}_{out}$ is the output of algorithm $\mathcal{A}$ and opt is the optimum set of disks with reduced radius covering the region of interest. In reduce radius setup, Funke et al. [11] proposed a 4-factor approximation algorithm for the RRC problem.

1.2   Our Results

We provide a PTAS (i.e., $(1 + \mu)$-factor approximation algorithm) for the LS-DUDC problem, which runs in $O(m^{3(1+\frac{1}{\mu})}n \log n)$ time $(0 < \mu \leq 1)$. Using this PTAS, we present an $(9 + \epsilon)$-factor approximation algorithm for the DUDC problem in $O(\max(m^{3(1+\frac{6}{\epsilon})}n \log n, m^6 n))$ time, where $0 < \epsilon \leq 6$. For the RRC problem, we describe an $(9 + \epsilon)$-factor approximation algorithm using the algorithm for the DUDC. We also propose an 2.25-factor approximation algorithm in the reduce radius setup, improving previous 4-factor approximation result [11].

2   PTAS for LSDUDC Problem

Let ℓ be a horizontal line. We use ℓ^+ and ℓ^- to denote the half-planes above and below ℓ respectively. The definition of the LSDUDC problem is as follows:

A set $\mathcal{P}$ of points and a set $\mathcal{D}$ of unit disks such that each point in $\mathcal{P}$ is in ℓ^- and center of each disk in $\mathcal{D}$ is in $\ell^+ \cup \ell^-$ and the union of the disks centered in ℓ^+ covers all points in $\mathcal{P}$ are given. The objective is to find minimum cardinality set $\mathcal{D}^* \subseteq \mathcal{D}$ such that union of disks in $\mathcal{D}^*$ covers $\mathcal{P}$.

Definition 1. *We use $\mathcal{U}$ and $\mathcal{L}$ to denote the set of disks in $\mathcal{D}$ with centers in ℓ^+ and ℓ^- respectively. For a disk $d \in \mathcal{D}$, its boundary arc and center are denoted by $\theta(d)$ and $\alpha(d)$ respectively. A disk $d \in \mathcal{U}$ is said to be lower boundary disk if there does not exist $X = \mathcal{U} \setminus \{d\}$ such that $d \cap \ell^- \subset (\cup_{D \in X} D) \cap \ell^-$. For a lower boundary disk $d \in \mathcal{U}$, we use the term lower region to denote the region $d \cap \ell^-$ and lower arc to denote the arc $\theta(d) \cap \ell^-$ (see Fig. 1). We use $D_\ell = \{d^1, d^2, \ldots, d^s\} \subseteq \mathcal{U}$ to denote the set of all lower boundary disks. We use B_{region} to denote the region covered by the disks in D_ℓ.*

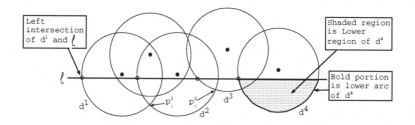

Fig. 1. Lower region, left intersection and lower boundary disks

Needless to mention that each disk in D_ℓ intersects the horizontal line ℓ and B_{region} contains all points in $\mathcal{P}$. Without loss of generality assume that $d^1, d^2, \ldots, d^s$ is the sorted order from left to right based on their left intersection point with the line ℓ (see Fig. 1). Since centers of the disks in D_ℓ are in ℓ^+ the number of intersection points (if any) of two disk arcs of D_ℓ in ℓ^- is one. For each disk $d^i \in D_\ell$ we define two points, namely p_l^i and p_r^i as follows:

p_l^i: If the disk d^i has intersection with d^{i-1} in ℓ^-, then p_l^i is the intersection point between $\theta(d^{i-1})$ and $\theta(d^i)$ in ℓ^-, otherwise p_l^i is the left intersection point between ℓ and $\theta(d^i)$.

p_r^i: If the disk d^i has intersection with d^{i+1} in ℓ^-, then p_r^i is the intersection point between $\theta(d^{i+1})$ and $\theta(d^i)$ in ℓ^-, otherwise p_r^i is the right intersection point between ℓ and $\theta(d^i)$.

Where d^0 and d^{s+1} are the two dummy disks having no intersection with d^1 and d^s respectively. For each $i = 1, 2, \ldots, s$ let $\mathcal{P}_i (\subseteq \mathcal{P})$ be the set of points lying between two vertical lines through p_l^i and p_r^i. Let e^i be the vertical line through the point p_r^i for $i = 1, 2, \ldots, s$. We use e^{i-} (resp. e^{i+}) to denote the region in

Algorithm 1. LSDUDC$(\mathcal{P}, \mathcal{D}, k, \ell)$

1: **Input:** Set $\mathcal{P}$ of points, set $\mathcal{D}$ of unit disks, a positive integer k and a horizontal line ℓ such that $\mathcal{P} \cap \ell^- = \mathcal{P}$ and union of the disks centered in ℓ^+ covers all the points in $\mathcal{P}$.

2: **Output:** Set $\mathcal{D}^* \subseteq \mathcal{D}$ of disks covering all the points in $\mathcal{P}$.

3: Set $\mathcal{D}^* \leftarrow \emptyset$

4: Find lower boundary disks set D_ℓ and arrange them from left to right as defined above. Let $D_\ell = \{d^1, d^2, \ldots, d^s\}$ be the lower boundary disks from left to right.

5: **for** $(i = 1, 2, \ldots s)$ **do**

6:    Compute the set $\mathcal{P}_i (\subseteq \mathcal{P})$

7: **end for**

8: Set $i \leftarrow 1$

9: **while** $(i \le s)$ **do**

10:    Find the maximum index j such that $\cup_{t=i,i+1,\ldots j} \mathcal{P}_t$ is covered by a set $\mathcal{D}_1 (\subseteq \mathcal{D})$ of disks with size k.

11:    $\mathcal{D}^* = \mathcal{D}^* \cup \mathcal{D}_1$, $i \leftarrow j + 1$

12: **end while**

13: Return $\mathcal{D}^*$

---

the left (resp. right) side of the vertical line e^i. Let $D^{i-}(\subseteq \mathcal{D})$ and $D^{i+}(\subseteq \mathcal{D})$ be the optimum cover of the points in $\mathcal{P} \cap e^{i-}$ and $\mathcal{P} \cap e^{i+}$ respectively.

Before proving approximation factor of the algorithm (Algorithm 1) for the LSDUDC problem, we first discuss some important properties of the disks in $\mathcal{U}$ and $\mathcal{L}$ separately.

Observation 1. *For two disks $d', d'' \in \mathcal{L}$, if $d', d'' \in D^{i-}$ and $d', d'' \in D^{i+}$, then both d' and d'' intersect e^i and both the intersections between $\theta(d')$ and $\theta(d'')$ lie in B_{region}.*

Proof. Both the disks d' and d'' intersect e^i because $d', d'' \in D^{i-}$ and $d', d'' \in D^{i+}$.

Since $d', d'' \in D^{i-}$ and $d', d'' \in D^{i+}$, then there exist points $p_0', p_0'' \in e^{i-} \cap \mathcal{P}$ and $p_1', p_1'' \in e^{i+} \cap \mathcal{P}$ such that $p_0', p_1' \in d'$ and $p_0'', p_1'' \in d''$ but $p_0', p_1' \notin d''$ and $p_0'', p_1'' \notin d'$ (see Fig. 2). Now, if d' and d'' do not intersect, then (i) either the distance of p_0' or p_1' from the line ℓ is greater than 1 (assuming $\alpha(d')$ is below $\alpha(d'')$) or (ii) either the distance of p_0'' or p_1'' from the line ℓ is greater than 1 (assuming $\alpha(d')$ is above $\alpha(d'')$), which leads to a contradiction because each point of $\mathcal{P}$ is covered by at least one disk centered above ℓ.

Now, if $\theta(d')$ and $\theta(d'')$ do not intersect in B_{region}, then either $\mathcal{P} \cap d' \subseteq \mathcal{P} \cap d''$ or $\mathcal{P} \cap d'' \subseteq \mathcal{P} \cap d'$ (Note: the centers of the disks d' and d'' lie below the line ℓ as $d', d'' \in \mathcal{L}$). Therefore, both d' and d'' cannot appear in the solutions D^{i-} and D^{i+}, which leads to a contradiction. Thus, $\theta(d')$ and $\theta(d'')$ intersect in B_{region}. $\square$

Definition 2. *A pair $(d', d'')(\in \mathcal{L} \times \mathcal{L})$ of disks is said to be weak (resp. strong) cover pair if $\theta(d')$ and $\theta(d'')$ intersect once (resp. twice) in B_{region}.*

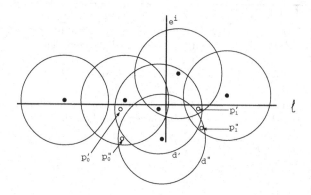

Fig. 2. Proof of Observation 1

Lemma 1. *For a weak cover pair $(d', d'')(\in \mathcal{L} \times \mathcal{L})$; $d', d'' \in D^{i-}$ and $d', d'' \in D^{i+}$ cannot happen simultaneously.*

Proof. On contrary, assume $d', d'' \in D^{i-}$ and $d', d'' \in D^{i+}$. By the definition of weak cover pair, $\theta(d')$ and $\theta(d'')$ does not intersect either in e^{i-} or e^{i+}. Therefore, either (i) $\mathcal{P}_{e^{i-}} \cap d' \subset \mathcal{P}_{e^{i-}} \cap d''$ or $\mathcal{P}_{e^{i-}} \cap d'' \subset \mathcal{P}_{e^{i-}} \cap d'$ or (ii) $\mathcal{P}_{e^{i+}} \cap d' \subset \mathcal{P}_{e^{i+}} \cap d''$ or $\mathcal{P}_{e^{i+}} \cap d'' \subset \mathcal{P}_{e^{i+}} \cap d'$, where $\mathcal{P}_{e^{i-}}$ (resp. $\mathcal{P}_{e^{i+}}$) is the set of points in $\mathcal{P}$ to the left (resp. right) of e^i (see Fig. 3). Thus, both the disks d', d'' cannot be in D^{i-} and D^{i+}. □

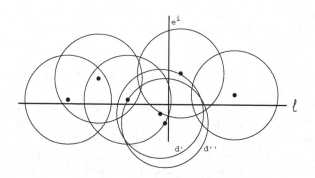

Fig. 3. Proof of Lemma 1

Lemma 2. *For a strong cover pair $(d', d'')(\in \mathcal{L} \times \mathcal{L})$, if $d', d'' \in D^{i-}$ (resp. $d', d'' \in D^{i+}$), then one intersection between $\theta(d')$ and $\theta(d'')$ lies in e^{i-} and other intersection lies in e^{i+}.*

Proof. If both the intersections between $\theta(d')$ and $\theta(d'')$ lies either in e^{i-} or e^{i+}, then from Observation 1 and the proof of Lemma 1 $d', d'' \in D^{i-}$ and

$d', d'' \in D^{i+}$ cannot happen simultaneously, which is a contradiction. Thus, the lemma follows.   □

Lemma 3. *For a strong cover pair* $(d', d'')(\in \mathcal{L} \times \mathcal{L})$ *with* $\alpha(d')$ *above* $\alpha(d'')$, *if the intersections of* $\theta(d')$ *and* $\theta(d'')$ *lie within lower boundary disks* d_x *and* d_y, *then either one intersection between* $\theta(d_x)$ *and* $\theta(d')$ *or* $\theta(d_y)$ *and* $\theta(d')$ *happens above the horizontal line* ℓ.

Proof. Without loss of generality assume that $\alpha(d')$ is above the intersection point of d' and d'' inside d_x. Let a and b be the two intersection points of $\theta(d)$ and $\theta(d')$. Therefore, $\alpha(d')$ should lie $\overline{\text{above at least one}}$ point among a and b. Let $\alpha(d')$ lies above a. By symmetry, $\overline{\alpha(d'), a}$ and $\overline{\alpha(d), b}$ are parallel (see Fig. 4). Thus, b must be above $\alpha(d)$ i.e., b must be above ℓ.   □

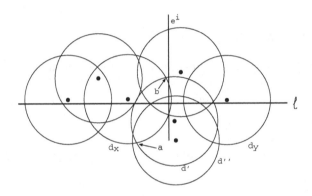

Fig. 4. Proof of Lemma 3

Lemma 4. $|D^{i-} \cap D^{i+} \cap \mathcal{L}| \leq 2.$

Proof. On the contrary, assume that $d_x, d_y, d_z \in D^{i-} \cap D^{i+} \cap \mathcal{L}$. Since $d_x, d_y, d_z \in D^{i-}$ as well as $d_x, d_y, d_z \in D^{i+}$, the disks d_x, d_y, d_z intersect each other in B_{region}. If any pair $(d, d') \in \Gamma = \{(d_x, d_y), (d_x, d_z), (d_y, d_z)\}$ do not form a weak cover pair nor a strong cover pair, then either $d \cap B_{region} \subseteq d' \cap B_{region}$ or $d \cap B_{region} \supseteq d' \cap B_{region}$, which contradict the fact that $d_x, d_y, d_z \in D^{i-} \cap D^{i+} \cap \mathcal{L}$. Again, from Lemma 1, no pair in Γ form a weak cover pair because $d_x, d_y, d_z \in D^{i-}$ as well as $d_x, d_y, d_z \in D^{i+}$. Therefore, each pair in Γ form a strong cover pair. Without loss of generality assume that $\alpha(d_x)$ is below $\alpha(d_y)$ and $\alpha(d_y)$ is below $\alpha(d_z)$ (see Fig. 5). If a is the intersection between $\theta(d_x)$ and $\theta(d_y)$ inside the lower boundary disk d (say) and below the horizontal line ℓ, then one intersection between $\theta(d_y)$ and $\theta(d_z)$ lies inside of d or d' (from Lemma 3). Therefore, $(d_x \cup d_y \cup d_z) \cap \mathcal{P} \cap e^{i-} \subseteq (d_x \cup d) \cap \mathcal{P} \cap e^{i-}$, which implies that D^{i-} is not optimum, leading to a contradiction. Thus, the lemma follows.   □

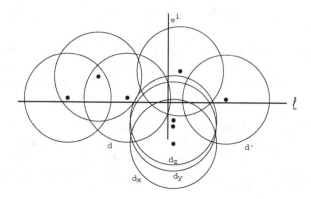

Fig. 5. Proof of Lemma 4

Lemma 5. $|D^{i-} \cap D^{i+} \cap \mathcal{U}| \le 1.$

Proof. Since the center of all the disks in $\mathcal{U}$ are in ℓ^+ and the points in $\mathcal{P}$ are in ℓ^-, two disks d_x, d_y in $\mathcal{U}$ can not intersect twice in ℓ^-. Therefore, $|D^{i-} \cap D^{i+} \cap \mathcal{U}|$ is at most 1. Thus, the lemma follows. $\qquad\square$

Lemma 6. $|D^{i-} \cap D^{i+}| \le 3.$

Proof. Follows from (i)$\mathcal{D} = \mathcal{U} \cup \mathcal{L}$ and (ii) Lemmas 4 and 5. $\qquad\square$

The following theorem says that the LSDUDC problem admits a PTAS.

Theorem 1. *Algorithm 1 produces $(1 + \frac{3}{k-3})$-factor approximation results in $O(m^k n \log n)$ time.*

Proof. For some integer t, let $j_1, j_2, \ldots, j_t$ the values of j in the while loop (line number 9) of the Algorithm 1. Let $\mathcal{Q}_v = \cup_{i=j_{v-1}+1, j_{v-1}+2,\ldots,j_v} \mathcal{P}_i$ for $v = 1, 2, \ldots, t$, where $j_0 = 0$. Algorithm 1 finds a covering for the sets $\{\mathcal{Q}_1, \mathcal{Q}_2, \ldots, \mathcal{Q}_t\}$ independently with each of size k (optimum size because in each iteration of the while loop in line number 9, Algorithm 1 finds maximum value of j's) except the covering of $\mathcal{Q}_t$. Let $\mathcal{D}^1, \mathcal{D}^2, \ldots, \mathcal{D}^t$ be the covering for $\mathcal{Q}_1, \mathcal{Q}_2, \ldots, \mathcal{Q}_t$ respectively. Lemma 6 says that $\mathcal{D}^i \cap \mathcal{D}^{i+1} \le 3$. Therefore, the total number of disks required to cover all the points by Algorithm 1 is $k(t-1) + |\mathcal{D}^t|$ whereas at least $(k-3)(t-1) + |\mathcal{D}^t|$ disks required in the optimum solution. Thus, the approximation factor of the Algorithm 1 is $(1 + \frac{3}{k-3})$.

The execution time to find lower boundary disks and to arrange them from left to right (line number 4) is $O(m \log m)$, where $m = |\mathcal{D}|$. To compute $\mathcal{P}_i$ for $i = 1, 2, \ldots s$ (for loop in line number 5) $O(n \log n)$ time is required, where $n = |\mathcal{P}|$. To implement while loop (line number 9), we first create set $\mathcal{P}^u (\subseteq \mathcal{P})$ of points such that $\mathcal{P}^u = \cup_{i=1,2,\ldots,u} \mathcal{P}_i$ for each $u = 1, 2, \ldots, s$, then for maximum j, we choose $j = 2^i$ for $i = 1, 2, \ldots, v$ such that $\mathcal{P}^{2^v}$ is not covered by k disks but

$\mathcal{P}^{2^{v-1}}$ is covered by k disks. Now, we need to perform a binary search among $[2^{v-1} + 1, 2^{v-1} + 2, \ldots, 2^v]$ for the maximum value of j. Therefore, the time complexity of the while loop (line number 9) is $O(m^k n \log n)$. Thus, the total time complexity of the Algorithm 1 is $O(m^k n \log n)$. □

2.1 An $(9 + \epsilon)$-Factor Approximation Algorithm for DUDC Problem

In this section, we wish to describe an $(9 + \epsilon)$-factor approximation algorithm for the DUDC Problem. Here a set $\mathcal{P}$ of n points and a set $\mathcal{D}$ of m unit disks are distributed in the plane; the objective is to choose minimum cardinality set $\mathcal{D}^*(\subseteq \mathcal{D})$ such that union of the disks in $\mathcal{D}^*$ covers $\mathcal{P}$. From Theorem 1, LSDUDC problem has an $(1 + \mu)$-factor approximation algorithm ($\mu = \frac{3}{k-3}$) and the running time of the algorithm is $O(m^{3(1+\frac{1}{\mu})} n \log n)$. Das et al. [10] proposed an approximation algorithm for DUDC problem using the algorithms for the LSDUDC and WSDUDC (with strip width $1/\sqrt{2}$) problems and proved that the approximation factor of the algorithm for the DUDC problem is

$6\times$ (approximation factor of an algorithm for the LSDUDC problem) + approximation factor of an algorithm for the WSDUDC (width $h = 1/\sqrt{2}$) problem.

Fraser and López-Ortiz [12] proposed a 3-factor approximation algorithm for WSDUDC (with width $h \leq 4/5$) problem in $O(m^6 n)$ time. Therefore, we have the following theorem for the DUDC problem.

Theorem 2. *The DUDC problem admits $(9 + \epsilon)$-factor approximation result in $O(\max(m^6 n, m^{3(1+\frac{6}{\epsilon})}) n \log n)$ time.*

Proof. The approximation factor of the algorithm for DUDC problem is ($6\times$ (approximation factor of an algorithm for the LSDUDC problem) + approximation factor of an algorithm for the WSDUDC (width $= 1/\sqrt{2}$) problem) [10]. Therefore, the approximation factor of the algorithm for the DUDC problem is $6 \times (1 + \mu) + 3 = 9 + \epsilon$, where $\epsilon = 6\mu$. The time complexity follows from time complexity of WSDUDC [12], and the complexity result stated in Theorem 1. □

3 Approximation Algorithms for RRC Problem

In the RRC problem, the inputs are (i) a set $\mathcal{D}$ of m unit disks and (ii) a rectangular region $\mathcal{R}$ such that $\mathcal{R} \subseteq \cup_{d \in \mathcal{D}} d$; the objective is to choose minimum cardinality set $\mathcal{D}^{**} \subseteq \mathcal{D})$ such that $\mathcal{R} \subseteq \cup_{d \in \mathcal{D}^{**}} d$. A *sector* f inside $\mathcal{R}$ is a maximal region inside $\mathcal{R}$ formed by the intersection of a set of disks. Thus each point within f is covered by the same set of disks. Let $\mathcal{F}$ be the set of all sectors (inside $\mathcal{R}$) formed by $\mathcal{D}$, and $|\mathcal{F}| = O(m^2)$. Now we construct a set of points $\mathcal{T}$ as follows: for each sector $f \in \mathcal{F}$ we add one arbitrary point $p \in f$ to $\mathcal{T}$. Therefore, covering the all sectors in $\mathcal{F}$ by minimum cardinality subset of $\mathcal{D}$ is equivalent to cover all the points in $\mathcal{T}$ by the minimum cardinality subset of $\mathcal{D}$). Thus, we have the following theorem:

Theorem 3. *The RRC problem has $(9+\epsilon)$-factor approximation algorithm with running time $O(\max(m^8, m^{6(1+3/\epsilon)} \log m)$.*

Proof. Consider an arbitrary point $p \in \mathcal{T}$. Let $f \in \mathcal{F}$ be the sector in which the point p lies. From the definition of sector, if a disk $d \in \mathcal{D}$ covers p, then the disk d also covers the whole sector f. Therefore, the instance $(\mathcal{R}, \mathcal{D})$ of the RRC problem is exactly same as the instance $(\mathcal{T}, \mathcal{D})$ of the DUDC problem. Note that $\mathcal{T} = O(m^2)$. Thus, the theorem follows from Theorem 2 by putting $n = m^2$. □

3.1   RRC Problem in Reduce Radius Setup

In this subsection we consider the RRC problem in reduce radius setup. In this setup, a set $\mathcal{D}$ of unit disks and a rectangular region $\mathcal{R}$ such that $\mathcal{R}$ is covered by the disks in $\mathcal{D}$ after reducing their radius to $(1 - \delta)$ are given. The objective is to choose minimum cardinality set $\mathcal{D}^{**}(\subseteq \mathcal{D})$ whose union covers $\mathcal{R}$. In the reduce radius setup an algorithm $\mathcal{A}$ is said to be β-factor approximation if $\frac{|\mathcal{A}_{out}|}{|opt|} \leq \beta$, where $\mathcal{A}_{out}$ is the output of $\mathcal{A}$ and opt is the optimum set of disks with reduced radius covering the region of interest. The reduce radius setup has many applications in wireless sensor networks, where coverage remains stable under small perturbations of sensing ranges and their positions. Here we propose an 2.25-factor approximation algorithm for this problem. The best known approximation factor for the same problem was 4 [11].

Observation 2. *Let $\nu = \sqrt{2}\delta$ and d be an unit disk centered at a point p. If d' is a disk of radius $(1 - \delta)$ centered within a square $\mathcal{S}$ of size $\nu \times \nu$ centered at p, then $d' \subseteq d$.*

Proof. Let c be the length of the diagonal of $\mathcal{S}$. Therefore, the maximum distance of any point within the square $\mathcal{S}$ of size $\nu \times \nu$ from the center point p is $c/2 = \delta$. Thus, the observation follows. □

Consider a grid with cells of size $\nu \times \nu$ over the region $\mathcal{R}$. Like Funke et al. [11] we also snap the center of each $d \in \mathcal{D}$ to the closest vertex of the grid and set its radius to $(1 - \delta)$. Let $\mathcal{D}'$ be the set of disks with radius $(1 - \delta)$ after snapping their centers. Let $\mathcal{R}'$ be a square of size 4×4 on the plane contained in $\mathcal{R}$. We define the regions *TOP, DOWN, LEFT, RIGHT, TOP-LEFT, TOP-RIGHT, DOWN-LEFT, DOWN-RIGHT* around $\mathcal{R}'$ as shown in Fig. 6. We now construct a set $\mathcal{D}_{RS}(\subseteq \mathcal{D}')$ such that any disk $d \in \mathcal{D}'$ and $d \notin \mathcal{D}_{RS}$ cannot participate to the optimum solution for covering the region $\mathcal{R}'$ by $\mathcal{D}'$. Note that, if a disk $d \in \mathcal{D}_{RS}$, then center of d is a grid vertex. The pseudo code for construction of $\mathcal{D}_{RS}$ is given in Algorithm 2.

Definition 3. *A disk $d \in \mathcal{D}'$ dominates another disk $d' \in \mathcal{D}'$ with respect to the region $\mathcal{R}'$ if $d \cap \mathcal{R}' \supseteq d' \cap \mathcal{R}'$.*

Lemma 7. *If $d \in \mathcal{D}'$ and $d \notin \mathcal{D}_{RS}$, then d cannot participate to the optimum solution for covering $\mathcal{R}'$ by minimum number of disks in $\mathcal{D}'$.*

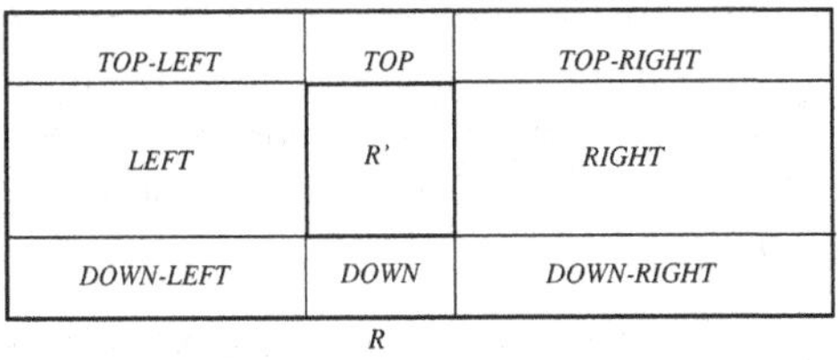

TOP-LEFT	TOP	TOP-RIGHT
LEFT	R'	RIGHT
DOWN-LEFT	DOWN	DOWN-RIGHT

R

Fig. 6. Definition of different regions

Algorithm 2. $Algorithm\_\mathcal{D}_{RS}(\mathcal{D}', \mathcal{R}', \nu)$

1: **Input:** Set $\mathcal{D}'$ of disks, a square region $\mathcal{R}'$ of size 4×4 and grid size ν.
2: **Output:** $\mathcal{D}_{RS}(\subseteq \mathcal{D}')$ such that no disk $d \notin \mathcal{D}_{RS}$ can participate to the optimum solution for covering the region $\mathcal{R}'$ by $\mathcal{D}'$.
3: Set $\mathcal{D}_{RS} \leftarrow \emptyset, \mathcal{D}_t \leftarrow \emptyset$
4: For each disk $d \in \mathcal{D}'$ having center in $\mathcal{R}'$, $\mathcal{D}_{RS} = \mathcal{D}_{RS} \cup \{d\}$
5: For each horizontal grid line segment h in *LEFT* add a disk $d \in \mathcal{D}'$ to $\mathcal{D}_{RS}$ if (i) $d \cap \mathcal{R}' \neq \emptyset$, (ii) center of d lies on h and (iii) center of d closest to $\mathcal{R}'$ than other disks having center on h. Similarly add disks to $\mathcal{D}_{RS}$ for the regions *RIGHT, TOP* and *DOWN*.
6: **for** (each horizontal grid line segment h in *TOP-RIGHT* from bottom to top) **do**
7:    Add a disk $d \in \mathcal{D}'$ to $\mathcal{D}_t$ if (i) $d \cap \mathcal{R}' \neq \emptyset$, (ii) center of d lies on h and (iii) there does not exits any disk $d' \in \mathcal{D}_t$ dominating d.
8: **end for**
9: $\mathcal{D}_{RS} = \mathcal{D}_{RS} \cup \mathcal{D}_t$
10: repeat steps 6-9 for *TOP-LEFT, DOWN-LEFT* and *DOWN-RIGHT*.
11: Return $\mathcal{D}_{RS}$

Proof. The center of d is in outside of $\mathcal{R}'$ as $d \notin \mathcal{D}_{RS}$ (see line number 4 of Algorithm 2). With out loss of generality assume that center of d is in *LEFT* and on the horizontal grid line segment h. By our construction of the set $\mathcal{D}_{RS}$, there exists a disk $d' \in \mathcal{D}_{RS}$ centered on h such that (a) $d' \cap \mathcal{R}' \neq \emptyset$, (b) center of d' lies on h and (c) center of d' closest to $\mathcal{R}'$ than other disks having center on h. Therefore, d' dominates d. Similarly, we can prove for other cases also. Thus, the lemma follows. □

Lemma 8. $|\mathcal{D}_{RS}| \leq \frac{16}{\nu^2} + \frac{20}{\nu}$.

Proof. The lemma follows from the following facts: (i) the maximum number of grid vertices in $\mathcal{R}'$ is $\frac{16}{\nu^2}$ and each of them can contribute one disk in $\mathcal{D}_{RS}$, (ii) the maximum number of horizontal grid line segment in the regions *TOP-LEFT, LEFT, DOWN-LEFT, DOWN-RIGHT, RIGHT* and *TOP-RIGHT* that can contribute a disk in $\mathcal{D}_{RS}$ is $\frac{12}{\nu}$ and (iii) the maximum number of vertical grid line segment in the regions *TOP* and *DOWN* that can contribute a disk in $\mathcal{D}_{RS}$ is $\frac{8}{\nu}$. Thus, the lemma follows. □

From Observation 2 and Lemma 8, we can compute a cover of $\mathcal{R}'$ by $\mathcal{D}''(\subseteq \mathcal{D})$ with minimum number of disks using brute-force method, where $\mathcal{D}''$ is the set of unit disks corresponding to the reduced-radius disks in $\mathcal{D}_{RS}$. The running time of the brute-force algorithm is $O(2^{\frac{16}{\nu^2}+\frac{20}{\nu}})$ (see Lemma 8). Although this worst-case running time of the brute-force algorithm is exponential in $\frac{1}{\nu^2}$, in practice, it is very small. We now describe approximation factor of our proposed algorithm for RRC problem in reduce radius setup.

Theorem 4. *In the reduce radius setup, the RRC problem has an 2.25-factor approximation algorithm with running time* $O(q2^{\frac{16}{\nu^2}+\frac{20}{\nu}})$, *where q is the minimum number of squares of size* 4×4 *covering* $\mathcal{R}$.

Proof. From the above discussion, for rectangle of size 4×4, we have optimum solution for the RRC problem. Note that the diameter of each disk of the RRC instance is 2. Therefore, we can apply shifting strategy described by Hochbaum and Maass [16] to solve the RRC problem and the approximation factor is $(1 + 1/2)^2 = 2.25$ and the running time of the algorithm is $O(q2^{\frac{16}{\nu^2}+\frac{20}{\nu}})$. Thus, the theorem follows.   □

Note that, Funke et al. [11] proposed a 4-factor approximation algorithm in $O(q2^{\frac{20}{\nu^2}})$ time for the RRC problem in reduce radius setup. Thus, our proposed algorithm is a significant improvement over the existing algorithm in literature.

4   Conclusion

In this paper, we have proposed a PTAS for the LSDUDC problem. Using this PTAS, we proposed an $(9 + \epsilon)$-factor approximation algorithm for the DUDC problem, improving previous 15-factor approximation result for the same problem [12]. The running time of our proposed algorithm for $\epsilon = 3$ (i.e., a 12-factor approximation of the DUDC problem) is same as the running time of 15-factor approximation algorithm. We have also proposed an $(9+\epsilon)$-factor approximation algorithm for the RRC problem, which runs in $O(\max(m^8, m^{4(1+3/\epsilon)} \log m)$ time. In the reduce radius setup, we proposed an 2.25-factor approximation algorithm. The previous best known approximation factor was 4 [11]. The running time of our proposed algorithm for the RRC problem in reduce radius setup is less than that of 4-factor approximation algorithm proposed in [11] for reasonably small value of $\delta(= \frac{\nu}{\sqrt{2}})$, where δ is the radius reduction parameter. Since the number of disks participating in the solution of 4×4 square is constant for fixed value of δ, the number of disks participating in the solution of $L \times L$ square is constant. Therefore, using the shifting strategy proposed by Hochbaum and Maass [16], we can design a PTAS for the RRC problem in reduce radius setup.

References

1. Agarwal, P.K., Procopiuc, C.: Exact and approximation algorithms for clustering. Algorithmica 33, 201–226 (2002)
2. Agarwal, P.K., Sharir, M.: Efficient algorithm for geometric optimization. ACM Comp. Surv. 30, 412–458 (1998)

3. Ambühl, C., Erlebach, T., Mihalák, M., Nunkesser, M.: Constant-factor approximation for minimum-weight (Connected) dominating sets in unit disk graphs. In: Díaz, J., Jansen, K., Rolim, J.D.P., Zwick, U. (eds.) APPROX 2006 and RANDOM 2006. LNCS, vol. 4110, pp. 3–14. Springer, Heidelberg (2006)
4. Brönnimann, H., Goodrich, M.: Almost optimal set covers in finite VC-dimension. Disc. and Comp. Geom. 14, 463–479 (1995)
5. Bose, P., Maheshwari, A., Morin, P., Morrison, J.: The grid placement problem. In: Dehne, F., Sack, J.-R., Tamassia, R. (eds.) WADS 2001. LNCS, vol. 2125, pp. 180–191. Springer, Heidelberg (2001)
6. Claude, F., Das, G.K., Dorrigiv, R., Durocher, S., Fraser, R., López-Ortiz, A., Nickerson, B.G., Salinger, A.: An improved line-separable algorithm for discrete unit disk cover. Disc. Math., Alg., and Appl. 2, 77–87 (2010)
7. Călinescu, G., Măndoiu, I., Wan, P.J., Zelikovsky, A.: Selecting forwarding neighbors in wireless ad hoc networks. Mob. Net. and Appl. 9, 101–111 (2004)
8. Carmi, P., Katz, M.J., Lev-Tov, N.: Covering points by unit disks of fixed location. In: Tokuyama, T. (ed.) ISAAC 2007. LNCS, vol. 4835, pp. 644–655. Springer, Heidelberg (2007)
9. Das, G.K., Das, S., Nandy, S.C.: Homogeneous 2-hop broadcast in 2D. Comp. Geom.: Theo. and Appl. 43, 182–190 (2010)
10. Das, G.K., Fraser, R., López-Ortiz, A., Nickerson, B.G.: On the discrete unit disk cover problem. Int. J. Comp. Geom. and Appl. 22, 407–420 (2012)
11. Funke, S., Kesseelman, A., Kuhn, F., Lotker, Z., Segal, M.: Improved approximation algorithms for connected sensor cover. Wir. Net. 13, 153–164 (2007)
12. Fraser, R., López-Ortiz, A.: The within-strip discrete unit disk cover problem. In: 24th Canadian Conference on Computational Geometry, pp. 61–66 (2012)
13. Fowler, R., Paterson, M., Tanimato, S.: Optimal packing and covering in the plane are NP-complete. Inf. Proc. Let. 12, 133–137 (1981)
14. Garey, M.R., Johnson, D.S.: Computers and Intractability: A Guide to the Theory of NP-Completeness. W.H. Freeman and Company, New York (1979)
15. Hwang, R., Lee, R., Chang, R.: The generalized searching over separators strategy to solve some NP-hard problems in subexponential time. Algorithmica 9, 398–423 (1993)
16. Hochbaum, D.S., Maass, W.: Approximation schemes for covering and packing problems in image processing and VLSI. J. of ACM 32, 130–136 (1985)
17. Mustafa, N.H., Ray, S.: Improved results on geometric hitting set problems. Disc. and Comp. Geom. 44, 883–895 (2010)
18. Narayanappa, S., Vojtechovsky, P.: An improved approximation factor for the unit disk covering problem. In: 18th Canadian Conference on Computational Geometry, pp. 15–18 (2006)
19. Yang, D., Misra, S., Fang, X., Xue, G., Zhang, J.: Two-tiered constrained relay node placement in wireless sensor networks: computational complexity and efficient approximations. IEEE Trans. on Mob. Comp. 11, 1399–1411 (2012)

Automated Extraction of Community Mobility Measures from GPS Stream Data Using Temporal DBSCAN

Sungsoon Hwang[1], Timothy Hanke[2], and Christian Evans[2]

[1] Department of Geography, DePaul University, Chicago, USA
shwang9@depaul.edu
[2] Physical Therapy Program, Midwestern University, Downers Grove, USA
{THANKE,CEvans}@midwestern.edu

Abstract. Inferring community mobility of patients from GPS data has received much attention in health research. Developing robust mobility (or physical activity) monitoring systems relies on the automated algorithm that classifies GPS track points into events (such as stops where activities are conducted, and routes taken) accurately. This paper describes the method that automatically extracts community mobility measures from GPS track data. The method uses temporal DBSCAN in classifying track points, and temporal filtering in removing noises (any misclassified track points). The result shows that the proposed method classifies track points with 88% accuracy. The percent of misclassified track points decreased significantly with our method (1.9%) over trip/stop detection based on attribute threshold values (10.58%).

Keywords: GPS track data, DBSCAN, trip detection, community mobility.

1 Introduction

The ability to get out and be physically active in one's life space is of significant importance to health and well-being. This is a particular challenge for the elderly and persons with physical disabilities. With Global Positioning System (GPS), one can determine when, where and how individuals move around in the community; that is, GPS can be used to measure community mobility of individuals. GPS-based community mobility measures have advantages over traditional measures [1-4]; they are more context-rich than measures based on laboratory testing, and more reliable than measures based on self-reporting completed retrospectively.

GPS tracks the position of properly equipped users by decoding and trilaterating signals transmitted from satellites of the U.S. Global Navigation Satellite System (GNSS). With GPS receivers, the movement of users can be recorded at a user-specified interval (e.g., every thirty seconds, every 10 meters). A collection of track points recorded at a certain interval during a specific time period is called GPS track data (or GPS stream data). Since this data contain spatiotemporal characteristics of movement in a fine resolution, it provides rich sources for tracking the movement of

B. Murgante et al. (Eds.): ICCSA 2013, Part II, LNCS 7972, pp. 86–98, 2013.
© Springer-Verlag Berlin Heidelberg 2013

individuals. For this reason, GPS has been used to complement travel survey in a transportation field [5, 6]. Recently, GPS track data has increasingly found its use in a health field.

An account of where these individuals visit when and how they get about (i.e., personal itinerary) has significance for quality of life because personal itinerary does not only represent levels of physical activity, but also reflects access to various opportunities and social interaction [7, 8]. As GPS technology has become increasingly accessible with location-aware devices [2], health practitioners can make more informed decisions about persons with disabilities' functioning and participation by making use of the personal itinerary extracted from GPS track data. Large volume and uncertainty inherent in GPS data, however, pose challenges in data management and analysis.

2   Challenges with Using GPS Data

Positional accuracy of GPS measurements varies although the accuracy has improved significantly over the last two decades [9]. The accuracy is affected by many factors, including ionospheric delay, geometry among satellites (known as Dilution of Precision), presence of building structure in surroundings, and receiver quality (such as precision of a clock built in GPS receivers, use of augmentation technology). Positional accuracy of GPS measurements can be assessed by comparing GPS measurements to measurements derived from reference data of independent source and higher accuracy (e.g., ground surveying, orthoimage). Performance of post-processing algorithms of GPS data depends on accuracy of GPS data, thus accuracy should be taken into account in developing a robust post-processing algorithm.

It is hard to obtain complete coverage of reliable GPS tracks over an extended period of time; it is mainly due to insufficient battery life and signal loss especially indoors. Large gaps in coverage render further analysis either impossible or infeasible. Recording interval of GPS track should be determined by judging the offset between battery life and desired temporal resolution. One can enforce rigorous users' protocol regarding battery replacement to increase coverage of GPS tracks. In addition, satellite signals do not penetrate building structure, making positional measurements indoors futile. Location fix is less accurate and reliable especially after a GPS receiver regains signal reception (for example, the moment that subjects exits a building) especially if warm-up time is long. GPS measurements can be complemented with accelerometers or smart home technology [4].

Human intelligence can be applied to post-processing GPS data with uncertainty on a case-by-case basis, but large volume of data makes manual classification tedious and subject to human errors. Tracking the movement of a subject during a week period with a 30 second record interval creates 20,160 track points. If this process can be automated, processing time can be reduced significantly. Further, automated post-processing of GPS track data is even more necessary as GPS data are increasingly available at an unprecedented rate through mobile devices. It is estimated that personal sensor data will increase from 10% of all stored information to 90% within the next decade according to the Chief Technology Officer of EMC. As GPS finds many application areas, it will be worthwhile developing automated algorithms that recognize patterns from a stream of GPS data accurately. Below we first consider post-processing methods of GPS data organized into data cleaning, event detection, and event characterization before we describe the new method we developed.

3   Review of GPS Post-processing Methods

The personal itinerary is an account of a journey for a certain period of time. The itinerary consists of two geographic entities/events: stops (or activities) and trips. A stop is defined as a place that a subject visits in order to participate in one or more activities for some duration. A trip is defined as a route from one stop (origin) to another stop (destination). Post-processing of GPS track data consists of data cleaning, event (trip/stop) detection, and event (trip/stop) characterization.

3.1   Data Cleaning

Data cleaning attempts to remove any anomaly present in data. Positional accuracy, completeness, and logical consistency of GPS measurements are major issues to address. Plotting GPS track points on top of data of higher accuracy in GIS can help gauge levels of positional accuracy. Track points with unrealistic attribute values (e.g., altitude, speed and acceleration) can be deleted for logical consistency. Some study considers smoothing positional values and attribute values as a means to handle uncertainty [6].

3.2   Event (Trip/Stop) Detection

Trip/stop detection is perhaps the most studied [5, 6] [9-12]. Since trips have higher speed than stops, trips can be detected based on speed thresholds [5]. Similarly, stops can be identified by little or no change in position of track points for a specified time duration [5]. Relying classification solely on thresholds (such as speed and change in position) is not a highly reliable method for trip/stop detection because many trips/stops apparently behave like its counterpart given the definition based on thresholds. For example, waiting for traffic light in driving can be misclassified into stops due to low speed although they are part of trips. GPS measurements after signal loss (e.g., existing a building) may be misclassified into trips because positional errors inflate values of change in position, and lead to false speed values.

Accuracy of threshold-based methods can be improved if more than one variable (e.g., speed, position in time, acceleration, heading change) is considered simultaneously in setting thresholds [6]. Smoothing or filtering can be also considered in detecting stops/trips; that is, instead of classifying track points one by one, track points can be grouped into temporal neighbors and overall characteristics of temporal neighbors (such as speed, change in position) can be considered in classification [6].

Kernel Density Estimation (KDE) has been proposed for identifying significant places visited [12]. KDE is typically used in calculating density from discrete point features. KDE scans the number and proximity of points within a specified bandwidth to regularly placed location, and transforms that information to kernel density function [13]. Frequently visited places can be identified using KDE because those places have unusually high density values. KDE, however, does not differentiate stops that are made in the same place but in different times. That is, KDE is for detecting spatial clusters, not spatiotemporal clusters such as stops.

Similarly, density-based spatial clustering algorithms (such as DBSCAN) can be considered in identifying significant places [12]. DBSCAN detects a group of dense spatial clusters by aggregating spatial clusters that are density-reachable [14]. DBSCAN begins with scanning the number of points within a specified bandwidth from any unvisited arbitrary data points (called seeds), and checks whether the number of points exceeds a pre-specified threshold value. If the abovementioned condition is met, the algorithm checks whether each point within the spatial cluster (identified above) forms another spatial cluster (i.e., expandable). An advantage of density-based clustering algorithms is that they are relatively robust to noise. Again, DBSCAN is to detect spatial clusters rather than spatiotemporal clusters. To detect stops, it is necessary to modify DBSCAN to temporal criteria so that spatial clusters can be disaggregated into stops based on when it occurred, and how long it lasted.

3.3   Event (Trip/Stop) Characterization

Trip/stop characterization assigns attributes to stops and trips. Location, arrival time, and departure time can be assigned to stops. Similarly, attributes like origin, destination, and a route taken can be assigned to trips. Additional attributes such as mode of navigation and purpose of trips can be extracted with further post-processing or/and with the aid of auxiliary data.

There have been growing interests in developing methods for detecting mode of navigation (or transportation) in recent years. Those methods typically consider multiple variables—speed, acceleration, heading change, total distance or duration, and proximity to transportation infrastructure (such as bus stop, bike racks, highway) [15, 16]. Artificial intelligence techniques, including Relational Markov Network, and Neural network, have been proposed [10, 11, 17]. The use of visual analytics tools can facilitate inference of navigation mode in a semi-automated way [18]. For example, graphing acceleration over time can help differentiate driving from walking because driving has a distinct acceleration profile. Purpose of trips is typically inferred by matching location of stops with land use data in a fine spatial resolution [19].

A few lessons from previous studies can be discussed in regard to developing the automated algorithm for extracting the personal itinerary from GPS track data. It is important to have uncertainty handling built in the algorithm since misclassification often results from uncertainty. Threshold-based methods alone cannot classify track points reliably. Much of GPS track data is filled with noise, and enforcing rigid thresholds without regard to context cannot escape misclassification. It appears that eclectic approaches that combine thresholds of appropriate variables, spatiotemporal clustering algorithms, and smoothing (or filtering) can produce reliable results.

4   The Method for Extracting Mobility Measures from GPS Data

We have developed a method that automates extraction of the personal itinerary from GPS track data. The method consists of four modules: (a) data cleaning, (b) stop/trip detection, (c) stop/trip characterization, and (d) mobility measures extraction.

4.1   Data Cleaning

After input data is geocoded and projected, data is checked for logical consistency and accuracy. Track points with duplicate temporal values are deleted for consistency; this problem (duplicate records) often occurs when input data is exported manually as a collection of files rather than one file for the time period monitored. Data is checked for completeness (that is whether there is any significant gap) due to signal loss or delay. A program deletes a track point following a significant gap. A significant gap is defined as temporal interval greater than 90 seconds and distance between two track points greater than 50 meters. This (deleting a track point after a significant gap) is necessary because signal loss or delay compromises positional accuracy of GPS measurements, leading to misclassification.

4.2   Event Detection

The stop/trip detection module consists of two components: (a) classifying track points into stops and trips, and (b) smoothing out classified track points. The unique aspect of the method described in this paper lies in using temporal DBSCAN for the first task (classification), and temporal filtering for the second task. The idea is that stops are equivalent to noisy spatiotemporal clusters, thus density-based spatial clustering algorithms that consider temporal information additionally are well suited to delineating stops. Trips are track points that are not stops. Further, whether a track point is stop or trip depends on whether temporally neighboring track points are stops or trips. Thus re-classifying track points over temporal neighbors can render classification less noisy.

DBSCAN is selected to detect stops over threshold-based methods described earlier because DBSCAN is robust to noise; this allows for classifying "track points that are scattered around stops but are not stops" into trips reliably. Instead of choosing seeds randomly for DBSCAN, the module narrows down the location of candidate stops using KDE. Then the location of extreme KDE values is fed into DBSCAN as seeds; this serves to reduce processing time for DBSCAN, and minimize possibility of missing seeds that might be stops due to random and incomplete search of the DBSCAN algorithm. To prevent the algorithm from being trapped in home (where vast majority of time is spent on), track points near home location are excluded for analysis. Spatial clusters identified from DBSCAN are checked to see if those clusters are consecutive for some duration of time. If the abovementioned requirement is met, spatial clusters are disaggregated into stops (as spatiotemporal clusters). So a place visited more than once are identified as multiple stops whose cardinality is the number of visits on the place. Track points that are not stops are classified into trips.

For KDE, pixel size is set to 30 meters, and bandwidth is set to 90 meters. Seeds are set to pixels where kernel density is greater than or equal to 900 per square kilometers. For DBSCAN, spatial search radius is set to 50 meters, minimum number of points is set to 5. For a spatial cluster to be qualified as a spatiotemporal cluster (that is stop), duration should be longer than 3 minutes, and track points consisting of a stop should be continuous.

Stops and trips identified above can still remain misclassified. In particular, data contain temporally noisy values (that is, trip/stop value is not assigned continuously, such as stop-trip-stop). To resolve these issues, the code calculates a majority value (the most common value among stop and trip) of consecutive track points whose total duration is 2 minutes and 30 seconds, and assigns a majority value to those track points as a final classification. For example, if temporally ordered track points have an array of values [stop, trip, stop, stop, stop] (assuming 30 second recording interval), then these track points are reclassified into stop. Majority filtering reclassifies track points by smoothing values of track points that might have been misclassified by clustering algorithms.

Unique identifiers (IDs) are assigned to stops and trips based on temporal order and a rule that a stop is followed by a trip, and vice versa. Thus a stop is a collection of track points that are spatially clustered and temporally continuous. A trip is a collection of track points that are not spatially clustered and temporally continuous. Track points that are classified into stops and trips with unique IDs are aggregated (or dissolved in GIS terms) into two geographic entities called stop centers and trip routes. *Stop centers* are mean centers of track points with a unique stop ID. *Trip routes* are polylines that link an origin stop to a destination stop.

4.3   Event Characterization

Once stop centers and trip routes with unique IDs are created, spatiotemporal characteristics are assigned. Stop centers have location, arrival time, departure time, duration of stop, and average speed. Street addresses of stop centers can be estimated using reverse geocoding in addition to latitude and longitude. Arrival time is derived from date and time of the first track point, and departure time is derived from date and time of the last track point that are classified into a stop with a unique ID. Trip routes include begin time, end time, length, and duration as attributes. Mode of navigation is determined on the basis of speed—driving if average speed is greater than 8 kilometers per hour, and walking otherwise. Other modes of navigation (e.g., bicycling, public mode of transportation) are not considered at this point. Trip routes also include IDs of both origin and destination stop centers.

4.4   Mobility Measures Extraction

This module calculates the number of stops, number of trips, duration of stops, duration of trips, length of trips, and number and length of walking trips. The module also calculates the percent of track points within multiple buffers from home at 50 meter (stay home), 1000 meter (neighborhood level), 5000 meter (town level), 20000 (out of

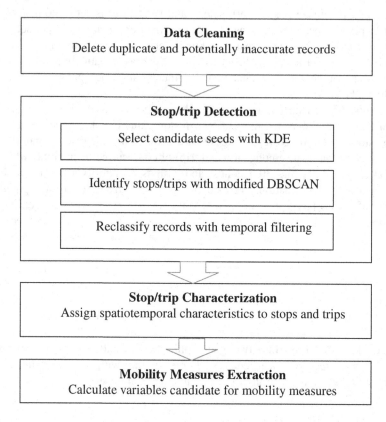

Fig. 1. Process of extracting mobility-related variables from GPS track data

town level), and beyond, respectively; this follows Life Space Assessment question-naire [4, 8]. To gather baseline data about how subjects participate in real-life situa-tions, subjects are asked to identify meaningful places (called targets) in one's envi-ronment, including personal, social, and/or occupational locations. The number (and percent) of targets reached is also reported to gauge the geographic extent of commu-nity participation of subjects. The Figure 1 depicts a simplified process of extracting variables useful in constructing mobility measures from GPS track data.

5   Results

The method described above was applied to extract personal itineraries of a healthy control subject during the period of week 1, week 5, week 9, and month 6 after treat-ment. The method was applied to a subject recovering from a major injury who com-pleted rehabilitation during the period of week 1, week 5, and week 9 after treatment. All GPS track data were collected in one week time period. Subjects agree to carry a GPS logger (DG-100 Data Logger) during waking hours of each week. Recording

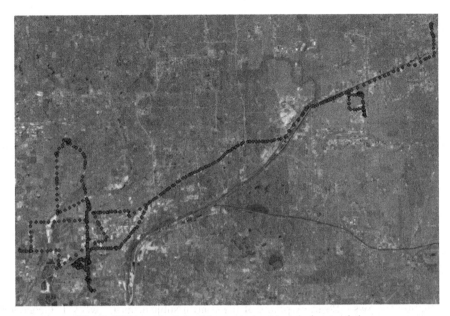

Fig. 2. GPS track points of the control subject during week 1

Fig. 3. Stop centers and trip routes of the control subject during week 1

interval was set to 30 seconds. A total of seven one-week track data from two subjects were analyzed. Each week data amounted to about 5,000-16,000 track points depending on the extent of gaps in the data. The entire process, from preprocessing to extraction of mobility-related variables, was automated.

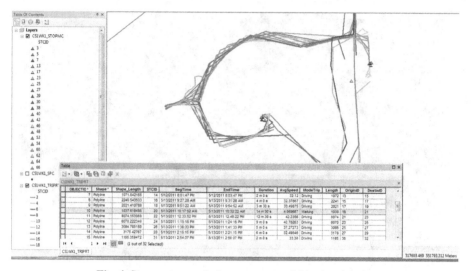

Fig. 4. Stop centers and trip routes at a neighborhood scale

Figure 2 shows GPS track points of the control subject during week 1. Geocoded track points are well aligned with orthoimagery of higher accuracy, indicating that data have good positional accuracy. Figure 3 shows stop centers and trip routes extracted from GPS track data shown in Figure 2; different routes are shown in different colors, and stop centers are shown as triangles. Figure 4 shows a screen shot of stop centers and trip routes at a neighborhood scale in ArcGIS, in which a walking trip is highlighted in a map and an attribute table.

Table 1 reports on how many stops and trips subjects made and how long those stops and trips were in duration. For example, the control subject made 7 stops during

Table 1. Number and duration of stops and trips, and length of walking trips

Subjects	Num stops	Dur stops	Num trips	Dur trips	Len walk
CS1WK1	21	9.50.48	32	8.0.31	1173
CS1WK5	7	3.4.43	19	4.27.30	524
CS1WK9	11	5.11.10	26	15.21.15	132
CS1MO6	6	3.21.34	19	6.9.5	551
RS3WK1	22	9.35.56	24	2.59.18	1305
RS3WK5	5	8.20.54	10	6.18.5	711
RS3WK9	8	11.19.59	14	8.11.40	3093

Num stops: the number of stops made
Dur stops: duration of stops in hour minute second format
Num trips: the number of trips made
Dur trips: duration of trips in hour minute second format
Len walk: length of walking trips in meters

week 5, and stayed at those stops for about three hours in total. During the same period, the same subject made 19 trips. Those trips took the subject approximately four hours and thirty minutes in total. The subject walked 524 meters during week 5, which is a decrease from 1.2 kilometers during week 1.

6   Evaluation

To evaluate how accurately the method described above classifies track points into stops and trips, we check classification errors against results of manual classification. For manual classification, we manually check all track points (totaling 61,557) to see whether they constitute stops or trips by looking through all track points superimposed over high-resolution imagery (ArcGIS 10.1 Map Service World Imagery).

Table 2 shows an error matrix compiled from seven data sets (from CS1WK1 to RS3WK9). The percent correctly classified is 94.2%, and Kappa index, a measure of classification accuracy, is 0.88. That is, data are correctly classified 88% of the time with the proposed method.

Table 2. Error matrix and classification accuracy of the proposed method

	trip	stop	Row Total
trip	133	11	144
stop	2	78	80
Column Total	135	89	224

Kappa index: 0.88

To examine effects of temporal DBSCAN and temporal filtering on classification accuracy, we compare performance of the test method to that of the threshold-based method (that is the control method). The control method classifies track points into trips if average speed in kilometer per hour is greater than 4.2, and into stops otherwise. All other aspects including data cleaning remain the same between two methods. Track points that constitute "staying at home" are excluded for the comparison. In other words, the extent to which track points are misclassified (i.e., misclassification rates) is calculated as the total number of misclassified track points out of the total number of track points outside of home.

Figure 5 graphs the percent of misclassified track points for seven data sets. On average, 10.58% of track points are misclassified with the control method, and 1.91% of track points are misclassified with the test method. This indicates that temporal DBSCAN along with temporal filtering significantly improves accuracy of differentiating between stops and trips from GPS track data.

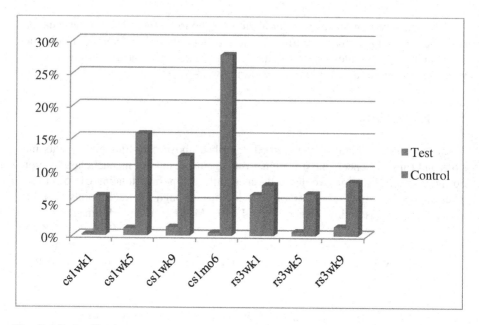

Fig. 5. Misclassification rates of the test method that use temporal DBSCAN and temporal filtering vs. those of the control method that does not

7   Conclusions

The method presented in this paper provides efficient, useful, and quantifiable data for community mobility measures within the constraints of the total time for GPS data collection. The proposed method utilizes temporal DBSCAN for detecting stops and temporal filtering for smoothing misclassified track points. The method classifies GPS track points into stops and trips with high accuracy (88%). This paper considers whether a clustering-based event detection method outperforms a threshold-based method as a predominant GPS post-processing technique. Results show that spatio-temporal clustering is effective in detecting stops and extracting the personal itinerary from GPS track data accurately.

Several limitations of this research can be discussed along with future research. The performance of the method needs to be evaluated against threshold-based me-thods in different variation. Results of the proposed method can be compared with self-reported information completed in real-time to evaluate the mtehod. Gaps in data need to be filled using other complementary methods of measuring mobility (e.g., accelerometer); accelerometry-based data can be synchronized with GPS data based on temporal information.

We live in the world where a stream of data on individuals' movement and activi-ties are fed into computer systems on a real time basis. Information extracted from GPS track data can be valuable for mining activity patterns of individuals, trajectories of social interaction, and community participation in geographic space [20]. Once these data are fused with environmental and demographic data, it is possible to extract patterns of spatial relationship; for example, one can examine how environmental factors (e.g., exposure to air pollution, pathogen) affect occurrence of diseases like cancer and asthma [21, 22, 23], and determine characteristics of the built environment (e.g., presence of sidewalk, disability-friendliness) that promote mobility of those with obesity or disability [24]. Applications of the automated post-processing of GPS data are wide ranging. It is hoped that this study demonstrates that robust analytic algorithms are at the core of realizing the potential of "Big Data".

References

1. Evans, C.C., Hanke, T.A., Zielke, D., Keller, S., Ruroede, K.: Monitoring community mobility with Global Positioning System technology after a stroke: a case study. J. Neurol. Phys. Ther. 36(2), 68–78 (2012)
2. Schenk, A.K., Witbrodt, B.C., Hoarty, C.A., Carlson Jr., R.H., Goulding, E.H., Potter, J.F., Bonasera, S.J.: Cellular telephones measure activity and lifespace in community-dwelling adults: proof of principle. J. Am. Geriatr. Soc. 59(2), 345–352 (2011)
3. Rainham, D., Krewski, D., McDowell, I., Sawada, M., Liekens, B.: Development of a wearable global positioning system for place and health research. Int. J. Health Geogr. 7(1), 59–75 (2008)
4. Tudor-Lock, C.: Assessment of enacted mobility in older adults. Top. Geriatr. Rehabil. 28(1), 33–38 (2012)
5. Srinivasan, S., Bricka, S., Bhat, C.: Methodology for converting GPS navigational streams to the travel-diary data format (2009), http://www.ce.utexas.edu/prof/bhat/ABSTRACTS/Srinivasan_Bricka_Bhat.pdf
6. Schuessler, N., Axhausen, K.W.: Processing raw data from global positioning systems without additional information. Transp. Res. Record 2105, 28–36 (2009)
7. Berke, E.M.: Geographic information systems (GIS): recognizing the importance of place in primary care research and practice. J. Am. Board Fam. Med. 23(1), 9–12 (2010)
8. Peel, C., Baker, P.S., Roth, D.L., Brown, C.J., Bodner, E.V., Allman, R.M.: Assessing mobility in older adults: the UAB Study of Aging Life-Space Assessment. Phys. Ther. 85(10), 1008–1019 (2005)
9. Stopher, P., FitzGerald, C., Xu, M.: Assessing the accuracy of the Sydney Household Travel Survey with GPS. Transportation 34(6), 723–741 (2007)
10. Liao, L., Fox, D., Kautz, H.: Extracting places and activities from GPS traces using hierarchical conditional random fields. Int. J. Robot. Res. 26(1), 119–134 (2007)
11. Liao, L., Patterson, D.J., Fox, D., Kautz, H.: Building personal maps from GPS data. Ann. N. Y. Acad. Sci. 1093(1), 249–265 (2007)
12. Schoier, G., Borruso, G.: Individual movements and geographical data mining: clustering algorithms for highlighting hotspots in personal navigation routes. In: Murgante, B., Gervasi, O., Iglesias, A., Taniar, D., Apduhan, B.O. (eds.) ICCSA 2011, Part I. LNCS, vol. 6782, pp. 454–465. Springer, Heidelberg (2011)
13. Carlos, H.A., Shi, X., Sargent, J., Tanski, S., Berke, E.M.: Density estimation and adaptive bandwidths: a primer for public health practitioners. Int. J. Health Geogr. 9(1), 39–46 (2010)

14. Ester, M., Kriegel, H.P., Sander, J., Xu, X.: A density-based algorithm for discovering clusters in large spatial databases with noise. In: Proceedings of the 2nd International Conference on Knowledge Discovery and Data Mining, KDD 1996, pp. 226–231. AAAI Press (1996)
15. Stenneth, L., Wolfson, O., Yu, P.S., Xu, B.: Transportation mode detection using mobile phones and GIS information. In: Proceedings of the 19th ACM SIGSPATIAL International Conference on Advances in Geographic Information Systems (ACMGIS 2011), pp. 54–63. ACM (2011)
16. Cho, G.H., Rodriguez, D.A., Evenson, K.R.: Identifying walking trips using GPS data. Med. Sci. Sports Exerc. 43(2), 365–372 (2011)
17. Gonzalez, P., Weinstein, J., Barbeau, S., Labrador, M., Winters, P., Georggi, N.L., Perez, R.: Automating mode detection using neural networks and assisted GPS data collected using GPS-enabled mobile phones. In: 15th World Congress on Intelligent Transportation Systems (2008)
18. Liao, B.: Anomaly detection in GPS data based on visual analytics. In: IEEE Symposium on Visual Analytics Science and Technology (2010)
19. Mavoa, S., Oliver, M., Witten, K., Badland, H.: Linking GPS and travel diary data using sequence alignment in a study of children's independent mobility. Int. J. Health Geogr. 10(1), 64 (2011)
20. Eagle, N., Pentland, A.: Reality mining: sensing complex social systems. Pers. Ubiquit. Comput. 10(4), 255–268 (2006)
21. Pentland, A., Lazer, D., Brewer, D., Heibeck, T.: Using reality mining to improve public health and medicine. Stud. Health Technol. Inform. 149, 93–102 (2009)
22. Betts, K.S.: Characterizing exposomes: tools for measuring personal environmental exposures. Environ. Health Persp. 120(4), a158 (2012)
23. Richardson, D.B., Volkow, N.D., Kwan, M.P., Kaplan, R.M., Goodchild, M.F., Croyle, R.T.: Spatial turn in health research. Science 339(6126), 1390–1392 (2013)
24. Duncan, M.J., Badland, H.M., Mummery, W.K.: Applying GPS to enhance understanding of transport-related physical activity. J. Sci. Med. Sport 12(5), 549–556 (2009)

Optimal Arc-Spline Approximation
with Detecting Straight Sections

Georg Maier, Andreas Schindler, Florian Janda, and Stephan Brummer

FORWISS, University of Passau,
Innstr. 43, 94032 Passau, Germany
{gmaier,schindler,jandaf,brummer}@forwiss.uni-passau.de
http://www.forwiss.uni-passau.de

Abstract. We present a method for approximating an open polygonal curve by a smooth arc spline with respect to a user-specified maximum tolerance. Additionally, straight sections of the polygon are detected resulting in a finite set of pairwise disjoint line segments. The proposed algorithm guarantees that the resulting arc spline does not exceed the tolerance and that sections of the pre-computed lines are part of the solution. Subject to these conditions we obtain the minimally possible number of circular arcs and line segments. Note that in contrast to existing approaches, we do not restrict the breakpoints of the arc spline to original points but compute them automatically.

Keywords: arc spline approximation, SMAP, minimum description length, line segment detection, polygon, simplification.

1 Introduction

Arc splines, which are curves composed of circular arcs and line segments (shortly: segments) have advantageous properties: exact offset and arc length computation, parameter free description and simple closest point calculation (cf. [1]). Therefore, research on this curve system has been very active in the past few years (e.g. [2–5]).

Various applications need methods guaranteeing that the resulting arc spline does not exceed a user-specified tolerance from the input data. Often, these algorithms follow geometric approaches: Suppose to approximate a simple polygonal curve, then the ε-tolerance zone is considered, which is given by the set of points which have an euclidean distance of at most ε. Heimlich and Held [6] introduced a method which generates tolerance zones using Voronoi diagrams. Regarding the restriction that breakpoints are original points, Drystale et al [7] proposed an algorithm for approximating an open polygonal curve with a minimal number of circular arcs and biarcs while remaining within a given tolerance region. This algorithm has a worst case complexity of $O(n^2 \log(n))$ for a n-vertex input polygon. The approach of Heimlich and Held [6] achieves a better runtime of $O(n \log(n))$, whereas the minimality of the segment number is lost.

B. Murgante et al. (Eds.): ICCSA 2013, Part II, LNCS 7972, pp. 99–112, 2013.

In order to construct an arc spline approximation with the actual minimal number of segments, the positions of the corresponding breakpoints have to be determined. The method presented by Maier [1] therefore computes an arc spline, which achieves the actual minimal number of segments for any open polygonal curve and user-specified tolerance. The currently best known implementation needs $O(n^2)$ time in worst case. Note that restricting the loci of the breakpoints can have considerable influence on the resulting segment number (cf. [8]).

2   Motivation for Line Segments

Practical applications demand the incorporation of preliminary knowledge. In some cases, line segments should appear at certain areas of the final approximation solution. For instance, such side conditions arise in the field of reverse engineering, the vectorization of pixel graphics or in the generation of digital maps. When using line segments instead of arcs with large radii numerical pitfalls can be avoided. Beyond that, line segments are actually used for road planning purposes, which justifies this approach.

We make extensive use of the presented algorithms to generate digital maps of high precision on behalf of the research activities in Ko-PER, which is part of the Ko-FAS research initiative (cf. [9]). A vehicle equipped with a calibrated camera and a highly accurate positioning unit records measuring points on all visible road markings. The extracted points are reconstructed in a global world coordinate system. Measuring points belonging to the same marking are grouped. For each group a tolerance zone is established. Within this zone we are searching for straight sections.

In this paper we present an algorithm for the detection of line segments that should be respected in the final arc spline approximation. That way, a curve representation of road markings including line segments is created. Its accuracy is controlled by a suitable tolerance zone while the number of curve segments is minimized which leads to the shortest possible description length (cf. [10]). Analogously, representations for individual lanes in the digital map are calculated.

Further, we integrate the constraint of respecting some line segments to our methodology by extending the original approach. Therefore, the next sections are devoted to some definitions and results from Maier [1] that we want to use subsequently. Some of the following proofs are shortened or even skipped, they can be found in Maier [1] too.

3   Tolerance Channels and Circular Visibility

The approach of Maier [1] deals with an approximation up to a given tolerance, which can vary locally. The approximation error is controlled by focusing on solutions staying inside a so-called *tolerance channel*.

Although tolerance channels can be formed by a comparatively broad class of bounding curves, for the sake of simplicity we focus on simple closed polygons within this scope. Generating an appropriate tolerance channel is a problem of its own that we do not treat here but we refer to Heimlich and Held [6]. For any n-vertex input a corresponding channel can be computed in $O(n \log(n))$ time.

Definition 1. *We call a triple (P, s, d) tolerance channel if P is a simple closed polygon and s and d are two disjoint edges of P denoting the start and the destination.*

An example is depicted in Fig. 2. The term polygon P is used to denote the union of both the boundary ∂P and the interior of the corresponding polygonal Jordan curve.

Definition 2. *Any smooth (G^1-continuous) arc spline staying inside the tolerance channel and connecting the start and destination segments with a minimum number of segments is called Smooth Minimum Arc Path (SMAP).*

Note that the breakpoints are not required to be original points but they are determined automatically. Obviously, this has considerably positive effects on the resulting segment number.

For the remaining part of this paper let (P, s, d) be a tolerance channel. To keep the notation as simple as possible, we assume that the two vertices of s are convex: A vertex v is convex if the interior angle at v is strictly smaller than $180°$. The definitions for the general case can be found in [1].

Definition 3. *A point $a \in P$ is said to be circularly visible (from s with respect to P) if there exists a segment γ in P that has its starting point on s and ends in a. The set of all circularly visible points from s with respect to P is denoted by V. An oriented arc γ, as above, is called visibility arc.*

4 Alternating Restrictions

The main instrument of the SMAP algorithm is the calculus of *alternating restrictions*. These are points visibility arcs and ∂P have in common, as indicated in Fig. 2. At the points a_1, a_2, a_3 the plotted visibility arc is alternately touched by ∂P from the left and from the right. Thus, we call them *left* and *right restriction points*:

Definition 4. *Let γ be a visibility arc and let a be a point on γ and on the boundary of P. The point a is called right restriction point (of γ with respect to s) if one of the following conditions holds:*

- *a is neither the starting nor the endpoint of γ and ∂P is locally right of γ at a.*
- *$a \in s$ is the starting point of γ and s is locally left of γ at a.*

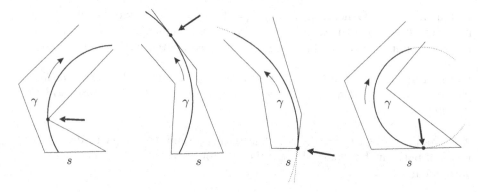

Fig. 1. Illustration of right restriction points. The straight arrows indicate that γ is restriction pointed from the right according to the corresponding orientation which is illustrated by the bended arrows.

At a right restriction point, the visibility arc cannot be moved to the right without either exceeding the tolerance boundary or violating the starting condition. The first two images in Fig. 1 illustrate the first case and the following two images show the second one. Likewise, we can define *left restriction points*.

Maier [1] showed that the boundary points in $\partial V \setminus \partial P$ lie on arcs, and the corresponding visibility arcs are denoted by *blocking arcs* if they are maximally extended with respect to inclusion in P. These arcs distinguish themselves from the other visibility arcs as they have at least three *alternating* restriction points a_1, a_2, a_3 ordered as a_1, a_2 and a_3 according to the orientation of the arc. They satisfy the following condition: Either a_1 and a_3 are left restriction points and a_2 is a right restriction point or a_1 and a_3 are right restriction points and a_2 is a left restriction point. Arcs passing through three alternating restriction points can be described in an efficient manner regarding to an algorithmic approach as they determine uniquely the three degrees of freedom an arc has.

Every connected component of $P \setminus V$ is separated from V by exactly one blocking arc. The blocking arc corresponding to the connected component including d is called the *window (with respect to s)*.

Theorem 1. *Let γ be a maximally extended visibility arc with endpoint on the left side of ∂P. Then γ is the window if and only if there are three alternating restriction points a_1, a_2, a_3 where a_3 is a right restriction point. Similar conditions hold if γ has its endpoint on the right side of ∂P.*

All possible configurations of left and right restrictions are given as follows: The corresponding arc

- passes through three vertices
- passes through two vertices and touches an edge
- passes through one vertex and touches two edges

We have now analyzed the circular visibility set V. The next step is to characterize the sets V^i of all points which can be reached by $i = 2, \ldots, k$ segments till V^k intersects d. Therefore, the following section gives a criterion for deciding if an oriented arc can be smoothly continued or not.

5   The Continuation Condition

Oriented arcs in P satisfying the so called *continuation condition (CC)* can be smoothly joined to a visibility arc. An oriented arc γ satisfies the CC with respect to an oriented arc η if either γ smoothly joins η or there are two intersection points x_1, x_2 of γ and η s.t. the orientation induced by γ and the one induced by η are equal. An illustration can be found in Fig. 2.

Theorem 2. *Let $x \in P \setminus V$ and let C be the connected component of $P \setminus V$ containing x. Then $x \in V^2$ if and only if there exists an oriented arc γ in $V \cup C$ ending in x and satisfying the CC with respect to the corresponding blocking arc.*

In this case we can even choose an arc γ that is extremal, i.e. having at least two alternating restrictions, which is fundamental for a constructive approach and hence for the algorithmic design.

Therefore, we are able to characterize the set V^2 by examining all oriented arcs γ satisfying the conditions above. However, we do not have to consider the whole set V^2, but only the component leading to d. This meets in elucidating a "modified" tolerance channel with the corresponding window as starting segment. The only differences to the kind of tolerance channels we have seen till now are the more complicated starting requirements given by Theorem 2. Maier [1] showed that the results presented in Section 3 hold for this kind of tolerance channel as well. Especially, the window of V^2 is characterized in the same manner. Recall that touching the start segment, which is here the window of V, is a left or right restriction point.

We are now able to use Theorem 2 inductively and exploit the properties of the sets V^k this way. Let us assume that the successively resulting windows have exactly three alternating restrictions. The general case requires some slight modifications, which cannot be treated within this scope. The outcome of this is a two step greedy algorithm traversing P from s to d in the forward step and back again from d to s in the backward step.

The Forward Step: After having found the first window ω_1 by identifying arcs with three alternating restrictions, the next windows ω_i can be computed such that the conditions of Theorem 2 and Theorem 1 are satisfied. In particular, ω_i has to satisfy the CC with respect to ω_{i-1}. The procedure is stopped when a point of d is circularly visible, and a visibility arc satisfying Theorem 2 and ending in d is computed. As it can be seen in Fig. 2, the windows do not represent a smooth arc spline. However, the computed windows are used in the backward step to generate a SMAP.

The Backward Step: The lastly calculated arc ω_k represents the last segment of the resulting SMAP. In particular, the minimum segment number is k. The

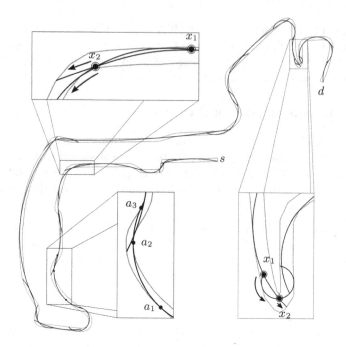

Fig. 2. Visualization of the forward step of the SMAP algorithm. In the bottom left zoom three alternating restriction points a_1, a_2, a_3 are marked with dots. In the remaining two zooms x_1 and x_2 indicate the intersection points of two subsequent windows.

predecessor segments are then determined by touching their successor and by two alternating restrictions. The whole procedure is finished when s has been reached. In Fig. 3 the backward step is visualized and the box shows a single backward step: γ_i joins its successor γ_{i+1} and satisfies the CC with respect to ω_{i-1} as the two emphasized dots indicate.

6   Generating Line Segments

As mentioned in Section 1 it is desirable to integrate pre-defined line segments that should be respected in the approximation solution. In general, there are various possibilities where these line segments come from: Preliminary knowledge, user interaction or heuristic approaches, like in the following which is based on *extremal line segments*: A maximally extended line segment in P is called extremal if it has two alternating restrictions a_1, a_2 such that either

- a_1 and the end point are on the left side and
- a_2 is on the right side

or vice versa. In particular, the start and end points are located on the boundary of the channel. That way, extremal line segments are related to the characterization of circular windows from Section 4. As we are only interested in "large"

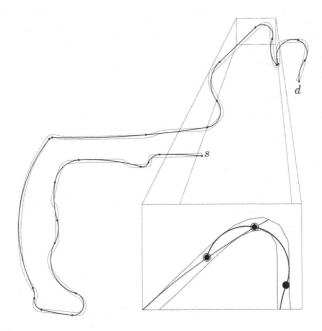

Fig. 3. Visualization of the backward step of the SMAP algorithm. The overview shows the SMAP result. A single backward step is shown in detail. The solid dot marks a breakpoint between the last and the current SMAP segment. In contrast the two emphasized dots mark intersections between the current SMAP segment and the next window. The Continuation Condition is fulfilled. Hence the next segment can be smoothly attached.

line segments, we consider only those candidates whose ratio between the length and ε is greater than a given threshold θ. The choice of θ depends significantly on the application. For our map generation purposes the threshold was chosen to $\theta = 10^4$ while the tolerance was $\varepsilon = 0.1$ m.

Naively, the extremal line segments can be determined by iterating over all combinations of restriction points on opposite sides. Though, this task is related to the calculation of the minimum link path (cf. [11]). Using techniques known from Chou and Woo [12], the extremal line segments can be computed in $O(n)$ complexity where n is the number of vertices in the input polygon. In order to determine some line segments that should be respected in the final approximation solution, many other heuristic approaches can be considered. However, we focus on the extremal line segments generated by Algorithm 1 as they are advantageous for the backward step of the SMAP algorithm. Fig. 4 shows some extremal line segment candidates and the largest segment which was chosen finally. All candidates have two alternating restrictions and end points on the opposite side of the last restriction point.

Algorithm 1. Compute line segments

Require: Tolerance channel P, tolerance ε, threshold θ
Ensure: $L = \{\lambda_1, \dots, \lambda_l\}$

  Compute all extremal line segments L' within P
  for all $l \in L'$ **do**
    if $\frac{length(l)}{\varepsilon} > \theta$ **then**
      insert l in C
    end if
  end for
  $i \leftarrow 1$
  while $C \neq \emptyset$ **do**
    $\lambda_i \leftarrow$ get largest line segment in C
    for all $l \in C, l \neq \lambda_i$ **do**
      if $\lambda_i \cap l \neq \emptyset$ **then**
        remove l from C
      end if
    end for
    insert λ_i in L
    $i \leftarrow i + 1$
  end while

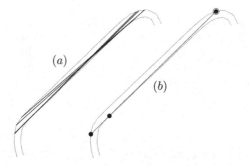

Fig. 4. Section of the racing course. (a): Candidate set of extremal line segments. (b): Selected extremal line segment with two alternating restriction points (dot) and end point (emphasized dot) on the opposite side of the channel.

7 Integration of Line Segments

Let $\lambda_1, \dots, \lambda_l$ be extremal line segments with tangent vectors v_i in P resulting from the strategy given above. In particular, these line segments are

- pairwise disjoint,
- ordered and oriented according to P and
- maximally extended in P.

Let $\mathfrak{S}$ denote the set of smooth arc splines in P starting at s and ending at d s.t. for all $\gamma \in \mathfrak{S}$ and all $i = 1, \ldots l$ we have

$$\forall x \in \lambda_i \cap \gamma \neq \emptyset : \tau_\gamma(x) = v_i,$$

where $\tau_\gamma(x)$ denotes the tangent unit vector of γ at x. We then search for a candidate $\gamma \in \mathfrak{S}$ having the minimally possible number of segments.

First, we consider $l + 1$ sub-problems, which we can divide into three types: The portion between s and λ_1, the one between each two line segments λ_i and λ_{i+1} ($i = 1, \ldots, l - 1$), and the one between the last line segment λ_l and d. If $\lambda_1 \cap s \neq \emptyset$ the first case lapses; in the same manner for $\lambda_l \cap d \neq \emptyset$ and the last case. Since Maier [1] treats the case of starting at a λ_i into direction v_i as well and the same results hold in this case, it is sufficient to focus on the first situation.

Let λ be an extremal line segment in P with tangent vector v and alternating points a_1, a_2 pointing into direction of the polygonal path from s to d. We then consider the set $\mathfrak{T}$ of smooth arc splines in P, which touch λ with direction v and start in s. Consequently, we search for a candidate $\gamma \in \mathfrak{T}$ having the minimum number of segments m. This number and a "modified" SMAP can be found using the results presented in the preceding sections since there exists a minimum number k s.t. $x \in V^k$.

Theorem 3. *Let $x \in V^k \setminus V^{k-1} \cap \lambda$ for some $k \geq 1$, where $V^0 := s$. Then we have $k \leq m := \min_{\gamma \in \mathfrak{T}} |\gamma| \leq k + 1$ with $|\gamma|$ denoting the segment number of γ. We can choose γ s.t. its m-th segment has at least two alternating restrictions.*

Proof. As shown by Maier [1] the set $V^k \setminus V^{k-1}$ can be characterized by considering the 1-visibility set of a suitably shrinked tolerance channel and the particular segments of a SMAP can be constructed by alternating restrictions. Without loss of generality we can assume that $k = 1$, i.e. $x \in V$. Hence we have to show that $1 \leq m \leq 2$. If there exists a γ ending at x with direction v, it is not hard to show, for continuity reasons, that we can choose γ with an additional restriction point.

Otherwise, we have $m \geq 2$. In any case, there is a visibility arc η ending at $x := \max(V \cap \lambda)$, where max is built with respect to the orientation of λ. Since we have assumed that γ has exactly three alternating restrictions and λ is maximally extended, $V \cap \lambda$ is infinite. It is easy to see that there is an arc γ_2 in V touching λ at x with direction v and satisfying the CC with respect to η. Hence we can show the existence of a visibility arc γ_1 touching γ_2 and having two alternating restriction points, when we apply Theorem 2. Therefore, the arc spline defined by γ_1 and γ_2 is an element of $\mathfrak{T}$ and we have $m = 2$. $\square$

This leads us to the following algorithmic strategy:

Forward Step: Assuming the situation above, we compute the corresponding window and check if it intersects λ. If such an arc cannot be found in the current step, we proceed with the SMAP algorithm. We continue in the same way till a point of λ is circularly visible. We then search for the next window, which touches λ. Thereafter we proceed as usual till the destination segment has been reached.

Backward step: When we have reached λ during the backward step, we have the situation of Theorem 3. Therefore, we first try to find a visibility arc touching λ at x, defined as in the proof. Again, we are able to constructively check the existence of a corresponding arc. There exists any valid visibility arc touching λ at x if and only if there exists one which has an additional restriction point (cf. box 1 in Fig. 5). Otherwise, we compute an arc γ_2 in the corresponding visibility set V^i touching λ at x and satisfying the CC of the predecessor window. We can choose γ_2 having an additional alternating restriction or touching some blocking arc. In any case, a valid γ_2 can be found in a constructive manner as well. Since γ fulfills the CC, the remaining part of our solution is computed according to the steps of the SMAP algorithm. Due to Theorem 3 we can be sure to achieve the minimally possible number of segments. In Fig. 6 an example is visualized.

Algorithm 2. Arc Spline Approximation with Line Segments

Require: Input points $p_1, \ldots, p_N$, tolerance ε, threshold θ
Ensure: Approximating arc spline
    Compute Tolerance Channel (P, s, d) for $p_1, \ldots, p_N$ and ε
    Compute extremal line segments $L = \{\lambda_1, \ldots, \lambda_l\}$ ordered according to P
    if $L = \emptyset$ **then**
        return SMAP of P
    end if
    if $\lambda_1 \cap s = \emptyset$ **then**
        $\gamma_1 \leftarrow$ modified SMAP starting at s and touching λ_1
        Insert γ_1 into G
    end if
    for $i = 2, \ldots, l$ **do**
        $\gamma_i \leftarrow$ modified SMAP touching λ_{i-1} and λ_i
        Insert γ_i into G
    end for
    if $\lambda_l \cap d = \emptyset$ **then**
        $\gamma_{l+1} \leftarrow$ modified SMAP touching λ_l and ending in d
        Insert γ_{l+1} into G
    end if
    return arc spline defined by G and L

To sum up, if we have the general situation formulated at the beginning of this section, we apply the strategy presented above to $l+1$ partial approximation problems, as already explained. This results in a set which contains arc splines and line segments $\gamma_1, \lambda_1, \ldots, \gamma_l, \lambda_l, \gamma_{l+1}$. We then restrict the line segments λ_i

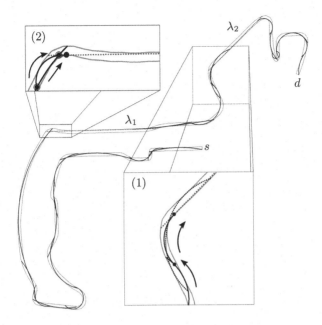

Fig. 5. The dotted lines λ_1 and λ_2 visualize two extremal line segments resulting from Algorithm 1. In (1) the dotted window is replaced by an arc which touches the following line segment. Breakpoints are drawn as solid points. In (2) such an arc does not exist and hence an additional one has to be inserted. The intersection points between the window and the inserted arc are visualized again as emphasized dots.

to the breakpoints x_i and y_i, where λ_i joins γ_i and γ_{i+1}, respectively. It remains to show that the resulting curve γ is simple. As the line segments λ_i are disjoint and the γ_i are simple, we only have to prove that $x_i \leq y_i$ with respect to the orientation of λ_i, $1 \leq i \leq l$ (cf. Fig. 7). For this purpose, let $1 \leq i \leq l$. Since λ_i has alternating restrictions a_1 and a_2 and the endpoint of λ_i lies on the opposite side of a_2 on ∂P and the simplicity of ∂P, the difference set $P \setminus \lambda_i$ has at least three connected components. Clearly, γ_i belongs to the one containing s and γ_{i+1} to the one containing d. Thus, γ_i touches λ_i at portion of λ_i between the start point $S(\gamma_i)$ and a_2, and γ_{i+1} between a_2 and the end point of λ_i. Hence we obtain $x_i \leq a_2 \leq y_i$. Note that for $x_i = a_2 = y_i$ the line segment λ_i vanishes. However, in this case $\lambda_i \cap \gamma$ is a singleton, where the G^1-continuity is satisfied as the according tangent vectors are equal to v_i. In any case, we have $\gamma \in \mathfrak{S}$, i.e. γ solves our approximation task as it yields the minimum number of segments. An overview of the whole approach is summarised in Algorithm 2.

Lemma 4. *The worst case complexity of Algorithm 2 is $O(n^2)$ for n input points and any tolerance ε.*

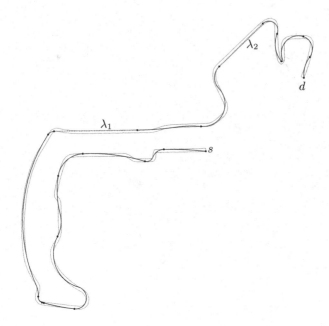

Fig. 6. The resulting SMAP with the two line segments λ_1 and λ_2

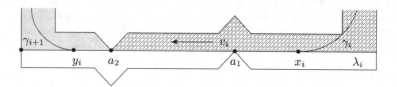

Fig. 7. Visualization of extremal arc λ_i with alternating restrictions a_1 and a_2. The grey portion indicates the connected component of $P \setminus \lambda_i$ containing d and the hatched one to the component which belongs to s.

Proof. Building the corresponding tolerance channel and computing extremal line segments $\lambda_1, \ldots, \lambda_l$ needs $O(n^2)$ time. Let us denote the sub-problems treating the portion between λ_{i-1} and λ_i by $S_1, \ldots, S_{l+1}$, where $\lambda_0 := s$ and $\lambda_{l+1} := d$. Clearly, to each sub-problem S_i belong n_i vertices of P s.t. $\sum_i n_i = O(n)$. For each S_i we firstly compute a SMAP γ_i from λ_{i-1} to λ_i resulting in $O(n_i^2)$ time. If $2 \leq i \leq l$, we have to check if we can replace the last segment of γ_i by one touching λ_i in direction v_i. If this is not the case, we insert an arc satisfying the CC as described above, which can be done in $O(n_i^2)$ time as well. Therefore, we have a worst case complexity of $\sum_{i=1}^{l+1} O(n_i^2) = O\left(\sum_{i=1}^{l+1} n_i^2\right) = O(n^2)$.

Although the worst case complexity for computing an extended SMAP does not differ from the usual case, integrating line segments usually speeds up the computational time significantly. The following example shall illustrate this issue: Suppose to have an input polygon with $k \cdot N$ points and the extended SMAP has k sup-problems with N points each. Then we have $\sum_{i=1}^{k} N^2 = k \cdot N^2 \ll k^2 \cdot N^2$ if k is considerably greater than 1.

8   Future Work

It should be possible to soften the side conditions of the approximation task in a way that not the line segments are fixed (expect a possible restriction) but only the areas in which they should appear. These areas could be defined by any strategy for detecting straight sections or user interaction, respectively. Whenever the algorithm runs into such a line area, an extremal line segment is searched satisfying the CC with respect to the last window. We are confident of being able to show that Theorem 2 holds for line segments and linear visibility as well. Then an optimal (new) line segment could be computed during the backward step, which should have positive effects on the overall segment number as the flexibility increases this way.

Acknowledgments. This work has been funded by the German Research Foundation (DFG - Deutsche Forschungsgemeinschaft) under grant number MA 5834/1-1.

References

1. Maier, G.: Smooth Minimum Arc Paths. Contour Approximation with Smooth Arc Splines. Shaker, Aachen (2010)
2. Meek, D.S., Walton, D.J.: Approximating smooth planar curves by arc splines. Journal of Computational and Applied Mathematics 59, 221–231 (1995)
3. Park, H.: Optimal Single Biarc Fitting and its Applications. Computer Aided Design and Applications 1, 187–195 (2004)
4. Piegl, L.A., Tiller, W.: Data Approximation Using Biarcs. Eng. Comput (Lond.) 18, 59–65 (2002)
5. Yang, X.: Efficient circular arc interpolation based on active tolerance control. Computer-Aided Design 34, 1037–1046 (2002)
6. Heimlich, M., Held, M.: Biarc Approximation, Simplification and Smoothing of Polygonal Curves by Means of Voronoi-Based Tolerance Bands. Int. J. Comput. Geometry Appl. 18, 221–250 (2008)
7. Drysdale, R., Rote, G., Sturm, A.: Approximation of an open polyonal curve with a minimum number of circular ars and biarcs. Computational Geometry 41, 31–47 (2008)
8. Maier, G., Pisinger, G.: Approximation of a closed polygon with a minimum number of circular arcs and line segments. Computational Geometry 46, 263–275 (2013)
9. Ko-FAS, research initiative (2012), http://www.ko-fas.de

10. Maier, G., Janda, F., Schindler, A.: Minimum Description Length Arc Spline Approximation of Digital Curves. In: Proceedings of IEEE 19th International Conference on Image Processing (ICIP), pp. 1869–1872. IEEE (2012)
11. Guibas, L.J., Hershberger, J., Leven, D., Sharir, M., Tarjan, R.E.: Linear Time Algorithms for Visibility and Shortest Path Problems Inside Simple Polygons. In: Symposium on Computational Geometry, pp. 1–13 (1986)
12. Chou, S.Y., Woo, T.C.: A linear-time algorithm for constructing a circular visibility diagram. Algorithmica 14, 203–228 (1995)

Identifying and Structuring Skeletal Noise

Thomas Delame, Céline Roudet, and Dominique Faudot

Le2i - Burgundy University, Dijon, France
tdelame@gmail.com,
{celine.roudet,dominique.faudot}@u-bourgogne.fr

Abstract. A skeleton is a thin centered structure within an object, which describes its topology and its geometry. The medial surface is one of the most known and used skeleton formulation. As other formulations, it contains noise, which complexifies its structure with useless parts. The connectivity of a skeleton is then unpredictable due to these useless parts. It can be a problem to segment the skeleton into logical components for example. We present here a technique whose purpose is to identify and structure such skeletal noise. It only requires a skeleton as input, making this work independent from any skeletonization process used to obtain the skeleton. We show in this paper that we significantly reduce the skeletal noise and produce clean skeletons that still capture every aspects of a shape. Those clean skeletons have the same local topology as the input ones, but with a clearer connectivity.

Keywords: skeleton, medial surface, skeletal noise, hairy pattern.

1 Introduction

A *skeleton* is a thin structure centered within an object, describing the topology and the geometry of this object. Such a skeleton could be then used as a shape representation model for every closed object. There exist several types of skeletons and we could divide them into two main categories:

1. curve skeletons; they are composed of curves and used for shape registration [5], mesh segmentation [4] and data reconstruction [18].
2. surface skeletons; they are composed of curves and surfaces, among them we can find medial surfaces [6], Midpoint loci [8] and PISA axes [16].

The work presented here is part of a process to make skeletons useful shape representation models. As stated in [11], surface skeletons better capture the geometry of objects than curve ones, making them best candidates for this purpose. Thus, we consider only surface skeletons, and in particular medial surfaces, because they are well defined and many algorithms exist to efficiently approximate them. A medial surface is made of *atoms*. An atom is a maximal inscribed ball lying inside the described object. Each atom a is connected to other atoms, called its neighbors $\mathcal{N}(a)$. Those links confer a topology on the skeleton.

B. Murgante et al. (Eds.): ICCSA 2013, Part II, LNCS 7972, pp. 113–128, 2013.

Skeletons are obtained from objects through a process called skeletonization. In such a process, a finite number of samples is used to capture an object $\mathcal{O}$. These samples cannot capture completely $\mathcal{O}$, and uncertainty about the object boundary arises: do a set of close samples capture a feature or a smooth part? In each case, atoms will be inserted in the skeleton. Atoms that do not capture any features produce the skeletal noise. Many methods have been proposed to remove unwanted components associated with this skeletal noise like [13] or [1]. A key goal of such methods is to preserve the topology of the skeleton during the process.

We noticed that the skeletal noise can be classified into two categories: Type 0 noise, also called clusters, and Type 1 noise, commonly referred to as hairy pattern. These noise types make the use of skeletons uneasy as shape representation models. For example, it is hard to estimate the tangent plane to an atom if one of its neighbors a belongs to Type 1 noise, since atoms such as a do not lie near the medial surface. Also connections are unnecessarily complex due to these noises. Connections get even worse when we look closely inside a cluster, locus of Type 0 noise, where the union of all its balls could be perfectly described by only one ball. The Figure 1 illustrates this classification.

In this paper, we propose new criteria to identify those skeletal noise. We also present a skeleton structure that contains a hierarchy, used to reflect the importance of atoms. Atoms are labeled by three number: 0 for Type 0 noise, 1 for Type 1 noise, and 2 otherwise. The hierarchy isolates skeletal noise from the remaining skeleton atoms: atoms from skeletal noise cannot have connections with relevant atoms. The neighbors are reorganized due to this isolation, in order to preserve the skeleton structure (i.e. logical components of the skeleton) thanks to structuring rules we created. There is no requirement about how the input skeleton was obtained, as our method take place apart from any skeletonization process. Thus, our work is general enough to be applied to every connected skeleton. More importantly, we show that our *cleaned skeleton* is simpler with a clearer connectivity, while not loosing any features or the original skeleton.

2   Previous Work

Since medial surfaces had always been noisy, removing this noise from it had motivated a lot of work. This is generally done during the skeletonization process: when the skeletonized object $\mathcal{O}$ is known. Thus, a noise removal technique is strongly linked to a skeletonization method. The most popular approaches for skeletonization are based on the Voronoi diagram or its dual Delaunay tetrahedralization. We review here such noise removal techniques.

Because such techniques are related to skeletonization, they must ensure that they provide a *good* skeleton. A good skeleton may be seen as a skeleton which converges to the medial surface when the sampling density of $\mathcal{O}$ tends to infinity. A very important result in 3D is that, unlike in 2D [9], the Voronoi vertices do not converge to the medial surface, as the sampling density tends to infinity [2].

An approach that guarantees convergence uses a subset of the Voronoi vertices, named the *poles* [1]. For a sample point p, the poles are the vertices of its

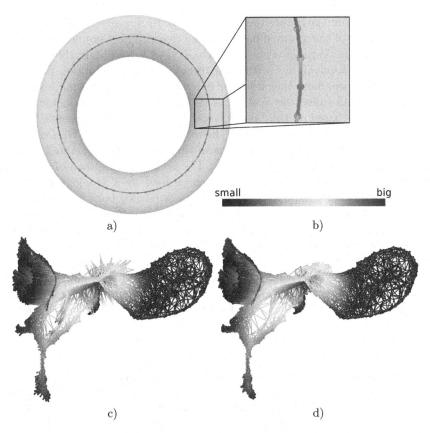

Fig. 1. Illustration of skeletal noise. a) This torus has 94% of its atoms belonging to Type 0 noise, such that the englarged view contains 68 atoms, while only 4 can be seen. b) To represent skeletons in this paper, we use a color code to transcript the radius of an atom: from blue for small values to hot colors for big values. c) The blender monkey model's skeleton contains hairy pattern, especially at the center of the skull. d) The cleaned skeleton of the monkey model obtained with our method.

Voronoi polyhedron that are the furthest away from p on the two sides of the surface. The balls centered in these poles (with radii equal to the distance to their samples) are called *polar balls*. The method is very robust and the produced skeleton, called the *Power Shape*, is visually reasonable. However, skeletal noise remains and many flat tetrahedra populate the skeleton. Also, this method was firstly intended to reconstruct a shape from a point cloud. Thus, it does not take advantage of any information contained in $\mathcal{O}$ except some sample positions.

Dey and Zhao [13] presented a method where (a subsequent part of) the output converge to the medial surface, by applying angle-based filter conditions to the Voronoi diagram. The filter parameters are scale and density independent. However the skeleton topology is ignored and some holes appear in many cases, inducing loss of information or topology changes in the described object.

In [10], the authors introduce the notion of *weak feature size* $\mathcal{F}(x)$ at a point x. This is the radius of the minimal ball enclosing closest object boundary points to x. If we remove the Voronoi vertices v with $\mathcal{F}(x) < \lambda$, we obtain the $\lambda - Medial$ *Axis*. The main disadvantage with such technique is the definition of λ for a shape: as λ increase, more skeletal noise is removed, as well as some features.

Another method called the *Scale Axis* [17] produces very nice looking skeletons. This work is based on the *Power Shape*, with useful enhancements. The input object is remeshed and sampled by the technique proposed in [7]. Polar balls are efficiently obtained from an input mesh thanks to a more adapted process. Those polar balls are scaled by a factor s. The skeleton, composed of the union of those scaled balls, is computed and then cleaned by a topology-preserving angle filtering. Finally, the balls in the skeleton are scaled back by a factor $1/s$. The scaling factor allows the skeletal noise removal by a spatially adaptive feature classification. With some constraints on the s parameter, the *Scale Axis* transform has been proven to have topological stability guarantees [14].

We propose in this paper to perform the noise removal apart from the skeletonization, working directly on the skeleton with no further information. As shown in the result section, our technique significantly remove the skeletal noise and keeps every details. Moreover, it maintains the topology of the skeleton.

Outline

In this paper, we first describe the model we have conceived to structure the skeleton (Section 3). We also present the atomic operations that can be realized on this model (Section 3.3). Then, we detail the methodology we used to identify skeletal noise of Type 0 (Section 4) and Type 1 (Section 5). We finally expose our validation process, and conclude.

3 Structuring Model

In this section, we present the skeleton model used to identify and structure the skeletal noise.

3.1 Atoms

Basically, atoms can be seen as balls whose union approximates an object. We consider them as nodes of a graph: they have a position, a radius, and links with other atoms. We add to this graph a hierarchy structure, reflecting the importance of the atoms. Each atom has then a rank that is equal to its level in the hierarchy: 0 for a useless atom, 1 for a unimportant atom, 2 for a more important and so on. The rank is the label of the atom node. Since we deal with a hierarchy, atoms may have also a father and sons. If an atom does not have a father, we call it *root*. Otherwise, we call it *sub-atom*. The Figure 2 shows a schematization of these concepts.

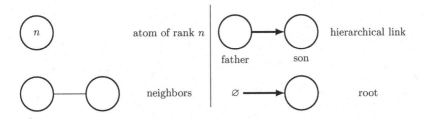

Fig. 2. Schematization of atoms

3.2 Skeleton

The skeleton has thus two structures: one composed of neighbor links (representing the topology), and another one composed of hierarchical links (reflecting the importance of atoms). In order to combine those two structures, we introduce the notion of n-hierarchy. A skeleton is said to be a n-hierarchy if:

  i) an atom is a root $\Longleftrightarrow$ an atom has rank n,
 ii) two atoms have different ranks $\Longrightarrow$ they cannot be neighbors,
iii) every sub-atom has a rank lower than the one of its father.

The structuring process starts with a 0-hierarchy skeleton $\mathcal{S}$: a graph where each node is 0 labeled. Then $\mathcal{S}$ is structured rank by rank, using criteria to detect when an atom a is less important than another atom b. The rank of such atom b is changed to reflect its importance relatively to a. Such an operation make impossible to keep the n-hierarchy property because there will be be roots at ranks n and $n + 1$ (See Figure 3). We hence release some of the constraints contained in the definition, to introduce the notion of n-consistent skeleton. This intermediary state specify how the skeleton should be before reaching a hierarchy of higher level. Such a skeleton meet the following requirements:

  i) roots can only be at the levels n and $n + 1$, and every atom of rank $n + 1$ is a root,
 ii) only two roots or two atoms with the same rank can be linked,
iii) every sub-atom has a lower rank than its father.

The roots of level $n + 1$ are called the *processed roots*, because they will remain roots of level $n + 1$ until the skeleton becomes a $n + 1$-hierarchy. Because of *ii)*, atoms corresponding to noise are not linked to important atoms. It is a way to isolate them. Items *i)* and *iii)* make the roots of a skeleton the most important atoms. If we consider only these roots we have a connected skeleton called in this paper *clean skeleton*, since the low ranks contain skeletal noise.

3.3 Structuring Operations

When an atom a is detected as less important than another atom b, we perform a structuring operation on the skeleton, called absorption and written $b \succ a$,

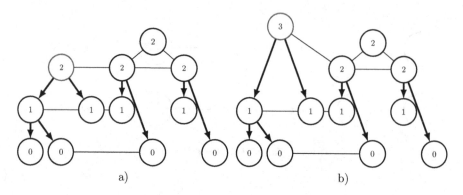

Fig. 3. Schematizations of skeletons. a) This skeleton is a 2-hierarchy. The atom in red is promoted to a higher rank because we detected it as important. We obtain the situation b), where this skeleton is 3-consistent.

to reflect this information. For each kind of addressed importance, a specific absorption is defined. In this paper, we deal only with importance relatively to noise. Thus we define one absorption for clusters and one for hairy pattern.

An absorption always sets the less important atom a as the son of a root atom c. After such action, c must update its rank r_c to have the same rank as the processed roots, i.e. $n + 1$. The rank r_c is then greater than the rank of a, in order to meet the constraint described in *iii)*. Then, every link between a sub-atom and c must be removed, to fulfill item *ii)*. The last two operations are called promotion and written `promote(c)`.

In order to maintain the skeleton topology, during the absorption, the links of a are transmitted to c. Moreover, we also need to remove the links between a and the other roots in the hierarchy (item *ii)*).

4   Identifying and Structuring Atom Clusters

This section deals with atom clusters and proposes a solution to identify them and structure them.

4.1   Observations about Atom Clusters

We know exactly the theoretical skeletons of simple objects, e.g. torus, sphere. When we observe the practical skeletons obtained for such objects, we can notice a huge difference between the number of perceived atoms and the real number of atoms. This difference is due to the presence of *clusters*: a high number of atom loci is contained in a very small spatial area. In such places, we perceive only one atom (at a reasonable scale), while in practice, there are so many, as shown in Figure 1 a).

Atoms are maximal inscribed balls. Thus no atom can be fully contained into other ones. If we suppose there is a difference of radii within a cluster, the biggest atom would contain at least one atom, as they are very close. This is a contradiction, and then every atom in a cluster has a similar radius. So, every atom in the cluster adds a very tiny piece of information (about the size of the machine precision). Consequently, a cluster should then be replaced by one of its atoms since it is useless to store a large number of atoms that add almost nothing.

Clusters are due to parts that are locally spherical. Basically, with a Voronoi diagram based skeletonization technique, each 4 cospherical samples produce an atom, located at the circumsphere. If there are more than 4 cospherical samples, duplicated atoms are created. Due to machine precision, duplicated atoms will not be at the same location, but very close and with very similar radii. Clusters could be then addressed in the skeletonization process by identifying samples that are cospherical, like in [19]. In the next section, we will show an identification criterion that is independent from the skeletonization technique.

4.2   Identification Criterion

Since atoms inside the same cluster are very close and have almost the same radius, the volume $\mathcal{V}_{a\setminus b}$ added by an atom a to another atom b of the same cluster is nothing compared to the volume $\mathcal{V}_b$. We define a test $\mathcal{F}_0(a,b)$ which indicates whether a and b belong to the same cluster and if a is less important than b. Here is the expression of the test $\mathcal{F}_0$:

$$\mathcal{F}_0(a,b) = (a \cap b \neq \emptyset) \bigwedge (\mathcal{V}_a < \mathcal{V}_b) \bigwedge (\mathcal{V}_{a\setminus b} < \kappa \cdot \mathcal{V}_b) . \tag{1}$$

This test allows to detect clusters with only one parameter, κ, which is easy to understand. Using a relative measure based on the volume of an atom makes this criterion insensitive to scale while being local. Thus, it suits models of any size while taking into account the local thickness of the shape, to avoid the labeling of small details as Type 0 noise. In our implementation, we noticed that $\kappa = 5\%_0$ is enough for all the tested objects.

There is no requirement on links between a and b, and we explain here why. First, there is no need to look at links to identify a cluster. Second, links inside a cluster are chaotic, we cannot rely on them for efficiency purpose. Finally, there exist isolated vertices inside a cluster for skeleton based on the *Power Shape* algorithm (due to the use of a regular tetrahedralization to define the links). Ignoring links when detecting clusters allows us to remove isolated vertices from the cleaned version of a skeleton.

4.3   Structuring Process

We process every atom of rank 0 such that there is no cluster in the *clean skeleton*, i.e. among the roots. For a cluster, only one of atom a will reach the rank 1, while every other atom will be the sons of a. Atoms that do not belong to a cluster are promoted to the next rank at the end of the process such that we obtain a 1-hierarchy.

To process an atom a, we use the test function $\mathcal{F}_0$ with already processed atoms $\mathcal{P}$. For $b \in \mathcal{P}$, if $\mathcal{F}_0(a, b) = true$, then $a \prec b$, otherwise if $\mathcal{F}_0(b, a) = true$, then $b \prec a$.

For clusters, the absorption is context dependent: the result is not the same depending whether input atoms are roots or not. We give a summary of structuring rules for this absorption in the Figure 4. For reasonable values of κ, this process keeps the skeleton topology because only very close atoms are removed (and links with removed atoms are transmitted to their fathers). Also, there is no need to impose a processing order, as every atom in a cluster is interchangeable.

5   Identifying and Structuring Hairy Pattern

This section deals with hairy pattern and propose a solution to identify them and structure them.

5.1   Observations about Hairy Pattern

The hairy pattern is one of the most known and recognizable skeletal noise. It consists in atoms that do not capture any feature. They are produced by circumspheres that have some of their 4 spherical sample points close from each other. This is why the *Lambda Axis* [10] prune atoms when distances between these samples are below a threshold.

When we explore the skeleton looking for Type 1 atoms, we notice something about their neighbors: they are located in a narrow cone, and most of them belong to the *stable skeleton*. The stable skeleton is the visually free of noise skeleton. This neighbor configuration gives to Type 1 atoms a spiky appearance, making them off-centered. Also, such atoms have a radius lower than their neighbors on the stable skeleton. So when we move from a Type 1 atom toward the stable skeleton, the radii of atoms get bigger.

The ends of thin skeleton parts, e.g. for fingers, are very similar to hairy pattern. The distinction is made with the length of such pattern: if this length is a small compared to surrounding parts, we have a hairy pattern. So, to deal with hairy pattern, we should limit the length of absorbed skeleton components, to avoid the classification of thin detail parts as noise.

5.2   Identification Criterion

Since Type 1 atoms are off-centered, we have built a simple criterion based on this. We check the location of the neighbors of an atom a by computing a sphere $\mathcal{BS}(a)$, which is the minimal sphere containing their positions. If a is not inside $\mathcal{BS}(a)$, we consider that a is off-centered and thus is a Type 1 atom. This parameter-free criterion is better suited than setting a threshold to control the thickness of the cone containing the neighbors.

To define $\mathcal{BS}(a)$, the atom a must have at least 2 neighbors. Otherwise, a has one neighbor b: it can either be noise or the termination of a thin curvilinear

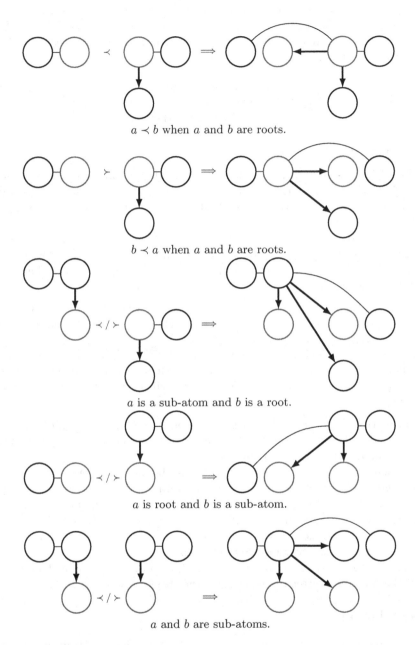

$a \prec b$ when a and b are roots.

$b \prec a$ when a and b are roots.

a is a sub-atom and b is a root.

a is root and b is a sub-atom.

a and b are sub-atoms.

Fig. 4. Schematization of cluster absorptions. The atoms concerned by this operation are depicted in red, a on the left that is the currently processed atom, and b on the right which had already been processed. For simplicity purpose, we do not schematize the operations to meet the constraints described by item *ii)* in the definition of a n-consistent skeleton.

component. In the latter case, a should not be absorbed. We distinguish those cases by counting the neighbors of b. If it is more than 2: b is not a part of a thin curvilinear component and a is not absorbed. This gives us the following test $\mathcal{F}_1$ to identify a Type-1 atom:

$$\mathcal{F}_1(a) = \begin{cases} a \in \mathcal{BS}(a) \,, \text{ if } \sharp\mathcal{N}(a) > 2 \\ \sharp\mathcal{N}(b) < 3 \,, \text{ if } \mathcal{N}(a) = \{b\} \\ \quad \text{false} \quad \,, \text{ in other cases} \end{cases} \tag{2}$$

5.3   Structuring Process

Initially, $\mathcal{S}$ is a 1-hierarchy. We have to test and structure the following set of atoms: $\mathcal{P} = Roots(\mathcal{S})$. Every atom $a \in \mathcal{P}$ is re-tested when a change (an absorption) is made in $\mathcal{P}$. We assume that $\mathcal{F}_1(a) = true$, otherwise we simply take the next atom in $\mathcal{P}$.

If a has only one neighbor b we realize the operation $a \prec b$. In the other case, we must choose among the neighbors of a, the atom b to use in the absorption. We propose to set b as the neighbor of rank 1 with the highest radius. This has two consequences:

- the absorption is made toward the stable skeleton because bigger atoms in a hairy pattern are closer to the stable skeleton than smaller atoms (see Section 5.1).
- choosing a rank 1 atom will limit the length of an absorbed part, since in combination with the identification criterion, there will be no further possible absorption. Thus thin detail skeleton parts will be protected, as shown in Figure 5.

Structuring rules for the hairy pattern absorption are detailed in the the Figure 6. Due to the identification criterion, atoms at crossings of skeleton components or inside a component are not detected as hairy pattern: their neighbors are all around them, thus they are inside the minimal bounding sphere. Only atoms in the boundary of components and hairy pattern are detected. As there is an effect which limits the length of the absorption, no components can completely disappear. Moreover, the links of an absorbed atom are transmitted to its father, without any loss of connectivity information. So, the topological structure of the *clean skeleton* remains the same after this step.

We impose to process $\mathcal{P}$ from the lowest radius to the biggest one. The result is then the same no matter the creation of the skeleton structure.

6   Validation

We validated our work with quantitative and qualitative comparisons. These comparisons were made with two different skeletons for some input shapes. The results express the "compression" realized by our technique on skeletons, while quantifying the modification in the geometric data. Also, the qualitative study shows the improvement in the skeleton connections and the conservation of the skeleton structure.

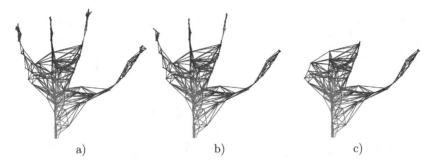

Fig. 5. Effect of the hairy pattern structuring process. The input skeleton of Homer hand is presented in a). With the method described here we obtain the result b). If we allow any neighbor of rank 1 to absorb an atom detected as noise, fingers disappear as shown by c).

6.1 Protocol

We took some input shapes, and we skeletonized them by two well-known algorithms that give theoretical guarantees about the results while reducing the skeletal noise: the *Power Shape* and the *Scale Axis*. Each of these skeletons was structured by the technique we presented previously, giving us four skeletons by input shape.

We first computed statistics about those skeletons, to quantify their complexity. We also measured the distance between the input shape and the shape described by a skeleton. To do so, we used the *Skin Surface* [15], a garbing algorithm which only considers the spheres. This way, we measure the possible loss of geometric data in the cleaned skeleton relatively to the input skeleton. If we had used a garbing algorithm that consider other primitives to enhance the garbing mesh, like [12] that removes surface noise in the garbing mesh, the loss of geometric data would have been hidden. We chose the Root Mean Square distance (RMS), and we computed the distance between the input shape and the skin surface of each skeleton by the *Mesh* software [3].

Then, we compared visually the quality of the skeletons. We examined the links between atoms, looked for remaining noise and checked that detail was not removed.

6.2 Quantitative Comparison

First, Figure 7 expresses that the skeletons have been reduced with our structuring process. In average, only 85% atoms of the *Power Shape* remain in the *clean skeleton*, and 53% of the *Scale Axis*. For the *Power Shape*, the majority of the skeletal noise is caused by hairy pattern. However, such a skeletonization technique was firstly designed to remove such noise type. Thus, a noise removal algorithm done apart from the skeletonization, like ours, is useful to produce clearer skeletons, as noise still remains.

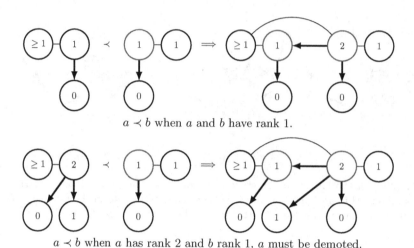

$a \prec b$ when a and b have rank 1.

$a \prec b$ when a has rank 2 and b rank 1, a must be demoted.

Fig. 6. Schematization of hairy pattern absorption. The atoms concerned by such operation are depicted in red, a is on the left and b on the right.

This statement is confirmed by the huge quantity of clusters contained in the *Scale Axis*. Such a skeleton is made of far more atoms than previous ones. A trend in skeletonization techniques is to sur-sample the input object, in order to produce more atoms. By doing so, we expect the skeleton to be more accurate and clear. As there are more atoms, they add fewer information to their neighbors, and our cluster identification criterion is triggered.

Table 1. RMS distances between original object and skin surfaces of both normal and cleaned skeletons. Results are given in percent relatively to the diagonal of the bounding box.

	Power Shape		Scale-Axis	
	Normal	Cleaned	Normal	Cleaned
armadillo	4.33	4.75	49.11	47.42
baby	7.15	7.23	41.17	40.35
bimba	7.36	7.41	82.24	80.94
boy	11.20	11.31	39.85	38.75
bunny	23.99	20.14	91.56	90.49
camel	16.12	15.67	64.73	63.78
dinopet	13.74	13.64	49.58	47.86
egea	4.87	5.79	134.65	133.23
fish	32.90	31.81	35.34	35.33
homer	10.16	10.43	65.71	63.47
horse	11.76	10.68	59.67	58.73

One could argue than such results are obtained because we remove details from the skeletons. However, Table 1 prove that we only removed skeletal noise from the skeletons. The geometric data loss with our *clean skeleton* obtained for

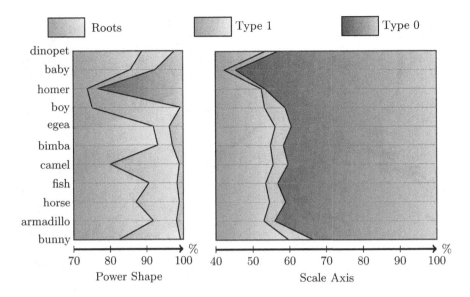

Fig. 7. Percentage of Type 0, Type 1 and roots in the cleaned *Power Shape* and the cleaned *Scale Axis*

the *Power Shape* is very small, about 1‰. For some cases this is even better, as skeletal noise can add material *outside* the original shape. For the *Scale Axis*, there is no loss of data reported: the RMS distance is always better. Thus our technique removed 47% of atoms in the *Scale Axis* without losing any detail compared to the input skeleton. This demonstrates that our technique efficiently remove skeletal noise from these skeletons.

The results also raised the question about the legitimacy of the actual trends in skeletonization: producing skeletons with more atoms does not improve the geometry captured, it mostly adds skeletal noise.

6.3   Qualitative Comparison

As clusters are by definition composed of atoms that cannot be distinguished visually, their absorptions make no visual difference. Also, we absorb the skeletal noise without any loss of detail, thus there is no missing parts in the skeleton. For these two reasons, there is not important visual differences between a skeleton and its cleaned version. We can only notice the removal of hairy pattern, erasing some small spiky component, and the disappearance of some atoms in skeleton boundaries as shown in Figure 8.

For *Scale Axis* skeletons, the noise of Type 1 – the only visible noise – is closer to the stable skeleton. Thus, the visual enhancement on the clean skeleton is less than for the *Power Shape*. We checked the visual enhancement on a *Power Shape* obtained with much more samples to raise the number of atoms. The produced skeleton is visually enhanced and the atom reduction is on average 53%, with a

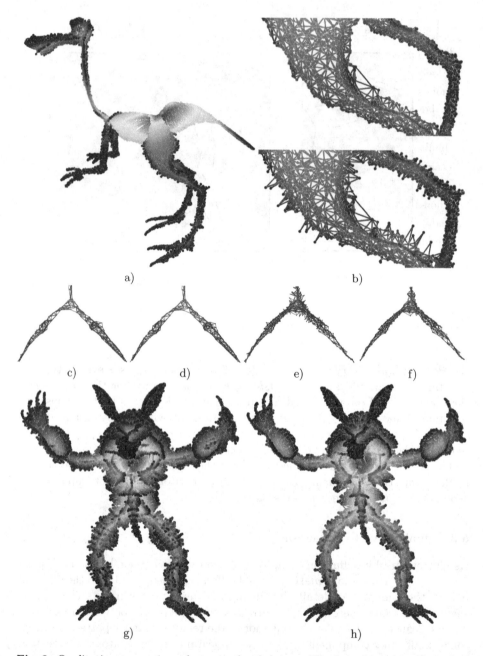

Fig. 8. Qualitative comparisons between the skeletons. The first three rows show the visual effects of our noise removal method on the dinopet model. a) is the *Scale Axis* for this model. b) shows the enhancement on the thigh from the input (top) to the cleaned skeleton (bottom). c), d), e) and f) propose a close view of one foot of this model for the *Power Shape* and its cleaned version, and for the *Scale Axis* and its cleaned version. A sur-sampled version of *Power Shape* in g) and its cleaned version in h).

relative proportion of Type 0 and Type 1 noise nearly the same as for the *Scale Axis*. Thus, even if raising the number of atom in a skeleton increase the Type 0 noise, the *Scale Axis* reduces the visual importance of Type 1 noise, while our method reduces both noises in terms of quantity.

6.4 Limitations

There are two limitations to our technique. The first one is due to the constraint we have on detail. Indeed, to use skeletons as shape representation models in computer graphics applications, they should be able to capture small features. Thus, the cleaning process let untouched some hairy patterns to not absorb small features because small features can also be detected as hairy pattern. This effect can be handled by two or more ranks of hairy pattern when needed, but it requires the user intervention to determine when to stop.

The second limitation comes from the definition of the κ value. If it remains small, the skeleton is perfectly clean and clear, no detail is loss. But with bigger values of κ, parts of the skeleton disappear, being iteratively absorbed. This is especially the case on models which have a lot of small features, like the Armadillo model, or on skeletons with highly dense atoms: an atom is very likely to add not enough material to its neighbors, and thus is absorbed. As the κ value is meant to handle only clusters, i.e. machine precision issues, we highly discourage the use of high values of κ: the skeleton will lose its structure and features will be missing.

7 Conclusion and Future Work

In this paper, we addressed the removal of skeletal noise, i.e. atom clusters and hairy pattern. Such noise produces unnecessarily complex skeleton, by raising its data size and disturbing its structure. Unlike the numerous methods that take place in a skeletonization algorithm, our technique can be directly used on any connected skeleton, requiring no additional data like the original shape or angle values stored with atoms. Another originality of our work, is the use of a hierarchy structure to process the skeletal noise. Depending on its contribution to the geometric data of a skeleton, an atom receive a rank, reflecting its importance. Less important atoms are isolated from more important ones by this structure.

We showed in this paper how to take advantage of this hierarchy to clean a skeleton, in order to use it as a true shape representation model and not only an intermediary processing model. Indeed, atom clusters are removed, and the majority of hairy pattern is erased, while the skeleton do not suffer from any loss of detail. Even on input skeletons produced by skeletonization algorithms that are known to remove skeletal noise, we significantly reduce the number of atoms. The components of the skeletons remain the same, and no "hole"appears.

We define a level of detail, by considering only atoms with a rank greater than a minimal one. Once we get rid of the skeletal noise, we can identify and structure the features of the skeleton the same way we dealt with noise.

This will build a multi-resolution shape representation model, allowing processing at different level of detail like rendering, segmenting, or interaction.

References

1. Amenta, N., Choi, S., Kolluri, R.K.: The power crust. In: 6th ACM Symposium on Solid Modeling and Applications, pp. 249–266. ACM, New York (2001)
2. Amenta, N., Choi, S., Kolluri, R.K.: The power crust, unions of balls, and the medial axis transform. Comput. Geom. Theory Appl. 19(2-3), 127–153 (2001)
3. Aspert, N., Santa-Cruz, D., Ebrahimi, T.: Mesh: Measuring errors between surfaces using the hausdorff distance. In: IEEE International Conference on Multimedia and Expo., vol. I, pp. 705–708 (2002)
4. Au, O.K.C., Tai, C.L., Chu, H.K., Cohen-Or, D., Lee, T.Y.: Skeleton extraction by mesh contraction. ACM Trans. Graph. 27, 44:1–44:10 (2008)
5. Aylward, S.R., Jomier, J., Weeks, S., Bullitt, E.: Registration and analysis of vascular images. International Journal of Computer Vision 55(2-3), 123–138 (2003)
6. Blum, H.: A Transformation for Extracting New Descriptors of Shape. In: Models for the Perception of Speech and Visual Form, pp. 362–380. MIT Press, Cambridge (1967)
7. Boissonnat, J.D., Oudot, S.Y.: Provably good sampling and meshing of surfaces. Graphical Models 67, 405–451 (2005)
8. Brady, M.J., Asada, H.: Smooth local symmetries and their implementations. International Journal of Robotic Research (1984)
9. Brandt, J.: Convergence and continuity criteria for discrete approximations of the continuous planar skeleton. Computer Vision, Graphics, and Image Processing: Image Understanding 59(1), 116–124 (1994)
10. Chazal, F., Lieutier, A.: Stability and homotopy of a subset of the medial axis. In: 9th ACM Symposium on Solid Modeling and Applications, pp. 243–248. Eurographics Association, Aire-la-Ville, Switzerland (2004)
11. Cornea, N.D., Silver, D., Min, P.: Curve-skeleton properties, applications, and algorithms. IEEE Trans. on Visualization and Computer Graphics 13, 530–548 (2007)
12. Delamé, T., Roudet, C., Faudot, D.: From a medial surface to a mesh. Computer Graphics Forum 31(5), 1637–1646 (2012)
13. Dey, T.K., Zhao, W.: Approximating the medial axis from the voronoi diagram with a convergence guarantee. Algorithmica 38(1), 179–200 (2003)
14. Giesen, J., Miklos, B., Pauly, M., Wormser, C.: The scale axis transform. In: 25th Annual Symposium on Computational Geometry, pp. 106–115. ACM, New York (2009)
15. Kruithof, N.G.H., Vegter, G.: Meshing skin surfaces with certified topology. In: 9th International Conference on Computer Aided Design and Computer Graphics, CAD-CG 2005, pp. 287–294. IEEE Computer Society (2005)
16. Leyton, M.: Symmetry–curvature duality. Computer Vision, Graphics, and Image Processing 38, 327–341 (1987)
17. Miklos, B., Giesen, J., Pauly, M.: Discrete scale axis representations for 3d geometry. ACM Trans. Graph. 29 (2010)
18. Tagliasacchi, A., Zhang, H., Cohen-Or, D.: Curve skeleton extraction from incomplete point cloud. ACM Trans. Graph. 28, 71:1–71:9 (2009)
19. Tam, R., Heidrich, W.: Shape Simplification Based on the Medial Axis Transform. In: IEEE Visualization, pp. 481–488. IEEE Computer Society, Washington, DC (2003)

GPU Integral Computations in Stochastic Geometry

Elise de Doncker and Rida Assaf

Western Michigan University, Kalamazoo MI 49008, USA
{elise.dedoncker,rida.assaf}@wmich.edu

Abstract. Stochastic geometry has applications in areas such as robotics, tomographic reconstruction with uncertainties, spatial Poisson-Vonoroi tessellations, and imaging from medical data using tetrahedral meshes. We examine numerical approaches for the computation of multivariate integrals for a family of problems where uniformly distributed points are picked as polyhedron vertices for tessellations, for example, tetrahedron vertices in a cube, tetrahedron or on a spherical surface. The classical cube tetrahedron picking problem yields the expected volume of a random tetrahedron in a cube, and helps furthermore assessing unsolved extremal problems (cf., A. Zinani, 2003). We demonstrate feasible numerical approaches including adaptive integration through region partitioning, quasi-Monte Carlo (based on a randomized Korobov lattice), and Monte Carlo techniques, which are the basic methods of our parallel integration package ParInt. We then describe our implementation of the Monte Carlo approach on GPUs (Graphics Processing Units) in CUDA C, and demonstrate its parallel performance for various stochastic geometry integrals.

Keywords: stochastic geometry, numerical integration, GPU.

1 Introduction

A family of problems, to determine the expected d-dimensional (dD) volume $E[V_n(K)]$ of the polyhedron formed by n points $X_1, \ldots, X_n$, uniformly distributed in the interior of a convex body K, is represented in [1] by

$$E[V_n(K)] = \frac{1}{|K|} \int_K \cdots \int_K \mathrm{conv}(X_1, \ldots, X_n)\, \frac{dX_1}{|K|} \cdots \frac{dX_n}{|K|}, \quad (1)$$

where $|K|$ denotes the d-dimensional volume of K and the integrand function is the volume $\mathrm{conv}(X_1, \ldots, X_n)$ of the convex hull generated by the n points.

J. J. Sylvester considered the plane case for a random triangle T in a convex set K and posed the problem to determine the shape of K for which the expected value $E[X]$ is minimal or maximal, for the variable $X = \mathrm{area}(T)/\mathrm{area}(K)$ (see, e.g., [2]). Note that the convex body K which maximizes the expected volume of a random d-dimensional simplex, is unknown for $d \geq 3$.

B. Murgante et al. (Eds.): ICCSA 2013, Part II, LNCS 7972, pp. 129–139, 2013.

As a direct application in $\mathbb{R}^3$, the mean volume of a tetrahedron whose vertices are uniformly distributed on a circular domain (*cap*) of the unit sphere is an essential element in the analysis of Poisson-Vonoroi and Delaunay tessellations [3]. Tessellations induce a partition into space-filling random polyhedra, and are used in such areas as molecular modeling, material science, pattern recognition and statistical data analysis. Tetrahedral meshes in general are found in applications including computer graphics, computer-aided design, robotics [4], electromagnetic simulations [5] and medical imaging [6–8].

Efron [9] determined the probability with which N points chosen at random in a convex region, are the vertices of a re-entrant polyhedron, where the last r selected points lie inside the convex hull of the first $N - r$ points. For example with $r = 1$, the probability that $N = 5$ points chosen at random in a convex region form the vertices of a re-entrant polyhedron (as related to (1)) is $N \times E[V_4(K)]$.

Results of the $E[V_n(K)]$ problem have been derived which cover specific cases for $d \geq 3$. For example, the problem to evaluate the expected volume of a random polytope in the interior of a sphere (ball), was considered early on in [10, 11]. Tetrahedron picking in a cube (*cube tetrahedron picking*) was handled, e.g., in [1, 2]. For tessellations on the surface of a sphere, S^2 in 3D, (*sphere-tetrahedron picking*) see [3]. The problem was solved for d-dimensional simplices, with $d + 1$ random vertices on the surface of S^{d-1} in $d \geq 2$ dimensions [12]. The solution of the problem for tessellations within a tetrahedron (*tetrahedron-tetrahedron picking*) [13, 1, 14] refutes a conjecture from 1991 that the expected volume would be a rational number [15, 16]. This as well as *cube-* and *octahedron-tetrahedron picking* contributes to the results with respect to tessellations in non-spherical convex bodies. A recurrence relation was established for $E[V_n(K)]$ with respect to arbitrary K in [17]. Overall, closed form solutions have been restricted to special cases and difficult to obtain.

In this paper we first examine approaches given by automatic numerical integration including adaptive partitioning of the domain, quasi-Monte Carlo (QMC) and Monte Carlo (MC) techniques, which have been implemented for *automatic integration* in sequential and parallel packages [18–22]. We choose these particular methods because they are at the basis of our ParInt package (developed further from ParInt 1.0 ©1999 by E. de Doncker, A. Gupta, A. Genz, R. Zanny). Over the years our interest in these techniques for automatic numerical integration has been sparked by applications in various fields, including many in statistics (e.g., multivariate normal, t- distributions), finite elements (e.g., automotive simulations), computational finance (financial derivatives, e.g., collateral mortgage obligations, mortgage backed security problems), high energy physics (e.g., interaction cross sections), computational chemistry (e.g., Fock matrix representing the electronic structure of an atom or molecule), and computational geometry.

A brief description of the basic methods is given in Section 2. In this paper they are mainly used for comparisons. A parallel Monte Carlo implementation for GPUs is outlined in Section 3. Section 4 gives results and addresses the parallel performance of the CUDA C implementation.

2   Automatic Numerical Integration

In this section we give a short description of the methods used to obtain some of the numerical results in this paper. Each of these techniques typically underlies an *automatic* numerical integration algorithm, set up as a *black-box* to which the user specifies the dimension, integrand function and integration domain, a limit on the number of function evaluations for termination, a tolerance for the error and possibly other parameters. The method may be *adaptive* if the course of its computations is governed by the behavior of the problem at hand, or *non-adaptive* if it evaluates a fixed sequence of integration rules until a termination criterion is satisfied.

The black-box algorithm yields an integral approximation

$$Q \approx I = \int_{\mathcal{D}} f(x) \; dx$$

and an absolute error estimate E, where it attempts to satisfy a criterion of the form $|Q - I| \leq E \leq$ tolerated error, within the allowed number of function evaluations; or indicates an error condition if the limit has been reached.

Adaptive methods are generally recommended for low to moderate dimensional problems (say, up to dimension 12), whereas Monte Carlo and number-theoretic type methods (lattice rules, QMC) can be used in higher dimensions. Apart from the adaptive and QMC algorithms described below, we will also include results from a crude Monte-Carlo algorithm (RCRUDE from the package MVNPACK [22]), with simple antithetic variates and a uniform (0,1) random number generator from [23].

2.1   Adaptive Region Partitioning

Many integration problems feature varying function behavior across the integration domain (e.g., singularities or peaks). An adaptive method based on region partitioning executes a number of iterations where, in each iteration, a subregion is selected, subdivided, and the integral and error estimates are updated accordingly. Region selection is based on the local error estimate which is obtained together with the local integral approximation over each region. By intensive partitioning in the vicinity of singularities or other hot spots, it is the goal to focus on the difficult areas and sample the integrand adaptively as needed.

The integration rules of the algorithm DCUHRE [18] were chosen for the adaptive methods in the PARINT parallel integration package [20, 21]. Some of the numerical results in Section 4 below have been obtained using this method.

2.2   Quasi-Monte Carlo

The Quasi-Monte Carlo (QMC) method [24] samples the integration function at the (N) points of a regular lattice, which are modified by a random vector.

Let the sequence of randomized (Korobov) lattice rule approximations be denoted by

$$K_N(\boldsymbol{\beta}) = \frac{1}{N} \sum_{i=1}^{N} f(\{\frac{i}{N}\mathbf{v} + \boldsymbol{\beta}\}),$$

where $\mathbf{v}$ is the lattice generator vector and $\boldsymbol{\beta}$ is a uniformly distributed random vector. By computing the rule for q random $\boldsymbol{\beta}$ vectors, the integral can be approximated as

$$I \approx \bar{K}_N = \frac{1}{q} \sum_{j=1}^{q} K_N(\boldsymbol{\beta}_j), \tag{2}$$

which allows for a standard error estimation

$$E_N^2 = \frac{1}{q(q-1)} \sum_{j=1}^{q} (K_N(\boldsymbol{\beta}_j) - \bar{K}_N)^2.$$

The program DKBVRC from the package MVNDST [19] is at the basis of the parallel QMC method in Parint [20]. The algorithm calculates the $\bar{K}_N$ values of (2) for successively larger numbers N of integration points until either an answer is found to the user-specified accuracy or the function count limit is reached.

3   Monte Carlo Simulation of Random Tetrahedron Picking on GPUs

3.1   Random Number Generation

The Monte Carlo approximation is obtained as the avarage of the function values at a set of N uniformly distributed random points. We rely on the CUDA pseudo-random number generator library (*curand*) to generate $N * d$ random floats for the coordinates of N d-dimensional points, directly on the GPU (device). A CUDA C program section allocating space for ndim $= N * d$ floats in an array a on the device, and calling the *curand* functions is shown below.

```
//  allocate ndim floats on device in array a
cudaMalloc (( void **) & a , ndim * sizeof ( float )) ;
//  create pseudo-random number generator
curandCreateGenerator (& gen , CURAND_RNG_PSEUDO_DEFAULT ) ;
//  set seed
curandSetPseudoRandomGeneratorSeed ( gen , 1234ULL ) ;
//  generate ndim random numbers on device
curandGenerateUniform ( gen , a , ndim ) ;
```

3.2   Device Functions

A part of the program to launch in parallel on the GPU is coded as a CUDA kernel. The CUDA runtime is then instructed to launch a number of parallel copies or *blocks*, and how many threads to launch per block. The integrand function evaluated at the random points from the CUDA kernel is referred to as a *device function*.

For tessellations on a spherical surface (*sphere-tetrahedron picking*), the vertices of the tetrahedron are random points on the unit spherical surface S^2. The integral (1) is 8-dimensional as it involves the integration for each of the four vertices over $K = S^2$ (which is 2-dimensional). S^2 can be described in parameter form by $(x = \sqrt{1 - u^2} \cos(\theta), \ y = \sqrt{1 - u^2} \sin(\theta), \ z = u)$, so that the integration is over $u \in [-1, 1]$ and $\theta \in [0, 2\pi]$. As suggested in [16], the dimension can be reduced from 8 to 5 by fixing the first vertex (x_1, y_1, z_1) at $(0, 0, 1)$ and taking the second (x_2, y_2, z_2) in the xz-plane $(y = 0)$ without loss of generality. The first vertex is thus obtained by $u_1 = 1$ and the second by $\theta_2 = 0$, so that the integration over the first sphere is removed and the second integration becomes one-dimensional. The integral obtained in this form corresponds to $\bar{V}(\pi)$ of [3],

$$
\bar{V}(\pi) = \frac{\int_{-1}^{1} du_2 \int_{-1}^{1} du_3 \int_{-1}^{1} du_4 \int_{0}^{2\pi} d\theta_3 \int_{0}^{2\pi} d\theta_4 \ |V_4|}{\int_{-1}^{1} du_2 \int_{-1}^{1} du_3 \int_{-1}^{1} du_4 \int_{0}^{2\pi} d\theta_3 \int_{0}^{2\pi} d\theta_4 \ 1}, \tag{3}
$$

where $|V_4|$ is the volume of the tetrahedron, given by the determinant

$$
V_4 = \frac{1}{6} \begin{vmatrix} x_1 & y_1 & z_1 & 1 \\ x_2 & y_2 & z_2 & 1 \\ x_3 & y_3 & z_3 & 1 \\ x_4 & y_4 & z_4 & 1 \end{vmatrix}.
$$

The integrand function for the 5D integral coded as a device function (to be called from the CUDA kernel) is given in Appendix A. The integration intervals $[-1, 1]$ and $[0, 2\pi]$ are transformed to $[0, 1]$. The implementation for the 8D integral can be given with dim $= 8$ and where the first and second vertex are computed similarly to the third and fourth.

Note that (3) corresponds to the special case of the expected volume $\bar{V}(\alpha)$ for $\alpha = \pi$ in [3], of the spherical *cap*

$$
S^2(\alpha) = \{ \ s(\phi, \vartheta) \mid ((\sin(\phi)\sin(\vartheta), \cos(\phi)\sin(\vartheta), \cos(\vartheta)); \ -\pi \le \phi \le \pi, \ 0 \le \vartheta \le \alpha \ \}.
$$

Thus the representation of S^2 above corresponds with $u = \cos(\vartheta)$ and $\alpha = \pi$. The expected volume of a random tetrahedron with vertices on S^2 is given by $\bar{V}(\pi) = \frac{4\pi}{105} \approx 0.11968$.

As another case, we compute the integral

$$
\begin{aligned}
E[V_4(T^3)] = 6^4 \int_0^1 dx_1 \int_0^{1-x_1} dy_1 \int_0^{1-x_1-y_1} dz_1 \int_0^1 dx_2 \int_0^{1-x_2} dy_2 \int_0^{1-x_2-y_2} dz_2 \\
\int_0^1 dx_3 \int_0^{1-x_3} dy_3 \int_0^{1-x_3-y_3} dz_3 \int_0^1 dx_4 \int_0^{1-x_4} dy_4 \int_0^{1-x_4-y_4} dz_4 \ |V_4|
\end{aligned}
$$

for the expected volume of a random tetrahedron within the 3D unit simplex. The integral is transformed as shown in the code of the device function in Appendix B, so all variables range over $[0, 1]$. The exact value is given by $E[V_4(T_3)] = \frac{13}{720} - \frac{\pi^2}{15015} \approx 0.017398$ (see [13, 1]).

For cube-tetrahedron picking, $E[V_4(C_3)]$ is given as an integral over the 12D unit cube, $[0, 1]^{12}$,

$$E[V_4(C^3)] = \prod_{i=1}^{4} \left[\int_0^1 dx_i \int_0^1 dy_i \int_0^1 dz_i \right] |V_4|, \tag{4}$$

and equals $E[V_4(T^3)] = \frac{3977}{21600} - \frac{\pi^2}{2160} \approx 0.0138428$ (see [1]).

4 Numerical Results and Parallel Performance

Table 1 compares the programs DCUHRE [18], DKBVRC [19] and RCRUDE [22] for the integral (4) of the *cube-tetrahedron* problem. Note that the problem is difficult in view of the absolute value function of the determinant integrated in 12 dimensions. The first column of Table 1 gives the number of integrand evaluations allowed. DKBVRC usually stays well under that, since it computes a sequence of rules each with a fixed number of points N_i at level i, and $N_i << N_j$ for $i < j$.

Table 1. Comparison for *cube-tetrahedron* problem ($E[V_4(C^3)]$)

# EVALS (M)	PROGRAM	RESULT	ABS. ERR.	ABS. ERR. Est.	TIME (s)
0.1	DCUHRE	1.381719e-02	3.36e-06	1.13e-02	0.3978e-02
	DKVBRC	1.394575e-02	1.03e-04	1.48e-04	0.5185e-02
	RCRUDE	1.363727e-02	2.06e-04	1.85e-04	0.2427e-01
1	DCUHRE	1.373211e-02	1.11e-04	6.96e-03	0.4372e-01
	DKVBRC	1.384961e-02	6.84e-06	3.67e-05	0.6687e-01
	RCRUDE	1.380387e-02	3.89e-05	5.90e-05	0.1214e-01
10	DCUHRE	1.373192e-02	1.11e-04	3.32e-03	0.4391e+00
	DKVBRC	1.384268e-02	9.37e-08	5.93e-06	0.8048e+00
	RCRUDE	1.384701e-02	4.24e-06	1.87e-05	0.4403e+00
30	DCUHRE	1.374961e-02	9.32e-05	2.36e-03	0.1323e+01
	DKVBRC	1.384218e-02	5.97e-07	4.13e-06	0.2729e+01
	RCRUDE	1.384282e-02	4.11e-08	1.08e-05	0.7280e+01
50	DCUHRE	1.381719e-02	2.56e-05	2.22e-03	0.2206e+01
	DKVBRC	1.384249e-02	2.88e-07	2.18e-06	0.4112e+01
	RCRUDE	1.384290e-02	1.27e-07	8.36e-06	0.1381e+02
100	DCUHRE	1.380235e-02	4.04e-05	1.85e-03	0.4410e+01
	DKVBRC	1.382480e-02	2.37e-08	1.28e-06	0.9232e+01
	RCRUDE	1.384405e-02	1.27e-06	5.91e-06	0.2427e+02
200	DCUHRE	1.380291e-02	3.99e-05	1.57e-03	0.8827e+01
	DKVBRC	1.384280e-02	2.37e-08	1.28e-06	0.9229e+01
	RCRUDE	1.384281e-02	3.50e-08	4.18e-06	0.4863e+02
	Exact:	1.384278e-02			

Table 2. GPU results for $E[V_4(S^2)]$, $E[V_4(C^3)]$ and $E[V_4(T^3)]$

# Ev. (M)	Problem	Result	Abs. Err.	Seq. Time (s)	Par. Time (s)	Speedup
0.1	$E[V_4(S^2)]$	1.196144e-01	6.53e-05	2.9787e-02	1.0117e-02	2.9442e+00
	$E[V_4(C^3)]$	1.383333e-02	3.57e-05	2.1131e-02	9.8808e-03	2.1385e+00
	$E[V_4(T^3)]$	1.743181e-02	3.36e-05	2.6249e-02	9.9260e-03	2.6445e+00
1	$E[V_4(S^2)]$	1.194591e-01	2.21e-04	2.9330e-01	1.0690e-02	2.7437e+01
	$E[V_4(C^3)]$	1.381283e-02	2.99e-05	2.0992e-01	1.1233e-02	1.8687e+01
	$E[V_4(T^3)]$	1.734215e-02	5.61e-05	2.5701e-01	1.1108e-02	2.3139e+01
10	$E[V_4(S^2)]$	1.196825e-01	2.74e-06	2.9632e+00	2.0068e-02	1.4374e+02
	$E[V_4(C^3)]$	1.384527e-02	2.50e-06	2.0518e+00	2.3818e-02	8.6146e+01
	$E[V_4(T^3)]$	1.740328e-02	5.04e-06	2.5688e+00	2.4165e-02	1.0630e+02
30	$E[V_4(S^2)]$	1.196864e-01	6.65e-06	8.8007e+00	4.2466e-02	2.0724e+02
	$E[V_4(C^3)]$	1.384298e-02	2.06e-07	6.1744e+00	5.1855e-02	1.1905e+02
	$E[V_4(T^3)]$	1.740763e-02	9.39e-06	7.7047e+00	5.2874e-02	1.4572e+02
50	$E[V_4(S^2)]$	1.196707e-01	9.03e-06	1.3848e+01	6.4795e-02	2.1372e+02
	$E[V_4(C^3)]$	1.384206e-02	7.11e-07	1.1376e+01	8.0151e-02	1.4193e+02
	$E[V_4(T^3)]$	1.739901e-02	7.76e-07	1.2827e+01	8.1835e-02	1.5675e+02
100	$E[V_4(S^2)]$	1.196731e-01	6.65e-06	2.9287e+01	1.1908e-01	2.4594e+02
	$E[V_4(C^3)]$	1.384286e-02	8.66e-08	2.0372e+01	1.5005e-01	1.3577e+02
	$E[V_4(T^3)]$	1.740003e-02	1.79e-06	2.5617e+01	1.5321e-01	1.6719e+02

The parameter *key* of Dcuhre is set to 4, which selects a cubature rule of polynomial degree 7. Dcuhre starts the test with good accuracy. However, the result is only considered as good as indicated by the estimated error, which is high and decreases only slowly. The (absolute) error and error estimate are listed in the fourth and fifth column, respectively. Dkbvrc performs very well for this problem, with respect to achieved accuracy and reliability. The accuracy of Rcrude starts out low, but does increase considerably, at the cost of computation time. These programs are written in Fortran (double precision). A version of the gfortran compiler was used, and the programs were run on a MacBook Pro with 3.06 GHz Intel Core 2 Duo processor and 8GB of memory. Elapsed execution times obtained with *etime* are given in seconds.

In Table 2 we give results from the GPU implementation, run on a cluster node with dual Intel Xeon E5-2670, 2.6GHz CPUs and 128GB of memory; and a M2090 GPU with 512 CUDA cores, 1.3GHz GPU clock rate and 5375 MB of global memory. The number of parallel blocks, and threads per block launched in the GPU runs was set to 16 and 512, respectively, for the tests described below. The problems addressed are $E[V_4(S^2)]$, $E[V_4(C^3)]$ and $E[V_4(T^3)]$, coded in CUDA C using single precision (float). For the timings in Table 2 we list the parallel time, and the speedup relative to a sequential computation which uses *erand48* for random number generation. The GPU computation is timed via the *cudaEventElapsedTime* mechanism. The speedup is the sequential time divided by the parallel time.

It emerges that for the small problem size of 100,000 sample points, the parallelization is not warranted. However, for larger sample sizes the speedups increase to values between 100 and 200, and up to 245 for the different problems. Indeed, as the number of samples increases by a factor of 10^3 (from 100,000 to 100 million), the sequential time increases by a factor of 10^3 but the parallel time increases only by a factor of about 15. Note that the sequential times are comparable to those of the sequential program RCRUDE in Table 1 for the *cube-tetrahedron* problem.

5   Conclusions

We tested adaptive, quasi-Monte Carlo (QMC) and Monte Carlo (MC) methods for the challenging stochastic geometry problem to determine the expected volume $E[V_4(K)]$ of a tetrahedron with uniformly distributed points in the interior of a cube (C^3), tetrahedron (T^3) and the surface of the 3D sphere (S^2). In the numerical tests we monitored the accuracy of the methods as the number of sample points increases. While the sequential tests are in favor of QMC for these problems, it emerges that MC is rendered a viable and versatile numerical candidate via its GPU implementation.

This work is part of efforts to port the PARINT package to a hybrid parallel environment, by adding efficient multi- and many-core capabilities (using OpenMP [25], CUDA [26] and OpenACC [27]), to its distributed (MPI [28, 29]) computations. In a future stage we intend to develop efficient kernels for QMC as well as MC on GPUs and (Xeon Phi) accelerators. We also plan on extending our application in stochastic computational geometry, for further approximations of the expected d-dimensional volume $E[V_n(K)]$ of the convex hull of n random points in a convex body.

Acknowledgement. The authors acknowledge support under grant number 1126438 by the National Science Foundation, and by NVIDIA for the CUDA Teaching Center awarded at Western Michigan University.

References

1. Zinani, A.: The expected volume of a tetrahedron whose vertices are chosen at random in the interior of a cube. Monatsh. Math. 139, 341–348 (2003), doi:10.1007/s00605-002-0531-y
2. Philip, J.: The expected volume of a random tetrahedron in a cube (2007), http://www.math.kth.se/~johanph/ETC.pdf
3. Heinrich, L., Körner, Mehlhorn, N., Muche, L.: Numerical and analytical computation of some second-order characteristics of spacial poisson-vonoroi tesselations. Statistics 31, 235–259 (1998)
4. Pach, J., Sharir, M.: Combinatorial Geometry and Its Algorithmic Applications: The Alcala Lectures. Amer. Mathematical Society (2009) ISBN-10: 0821846914, ISBN-13: 978-0821846919
5. Dardenne, J., Siauve, N., Valette, S., Prost, R., Burais, N.: Impact of tetrahedral mesh quality for electromagnetic and thermal simulations. In: COMPUMAG, Florianopolis, Brazil, pp. 1044–1045 (2009)

6. Sitek, A., Huesman, R.H., Gullberg, G.T.: Tomographic reconstruction using an adaptive tetrahedral mesh defined by a point cloud. IEEE Transactions on Medical Imaging 25(9), 1172–1179 (2006)
7. Fedorov, A., Chrisochoides, N.: Tetrahedral mesh generation for non-rigid registration of brain mri: Analysis of the requirements and evaluation of solutions. In: Garimella, R.V. (ed.) Proceedings of the 17th International Meshing Roundtable. Springer, Heidelberg (2008), doi:10.1007/978-3-540-87921-3
8. Zhang, Y., Wang, W., Liang, X., Bazilevs, Y., Hsu, M.C., Kvamsdal, T., Brekken, R., Isaksen, J.: High-fidelity tetrahedral mesh generation from medical imaging data for fluid-structure interaction analysis of cerebral aneurysms. Computer Modeling in Engineering & Sciences (CMES) (2), 131–148 (2009)
9. Efron, B.: The convex hull of a random set of points. Biometrica 52, 331–343 (1965)
10. Buchta, C., Müller, J.: Random polytopes in a ball. J. Appl. Prob. 21, 753–762 (1984)
11. Affentranger, F.: The expected volume of a random polytope in a ball. J. Microscopy 151, 277–287 (1988)
12. Miles, R.E.: Isotropic random simplices. Adv. Appl. Probability 3, 353–382 (1971)
13. Buchta, C., Reitzneri, M.: The convex hull of random points in a tetrahedron: Solution of Blaschke's problem and more general results. J. Reine Angew. Math. 536, 1–29 (2001)
14. Philip, J.: The average volume of a random tetrahedron in a tetrahedron. TRITA MAT 06 MA 02 (2006), http://www.math.kth.se/~johanph/ev.pdf
15. Croft, H.T., Falconer, K.J., Guy, R.K.: Unsolved problems in geometry, pp. 54–57. Springer, New York (1991), B5. Random Polygons and Polyhedra
16. Weisstein, E.W.: Sphere tetrahedron picking. In: MathWorld – A Wolfram Web Resource (March 2013), http://mathworld.wolfram.com/topics/RandomPointPicking.html
17. Buchta, C.: Distribution-independent properties of the convex hull of random points. J. Theor. Prob. 3, 387–393 (1990)
18. Berntsen, J., Espelid, T.O., Genz, A.: An adaptive algorithm for the approximate calculation of multiple integrals. ACM Trans. Math. Softw. 17, 437–451 (1991)
19. Genz, A.: MVNDST (1998), http://www.math.wsu.edu/faculty/genz/software/fort77/mvndstpack.f
20. de Doncker, E., Kaugars, K., Cucos, L., Zanny, R.: Current status of the ParInt package for parallel multivariate integration. In: Proc. of Computational Particle Physics Symposium (CPP 2001), pp. 110–119 (2001)
21. Li, S., Kaugars, K., de Doncker, E.: Distributed adaptive multivariate function visualization. International Journal of Computational Intelligence and Applications (IJCIA) 6(2), 273–288 (2006)
22. Genz, A.: MVNPACK (2010), http://www.math.wsu.edu/faculty/genz/software/fort77/mvnpack.f
23. L'Equyer, P.: Combined multiple recursive random number generators. Operations Research 44, 816–822 (1996)
24. Cranley, R., Patterson, T.N.L.: Randomization of number theoretic methods for multiple integration. SIAM J. Numer. Anal. 13, 904–914 (1976)
25. OpenMP website, http://www.openmp.org
26. CUDA NVIDIA Developer Zone, https://developer.nvidia.com/category/zone/cuda-zone
27. OpenACC website, http://www.openacc-standard.org
28. Open MPI website, http://www.open-mpi.org
29. MPI website, http://www-unix.mcs.anl.gov/mpi/index.html

Appendix

A   CUDA C Device Function (Integrand Evaluation) for Spherical Surface Tessellation

```
const int dim = 5;
struct point {
    float coordinates[dim];
};
__device__ float f( point p ) {
    float u2,u3,u4,t3,t4;
    float x1,y1,z1,x2,y2,z2,x3,y3,z3,x4,y4,z4;
    float d1,d2,d3,d4,f0;
    const float pi2 = atan(1.0)*8;

    u2 = 2*p.coordinates[0]-1;
    u3 = 2*p.coordinates[1]-1;
    u4 = 2*p.coordinates[2]-1;
    t3 = p.coordinates[3]*pi2;
    t4 = p.coordinates[4]*pi2;
    x1 = 0;
    y1 = 0;
    z1 = 1;
    x2 = sqrt(1.0-u2*u2);
    y2 = 0;
    z2 = u2;
    x3 = sqrt(1.0-u3*u3)*cos(t3);
    y3 = sqrt(1.0-u3*u3)*sin(t3);
    z3 = u3;
    x4 = sqrt(1.0-u4*u4)*cos(t4);
    y4 = sqrt(1.0-u4*u4)*sin(t4);
    z4 = u4;

    d1 = x2*y3*z4+y2*z3*x4+z2*x3*y4-z2*y3*x4-x2*z3*y4-y2*x3*z4;
    d2 = x1*y3*z4+y1*z3*x4+z1*x3*y4-z1*y3*x4-x1*z3*y4-y1*x3*z4;
    d3 = x1*y2*z4+y1*z2*x4+z1*x2*y4-z1*y2*x4-x1*z2*y4-y1*x2*z4;
    d4 = x1*y2*z3+y1*z2*x3+z1*x2*y3-z1*y2*x3-x1*z2*y3-y1*x2*z3;

    f0 = d1-d2+d3-d4;
    return fabs(f0)/6.0;
}
```

B   CUDA C Device Function (Integrand Evaluation) for Tessellation in Tetrahedron

```
const int dim = 12;
struct point {
    float coordinates[dim];
```

```
};
__device__ float f( point p ) {
    float x1,y1,z1,x2,y2,z2,x3,y3,z3,x4,y4,z4;
    float d1,d2,d3,d4,f0;

    x1 = p.coordinates[0];
    y1 = (1.0-x1)*p.coordinates[1];
    z1 = (1.0-x1-y1)*p.coordinates[2];
    x2 = p.coordinates[3];
    y2 = (1.0-x2)*p.coordinates[4];
    z2 = (1.0-x2-y2)*p.coordinates[5];
    x3 = p.coordinates[6];
    y3 = (1.0-x3)*p.coordinates[7];
    z3 = (1.0-x3-y3)*p.coordinates[8];
    x4 = p.coordinates[9];
    y4 = (1.0-x4)*p.coordinates[10];
    z4 = (1.0-x4-y4)*p.coordinates[11];

    d1 = x2*y3*z4+y2*z3*x4+z2*x3*y4-z2*y3*x4-x2*z3*y4-y2*x3*z4;
    d2 = x1*y3*z4+y1*z3*x4+z1*x3*y4-z1*y3*x4-x1*z3*y4-y1*x3*z4;
    d3 = x1*y2*z4+y1*z2*x4+z1*x2*y4-z1*y2*x4-x1*z2*y4-y1*x2*z4;
    d4 = x1*y2*z3+y1*z2*x3+z1*x2*y3-z1*y2*x3-x1*z2*y3-y1*x2*z3;

    f0 = fabs(d1-d2+d3-d4);
    f0 = f0*(1.0-x1)*(1.0-x1-y1)*(1.0-x2)*(1.0-x2-y2);
    f0 = f0*(1.0-x3)*(1.0-x3-y3)*(1.0-x4)*(1.0-x4-y4);
    return f0*1296;
}
```

Integrated Random Local Similarity Approach for Facial Image Recognition

Henry H.M. Huang[1] and Marina L. Gavrilova[2]

[1] Guangxi Normal University, Guilin, Guangxi, China
huanghm@gxnu.edu.cn
[2] University of Calgary, Calgary, AB, Canada
mgavrilo@ucalgary.ca

Abstract. Face recognition is a fundamental capability of humans to recognize each other, which predominantly made its way into computing domain. The demand for fast and highly reliable face recognition methods is as high as ever. This paper proposes one solution based on a novel similarity measure for frontal facial image recognition, which can be computed rapidly while maintaining a high recognition performance. The new proposed similarity measure is named Integrated Random Local Similarity (*IRLS*), based on an appropriate combination of holistic similarity of facial prominent points expressed in binary image and pixel-wise local similarity of local regions in original gray-level image. The holistic similarity is estimated by the ratio of intersection of the candidate image and the gallery image to the sum of the candidate images prominent points in original gray-level image spatial domain. Experiments have been conducted on AR database. The preliminary experiment results shows that *IRLS* is a very robust approach which maintains high recognition performance, and deserves to be investigated with larger dataset.

Keywords: Local Similarity, Kullback-Leibler Distance, Integrated Filter, Face Recognition, Biometrics.

1 Introduction

Considerable efforts have been dedicated over last decade to face recognition [1, 2] in both academic and government sectors. However, face detection, segmentation, localization and recognition under varying conditions, illumination and occlusion is still a challenging problem [4, 6, 11, 13]. Almost all mainstream algorithms for face recognition need to construct some "models" before estimating the similarity between a pair of facial images.

The needing for more additional prior information to build such models severely hinders the application and performance of the corresponding facial recognition method. Furthermore, if the training set changes by adding or removing facial images, facial models such as some probability distribution functions need to be re-estimated, the similarity value may vary accordingly, prone to some object(s) to be reclassified and increasing training time.

B. Murgante et al. (Eds.): ICCSA 2013, Part II, LNCS 7972, pp. 140–149, 2013.
© Springer-Verlag Berlin Heidelberg 2013

The goal of this paper is to develop a robust similarity measure which only relies on the information of the pairing images, none of any other images in database being required. We make an attempt to establish a similarity measure, which is suitable to allow recognizing faces even under partially occluded conditions, varied expressions and inconsistent illumination. Sparse and redundant representations [3] of face images can achieve high recognition rates [4] on occluded faces with different expressions and illumination. However, this approach fails when only a single image sample is available for per person, that is a common practical scenario. In addition, those methods suffer from the heavy time consumption for decomposing each test image to its sparse linear representation on the whole learning sample set. When a new image is added to learning sample set, the training must be repeated and corresponding decompositions must be iterated.

A variety of face recognition methods have been developed in the past, including those based on geometry [14], appearance [15], or fusion methods [16]. The methodology started to make its way into novel application areas, such as virtual reality domain [17] or a new generation future cognitive security systems [18]. In this paper, we propose a radically new idea of integrating holistic image feature and local features. Both of the holistic feature and local feature are calculated in spatial domain. The holistic feature is the result of a serial of linear filtering of original image. The local feature is calculated at pixel-wise and based on Kullback-Leibler [7] distance.

This approach is quite simple and also fast, with the benefits listed below:

- The estimation of the proposed similarity – IRLS - only needs the information of two pairing images, none of any other images;
- Only one image per person in learning sample set is sufficient for the proposed method. On contrary, performance of other methods highly depends on a large number of varied expressions and varied illuminations images [4].
- The method allows to achieve a comparable or a better   recognition rate when compared with Gaussian mixture distribution model on eigen-representation [6], and also outperforms the method presented in [12] for   occluded images.
- The proposed approach is faster, simpler and more intuitive when compared with Hidden Markov Models [11], while it maintains nearly equivalent high recognition rate.

We name the measure as Integrated Random Local Similarity (IRLS). The results of a series of experiments conducted on AR Face database [9]   show that the IRLS similarity can outperforms the previously proposed measures, such as PCA , LDA and FaceIt methods [8]; and is better than HHM for certain expression types while maintaining comparable recognition rates for other cases[11].

These preliminary results suggest that IRLS similarity may be an excellent measure for frontal facial image for face recognition, with very fast recognition rate and good generalization capabilities. IRLS deserves to be investigated with more images and more facial expressions.

2   The Methodology

The integrated random local similarity measure (IRLS), proposed in this paper, is a measure that depends on the pairing of two compared images, without involving training stage or any other images to establish some model or estimate the value of the similarity. It only relies on the weighted holistic similarity and the combined local similarity. The holistic similarity is estimated by pairing of two binary images. Binary facial images are actually facial prominent points which being generated from gray-level image by linear filtering. The combined local similarity is formed by the pixel-wise similarity, and the latter is estimated by Kullback-Leibler (KL) distance between paired sections in the spatial domain. The paired sections can be distinct parts of original image or can be overlapped random sections, even can be as small as a single line of pixels or a single pixel point in an extreme case.

2.1   Binary Image of Facial Prominent Points

The binary image of facial prominent points is a binary image where only partial discriminatory and distinguished   points of original image x are set "on" (the pixel value of corresponding position is 1), all other points are set "off" (the pixel value of corresponding position is 0). Such binary image is a product of original image. Ideally, all background pixels and non-facial pixels such as facial obstructions (scarf, sunglasses) shall be assigned to "0" values (set "offs") as they cannot contribute to facial recognition.

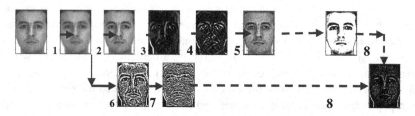

Fig. 1. Process of Generating Facial Prominent Points Binary Image

The steps of 1st serial filtering process: (**1**)6x6 averaging,(**2**)3x3 flat filtering-out, (**3**)3x3 vertical edge detecting, (**4**)3x3 horizontal edge detecting, (**5**)3x3 special "non-sharp" filtering;

The steps of 2nd serial filtering process: (**1**)6x6 averaging, (**6**)5x5 flat filtering-out, (**7**)3x3 horizontal edge detecting; (**8**)post-filter processing: normalization, binarization.

The facial prominent points binary image is generated by a combination of two series of linear filtering processes, displayed in Figure 1. The 1st series are comprised of five linear filters: a 6x6 averaging filter, a 3x3 flat filtering-out filter, a 3x3 vertical edge detecting filter, a 3x3 horizontal edge detecting filter, and a 3x3 special "non-sharp" filter. The 2nd series involve three linear filters: a 6x6 average filter, a 5x5 flat filtering-out filter and a 3x3 horizontal edge detecting filter.

Some typical binary images of facial prominent points are shown in Figure 2. The original gray-level facial images in AR database are also displayed to the left of the corresponding binary images.

Fig. 2. Some Binary Images of Facial Prominent Points

2.2 Holistic Similarity

After suitable alignment and scaling, the holistic similarity between two pairing images can be estimated in the spatial or frequency domains. We estimate the holistic similarity in spatial domain by comparing two binary images of facial prominent points: one is candidate image and another is   gallery image.

Suppose the two binary images are IG and IT, IG is a gallery binary image, and IT is a candidate binary image. These binary images are generated by "facial prominent points" as described in former section. Before estimating holistic similarity, if the sizes of IG and IT differ, the larger one is truncated to the size of the smaller one. The holistic similarity (S_{Hol}) is then computed as:

$$S_{Hol} = \frac{sum(sum(IG \cap IT))}{sum(sum(IT))} \tag{1}$$

where $IG \cap IT$ is a   matrix-wise logical AND; sum means summation; sum(sum(IT)) means adding all the elements of matrix IT, which resulting in the total count of "on" elements of IT; and sum(sum($IG \cap IT$)) means the total count of value '1' elements in the intersection of matrix IG and matrix IT.

The holistic similarity S_{Hol} is a measure which expresses the degree of overlapping in a candidate binary image, being compared to the gallery binary image. The range of S_{Hol} is [0, 1], the larger of the S_{Hol} value, the more similar between the candidate image and the gallery image. If the binary image IT is identical to the binary image IG, $S_{Hol} = 1$.

2.3 Kullback-Leibler Distance

The Kullback-Leibler distance [7] measures the difference of two distribution probabilities $p(x)$ and $q(x)$, where x is a random variable. If x is a continuous random variable, the Kullback-Leibler distance can be defined as [7]:

$$D_{KL}(p \parallel q) = \int_{-\infty}^{\infty} p(x) \ln \frac{p(x)}{q(x)} dx \tag{2}$$

If x is a discrete random variable, the two distribution probabilities p and q are also discrete variables, then the Kullback-Leibler distance is defined as [7]:

$$D_{KL}(p \parallel q) = \sum_i p(i)\ln\frac{p(i)}{q(i)} \qquad (3)$$

In probability and information theory, the Kullback-Leibler distance is also named Kullback-Leibler divergence (information divergence, information gain, relative entropy, or KLIC).

The Kullback-Leibler distance between candidate image IT and gallery image IG is defined as:

$$D_{KL} = \sum_{i=1}^{N}\left(IT(i)\ln\frac{IT(i)}{IG(i)} + IG(i)\ln\frac{IG(i)}{IT(i)} \right) \qquad (4)$$

where N is the total count of pixels in candidate or gallery image, $IT(i)$ is the vectorized original gray-level candidate image or its section(s), $IG(i)$ is the vectorized original gray-level gallery image or its section(s).

2.4   Local Similarity

The original candidate image or gallery image can be partitioned into a serial of sections. These sections can be distinct parts with same size of original image, with several rows and several columns (see Fig.3.), and also can be sections (subimages) of varied size, even can be a single line.

After the Kullback-Leibler distances of all pairing sections have been calculated according to formula (4), the local similarities(S_{Loc}) can be estimated for each pairing sections between a candidate image and a gallery image. For the sample partition in Figure 3, we can get nine values of the D_{KL}:

$$D_{KL}^1, \ D_{KL}^2, \ldots\ldots, \ D_{KL}^C$$

where C is the count of sections in any of pairing images i.e. in this example $C = 9$. C can also be the number of rows(m) or columns(n) in one image pairing, if section is a horizontal line or a vertical line. It can even be the total pixels number ($N = m \times n$) in some image pairing, when section is just a pixel. In this case, the pixel pairing

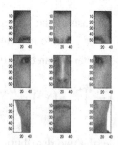

Fig. 3. Partitioning into Sections of Original Gray Image

represents the same positions in the candidate as in the original image, with different features assigned to it. If sections are generated by random strategy, C can be larger than N, due to the sections, which is corresponded to the same image positions, being generated in sequential partitions, can be overlapped .

Now, the local similarity S_{Loc} measure between each of the paired sections of the candidate image and the gallery image can defined as the follows:

$$S_{Loc}(i) = e^{-D_{KL}^i}, \quad i = 1,2,...,C \tag{5}$$

where $i = 1,2,...,C$, and C is the count of sections in a set of paired images.

If two paired sections are identical, then the corresponding Kullback-Leibler distance $D_{KL} = 0$, according to (5), then $S_{Loc} = 1$. The more difference is there between the two paired sections, the larger the value of D_{KL} will be. The range of D_{KL} is $[0, \infty)$, and in this range, the exponential function is monotonously descending with the maximum $S_{Loc} = 1$ at $D_{KL} = 0$. Thus, the local similarity S_{Loc} is compatible and consistent with the holistic similarity S_{Hol}, as their range is [0, 1]. The larger the value (S_{Loc}, or S_{Hol}), the more similar the candidate image and the gallery image.

2.5   Random Local Similarity

Random local similarity is calculated similarly to S_{Loc}. However, the pairing sections are not predefined, but being generated randomly. The position (x, y) and size (w, h) of section is randomly generated from original image (the candidate or the gallery image). For pairing those sections, the positions of sections within candidate image and within gallery image can be different, however, their sizes must be the same.

2.6   Integrated Random Local Similarity

There are two steps to reach the final similarity - integrated random local similarity (IRLS) - S_{IRLS}. First, a combined random local similarity S_{CLoc} is calculated from all estimated random local similarities.

Since paired sections are randomly generated, sequential pairing sections may occupy the same part of the original image. In order to avoid repeated summation of local similarity S_{Loc}, the value of S_{Loc} is replicated and assigned to each pixel in the two pairing sections. Thus, there exist two matrixes corresponding to the candidate image and the gallery image, with the same sizes as the corresponding images. They serve as the storage of the maximal S_{Loc}, which re-assigned to the pixels corresponding pairing sections at each iteration of random local similarity calculation.

Adding up all the pixel-wise similarities in the candidate similarity matrix and gallery similarity matrix, and then divided by corresponding total number of image pixels, we obtain two quantities - S_{CLoc}^T, S_{CLoc}^G:

$$S_{CLoc}^T = \frac{1}{N_T}\sum_{i=}^{m_T}\sum_{j=1}^{n_T}S_{Loc}^T(i,j)$$

$$S_{CLoc}^G = \frac{1}{N_G}\sum_{i=}^{m_G}\sum_{j=1}^{n_G}S_{Loc}^G(i,j) \quad\quad (6)$$

where T means candidate image; G means gallery image; m_T, m_G - the row numbers; n_T, n_G - the column numbers; $N_T = m_T \times n_T, N_G = m_G \times n_G$ - total number of pixels in candidate image and gallery image, respectively.

Then, the first stage of integrated random local similarity S_{CLoc} is the lower of S_{CLoc}^T and S_{CLoc}^G:

$$S_{CLoc} = \min(S_{CLoc}^T, S_{CLoc}^G) \quad\quad (7)$$

In the end, the final integrated random local similarity S_{IRLS} is defined as the weighted sum of the holistic similarity and the combined local similarity:

$$S_{IRLS} = \alpha S_{Hol} + (1 - \alpha)S_{CLoc} \quad\quad (8)$$

where α is an adjustable parameter within [0,1].

3   Experimentation

Facial images from AR database [9] are used as the experimental dataset to validate the proposed similarity measure performance and computational speed. The sample images in AR database were captured twice per subject over two weeks interval. During each session, 13 images with 4 varying facial expressions: neutral, smiling, angry, screaming, three varying illumination and two types of occlusion (sunglasses and scarves), were captured. These images belong to one hundred subjects, with twenty six images for each subject.

These images in AR database have been calibrated, warped and normalized to the size of 165x120 pixels. The 26 captured images are arranged as sub-datasets referred to as AR01,AR02, AR03,..., AR26. Each of these 26 sub-datasets contains 100 images, one per subject, for each of 100 subjects. For example, AR01, AR14 are neutral expressions sub-datasets corresponding to 1st or 2nd session; AR02, AR15 are smile expressions sub-datasets; AR04, AR17 are scream expressions sub-datasets; AR05, AR18 are left-directional illumination with neutral expressions sub-datasets; AR08, AR21 are sub-datasets of sunglasses occlusions with normal illumination; AR11, AR24 are sub-datasets of scarves occlusions with normal illumination.

Table 1. Recognition Results of Only One Image per Person in Learning Set(AR01)

(a)

	Expression						Illumination					
	AR02	AR15	AR03	AR16	AR04	AR17	AR05	AR18	AR06	AR19	AR07	AR20
PCA	87	71	86	69	39	30	81	61	79	59	82	62
LDA	96	78	89	67	60	40	87	54	82	60	86	60
FaceIt	96	86	92	79	76	48	96	87	93	81	86	**69**
HMM	99	88	99	**90**	**99**	68	99	91	98	84	**97**	65
IRLS	**99**	**86**	**97**	83	**88**	58	74	66	82	66	**89**	50

(b)

	Occlusion + Illumination											
	AR08	AR21	AR09	AR22	AR10	AR23	AR11	AR24	AR12	AR25	AR13	AR26
PCA	48	35	26	25	21	13	27	14	21	14	11	09
LDA	45	25	31	18	27	17	44	32	33	25	31	20
FaceIt	10	10	09	10	06	04	81	65	72	50	72	48
HMM	**99**	**90**	**95**	**68**	**95**	**65**	**94**	72	**85**	55	**83**	54
IRLS	98	80	64	56	78	57	87	**70**	51	46	69	**65**

Note: * PCA, LDA, FaceIT[8] and HMM resultant data are excerpted from [11].
 ** IRLS -- Integrated Random Local Similarity proposed in this paper.

As can be seen from Table 1, IRLS if compared with PCA, LDA and FaceIt, the recognition algorithm proposed in this paper is clearly superior for most of the sub-datasets. Due to S_{IRLS} being computed in spatial domain, it's performance is only hindered on varied illumination samples. The computational speed for calculating the holistic similarity S_{Hol}, and the local similarity S_{Loc}, is very fast, only on the scale of 0.01 - 0.1 second for a pairing calculation on a Acer Notebook (ASPIRE 5542G, 2.0G HZ AMD Athlon CPU, 2 GB Memory). However, the calculation of random local similarity S_{CLoc} is rather time-consuming, typically on several seconds scale, but being adjustable by adopting suitable randomizing strategy, and obviously better than the decomposition of sparse representation[4].

4   Discussion

The newly proposed similarity measure S_{IRLS} is computed in the spatial domain. It requires only two paired images – one is the candidate image and another is the gallery image. The value of S_{IRLS} is in the range of $[0, 1]$. The larger the value of S_{IRLS}, the more similar between the two pairing images is. If the candidate image is compared one-to-one against each image in the gallery (the dataset), similarity is computed for each pair, the class identity of the gallery image with the maximum similarity value is used for image recognition. If the candidate image is compared with only one specified image in the gallery, only one value of S_{IRLS} is computed and can be used to estimate image similarity.

It is very rare if the candidate image is identical to the specified gallery image. The value of S_{IRLS} is often less than 1. In such a case, some threshold S_{IRLS}^T shall be specified. If $S_{IRLS} \geq S_{IRLS}^T$, then the candidate image is considered to belong to the same class of the paired gallery image.

HMM, LDA, PCA and other algorithms need large collection of images to establish some models and conduct learning processes. The proposed   IRLS method only relies on two images-- one candidate image and one gallery images, conducts a single image pairing to evaluate the similarity of the two compared images.

However, if the candidate image is not calibrated, warped or normalized, the value S_{IRLS} may be low ever if the pairing image actually corresponds to the same subject. Thus, normalization should be used as the first step for this method. Future works shall include expanding the method to other similarity measures.

Acknowledgment. The authors would like to express heartedly appreciation to A.M. Martinez for permission to use the AR face database and for this research. They are thankful for partial support of NSERC, Canada and Chinese Post-Doctoral Scholarship (GED sponsored Guangxi Scholars), China.

References

1. Li, S.Z., Jain, A.K. (eds.): Handbook of Face Recognition, 2nd edn. Springer-Verlag London Limited (2011)
2. Delac, K., Grgic, M. (eds.): Face Recognition. I-TECH Education and Publishing, Croatia (2007)
3. Elad, M.: Sparse and Redundant Representations From Theory to Applications in Signal and Image Processing. Springer Science+Business Media, LLC (2010)
4. Wright, J., Yang, A.Y., Ganesh, A., Shankar Sastry, S., Ma, Y.: Robust Face Recognition via Sparse Representation. IEEE Trans. on Pattern Analysis and Machine Intelligence 31(2), 210–227 (2009)
5. Fidler, S., Skocaj, D., Leonardis, A.: Combining Reconstructive and Discri-minative Subspace Methods for Robust Classification and Regression by Subsampling. IEEE Trans. Pattern Analysis and Machine Intelligence 28(3), 337–350 (2006)
6. Martinez, A.M.: Recognizing Imprecisely Localized, Partially Occluded, and Expression Variant Faces from a Single Sample per Class. IEEE Trans. Pattern Analysis and Machine Intelligence 24(6), 748–763 (2002)
7. http://en.wikipedia.org/wiki/Kullback%E2%80%93Leibler_divergence
8. FaceIt® SDK: http://www.l1id.com/pages/101-faceit-sdk
9. Martinez, A.M., Benavente, R.: The AR face database. CVC Technical Report #24 (June 1998)
10. Bhuiyan, A.A., Liu, C.H.: On Face Recognition using Gabor Filters. Proceedings of World Academy of Science, Engineering and Technology 22, 51–56 (2007)
11. Le, H.-S., Li, H.: Recognizing Frontal Face Images Using Hidden Markov Models with One Training Image per Person. In: Proceedings of the 17th International Conference on Pattern Recognition (ICPR 2004), pp. 1051–4651 (2004)
12. Zhang, W., Shan, S., Gao, W.: Local Gabor Binary Patterns Based on Kullback–Leibler Divergence for Partially Occluded Face Recognition. IEEE Signal Processing Letters 14(11), 875–878 (2007)
13. Beymer, D., Poggio, T.: Face Recognition from One Example View. MIT Artificial Intelligence Lab., A.I. Memo No. 1536, C.B.C.L. Paper No. 121, pp. 500–507 (September 1995)
14. Yuan, L., Gavrilova, M., Wang, P.: Facial metamorphosis using geometrical methods for biometric applications. IJPRAI 22(3), 555–584 (2008)

15. Yanushkevich, S., Gavrilova, M., Wang, P., Srihari, S.: Image Pattern Recognition: Synthesis and Analysis in Biometrics. Series in Machine Perception and Artificial Intelligence, vol. 67. World Scientific Publishers (2007)
16. Monwar, M., Gavrilova, M.L.: A Multimodal Biometric System Using Rank Level Fusion Approach. IEEE Transactions on Man, Systems and Cybernatics TMSC Part B 39(5), 867–878 (2009)
17. Yampolskiy, R., Gavrilova, M.: Artimetrics: Biometrics for Artificial Entities. IEEE Robotics and Automation Magazine (appeared on-line December 2012)
18. Wang, Y., Berwick, R., Haykin, S., Pedrycz, W., Kinsner, W., Baciu, G., Zhang, D., Bhavsar, V., Gavrilova, M.: Cognitive Informatics and Cognitive Computing in Year 10 and Beyond. International Journal of Cognitive Informatics and Natural Intelligence 5(4), 1–21 (2011)
19. Bui, L., Tran, D., Huang, X., Chetty, G.: Novel Metrics for Face Recognition Using Local Binary Patterns. In: König, A., Dengel, A., Hinkelmann, K., Kise, K., Howlett, R.J., Jain, L.C. (eds.) KES 2011, Part I. LNCS (LNAI), vol. 6881, pp. 436–445. Springer, Heidelberg (2011)

A Gabor Filter-Based Approach to Leaf Vein Extraction and Cultivar Classification

Dominik Ludewig Michels and Gerrit Alexander Sobottka

Institute of Computer Science II, University of Bonn
Friedrich-Ebert-Allee 144, 53113 Bonn, Germany
michels@uni-bonn.de,
sobottka@cs.uni-bonn.de

Abstract. We devise a new algorithm for the extraction of vine leaf veins. Our method performs a directional edge tracing on the responses of appropriate adaptive Gabor filters in order to extract the network of the main veins. The respective curvature vectors are used for the classification of different cultivars using support vector machines. We evaluate the advantageous behavior and the robustness of our approach on a test set consisting of 150 light transmitted images of different vine leaves.

Keywords: Leaf Classification, Feature-Based Classification, Feature Extraction, Gabor Filter, Edge Tracing, Support Vector Machines.

1 Introduction

Population growth, climate change, and the shortage of resources have caused an increased global interest of the agricultural community in intelligent farming methods. As a result, the integration of agricultural concepts and modern IT has paved the way for tremendous crop yield increases over the last decade. Dedicated robots for farm working, usually four-wheeled vehicles with robot manipulators, have been developed as part of the smart farming process. Equipped with recent satellite and sensor technologies they are able to autonomously navigate through vineyards, corn- or strawberry fields and at the same time take over the sowing and harvesting work of the agricultural laborer. Exhausted, depleted, and pesticide contaminated soils, on the other hand, reveal the disastrous consequences of the long-term use of classical monocrops and force modern agriculture more and more to embark on whole system approaches like sustainable agriculture, integrated farming, and permacultures, i.e. well-designed agriculturally productive ecosystems with the diversity, long-term stability, and synergistic properties of natural ecosystems. This poses new requirements for the autonomous harvesting vehicles: it is mandatory to furnish them with robust identification mechanisms that ensure a correct plant recognition by means of their phenotypic characteristics. Consequently, non-destructive approaches to the problem of computer aided analysis and screening of plant phenotypes have experienced a growing interest in the crop science community over the last decade. Matured computer

B. Murgante et al. (Eds.): ICCSA 2013, Part II, LNCS 7972, pp. 150–159, 2013.
© Springer-Verlag Berlin Heidelberg 2013

Fig. 1. Illustration of different level veins colored in blue (level 0), red (level 1), and yellow (level 2)

vision techniques are employed for an automated plant recognition by means of the vein networks of their leaves which act as a unique classifier–a kind of fingerprint–for a specific cultivar.

In our contribution we tackle the challenging problem of extracting the main veins of vine leaves by means of a Gabor filter approach. The reason for focussing on vine leaves is that the perennial vinegrape is one of the oldest and most important crop plants in human history and the automated classification of cultivars of vinegrapes is still an important and challenging topic. The extracted vein data is a unique classifier for a certain cultivar and can thus be used as input for support vector machines to perform the final classification. We evaluate the robustness of our method on a test set consisting of 150 images of vine leaves with different color patterns.

2 Related Work

Leaf vein extraction and cultivar classification is a well-established field in the computer vision and machine learning community. In this section we give a brief overview over the recent achievements.

In [9] the authors present a leaf vein extraction method based on the gray-scale morphology. An independent component analysis is used in [5] to realize a robust vein extraction method.

A leaf recognition algorithm based upon probabilistic neural networks is presented in [8]. Their approach allows for a robust plant classification. In contrast, the method described in [4] makes use of region-based features. A rather exotic approach is presented in [1] where the authors discuss an ant colony algorithm.

In contrast to these methods we aim to exploit knowledge from the position space on the first hand and extend it with information from the frequency domain. This is achieved on the firm basis of appropriate Gabor filters that have originally been developed by Dennis Gabor in the last century.

In order to find the point of intersection of the five main veins, the configuration of a starting template is determined by means of a principal component analysis (cf. [6]) and a subsequent optimization step based on the simulated annealing method that has been introduced in [3].

2.1   Problem Setting

Our goal is to automatically extract the network of main veins in vine leaves, the so called veins of level zero. A vein emanating from a level n vein is of level $n+1$, cf. Fig. (1). It is important to note that we make intensive use of the fact that the phenotypical appearance of vine leaves always shows five level zero veins. These veins have a common start point–the so called central point of the leaf– and a well-defined endpoint. The extraction shall be performed automatically on a input image. The desired output is a vectorized representation of the veins of level zero as a pixel sequence or as a so called *chain code*.

We demonstrate the efficiency of our approach on a test set containing 150 images of vine leaves of different cultivars, specifically Kerner (35 images), Müller-Thurgau respectively Rivaner (38 images), Riesling (40 images), and Scheurebe (37 images). The images show diverse characteristics like overlapping parts, handwritten labels, and severe discolorations, e.g. due to leaf diseases, cf. Fig. (2).

One of the major problems we have to deal with is that the difference in the intensities of the level zero veins and their higher order branches is not large enough to prevent a standard vein detection algorithm from spuriously changing its tracing direction from the principal to a secondary direction at the branch-offs. Beside this level zero veins typically tend to become smaller with increasing distance from the center point. As a consequence, it is difficult to distinguish between veins of different levels. This in turn promotes erroneous classifications of the vein levels. Here, our adaptive Gabor filter-based approach comes into play.

The choice to take veins of level zero is well founded due to the fact, that it has become an established phenotypic feature. For example it is recorded under the number OIV 070-1 by the International Plant Genetic Resources Institute (cf. [2]).

3   A Gabor Filter-Based Approach

In order to overcome the aforementioned problems we aim at combining knowledge from the position space as well as the frequency domain. This is achieved with so called Gabor filters that are closely related to short-time Fourier transforms. In order to allow the reader to acquire a deeper understanding of how our

Fig. 2. Some leaves from our test set. Diverse characteristics, like overlapping parts, handwritten labels, and severe discolorations can be identified.

approach works we give a brief overview of the mathematical concepts behind Gabor filters and explain how they can be effectively applied in the context of leaf vein detection.

3.1   Gabor Filter

Appropriate transformations to the frequency domain allow us to study the frequency information of a time dependent phenomenon. In the one-dimensional, continuous case such a transformation is given by the Fourier transform

$$\mathcal{F}[f](\omega) = \int_{\mathbb{R}} f(t)\exp(-2\pi i\omega t)\,dt,$$

which maps the signal f in the time domain to the Fourier transformed version $\mathcal{F}[f]$ in the frequency domain. This transformation is represented by a scalar product in the functional space with the arguments f and a complex exponential with frequency ω. Hence, $\mathcal{F}[f](\omega)$ can be considered as the complex amplitude of the occurrence of the fundamental oscillation with frequency ω in the signal f. In other words, the graph of $\mathcal{F}[f]$ shows how much of the signal f lies within each given frequency.

But there is no time domain information available in $\mathcal{F}[f]$. Since we are interested in a combination of time and frequency information, it is a common method to add a τ-shifted time domain window function g to the first argument of the scalar product in the sense of a multiplication with the signal f. Therefore, the resulting so called windowed or short-time Fourier transform

$$\mathcal{F}_g[f](\omega, \tau) = \int_{\mathbb{R}} f(t) \, g(t - \tau) \exp(-2\pi i \omega t) \, dt,$$

describes the frequency behavior of the signal f in the time domain neighborhood of τ.

The transformation $\mathcal{G}[f] := \mathcal{F}_{g_\sigma}[f]$ with the Gaussian window function

$$g_\sigma(t) = \frac{1}{\sqrt{2\pi\sigma^2}} \exp\left(-\frac{t^2}{2\sigma^2}\right),$$

with variance σ is known as the Gabor transformation of the signal f and shows how much of the signal f limited to $t \in [\tau - \sigma, \tau + \sigma]$ matches a given frequency. An arbitrarily close resolution in the time as well as in the frequency domain is not possible at the same point in time and limited by Heisenberg's Uncertainty principle. One can show that the Gabor transformation is a Fourier transformation with minimal uncertainty.

The Gabor filter G_θ with orientation θ used in this approach is formally given by

$$G_\theta(\mathbf{x}, \lambda, \sigma, \psi, \gamma) = \exp\left(-\frac{\hat{x}_1^2 + \gamma^2 \hat{x}_2^2}{2\sigma^2}\right) \exp\left(i\left(\frac{2\pi}{\lambda}\hat{x}_1 + \psi\right)\right), \tag{1}$$

at the point $\mathbf{x} = (x_1, x_2)^{\mathsf{T}}$. It can be regarded as an oriented two-dimensional discrete version of the Gabor transformation $\mathcal{G}$.

The left factor in Eqn. (1) describes a two-dimensional elliptic Gaussian in which the spatial aspect ratio γ denotes the ellipticity. The variance is again given by σ and the direction is determined by the angle θ of the normal direction that influences the vector $\hat{\mathbf{x}} = \mathbf{R}_\theta^{\mathsf{T}} \cdot \mathbf{x}$. The matrix $\mathbf{R}_\theta$ describes the two-dimensional mathematically positive rotation by the angle θ, which is given by

$$\mathbf{R}_\theta = \begin{pmatrix} \cos\theta & -\sin\theta \\ \sin\theta & \cos\theta \end{pmatrix}.$$

The right factor in Eqn. (1) analogously describes a complex exponential with phase shift ψ and frequency $\omega = 1/\lambda$, in which λ denotes the wavelength.

With the use of Euler's formula we can split the complex Gabor filter Eqn. (1) in a real and an imaginary part. Only the real part is evaluated in our algorithm and given by

$$G_{\text{real}_\theta}(\mathbf{x}, \lambda, \sigma, \psi, \gamma) = \exp\left(-\frac{\hat{x}_1^2 + \gamma^2 \hat{x}_2^2}{2\sigma^2}\right) \cos\left(\frac{2\pi}{\lambda}\hat{x}_1 + \psi\right). \tag{2}$$

The Gabor filter G_{real_θ} is applied to an image F by using the two-dimensional convolution $F \otimes G_{\text{real}_\theta}$, pointwise defined by

$$F(x_1, x_2) \otimes G(x_1, x_2) = \sum_{\tau_1} \sum_{\tau_2} F(\tau_1, \tau_2) G(x_1 - \tau_1, x_2 - \tau_2).$$

The main idea behind our approach is that for a given angle θ a θ-oriented vein of order zero will always dominate the Gabor filter response in contrast to the higher order veins that branch off. We exploit this fact by tracing the course of the veins on $F \otimes G_{\mathrm{real}_\theta}$.

3.2 The Algorithm

This motivates the following algorithm: In a precomputing step the central point of the leaf is detected with a template-based matching strategy as illustrated in Fig. (3). The image template is rotated so that its direction fits the orientation

Fig. 3. Illustration of the template-based strategy to detect the central point. The orientation is found by a principal component analysis. The directions correspond to the maximal and minimal variance respectively. Because the orientation is determined in such a way, the simulated annealing algorithm has to minimize a scalar fitness function of only two variables, which depends on the shift of the template image.

of the leaf which in turn is found by means of a principal component analysis (cf. [6]) on the binarized image. The image template is then shifted along the main directions until the color distance to the leaf pixels reaches a minimum. This process is controlled by a simulated annealing strategy, cf. [3]. In order to reduce the influence of minor variations, the average of the squared color distances is used as fitness function. The barycenter is a good initial guess to start from. The image template is created from the average image of the intensity normalized center regions of twelve leaves from our test set.

Fig. 4. Illustration of a single filter step of our vein tracing method. It is not possible to decide which one of the two colored parts belongs to the level zero vein (left) by just comparing pixel intensities. After applying G_{real_θ} to the red channel F of the image with angle θ computed from prior curvature values the obtained filter response $F \otimes G_{\mathrm{real}_\theta}$ sheds a light on the actual course of the vein because it masks all other directions (right).

Starting at the central point found by the procedure described above, the algorithm traces the course of the veins by comparing the intensity values on the red channel. Here the contrast between the veins and the other parts of the leaf is significantly improved, cf. Fig. (4). The reason for this is that the leaf pigments consist of the green chlorophyll and the different reddishbrown carotenes.

During the tracing procedure the algorithm keeps track of the curvature of the vein. The curvature is defined as slope of the secant of the last five percent of the main vein with respect to the estimated total length of the detected vein. We use a simple heuristic to detect when the algorithm is about to leave the exact course of the main vein: if the ratio of the difference of the current and the last curvature and that of the last three curvature differences is greater than ten percent the algorithm jumps back to the location where the curvature has changed and applies the real Gabor filter Eqn. (2) to the red channel F of the image. Since we feed the Gabor filter with the angle θ–computed from the curvature values–of the direction of the current subsection of the vein it is guaranteed that all subsections of veins with this particular direction will be visible in the filter response while other directions are masked. Therefore, the course of the vein is traced on the filter response $F \otimes G_{\mathrm{real}_\theta}$ and not on the original image. It should be clear that when the curvature and therefore the angle of the current subsection of the vein under consideration changes "too much" the filter response must be recomputed. Our tracing algorithm stops when the border of the leaf is reached. This is easily detected because of the high contrast between the background and the leaf image.

The main veins typically become thinner as their distance from the central point increases. We account for this fact by adapting the wavelength and the variance parameters λ and σ of the Gabor filter through interpolation from the intensity values of the original image in the area where–based on the last curvature values–the next part of the vein is expected. In contrast to the wavelength and the variance, the influence of the spatial aspect ratio γ and the phase shift ψ can be regarded as global. Thus, we work with a constant aspect ratio and ignore the phase shift.

4   Cultivar Classification

Our Gabor filter-based algorithm has been implemented in a vine leaf classification system written in Java. The different steps carried out for the final classification of a given vine leaf are illustrated in Fig. (5). The curvature values of the main veins are stored in a vector and fed into a support vector machine. In our implementation we embark on the WEKA SVM (cf. [7]) and use a simple linear kernel function. The training set contains 20 images of the test set: five leaves of each of the four white vine cultivars.

5   Evaluation and Results

We apply the devised Gabor filter-based algorithm to the test set consisting of 150 images of vine leaves of different cultivars with a broad spectrum of color characteristics, cf. Sec. (2.1). The set therefore contains 750 different zero order veins. The vein extraction algorithm failed in only 13 of the 750 cases ($< 2\%$). This was due to the fact that there has been either a hole in the outer leaf lobes or a shadow cast prevented the recognition of the correct course of the vein.

The cultivars of the untrained 130 images in the test set have been classified with the following success rates: Kerner 93% (28 of 30), Müller-Thurgau respectively Rivaner 97% (32 of 33), Riesling 94% (33 of 35), and Scheurebe 100% (32 of 32). Hence, all classifications were performed with usable success rates significantly above the 93% threshold. The classification of a single leaf took about 5 sec on a 3.20 GHz Intel Core i7-3930K, 32 GB RAM, under Microsoft Windows 7. It comprises all the processing steps described in Fig. (5). The input images have been used in a downscaled resolution of 400×400 pixels, because there was no significant difference in the success rate compared to the larger scaled images.

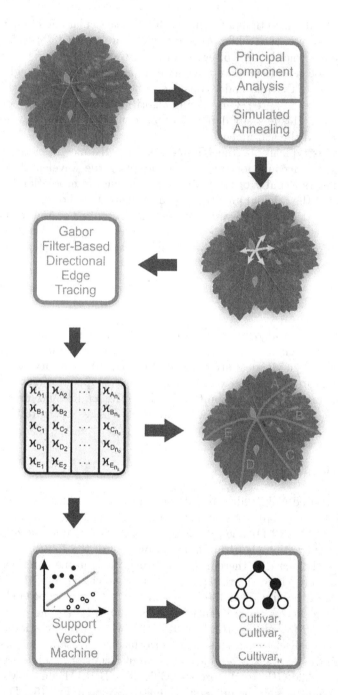

Fig. 5. Illustration of our full-fledged vine leaf classification system: the final classification is based on the curvature vectors that act as a kind of fingerprint using a support vector machine

5.1 Conclusion and Future Work

We have demonstrated that working on appropriate Gabor filter responses allows us to accurately extract the main veins of vine leaves. Our approach finds the correct paths in almost all test cases. At the same time it is able to handle challenging problems like discolorations and even in the case of equally oriented handwritten labels it stays on the correct path because the Gabor filter masks all differently oriented parts of the label. However, problems can still occur in the case of holes caused by pest infestation or shadow casts. Our future work focuses on a robust method that detects such holes and applies a gap filling algorithm, e.g. as part of the precomputing step of our leaf classification system. Moreover, we aim at generalizing the presented approach to other cultivars with different phenotypical characteristics in the leaves to overcome the current restriction to five veined vine leaves.

References

1. Cope, J.S., Remagnino, P., Barman, S., Wilkin, P.: The extraction of venation from leaf images by evolved vein classifiers and ant colony algorithms. In: Blanc-Talon, J., Bone, D., Philips, W., Popescu, D., Scheunders, P. (eds.) ACIVS 2010, Part I. LNCS, vol. 6474, pp. 135–144. Springer, Heidelberg (2010)
2. International Plant Genetic Resources Institute: Descriptors for Grapevine: (Vitis Spp.). Descriptors IBPGRI. International Plant Genetic Resources Institute (1997)
3. Kirkpatrick, S., Gelatt, C.D., Vecchi, M.P.: Optimization by simulated annealing. Science 220, 671–680 (1983)
4. Lee, C.-L., Chen, S.-Y.: Classification of leaf images. International Journal of Imaging Systems and Technology 16(1), 15–23 (2006)
5. Li, Y., Chi, Z., Feng, D.D.: Leaf vein extraction using independent component analysis. In: IEEE International Conference on Systems, Man and Cybernetics, SMC 2006, vol. 5, pp. 3890–3894 (2006)
6. Pearson, K.: On lines and planes of closest fit to systems of points in space. Philosophical Magazine 2, 559–572 (1901)
7. WEKA 3. Data Mining with Open Source Machine Learning Software in Java
8. Wu, S., Bao, F., Xu, E., Wang, Y.-X., Chang, Y.-F., Xiang, Q.-L.: A leaf recognition algorithm for plant classification using probabilistic neural network. In: 2007 IEEE International Symposium on Signal Processing and Information Technology, pp. 11–16 (December 2007)
9. Zheng, X., Wang, X.: Leaf vein extraction based on gray-scale morphology. International Journal of Image, Graphics and Signal Processing 2, 25–31 (2010)

Economical Assessment of Large-Scale Photovoltaic Plants: An Italian Case Study

Enrico Maria Mosconi[1], Maurizio Carlini[1], Sonia Castellucci[1], Elena Allegrini[1], Luca Mizzelli[1], and Michelangelo Arezzo di Trifiletti[2]

[1] University of Tuscia, via San Camillo de Lellis, Viterbo, Italy
{enrico.mosconi,maurizio.carlini,sonia.castellucci,
allegrini.e}@unitus.it, lucamizzelli@libero.it
[2] Embassy of the United States of America
arezzomf@yahoo.it

Abstract. Photovoltaic power is characterized by higher costs than coal-fired technologies. Thus, photovoltaic systems require policy support to become more affordable. This is even more important if the goals -set by Directive 2009/28/EC- must be achieved by 2020. The present paper carefully describes the economical feasibility of a large scale photovoltaic plant, with specific regard to the Italian situation. An Italian case study is investigated and attention is paid to an extremely important tool, i.e. Project Financing. In order to better understand the economical assessment, recent public incentives and supporting measures are taken into account. After analyzing the cash flows and evaluating the Net Present Value (NPV) and the Internal Rate of Return (IRR), the results - obtained by the simulation for different Feed-In Tariff (FIT)s- show that the income is influenced by the date of commissioning and by the amount of incentives.

Keywords: photovoltaic, economical, feasibility, large-scale plant.

1 Introduction

The Directive 2009/28/EC set many climate and energy targets in order to reach and ensure a clean and sustainable future for the forthcoming generations. It is also well known as "Directive 20-20-20" since its aim is to reduce greenhouses gases emissions by 20%, to produce 20% of energy from renewable sources and to decrease the consumption by 20% improving the energy efficiency. These goals must be achieved by 2020 [1-2].

Each EU Member State implemented the above-mentioned regulation. To ensure that the mandatory national overall targets are correctly reached, each Member State worked towards an indicative trajectory which paths the achievement of the final goals. According to that, a national renewable energy action plan must be adopted by each Member State, focusing on the share of energy from renewable sources

B. Murgante et al. (Eds.): ICCSA 2013, Part II, LNCS 7972, pp. 160–175, 2013.
© Springer-Verlag Berlin Heidelberg 2013

consumed in transport, electricity, cooling and heating. With specific regard to Italy, the target for share of energy from renewable resources in gross final consumption in 2020 would be at least 17% [1-2].

In this scenario, solar energy could play a significant role. The amount of energy produced by photovoltaic panels, in the future, could overtake that produced by fossil fuels. This could lead countries from being predominantly fossil fuelled to being fuelled by locally available sources [3-10].

Moreover, an important aspect has to be further investigated: PhotoVoltaic (PV) systems clearly require economical feasibility in order to become more affordable. However, it is basic to take into account the incentives for the production of electricity by PV systems. Campoccia *et al.* investigated and compared the different supporting measures adopted in several European countries, namely France, Germany, Italy and Spain. According to this study, subsidies and prices of energy strongly vary so that it is difficult to find out the most effective option [11-15].

In 2010, PV installations strongly increased in terms of number of plants (+ 215%) and installed power (+ 324%) if compared with the values reported at the end of 2009. Moreover, the European Photovoltaic Industry Association stated that Italy might reach "Grid Parity" (GP) by the end of 2013, considering the growing trend of the past years. GP occurs when an alternative energy source can generate electricity at level generalized costs that is less than or equal to the price of purchasing power from the electricity grid. Thus, it corresponds to the point when PV-generated electricity becomes competitive with the retail rate of grid power [16].

Table 1 shows the number of PV installations in Italy -classified by size into three different categories (i.e. less than 20kW, 20÷50 kW, more than 50 kW)- and the power produced by each typology [17]:

Table 1. PV plants

Size of PV plants	Number of installations	Power (MW)
<20 kW	426700	2553
20÷50 kW	16037	632
>50 kW	39086	13202

According to Atlasole –which is the geographic information system published by GSE (in Italian "Gestore dei Servizi Elettrici")- the total number of photovoltaic plants in Italy is more than 480 000 (update on the 6th of February 2013) and the produced power is up to 16 387 MW. Most of the plants have small sizes (less than 20 kW), as shown in figure 1, and represent almost 90% of the total installations. Furthermore, only the 8% of the PV systems overcome 50 kW: however, this latter category produces more than 13 200 MW representing the 80% of the total power, as it can be seen in figure 2 [17].

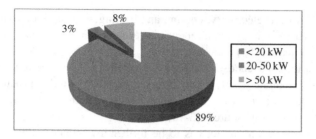

Fig. 1. Size of photovoltaic plants in Italy

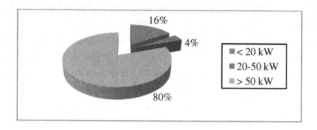

Fig. 2. Power of PV plants in Italy

Figure 3 and 4 show the number of installations and the produced power per region respectively. With specific regard to Latium, it can be seen that the total installed power is equal to 1 066.7 MW corresponding to 27 045 PV plants. Furthermore, Latium reaches the seventh rank in term of both aspects if compared with the other Italian regions. As shown in figure 3, the highest number of installations occur in Lombardia and Veneto. Nevertheless, the most significant amount of produced power (figure 4) is generated in Puglia and Lombardia in second position [17].

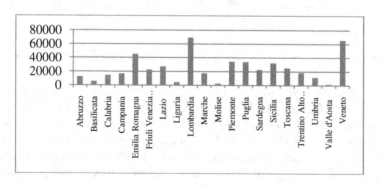

Fig. 3. Number of installation for each region

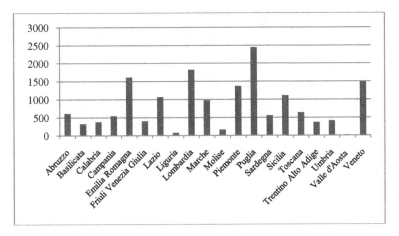

Fig. 4. Produced power for each region (MW)

In order to evaluate the economical feasibility of large-scale plants in Italy, the incentives system has to be further investigated. The current regulation is the so-called "Fifth Feed-in Scheme", which was approved by the Ministry of Economy and Finance on the 5th of July 2012 and came into force on the 27th of August 2012 [17].

The aim of the present paper is to describe the economical feasibility of large-scale PV plants, with specific regard to the Italian situation. More precisely, an Italian case study will be further investigated, focusing the attention on FIT and supporting measures. In order to better understand the economical assessment, the cash flows, the Net Present Value (NPV) and the Internal Rate of Return (IRR) will be analyzed. The comparison between the results is obtained by the simulation for two different supporting measures (namely "Fourth" and "Fifth Feed-in Scheme") and shows that the income decreases as the incentives are reduced, since the specific cost per PV module is having a decreasing trend.

2   Material and Methods

The so called "Feed-in Scheme" is the government tool which grants incentives for electricity generated by photovoltaic plants connected to the grid. Italy introduced these supporting measures in 2005. PV plants -with a minimum capacity of 1 kW- may benefit from the tariff which varies depending on the capacity and on the category of the plant, and is granted over a period of 20 years [17].

According to AEEG (which stands for "Autorità per l'Energia Elettrica e il Gas"), the indicative yearly cumulative cost of incentives has reached € 6 billion. The 5th feed-in scheme will cease to have effect 30 calendar days after reaching an indicative cumulative cost of incentives equal to € 6,7 billion per year [18].

Three different typologies of PV plants are effected by the present Decree [17]:

- PV plants on buildings which respect all the installation requirements reported in Annex II of the present regulation, namely Building-Integrated Photovoltaic plants (BIPV) with innovative features;
- Concentrating Photovoltaic Plants (CSP);
- other PV plants, not belonging to the above described categories and including ground-mounted solutions as long as they fulfill the requirements specified in articles 7, 8 and 9 in order to benefit from the FIT.

The present scheme provides two different mechanisms to access incentives, depending on the type of installation and on the nominal capacity of the plant, namely [17]:

1. direct access for:

- PV plants (up to 50 kW) installed on buildings: the modules must replace roofs/covers from which asbestos has been completely removed;
- PV plants, not exceeding 12 kW and including those with an increase in capacity less than 12 kW;
- BIPV plants with innovative features until reaching an indicative cumulative cost of incentives equal to € 50 million;
- CPV plants until reaching an indicative cumulative cost of incentives equal to € 50 million;
- PV plants built by public administrations until reaching an indicative cumulative cost of incentives of € 50 million;
- PV plants applying for a tariff which is 20% lower than the one the same plants could get via Registry.

2. via registry:

All the plants not belonging to the above-listed categories may access the incentives by enrolling into appropriate electronic Registries held by GSE. However, in order to benefit from the FIT, their position in the relevant ranking list should not exceed the applicable cost limit. GSE will issue the calls for applications every 6 months which must be submitted within the next 60 days.

The 5th feed-in scheme grants an all-inclusive feed-in tariff to the share of net electricity injected into the grid and a premium tariff to the share of net electricity consumed on site. Thus, it is different from the previous support schemes. Table 2 defines the current incentives for PV plants according to the Fifth Feed-in scheme, with specific regard to those plants working in the first semester of application of the present regulation. It can be easily seen that incentives decrease as the nominal power becomes higher, as shown in figure 5. Moreover, a lower support measure is provided to the category "Other PV Plants" if compared with "PV Plants on Buildings" (figure 6) [17].

Table 2. Incentives for PV plants according to the current feed-in scheme

Power (kW)	PV plants on buildings		Other PV plants	
	All-inclusive tariff (€/MWh)	Premium tariff (€/MWh)	All-inclusive tariff (€/MWh)	Premium tariff (€/MWh)
1÷3	182	100	176	94
3÷20	171	89	165	83
20÷200	157	75	151	69
200÷1000	130	48	124	42
1000÷5000	118	36	113	31
>5000	112	30	106	24

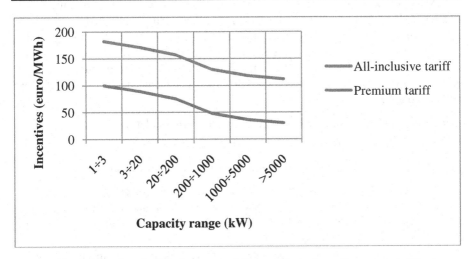

Fig. 5. Incentives for PV plants on buildings: all-inclusive and premium tariff

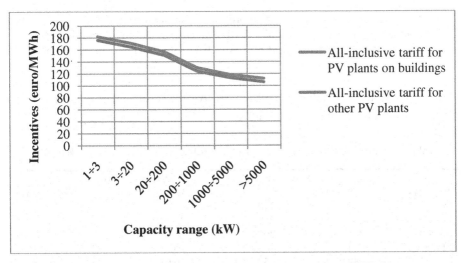

Fig. 6. All-inclusive tariff fro PV plants on buildings and other PV plants

The Fourth Feed-in scheme will continue to be valid to [17]:

- Small photovoltaic plants, building integrated plants with innovative features and concentrating PV plants, commissioned before 27 August 2012;
- Large PV plants whose position in the Registry does not exceed the cost limit and which submit to be completed within 7 months (or 9 months if plant capacity is more than 1 MW) of the publication of the ranking list;
- Plants installed on public buildings and in those areas owned by Public Administration, commissioned before 31 December 2012.

The Fourth Feed-in scheme was published on 12 May 2011 and takes into account all the PV plants with a nominal power at least equal to 1 kW and commissioned between 1 June 2011 and 31 December 2016. The tariff introduced by the past support scheme consists of two main components: the premium and the price paid for the electricity produced by the system. Thus, the tariff included both the incentives and the value of the electricity fed into the grid. Moreover, a specific support will be applied to the self-consumed electricity. PV plants are considered eligible if they belong to one of the below-listed categories [17]:

- "building -integrated PV Plants" according to the criteria defined in Annex 2;
- "Other PV Plants", including ground-mounted PV systems.

The decree defines two types of PV plants depending on their capacity:

- small plants, namely BIPV with a nominal power lower than 1 MW; other PV systems with a nominal power up to 1 MW and operating under net metering; PV systems installed on buildings and in those areas belonging to the Public Administration;
- large plants, which include those systems which do not meet the above-mentioned criteria.

PV plants must be enrolled into an appropriate electronic registry in order to be eligible and access the incentives: only small plants can avoid this step. It is important to underline that the registry is held by GSE [14]. Table 3 shows the tariffs for PV plants commissioned in 2012: it is extremely clear in figure 7 that the incentives decrease as the nominal power grow and passing from the first to the second semester of the year. Moreover, "other PV plants" receive lower support if compared with BIPV, as it can be seen in figure 8 [17].

Table 3. Incentives (€/MWh) for PV plants according to the Fourth Feed-in scheme

Power (kW)	1st Semester 2012		2nd Semester 2012	
	BIPV plants	Other PV plants	BIPV plants	Other PV plants
1÷3	274	240	252	221
3÷20	247	219	227	202
20÷200	233	206	214	189
200÷1000	224	172	202	155
1000÷5000	182	156	164	140
>5000	171	148	154	133

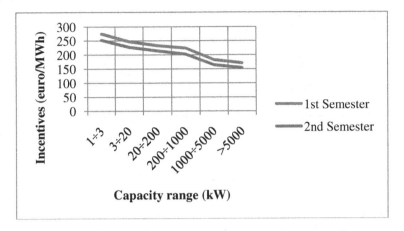

Fig. 7. Incentives for BIPV plants in 2012: comparison between the first and second semester

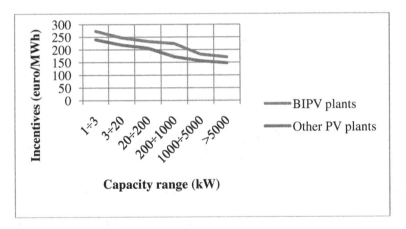

Fig. 8. Incentives for "BIPV Plants" and "Other PV Plants" for the first semester of 2012

In order to evaluate the economical feasibility of a large-scale PV plant and its convenience, some economical factors need to be further defined. Actually they are strongly related to:

- efficiency decrease of PV panels every year;
- maintenance and management costs;
- yearly increasing of electricity prices;
- inflation.

All the above-listed parameters are connected to the cash flows (C_t^*) which is obtained by adding all the costs $(C_{j,t})$ and all the profits $(P_{j,t})$ related to the generic t-th year, as shown in the following expression [8]:

$$C_t^* = \sum_j P_{j,t} - \sum_j C_{j,t} \tag{1}$$

The cash flows are successfully annualized by the expression [3]:

$$C_t = \frac{C_t^*}{(1+i)^t} \tag{2}$$

where i represents the Weighted Average Cost of Capital (WACC). It refers to the index which defines the average expected return considering the assets of the plant's owner. The evaluation of the Net Present Value (NPV) and the Internal Rate of Return (IRR) lead to assess the effectiveness of installing a PV plant and are given by [3]:

$$NPV = \sum_{t=1}^{N} \frac{C_t^*}{(1+i)^t} - C_o \tag{3}$$

$$C_0 = \sum_{t=1}^{N} \frac{C_t}{(1+IRR)^t} = 0 \tag{4}$$

respectively. N is the lifetime of the investment and C_0 represents the initial investment cost.

An economical simulation is carried out in order to evaluate how the income has changed passing from the Fourth to the Fifth Feed-in scheme. The feasibility is investigated for a specific Italian case study which consists of a large-scale PV plant. Three different solutions are implemented:

1. ground-mounted PV plant according to the Fifth FIT (called "Scenario 1");
2. ground-mounted PV plant according to the Forth FIT(called "Scenario 2");
3. building-integrated PV plant according to the Fifth FIT (called "Scenario 3");

The assumptions shown in table 4 have been adopted in the simulation. From the fiscal point of view, the "Worst Case" is represented by a PV plant located on a commercial activity: this condition is considered in the implementation since it is the most precautionary solution. With regard to scenario 1 and 2, the large PV plant will

Table 4. Technical parameters of the PV plant

Parameters	
Nominal power	9 872.52 kWp
Location	Province of Viterbo, Italy
Latitude	42°25'00"
Longitude	12°06'00"
Altitude	326 m
Climatic data	UNI 10349
Albedo coefficient	0.20
Tax system	"the Worst Case"
Self-consumption	100%
Module	Polycrystalline silicon
Power generated by the module	245 W_p
Total number of modules	40 296
Total surface occupied by the PV plant	67 294.32 m^2
Installation	fixed
Balance of system	80%

Table 5. Description of the three different scenarios: technical features

	Scenario 1	Scenario 2	Scenario 3
Electricity (kWh)	13 010 830.6	13 037 686.6	13 010 830.6
PV tilt (°)	32	32	6
Saved TOE	2 992.49	2 998.67	2 758.60

Table 6. Description of the three different scenarios: economical features

	Scenario 1	Scenario 2	Scenario 3
Total costs (€)	11 847 024	11 847 024	11 847 024
Specific cost (€/kW)	1 200	1 200	1 200
Insurance	94 776.19	94 776.19	94 776.19
Policy support	5th FIT	4th FIT	5th FIT
Type of incentive	"Other PV Plants"	"Other PV Plants"	BIPV
Amount of incentive (€/MWh)	106	133	112
Self-consumption tariff (€/MWh)	2.4	0	3

be installed on a land belonging to the Ministry of Defense and on an industrial cover respectively. This choice is due to the fact that the current regulation (Ministerial Decree 24 March 2012) does not allow the construction of PV plants on agricultural lands. Exceptions to the law are represented by those agricultural areas which belong to Army or achieved the authorization for the plant itself before 24 March 2012 [17]. Moreover, the energy produced by the PV plant is self-consumed. Tables 5 and 6

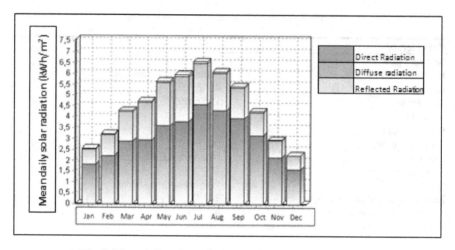

Fig. 9. Mean daily solar radiation on the plane of PV module

carefully describe the main features for each scenario. Figure 9 represents the mean daily solar radiation on the plane of the PV module (kWh/m^2) for the chosen location.

3   Results and Discussion

The simulation has been implemented using a specific software for the technical and the economical feasibility of PV plants. It leads to calculate the main economical parameters, i.e. the cumulative cash flow, the net present value and the internal rate of return for each scenario (figure 10-18).

SCENARIO 1:

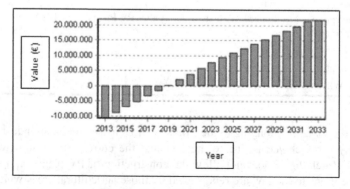

Fig. 10. Cumulative cash flow

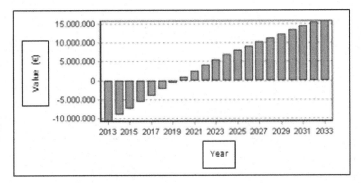

Fig. 11. Net present value

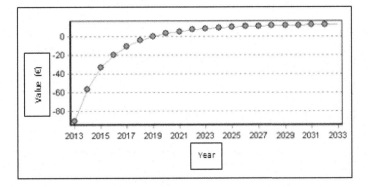

Fig. 12. Internal rate of return

SCENARIO 2:

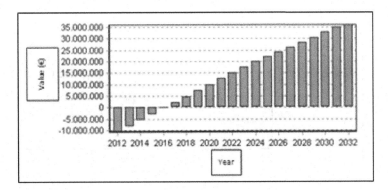

Fig. 13. Cumulative cash flow

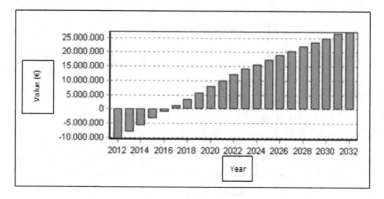

Fig. 14. Net present value

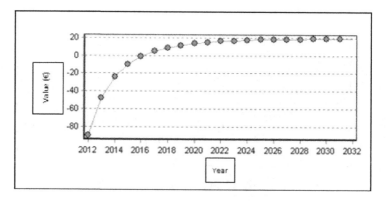

Fig. 15. Internal rate of return

SCENARIO 3:

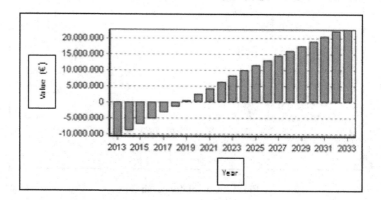

Fig. 16. Cumulative cash flow

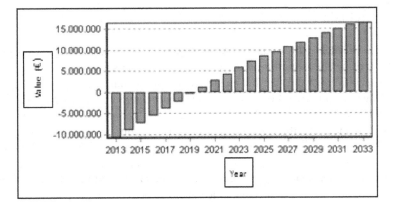

Fig. 17. Net present value

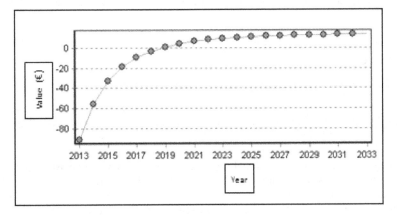

Fig. 18. Internal rate of return

If the three different scenarios are compared, it can be seen that the cumulative cash flow becomes positive in 2019 for the scenario 1, in 2016 for the scenario 2, and in 2019 for the scenario 3. The second case leads to a positive value of the cash flow earlier than the other two situations, since the incentive amount is higher. Moreover, the first and the third scenario start to have an income after 6 years, although they receive different policy support and are commissioned in the same period. This is due to the fact that the scenario 3, which gains the highest support measure, is subject to a stronger fiscal system than the scenario 1.

After analyzing the cash flows and evaluating the Net Present Value (NPV) and the Internal Rate of Return (IRR), the results -obtained by the simulation for different Feed-In Tariff (FIT)s- show that the income strongly depends on the date of commissioning and the amount of incentives.

4 Conclusion

In the present paper, the economic feasibility of a large scale PV plant has been studied, focusing the attention on the policy support (the so called "Feed-in Tariff") and with specific regard to the Italian situation. Photovoltaic technology has actually an increasing importance and is becoming even more attractive if the ambitious goals set by Directive 2009/28 must be achieved by 2020. The economical feasibility has been implemented for three different scenarios: it can be seen that the income decrease as the incentive amount becomes lower and passing from the 4th to the 5th FIT. The reduction of incentives is due the decreasing trend of PV module cost. Moreover, it is clear that a ground-mounted PV plant is more convenient if it gains measure supports from the 4th FIT since the incentive is higher. Nevertheless, if scenario 1 and 3 are compared considering the same FIT, it can be seen that the second case is more convenient although the incentive is higher. This is due to a harder fiscal system. It is expected that as far the incentives decrease according to the reduction of the PV module costs, the income is lower and thus affecting the pay-back time.

References

1. European Directive 2009/28
2. Cellura, M., Di Gangi, A., Longo, S., Orioli, A.: Photovoltaic Electricity Scenario Analysis in Urban Contests: an Italian Case Study. Renewable and Sustainable Energy Reviews 16, 2041–2052 (2012)
3. Al-Salaymeh, A., Al-Hamamre, Z., Sharaf, F., Abdelkader, M.R.: Technical and Economical Assessment of the Utilization of Photovoltaic Systems in Residential Buildings: the Case of Jordan. Energy Conversion and Management 51, 1719–1726 (2010)
4. Marucci, A., Monarca, D., Cecchini, M., Colantoni, A., Manzo, A., Cappuccini, A.: The Semitransparent Photovoltaic Films for Mediterranean Greenhouse: A New Sustainable Technology. Mathematical Problems in Engineering, art. no. 4519354 (2012)
5. Petroselli, A., Biondi, P., Colantoni, A., Monarca, D., Cecchini, M., Marucci, A., Cividino, S.: Photovoltaic Pumps: Technical and Practical Aspects for Applications in Agriculture. Mathematical Problems in Engineering, art. no. 342080 (2012)
6. Rottenberg, F., Lucentini, M., Di Palma, D.: Trigeneration Plants in Italian Large Retail Chains: a Calculating Model for the TPF Projects with the Evaluation of All the Incentivizing Mechanisms. SDEWES 2011, to appear on Journal of Sustainable Development of Energy, Water and Environment Systems (2013)
7. Cocchi, S., Castellucci, S., Tucci, A.: Modeling of an Air Conditioning System with Geothermal Heat Pump for a Residential Building. Mathematical Problems in Engineering, art. no. 781231 (2013)
8. Villarini, M., Limiti, M., Abenavoli, R.I.: Overview and Comparison of Global Concentrating Solar Power Incentives Schemes by Means of Computational Models. In: Murgante, B., Gervasi, O., Iglesias, A., Taniar, D., Apduhan, B.O. (eds.) ICCSA 2011, Part IV. LNCS, vol. 6785, pp. 258–269. Springer, Heidelberg (2011)

9. Bocci, E., Villarini, M., Bove, L., Esposto, S., Gasperini, V.: Modelling Small Scale Solar Powered ORC Unit for Standalone Application. Mathematical Problems in Engineering, art. no. 124280 (2012)
10. Carlini, M., Villarini, M., Esposto, S., Bernardi, M.: Performance Analysis of Greenhouses with Integrated Photovoltaic Modules. In: Taniar, D., Gervasi, O., Murgante, B., Pardede, E., Apduhan, B.O. (eds.) ICCSA 2010, Part II. LNCS, vol. 6017, pp. 206–214. Springer, Heidelberg (2010)
11. Suri, M., Huld, T.A., Dunlop, E.D., Ossenbrik, H.A.: Potential of Solar Electricity Generation in the European Union Member States and Candidate Countries. Solar Energy 81, 1295–1305 (2007)
12. Robinson, D., Stone, A.: Solar Radiation Modeling in the Urban Context. Solar Energy 77, 295–309 (2004)
13. Marucci, A., Carlini, M., Castellucci, S., Cappuccini, A.: Energy Efficiency of a Greenhouse for the Conservation of Forestry Biodiversity. Mathematical Problems in Engineering, art. no. 768658 (2013)
14. Carlini, M., Castellucci, S., Allegrini, E., Tucci, A.: Down-hole Heat Exchangers: Modelling of a Low-Enthalpy Geothermal System for District Heating. Mathematical Problems in Engineering, art. no. 845192 (2012)
15. Carlini, M., Cattani, C., Tucci, A.O.M.: Optical Modelling of Square Solar Concentrator. In: Murgante, B., Gervasi, O., Iglesias, A., Taniar, D., Apduhan, B.O. (eds.) ICCSA 2011, Part IV. LNCS, vol. 6785, pp. 287–295. Springer, Heidelberg (2011)
16. European Photovoltaic Industry Association, http://www.epia.org
17. Gestore dei Servizi Elettrici, http://www.gse.it
18. Autorità dell'Energia Elettrica e del Gas, http://www.autorita.energia.it

Modelling and Experimental Validation of an Optical Fiber for Solar Devices

Maurizio Carlini and Andrea O.M. Tucci

DAFNE, University of Tuscia, Viterbo. Italy
andrea.tucci@ino.it

Abstract. In this paper it has been presented a numerical model of an optical fiber used in the solar applications. We discussed the overall thermo-mechanical design and the behaviour of the system, composed of the fiber and its holder. Starting from the geometrical choices and from the analyses of the selected fiber, the response of the device is computed via numerical simulations. The geometry shape and characteristics of the optical components are selected among commercially available optical fibers while the shaping of the aluminium component used as clamper is chosen in order to easily set up the laboratory measures. The calculated model has been validated by means of an experimental setup. The results are well suitable to design a device for the transmission of concentrated solar energy via an optical fiber.

Keywords: FEA, optical fiber, solar concentrator, modeling of an optical fiber.

Nomenclature

Abbreviation

NA	Numerical Aperture
DOF	Degree of Freedom
CR	Concentration Ratio inlet power
AR	Anti reflection

Symbols

p	Pressure [bar]
T	Temperature [K]
V	Velocity [m/s]
F	Focal length [mm]
Ø	Diameter of lens [mm]
Re	Reynolds Number
J	Radiosity [W/m^2]
P	Density [kg/m^3]
u	Velocity vector [m/s]
μ	Dynamic viscosity [Pa·s]
Cp	Specific heat capacity at constant pressure [J/(kg·K)]
q	Heat flux by conduction [W/m^2]
σ	Constant of Stefan-Boltzmann [W/(m^2·k^4)]
DP	Total displacements [μm]

B. Murgante et al. (Eds.): ICCSA 2013, Part II, LNCS 7972, pp. 176–191, 2013.
© Springer-Verlag Berlin Heidelberg 2013

1   Introduction

With the rise of environmental consciousness in the recent years the green energy industries is gradually emerged and the efforts made to improve the technologies based on energy from renewable sources have been increased.

The shortage of fossil fuel resources and the increasing of the emissions of pollutants like CO, SOx, NOx makes the sun, a source of inexhaustible, clean energy, a competitive resource.

Among the solar devices based on the exploitation of the radiant energy, the ones that focus the sun image and the high density radiation through a collector by means a waveguide are suitable of many applications. Their principle of operation is based on the transmission of the concentrated solar energy via optical fibers. Such a principle has been applied in several fields. Besides the photovoltaic and the heating plants, other applications dealt with the indoor illumination.

Increasing in the use of natural lightning for lightning purposes in the interior spaces saves both the amount of electric energy used and the quality of the internal environment [1],[2],[3],[4]. A class of photo-bioreactors exploits this technique, too: their efficiency is related to the luminous flux and its distribution. Coupling the waveguide with the concentrating equipments improves the efficiency of the system [5],[6],[7]. Moreover, the concentrated solar power transmitted with optical fibers could be suitable to replace the solar surgery, as validated by some experimental results [8],[9],[10] as well as in some other applications [11],[12]. The efficiency of these systems strongly depends on the coupling between the fiber and the collector then a thermo-mechanical analysis [13] of the waveguide help to help to increase the performances of the whole device.

The topic of this paper is the modelling and analysis of the response of an optical fiber which may be used by these class of devices. An experimental setup has been built to validate the numerical results.

2   Material and Methods

This section describes both the numerical model and the tests discussed.

2.1   Numerical Model

Geometry and Physical Properties

The proposed device has been numerically modelled in order to simulate a naked optical fiber along with its mounting, surrounded by the air. The environment is represented by a cylinder large enough to describe the air fluxes. The cylinder top is the inlet zone and the cylinder bottom the outlet one. The power is injected into the top end of the optical fiber through an area of 0.78mm2 — the sun image focused by the collector.

The model has been analyzed by means of the Finite Elements Analysis (FEA) method. The geometry of the device allows to implement an axisymmetric model, which dramatically reduces the DOF's and the computational time.

The adopted scheme, shown in Figure 1, consists of two cylinder — the fused silica core and its silicon cladding, respectively (F and H in the Figure) —, a frame — the mechanical mounting (G in the Figure) — and a big cylinder— the air environment. The dimensions of the air are much larger than the ones of the other components in order to avoid any edge effect. The ambient air is assumed to be equal to 293K at air outer boundaries.

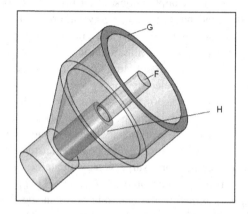

Fig. 1. 3D view of the assembly: F represents the optical fiber, H the cladding and G the mechanical mounting

The geometry is sketched in Figure 2. The red dash line represents the axis of symmetry, and A, B, C and D, respectively, are the sample points where the temperature is sensed. The physical characteristics and the dimensions of the analyzed layout are summarized in the Table 1.

The physical proprieties assumed for the air and the aluminium used for the mounting are taken from [14] and [15], respectively.

Table 1. Geometrical dimensions

Component	Width; Height [m]
Fused silica core	5×10^{-4}; 15×10^{-3}
Silicon cladding	5×10^{-4}; 5×10^{-3}
Air box	0.24; 0.48

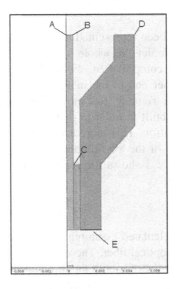

Fig. 2. Axial section of the model. The sample points are placed on the optical fiber, C on the cladding, and D on the mechanical mounting. The point E is the three axes constraint used in the mechanical analysis.

Table 2. Fused silica physical proprieties

Property	Unit	Value
Coefficient of thermal expansion	1/K	4.6×10^{-6}
Heat capacity	J/(Kg·K)	728
Thermal conductivity	W/(m·K)	1.35
Young's modulus	Pa	73.6×10^{9}
Poisson's ratio		0.17
Density	Kg/m^3	2210
Emissivity		0.96
Absorptivity		0.02

Table 3. Silicon cladding physical proprieties

Property	Unit	Value
Coefficient of thermal expansion	1/K	3.4×10^{-6}
Heat capacity	J/(Kg·K)	810
Thermal conductivity	W/(m·K)	0.2
Young's modulus	Pa	5.79×10^{9}
Poisson's ratio		0.27
Density	Kg/m^3	1000
Emissivity		0.90

Meshes

The proposed geometry has been represented by two different unstructured meshes. The first mesh is used to depict the whole system assuming the environment in laminar condition and it is composed of about 2700 free triangular elements. A refinement of the optical fiber and of the mounting have been applied to improve the spatial resolution when running the model. The second mesh, with about 100,000 elements, has been built in order to represents the model assuming the environment in turbulent condition. This more accurate discretization allows to properly represent the behaviour of the air flux and consequently the contribute of the convective fluxes. The figure 3 shows the refinement applied to the optical fiber and to the mechanical frame.

Input Data

The numerical models are calculated assuming three possible CR between the solar collector and the head of the optical fiber. These ratios are about 750, 1500 and 3000. The other values of the input data for both the turbulence model simulations and the laminar ones are summarized in Table 4. The boundaries of the boxes which represent the air are considered thermally open and the ambient temperature is assumed equal to 293K.

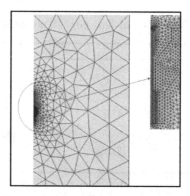

Fig. 3. Refinement of the mesh applied to the optical fiber and to the mounting

Table 4. Input data values

Data	Unit	Value (laminar/turbulent)
Wind velocity	m/s	0.005/10
Average irradiance	W/m^2	1000
To ambient	K	293
Po ambient	bar	1

Experimental Setup

The experimental setup simulates a scaled operating condition with respect to the concentrated sun inlet power. The system adopts as source a 150W xenon lamp (SVX 1530 of Muller Elektronik-Optik), focused by an optical reimager and spatially filtered by a 1000μm pinhole located in the focus. This focus is re-imagined on the top end of the fiber using two AR coated doublets with f=100m and Ø=50.8mm.

The diameter of the focus is 1.2mm and the supplied power has been measured using a power meter (Laser Star produced by OPHIR): the value is equal to $130 \cdot 10^{-3}$W, corresponding to a CR of about 95.

The used fiber, M30L02 distributed by Thorlabs, is a commercial one which a nominal core of 1000μm and its NA is 0.39. It has been adapted to reproduce the simulation model: the protection jacket and the cladding have been removed, taking the fused silica core in direct contact with the air.

This component has been mounted onto an aluminium frame and locked via a conic Teflon ring to minimize the stresses.

The PT100 sensors selected for the temperature measurements are the model PT111 manufactured by Lakeshore Inc.: they are positioned by plastic rods locked via clamps. The mechanical mounting has also two grooves suitable to house the chosen sensors to test the cladding temperature.

Two additional PT-100-based thermal resistances are used to detect the temperatures of the D point on the mechanical frame (see Figure 2) and the air temperature.

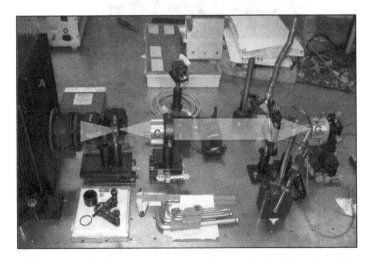

Fig. 4. The experimental setup is composed starting from a 150W Xenon lamp (A) focused by an optical reimager. A pinhole (B), exploited as spatial filter, was placed in the focus. This focus is re-imagined using two AR doublets (C and F) on the top of the optical fiber which has been mounted via a mechanical frame placed on a θXYZ stage (G). The projected spot presents an area of 1.13mm^2 and over this area supplies a power of 130mW. The temperature measurement has been performed using PT 100 sensors leaned on the fiber core via plastic rods locked by two clamps (D,E).

The optical fiber is placed on a small prism to rest the temperature sensors using an appropriate pressure value while keeping the alignment. A notch placed on of the support provides the minimization of the contact area between the fiber core and the backing prism. The whole assembly is joined to a θXYZ stage in order to allow a fine alignment of the fiber with respect to the power source. The experimental setup is shown in Figure 4.

The arranged experimental setup, it has been used also to evaluate the efficiency (η) of the chosen optical fiber. Using a variable aperture between the two optical doublets C and F (figure 4) it has been measured the transmission at different aperture(f/#). The maximum tramission is about 0.95, and it has been obtained at the minimum f-number aperture. This scheme has been used for all measurement.

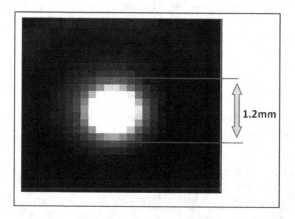

Fig. 5. Diameter of the focus

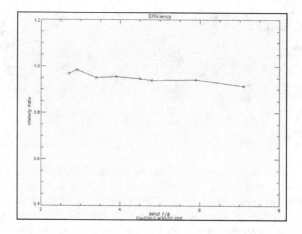

Fig. 6. Efficiency of transmission of the chosen optical fiber related to the f/#

3   Theoretical Background and Calculation Approach

The numerical models are based on a multiphysics approach which considers both the rates of exchanged radiant, convective and conductive energy transmissions and the mechanical effects induced by the heating.

The calculation approach applied to those models depends on the Reynolds number [16].

3.1   Laminar Flow

These simulations take into account a non-isothermal modelling of the air, considered as compressible. The behaviour of the fluid and the heat transfer are described by means of the following equations [17]:

$$\frac{\partial \rho}{\partial t} + \nabla \cdot (\rho \mathbf{u}) = 0 \tag{1}$$

$$\rho \frac{\partial \mathbf{u}}{\partial t} + \rho \mathbf{u} \cdot \nabla \mathbf{u} = -\nabla p + \nabla \cdot \left(\mu (\nabla \mathbf{u} + (\nabla \mathbf{u})^T) - \frac{2}{3} \mu (\nabla \cdot \mathbf{u}) \mathbf{I} \right) + \mathbf{F} \tag{2}$$

Equations 1 represent the continuity equation and the momentum equations, where F is the body force vector. Equation 2 describes the simplified heat equation for a fluid, where the pressure work term and the viscous heating term have been neglected.

$$\rho C_p \left(\frac{\partial T}{\partial t} + (\mathbf{u} \cdot \nabla) T \right) = -(\nabla \cdot \mathbf{q}) + Q \tag{3}$$

The heat transfer is governed by

$$\rho C_p \frac{\partial T}{\partial t} = -(\nabla \cdot \mathbf{q}) + Q \tag{4}$$

The radiant energy between the source and the fiber core is described by the following equation

$$J = \rho G + \varepsilon \sigma T^4 \tag{5}$$

where ε is the surface emissivity, ρ is the absorptivity, G the irradiation and J the radiosity.

A linear mechanical analysis has been run using the output of the thermal analysis and restraining the model only at the point E (Figure 2) about all its DOF's. Both the steady-state and the time-step models have been solved via the direct solver MUMPS [18], assuming as dependent variables the air velocity field $\mathbf{u}(u,v,w)$, the pressure P,

Table 5. CPU time and DOF laminar simulations

CR	750	1500	3000
CPU time [s] (Steady state/ Time dependent)	41/342	41/78	41/94
DOF	8700	8700	8700

the radiosity J, and the field mechanical displacement **u2** (u2,v2,w2). The number of DOF's and the CPU time are summarized in Table 5 as a function of the CR.

In order to evaluate the time constant, the time-step simulations are stopped when the difference of the mean temperatures of the areas F and G are lower than 1mK. The CR time-dependent simulations take more time because of the thermal stability is reached after many   steps.

3.2   Turbulent Flow

These steady-state simulations take into account the turbulent flows by means of the k-ε model [19] and described the turbulent conductivity using the Kays-Crawford one [20].

$$
Pr_T = \left(\frac{1}{2Pr_{T\infty}} + \frac{0.3}{\sqrt{Pr_{T\infty}}} \cdot \frac{c_{p\mu_T}}{\lambda} - \left(0.3 \frac{c_{p\mu_T}}{\lambda} \right)^2 \left(1 - e^{-\frac{\lambda}{0.3 c_p \mu_T \sqrt{Pr_{T\infty}}}} \right) \right)^{-1}
$$

(6)

Equation 5, where λ is the thermal conductivity, defines   the Prandtl number assuming PrT∞ =0.85.

A wall function [21], which locally modifies the K-ε model, is added to solve the thin region close to the wall. The backward-facing step model ,[22],[23], shows how to refine the mesh and how to model the inlet region.

The turbulent kinetic energy k and the rate of dissipation of turbulent kinetic energy ε are added to the other   dependent variables used for the laminar simulations, whereas the direct solver PARDISO [25] is used to solve the model.

The number of DOF's and the CPU time are summarized in Table 6 as a function of CR.

Table 6. CPU time and DOF turbulent simulations

CR	750	1500	3000
CPU time [s]	120735	12125	12576
DOF	360000	360000	360000

4   Results

This section describes both the numerical results and the experimental outcomes.

The data of the simulations which assume the environment in a laminar condition are shown in two different subsections. The former describes the model assuming the inlet power equal to the concentrated sun irradiance, the latter simulates a CR corresponding to the experimental setup. The analysis of the mechanical deformation caused by this inlet power has not been performed because the displacements induced by the thermal stresses are negligible.

4.1 Numerical Results

Laminar Simulation

The temperatures of A, B, C and D points and their total mechanical displacements (see Figure 2) are summarized in   Table 7 as a function of CR.

The time-step simulations took about 274, 472, 525 seconds for the three cases defined in section 2.1.3. Figure 7 shows the time responses of the selected sample points to   CR=1500. The other CR values give similar trends.

Table 7. Temperatures of the selected control points and total displacements

CR	A T/DP	B T/DP	C T/DP	D T/DP
750	306.0k/0.19	304.1/0.19	293.7k/0.05	293.6k/0.22
1500	318.9k/0.39	315.1/0.38	294.4/0.18	294.2k/0.44
3000	343.9k/0.77	336.2/0.76	295.7k/0.21	295.4k/0.88

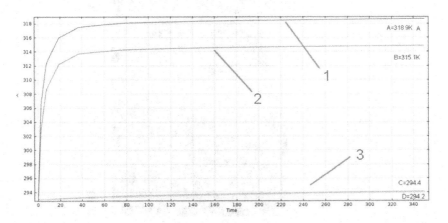

Fig. 7. Thermal time evolution of the selected sample points for a CR=1500. The assembly needs about   340 seconds to reach the thermal stability. The curves 1,2 and 3 represent the time thermal response of the top end of the optical fiber, the cladding and the mechanical frame, respectively, calculated on the sample points defined Figure 2. Note that the curves which describes the thermal behaviour of the C and D points are overlapped.

Laminar Simulation CR95
This numerical simulation assumes an initial temperature of 297.4K, corresponding to the measured laboratory temperature, and a value of the inlet power of $115 \cdot 10^{-3}$ W/mm^2. The numerical results of the sample points are shown in Table 8.

Turbulent Simulation
The temperatures of A, B, C and D points and their total mechanical displacements (see Figure 2) are summarized in Table 9 as a function of CR. Figure 8 and 9 show the temperature distribution in the assembly and the behaviour of the fluid near the boundaries, respectively.

Table 8. Temperatures of the selected control points

CR	A	B	C	D
95	299.1k	299.1k	297.4k	297.4k

Table 9. Temperatures of the selected control points and total displacements for turbulent flow

CR	A T/DP	B T/DP	C T/DP	D T/DP
750	300.6k/0.16	299.5/0.16	293.7k/0.02	293.2k/0.09
1500	306.9k/0.31	303.5/0.30	293.9/0.02	293.2k/0.09
3000	330.5k/0.38	312.3/0.4	295k/0.02	294.8k/0.11

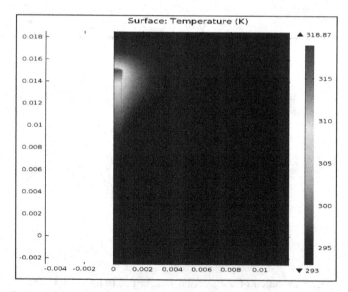

Fig. 8. Distribution of the temperature in the whole assembly. Only the optical fiber is affected by a temperature increasing. The optical fiber has a fused silica core with a nominal core of 1000μm. All computed simulations have a temperature distribution like the depicted one.

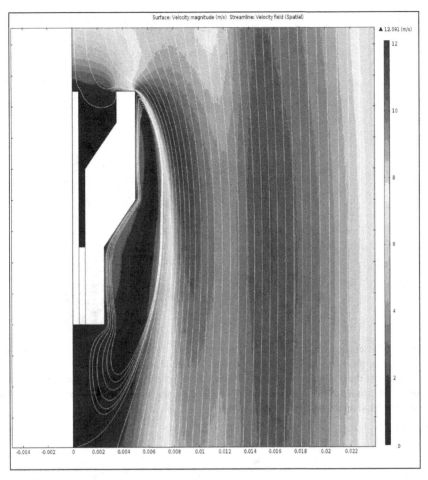

Fig. 9. Air velocity distribution in the turbulent cases. The turbulent simulations have been computed assuming an inlet air velocity of about 10m/s in order to represent a fully turbulent motion of the air within the considered box. The stream function plot shows   the recirculation zone (N) which occurs near the boundaries of the mechanical mounting (M) .

Experimental Outcomes

The results of the thermal analysis are summarized in the following table.

Table 10. Temperatures of the selected control points

CR	A	B	C	D
95	299.2k	299.2k	297.4k	297.4k

Figure 10 and Figure 11 show the PT111 sensor used to test the head of the optical fiber core and the arrangement used to measure both the spot size and the inlet power.

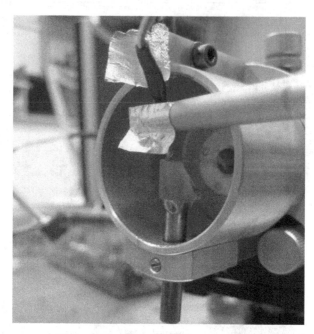

Fig. 10. PT111 sensor over the top end of the optical fiber. A small prism (R) is placed on the opposite side of the fiber core (S) to compensate the displacements caused by the probe force and allow to rest the temperature sensor (Q) using the appropriate pressure value while keeping the optical fiber alignment. A silver film (P) sticked on the plastic rod (O) has been used to minimize the irradiance contribute on the sensor.

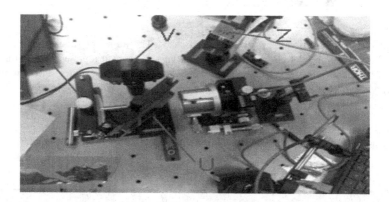

Fig. 11. . Measurements of the dimension of the focus and of its power. The rays, focused by doublet T, are folded via a mirror (U) and projected over the sensor of the power meter V (Laser Star Product by Ophir). The measurement of dimension of the focus has been made by a CCD UI-1120SE-M manufactured by IDS (Z) using the same optical layout.

5   Discussion

The results show that only the fiber core is affected by a relevant temperature in-creasing. The temperatures of the sample points located on the cladding and on the mechanical mounting, both in turbulent and in laminar conditions, are quite close to

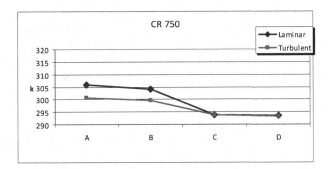

Fig. 12. A comparison between the temperatures of the sample points in laminar and turbulent condition at CR 750

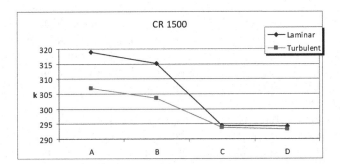

Fig. 13. A comparison between the temperatures of the sample points in laminar and turbulent condition at CR 1500

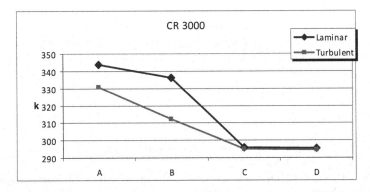

Fig. 14. A comparison between the temperatures of the sample points in laminar and turbulent condition at CR 3000

the initial temperature, as highlighted in Tables 7 and 9. These results have also been confirmed by the curves shown in Figure 7, where the time thermal responses of C and D points at the stop condition assume values close to the ambient one. The small displacements in Tables 7 and 9 show that thermal distortions are negligible. The following figures show the thermal comparison of the selected points at various CR considering both laminar and turbulent condition.

At 1500 and 3000 CR the temperature surface of the fiber near the top is lower respectively of about 12 and 24k. These differences due to the forced convection may be used as a starting point to design a possible required cooling system for high concentration devices using commercial fiber whose cladding may be damaged by high temperature . Moreover the fluid dynamics analysis in the turbulent simulations, as the one plotted in the Figure 9, may be used to design a mechanical mounting which creates a recirculation zone to improve the dissipation above the top of the fiber. The comparison between the experimental outcomes and the numerical results of the simulation also validates the created model. The differences between the measurements and the theoretical values are due to the measurements errors.

6   Conclusions

A numerical model of an optical fiber used for solar energy applications has been investigated by means of FEA in order to analyze its thermo-mechanical response. Various inlet powers take into account the required range of the amount of energy to be supplied. The calculations have been performed considering both laminar and turbulent environments and assuming both the steady-state condition and the time evolution of the system.

An experimental setup has been arranged to validate the model using a scaled power source in laminar conditions: the comparison between the computed data and the experimental outcomes shows that the numerical model simulates correctly the actual device — the experimental outcomes well match the numerical results. The few percent differences are very close to the measurement errors.

The proposed numerical model enhances the design of the systems with high concentration ratios. The simulation allows to optimize the cooling system and to avoid misalignments due to the thermal stresses.

As a conclusion, the suggested numerical approach is suitable for further, still unexplored solutions and allows to enhance the systems conveying the solar power by means of optical fibers.

References

1. Hana, H., Kimb, J.T.: Application of high-density daylight for indoor illumination next term. Energy 35(6), 2654–2666 (2010)
2. Kima, J.T., Kimb, G.: Overview and new developments in optical daylighting systems for building a healthy indoor environment. Building and Environment 45(2), 256–269 (2010)

3. Benton, C.C.: Daylighting can improve the quality of light and save energy. Architectural Lighting, 46–48 (1986)
4. Han, H.J., Jeon, Y.I., Lim, S.H., Kim, W.W., Chen, K.: New developments in illumination, heating and cooling technologies for energy-efficient buildings. Energy 35(6), 2647–2653 (2010)
5. Guoa, C.-L., Zhua, X., Liao, Q., et al.: Enhancement of photo-hydrogen production in a biofilm photobioreactor using optical fiber with additional rough surface. Bioresource Technology 102(18), 8507–8513 (2011)
6. Chen, C.Y., Saratale, G.D., Lee, C.M., Lee, P.C., Chang, J.S.: Phototrophic hydrogen production in photobioreactors coupled with solar-energy-excited optical fibers. International Journal of Hydrogen Energy 33(23), 6886–6895 (2008)
7. An, Y., Kim, B.W.: Biological desulfurization in an optical-fiber photobioreactor using an automatic sunlight collection system. Journal of Biotechnology 80(1), 35–44 (2000)
8. Feuermann, D., Gordon, J.M., Huleihil, M.: Solar fiber-optic mini-dish concentrators: first experimental results and field experience. Solar Energy 72(6), 459–472 (2002)
9. Gordon, J.M., Feuermann, D., Huleihil Laser, M.: surgical effects with concentrated solar radiation. J. Appl. Phys. 93, 4843–4851 (2003)
10. Feuermann, D., Gordon, J.M.: Gradient-index rods as flux concentrators with applications to laser fibre optic surgery. Opt. Eng. 40, 418–425 (2001)
11. Barlev, D., et al.: Innovation in concentrated solar power. Solar Energy Materials and Solar Cells 95(10), 2703–2725 (2011)
12. Ciamberlini, C., Arancini, F., et al.: Solar system for exploitation of the whole collected energy. Optics and Lasers in Engineering 39(2), 233–246 (2003)
13. Carlini, M., et al.: International Conference on applied Energy, Suzhou (2012)
14. Comsol chemical reaction engineering module user guide v4.2a, 275–278 (2011)
15. Juvinall – Marschek: Fondamenti della progettazione di macchine, p. 821, Edizioni ETS
16. Pope, S.B.: Turbulent Flow, pp. 264–266. Cambridge Press (2000)
17. Comsol Heat transfer module users guide v4.2a, 238–239 (October 2011)
18. Comsol Multiphysics reference guide v4.2a, 521 (October 2011)
19. Pope, S.B.: Turbulent Flow, pp. 373–383. Cambridge Press (2000)
20. Comsol Heat transfer module users guide v4.2a, 242 (October 2011)
21. Comsol Heat transfer module users guide v4.2a, 243–244 (October 2011)
22. Armalyt, B.F., Durst, F., Pereira, J.C.F., Schonung, B.: Experimental and theoretical investigation of backward-facing step flow. J. Fluid Mech. 127, 473–496 (1983)
23. 1st NAFEMS Workbook of CFD Examples. Laminar and Turbulent Two-Dimensional Internal Flows, pp. 38–49. NAFEMS (2000)
24. Comsol Multiphysics reference guide v4.2a, 522 (October 2011)

Characterization of Biomass Emissions and Potential Reduction in Small-Scale Pellet Boiler

Daniele Dell'Antonia[1], Gianfranco Pergher[1], Sirio R.S. Cividino[1], Rino Gubiani[1],
Massimo Cecchini[2], and Alvaro Marucci[2]

[1] DISA, University of Udine. Italy
{daniele.dellantonia,gianfranco.pergher,rino.gubiani}@uniud.it
[2] DAFNE, University of Tuscia, Viterbo. Italy
{cecchini,marucci}@unitus.it

Abstract. In recent years it has been proved that residential biomass combustion has a direct influence on ambient air quality, especially in the case of cereals. The aim of this study is the characterization of the emissions in small-scale fixed-bed pellet boiler (heat output of 25 kW) of beech and corn, and of its potential reduction to an addition of calcium dihydroxide. In the biomass combustion test 7 fuel mixtures were investigated with regard to the particulate content (PM_{10}), gaseous emissions and combustion chamber deposit.

The corn kernels tanned with calcium dihydroxide determined a decrease in particulate emissions (54 ± 13 mg MJ^{-1}) in comparison to corn, whereas in the combustion of corn pellet with 1% calcium dihydroxide high emissions were observed (193 ± 21 mg MJ^{-1}). With regard to SO_2 emissions, the combustion of corn with the additives make a reduction in comparison to additive-free corn.

Keywords: Combustion, Emissions, Particulate matter, Beech, Corn, Pellet boiler.

1   Introduction

On average, biomass contributes some 9-10% to total energy supply in the world [1], but in European countries, biomass energy is 5-6% of total energy demand [2]. In Italy, a consumption of about 19 million tons of wood for domestic heating has been estimated (about 20% total energy consumed for residential heating) [3]. Wood is the most common and cheapest source of heat in many mountain areas during the winter, and it is burned in simple, traditional wood stoves (with lower thermal efficiency) as well as in modern boilers (electronic control with the parameters). In recent years, however, a different possibility has become of interest, especially in rural areas (i.e. the use of agricultural residues such as straw, pruning of vines and fruit tree, residues of food processing, etc.).

Biomass combustion is regarded "CO_2 neutral", but at the same time is an important source of both gaseous and particulate pollutants (Table 1), such as particulate

B. Murgante et al. (Eds.): ICCSA 2013, Part II, LNCS 7972, pp. 192–206, 2013.

matter, carbon monoxide (CO), nitrogen oxides (NO_X), sulphur dioxide (SO_2) and hydrocarbons. The combustion emissions of biomass are associated with adverse health effects, such as pulmonary and cardiovascular symptoms [4, 5, 6]. Exposure to particulate matter has been estimated to cause an average loss of life expectancy of 9 months in EEA-32 countries (EU-25) [7]. As pointed out by Smith (1994), the products of incomplete combustion have three major adverse effects: energy loss, impact on human health and impact on the environment. He estimated that the use of biomass fuels contributed 1 to 5% of all CH_4 emissions, 6 to 14% of all CO emissions, 8 to 24% of all total non-methane organic compounds (TNMOC) emissions and thus 1 to 3% of all human induced global warming [8].

According to recent studies, small scale biomass combustion is the most important emission source of particulate matter in the cold season in Europe and in the USA. For example, particle emissions have been reported to be 50 times higher from a wood stove or a fireplace than from more controlled devices [9]. A survey conducted in five western Montana valley communities during 2006/2007 showed that wood smoke (likely from residential wood stoves) was the major source of $PM_{2.5}$ in each one of the communities, contributing from 56% to 77% of the measured wintertime $PM_{2.5}$ [10]. In 2010, an investigation in a residential site (Dettenhausen, Germany) concluded that during winter months the contribution from biomass combustion in the heating system to ambient particulate matter pollution was around 59% [11, 12]. Thus, it is evident that residential wood combustion has a direct influence on ambient air quality.

Fine particles originated in small-scale wood combustion mainly consist of ash, elemental carbon and organic material. Ash particles are formed when ash species volatilize in the hot combustion chamber and form particles when the flue gas cools down. In small-scale combustion, the most common ash particles are alkali metals, such as K_2SO_4, KCl, K_2CO_3 and KOH [6].

Moreover, the combustion processes with biomass, especially with herbaceous biomasses which have a high content of alkaline elements, are prone to experience slagging, fouling and corrosion in the boiler plant. The biomass ash sintering on heat exchangers causes a decrease of heat transfer capacity and difficulty in cleaning the deposited ash. In relation to the heat exchangers, elements such as K, S and Cl are enriched in the ash deposits [13].

Recently, different studies have used additives into the biomass, in order to reduce the slagging phenomena in the plant and the particulate matter in the flow emission. In Denmark it is commonly recommended to add 1-2% of limestone ($CaCO_3$) to avoid sintering [14]. Indeed, the combustion of cereal grains with the CaO in residential appliances (barely, rye and wheat) was shown to reduce slagging in a burner by formation of high-melting calcium/magnesium potassium phosphates. Thus, the formation of potassium fly-ash particles can be prevented by incorporation of potassium in calcium compounds they stay in the bottom ash [15]. Coda (2001) added some limestone to the straw combustion in a fluidised bed combustor, and observed that lime reacts with HCl, resulting in an enhanced share of Cl in the larger fly-ash particles, that was not bound as alkali chloride, followed by a corresponding reduction of gaseous HCl in the flue gas [16]. Subsequently, Wolf (2005) used Ca-based additives, including limestone, in high alkaline biomass combustion and found that emissions of gaseous SO_2 could be reduced with up to 25% [17]. It was suggested that limestone addition may produce a shift from alkali chlorides to alkali sulphates in the fine fly-ash particles. Recently, Bäfver (2009) concluded that the emission of particle mass can be lowered by supplying kaolin from residential combustion of oat grain (2% kaolin added to the fuel decreased the emission of particle with 31%, and 4% lowered the emission by 57%). On the other hand, however, the particle mass was not affected by addition of limestone, although limestone decreased HCl (Cl increased in the bottom ash) and SO_2 in the flue gas was unchanged because the high concentration of phosphorous in the fuel hindered SO_2 captured by limestone [15].

The objective of the present work was to investigate the effects of the biofuels mixture (corn with beech and corn with calcium dihydroxide ($Ca(OH)_2$)) on the gaseous and solid emission in small appliances suitable for residential use. For the gaseous emissions were measured the total particulate matter and the concentration of the main pollution gases. The solid emissions were assessed in relation of the deposit in the combustion chamber.

Table 1. Range of emissions in the atmosphere from the combustion of various biomass, expressed as mass per energy [4, 5, 9]

Boiler/burner	Fuels	Power (kW)	CO_2 (%)	CO (mg MJ^{-1})	CH_4 (mg MJ^{-1})	NO_X (mg MJ^{-1})	PM (mg MJ^{-1})
Old-type boilers [1] [9]	Wood	6-24	4.7-8.4	4,100-16,400	610-4,800	28-72	87-2,200 [5]
Modern boilers [2] [9]	Wood	12-34	5.1-11.5	507-3.781	0.8-73	60-125	18-89 [5]
Modern boilers [2] [9]	Wood pellet	3-22	3.7-13	30-1.100	0-14	62-180	12-65 [5]
Pellet boiler [5]	Wood pellet	25	—	80 ± 67	0.3 ± 0.2	49 ± 7	19.7 ± 1.6 [6]
Stove [3] [4]	Beech	6-6.5	—	2,472-2,779	151-304.5	155-156	111.4-131.3 [7]
Stove [3] [4]	Oak	6-6.5	—	2,948-3,074	123.5-223.5	163-166	107.3-121.8 [7]

(1) Water-cooled multi fuel boiler, up-draught combustion
(2) Ceramic wood boiler with fuel gas fan, down-draught combustion
(3) Logwood stove with manual operation
(4) Improved biomass stove for space heating in developing countries
(5) Mass of particles below aerodynamic diameter of 2.5 μm
(6) Mass of particles below aerodynamic diameter of 1 μm
(7) Mass of particles below aerodynamic diameter of 10 μm

Nomenclature

Feedstock

C	Corn grain
CL	Corn grain tanned with 1% calcium dihydroxide
CB12	Pellet of corn (87,5%) and beech (12,5%)
CB25	Pellet of corn (75%) and beech (25%)
CB50	Pellet of corn (50%) and beech (50%)
CFL	Pellet of corn (99%) and calcium dihydroxide (1%)
B	Pellet of beech

Symbols

Mm/v	Mass pollution to volume concentration (mg m^{-3})
Mv/v	Volume concentration (ppm)
Mm/e	Mass pollution to biomass energy (mg MJ^{-1})
Mm	Particulate mass on the filter sampling (mg)
F	Flow of the gas emission (m^3 s^{-1})
P	Nominal heat input (W)
FM	Formula weight (g kmol^{-1})
B	Burning rate (kg h^{-1})
U	Biomass moisture content (%)
TE	Temperature of the flue gas emission (°C)
TC	Temperature in the combustion chamber (°C)
LHV	Lower heating value (MJ kg^{-1}d.b.)
EC	Energy content (MJ kg^{-1}w.b.)
PM	Particulate matter in the flow emission (mg MJ^{-1})
SP	Surface of steel plates (m^2)
Dm	Mass of the deposit (mg)
De	Mass deposit in relation to biomass energy (mg m^{-2} MJ^{-1})
daf	dry basis, ash-free

2 Materials and Methods

The experiments were carried out to assess the emissions of the main flue gases (O_2, CO_2, CO, NO, NO_2, NO_X and SO_2) and particulate matter in a fixed-bed boiler with a maximum heat output of 25 kW.

The experiments were performed with seven combinations of biomass samples from agro-forestry products (Table 2) and additives, added to improve the combustion and reduce emissions of pollutants in the atmosphere. For each sample were made three tests. The feedstocks included:

- corn grain and beech pellet;
- pellet of corn and beech (12.5%, 25% and 50%);
- pellet of corn and 1% calcium dihydroxide ($Ca(OH)_2$);
- corn grain tanned with calcium dihydroxide ($Ca(OH)_2$).

The pellet samples were obtained by mixing corn flour with ground beech and/or calcium dihydroxide additive. The mixture was pelletized in a professional installation (Costruzioni Nazzareno s.r.l., Breda di Piave (TV) - Italy). The tanned corn samples were obtained by mixing whole corn with the additive (calcium dihydroxide), and then manually sieving the mixture to remove the excess additive. The percentage of additive was determined base on the weight difference before and after treatment (0.87%). The test did not include pellet samples made with pure corn flour since, because of technical and economical aspects, it is hardly convenient to pelletize the flour instead of using the whole corn directly.

Table 2. Characteristics of biomass samples used in the tests, and operating parameters of the appliances

| Biomass | Symbol | Shape | Biomass | | | | | |
			U (%)	LHV (MJ kg^{-1}$_{d.b.}$)	EC (MJ kg^{-1}$_{w.b.}$)	P (kW)	B (kg h^{-1})	F (m^3 s^{-2})
Corn grain	C	Grain	10.1	17.7	15.7	20.7	4.7	1.8
Corn grain + 0.87% hydrated lime	CL	Grain	9.4	17.7	15.8	21.7	4.9	1.7
Corn flour (87.5%) + Beech (12.5%)	CB$_{12}$	Pellet	6.4	17.8	16.5	21.6	4.7	1.7
Corn flour (75%) + Beech (25%)	CB$_{25}$	Pellet	6.7	17.9	16.5	21.2	4.6	1.7
Corn flour (50%) + Beech (50%)	CB$_{50}$	Pellet	7.5	18.0	16.5	20.0	4.3	1.6
Corn flour (99%) + 1% hydrated lime	CFL	Pellet	7.3	17.3	15.9	21.0	4.7	2.0
Beech	B	Pellet	6.2	19.6	18.2	17.5	3.4	1.1

U, in %, is the biomass moisture content
LHV, in MJ kg^{-1}d.b, is the lower heating value on dry basis
EC, in MJ kg^{-1}w.b., is the energy content on wet basis
P, in W, is the nominal heat input
B, in kg h^{-1}, is the burning rate
F, in m^3 s^{-1}, is the flow of the gas emission

The lower heating value (LHV) of C and B was determined with a calorimetric bomb (C200, Ika, Staufen Germany); for the other sample feedstocks, the energy content was calculated based on the percentage of the different components. Afterwards, the energy content was calculated taking into account the latent heat of the water absorbed in the structure of the biomass. The following formula was used to calculate the energy content [19]:

$$EC = \frac{LHV \cdot (100 - U) - 2.44 \cdot U}{100} \quad (1)$$

where:
EC, in MJ kg^{-1}w.b., is the energy content;
LHV, in MJ kg^{-1}d.b., is the lower heating value;
U, in %, is the moisture content of the biomass;
0.025, in MJ kg^{-1}, is the heat of water vaporization.

Table 2 shows the analysis of moisture and energy content of the feedstocks. The combustion equipment was similar to most commercial pellet appliances (Granola 25R, Arca Caldaie, San Giorgio di Mantova - Italy) with slight differences in the hydraulic and heat dissipation systems (Fig. 1).

The small-scale pellet boiler represented a modern continuous combustion technology with a nominal power output of 25 kW and can be operated with any kind of wood and herbaceous feedstock, either in form of pellets or grain. In this combustion equipment there is not a hot water cylinder (buffer tank) as a heat storage and the heat is directly dissipated by a fan heater.

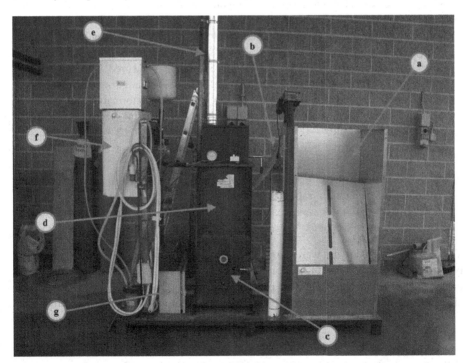

Fig. 1. Combustion equipment: a) fuel tank, b) screw feeding, c) combustion chamber, d) turbulators for heat exchangers, e) steel chimney (12 mm), f) fan heater, g) boiler control unit

The feedstock is supplied to the combustion chamber from an external fuel storage by a transport screw. The chamber combustion had a square shape with a pig iron plate for gas circulation. Its size was 520 mm, 550 mm, 590 mm (width, length, height). The fumes pass to the heat exchanger of the boiler and subsequently in the steel chimney, where the sampling probes were placed.

Biofuel boilers typically present a very long period of start-stop, and it takes them a long time to work at full performance. For this reason, the biomass plant included a heat sink, whose function was a thermic dissipation, that allows to maintain a constant power of the combustion. The biomass combustion was corrected with the operation time of the screw feeding and the flow air with the electronics unit (Termotre 32.10, Tecnosolar, Villimpenta (MN) - Italy). In this way, it was possible to obtain a constant power (about 20 kW) and determine the fuel consumption for the all test (Table 2). The tests performed during biomass combustion have provided the following samplings (Fig. 2):

- the flue gas emission was measured with a multi-gas analyzer (Vario plus industrial, MRU air Neckarsulm-Obereisesheim; DIN EN 50379-1+2). Flue concentrations of CO_2 and CO were measured with an infrared bench, while O_2, NO, NO_2 and SO_2 were measured electrochemically with a sensor with three electrodes. The NO_X gas was calculated as the sum of NO and NO_2;
- the particulate matter was assessed with a gravimetric method, which required the isokinetic conditions to collect the particulate (UNI EN 13284-1 and UNI 10169). The gas emission was collected with a constant flow rate sampler (ZB1 battery, Zambelli, Milano - Italy). The speed of flue gas was measured with a Pitot tube connected to the multi-gas analyzer and the flow of the sampling probe was corrected (gas speed per 6 mm diameter probe). The particulate matter was collected on quartz fiber filters with a 47 mm in diameter (FAFA quartz filter, Millipore, Billerica Massachusetts – USA). Its efficiency is >99.998% (0.3 micron particles) and it is considered the most penetrating particle in an air flow stream (maximum temperature of 950 ° C);
- the temperature in the combustion chamber was measured using a thermocouple (Tastotherm MP 2000, Infrapoint, Saalfeld – Germany);
- the deposit in the boiler was measured with a three steel plates in the combustion chamber, before the heat exchanger (40 mm x 25 mm x 8 mm).

The experimental analyses were carried out in the summer when the ambient humidity was lower than 65% and temperatures varied between 20 and 32°C. For all tested boilers, the same methodology was performed. Initially, the boiler was turned on and after 30 minutes and not before the chamber temperature was stable, the emissions were recorded for 60 minutes. The gas emission values were measured in parts per million (ppm, volume to volume) and then expressed as a mass of pollutant per content energy of the fuel (mg MJ^{-1}). The following formula was used to calculate the pollution mass to volume concentration:

$$C_{m/v} = C_{v/v} \frac{FM}{22.4} \quad (2)$$

where:

$C_{m/v}$, in mg m^{-3}, is the mass to volume concentration;

$C_{v/v}$, in ppm, is the concentration of volume to volume;

FM, in g kmol^{-1}, is the molecular weight;

22.4, in m^3 kmol^{-1}, is the specific volume of a gas at 20°C temperature and 101.325 kPa (Avogadro's law).

The following formula was used to calculate mass pollution to biomass energy:

$$C_{m/e} = \frac{C_{m/v} \cdot F}{P \cdot 10^6} \quad (3)$$

where:

$C_{m/e}$, in mg MJ^{-1}, is the mass per unit of biomass energy;

$C_{m/v}$, in mg m^{-3}, is the mass to volume concentration;

F, in m^3 s^{-1}, is the flow rate of the gas emission (gas speed per 12 mm steel chimney diameter);

P, in W, is the nominal heat input.

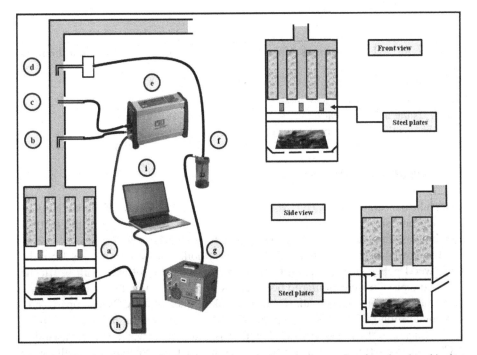

Fig. 2. Schematic of the experiment set-up: a) steel plates in the combustion chamber, b) pitot tube; c) instrument probe for combustion gas sampling, d) probe for dust sampling, e) multi-gas analyzer; f) silica gel to remove moisture from the flue gas; g) dust sampler; h) instrument to measuring the temperature in the combustion chamber; i) computer for data collection

For the particulate matter, the quartz filters were collocated after the analysis in a heater at 105°C, in order to remove residual moisture. The filters were subsequently weighed to determine the total particulate matter collected. The following formula was used to calculate the PM to biomass energy:

$$PM = \frac{M_m}{P \cdot 3,600 \cdot 10^6} \quad (4)$$

Where:
PM, in mg MJ^{-1}, is the particulate mass per unit biomass energy;
M_m, in mg, is the mass of particulate matter collected by the filter;
P, in W, is the nominal heat input;
3,600, in s, is the duration of the test.

The plates were weighed before and after the experimental test, in order to determine the amount of dust deposit. The weight was related to the mass of deposit per unit energy (mg m^{-2} MJ^{-1}) in the fuel (Fig. 2). The following formula was used to calculate the mass deposit to biomass energy:

$$D_e = \frac{D_m}{SP \cdot 3,600 \cdot 10^6} \quad (5)$$

where:
De, in mg m^{-2} MJ^{-1}, is the mass deposit per unit sampling area in relation to biomass energy;
D_m, in mg, is the total mass deposit on the plates;
SP, in m^2, is the surface of steel plates;
P, in W, is the nominal heat input;
3,600, in s, is the duration of the test.

3   Results and Discussion

The results concern O_2, CO_2, CO, NO, NO_2, NO_X, SO_2, PM and De. The total gaseous and solid emissions (amount of emitted substance per MJ of energy in the fuel) of these biomass fuels are also given in Table 3 and in Table 4.

Data showed higher emission of CO (unburnt pollution) in the flue gas from the beech wood pellet, in comparison to the corn. As for pellet of corn with beech the unburned emission were similar to beech combustion. In relation to the additive utilization the corn grain tanned with ($Ca(OH)_2$) make a decrease of CO in comparison to corn combustion (-32%) while the pellet combination with corn and 1% ($Ca(OH)_2$) gave an increase of CO (+158%) (Table 4). The analysis showed a strong variability of CO emissions (Fig. 3). It has been observed that CFL combustion results were more variable than all the others analyzed samples. Perhaps, calcium adsorbs the heat in the combustion chamber and causes a strong variability in the combustion temperature (795°C ± 10) [22-28].

Table 3. Emissions of O_2, CO_2, PM, $C_{d/v}$, TE and TC of fuel combustion (the standard error is reported to the average of the individual repetitions for each sample)

Feedstock	Symbol	O_2 (%)	CO_2 (%)	PM (mg MJ^{-1})	Deposit (mg cm^{-2} m^{-3})	TE (°C)	TC (°C)
Corn grain	C	15 ± 0.1	5.7 ± 0.1	74 ± 9	12 ± 8	140.0 ± 0.3	734 ± 6
Corn grain + 1% hydrated lime	CL	14.6 ± 0.1	6.1 ± 0.1	54 ± 13	108 ± 21	151 ± 0.6	613 ± 8
Corn flour (87.5%) + Beech (12.5%)	CB$_{12}$	14.9 ± 0.0	6.1 ± 0	72 ± 1	13 ± 6	152.8 ± 0.3	680 ± 7
Corn flour (75%) + Beech (25%)	CB$_{25}$	14.6 ± 0.0	6.3 ± 0.1	65 ± 4	14 ± 5	155.7 ± 0.2	726 ± 7
Corn flour (50%) + Beech (50%)	CB$_{50}$	15.6 ± 0.1	5.1 ± 0.1	66 ± 9	8 ± 5	153.6 ± 0.5	804 ± 8
Corn flour (99%) + 1% hydrated lime	CFL	14.4 ± 0.0	6.3 ± 0	193 ± 21	62 ± 2	162.2 ± 0.3	795 ± 10
Beech	B	15 ± 0.1	5.6 ± 0.1	164 ± 11	15 ± 1	158.0 ± 0.2	751 ± 5

PM is the particulate matter
TE is the temperature of the flue gas emission
TC is the temperature in the combustion chamber

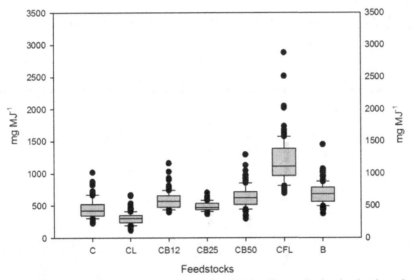

Fig. 3. Box plot of the variability in the CO emission for the all tests (in the simplest box plot the central rectangle spans the first quartile to the third quartile; the segment inside the rectangle shows the median and above and below the box show the locations of the minimum and maximum).

One of the main environmental impacts of solid biofuel combustion is caused by NOX emissions. The main mechanism of NOX formation in the fuel biomass, when the temperature of combustion is between 800-1,200°C, is the oxidation of the nitrogen contained in the fuel. The determining factors are the amount of fuel bound nitrogen in the ultimate analyses, the O2 concentration in the flame area and in a lower degree its temperature [22]. Furtermore the geometric of furnace and the type of combustion technology applied influencing variables for NOX formation [21]. According to Table 4, beech wood combustion have the lowest nitrogen oxides emissions (145±2 mg MJ-1), while higher concentrations are found in corn (447±4 mg MJ-1) and in corn mixed with beech (about 414 mg MJ-1). The emission data confirm the higher nitrogen content in the fuel, the higher NOX emissions from the biomass combustion.

Table 4. Emission of CO, NO, NO_2, NO_X, and SO_2 of fuel combustion (the standard error is reported to the average of standard error of the individual repetitions for each sample)

Feedstock	Symbol	CO (mg MJ⁻¹)	NO (mg MJ⁻¹)	NO_X (mg MJ⁻¹)	NO_2 (mg MJ⁻¹)	SO_2 (mg MJ⁻¹)
Corn grain	C	457 ± 18	279 ± 3	447 ± 4	19.5 ± 0.8	67.3 ± 1.9
Corn grain + 1% hydrated lime	CL	309 ± 11	259 ± 6	412 ± 9	13.9 ± 0.8	46.4 ± 2.2
Corn flour (87.5%) + Beech (12.5%)	CB_{12}	588 ± 16	261 ± 2	402 ± 3	1.5 ± 0.1	19.5 ± 0.7
Corn flour (75%) + Beech (25%)	CB_{25}	488 ± 9	305 ± 3	471 ± 4	3.9 ± 0.1	45.9 ± 1
Corn flour (50%) + Beech (50%)	CB_{50}	633 ± 17	237 ± 4	368 ± 5	4.9 ± 0.4	9.1 ± 0.7
Corn flour (99%) + 1% hydrated lime	CFL	1181 ± 36	369 ± 5	566 ± 7	0 ± 0	35.0 ± 1.3
Beech	B	691 ± 19	94 ± 1	145 ± 2	0 ± 0	0 ± 0

HC is the hydrocarbon combustion expressed as CH_4

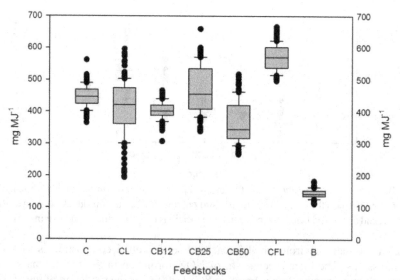

Fig. 4. Box plot of the variability in the NO_X emission for the all tests

The emissions of NOX for the combustion of corn grain tanned with calcium dihydroxide (412±9 mg MJ-1) resulted the same of corn combustion whereas the pellet combination with corn and 1% calcium dihydroxide has resulted in an increase of NOX emissions (566±7 mg MJ-1). The variability of NOX emissions resulted to be very high for CL, CLF, CB25 and CB50 (Fig. 4).

The sulphur contained in the solid biofuel forms mainly gaseous SO2. Due to the sub-sequent cooling of the flue gas in the boiler section of the combustion plant, SOX forms sulphates and condenses on the heat exchanger surfaces or reacts directly with

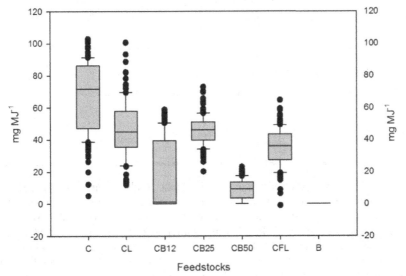

Fig. 5. Box plot of the variability in the SO$_2$ emission for the all tests

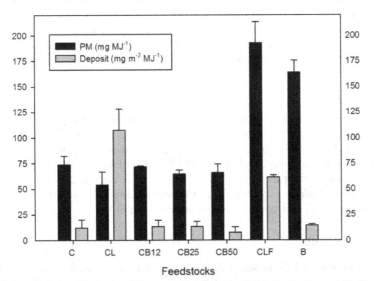

Fig. 6. Emissions of particulate matter (PM) and deposit (Cd/e) in combustor chamber for the feedstock combustion (reported the standard error of the repetitions for each samples)

Table 5. List of the equation in the text

Symbol	Unit	Description	Equation	
CE	MJ kg^{-1}$_{d.b}$	Energy content was calculated taking into account the latent heat of the water absorb in the structure of the biomass	EC=(LHV·(100-U)-2.44·U)/100	(1)
C$_{m/v}$	mg m^{-3}	Mass to volume concentration	C(m/v)=C(v/v)·FM/22.4	(2)
C$_{m/e}$	mg MJ^{-1}	Mass per unit of biomass energy	C(m/e)=(C(m/v)·F)/(P·106)	(3)
PM	mg MJ^{-1}	Particulate mass per unit biomass energy	PM=Mm/(P·3,600·106)	(4)
D$_e$	m^{-2} MJ^{-1}	Mass deposit per unit sampling area in relation to biomass energy	De=Dm/(SP·3,600·106)	(5)

fly ash particles deposited on heat exchanger surfaces (sulphation). The average emission of sulphur dioxide in corn combustion were 67 ± 2 mg MJ-1 while beech wood emissions were lacking because these fuels has a low concentration (Table 4)

The emission for corn with beech was smaller in comparison to corn (-72% for CB12, -31% for CB25 and -87% for CB50). The efficiency of sulphur fixation in the ash depends on the concentration of alkali and earth-alkali metals (especially Ca) in the fuel (fuels like wood chips and bark can have high Ca contents and cause therefore a high S fixation) [20, 21]. This consideration was confirmed by the result of the lime utilization; the emission were lower by 31% for CL and 48% for CFL.

The mass concentration of particulate matter from combustion of corn and pellet corn with beech ranged from 65 to 74 mg MJ^{-1}. The highest concentration occurred during the combustion of beech wood with 164 ± 11 mg MJ^{-1}. The utilization of lime as an additive in the corn pellets resulted in increased emissions of particulate matter (193 ± 21 mg MJ^{-1}) while the use of lime tanned with corn resulted in a decrease of emissions 54 ± 13 mg MJ^{-1}). Probably this decrease in emissions was determined by the dust of calcium dihydroxide in the combustion chamber that captured particulates matter (Fig. 6). The deposit in combustion chamber was the same for the feedstocks without the additive although there was tendency to lower emission for the CB50 (-33% in comparison to corn combustion) [29-31]. In relation to the additive utilizations, shows a significant increase of the deposits in the combustion chamber caused by the presence of calcium compounds formed during combustion (+800% for the CL and +417% for the CFL in comparison to corn combustion).

4   Conclusions

The results presented in this study clearly show that substantial differences in the combustion emissions were produced by the different feedstock from small-scale appliances. The unburned (CO and PM) emitted from combustion of corn was smaller than the emission from beech combustion, but the NO_X emission was much higher for corn utilization because there is higher nitrogen content in the feedstock. The calcium dihydroxide added to the fuel in the corn pellets reduces the SO_2 emissions. Conversely the additive increase the emissions of particulate matter and NO_X emissions from corn feedstock.

The calcium dihydroxide may be used to reduce the corrosive gasses, but the operation of the boiler and the method of the additive supplied play an important role for the result. In the future experiments, more focus should be put on the residence times of the fuel in the combustion chamber and the reaction with the bottom ash. In additional, a set of chemical analysis is recommended to be performed the particulate emissions and the bottom ash in order to provide more specific data to aid in the evolution of reduction emissions to small-scale biomass combustion.

References

1. ENEA, Agenzia Nazionale per le Nuove Tecnologie, l'Energia e lo Sviluppo Economico Sostenibile, Lungotevere Thaon di Revel, 76 Roma. Le fonti rinnovabili, ricerca e innovazione per un futuro low-carbon (2010)
2. Eurostat, http://epp.eurostat.ec.europa.eu/p (accessed: January 10, 2011)
3. APAT, Agenzia per la Protezione dell'ambiente e dei Servizi Tecnici. Stima dei consumi di legna da ardere per riscaldamento ed uso domestico in Italia (2008)
4. Schmidl, C., Luisser, M., Padouvas, E., Lasselsberger, L., Rzaca, M., Santa Cruz, C.R., Handler, M., Peng, G., Bauer, H., Puxbaum, H.: Particulate and gaseous emissions from manually and automatically fired small scale combustion systems. Atmospheric Environment 45, 7443–7454 (2011)
5. Lamberg, H., Nuutinen, K., Tissari, J., Ruusunen, J., Yli-Pirilä, P., Sippula, O., Tapanainen, M., Jalava, P., Makkonen, U., Teinilä, K., Saarnio, K., Hillamo, R., Hirvonen, M.R., Jokiniemi, J.: Physicochemical characterization of fine particles from small-scale wood combustion. Atmospheric Environment 45, 7635–7643 (2011)
6. Sippula, O.: Fine particle formation and emissions in biomass combustion. Report series in aerosol science n:o 108 (2010)
7. European Environment Agency (EEA) Report, n. 2/2007. Air pollution in Europe 1990-2004 (2007)
8. Smith, K.R.: Health, energy, and greenhouse-gas impacts of biomass combustion in household stoves. Energy for Sustainable Development 1(4), 23–29 (1994)
9. Johansson, L.S., Leckner, B., Gustavsson, L., Cooper, D., Tullin, C., Potter, A.: Emission characteristics of modern and old-type residential boilers fired with wood logs and wood pellets. Atmospheric Environment 38, 4183–4195 (2004)
10. Ward, T., Lange, T.: The impact of wood smoke on ambient PM2.5 in northern Rocky Mountain valley communities. Environmental Pollution 158, 723–729 (2004)
11. Bari, A., Baumbach, G., Kuch, B., Scheffknech, G.: Wood smoke as a source of particle-phase organic compounds in residential areas. Atmospheric Environment 43, 4722–4732 (2009)
12. Bari, A., Baumbach, G., Kuch, B., Scheffknech, G.: Institute Temporal variation and impact of wood smoke pollution on a residential area in southern Germany. Atmospheric Environment 44, 3823–3832 (2010)
13. Fernandez, L.M.J., Murillo, L.J.M., Escalada, C.R., Carrasco, G.J.E.: Ash behaviour of lignocellulosic biomass in bubbling fluidised bed combustion. Fuel 85, 1157–1165 (2006)
14. Ronnback, M., Johansson, L., Claesson, F., Johansson, M., Tullin, C.: Emission from small-scale combustion of energy grain and use of additives to reduce particle emissions. In: Proceeding of 16th European Biomass Conference and Exhibition, Valencia, pp. 1429–1435 (2008)

15. Bäfver, L.S., Rönnbäck, M., Leckner, B., Claesson, F., Tullin, C.: Particle emission from combustion of oat grain and its potential reduction by addition of limestone or kaolin. Fuel Processing Technology 90, 353–359 (2009)

16. Coda, B., Aho, M., Berger, R., Hein, K.R.G.: Behavior of chlorine and enrichment of risky elements in bubbling fluidized bed combustion of biomass and waste assisted by additives. Energy Fuels 15, 680–690 (2001)

17. Wolf, K.J., Smeda, A., Müller, M., Hilpert, K.: Investigations on the influence of additives for SO_2 reduction during high alkaline biomass combustion. Energy Fuels 19, 820–824 (2005)

18. Koyuncu, T., Pinar, Y.: The emissions from a space-heating biomass stove. Biomass and Bioenergy 31, 73–79 (2007)

19. Hartmann, H., Böhm, T., Maier, L.: Naturbalassene biogene Festbrennstoffe – umweltrelewante Eigenschaften und Einflussmöglichkeiten. Serie Umwelt & Entwicklung. Bayerisches Staatsministerium für Landesentiwicklung und Umweltfragen, vol. 154 (2000)

20. Obernberger, I., Theka, G.: Physical characterisation and chemical composition of densified biomass fuels with regard to their combustion behaviour. Biomass and Bioenergy 27, 653–669 (2004)

21. Obernberger, I., Brunner, T., Barnthaler, G.: Chemical properties of solid biofuels—significance and impact. Biomass and Bioenergy 30, 973–982 (2006)

22. Teixeira, F.N., Lora, E.S.: Experimental and analytical evaluation of NOX emissions in bagasse boilers. Biomass and Bioenergy 26, 571–577 (2004)

23. UNI CEN/TS 1496. Solid biofuels: fuel specifications and classes

24. Dell'Antonia, D., Gubiani, R., Maroncelli, D., Pergher, G.: Gaseous emissions from fossil fuels and biomass combustion in small heating appliances. Journal of Agricoltural Engineering 4, 1–10 (2010)

25. Monarca, D., Cecchini, M., Colantoni, A.: Plant for the production of chips and pellet: technical and economic aspects of an case study in the central italy. In: Murgante, B., Gervasi, O., Iglesias, A., Taniar, D., Apduhan, B.O. (eds.) ICCSA 2011, Part IV. LNCS, vol. 6785, pp. 296–306. Springer, Heidelberg (2011)

26. Monarca, D., Colantoni, A., Cecchini, M., Longo, L., Vecchione, L., Carlini, M., Manzo, A.: Energy characterization and gasification of biomass derived by hazelnut cultivation: analysis of produced syngas by gas chromatography. Mathematical Problems in Engineering, art. no. 102914 (2012)

27. Carlini, M., Castellucci, S.: Efficient energy supply from ground coupled heat transfer source. In: Taniar, D., Gervasi, O., Murgante, B., Pardede, E., Apduhan, B.O. (eds.) ICCSA 2010, Part II. LNCS, vol. 6017, pp. 177–190. Springer, Heidelberg (2010)

28. Colantoni, A., Giuseppina, M., Buccarella, M., Cividino, S., Vello, M.: Economical analysis of SOFC system for power production. In: Murgante, B., Gervasi, O., Iglesias, A., Taniar, D., Apduhan, B.O. (eds.) ICCSA 2011, Part IV. LNCS, vol. 6785, pp. 270–276. Springer, Heidelberg (2011)

29. Monarca, D., Cecchini, M., Colantoni, A., Marucci, A.: Feasibility of the electric energy production through gasification processes of biomass: technical and economic aspects. In: Murgante, B., Gervasi, O., Iglesias, A., Taniar, D., Apduhan, B.O. (eds.) ICCSA 2011, Part IV. LNCS, vol. 6785, pp. 307–315. Springer, Heidelberg (2011)

30. Monarca, D., Cecchini, M., Guerrieri, M., Colantoni, A.: Conventional and alternative use of biomasses derived by hazelnut cultivation and processing. Acta Horticulturae 845, 627–634 (2009)

31. Marucci, A., Monarca, D., Cecchini, M., Colantoni, A., Manzo, A., Cappuccini, A.: The Semitransparent Photovoltaic Films for Mediterranean Greenhouse: A New Sustainable Technology. Mathematical Problems In Engineering, art. no. 451934, 1–16 (2012)

Use of Hydro Generator on a Tanker Ship:
A Computer-Generated Simulation Study

Wilfredo Yutuc

Malaysian Maritime Academy (ALAM)
Batu 30, Tanjung Dahan
Kuala Sungai Baru, Melaka
Malaysia 78200
wilfredo.yutuc@alam.edu.my

Abstract. Considering ways of utilizing renewable energy sources in the marine environment, the possibility of fitting a ducted hydro generator on the hull of an existing 157 209 gross tonnage tanker ship was investigated. A three-dimensional (3D) model of the hydro generator was created and, using computational fluid dynamics (CFD) analysis, flow simulations were conducted to determine the additional hull resistance when a hydro generator is virtually fitted on the ship's hull. The additional hull resistance converted to resistance power was necessary to find out the net power of the hydro generator and determine its effectiveness and impact on the ship's fuel consumption. Using the characteristics of the tanker ship in study, an estimated 3.46% in fuel savings was obtained with the use of the hydro generator.

This study showed the design concept and computer-generated simulation carried out to analyze the performance of the hydro generator when integrated to the tanker ship running at full-speed sea condition of 15.5 knots with a head and tail water current velocity range of up to 2.5 m/s.

Keywords: computational fluid dynamics, hydro generator, renewable energy.

1   Introduction

The marine environment, mainly composed of vast oceans, can be considered as one of the largest unexploited renewable energy sources on our planet. Finding means of harvesting the available energy from these oceans can greatly contribute in addressing the current environmental issues on global warming and sustainability [1].

For navigating ships, considering the kinetic energy around the hull as it moves, predictable amount of wave energy can be extracted from the water. This can be made possible by a hydro generator, which is also termed "tidal generator". It operates under the principle of capturing and converting the kinetic energy and flow-pressure energy (if ducted) of the moving water into electrical energy. This energy, converted

B. Murgante et al. (Eds.): ICCSA 2013, Part II, LNCS 7972, pp. 207–219, 2013.
© Springer-Verlag Berlin Heidelberg 2013

into power, can be directed to the ship's main bus bar to support power generation onboard.

This paper investigated the possibility of installing a hydro generator on an existing tanker ship, in a manner as shown in Fig. 1, having the following characteristics:

Length, Overall	329.99 m
Length, Between Perpendiculars	316.00 m
Breadth, Moulded	60.00 m
Depth, Moulded	29.70 m
Draft Loaded, Moulded	19.20 m
Deadweight	259 994 ton
Gross Tonnage	157 209
Net Tonnage	99 808
Speed	15.50 knots
Main Engine (ME) Output at	
Maximum Continuous Rating (MCR)	25 090 kW
ME Speed	78.60 rpm
ME Specific Fuel Consumption (SFC)	
at MCR	168.50 g/kW-h
Diesel Generator (DG) Rated Output	1 020 kW x 3
DG SFC	195.00 g/kW-h

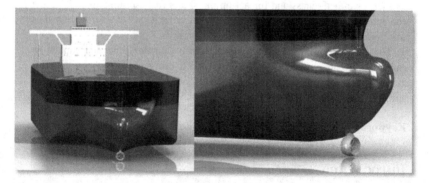

Fig. 1. Graphic conceptualization of the hydro generator fitted on the bow of the tanker ship in study

It is intended to provide the ship's electrical power requirement when the ship is running at full speed. This will reduce fuel consumption required to generate electrical power onboard, aside from the advantages brought about by the use of renewable energy.

Using a computer-aided design (CAD) software and the best dimension and scale available for a ducted hydro generator, a 3D model was created. Thereafter, CFD analysis was used to determine the increase in ship's hull resistance due to the fitting of the hydro generator. The additional hull resistance, converted to resistance power was necessary to examine the generator's estimated net power and its effect on the ship's fuel consumption.

2 Theory

Conversion of the linear movement of a flowing fluid into a useful rotational movement is usually done using a turbine. The power, P_T, in W harnessed by the turbine, is decided by its sweep area, A, in m^2, according to

$$P_T = \frac{1}{2} C_p A \rho_{fluid} V^3 \tag{1}$$

where C_p is the power coefficient, ρ_{fluid} is the fluid density, in kg/m^3, and V is the fluid velocity, in m/s.

Not all of the extracted energy from the fluid is converted to other energy forms. In this case, wherein electrical energy is desired, the non-conversion is due to the losses in the system, as shown in Fig. 2.

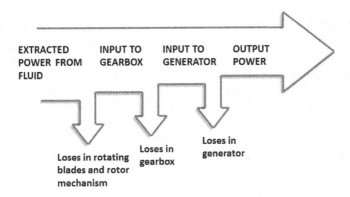

Fig. 2. Power flow of a turbine energy extractor

To find the turbine's generated electrical power output, P_{out} in W, the power, P_T, is multiplied by the blade efficiency, η_{blade}, gearbox efficiency, η_{gear}, and generator efficiency, η_{genr}, which can be written as

$$P_{out} = P_T \left(\eta_{blade} \eta_{gear} \eta_{genr} \right). \tag{2}$$

For output power estimation, blade efficiency of 90%, gearbox efficiency of 95% and generator efficiency of 70% can be assumed [2].

There are different turbine types and designs that can be used to extract energy from a moving fluid [3]. In view of the benefits of putting a duct or shroud around a turbine, a ducted water turbine was considered in this investigation.

A duct surrounding a turbine serves as a convergent-divergent diffuser which creates a drop in fluid pressure behind (downstream) the rotor blades, allowing increased fluid flow through the turbine, thereby increasing the power. In addition, a ducted or "diffuser-augmented" turbine eliminates tip losses on axial flow turbine blades and improves efficiency. It is not subject to the Betz limit which defines an upper limit of 59.3% of the incident kinetic energy that can be converted to shaft power by a single actuator disk turbine in open flow [4]. Increase in theoretical maximum power coefficient for a diffuser-augmented turbine of up to 3.3 times higher than the Betz limit of 59.3% has been claimed [5]. Experimentation with wind tunnel models even reported a power augmentation factor of 4.25 for a turbine with a diffuser, producing 4.25 times as much power than the same turbine in open flow [6]. Actual test data have also shown increase in performance by a factor of about 3 when a duct was added to a water turbine [4].

The seeming violation of the Betz law by ducted turbines in extracting more energy from the fluid than what is available arises from the C_p formula, which can only calculate the percentage of kinetic energy extracted from the fluid. Ducted turbines capture not only the kinetic energy but also certain percentage of the flow-pressure energy. In this study, an augmentation factor of 3 or a power coefficient, C_p, of 1.77 was used.

To calculate the hydro generator's generated power, relative water current velocities with respect to the ship were considered. While the ship is running in the open sea, the water current may flow from different directions with different magnitudes. This can greatly influence the power produced by an open flow hydro generator. However, this might not hold true for a ducted hydro generator. Due to the duct surrounding the turbine, only the head and tail water current will have considerable effect on energy extraction and power conversion. Also, the ship's cruising speed of 15.5 knots or 7.97 m/s is expected to ensure positive power in most conditions. In this investigation, a head and tail water current velocity range of up to 2.5 m/s was considered [7-8].

3   Method

The results and findings of this computer-generated simulation study were based on the calculation of (1) generated power, (2) resistance power, (3) net power, (4) fuel consumption – taking into consideration the additional resistance power and (5) fuel savings – obtained by comparing the consumption when using the diesel generator with the hydro generator.

3.1   Generated Power

The hydro generator's generated power was calculated using Equations (1) and (2). A 2.5-m turbine diameter was initially selected for the hydro generator with a hub of

20% or 0.5 m. The effect of water velocity from two directions (head and tail current flow) on the generated power was examined with the ship running on full speed sea condition. The generated output power is shown in Table 1.

Table 1. Generated power of the hydro generator on up to 2.5 m/s water current

Turbine Diameter (m)	Water Current (m/s)	Relative Velocity, V_R (m/s)		Generated Power (kW)		Average Generated Power (kW)
		Against the Current	With the Current	Against the Current	With the Current	
2.50	0.00	7.97	7.97	1294.57	1294.57	1294.57
	0.50	8.47	7.47	1553.83	1065.89	1309.86
	1.00	8.97	6.97	1845.57	865.87	1355.72
	1.25	9.22	6.72	2004.22	776.00	1390.11
	1.50	9.47	6.47	2171.71	692.57	1432.14
	2.00	9.97	5.97	2534.18	544.10	1539.14
	2.50	10.47	5.47	2934.89	418.52	1676.71

3.2   Resistance Power

Drag resistance plays a vital role in considering the net power benefitted from the hydro generator. The main engine needs to burn more fuel and produce more power to overcome this resistance in order to propel the ship and maintain its speed.

To determine the resistance power, a 3D model of the hydro generator was created using a CAD software, as seen in Fig.3, with the best dimension and scale available for ducted tidal turbines.

Fig. 3. Hydro generator 3D model

CFD analysis (Fig.4) was then used to estimate the drag brought about by the integration of the hydro generator to the ship. "Flow Simulation", a CFD analysis program embedded in SolidWorks® was used. It solves the Navier-Stokes equations, which are formulations of mass, momentum and energy conservation laws for fluid flows. It employs one system of equations to describe both laminar and turbulent flows.

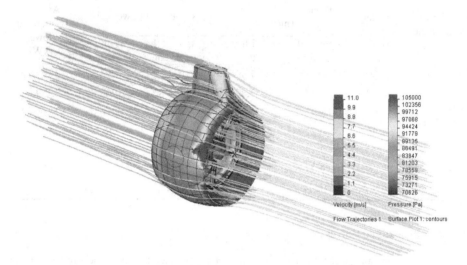

Fig. 4. CFD analysis on the hydro generator with bracket to estimate drag

In fluid dynamics, drag, sometimes called fluid resistance, refers to forces that oppose the relative motion of an object through a fluid, which is, in this case, water. Drag forces act in a direction opposite the oncoming flow velocity. Unlike other resistive forces such as dry friction, which is nearly independent of velocity, drag forces depend on velocity. For a solid object moving through a fluid, the drag is the component of the net aerodynamic or hydrodynamic force acting opposite to the direction of the movement [9]. Therefore, drag brought about by the integration of the aero and hydro generators to the ship opposes its motion and has to be overcome by additional propeller thrust. The power, P_R, in W to overcome fluid resistance or drag is equivalent to

$$P_R = F_D V \tag{3}$$

where F_D is the drag force, in N, and V is the fluid velocity, in m/s.

From the CFD analysis conducted, drag results were converted to equivalent resistance power using Equation (3) and are shown in Table 2.

Table 2. Drag and resistance power of the hydro generator on up to 2.5 m/s water current.

Water Current (m/s)	Relative Velocity V_R (m/s)		Drag (kN)		Resistance Power (kW)		Average Resistance Power (kW)
	Head Water Current	Tail Water Current	Head Water Current	Tail Water Current	Head Water Current	Tail Water Current	
0.00	7.97	7.97	33.02	33.02	263.17	263.17	263.17
0.50	8.47	7.47	37.23	28.99	296.72	231.05	263.89
1.00	8.97	6.97	41.82	25.15	333.31	200.45	266.88
1.25	9.22	6.72	44.17	23.42	352.03	186.66	269.35
1.50	9.47	6.47	46.51	21.70	370.68	172.95	271.82
2.00	9.97	5.97	51.54	18.62	410.77	148.40	279.59
2.50	10.47	5.47	56.87	15.66	453.25	124.81	289.03

3.3 Net Power

In this study, the net power is the ultimate beneficial power obtained from the hydro generator, measured as the generated power less the resistance power. A summary of the powers obtained, including the net power for the hydro generator is shown in Table 3.

Table 3. Power summary

Water Current (m/s)	Generated Power (kW)		Resistance Power (kW)		Net Power (kW)		Average Net Power (kW)
	Head Water Current	Tail Water Current	Head Water Current	Tail Water Current	Head Water Current	Tail Water Current	
0.00	1294.57	1294.57	263.17	263.17	1031.41	1031.41	1031.41
0.50	1553.83	1065.89	296.72	231.05	1257.10	834.84	1045.97
1.00	1845.57	865.87	333.31	200.45	1512.26	665.42	1088.84
1.25	2004.22	776.00	352.03	186.66	1652.18	589.34	1120.76
1.50	2171.71	692.57	370.68	172.95	1801.03	519.62	1160.32
2.00	2534.18	544.10	410.77	148.40	2123.41	395.69	1259.55
2.50	2934.89	418.52	453.25	124.81	2481.64	293.71	1387.67

3.4 Fuel Consumption

The tanker ship considered in this study was equipped with 3 diesel generator engines to generate electrical power supply. While running on full-speed sea condition, a 960 kW diesel generator engine supplied all the necessary electrical power requirement.

To maintain the ship's speed, the main engine needed to overcome the resistance brought about by the hydro generator's integration. This meant that the main engine had to produce more thrust and thus, would require additional fuel consumption. Considering the main engine's SFC at maximum continuous output of 168.5 g/kW-h, the additional main engine fuel consumption to develop more thrust to overcome the resistance of the hydro generator is shown in Table 4. This will also give an average additional fuel oil consumption of 45.8 kg/h.

Table 4. Additional fuel oil consumption to overcome hydro generator average resistance

Water Current (m/s)	Drag (kN)		Resistance Power (kW)		Average Resistance Power (kW)	Additional Fuel Consumption to Overcome Average Resistance (kg/h)
	Head Water Current	Tail Water Current	Head Water Current	Tail Water Current		
0.00	33.02	33.02	263.17	263.17	263.17	44.34
0.50	37.23	28.99	296.72	231.05	263.89	44.46
1.00	41.82	25.15	333.31	200.45	266.88	44.97
1.25	44.17	23.42	352.03	186.66	269.35	45.38
1.50	46.51	21.70	370.68	172.95	271.82	45.80
2.00	51.54	18.62	410.77	148.40	279.59	47.11
2.50	56.87	15.66	453.25	124.81	289.03	48.70

To sustain the ship's electrical power requirement at full speed sea conditions, the hydro generator considered in this study has to generate 960 kW of electrical power, equivalent to one diesel generator engine which supplied all the electrical power requirement when the ship was running at full speed.

In the condition wherein the diesel generator engine was running to supply the ship's electrical power requirement on full-speed sea condition, the fuel consumption was obtained, taking the maximum rated output of both main and diesel generator engines. Table 5 reflects the data.

Table 5. Main and diesel generator engine fuel consumption at full speed sea condition

	Specific F.O. Cons. (g/kW-h)	Max. Continuous Output (kW)	Fuel Cons. (kg/h)
Main Engine	168.50	25,090	4,227.70
Diesel Generator Engine	195.00	1,020	198.90
(AC Generator)		(960)	
Total			4,426.60

4   Results

The 2.5-m turbine diameter hydro generator had a calculated average generated power which met the 960 kW power requirement of the ship under study, regardless of water current direction, as shown in Fig.5.

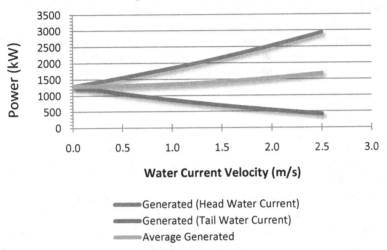

Fig. 5. Generated and average generated power of the hydro generator at different water current velocities (ship at 15.5 knots)

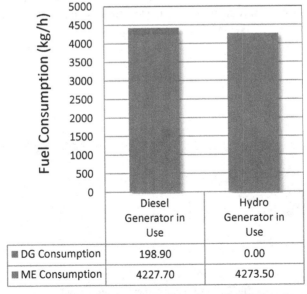

Fig. 6. Comparison of the calculated fuel consumption using a diesel generator and hydro generator at full speed sea conditions

As the hydro generator supplies the electrical power requirement when the ship is running at full speed, no diesel generator engine is operated. This reduces the ship's total fuel oil consumption by eliminating the need to operate a diesel generator engine for generating electrical power.

A comparison of the calculated fuel consumption using a conventional diesel generator and a hydro generator - taking into account the additional main engine fuel consumption to overcome resistance due to its integration, is shown in Fig.6. A reduction in fuel consumption of 153.1 kg/h can be obtained. This was equivalent to 3.46% of the 4 426.6 kg/h combined rated main and diesel generator engine fuel consumption at full speed sea condition. This meant savings of 3.46%.

5   Discussion

The hydro generator arrangement can be integrated to the ship's electrical power plant similar to shaft generator systems, which are already in existence. It is connected in line when the ship is running at its rated rpm. For safety reasons, one stand-by diesel generator should be readily available whenever the hydro generator is in use. Since its operation is affected by the water current's direction and velocity, its variable frequency and voltage output have to be converted into fixed voltage and frequency to match with machinery requirement. The use of a rectifier/inverter module can help address this concern, which operates by rectifying the variable frequency into direct current (DC) which is later inverted into fixed frequency alternating current (AC). The power management system onboard ships may provide a central place to make efficient utilization of all the electrical power as shown in Fig. 7. It usually includes the mode controllers, power flow meters, transfer switches and protection circuit breakers, and battery charge and discharge regulators.

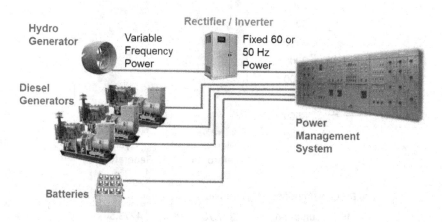

Fig. 7. Power management system layout

Then again, unlike tidal turbine generators which operate at lower water velocities, the hydro generator integrated in the ship's hull should be designed and constructed to operate at higher water velocities and withstand the sea forces and vibration experienced by the ship during adverse weather condition.

For the protection of marine life, a hollow turbine center with shrouded blade tips, non-usage of hydrocarbon-based lubricants and very low operating frequency sound are only some of the safety features available in some hydro generator designs [10-11].

On maintenance issues and concerns, there are hydro generators wherein maintenance schedule may be made to coincide with the ship's drydocking period. For instance, Clean Current's tidal turbine generators are constructed with a bearing system seal which is scheduled for replacement every 5 years and with the generator overhaul scheduled every 10 years. Moreover, these tidal turbine generators are designed for a service life of 25 to 30 years [10], which can be considered even longer than a ship's average service life.

In 2005, the unit capital cost of tidal turbine generators ranged from USD 1 700 to 2 000 per kW [12-13]. To date, it is estimated that the market is at USD 2 000 to 3 000 per kW [10]. Nonetheless, there are some makers who claim to achieve a target capital cost of less than USD 1 610 per kW [14]. Taking the average of these figures, which is equivalent to USD 2 055 per kW, a 960 kW hydro generator may be estimated to incur a capital cost of USD 2 million, with a total annual operation and maintenance cost estimated at USD 82 000 [12]. Taking this into consideration and with a zero salvage value, a payback period of 5 years can be expected.

With an average annual sea time of 264 days or 6 336 h for the ship in study [15] and with a bunker price of IFO 380 at USD 505 per ton (991 kg/m3 density) [16], an estimated annual savings of USD 489 871 can be projected with the use of a hydro generator.

6   Conclusion

Theoretical data obtained in this study approves the use of a hydro generator as another means to supply the tanker ship's electrical power requirement at full-speed sea conditions using renewable energy source. With the appropriate size of turbine diameter, it can theoretically generate the required power, regardless of water current velocity and direction. Also, taking into consideration the estimated savings in fuel consumption, capital and operational cost recoveries and amortization benefits can be realized within the first quarter of the equipment life cycle. Since the benefits in using hydro generators can be enjoyed during full-speed sea conditions, tanker ships, which spend most of their time at sea, can be expected to benefit more from such applications.

7   Recommendation

It is recommended that further studies be conducted on the validation of the simulation results presented in this paper. This is to determine the degree to which the model accurately represented the real world from the perspective of its intended use. Tests using scaled model on a towing tank may also be considered. Studies should include identification of the best efficient location on the hull where to affix the unit, including the design of the hydrodynamic mounting or means of attachment that will integrate it to the ship.

One issue that is of concern is the protection of marine life. Although there are tidal turbines designed with a hollow center, means of protection should still be installed to protect sea creatures from being caught by the rotating turbine blades.

In terms of design and construction, the hydro generator integrated in the ship's hull should be able to operate at higher water velocities. This is contrary to tidal turbine generators which are driven by tidal currents and operate at lower water velocities. The hydro generator's strength, durability and reliability, being located underwater, should also be taken into consideration to withstand pounding and slamming in harsh weather conditions. Vibration that could be brought about by its integration should also be looked into. Safety of Life at Sea (SOLAS) requirements and classification society rules on the use of hydro generators on ships are other important areas of further research.

As regard the ship's characteristics, the effects on hull center of buoyancy, trim, dynamic stability, seakeeping at all headings including the ease of maneuvering are some other things which should be further looked into. Drag effects to the ship running at a lower rpm and during maneuvering are other concerns.

Possibility of installing hydro generators on other ship types could also be explored.

References

1. Brook, J.: Wave Energy Conversion. Elsevier Ocean Engineering Book Series, vol. 6, pp. 1–6. Elsevier Science Ltd., UK (2003)
2. El-Sharkawi, M.A.: Electric Energy: An Introduction, 2nd edn., p. 107. CRC Press, Boca Raton (2009)
3. Streeter, V.L.: Handbook of Fluid Dynamics. McGraw-Hill Company, Inc., New York (1961)
4. Kirke, B.: Developments in Ducted Water Current Turbines (2006), http://www.cyberiad.net/library/pdf/bk_tidal_paper25apr06.pdf (retrieved: February 4, 2011)
5. Riegler, G.: Principles of Energy Extraction from a Free Stream by Means of Wind Turbines. Wind Engineering 7(2), 115–126 (1983)
6. Gilbert, B.L., Foreman, K.M.: Experiments with a Diffuser-Augmented Model Wind Turbine. J. Energy Resources Technology, Trans. ASME 105, 46–53 (1983)
7. Adams, J., et al.: Ocean Currents. Microsoft Encarta. CD-ROM, vol. 2. Microsoft, Redmond (1999)
8. Mason, K.: Composite Tidal Turbine to Harness Ocean Energy. Composites Technology (2005), http://www.compositesworld.com/articles/composite-tidal-turbine-to-harness-ocean-energy (accessed: February 1, 2011)
9. Wikipedia, http://en.wikipedia.org/wiki/Drag_(physics) (accessed: February 6, 2013)

10. Clean Current, `http://www.cleancurrent.com/` (accessed: December 30, 2011)
11. OpenHydro Tidal Technology, `http://www.openhydro.com/` (accessed: February 8, 2011)
12. Previsic, M., Polagye, B., Bedard, R.: System Level Design, Performance, Cost and Economic Assessment – Maine Western Passage Tidal In-Stream Power Plant. EPRI – TP – 006 – ME (2006), `http://archive.epri.com/oceanenergy/attachments/streamenergy/reports/006_ME_RB_06-10-06.pdf` (accessed: February 2, 2011)
13. Hagerman, G.: Tidal Stream Energy in the Bay of Fundy. Paper presented at the Energy Research and Development Forum, Antigonish, Nova Scotia (2006), `http://ns.energyresearch.ca/files/George_Hagerman.pdf` (retrieved: February 2, 2011)
14. Neptune Renewable Energy, `http://www.neptunerenewableenergy.com/index.php` (accessed: February 5, 2011)
15. Abstract of Deck Log, Bunga Kasturi 5, January 2010 to December 2010. American Eagle Tankers (AET) Shipmanagement, Malaysia
16. MER: Bunker Prices. Marine Engineers Review, December/January 2011 Issue, p. 13 (2010)

Wind – Solar Hybrid Systems in Tunisia: An Optimization Protocol

Karemt Boubaker[1], Andrea Colantoni[2,*], Leonardo Longo[2],
Simone Di Giacinto[2], Giuseppina Menghini[2], and Paolo Biondi[2]

[1] Unité de physique des dispositifs à semi-conducteurs,
Tunis EL MANAR University, 2092 Tunis, Tunisia
mmbb11112000@yahoo.fr
[2] Department of Agriculture, Forest, Nature and Energy (DAFNE), University of Tuscia,
Via S. Camillo de Lellis snc, 01100 Viterbo, Italy
{colantoni,leonardolongo,s.digiacinto,biondi}@unitus.it

Abstract. In this work, potentials, state-of-the-art and development of hybrid
wind-solar plants in the eastern-North Africa zone have been studied. Since the
use of the renewable energy sources requires an accurate evaluation and planning, an optimization procedure has been adopted: the protocol exploits data -
such as solar radiation and cumulative mean wind speed- which are available
for each node of the grid representing the Tunisian territory. The aim of this paper is to identify several optimal locations which can host a hybrid system
based on solar and wind technologies.

Keywords: Wind-solar hybrid systems, Optimization, Renewable energy.

1 Introduction

Renewable energy sources (RES) play an important role in targeting energy security,
sustainable development and environment preservation. Nowadays, wind and solar
sources have been recognized as the most promising clean technologies because of
the decreasing trend of costs, oppositely to other energies [1-3]. In this scenario the
disadvantages related to the use of fossil fuels (such as environmental impacts) have
forced public administration and stakeholders to move towards RES [30-33]. Solar
and wind are inexhaustible and fuel-free sources which do not cause pollution in
electricity production. They are also able to yield energy with minimal transmission
loss due to the ease of their implementation in both urban and rural areas [4-5].

The eastern North-Africa zone is situated between Europe and Africa and can be
seen as a bridge beyond the Mediterranean sea. This area -if compared with the neigh-
borhood- has a low amount of fossil fuels and lacks of natural gas reserve [6-9]. In the
last three decades, energy consumption has increased parallel to the technological development in the whole region. In this region, most of the electricity is generated by
thermal power plants exploiting fossil fuels and gas. As a consequence, wind and solar
energy can be an alternative to their use and must be investigated very seriously.

---

*Corresponding author.

B. Murgante et al. (Eds.): ICCSA 2013, Part II, LNCS 7972, pp. 220–230, 2013.
© Springer-Verlag Berlin Heidelberg 2013

At the end of 2005 the installed power station capacity in Eastern-North African zone totaled roughly 2,900 GW. The mean annual peak load along this period was approximately 1,890 MW. The contribution of hydroelectric power plants and wind power installations did not exceed 60 MW and 10 MW, respectively. The remaining capacity was wholly accounted by thermal power stations [10]. In this region, the commercial use of wind power for generating electricity is still in its infancy. Since 1980, some small wind energy plants have been used on a decentralized basis, e.g. for the desalination of brackish water or pumping water as part of field irrigation schemes in isolated zones [11]. According to some recent studies [12] wind potential may reach 1,000 MW, with a possibility of implementing 250 MW-wind farms in the northern part of the zone.

The aim of the present paper is to identify several optimal locations which can host a hybrid system based on the use of solar and wind technologies. This goal is success-fully reached by an optimization protocol which includes the construction of a pondered grid on the territory and the Boubaker Polynomial Expansion Scheme (BPES). The study shows that the integrated use of wind and solar energy has a strong potential and requires an accurate planning in order to decrease the dependence on fossil fuels.

2   Material and Methods

2.1   Potentials and Resources of Wind Energy

The Meteorology National Institute (INM) provided 10-year mean records of synoptic observations for several locations in Tunisia [13]. Wind speed has been assumed to be

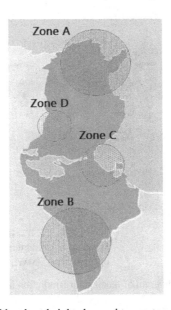

Fig. 1. Global wind speed levels at heights beyond ten meters within four zones [14]

stationary within each month with a maximum of 5.0 m/s at heights beyond ten meters. Records show that wind mean speed in Tunisia varies between 2.0 and 5.0 m/s.
Four major zones (Figure 1) were highlighted: ZA (Bizerte, Tunis, Klebia, Tabarka,
etc.), ZB (Elborma, Remada, etc.), ZC (Gabes, Djerba,   Sfax, Medenine, etc.)   and
ZD (Thala,   Gafsa, Sidi-Bouzid, etc.) according to the relevance of measurements
along with the availability of active meteorological stations.

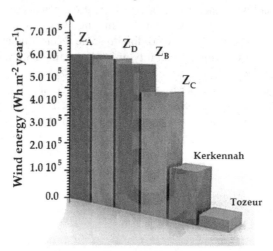

Fig. 2. Cumulative mean annual data for wind energy in the main zones [15]

Figure 2 shows the cumulative mean annual data of wind energy in the above-
described areas. It can be easily seen that zone A has the highest values.

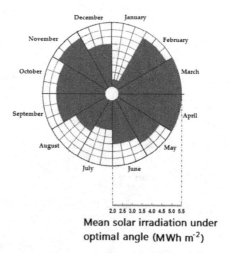

Mean solar irradiation under
optimal angle (MWh m^{-2})

Fig. 3. Rose diagram of mean solar irradiation in Tunisia [15-16]

Nowadays, a single wind farm has been built, in Sidi Daoud (Northen part) near Cap Bon. It has been working since 2000. Average annual wind velocity at this location is 8.4 m/s at 30 m a.g.l. . The project was put out to tender in 1996 on the basis of a feasibility study which was drawn up between 1990 and 1992.

The wind farm is equipped with 32 turbines, 330 kW each. In 2002 it generated 30 GWh of electricity and later in 2003 was expanded by the addition of 12 turbines which reached 8.7 GWh. According to STEG (Société Tunisienne d'Electricité et de Gaz), the estimated cost benefits are up to 26% in terms of avoided fuel costs [14].

2.2 Potentials and Resources of Solar Energy

Tunisia has an important solar potential, boasting mean annual irradiation rates between 1.5 and 5.5 MWh/m^2 (Figure 3).

Such rates encouraged the use of solar panels in commercial and residential installations through important official subsidies and assistance. The unique implemented high scale plant is in El Borma at the Algerian southern frontiers, with a nominal power of 2.1 GW [17].

Solar radiation chart is shown in Figure 4, along with existing plants in 2012. It can be seen that the annual solar radiation is more than 1.7 MWh/m^2 in the whole country and the highest values occur in the southern areas (yellow in the map in Figure 4).

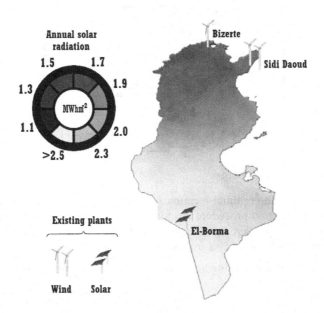

Fig. 4. Existing plants in 2012 along with annual solar radiation repartition [15-16]

2.3   Hybrid Elementary Units Plan

The recent energy plants are aimed at increasing the share of RES from 4% -in terms of total electrical installed power in 2010- to 16% and 40% in 2016 and 2030, respectively. With specific regard to power demand, this target corresponds to the implementation of approximately 5.4 GW throughout the territory. If all the resources and the projected plants (53-90 MW) are taken into account, the amount of annual required installation should be equal to 0.3 GW, which is equivalent to the installation of eight plants per year, for a total period of 20 years.

According to the disposition given in Figures 3 and 4, optimal loci for different kinds of plants do not match along the territory due to resources disparity. This constraint is the main reason for proposing medium-sized hybrid elementary units whose scheme is presented in Figure 5.

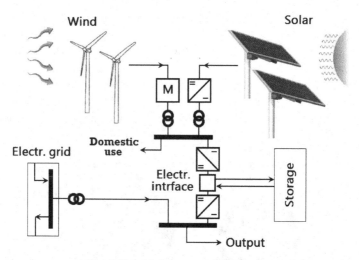

Fig. 5. Hybrid elementary unit scheme

2.4   Optimization Protocol

In order to find out the optimal location for the above described hybrid plant in the territory, an optimization procedure is adopted. The optimization protocol is based on two major items: the pondered grid and the Boubaker Polynomials Expansion Scheme BPES. The pondered grid is constructed through the data given in (2) and (3). Each node in the map is indexed using an integer index $k\big|_{k=1..M_0}$ and then introduced within a continuum coordinate system (Figure 6). M_0 depends on the size of the grid, for the optimization procedure a single mesh is 0.5 km^2 and the numeric value of M_0 is $78 \cdot 10^4$.

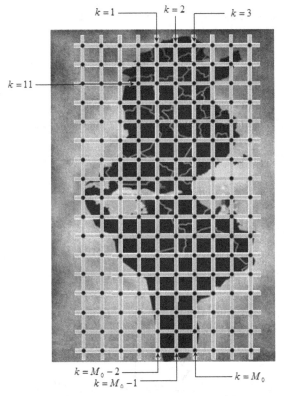

Fig. 6. Optimization protocol grid

A preset number N_0 of aggregates $z_i|_{i=1..N_0}$ affects each point. The positive aggregates involve several parameters including: irradiation level, wind performance, recovery of investment, gain of cost versus conventional energy supply scenario and delivery cost, area availability, installation history and mean costs of different options. For this simulation, the numeric value of N_0 is 11. Each aggregate $z_i|_{i=1..N_0}$ varies inside the range $[z_i^{\min}, z_i^{\max}]$. The range of variation for some aggregates used in the model is shown in Table 1.

Table 1. Range of the aggregates

z_i	Aggregate	Range [%]
z_1	Wind performance	0-100
z_2	Irradiation level	0-100
z_3	Accessibility	0-100
z_4	Mean costs	0-100

In a first step, and for standardizing purposes, each aggregate $z_i|_{i=1..N_0}$ is normalized using (1):

$$\tilde{z}_i = \left[\frac{z - z_i^{min}}{z_i^{max} - z_i^{min}} \right] \tag{1}$$

Consecutively, the optimization is carried out through the Boubaker Polynomials Expansion Scheme BPES [18-29] a standardized weight function set as:

$$\xi(N_0, M_0) = \frac{1}{M_0} \sum_{i=1}^{M_0} \left[\frac{1}{2N_0} \sum_{k=1}^{N_0} \lambda_k \times B_{4k}(r_k \tilde{z}_i) \right] \tag{2}$$

where:

- B_{4k} are the 4k-order Boubaker polynomials;
- r_k are B_{4k} minimal positive roots;
- $\lambda_k|_{k=1..N_0}$ are unknown pondering real coefficients.

The BPES protocol ensures the validity of the optimizing test thanks to Boubaker polynomials first derivatives properties:

$$\begin{cases} \sum_{q=1}^{N} B_{4q}(x) \Bigg|_{x=0} = -2N \neq 0; \\ \sum_{q=1}^{N} B_{4q}(x) \Bigg|_{x=r_q} = 0; \end{cases} \tag{3}$$

$$\begin{cases} \sum_{q=1}^{N} \frac{dB_{4q}(x)}{dx} \Bigg|_{x=0} = 0 \\ \sum_{q=1}^{N} \frac{dB_{4q}(x)}{dx} \Bigg|_{x=r_q} = \sum_{q=1}^{N} H_q \end{cases} \tag{4}$$

$$\text{with}: H_n = B'_{4n}(r_n) = \left(\frac{4r_n[2 - r_n^2] \times \sum_{q=1}^{n} B_{4q}^2(r_n)}{B_{4(n+1)}(r_n)} + 4r_n^3 \right)$$

The BPES solution is obtained through four steps:

- Determining the set of the pondering real coefficients, $\lambda_k|_{k=1..N_0}$ which maximizes the standardized weight function $\xi\ (N_0, M_0)$.

- Testing the rank $i|_{i\in\{1,..M_0\}}$ which gives minimal value of the first derivatives of $\xi\ (N_0, M_0)$.

- Affecting the adequate hybrid plant prototype to the location of this rank and hence eliminating it from the grid.

- Continuing the process after decrementing M_0 and adjusting N_0 at each step if necessary.

3   Results and Discussion

The result of this algorithm consists of an optimized and well scheduled planning in order to implement the net of hybrid elementary units plan under the given constraints (Figure 7). The study –carried out by the useful optimization procedure- leads to the definition of five different locations which can host the hybrid systems. The plants have different capacities (Table 2), from 38 MW (eastern-south Tunisia) to 182 MW (middle part of Tunisia), depending on the wind and solar potential of that area. A different time of creation is provided for each plant, the forecast is that the systems will be completed by 2030 with a total capacity of just over 450 MW.

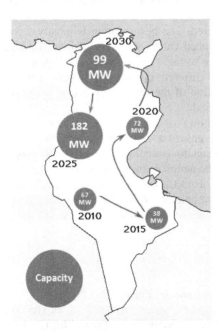

Fig. 7. Result of the algorithm

Table 2. Solar and wind contribution in total capacity

Site	Solar contribution (%)	Wind contribution (%)	Total capacity (MW)
1	45	55	182
2	30	70	99
3	77	23	72
4	70	30	67
5	75	25	38

Tunisia is also looking at the possibility of a transmission line running from Egypt to Morocco, whereby a feasibility study has already begun. A link between Tunisia and Italy is also being considered so that opportunities of commercializing energy can be developed between the two countries.

Transmission network in Tunisia involves 60 HV substations and about 5000 km of HV lines. The interconnection network is connected to Europe through networks in Algeria and Morocco. Interconnection with Libya will help to extend the synchronous zone to Machrek countries, which means that interconnection from Syria through Libya, Egypt and Jordan would take place.

4   Conclusion

The protocol optimization is an important tool to evaluate the possible use of RES – such as hybrid systems based on solar and wind technologies- and to identify the optimal locations of these plants. In order to reach this goal, the potential of wind and solar energy is considered. The best solution is mostly a compromise between solar radiation, mean wind speed values, recovery of investment, gain of cost versus conventional energy supply scenario and delivery cost, area availability, installation history and mean costs of different options. By means of the simulation several sites - suitable for the realization of the wind-solar hybrid systems with a total capacity of 450 MW – were identified.

The present paper shows an opportunity to reduce the dependence on fossil fuels and, as a consequence, ensure a clean and sustainable future in agreement with the energy policy of Tunisia to encourage the production of energy from renewable sources and to reduce the national energy consumption.

Nevertheless, some aspects have to be considered: the development of hybrid plants need incentives to compete with fossil fuels, which still covers most of energy demand. According to that, research can be addressed to the analysis of the best available technologies.

References

1. Bahri, Y.: Revue Tunisienne de l'Energie 24, 37 (1991),
   http://www.industrie.gov.tn/fr/doc.asp?mcat=20&mrub=91
2. Hadj Sassi, B., Gattoufi, B.: Revue Tunisienne de l'Energie 14, 53 (1988),
   http://www.industrie.gov.tn/fr/doc.asp?mcat=20&mrub=91

3. Societe Tunisienne de l'Electricite et du Gaz (STEG): Large scale integration of solar and wind power in Mediterranean countries. MED 2010 Project, Contrat ENK5-CT-2000-00307, Direction of the Studies and Planning, Tunisia, http://www.umc.edu.dz/vf/proceeding/sigcle2010/pages/themes/ energies_renouvelables/SessionV/043ElAmouriSess5.pdf

4. Czisch, G.: Scenarios for a Future Electricity Supply: cost-optimized variations on supplying Europe and its neighbours with electricity from renewable energies. The Institution of Engineering and Technology (IET), 640 (2011)

5. Czisch, G.: Low Cost but Totally Renewable Electricity Supply for a Huge Supply Area, - a European/Trans-European Example. University of Kassel, Bath, UK, Claverton Energy Conference (2008)

6. Khemiri, A., Hassairi, M.: Development of energy efficiency improvement in the Tunisian hotel sector: a case study. Renewable Energy 30, 903–911 (2005)

7. Trieb, F., Müller-Steinhagen, H., Kernb, J.: Financing concentrating solar power in the Middle East and North Africa-Subsidy or investment? Energy Policy 39, 307–317 (2011)

8. Trieb, F., Müller-Steinhagen, H.: Concentrating solar power for seawater desalination in the Middle East and North Africa. Desalination 220, 165–183 (2008)

9. Trieb, F., Nitsch, J.: Recommendations for the market introduction of solar thermal power stations. Renewable Energy 14, 17–22 (1998)

10. Marucci, A., Monarca, D., Cecchini, M., Colantoni, A., Manzo, A., Cappuccini, A.: The Semitransparent Photovoltaic Films for Mediterranean Greenhouse: A New Sustainable Technology. Mathematical Problems in Engineering, art. no. 451934 (2012)

11. Bourouni, K., Chaibi, M.: Solar energy for application to desalination in Tunisia: description of a demonstration project. Renewable Energy in the Middle East, 125–149 (2009)

12. Elamouri, M., Ben Amar, F., Trabelsi, A.: Vertical characterization of the wind mode and its effect on the wind farm profitability of Sidi Daoud-Tunisia. Energy Conversion and Management 52, 1539–1549 (2011)

13. Brand, B., Zingerle, J.: The renewable energy targets of the Maghreb countries: Impact on electricity supply and conventional power markets. Energy Policy 39, 4411–4419 (2011)

14. Elamouri, M., Ben Amar, F.: Wind energy potential in Tunisia. Renewable Energy 33, 758–768 (2008)

15. Ghezloun, A., Oucher, N., Chergui, S.: Energy policy in the context of sustainable development: Case of Algeria and Tunisia. Energy Procedia 18, 53–60 (2012)

16. Khalfallah, E.: Les énergies renouvelables en Tunisie: enjeux et perspectives. In: Energie-Francophonie 71- deuxième trimestre (2006)

17. De Oliveira, W.S., Fernandes, A.J.: Global Wind Energy Market, Industry and Economic Impacts. Energy and Environment Research 2, 79–97 (2012)

18. Ghanouchi, J., Labiadh, H., Boubaker, K.: An attempt to solve the heat transfer equation in a model of pyrolysis spray using 4q-order m-Boubaker polynomials. Int. J. of Heat and Technology 26, 49–53 (2008)

19. Slama, S., Bessrour, J., Boubaker, K., Bouhafs, M.: A dynamical model for investigation of A3 point maximal spatial evolution during resistance spot welding using Boubaker polynomials. Eur. Phys. J. Appl. Phys. 44, 317–322 (2008)

20. Slama, S., Bouhafs, M., Ben Mahmoud, K.B.: A Boubaker Polynomials Solution to Heat Equation for Monitoring A3 Point Evolution During Resistance Spot Welding. International Journal of Heat and Technology 26, 141–146 (2008)

21. Lazzez, S., Ben Mahmoud, K.B., Abroug, S., Saadallah, F., Amlouk, M.: A Boubaker polynomials expansion scheme (BPES)-related protocol for measuring sprayed thin films thermal characteristics. Current Applied Physics 9, 1129–1133 (2009)

22. Barry, P., Hennessy, A.: Meixner-Type results for Riordan arrays and associated integer sequences, section 6: The Boubaker polynomials. Journal of Integer Sequences 13, 1–34 (2010)

23. Agida, M., Kumar, A.S.: A Boubaker Polynomials Expansion Scheme solution to random Love equation in the case of a rational kernel. El. Journal of Theoretical Physics 7, 319–326 (2010)

24. Yildirim, A., Mohyud-Din, S.T., Zhang, D.H.: Analytical solutions to the pulsed Klein-Gordon equation using Modified Variational Iteration Method (MVIM) and Boubaker Polynomials Expansion Scheme (BPES). Computers and Mathematics with Applications 59, 2473–2477 (2010)

25. Kumar, A.S.: An analytical solution to applied mathematics-related Love's equation using the Boubaker Polynomials Expansion Scheme. Journal of the Franklin Institute 347, 1755–1761 (2010)

26. Fridjine, S., Amlouk, M.: A new parameter: An ABACUS for optimizig functional materials using the Boubaker polynomials expansion scheme. Modern Phys. Lett. B 23, 2179–2182 (2009)

27. Milgram, A.: The stability of the Boubaker polynomials expansion scheme (BPES)-based solution to Lotka-Volterra problem. J. of Theoretical Biology 271, 157–158 (2011)

28. Benhaliliba, M., Benouis, C.E., Boubaker, K., Amlouk, M., Amlouk, A.: A New Guide To Thermally Optimized Doped Oxides Monolayer Spray-grown Solar Cells: The Amlouk-boubaker Optothermal Expansivity ψab in the book. Solar Cells - New Aspects and Solutions, 27–41 (2011)

29. Rahmanov, H.: A Solution to the non Linear Korteweg-De-Vries Equation in the Particular Case Dispersion-Adsorption Problem in Porous Media Using the Spectral Boubaker Polynomials Expansion Scheme (BPES). Studies in Nonlinear Sciences 2, 46–49 (2011)

30. Villarini, M., Limiti, M., Abenavoli, R.I.: Overview and Comparison of Global Concentrating Solar Power Incentives Schemes by Means of Computational Models. In: Murgante, B., Gervasi, O., Iglesias, A., Taniar, D., Apduhan, B.O. (eds.) ICCSA 2011, Part IV. LNCS, vol. 6785, pp. 258–269. Springer, Heidelberg (2011)

31. Bocci, E., Villarini, M., Bove, L., Esposto, S., Gasperini, V.: Modelling Small Scale Solar Powered ORC Unit for Standalone Application. Mathematical Problems in Engineering, art. no. 124280 (2012)

32. Cocchi, S., Castellucci, S., Tucci, A.: Modeling of an Air Conditioning System with Geothermal Heat Pump for a Residential Building. Mathematical Problems in Engineering, art. no. 781231 (2013)

33. Carlini, M., Castellucci, S.: Modelling the vertical heat exchanger in thermal basin. In: Murgante, B., Gervasi, O., Iglesias, A., Taniar, D., Apduhan, B.O. (eds.) ICCSA 2011, Part IV. LNCS, vol. 6785, pp. 277–286. Springer, Heidelberg (2011)

Use of Semi-transparent Photovoltaic Films as Shadowing Systems in Mediterranean Greenhouses

Alvaro Marucci, Danilo Monarca, Massimo Cecchini,
Andrea Colantoni, Elena Allegrini, and Andrea Cappuccini

DAFNE, University of Tuscia, Via San Camillo de Lellis s.n.c., 01100 Viterbo, Italy
{marucci,colantoni,monarca,cecchini,
cappuccini,allegrini}@unitus.it

Abstract. In Mediterranean greenhouses, active and passive cooling systems are almost always needed due to high values of solar radiation (nearly 1000 Wm^{-2}) -especially during summer season- and high values of air temperature (near 40°C). Nevertheless, the use of the above-mentioned systems imply the increase of the operating costs of greenhouses.

The aim of the present study is to investigate the possible use of semi-transparent photovoltaic covers as shadowing systems in Mediterranean greenhouses. In order to reach this goal, the energy transfers have been calculated and the energy balance for a greenhouse has been determined. More precisely, three cases have been taken into account: traditional cover with EVA (Ethylene vinyl acetate)film, cover with double film, namely EVA for the internal part and polyethylene for the external one, and cover with double film, namely EVA for the internal part and semi-transparent photovoltaic film for the external one.

Keywords: greenhouse, photovoltaic film, shading, renewable sources.

1 Introduction

The use of plastic film covers has greatly contributed to the development of the greenhouses in Italy and at the same time to the generation of a new class of greenhouses, different from those of Central and Northern Europe, the so called "Mediterranean Greenhouse". It is characterized by a certain structural simplicity, cheap covers and lack of fixed winter conditioning systems as the available solar energy is-broadly speaking-more than sufficient to meet the energy requirements [1].

During the hot season, however, some problems occur within the greenhouse due to intense solar radiation and high air temperature too. In order to overcome this situation, cooling systems – namely active and passive solutions- have to be taken into account.

Choosing cooling systems strongly system depends on many factors e.g. climate, technology and available resources [2]. Growers in cold climate countries, with advanced technology, abundant water and low solar radiation have generally moved towards the use of active methods of cooling, such as mechanic ventilation and water evaporation [3-5]. On the other hand, in countries with warmer climates, less

B. Murgante et al. (Eds.): ICCSA 2013, Part II, LNCS 7972, pp. 231–241, 2013.
© Springer-Verlag Berlin Heidelberg 2013

sophisticated technologies, scarcer water supplies and higher solar radiation, passive methods of cooling dominate, as they require little initial investment, have lower energy costs and use less water.

In the latter case, the conversion of solar energy into electric energy by photovoltaic effect could be a viable solution allowing you to take advantage of the solar energy excess with economic and productive benefits.

The traditional photovoltaic silicon-based panels are not transparent and do not let the solar radiation penetrate inside the greenhouse. Thus, plants cultivation becomes problematic and it is difficult to create greenhouse effect necessary to develop the microclimatic conditions for crops.

In recent years, research has focused the attention on the production of photovoltaic films [6-8].

In some periods of the year, in Mediterranean areas, the solar radiation exceeds the needs of the crops cultivated in greenhouses [9]. This has led to consider the use of these semi-transparent photovoltaic films not only for energy production but as a passive cooling system too.

This application would ensure the grower an advantage economically by selling electricity and in terms of production since energy and thermal excess –harmful to plants cultivation in greenhouse- would be reduced.

The use of the double covering in greenhouses increases the cover opacity to the long wave infrared radiation and the cost of the investment and moreover leads to decrease transparency at Photosynthetically Active Radiation (PAR), generating production problems. When a transparent layer is added to the cover, at least the 10% of the light is lost. This value varies depending on material and angle of incidence [10].

The aim of the present work is to evaluate the possible use of innovative semi-transparent photovoltaic films as shading systems for Mediterranean greenhouses. In order to reach this goal, the inputs and outputs of energy have been calculated and the energy balance for a greenhouse has been determined. More precisely, three cases have been taken into account: traditional cover with EVA film, cover with double film, namely EVA for the internal part and polyethylene for the external one and cover with double film, namely EVA for the internal part and semi-transparent photovoltaic film for the external one.

Climatic conditions of Viterbo are taken into account.

The energy balance has lead to evaluate the possible use of semi-transparent photovoltaic films as passive cooling systems and for energy production from Renewable Energy Sources (RES).

2   Materials and Methods

The present research has been carried out with specific regard to a greenhouse assuming different types of cover materials:

EVA (Ethylenevinylacetate)
Double covering: EVA-Polyethylene

Double covering: EVA- semi-transparent photovoltaic film.
Here is the size of the greenhouse:
Length = 50 m
Width = 12 m
Eaves height = 3.5 m
Slope of the pitches = 22°
Thickness of the films = 0.18 mm

The external climatic conditions are referred to the Educational and Experimental Farm "Nello Lupori" of the University of Tuscia in Viterbo (λ = 42°25'7"68 N; φ = 12°6'34"20 E; Altitude= 326 m). Climatic data were obtained from elaboration of historical series of data (2000-2012) measured by sensors and stored by a data acquisition system with data logger Campbell CR 10. In order to evaluate the effect of different covers on the internal greenhouse microclimate, the energy balance - between input (i.e. solar radiation) and energy losses due to trasmission, ventilation and radiation on an average clear day of each month- has to be calculated [13, 14]. Furthermore, the energy balance is related to the occurring energy flows, as shown in figure 1.

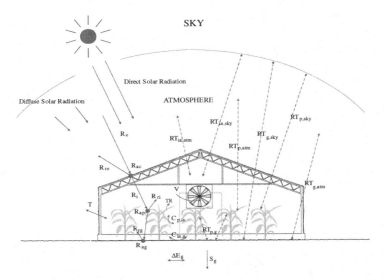

Fig. 1. Energy flows in greenhouse

The energy balance equation is given by (1):

$$R_i = RT_{g,sky} + RT_{g,atm} + RT_{p,sky} + RT_{p,atm} + RT_{ia,sky} + RT_{ia,atm} + T + V + \Delta E_g + S_g \qquad (1)$$

Nomenclature
Alphabetic symbols

R = global solar radiation [MJ m^{-2}d^{-1}];
RT = energy lost by radiation [MJ m^{-2}d^{-1}];
C = energy exchanged by convection [MJ m^{-2}d^{-1}];
T = energy lost by transmission [MJ m^{-2}d^{-1}];
V = energy lost by ventilation [MJ m^{-2}d^{-1}];
TR = plant transpiration [MJ m^{-2}d^{-1}];
S = energy transferred by conduction in the deep layers [MJ m^{-2}d^{-1}];
E = thermal storage [MJ m^{-2}d^{-1}].

Subscripts

e = external;
re = reflected external;
i = internal;
ri = reflected internal;
ap = absorbed by the plants;
rg = reflected from the ground;
ag = absorbed by the ground;
ac = absorbed by the cover;
p = plants;
g = ground;
ia = indoor air;
atm = atmosphere;

The global solar radiation (direct and diffuse) inside the greenhouse in clear sky conditions was calculated using the equation (2):

$$R_i = R_e \left[\tau_b \left(1 + C_d\right) + \tau_d \right] \cos \theta_i \cdot \tau_c \qquad (2)$$

Where:

$$R_e = 1367 \left[1 + 0.033 \cos \left(\frac{360 \cdot n}{365} \right) \right]$$

n = julian day;

$\cos \theta_i = \mathrm{sen}\lambda \mathrm{sen}\delta \cos \theta_z - \cos\lambda \mathrm{sen}\delta \mathrm{sen}\theta_z \cos \psi + \cos\lambda \cos\delta \cos\omega\cos\theta_z +$
$+ \mathrm{sen}\lambda \cos\delta \cos\omega\mathrm{sen}\theta_z \cos \psi + \cos\delta \mathrm{sen}\omega\mathrm{sen}\theta_z \mathrm{sen} \psi$

$\cos\theta_z = \mathrm{sen}\phi \, \mathrm{sen}\delta + \cos\phi \, \cos\delta \, \cos\omega;$

ϕ = latitude (°);

$$\delta = 23.45 \cdot \mathrm{sen} \cdot \left[\frac{360}{365} \cdot (284 + n) \right]$$

(°);

$$\omega = \frac{360}{24} \cdot (12 - h)$$
$(°)$;

ψ = angle between the horizontal projection of the surface normal and the direction south $(°)$

h = hour days   (h);

A = altitude   (km);

τ_b = transmissivity of the atmosphere to the direct radiation= $A_0 + A_1 \times e^{\frac{-k}{\cos\theta_z}}$;

τ_d = transmissivity of the atmosphere to the diffuse radiation = $0.271 - 0.294\, \tau_b$;

$A0 = [0.4237 - 0.00821\ (6-A)2]\ [1 + 0.03\ \text{sen}\ (\pi\frac{91+n}{182})]$;

$A_1 = [0.5055 + 0.00595\ (6.5-A)^2]\ [1 + 0.01\ \text{sen}\ (\pi\frac{91+n}{182})]$;

$K = [0.2711 + 0.01858\ (2.5-A)^2]\ [1.01 - 0.01\ \text{sen}\ (\pi\frac{91+n}{182})]$.

C_d = Diffusivity coefficient of the plastic film cover indicating the direct radiation that is converted into diffuse one in the internal environment from the cover).

Trasmissivity (τ_c) of the cover and the walls to solar radiation was determined using the model described by Marucci et al. [14].

The energy lost by transmission from walls and the cover was calculated with the relation (3):

$$T = K\ S\ (T_i - T_e)$$ (3)

K = thermal trasmittance (W m^{-2} K^{-1});

S = surface (m^2);

T_i = internal temperature (K);

T_e = external temperature (K).

The energy lost by ventilation was calculated with the relation (4):

$$E_v = V\ (H_i - H_e)$$ (4)

E_v = energy lost by ventilation (KJ h^{-1});

V= flow rate of ventilation (KgDryAir h^{-1});

H_i = internal air enthalpy (KJ KgDryAir^{-1});

H_e = external air enthalpy (KJ KgDryAir^{-1}).

The enthalpy H (KJ KgDryAir^{-1}) of a Kg of air, at temperature t (°C) and with a water content, in the form of water, equal to x (Kg KgDryAir^{-1}) is obtained from (5):

$$H = 1.005\ t + x\ (2499.5 + 2.005\ t)$$ (5)

x (Kg KgDryAir^{-1}) represents the water vapor contained in the air and can be determined through the psychrometric diagram or with the following formula (6):

$$x = 0.6215 \frac{P_{H_2O}}{p - P_{H_2O}} \tag{6}$$

$p = 1.013$ (Kg cm^{-2}) atmospheric pressure;

P_{H_2O} = vapor pressure of the air, expressed in Kg cm^{-2} at the temperature T (K) and at relative humidity UR (%) (7);

$$P_{H_2O} = 1.41 \times 10^{10} \, e^{\left(\frac{-3928.5}{T-41.5}\right)} \times UR \tag{7}$$

The energy lost by radiation was calculated with the relation (8):

$$RT_{12} = \sigma \, \varepsilon_{1,2} F(T_1^4 - T_2^4) \tag{8}$$

RT_{12} = Energy lost by radiation (W m^{-2});

σ = Stefan–Boltzmann constant (W m^{-2} K^{-4});

$\varepsilon_{1,2}$ = emissivity;

F = view factor;

T = absolute temperature (K) [11-13].

The terms in the energy balance equation lead to gain useful information on the possible use of semi-transparent photovoltaic films to convert solar radiation surplus into electric energy. This becomes even more important in those areas where intense solar radiation occurs.

3   Results and Discussion

If solar radiation inside the greenhouse and energy losses in clear sky conditions are compared considering different covering materials, a good agreement between inputs and outputs emerges.

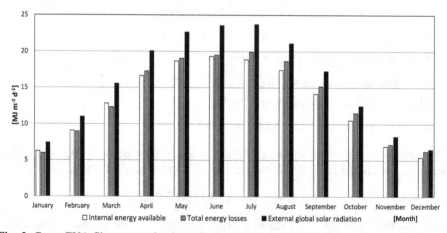

Fig. 2. Cover EVA film: external solar radiation, internal solar radiation and energy losses in clear sky conditions

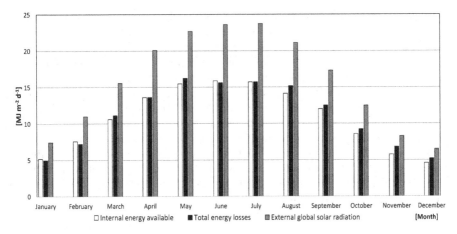

Fig. 3. Double covering EVA-Polyethylene: external solar radiation, internal solar radiation and energy losses in clear sky conditions

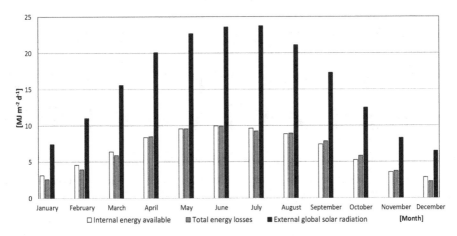

Fig. 4. Double covering EVA- photovoltaic film: external solar radiation, internal solar radiation and energy losses in clear sky conditions

The internal solar radiation –available for plants growth- strongly varies depending on the greenhouse cover material. Considering the EVA cover and the specific location of the present study, the internal solar radiation is approximately equal to 20 MJ $m^{-2}d^{-1}$ in the period May-August: this value greatly exceeds the needs of the most demanding horticultural crops (es. tomato, *Solanum lycopersicum*) and forces the use of use active and passive cooling systems with very high energy consumption.

In the case of the double covering (EVA-Polyethylene) (Fig. 3) the inside solar radiation available for the plants and the related energy losses decrease by about 25%. However it is still necessary to resort to cooling systems requiring less energy than the previous case.

Instead of dissipating the energy surplus with cooling systems and the consequent costs, Photovoltaic elements can be used to convert the solar radiation surplus into electric energy and, moreover, to shade the greenhouse itself.

Photovoltaic panels are already used for this purpose but hardly ensures partial and uniform shade which is extremely important for plant growth and productivity.

The semi-transparent PV elements have recently widespread: they let the solar radiation – needed by plants- penetrate inside the greenhouse. Considering that plants require almost 50% of the external solar radiation during warm periods, the remaining part can be used to produce electricity.

In the event combined with an EVA film a semi-transparent photovoltaic film that passes 50% [1] of the incident solar radiation (Fig. 4), the internal solar radiation, from May to August, reaching values slightly less than 10 MJ $m^{-2}d^{-1}$ which should be enough for a normal growing major vegetable crops.

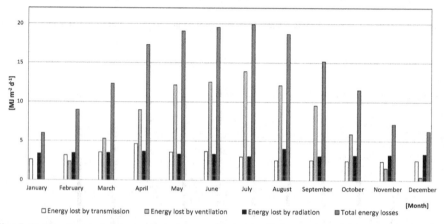

Fig. 5. Energy losses due to transmission, ventilation, radiation and total for greenhouse covered with EVA film

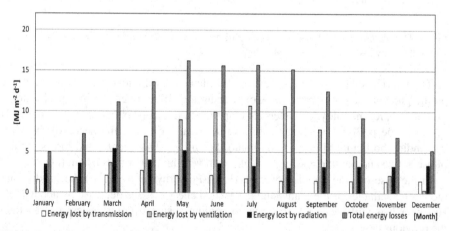

Fig. 6. Energy losses due to transmission, ventilation, radiation and total for greenhouse with double covering (EVA- Polyethylene)

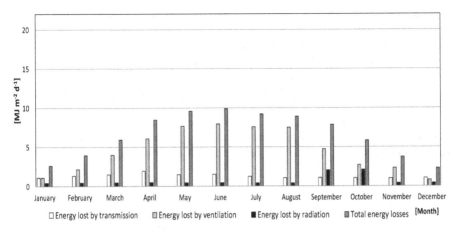

Fig. 7. Energy losses due to transmission, ventilation, radiation and total for greenhouse with double covering (EVA-photovoltaic film)

In case of totally or partially cloudy sky, photovoltaic shading films should be removed and, moreover, in different climatic conditions, transparency of PV films should change.

Energy losses have to be taken into account in the three above-described greenhouses since they strongly vary, especially those due to irradiation. Actually PV semi-transparent films are almost opaque in the long wave infrared: in the case of the latter greenhouse cover, indeed, the losses due to radiation represent only 12-15% of those in EVA covers.

With the latter greenhouse cover, in fact, the losses by radiation are just 12-15% of those with EVA cover.

In the greenhouse double cover EVA-Polyethylene, however, since the latter is slightly opaque in the long wave infrared, the radiation losses are similar to those under EVA.

The losses by transmission from the walls and the cover significantly decrease in both cases if double covering is due to the drastic reduction in the coefficient of thermal transmittance.

The energy lost by transmission from the greenhouse with double covering EVA Polyethylene is about 60% of that with only EVA film. The greenhouse double cover EVA- photovoltaic film shrunk down to about 42% of the EVA film: this is due to the low thermal transmittance of this covering material.

The energy lost by ventilation decreases gradually passing from single to double film with polyethylene. Lower values will occur with the shading photovoltaic films. The greenhouse double cover EVA-polyethylene reduces energy losses due to ventilation less than 80% EVA film, while the greenhouse double cover EVA-photovoltaic film these energy losses amount to about 65% of those with single greenhouse cover.

The above-described considerations are referred to completely clear sky and to a specific location.

The performed experiments lead to interesting perspectives on the possible use of semi-transparent PV elements as shading systems for greenhouses and as an efficient solution to convert solar radiation surplus into electricity, especially during summer and in warmer areas.

If plants needs are analyzed during the whole year, the solar energy excess –with specific regard to the chosen location- can be exactly determined and, as a result, the transparency of the PV films can be defined.

The choice of crops can be even optimized to increase the efficiency to a maximum value.

In addition to the need of increasing conversion efficiency of semi-transparent PV elements –which nowadays is approximately 2%- another problem has to faced: films anchoring and handling must not damage efficiency and duration.

4   Conclusions

The present paper focuses the attention on the possible use of innovative semi-transparent PV films for shading greenhouses and to produce electricity by converting the solar radiation surplus which exceeds crop needs. The research has been carried out with regard to different types of greenhouse covers: EVA film, double film EVA-Polyethylene and double film EVA-photovoltaic film.

Considering the climatic conditions of the Mediterranean areas and with specific regard to clear skies, the results show that a large amount of energy surplus occurs if compared with the crop needs. Thus, the solar radiation excess can be converted into electricity by PV elements. However, the development of flexible and semitransparent PV elements is needed.

References

1. Marucci, A., Gusman, A., Pagniello, B., Cappuccini, A.: Limits and prospects of photovoltaic covers in mediterranean greenhouse. Journal of Agricultural Engineering XLIV(e1), 1–8 (2013) (in press)
2. Garcıa, M.L., Medrano, E., Sanchez-Guerrero, M.C., Lorenzo, P.: Climatic effects of two cooling systems in greenhouses in the Mediterranean area: External mobile shading and fog system. Biosystems Engineering 108, 133–143 (2011)
3. Bailey, B.J.: The environment in evaporatively cooled greenhouses. Acta Horticulturae 287, 59–66 (1990)
4. Boulard, T., Baille, A.: A simple greenhouse climate control model incorporating effects of ventilation and evaporative cooling. Agricultural and Forest Meteorology 65, 145–157 (1993)
5. Jain, D., Tiwari, G.N.: Modeling and optimal design of evaporative cooling system in controlled environment greenhouse. Energy Conversion and Management 43, 2235–2250 (2002)
6. Hua, H., Kung, S.-C., Yang, L.-M., Nicho, M.E., Penner, R.M.: Photovoltaic devices based on electrochemical–chemical deposited CdS and poly3-octylthiophene thin films. Solar Energy Materials & Solar Cells 93, 51–54 (2009)
7. Shin, G.H., Allen, C.G., Potter Jr., B.G.: RF-sputtered Ge–ITO nanocomposite thin films for photovoltaic applications. Solar Energy Materials & Solar Cells 94, 797–802 (2010)

8. Zhou, Y., Taima, T., Shibata, Y., Miyadera, T., Yamanari, T., Yoshida, Y.: Controlled growth of dibenzotetraphenylperiflanthene thin films by varying substrate temperature for photovoltaic applications. Solar Energy Materials & Solar Cells 95, 2861–2866 (2011)
9. Marucci, A., Monarca, D., Cecchini, M., Colantoni A., Cappuccini, A.: The semitransparent photovoltaic films for mediterranean greenhouse: A new sustainable technology. Mathematical Problems in Engineering, 14 pages (2012)
10. Hurd, R.G.: Energy saving techniques in greenhouses and their effects on the tomato crops. Scientia Horticulturae 33, 94–101 (1983)
11. Campiglia, E., Colla, G., Mancinelli, R., Rouphael, Y., Marucci, A.: Energy balance of intensive vegetable cropping systems in central Italy. Acta Horticulturae 747, 185–191 (2007)
12. Marucci, A., Campiglia, E., Colla, G., Pagniello, P.: Environmental impact of fertilization and pesticide application in vegetable cropping systems under greenhouse and open field conditions. Journal of Food, Agriculture and Environment 9(3-4), 840–846 (2011)
13. Marucci, A., Pagniello, B.: Simulation of the growth and the production of the tomato in typical greenhouses of the Mediterranean environment. Journal of Food, Agriculture and Environment 9(3-4), 407–411 (2011)
14. Marucci, A., Carlini, M., Castellucci, S., Cappuccini, A.: Energy Efficiency of a Greenhouse for the Conservation of Forestry Biodiversity. Mathematical Problems in Engineering, 7 pages (2013)

Waste Wood Biomass Arising
from Pruning of Urban Green in Viterbo Town:
Energy Characterization and Potential Uses

Maurizio Carlini[1], Sonia Castellucci[2], Silvia Cocchi[1], and Alberto Manzo[3]

[1] DAFNE, University of Tuscia, Viterbo, Italy
{maurizio.carlini,silvia.cocchi}@unitus.it
[2] CIRDER, University of Tuscia, Viterbo, Italy
sonia.castellucci@unitus.it
[3] Ministry of Agriculture, Food and Forestry, Rome, Italy
a.manzo@mpaaf.gov.it

Abstract. The use of residual biomass arising from urban green pruning for energy production is a current and interesting subject for three main reasons: to achieve the aims of the Kyoto Protocol, to reduce the reliance on fossil fuels and to manage the urban green in a sustainable way.

The aim of this study is a qualitative and quantitative analysis of waste wood biomass from urban green pruning in Viterbo in order to use it for energy production. Moisture, ash, C, H, N contents and Calorific Value analysis are carried out. Starting from the results analysis, two energy uses are proposed: wood-chips boiler and gasification.

Keywords: biomass, waste wood biomass, energy characterization, wood-chips boiler.

1 Introduction

Renewable Energy Sources (RES) have lately become a priority for many countries because of the need to reduce the greenhouse gases emissions and consequently the global warming problem [1-3]. Thus, they can be seen as a positive alternatives to the use of fossil fuels and their derivatives [4]. Biomass encompasses all organic matter of vegetable or animal origin and represents a promising RES. More precisely biomass from plant is a product of photosynthesis, and can be considered carbon neutral since the amount of the released carbon, during its energetic conversion process, is similar to the quantity absorbed during its life time [5].

Biomass is more equally distributed over earth's surface than fossil fuels and therefore allows to diversify and decentralize the energy supply [6].

Biomass can be classified as follows [7]:

- agricultural and forestry residues,
- herbaceous crops,

B. Murgante et al. (Eds.): ICCSA 2013, Part II, LNCS 7972, pp. 242–255, 2013.

- aquatic and marine biomass,
- organic wastes (municipal solid wastes, municipal sewage sludge, animal wastes, etc.).

In agriculture a large quantity of residual and not used biomass– e.g. pruning of fruit trees, grapevines, olive trees and straw - remains [8-9]. Towns are characterized by the generation of an amount of residual biomass which can be exploited, such as wastes arising from urban green pruning.

The use of residual biomass for energy production is becoming an extremely current subject [10]. In particular, the use of waste biomass arising from urban green pruning leads to extremely positive advantages:

- to achieve the aims of the Kyoto Protocol,
- to reduce the reliance on fossil fuels,
- to manage the urban green in a sustainable way,
- to reduce the environmental impact of waste disposal,
- to reduce costs.

In order to produce energy, biomass can be exploited by the use of several processes, namely thermo-chemical conversion, bio-chemical conversion, mechanical extraction. Direct combustion, gasification technology and pyrolysis belong to the first category as shown in figure 1. Anaerobic digestion, alcoholic fermentation and aerobic digestion are included in the second group, while biodiesel production comes from the last process conversion [11].

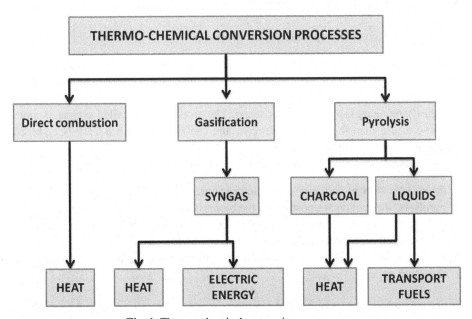

Fig. 1. Thermo-chemical conversion processes

The most suitable use for each biomass strongly depends on biomass properties: calorific value, C/N ratio, moisture, ashes, and volatile matter contents [12]; in particular, C/N ratio and moisture content play a fundamental role in order to define the most appropriate conversion technology. Broadly speaking waste biomass arising from urban green pruning consists of wood and leaves, and thus can be successfully exploited in thermo-chemical processes.

In the present work the amount of the biomass coming from urban green pruning in Viterbo has been estimated and some samples have been analyzed in order to determine their characteristic properties. Two different scenarios can be taken into account and further investigated:

- energy conversion in a wood-chips boiler for heat production (namely combustion),
- use in a gasification plant for heat and electricity production.

At the end of the paper, an economical assessment is carried out in order to choose the most appropriate solution among the above-mentioned possibilities.

Combustion includes a series of chemical reactions (1) which mostly leads to CO_2 and H_2O formation [13]:

$$C_{42}H_{60}O_{28} + 43O_2 \rightarrow 42CO_2 + 30H_2O \tag{1}$$

Heat represents the main product. These steps can be distinguished during the process:

- drying (moisture evaporation at low temperature),
- pyrolysis (thermal degradation of the biomass with formation of tars, chars and gases),
- combustion (complete oxidation of the biomass).

Gasification technology consists of a partial thermal oxidation producing [14-15]: gases (CO_2, water, CO, H_2 and gaseous hydrocarbons), chars (small quantities), ashes, tars and oils.

The gasifying agent is air, oxygen or steam. The produced gas, called syngas, is more versatile to use than the original biomass, and can be used with a CHP (Combined Heat and Power) system [16].

The phases of gasification process are the following [17]:

- drying (moisture evaporation at low temperature),
- pyrolysis (thermal decomposition of the biomass in absence of oxygen or air),
- oxidation (reaction between solid carbonized biomass and oxygen; a large amount of heat is released with oxidation of carbon and hydrogen),
- reduction (in substoichiometric presence of oxygen, some endothermic reactions of reduction occur).

2 Materials and Methods

The data concerning the urban green in Viterbo were collected and later processed starting from urban green census. For each species, an evaluation of the amount of the biomass from pruning was carried out for a period of 5 years.

In Viterbo, linden trees (*Tilia cordata*) are widespread and nearly represent the 50% of total plants, and form the tree-lined road of the town together with plane trees (*Platanus*). Every year the pruning is necessary to prevent problems especially on roads.

Table 1 shows the amount of the various species of the biomass for a period of 5 years: the total amount of yearly biomass from pruning is equal to 625 200 kg/year [18].

Biomass needs to be minced into small pieces for its use in gasification or boiler, a wood-cheaper can be used.

Table 1. Trees of urban green in Viterbo and evaluation of the biomass quantity for a period of 5 years

Scientific name	Number of plants	[%]	Biomass [kg]	[%]
Tilia cordata	1066	48.21	1 599 000	51.15
Platanus spp.	265	11.99	530 000	16.95
Robinia pseudoacacia	153	6.92	153 000	4.89
Ulmus spp.	121	5.47	121 000	3.87
Cercis siliquastrum	112	5.07	112 000	3.58
Pinus pinea	103	4.66	154 500	4.94
Ligustrum lucidum	75	3.39	37 500	1.20
Acer negundo	55	2.49	82 500	2.64
Prunus cerasifera	53	2.40	79 500	2.54
Populus nigra	52	2.35	78 000	2.50
Quercus spp.	51	2.31	51 000	1.63
Ailantus altissima	32	1.45	32 000	1.02
Quercus ilex	25	1.13	37 500	1.20
Prunus cerasus	12	0.54	18 000	0.58
Lagerstroemia spp.	9	0.41	9 000	0.29
Cedrus spp.	8	0.36	8 000	0.26
Acer spp.	4	0.18	6 000	0.19
Cupressus semprevirens	3	0.14	1 500	0.05
Olea europea	3	0.14	4 500	0.14
Paulownia tomentosa	3	0.14	4 500	0.14
Acer platanoides	2	0.09	3 000	0.10
Ilex aquifolium	1	0.05	500	0.02
Laurus nobilis	1	0.05	1 500	0.05
Fraxinus spp.	1	0.05	1 500	0.05
Mespilus germanica	1	0.05	500	0.02
TOTAL	**2211**		**3 126 000**	

Table 2. Amount of available biomass from pruning of urban green

Biomass		
Annual total biomass	625 200	kg/year
Annual total biomass after wood chipper phase	606 444	kg/year

A loss of 3% was taken into account to avoid overestimating the real available quantity after the wood chipper phase (Table 2) [19].

Moreover, it has to be underlined that the storage phase determines a weight reduction too in the biomass but is due to moisture loss. This becomes even more useful if the energy conversion process is considered since a specific low moisture content is required.

2.1 Biomass Characterization

In order to determine the most suitable conversion process for a particular biomass, characteristic properties of the biomass are needed, and the following ones were determined in the laboratory of CIRDER (University of Tuscia):

- moisture content (on dry and wet basis),
- ash content,
- High and Low Calorific Value (HCV, LCV),
- C, H, N contents.

Sampling was the first phase of this study, where waste wood biomass arising from pruning of urban green were taken according to EN 14778:2011 [20]. The samples were prepared according to EN 14780:2011 [21].

Leco CHN-2000 machinery was used in order to analyze the content of carbon, hydrogen and nitrogen. All the tests were carried out in accordance to the technical standard EN 15104:2011 [22].

The moisture content was determined following the EN 14774-1:2009 [23] and EN 14774-3:2009 [24] technical standards. Each sample was subject to the determination of total moisture content, in order to know the actual humidity of the sample at the time of sampling. The sample was dried in an oven at a temperature of 105 ± 2 °C. To prevent the loss of the volatile substances, the drying time should, normally, not exceed 24 hours. The total contents of moisture on wet basis M_{ar} and on dry basis U_d were calculated using the formula provided by the technical standards. Other moisture tests were performed using the sample crushed and sieved to 1 mm during the determination of calorific value and ash content in order to use this value to correct the ash content and the calorific value. The moisture content in the analysis sample M_{ad} was calculated using the formula provided by the technical standard.

The procedure described in the EN 14775:2009 technical standard was used in order to estimate ash content [25]. The ash content was determined by calculation from the mass of the residue remaining after the sample was heated to a controlled temperature of 550 ± 10 °C. The ash content on dry basis A_d of the sample, expressed as a percentage by mass on a dry basis, was calculated using the formula provided by the standard. The result is expressed as the mean of two determinations.

Gross calorific value was determined experimentally using a calorimeter Parr 6200 and according to EN 14918:2009 [26]. Three tests were performed for each sample. The final result is given by the weighted average of the obtained three values.

The High Calorific Value (HCV) was rectified with the M_{ad} value and the HCV on dry basis was calculated. Low Calorific value (LCV) on dry basis was calculated with a formula depending on hydrogen content, according to technical standard.

Properties

The moisture of the biomass from urban green pruning was determined both on the sample as soon as the pruning and on the sample after storage. The moisture content of the sample as soon as the pruning was 45% on wet basis.

The sample after storage was processed to all the analysis because it represents the real biomass to use in energy conversion processes. The properties of the analyzed biomass are shown in Table 3.

The moisture content on dry basis is about 14% and is suitable in thermo-chemical conversion processes. The C/N ratio is high enough for thermo-chemical process. The N content is not very low because the sample of pruning is composed of leaves and wood. Also the Calorific Value is high and therefore is appropriated for thermo-chemical technology.

2.2   Energy Uses

In this work two energy conversions technology have been considered:

- combustion technology for wood-chips boiler for heat production;
- gasification technology within a gasifier for heat and electricity production.

In this study, the use of a wood-chipper is involved. The wood-chips can be used both in wood-chip boiler for heat production and in gasification plant with CHP.

Table 3. Properties of the pruning from urban green sample (after storage)

Properties		
Moisture on wet basis M_{ar}	13.99	%
Moisture on dry basis U_d	16.27	%
Ash content	10.61±1.13	%
HCV on dry basis	19.98	MJ/kg
LCV on dry basis	18.64	MJ/kg
C content	46.22±0.48	%
H content	5.86±1.19	%
N content	2.25±0.18	%
C/N ratio	20.54	-

Table 4. General information about the available biomass

Input information		
Total biomass	625 200	kg/year
Biomass (moisture = 45%) after chipping	606 444	kg/year
Biomass (moisture = 14%) after chipping	387 842	kg/year
LCV (on dry basis)	18.64	MJ/kg
LCV (for moisture = 14%)	16.19	MJ/kg
Density	300	kg/m³
LCV (on dry basis) [kWh/m³]	1 554	kWh/m³
LCV (for moisture = 14%) [kWh/m³]	1 349	kWh/m³
Biomass (moisture = 14%) after chipping [m³]	1 292.81	m³/year

In order to size the plants, the following data – summarized in Table 4 - had to be considered:

- amount of annual total biomass from pruning;
- amount of annual biomass after chipping phase (B_i);
- amount of annual biomass after chipping with moisture on wet basis of 14% (B_f), calculated with equation (2)

$$B_f = B_i \left(1 - \frac{M_i - M_f}{100 - M_f}\right) \qquad (2)$$

where M_i is the initial moisture and M_f is the final moisture [27];
- density of the wood-chips;
- Low Calorific Value (LCV).

The most important benefit of the wood-chip boiler is due to its automation. A storage site for the biomass is necessary and can be a silo for small plants or a barn for big plants. The biomass is transported to the boiler through a cochlea. The combustion process and the biomass supply are controlled by a microprocessor. A lambda sensor analyzes the exhaust gases and communicates with the microprocessor. The data of temperature and pollutants are compared by software which automatically it doses the biomass supply and the combustive agent (air).

The wood chip plant becomes cost-effective if power increases and therefore it is suitable for heating of condominium, big building used as offices or schools, big industrial buildings and shopping centers.

In order to determine the heating period, the Heating Day map has to be considered. According to the value of Heating Degree Days (HDDs), Italy is divided into six different zones: the zone A is the hottest one, the zone F is the coldest one. HDD increases when the climate becomes colder. Viterbo belongs to zone D (Fig. 2) [28]. In the zone D the heating period is from the 1st of November to the 15th of April for maximum 12 hours/day.

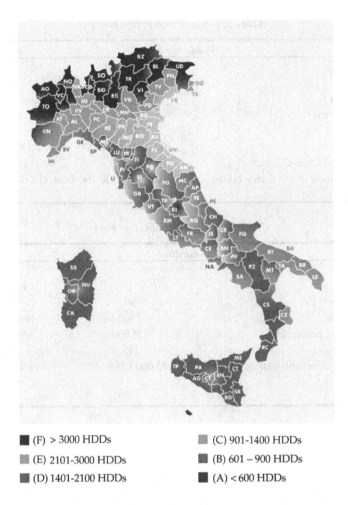

■ (F) > 3000 HDDs	▨ (C) 901-1400 HDDs
▨ (E) 2101-3000 HDDs	■ (B) 601 – 900 HDDs
■ (D) 1401-2100 HDDs	■ (A) < 600 HDDs

Fig. 2. Heating degree day map in Italy [29]

The accessible amount of biomass allows to install 1 095 kW of power (Table 5), since a efficiency value of 0.9 has been considered [30].

Therefore five wood-chips boiler of about 200 kW can be installed, heating five public buildings, in particular schools or public offices with a surface of about 3 500 to 4 000 m^2 each.

These parameters are favorable for building belonging to "class C" energy consumption: 58-85 kWh/m^2.

The second studied possibility has been the use of a gasification plant. The sizing of the plant has been carried out considering a specific consumption (C_s) of biomass equal to 1 kg/kWh [31]. The use of biomass leads to reach 50 kW$_e$ and 115 kW$_t$ (Table 6).

Table 5. Sizing of wood-chips boiler

Wood-chip boiler		
Power	1 095	kW
Heating period	1 November – 15 April	
Max heating hours	10	h/day
Efficiency	0.9	
Energy	1 500 000 - 1 650 000	kWh/year
Heating demand (class C building)	58 - 87	kWh/m^2
Number and power of wood-chip boilers	5 plants of 200 kW for about 3500 - 4000 m^{2}each	

Table 6. Sizing of gasification plant

Gasifier		
C_s	1	kg/kWh
Electric Power	50	kW
Operating time	7 500	hours/year
Electric energy production	375 000	kWh/year
Heat Power	115	kW
Heat energy production (during heating period)	165 000 - 189 150	kWh/year

The amount of heat and energy production for the two cases is very different, in favor of the wood-chips boiler, and not to be compared, though the gasification plant produces electricity. Furthermore the need of the Municipality of Viterbo is mostly related to thermal energy and the wood-chips boilers solution permits to heat around five big public buildings.

3   Results

In order to study the advantages coming from the use of wood-chips boiler plants, a cost-benefits analysis is necessary.

The economic assessment is carried out for a 200 kW single plant. The following parameters should be considered:

- Capital Costs (Boiler, Wood-Chipper),
- Operating costs (amortization, labor),
- Profits (avoided costs of diesel or methane, possible subsidies).

For each parameter, average costs have been taken into account as summarized in table 7.

The main profit is represented by the avoided costs of methane or diesel and moreover depends on the type of the replaced plant.

The total investment has been studied with the following main financial indicators (Table 8):

- Net Present Value (NPV)
- Payback period (PB)
- Internal Rate of Return (IRR)
- Benefit /Cost ratio (B/C)

The used discounted rate was 3%.

Table 7. Costs for a wood-chips boiler plant (200 kW)

Costs		
Capital Costs		
Wood-chips boiler	69 000	€
Wood-chipper	13 200	€
Total	**82 200**	**€**
Operating Costs		
Amortization Payment	7 919	€/year
Labor costs	1 500	€/year
Total	**9 419**	**€/year**
Profits		
Avoided methane consumption	30 612	m^3/year
Avoided costs of methane	22 194	€/year
or		
Avoided diesel consumption	30 000	l/year
Avoided costs of diesel	32 850	€/year

Table 8. Main considered financial indicators

	Financial Indicators	
	Wood-chips boiler vs. methane boiler	Wood-chips boiler vs. diesel boiler
NPV	1 603 583	3 009 810
PB	4	3
IRR	49 %	71 %
B/C	2.21	3.27

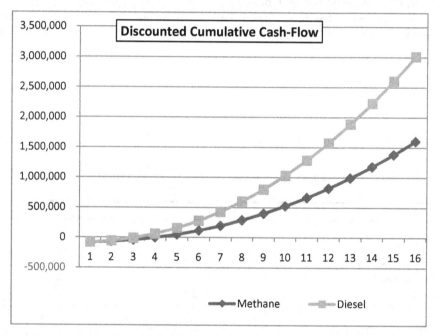

Fig. 3. Discounted Cumulative Cash-flow for the cases of methane or diesel boiler substitution for a wood-chips boiler

In figure 3 a comparison between the discounted cumulative cash flow of the substitution cases of methane or diesel boilers for a wood-chips boiler is shown.

The cash flows (C_t^*) is obtained by adding all the costs ($C_{j,t}$) and all the profits ($P_{j,t}$) related to the generic t-th year, as shown in the following expression (3):

$$C_t^* = \sum_j P_{j,t} - \sum_j C_{j,t} \qquad (3)$$

In figure 4 a comparison between the discounted cash inflows and outflows of the substitution cases of methane or diesel boilers for a wood-chips boiler is reported.

The project is evidently profitable because of the avoided costs for methane or diesel, in favour of wood-chips, and determines a recovery of the investment cost after 3 or 4 years.

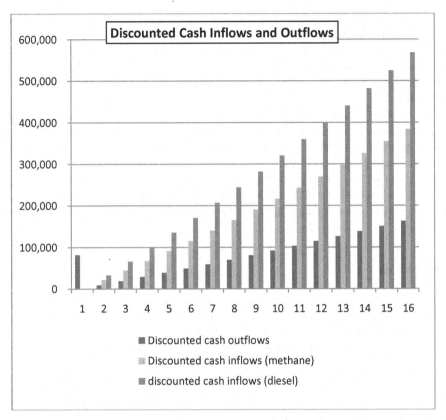

Fig. 4. Discounted cash inflows and outflows for the cases of methane or diesel boiler substitution for a wood-chips boiler

4   Conclusion

The present paper investigates the use of residual biomass arising from urban green pruning for energy production in Viterbo town.

The suitable amount has been evaluated and some samples have been analyzed in laboratory in order to determine the main characteristics of the biomass.

The total amount of the biomass is equal to 625 200 kg/year.

The high C/N ratio (equal to 20.54) and the low moisture content (equal to 14 %) prove that the analyzed biomass can be successfully and efficiently exploited in thermo-chemical conversion processes. Two possibilities for energy conversion have been estimated: use in wood-chips boiler for heat production and use in a gasification plant with CHP system.

The first scenario allows to install a total capacity of 1 095 kW to use with a wood-chips boiler. Because of the high power, the solution of 5 plants - 200 kW each - has been chosen and leads to heat 5 big public buildings, e.g. schools or offices whose area vary from 3 500 to 4 000 m^2.

The second opportunity allows to install a 50 kW_e and 115 kW_t gasification plant. This solution is certainly less favorable because the heat energy production is much lower than the wood-chip boiler one, although electricity can be produced in gasification. Public buildings heating requires the highest energy supplies: the wood-chips boiler solution is the best choice decreasing costs and reducing greenhouse gases emissions.

The solution of a 200 kW wood-chips boiler has been compared with two cases: diesel and methane boiler, both with the same power. The project is evidently profitable because of the avoided costs for methane or diesel, to advantage of wood-chips, and determines a recovery of the investment cost after 3 or 4 years.

Therefore starting from the month of January 2013 in Italy some subsidies are provided by the national law "DM 28 December 2012". The incentives cover the 40% of the total investment costs and make the wood-chips boiler even more favorable and attractive.

References

1. Lineback, N., Dellinger, T., Shienvold, L.F., Witcher, B., Reynolds, A., Brown, L.E.: Industrial Greenhouse Gas Emissions: Does CO2 from Combustion of Biomass Residue for Energy Really Matter? Clim. Res. 13, 221–229 (1999)
2. Bocci, E., Zuccari, F., Dell'Era, A.: Renewable and Hydrogen Energy Integrated House. International Journal of Hydrogen Energy 36, 7963–7968 (2011)
3. Villarini, M., Limiti, M., Abenavoli, R.I.: Overview and Comparison of Global Concentrating Solar Power Incentives Schemes by Means of Computational Models. In: Murgante, B., Gervasi, O., Iglesias, A., Taniar, D., Apduhan, B.O. (eds.) ICCSA 2011, Part IV. LNCS, vol. 6785, pp. 258–269. Springer, Heidelberg (2011)
4. Cocchi, S., Castellucci, S., Tucci, A.: Modeling of an Air Conditioning System with Geothermal Heat Pump for a Residential Building. Mathematical Problems in Engineering article no. 781231 (2013)
5. Kobayashi, N., Fan, L.S.: Biomass Direct Chemical Looping Process: a Perspective. Biomass Bioenergy 35, 1252–1262 (2011)
6. Naik, S., Goud Vaibhav, V., Rout, P.K., Jacobson, K., Dalai, A.K.: Characterization of Canadian Biomass for Alternative Renewable Biofuel. Renewable Energy 35, 1624–1631 (2010)
7. Panwar, N.L., Kothari, R., Tyagi, V.V.: Thermochemical Conversion of Biomass – Eco Friendly Energy Routes. Renewable and Sustainable Energy Reviews 16, 1801–1816 (2012)
8. Dodić, S.N., Vasiljević, T.Z., Marić, R.M., Radukin Kosanović, A.J., Dodić, J.M., Popov, S.D.: Possibilities of Application of Waste Wood Biomass as an Energy Source in Vojvodina. Renewable and Sustainable Energy Reviews 16, 2355–2360 (2012)
9. Bocci, E., Di Carlo, A., Marcelo, D.: Power Plant Perspectives for Sugarcane Mills. Energy 34, 689–698 (2009)

10. Mc Kendry, P.: Energy Production from Biomass (part 2): Conversion Technologies. Bioresource Technology 83, 47–54 (2002)
11. Monarca, D., Cecchini, M., Guerrieri, M., Colantoni, A.: Conventional and Alternative Use of Biomasses Derived by Hazelnut Cultivation and Processing. Acta Horticulturae 845, 627–634 (2009)
12. Mc Kendry, P.: Energy Production from Biomass (part 1): Overview of Biomass. Bioresource Technology 83, 37–46 (2002)
13. Küçük, M.M., Demirbas, A.: Biomass Conversion Processes. Energy Conversion Manage 38(2), 151–165 (1997)
14. Monarca, D., Colantoni, A., Cecchini, M., Longo, L., Vecchione, L., Carlini, M., Manzo, A.: Energy Characterization and Gasification of Biomass Derived by Hazelnut Cultivation: Analysis of Produced Syngas by Gas Chromatography. Mathematical Problems in Engineering, art. no. 102914 (2012)
15. Monarca, D., Cecchini, M., Colantoni, A., Marucci, A.: Feasibility of the Electric Energy Production through Gasification Processes of Biomass: Technical and Economic Aspects. In: Murgante, B., Gervasi, O., Iglesias, A., Taniar, D., Apduhan, B.O. (eds.) ICCSA 2011, Part IV. LNCS, vol. 6785, pp. 307–315. Springer, Heidelberg (2011)
16. Puig-Arnavat, M., Bruno, J.C., Coronas, A.: Review and Analysis of Biomass Gasification Models. Renewable and Sustainable Energy Reviews 14, 2841–2851 (2010)
17. Di Carlo, A., Bocci, E., Naso, V.: Process Simulation of a SOFC and Double Bubbling Fluidized Bed Gasifier Power Plant. International Journal of Hydrogen Energy 38, 532–542 (2013)
18. La Marca, O.: Elementi di dendrometria. Patron, Bologna (1999)
19. Monarca, D., Cecchini, M., Colantoni, A.: Plant for the Production of Chips and Pellet: Technical and Economic Aspects of an Case Study in the Central Italy. In: Murgante, B., Gervasi, O., Iglesias, A., Taniar, D., Apduhan, B.O. (eds.) ICCSA 2011, Part IV. LNCS, vol. 6785, pp. 296–306. Springer, Heidelberg (2011)
20. EN 14778:2011. Solid Biofuels: Sampling (2011)
21. EN 14780:2011. Solid Biofuels: Sample Preparation (2011)
22. EN 15104:2011. Solid Biofuels: Determination of Total Content of Carbon, Hydrogen and Nitrogen. Instrumental methods (2011)
23. EN 14774-1:2009. Solid Biofuels. Determination of Moisture Content, Oven Dry Method. Part 1: Total Moisture, Reference Method (2009)
24. EN 14774-3:2009. Solid Biofuels. Determination of Moisture Content, Oven Dry Method. Part 3: Moisture in General Analysis Sample (2009)
25. EN 14775:2009. Solid Biofuels: Determination of Ash Content (2009)
26. EN 14918:2009. Solid Biofuels: Determination of Calorific Value (2009)
27. Bernetti, G., De Favero, R., Pividori, M.: Selvicoltura produttiva. Manuale Tecnico, Il Sole 24 Ore Edagricole (2012)
28. Carlini, M., Castellucci, S., Allegrini, E., Tucci, A.: Down-hole Heat Exchangers: Modeling of a Low-enthalpy Geothermal System for District Heating. Mathematical Problems in Engineering, art. no. 845192 (2012)
29. Legislative Decree 29/12/2006 n. 311 published on Gazzetta Ufficiale 01/02/2007 no. 26
30. Chandrasekaran, S.R., Laing, J.R., Holsen, T.M., Raja, S.: Emission characterization and efficiency measurements of high-efficiency wood boilers. Energy and Fuels 25(11), 5015–5021 (2011)
31. Caputo, C.: Gli Impianti Convertitori d'Energia, vol. I. CEA, Milano (1997)

Technical-Economic Analysis of an Innovative Cogenerative Small Scale Biomass Gasification Power Plant

Enrico Bocci[1,*], Andrea Di Carlo[2], Luigi Vecchione[3], Mauro Villarini[3], Marcello De Falco[4], and Alessandro Dell'Era[1]

[1] G. Marconi University, Energy Department, Rome, Italy
{e.bocci,a.dellera}@unimarconi.it
[2] Sapienza University, Mechanic Department, Rome, Italy
andrea.dicarlo@uniroma1.it
[3] Tuscia University, Interdepartmental Centre for Research and Dissemination of Renewable Energy, Orte (VT), Italy
{l.vecchione,m.villarini}@unitus.it
[4] Bio-medical Campus University, Engineering Faculty, Rome, Italy
m.defalco@unicampus.it

Abstract. The use of biomass waste in high efficient low pollutants emissions micro-cogeneration plants overcomes the main biomass barriers: competition with the food and material uses, dispersion of a low energy density fuel and high emissions.

This paper is focused on a small (100 kW_{th}) steam gasification fluidized bed and hot gas conditioning system. In fact, the gasification without air leads to a high calorific value gas; the hot gas conditioning allows reducing pollutants by converting more gas; the fluidized bed allows a better process and heat transport management.

Beside the design analysis, a technical and economic analysis is proposed. In particular, the feasibility study has been carried out through the main methods of NPV (Net Present Value) and PBP (Pay Back Period) assessment. The study highlights the economic viability of the proposed system, which has always a acceptable PBT and a positive NPV despite the small plant size.

Keywords: biomass, gasification, hot gas conditioning, power plant, economic analysis.

1 Introduction

This work lies within the framework of a research aimed at the diffusion of sustainable development model with the concept of closed-cycle, in which resources are not consumed but used and reused over time.

The implementation of this model proceeds through the identification and use of renewable energy sources, the use of appropriate energy carriers to make available the

---

[*] Corresponding author.

B. Murgante et al. (Eds.): ICCSA 2013, Part II, LNCS 7972, pp. 256–270, 2013.

energy sources where and when they are necessary, the use of different sources to enhance the unique characteristics that each location can offer and to encourage the development of a more decentralized energy production [1].

According to these criteria, this work analyses the technical and economic feasibility of a small power plant based on the gasification of biomass residues (lignocellulosic from forest areas, such as coppice woodland, agricultural, such as straw, and industrial waste as of wood industry). Biomass is the most common form of renewable energy and the largest reservoir of solar energy, but the energy use of the organic substances is limited by their low energy density, complexity of the supply chain (often in competition with the main uses of organic matter, as food and materials) and high local emissions of pollutants [2]. Using organic wastes as feedstock for high efficient micro-cogeneration plants would solve all the drawbacks associated to biomass utilization as energy source, since the small size plants can be installed near the point of biomass production (short chain) and the high conversion efficiency would reduce the local pollutants emissions. One of the major limitations associated with the current use of the large energy potential of waste (e.g. the Italian territory amounted to about 30 million metric tons / year [3–5]) is their dispersion in large area. For obvious reasons of transport, this fuel with a low energy density and perishable has so far found only limited use in large power plants, for the need of large territories with feedstock availability [6, 7]. The use of small scale and efficient systems permits the efficient use of biomass in most territories, enabling a larger and more sustainable waste energy exploitation.

One of the innovative aspects of the system analysed is the application of the technology of fluidized bed steam gasification with hot gas conditioning applied at small scale in the order of 100-500 kW$_{th}$. In fluidized bed gasifiers, the concentration of TAR and particulate matter (PM) ranges between 5 and 100 g/Nm^3 of the produced gas. Moreover, corrosive and pollutant characteristics of TAR compounds prohibit direct utilization of the produced gas stream. Catalytic steam reforming seems the best way to eliminate TAR compounds, converting these into useful syngas while high temperature ceramic filters can completely remove the particulate from the gas [8]. The clean fuel gas could be thus delivered at temperatures as high as those required to exploit it in high efficiency power generation devices like SOFC or MCFC [9–14] in future applications.

2   Gasification Technology

The technologies usually applied for small plants are fixed bed (in particular the downdraft technology). These technologies have operability and durability limitations associated with the high temperatures (there are isolated hot spots that cause heat loss and malfunctioning of the process), furthermore they have efficiency limitations associated with the use of air as gasification medium (which results in a low quality of the syngas obtained in terms of composition and heating value)[15, 16].

In this work, the proposed idea is to develop a small gasification fluidized bed (100-500 kW$_{th}$) operating with air and steam in single dual-chamber fluidized reactor to produce a syngas with a high calorific value, which can be exploited in internal combustion engine for the production of electricity.

Fluidized-bed reactors are common in those processes where catalysts must be continuously regenerated, also facilitating heat transfer, temperature uniformity, and higher catalyst effectiveness factors. In particular, the fluidization state allows a steady circulation of the bed material among different reactors. In this way, it is possible to exploit the bed material as thermal carrier to enhance heat exchange among different reactors [17].

The gasifier analysed in this work is constituted by a single reactor, in which there are two distinct chambers communicating with each other, as shown in Figure 1.

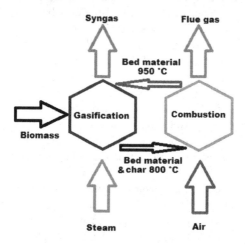

Fig. 1. Scheme of indirect heating

In detail, the fuel is fed into the gasification zone where it pyrolyses and is gasified at 850 °C with only steam (steam inlet temperature 500 °C). The main reactions occurring in this zone are:

$$C + H_2O \rightarrow CO + H_2 \tag{1}$$

$$C + CO_2 \rightarrow 2CO \tag{2}$$

$$C + 2H_2 \rightarrow CH_4 \tag{3}$$

$$CH_4 + H_2O \leftrightarrow CO + 3H_2 \tag{4}$$

$$CO + H_2O \leftrightarrow CO_2 + H_2 \tag{5}$$

The bed material (olivine sand), together with residual charcoal from (1), (2) and (3), circulates to the combustion zone. This zone is fluidized with air and here the charcoal is burned, heating the bed material at higher temperature (950 °C) than at the entrance.

The hot bed material from combustor, circulated again to the gasifier, supplies the thermal power for the gasification reactions. With this concept, the two reaction chambers (combustion with air and gasification with only steam) are physically separated and it is possible to get a high-grade product gas, not diluted with the N_2 of the air, without use of pure Oxygen. The steam to biomass ratio has an optimal value of about 0.5. The result is the production of a gas with a high calorific value (greater than 10 MJ/Nm^3) with a high content of hydrogen, followed by carbon monoxide, dioxide, and steam and with small percentages of nitrogen and light hydrocarbons such as CH_4 and traces of heavy hydrocarbons, such as tar.

The average composition of the syngas referred to experimental tests carried out with circulating fluidized bed technology is reported in Table 1 [18].

Table 1. Average composition of the syngas

Species	Comp. molar wet (%)	Comp. molar dry (%)
H_2	38	46.0
CO	17	21.0
CH_4	7	8.5
CO_2	15	18.0
N_2	5	6.5
H_2O	18	0.0

The gas produced can be exploited both in conventional machine, such as Internal Combustion Engine (ICE) or micro Gas Turbine (mGT), and in Fuel Cells (FC). Unfortunately, although the fuel cells (and particularly those at high temperature) represent a technology more efficient and with lower emissions than combustion machines, actually they still have a cost per kW much higher.

3   Process Description

The system mainly consists of the gasification unit above described and the hot gas conditioning & cleaning system. Finally, the ICE will convert the syngas with a good compromise between a high overall efficiency (electrical and thermal) and a low cost.

Table 2. Main technical specifications of the system studied

Parameter	Value	Unit
Biomass flow	24	kg/h
Biomass LHV	13	MJ/kg
Biomass power	94	kW$_{th}$
Produced syngas flow	1.5	Nm^3/kgbio
Syngas LHV	10-11	MJ/Nm^3
Produced electrical power	26	kW$_e$
Produced thermal power	50	kW$_{th}$

Table 2 shows the main technical system specifications for 94 kW_{th} of fed biomass. The system, in this case, can process up to 24 kg/h of biomass (with a Low Heating Value of 13 MJ/kg, i.e. 30% moisture woodchips, giving an input power of 94 kW_{th}), producing approximately 26 kW_e and 51 kW_{th} (using, for the performance of the Internal Combustion Engine, the data of the MAN E0384 [19]).

The power plant flow sheet is shown in Figure 2. The system work at ambient pressure. The air and the steam are pre-heated at a temperature of 300°C and 250°C before being injected into the gasifier. Water is heated and vaporized through the heat exchanger (1), using part of the heat contained in the syngas outgoing the TAR Cracker. Air is heated through the heat exchanger (2), using the heat provided by the products exiting from the combustion chamber of the gasifier. The gasifier is fed with biomass via a dual screw feeder: a dosing screw feeder and a fast screw feeder; the use of the fast screw permits to limit the residence time of the biomass in the feeding system, so to avoid that the process of pyrolysis could start before biomass enters in the reactors.

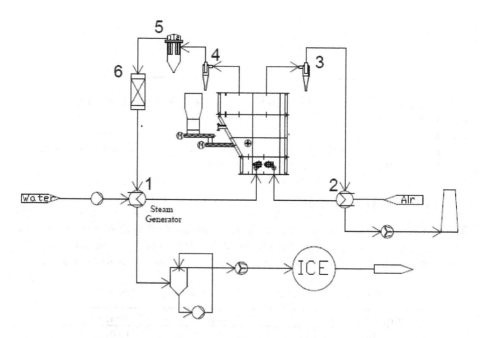

Fig. 2. Power plant flow sheet

The outlet streams of the gasifier are a syngas flow at about 800° C and the exhausted products of combustion at approximately 950°C. The latter passes through the first cyclone (3) to remove the coarse particulate and subsequently in the heat exchanger (2), for the thermal recovery before being released into the atmosphere.

The syngas instead, passes through the cyclone (4) where the coarsest particles are retained and subsequently at a temperature of about 750° C, through the high temperature particulate filter (5) capable of remove fine particles (≥ 0.3 μm). The syngas finally passes through the TAR Cracker (6) to completely remove the condensable TAR; the sensible heat of the gas, now free from tars and particulates, is available to be completely recovered. Downstream of the TAR Cracker, are placed the heat exchanger (1) used to vaporize the water to be injected into the gasifier and a scrubber to remove any trace of residual tar. The main components of the plant as well their dimensioning are described below. It is not currently possible to describe the gasifier in more detail since it is the subject of a patent.

4   Cyclone

To estimate the separation efficiency of the two cyclones, the granulometric distribution of the elutriated particulate are assessed. It was hypothesized that the amount of entrained particulate was equal to 20 g/Nm^3 with a density of 600 kg/m^3 [7].

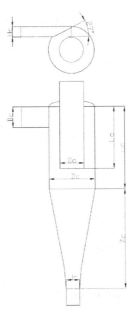

Fig. 3. Blueprint of cyclones

To determine the removal efficiency the law of Theodore and De Paola [20] has been adopted. Figure 3 shows a schematic drawing of the two cyclones and in Table 3 the main dimensions are reported.

Table 3. Size of the different sections of cyclones

Section	Size
Dc	110 mm
Bc	50 mm
Hc	25 mm
Jc	25 mm
Zc	242 mm
Lc	200 mm
Lo	150 mm
Do	60 mm

The Figure 4 shows the particle size distribution before and after the cyclone. The concentration of total particulate matter will be reduced to 1.6 g/Nm^3. The pressure drops should be around 10 mbar.

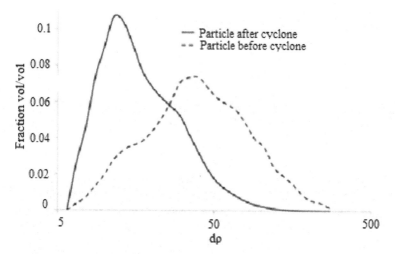

Fig. 4. Particle size distribution

5   Particulate Filter Ceramic High T

To completely remove the fine particles, ceramic filters, DIA-SCHUMALITH DS 10-20 [21], for high temperature have been installed. The technical characteristics of these filters are shown in following Table 4.

Table 4. Technical characteristics of ceramic filters

Parameter	Value
Filtration fineness [μm]	0.3
Operating temperature °C	750
Support material	SL 20
Membrane type	DIA 10
Membrane material	Mullite grains
Mean pore size [μm]	50
M. pore size (membrane) [μm]	10
Porosity support material [%]	38
Material density [g/cm3]	1.90
Permeability 10-13m2	55(air at 20 °C)
Dimension cm (H/Do/Di):	100/6/4

The candle filters are hollow cylinders (closed on one end) typically with a diameter of about 6 cm and a length of 1–1.5 m; the outer surface is made of a thin, micro porous layer that forms the actual filtering surface. The dusty feed gas flows outside such cylinders and percolates through their porous structure driven by a differential pressure: fine particles accumulate on the filtering surface building up a cake, while the clean gas is collected in the hollow space. A filter vessel for power plant requires a certain number of candle filters arranged in several clusters. Due to accumulation of particulate on the filter (cake), the equivalent permeability will tend to decrease thereby increasing the pressure drop along the filter: it is therefore necessary to remove this cake. Therefore, each cluster is periodically cleaned by an instantaneous reverse-gas flow back-pulse procedure (for approximately 250 ms) to remove the dust cake that builds up on the filtering surface, during the operation of the remaining cluster. In this way, it is possible to ensure continuity of the process. The regeneration is never complete because there is a fraction of a cake that cannot be removed and which will gradually decrease the initial permeability of the filters. Such permeability reaches an asymptotic value, about one half of the initial [22], after several cycles that can be considered as the basis for the calculation of the pressure drops. To limit the number of filters but at the same avoiding excessive pressure drops the system is dimensioned for a superficial velocity of 90 m/h. From the results obtained by Di Carlo and Foscolo [8], it can be assumed that the pressure drop changes over time with the following law:

$$\Delta p = 3.64 \cdot e^{11} + 3.02 \cdot e^{7} \cdot t \qquad (6)$$

where t is expressed in seconds and Δp in Pa.

The Figure 5 shows a typical pressure drop due to the formation of the cake on the filters, obtained with the law shown above, and for the regeneration back pulse in 250 ms, for 3 cycles of operation.

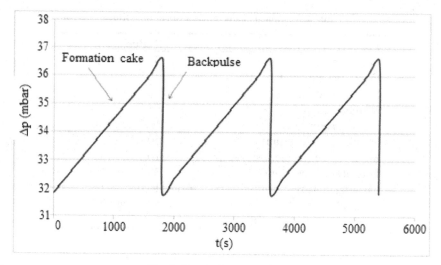

Fig. 5. Trend of pressure drop

With the above considerations, it was decided to divide the system into three sectors each 3 ceramic filters. Regarding the soot formation and emission see, e.g. Bocci [23].

6   Tar Reformer

Corella et al [24] have shown that at 800 ° C, with a fraction of steam close to 17% and a Space Velocity (SV) of 26600 h⁻¹, with a gas obtained by gasification of biomass in fluidized bed, it is possible to convert 98-100% of the tar using the Ni-based catalyst of Topsoe (RK-67 [25]). As suggested by the Topsoe [25, 26] in order to limit the formation of coke on the heavy hydrocarbons reforming catalyst, it is preferable to use a mixture of catalysts RK-67-7H (50%), RK-202 (35%), RK-212 (15%) the last with an high content of K_2O (> 1%). This is an alkaline compound, which restricts the acidity of the Al_2O_3 support of the catalyst, inhibiting the formation of coke. By using the data reported in Corella et al [24], the tar reformer dimensioning has been performed: it is a cylinder of 125 mm inner diameter and 500 mm long, so as to find a SV of 26600 h⁻¹. Tar concentration in the syngas outgoing the gasifier, using olivine as bed material varies between 2 and 10 g/Nm³. At the exit of the reformer a concentrations within the range 0-200 mg/Nm³ can be thus expected. The pressure drop in the tar cracker can be neglected.

7   Steam Generator and Heat Exchanger Recovery

It is hypothesized to generate steam at 250 °C and preheat air at 300 °C approximately exploiting the thermal recovery of the syngas effluent from tar reforming and exhausted from the burner, respectively. Then two heat exchangers type BEM are dimensioned using the tool ASPEN HTFS + 2006.

Given the high temperature differences, between hot fluid and cold it was decided to insert a compensator for the axial expansion welded on the shell side.

The Figure 6 shows a schematic drawing of the two heat exchangers, while in Tables 5 and 6 the main results of the sheet corresponding THEME are reported.

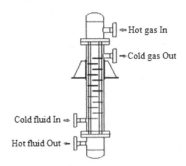

Fig. 6. Heat exchanger type BEM

Table 5. TEMA sheet of the steam generator

Parameter	Value	
Size [mm]	135/1530	
Surface [m^2]	1.4	
Heat exchanged [kW]	7.7	
Transfer rate [W/m^{2}K]	34.1	
Tube external diameter [mm]	20	
Shell external diameter [mm]	141.3	
Fluid	*Steam*	*Syngas*
Flow [kg/h]	10	26
Velocity [m/s]	1.66	7.02
Temperature In/Out [°C]	25/250	628/110
Pressure drop [mbar]	0.27	3.89

Table 6. Sheet TEMA exchanger recovery

Parameter	Value	
Size [mm]	135/860	
Surface [m^2]	0.8	
Heat exchanged [kW]	8	
Transfer rate [W/m^{2}K]	30	
Tube external diameter [mm]	20	
Shell external diameter [mm]	141.3	
Fluid	*Air*	*Exhaust*
Flow [kg/h]	100	110
Velocity [m/s]	9	21
Temperature In/Out [°C]	25/303	630/405
Pressure drop [mbar]	24	10

8   Suction Blowers

The gasification system must work at negative pressure to avoid leakages of gas to the outside of the plant. The pressure drops mainly derive from the high temperature filtration system.   It has also to be considered that syngas has high concentration of H_2. For both these reasons, there is a tendency to use brushless type aspirators whose speed of rotation (and therefore the suction flow rate) can be controlled via a voltage signal 1.5-10 VDC.

Then the following models of Nothland have been selected:

- BBA14-221HEB-00 side syngas able to suck up to 290 Nm^3/h of air ($\approx$ half of syngas) with a maximum vacuum of 140 mbar.
- BBA14-213HEB-00 side fumes able to suck up to 150 Nm^3/h of air ($\approx$ fumes) with a maximum vacuum of 175 mbar.

More details and characteristics of vacuum cleaners are available directly on the site of the Northland [27].

9   Economic Analysis

Doing a technical and economic analysis of an innovative power plants it is always a rough assessment due to the lack of experimental and verified data, as showed by the authors in other innovative power plants analysis [28–30]. In this case, where the system is patent pending and testing, even if still remain an uncertainty on the energy data (not only efficiency but especially reliability, e.g. annual equivalent hours) it is more easy to calculate the global equipment cost at around 100000 € and make an economic analysis of the global investment and operating cost and revenues [30]. In particular, the cost of the drying system and fuel loading, of the gasifier and the air conditioning system and cleaning is equal to about 40000 € (i.e. about 400 €/kW) while the cost of the MCI is equal to about 35000 (1250 €/kW approximately). If these costs are added to control costs, the total installation cost are about 104000 €, which is the cost of production of the prototype in university laboratories. Moreover, have to be considered: the financing cost (around 45000 €, 13 years loan at 6% rate), the biomass purchasing cost (70 €/t), the general maintenance and operating costs, set equal to 5% of the plant cost. The annual revenue comes from the electricity sale to the grid (price of 0.28 €/kWh, as Italian all-inclusive tariff for the production of electricity from biomass for 6500 annual equivalent hours) and the thermal energy used (instead of generated at price of 0.06 €/kWh for 4000 annual equivalent hours).

Thus the 94 kW_{th} inlet power plant ensures a production of electricity of about 26 kW_e (27.7% energy efficiency) and a production of thermal power of 45 kW_{th} (thermal efficiency of 47.8%; cogeneration efficiency of 75.5%). With these data, it is possible to obtain the cash flows for the case of the sale of electricity and the case of sale of both electricity and thermal energy, as illustrated in the following figures.

It can be seen that in both cases the cash flows are extremely positive, obtaining a return on investment equal to 8 and 5 years.

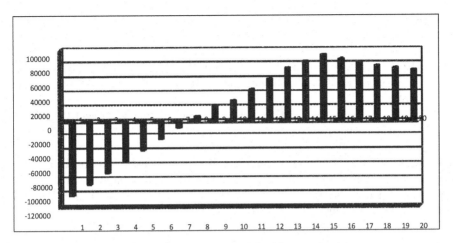

Fig. 7. Cash flows only electricity (€ versus year)

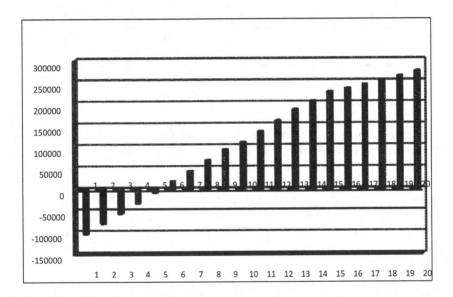

Fig. 8. Cash flows in cogeneration (€ versus year)

The payback time depends mainly on annual equivalent hours of system operation, biomass cost, investment costs, electricity and heat sale prices.

Revenues can be considered fixed, given the all-inclusive tariff for 15 years, and then the average selling price of electricity of 0.10 €/kWh and the average selling price of thermal energy in 0.06 €/kWh. Operating costs also can be considered fixed due to the small size system, so from one side a 24h/24 staff is not required (otherwise the cost of 5 annual people make uneconomic investment) and, on the other hand, the power plant can easily be automated. From these considerations, an operating costs equal to 5% of the cost of the plant has been assumed.

Instead, for annual equivalent hours and for biomass cost the following assumptions can be made:

- Hours of actual work per year: within the range 3500 (average hours per year of biomass in 2009, GSE data) and 6500 hours (the operating hours for the project);
- Cost of biomass: from 5 €/ton (average price of scrap wood) and 15 €/ton (retail price of pellets).

Whereas the best combination (more hours of operation and lower cost of biomass) and the worst (fewer hours of operation and higher cost of biomass) are obtained, always considering the sale of thermal energy as shown, the return times in years from investment are illustrated in the following table.

Table 7. PBT as a function of hours and cost biomass

Equivalent annual operating hours	biomass cost (€/t)	Pay-back Time
3500	150	15
6500	50	4

10   Conclusions

The proposed system has a payback time of 4 - 15 years, after those periods would gain each year from 10000 to 30000 €. This scenario allows understanding that the power plant, limiting the capital cost at 100000 euro, which is a feasible production cost for several units, is economically competitive even without any incentives if the biomass cost is low and the annual equivalent hours are high. In the evaluations made, the sale of thermal energy is crucial for the economic return on investment. Also in the case of residual biomass of industrial processing or forestry (such as, for example, shells of nuts machining confectionery), the cost of biomass is greatly reduced and it can be considered zero in case of no expenditure due to the disposal of waste.

Acknowledgements. The authors acknowledge the financial support of European Contract ID 299732 (regarding the research project UNIfHY, UNIQUE For Hydrogen production, under the topic SP1-JTI-FCH.2011.2.3: Biomass-to-hydrogen (BTH) thermal conversion process) funded by the European Union, Seventh Framework Programme, Fuel Cells and Hydrogen Joint Technologies Initiative (FP7, FCH-JU).

References

1. Marucci, A., Carlini, M., Castellucci, S., Cappuccini, A.: Energy Efficiency of a Greenhouse for the Conservation of Forestry Biodiversity. Mathematical Problems in Engineering (2013)

2. Bocci, E., Villarini, M., Caffarelli, A., D'Amato, A.: Sistemi A Biomasse: Progettazione E Valutazione Economica. Maggioli Editore, Torino (2011)
3. Orecchini, F., Bocci, E.: Biomass to hydrogen for the realization of closed cycles of energy resources. Energy 32, 1006–1011 (2007)
4. Orecchini, F., Bocci, E., Zuccari, F., Santiangeli, A.: Elettricita da fonti energetiche rinno-vabili. Le potenzialita dell'Italia. Termotecnica-Milano 58, 81–88 (2004)
5. Pari, L.: Energy production from biomass: the case of Italy. Renewable Energy 22, 21–30 (2001)
6. Bocci, E., Di Carlo, A., Marcelo, D.: Power plant perspectives for sugarcane mills. Energy 34, 689–698 (2009)
7. Monarca, D., Cecchini, M., Colantoni, A.: Plant for the production of chips and pellet: technical and economic aspects of an case study in the central Italy. In: Murgante, B., Gervasi, O., Iglesias, A., Taniar, D., Apduhan, B.O. (eds.) ICCSA 2011, Part IV. LNCS, vol. 6785, pp. 296–306. Springer, Heidelberg (2011)
8. Di Carlo, A., Foscolo, P.U.: Hot syngas filtration in the freeboard of a fluidized bed gasifier: Development of a CFD model. Powder Technology 222, 117–130 (2012)
9. Toonssen, R., Sollai, S., Aravind, P.V., Woudstra, N., Verkooijen, A.H.M.: Alternative system designs of biomass gasification SOFC/GT hybrid systems. International Journal of Hydrogen Energy 36, 10414–10425 (2011)
10. Di Carlo, A., Bocci, E., Dell'Era, A.: Comparison by the use of numerical simulation of a MCFC-IR and a MCFC-ER when used with syngas obtained by atmospheric pressure biomass gasification. International Journal of Hydrogen Energy 36, 7976–7984 (2011)
11. Orecchini, F., Bocci, E., Di Carlo, A.: Process simulation of a neutral emission plant using chestnut's coppice gasification and molten carbonate fuel cells. Journal of Fuel Cell Science and Technology 5 (2008)
12. Di Carlo, A., Bocci, E., Naso, V.: Process simulation of a SOFC and double bubbling fluidized bed gasifier power plant. International Journal of Hydrogen Energy 38, 532–542 (2013)
13. Colantoni, A., Giuseppina, M., Buccarella, M., Cividino, S., Vello, M.: Economical analysis of SOFC system for power production. In: Murgante, B., Gervasi, O., Iglesias, A., Taniar, D., Apduhan, B.O. (eds.) ICCSA 2011, Part IV. LNCS, vol. 6785, pp. 270–276. Springer, Heidelberg (2011)
14. Di Carlo, A., Borello, D., Bocci, E.: Process simulation of a hybrid SOFC/mGT and enriched air/steam fluidized bed gasifier power plant. International Journal of Hydrogen Energy 38, 5857–5874 (2013)
15. Warnecke, R.: Gasification of biomass: comparison of fixed bed and fluidized bed gasifier. Biomass and Bioenergy 18, 489–497 (2000)
16. Athanasiou, C., Vakouftsi, E., Coutelieris, F.A., Marnellos, G., Zabaniotou, A.: Efficiencies of olive kernel gasification combined cycle with solid oxide fuel cells (SOFCs). Chemical Engineering Journal 149, 183–190 (2009)
17. Chao, Z., Wang, Y., Jakobsen, J.P., Fernandino, M., Jakobsen, H.A.: Numerical Investigation of the Sorption Enhanced Steam Methane Reforming in a Fluidized Bed Reactor. Energy Procedia 26, 15–21 (2012)
18. Monarca, D., Colantoni, A., Cecchini, M., Longo, L., Vecchione, L., Carlini, M., Manzo, A.: Energy Characterization and Gasification of Biomass Derived by Hazelnut Cultivation: Analysis of Produced Syngas by Gas Chromatography. Mathematical Problems in Engineering (2012)
19. MAN Truck & Bus: Power Gas EN 121017 (2011), http://www.man-engines.com/media/content_medien/doc/Power_Gas_EN_121017_screen.pdf

20. Theodore, L., Paola, V.D.: Predicting cyclone efficiency. Journal of the Air Pollution Control Association 30, 1132–1133 (1980)
21. Pall Corporation: Data Sheet PIDIASCHUMALEN (2009),
    http://site.pall.com/pdf/PIDIASCHUMALEN.pdf
22. Simeone, E., Nacken, M., Haag, W., Heidenreich, S., de Jong, W.: Filtration performance at high temperatures and analysis of ceramic filter elements during biomass gasification. Biomass and Bioenergy 35, S87–S104 (2011)
23. Bocci, E., Rambaldi, L.: Soot emission modelization of a diesel engine from experimental data. In: Murgante, B., Gervasi, O., Iglesias, A., Taniar, D., Apduhan, B.O. (eds.) ICCSA 2011, Part IV. LNCS, vol. 6785, pp. 316–327. Springer, Heidelberg (2011)
24. Corella, J., Orio, A., Toledo, J.-M.: Biomass gasification with air in a fluidized bed: Exhaustive tar elimination with commercial steam reforming catalysts. Energy & Fuels 13, 702–709 (1999)
25. HALDOR TOPSØE: Data Sheet RK-67-7H, http://www.topsoe.com/
    business_areas/synthesis_gas/~/media/PDF%20files/
    Steam_reforming/Topsoe_steam_reforming_cat_r%2067%207h.ashx
26. HALDOR TOPSØE: Data Sheet RK-200 SERIES, http://www.topsoe.com/
    business_areas/synthesis_gas/~/media/PDF%20files/Steam_
    reforming/Topsoe_steam_reforming_cat_rk%20200_series.ashx
27. Northland Motor Technologies: Data Sheet BBA14-22, BBA14-21,
    http://www.northlandmotor.com/UI/HiResGallery.aspx
28. Orecchini, F., Bocci, E., Di Carlo, A.: MCFC and microturbine power plant simulation. Journal of Power Sources 160, 835–841 (2006)
29. Rambaldi, L., Bocci, E., Orecchini, F.: Preliminary experimental evaluation of a four wheel motors, batteries plus ultracapacitors and series hybrid powertrain. Applied Energy 88, 442–448 (2011)
30. Bocci, E., Zuccari, F., Dell'Era, A.: Renewable and hydrogen energy integrated house. International Journal of Hydrogen Energy 36, 7963–7968 (2011)
31. Monarca, D., Cecchini, M., Colantoni, A., Marucci, A.: Feasibility of the electric energy production through gasification processes of biomass: technical and economic aspects. In: Murgante, B., Gervasi, O., Iglesias, A., Taniar, D., Apduhan, B.O. (eds.) ICCSA 2011, Part IV. LNCS, vol. 6785, pp. 307–315. Springer, Heidelberg (2011)

Technical-Economic Analysis of an Innovative Small Scale Solar Thermal - ORC Cogenerative System

Mauro Villarini[1], Enrico Bocci[2], Andrea Di Carlo[3], Danilo Sbordone[4],
Maria Carmen Falvo[4], and Luigi Martirano[4]

[1] University of Tuscia, Interdepartmental Renewable Energy Research Centre, Orte (VT), Italy
mauro.villarini@unitus.it
[2] G. Marconi University, Energy Department, Rome, Italy
e.bocci@unimarconi.it
[3] Sapienza University, Mechanic Department, Rome, Italy
andrea.dicarlo@uniroma1.it
[4] Sapienza University, Electric Department, Rome, Italy
{danilo.sbordone,mariacarmen.falvo,luigi.martirano}@uniroma1.it

Abstract. The generation of electricity from solar thermal source has continued to garner more attention due to the very attractive environmental performance, the applicability to small distributed users and as alternative to the fossil fuels power generation. Indeed, solar thermal energy, through the coupling with ORC systems, becomes fit for electric power generation as well and, with a proper thermal storage, can deliver more equivalent hours than other residential systems. The paper analyses, by means of system modelling simulation, a small solar power plant composed by a CPC heat pipe solar collector device feeding a thermal storage, an ORC and an absorber unit. Beside the analysis of configuration and design, this paper proposes an economic analysis taking into consideration the applicable incentive. The evaluation highlights the economic viability of the proposed system and tries to define a roadmap to optimize results consisting in acceptable PBT and positive NPV.

Keywords: solar thermal, small scale power plant, ORC, Absorber, Economic Analysis, EFPH, NPV, PBT.

1 Introduction

The present paper illustrates the analysis of a solar powered Organic Rankine Cycle (ORC) system implemented according the output of the model calculations obtained by a TRNSYS platform. This study is the preliminary part of the STS (*Solar Trigeneration System*) research project which includes, after a prediction and evaluation activity, the collection of performance data from a 3 kW prototype under construction at University of Tuscia (Viterbo, Italy).

The background to the small-scale power generation by solar thermal systems has been the application of solar cooling [1, 2] which has been a way to optimize the use of solar energy in comparison with plain domestic hot water production or domestic heating.

B. Murgante et al. (Eds.): ICCSA 2013, Part II, LNCS 7972, pp. 271–287, 2013.
© Springer-Verlag Berlin Heidelberg 2013

The cooling use of solar energy allows to harness the summery peak of solar radiation. Solar cooling application generates the following benefits [1,2]:

- reduction of power peak on electric grid in the summer and of consequent summery energy black-out;
- reduction of losses on electricity distribution grid;
- less energy request during a period of the year when the big power plants suffer derating due to climate conditions.

These main advantages would have a major benefit in case the use of solar energy could have a further application: electric power generation. This extension of the solar system would permit to decide what to do by the solar energy input: not only the production of hot water but also answer to the heating, cooling and also electric demands. The object of present search is precisely the development of a plant which aims to produce all the electric and thermal energy needed for a 150 square meter house and, even if fed by aleatory energy source, it allows to direct and manage the energy flows better than other uncertain and uncontrollable renewable energy systems in virtue of a plethora of use options and of the thermal storage. The option to be able to decide how to use energy in distributed generation increases their efficiency, usefulness and availability. This is one of the most relevant limit of photovoltaic systems nowadays. Grid connected photovoltaic plant production needs to be consumed immediately where it is generated or to be put it into the grid [1–5].

A solar ORC cogenerative system allows to store the thermal energy or to use the energy through a plethora of ways.

There are still several aspects to optimize:

- thermal and electrical efficiency (minimization of the components heat losses, maximization of the ORC efficiency, quality of the storage components, etc.);
- overall system cost;
- encumbrance;
- system complication.

The system described in the present paper has been designed and analysed in order to determine the best economical trade-off conditions varying some selected parameters as the solar CPC plant area beside the environmental impact of the power system and of its other components. The TRNSYS model supports the analysis allowing the user to interact with the numerous parameters involved and obtaining different scenarios and results [6–10].

2   System Configuration

The system, as showed in Figure 1, mainly consists of a solar collector field feeding a heat storage tank which feeds an ORC system which feeds a low temperature heat storage tank which feeds the house heating and cooling (via absorber) loads. The heat contained by the tank can be managed and directed according to thermal levels available (temperature) and final user requirements. Water mixed with an additive (60% of Glycol) is the thermal vector fluid of the primary loop of the plant. The solar collector field has an area of approximately 50 square meters with which corresponds a thermal power of 35 kW while storage size is of 1000 litres. The water system temperature considered is of 15 °C which is an annual mean value for the water as distributed in central Italy.

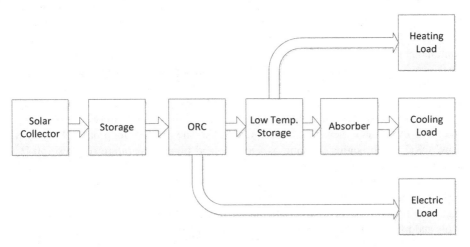

Fig. 1. Energy flow system

After single components description, overall system configuration is broached in order to comprehend components interaction. The Table 1 below summarizes the main power plant components parameters.

Table 1. Power plant components parameters

Component	Parameter	Value
Solar collectors	Square meters	50 m^2
Pumps	Flow	4000 kg/h
HT and LT Storage	Cubic meters	1 m^3
ORC	Nominal Electric Power	3 kW
Absorber	Nominal Cooling Power	10 kW
House	Square meters	150 m^2

In the present analysis sun irradiation is the main energy source whose integration with other renewable energy resources [11–14] will be assessed in following modelling and experimental activities. Further option consists in provisioning ORC generators deriving from recovery of exhaust and coolant heat both in stationary and mobile systems [15]. In simple solar case, energy contribution is processed by solar CPC collectors. Correct energy flow management is unavoidable for efficiency optimization. Solar fraction of solar thermoelectric is a basic parameter which define the potential of solar system. Solar power has to be compared with nominal thermal power necessary to feed the ORC evaporator. Usually, in major size plant, solar power is bigger than nominal input power requested by evaporator. This option is chosen in order to guarantee adequate thermal power at a temperature suitable to ORC and to compensate night and cloudy hours. Considering our system configuration, solar power should be twice or one time and a half the nominal thermal power. According to urban encumbrance requirements, consisting in occupation of a moderate area by solar field, the choice is restricted to a bit more than nominal power. Nominally, 3 kW_{el} ORC unit should be fed by a 30 kW_{th} solar thermal generator. For instance, a 90 kW_{th} solar field would allow to continuously produce electric power. Nevertheless, continuous generation would not justify overall system costs and field encumbrance. That is why bigger than 35 kW solar plant, occupying 50 square meters, would difficultly be possible to supply energy to an apartment. The size established in our analysis is a 35 kW solar plant feeding the ORC evaporator heat exchanger which hydraulically separates solar fluid loop and R235fa ORC loop. Between solar loop and ORC evaporator there is a heat storage tank. The primary loop has a 4000 kg/h flow. ORC pump begins working as soon as feed pipe temperature reaches 150 °C and stops as soon as it goes down to 100 °C. After this first conversion step the secondary loop feeds ORC evaporator with a flow rate of 800 kg/h and a temperature variation between go and return next to 15°C. The organic fluid is heated till vapour stage and transmits energy to an alternative device. During expansion, organic fluid cools and by the condenser comes back to liquid stage.

2.1   Solar Thermal Collector and HT Storage

The solar thermal collectors have been developed by the companies K-Engineering and Kloben Sud partners in the STS project and have been patented. By a meticulous design activity, it has been accomplished a collector capable to produce water at temperature bigger than 100°C. The patent obtained is an upgrade of the technology aimed to produce hot water at medium temperatures to feed a solar cooling system. The sealing material is made of copper heat pipe to the copper header tube and the mounting. Reflector material is EN AW 1085 EN 573-3 anodized as superficial treatment. Heat pipe specifications include high vacuum of $10^{-1}/10^{-2}$ [mbar] inside the copper tubes. Transfer fluid is water in quantity 5 to 8 g inside the copper tubes. Water has been mixed with additive in order to elevate fluid vaporization temperature and to not spawn an over pressure level inside the loop. Insulation has been obtained by high vacuum between inner and outer tubes of the glass evacuated. Insulation over the header copper collector is in mineral wool. The absorber of solar collector has a Al–N/Al selective graded surface and his absorptance coefficient α is not lower than 0.92 while the emittance coefficient ε is not higher than 0.065. At present, research group is reckoning to apply an optimization system [16] developed for solar system in order to further improve collector performances. A suitable heat storage system is required to maximize the productivity of the solar plant and to provide solar heat at the desired rate regardless the instantaneous solar radiation availability and the thermal needs [17].

2.2   ORC Unit and LT Storage

The ORC unit considered in the simulation is connoted by same technical specifications of the unit under installation and testing at university La Tuscia. The selected generator, called "Piglet", is produced in Italy and is an alternative engine with a declared yield included in the range 8%-10%. There are only a few producer capable to manufacture such a small size ORC generator. The engine producer datasheet requires a 20 square meters area of solar collectors to generate almost 2 kW. Nevertheless, transitory condition must be taken into account and choose a precautionary overestimating the solar collectors field area allows, during the test stage, to vary the number of operative solar collector strings analysing the system feedback to different thermal powers input. In the early system configuration, ORC evaporator receive heat by hot fluid coming from the energy storage interposed between solar primary loop and ORC evaporator loop. In a following stage, will be assessed the convenience of a second branch connecting directly solar thermal generator with ORC evaporator. When it is not necessary making solar energy flow to pass through the heat storage, use an eventual direct connection with ORC evaporator is possible in order to eliminate many losses.

Fig. 2. Internal details vision of piglet ORC

The temperature difference between evaporator and condenser is at minimum 35°C.

2.3   ORC Internal Loop Working Fluid

The working fluids that are suitable for low temperature ORC cycles are well known in the literature [18].

The choice of the fluid depending on the following main factors [6]:

* the critical point of the working fluid (in this case of small solar power plant the maximum temperature can be within 100 and 200 °C),
* the specific volume ratio over the expander must be low in order to reduce the size, and, thus, the cost (in general the higher the critical temperature, the higher the specific volume ratio over the expander),
* the working fluid should have a null Ozone Depleting Potential (Montreal Protocol and, for EC, Regulation 2037/2000),
* the fluid must fulfil other conditions such toxicity, cost and flammability,
* if scroll compressors is used for the expander the temperature must be below 150 °C because refrigeration compressors are not designed for temperatures higher than 150 °C.

We have chosen the fluid R245fa (1,1,1,3,3-penta-fluoropropane), because:

* • its critical temperature is greater than 150 °C (temperature that can be reached in a boiler feed by vacuum solar panels),
* • it shows very advantageous swept volumes, no toxicity and low cost,
* • it has a null Ozone Depleting Potential,
* • it is the most often fluid used in small ORC applications [19],

The data from the NIST Chemistry WebBook [20] indicates the thermodynamic properties of the fluid chosen, R245fa, varying according to the different working conditions. Thus it is necessary to develop a mathematical model of the fluid in order to have the thermodynamic properties of the fluid inside the global power plant simulation. The model developed is based on the following assumptions. Working fluid properties are shown in Table 2.

Table 2. R245fa properties

Molecular weight (g/mol)	134.05
Normal boiling point (°C)	15.14
Critical pressur (Mpa)	3.65
Critical temperature (°C)	154.01
Ozone Depleting Potential	0
Global warming protection	950
Safety group classification	B1

2.4 Size of Generator Main Components

For sure, coupling of solar collectors with ORC generator entails a relevant encumbrance nearby the user house or more generally building. This aspect has to be analysed in order to optimize the room occupation inside a building and over its roof. The solar collectors foreseen in the actual early design stage occupy 50 square meters and can be mounted on roof even if flat.

The other component distinguished by remarkable size is undoubtedly the heat storage. Nevertheless, this and other components like pumps, two or three way valves, and simple valves are part of a normal plumbing and heating system. The room over occupied in comparison with a traditional thermohydraulic plant is due to solar collectors and ORC engine and distribution piping associable with them. Not even absorption chiller can be considered as it substitutes traditional heat pump system. On the other hand, evaporative cooling tower or dry cooler has to be associated to the absorption chiller and it is a component that would not be present in a traditional plant. The cooler associated to the absorption chiller has to be placed outside to permit heat exchange and has the following sizes: 2300 mm width, 1150 mm length, and 950 mm height. While the other relevant component of the system is the ORC piglet with the following size: 1650 mm width, 1550 mm length, and 1450 mm height.

3   Model and Simulation

The plant model has been developed in TRNSYS 17 that is a powerful software using FORTRAN subroutines and able to simulate energy patterns with time dependent inputs. In this way is possible to include the fluctuant and variable sun irradiation as input and to monitor the energy fluxes linked to it. Each FORTRAN subroutine, linked together in the TRNSYS environment, represents a component of the system, and a mathematical model simulates his functions having inputs, outputs, variable with time, and parameters, constant for all the duration of the simulation. The TRNSYS library has a wide of already tested subroutines ("type") for the simulation of solar collectors, thermal storage, and piping. The subroutine for the simulation of the Organic Rankine Cycle has been developed to include this component in the simulation. Through the graphic environment of Studio has been drawn the model shown in Fig. 2 used to run the TRNSYS simulation. The system simulated is composed by components showed in the model layout.

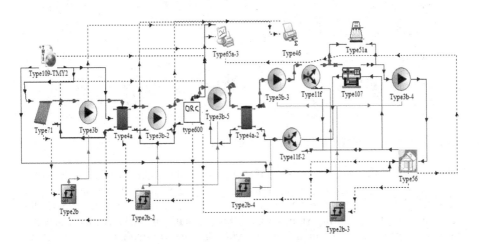

Fig. 3. Power plant model layout

As indicated in the Fig. 3 the TRNSYS model starts with solar radiation and environmental values, Type109, set with Rome irradiation data. The Type109 gives the irradiation values to the solar collector, modelled by Type71. The Type 71 is connected to the storage (Type4a) via the pump (Type3b) that is activated when the solar collector temperature fluid is greater than 10°C and it is stopped when the solar collector temperature fluid is lower than 2°C always of the average tank temperature. The type2b has been always used to implement the control system that gives the command to the different pumps (Type3b).

The storage is connected to the ORC that has been developed through an imported Fortran model [6] represented with type600. The relative pump (Type3b-2) is activated when the higher storage temperature fluid is greater than 150°C and it is stopped when the higher storage temperature fluid is lower than 80°C. The ORC is connected to the absorber system (type107) and to the thermal load (type56). The system decide to direct the ORC waste power to the absorber or to the thermal load in base of load demand (cooling or heating) through the three-way valves (type11f and type 11f-2). The absorber waste is re-cooled via the evaporative tower (type51a).

4   Model Outputs

The results of the simulations are showed in the next figures. The simulation step has been always 15 minutes and always on the left there are power and flow values, on the right there are temperature values. The Figure 4 and 5 below show the annual, simulation time 8760 hours, results of the simulation regarding the solar and the ORC power plant sections.

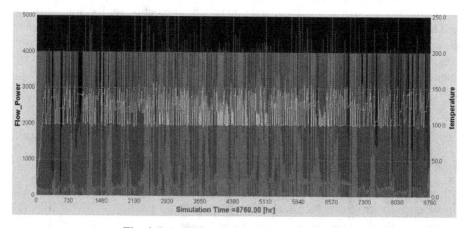

Fig. 4. Solar power plant section annual results

From Figure 4 it is possible to see that the total solar irradiation varies from 0 to 3800 kJ/hm^2, the average tank temperature after few days reach 150°C and then varies, as fixed, between 150 and 100 °C, the CPC temperature varies between 5°C, during winter, or 15 °C, during summer, and 150°C (with little peaks that reach also 250°C), the pump flow is functioning at 4000 kg/h and depending on the difference between CPC and average tank temperatures as fixed.

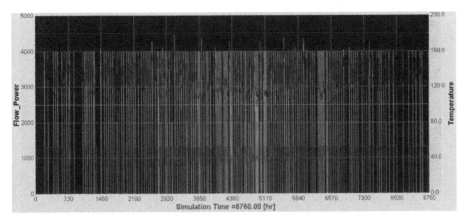

Fig. 5. ORC power plant section annual results

From Figure 5 it is possible to see that the total power varies from 0 to 3000 W (with peaks that reach also 4500 W during summer), the tank to ORC temperature varies between 100°C and 150°C, the evaporator temperature varies between 0°C and 120°C, the condenser temperature is always over 60 °C (in order to have the necessary temperature for the heating or absorber demand), the pump flow is functioning always at 4000 kg/h and depending on the difference between tank to ORC and the 100°C stationary temperature as fixed.

The Figures 6 and 7 below show the results of the simulation regarding the solar, and the ORC respectively during the first 5 days (thus first January days).

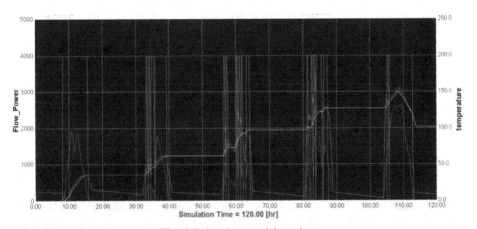

Fig. 6. Solar plant model results

From Figure 6 it is possible to see the variation of the CPC temperature corresponding to the variation of the total solar radiation and thus the relative activation (and the flow, fixed at 4000 kg/h) of the pump. It is easy to see that after five days of solar radiation the average tank temperature reach the fixed ORC starting temperature (150 °C). From the global simulation it is possible to see that 150°C will be reached again, on average, after 1-2 days in winter and after 0-1 days during summer.

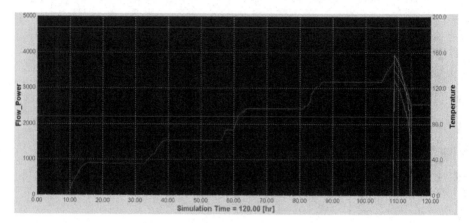

Fig. 7. ORC plant model results

From Figure 7 it is possible to see the starting point of the ORC when the average tank temperature reach 150°C. The ORC pump flow (fixed at 4000 kg/h), the evaporator temperature, the electric power produced and the ORC to tank temperature are also showed.

The simulation output provides a current performance varying between 2 and 4 kW depending on evaporator and thus tank temperature that depends on solar irradiation. The annual electricity production is approximately of 5000 kWh keeping mean ORC efficiency included in the range 4%–9%, in accordance with existing references for similar systems [21]. This meaning that the ORC Equivalent Full Power Hours, considering that this ORC have a nominal power of 3 kWe, are 1666 hours. On the other hand, higher thermal levels at solar loop allows to reach almost 10% instantaneous efficiency [22]. The multi-generator also produces 20000 kWh thermal energy and approximately 10000 kWh for refrigeration. The thermal energy output obtained counts on use of low temperature heating distribution plant. By the way, the concept to emphasize is the conflict between efficient energy production system and ineffective final user devices. That is the reason why it is necessary a right combination of such an efficient and zero emission energy generation system, an effective distribution loop and low temperature heating terminals. Then, the coupling of a STS generator and ineffective heaters would be a nonsense.

One of the most important targets to reach by the present research activity is to improve the EFPH (acronym of Equivalent Full Power Hours). The EFPH parameter is given by energy/power ratio that is the ratio between the energy produced in one typical year and the nominal power of the energy system. In fact, the meaning of energy systems power has a partial significance in order to evaluate the energy generation potential. Deliver the only power, as an input data, is not a complete information. In particular case of renewable energies, different sources can generate very different quantities of energy even though having identical power. The EFPH ratio is indicative of the energy potential resident in each kW of the power system.

5 Economic Analysis

Considering the particular characteristics to assess the opportunity to finance this kind of application, the following aspects have to be considered:

- a low level of technological risk in plant construction;
- a strong captive market, created by means of long-term contracts signed with solid counterparties (i.e. off-takers) with the aim to mitigate the market risk both in terms of pricing and volume on the sell side;
- a reliable supply market, in case of production initiative driven by volatile raw materials ' markets, created by means of long-term contracts signed with solid counterparties (i.e. suppliers) with the aim to mitigate the market risk both in terms of pricing and volume on the supply side.

Main financial parameters, used in the present analysis, are:

- the NPV (acronym of Net Present Value) which is the net present value (NPV) of a time series of cash flows, both incoming and outgoing, and is defined as the sum of the present values of the individual cash flows;
- PBT or PBP (acronym of payback time or period) in capital budgeting refers to the period of time required for the return on an investment to "repay" the sum of the original investment. It is also widely used in other types of investment areas, often with respect to energy efficiency technologies, maintenance, upgrades, or other changes.

5.1 Concentrating Solar Power Market and Key of Success in Italian Market

In the following paragraphs, is shown, by statistics pertinent to two important European countries markets, a big plants focused solar thermoelectric sector [23]. Spain leadership shows a relevant application of power electric production by solar thermal source. In virtue of the incentive introduction law occurred in 2008, Italy had a good onset of solar thermoelectric application but biggest efforts have been addressed to the other technology to produce electricity by solar irradiation: the photovoltaic.

The reasons for Spanish success and Italian failure in CSP development are related to the efficient way to manage the theoretical key drivers for success. Firstly, Spain is far more apt for CSP technology than Italy from a geomorphologic point of view. Spain has far more deserted flat areas that would not have any alternative use, and is less exposed to earthquakes than Italy. In addition to this as Italian territory is mostly developed along its longitude whereas Spain is equally developed along its longitude and its latitude, Spain has far larger areas with high solar radiation than Italy. Secondly, Italian entrepreneurs and industrial operators are less risk takers than the Spanish ones and this impeded the success of CSP due to its lack of positive track records. Moreover timing for the introduction the CSP Governmental incentive schemes was far more favourable for the Spanish industry. Indeed the Spanish Government introduced the first CSP feed-in-tariff in 1998, at the beginning of its economic boom which ended with the financial crisis of 2008. On the other hand, Italy introduced the current CSP incentive scheme in April 2008 and this not only slowed down investments proposal from the industry but also undermined the financial support of banks.

Banks and financial institutions attitude to taking risk and their size also affected the development of CSP in Italy.

Least, the authorisation process in Italy is far more complicated and longer than in Spain as it is depends on approval from many local authorities (municipalities, provinces and regions). This made international investors reluctant to finance CSP projects in Italy as the delay to obtain the authorisations would increase the risk to see the technology turn obsolete.

Many of the obstacles connected with Italian CSP failure can be obviated by the evolution of small size plants because, as mentioned, residential application of technologies have a relevant role in the renewable energy propagation scenario because of the lack of boundless surfaces where to install huge extensive solar power plants contrary to Spain territory features. Furthermore, the diffusion on the territory of small scale systems working nearby the energy user constitutes the vocational guidance of distributed generation concept.

5.1.1   The Spanish Case

The first step of Spain to promote the CSP started in 1998 by Royal Decree 2818/1998 which promoted the development of solar technologies (non-hybrid).

After that, a series of other Royal Decrees introduced changes in 2002 and 2004 (Royal Decree 841/2002, Royal Decree 436/2004), until the latest Royal Decree 661/2007, introduced on 25 May 2007, that established a new incentive arrangements which introduced the annual adjustment of the tariff on the basis of inflation.

Spain is currently the world leader in CSP with an installed capacity of over 700 MW even higher than the USA. The Table 3 below shows the data of the main CSP Plants.

Table 3. CSP Plant

Plant	MW	Developer	Commercial Operation Date
PS 10	11	Abengoa Solar	2007
Andasol 1	50	ACS-Cobra Energy	2008
PS 20	20	Abengoa Solar	2009
Puertollano	50	Iberdrola	2009
Puerto Errado	1.4	Novatec Solar España S.L.	2009
Andasol 2	50	ACS-Cobra Energy	2009
Solnova 1	50	Abengoa Solar	2009
Solnova 3	50	Abengoa Solar	2009
Solnova 4	50	Abengoa Solar	2009
Alvarado 1	50	Acciona	2009
La Florida	50	Renovables SAMCA	2010
Extresol 1	50	ACS-Cobra Energy	2010
Majada 1	50	Acciona	2010
Palma del Rio II	50	Acciona	2010
La Dehesa	50	Renovables SAMCA	2011
Manchasol 1	50	ACS-Cobra Energy	2011
Extresol 2	50	ACS-Cobra Energy	2010

The Spanish market is ruled by large companies which usually have the expertise and the capabilities to play all the roles involved in project finance being the developer, the sole owner, the EPC contractor and the operator.

5.1.2   The Italian Case

The Ministerial Decree (MD) of 11 April 2008, published in May 2008, laid down criteria and procedures for supporting electricity generation from the solar source via plants that uses thermodynamic cycles.

Feed-in tariffs are recognised to solar electricity generated by solar plants or by hybrid plants over a period of 25 years as a fixed value without any inflation adjustment and can be accumulated with the selling price on the grid. The incentive values vary from 0.22 to 0.28 €/kWh depending on the value of the solar share.

The first concentrating solar power plant constructed and operating in Italy is the 5 MW Archimede plant in Priolo Gargallo (Syracuse, Sicily) owned by Enel and inaugurated on 15 July 2010. This plant has a high technological and innovative value as it is the first in the world to be integrated with a combined-cycle plant and use the Enea molten-salt technology instead of a conventional synthetic oil technology. For this reason it can be considered as an R&D investment and it has not been financed in project finance by banks but via corporate finance solutions by Enel. After that, no further plants are under construction nor development in Italy. The incentive scheme was conceived to favour growth and optimization of some technologies developed in Italy in the last years but beyond the Sicily solar power station no further investments and construction have been made.

A new event changed the history of solar thermoelectric sector: the incentive structure has been modified by the Ministerial Decree (MD) of 6th July 2012. The new law modifies the previous one and does not fix a lower limit for the square meters acceptable. From 1st of January 2013 small plants with less than 10000 square meters can be admitted to the feed-in-tariff promotion. While in plants made of more than 10000 square meters of solar collectors, thermal energy capacity of the storage tank must be greater than a value of kWh per square meter (for instance a 0.4 kWh/m^2 lower limit regards plants with more than 10000), residential storage systems do not have any kind of capacity limit to respect.

Then, plants connected to the electric grid from 1st of January 2013 until 31st December 2015 composed of less than 2500 square meters can gain access to 0.36 euros per kWh produced. Even in this case, feed-in tariffs are recognised to solar electricity generated by solar plants over a period of 25 years as a fixed value without any inflation adjustment and can be accumulated with the selling price on the grid. Furthermore, lower incentives can be recognized to solar thermal power generators with an integration fraction of energy produced by biomasses.

5.1.3   Applicability of Italian Incentive Scheme to Small Scale System

As shown, until nowadays, market of solar thermoelectric application has only been focused on the big scale power generators. Nevertheless, for instance in Italy, unused incentives can be apply also to small scale systems with a consequent exploitation of economic resources foreseen by the Italian government to be applied to big scale ones. The previous inquiry passes through the economical profit of this application for the Italian energy users.

The occurred modification solar thermodynamic incentive have been modified as illustrated in the previous paragraphs and a new future could be opening up for it.

5.2   Economic Evaluation

It is possible to make a rough assessment of the economic return associated with the construction of the solar plant coupled to a 3 kW power Organic Rankine Cycle generator. In particular, the feasibility study was carried out through the main methods of NPV (Net Present Value) and PBP (Pay Back Period) assessment. Considering an implementation cost of around € 55200 (the system is still in testing and will be patent pending as soon as possible). Indeed, the cost of the solar thermal plant with the primary loop, is equal to € 26200, the cost of the ORC is about € 10000, and the cost of the absorption refrigerator is equal to € 7000. Adding these costs to control and installation costs, a total amount € 60200 is reached. These amount are similar to the cost of construction of the prototype at university laboratories. To these costs must be added the cost of maintenance and management. This kind of costs has been assessed but lack of an appropriate follow-up period with particular reference to small size ORC is noteworthy.

Revenue consists of the electricity consume savings, with a cost ever and quickly growing, government incentive (0.36 €/kWh), thermal energy savings (0.60 €/m^3) and further electricity savings due to refrigeration.

From the figures it can be seen that the cash flows are slowly positive, obtaining a return on investment equal to 13 years.

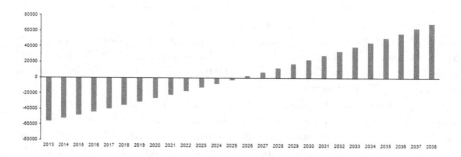

Fig. 8. Cash flows in cogeneration

The payback time depends mainly on: annual hours of work of the system, cost of electricity and heat and, especially, investment cost.

Part of revenues can be considered fixed, given the feed-in-tariff for 25 years. Operating costs also can be considered as being a system of small size and, on the other side, they can easily be automated.

The other relevant parameters calculated are NPV and IRR. The first one is positive (a bit more than € 11000) while the second one is equal to 7%.

6   Conclusions

As shown the system designed would have a payback period of 13 years. The total amount of income during the 25 years period is about 155000 Euro and the yearly gain, after the payback time, would amount to a value between € 6200 and € 7900. The evaluation shows that, without the "conto energia" incentives, the payback time would be of 18 years with an IRR of 4% and a negative NPV. In order to reach the grid parity conditions an investment cost reduction is indispensable and the road is long and winding. The evaluation shows that standardization of system do not have a so relevant impact. The mentioned cost reduction involves more specifically other items of expenditure as solar plant (particularly solar collectors) and ORC generator or absorption refrigerator. Currently, the ORC generator requires a 5,000 € /kW expenditure which is far from an affordable investment as showed by authors for other new technology small scale power plants [24–27]. Costs could benefit from enlargement of these components production scale, harnessing bigger size plants whose attraction is much more effective from the point of view of the investment [28–31].

Furthermore, absorption chiller can be numbered between the devices and components part of the solar generator but reckoning to provide a new building or house or, more generally, a new construction with such a solar power system or considering the exigency to update or replace the apartment chiller, it would not be correct to ascribe the refrigerator to the sun generator items of expenditure list.

Nevertheless, the advantages of the system do not concern only the private user of the multi energy generator but also the national energy distribution system: the grid. This system do not require the energy grid reconstruction as it is happening owing to myriad of territory distributed photovoltaic and wind energy installations. In fact, in virtue of system controllability, energy received by aleatory solar irradiation can be used with a delay in comparison with time of collection. That means that the cost load (direct and indirect) of this kind of system is not comparable to other renewable technologies by now widely employed.

Acknowledgements. The authors acknowledge the financial support of Italian Contract SEC-DEC-2011-0000564 (regarding the research project STS, Solar Trigeneration System) funded by the Italian Ministry for the Environment, Land and Sea (Ministero dell'Ambiente e della Tutela del Territorio e del Mare).

References

1. I Sistemi Solari Termici Per La Climatizzazione - Villarini Mauro, Germano', Domenico, Fontana Francesco, Limiti Maurizio - Maggioli Editore - Libro + Cd-Rom - Hoepli.it, http://www.hoepli.it/libro/i-sistemi-solari-termici-per-la-climatizzazione-/9788838754388.html
2. Germanò, D., Gasparro, M., Villarini, M.: Principi di progettazione degli impianti solari termici. Maggioli Editore (2011)

3. Eltawil, M.A., Zhao, Z.: Grid-connected photovoltaic power systems: Technical and potential problems—A review. Renewable and Sustainable Energy Reviews 14, 112–129 (2010)
4. Falvo, M.C., Martirano, L., Sbordone, D.: Sustainable Energy Microsystems for a Smart Grid. In: Apolloni, B., Bassis, S., Esposito, A., Morabito, F.C. (eds.) Neural Nets and Surroundings. SIST, vol. 19, pp. 259–269. Springer, Heidelberg (2013)
5. Falvo, M.C., Martirano, L., Sbordone, D., Bocci, E.: Technologies for Smart Grids: a brief review. In: 12th IEEE-EEEIC International Conference on Environment and Electrical Engineering. IEEE Press, New York (2013)
6. Bocci, E., Villarini, M., Bove, L., Esposto, S., Gasperini, V.: Modeling Small Scale Solar Powered ORC Unit for Standalone Application. Mathematical Problems in Engineering (2012)
7. Carlini, M., Villarini, M., Esposto, S., Bernardi, M.: Performance analysis of greenhouses with integrated photovoltaic modules. In: Taniar, D., Gervasi, O., Murgante, B., Pardede, E., Apduhan, B.O. (eds.) ICCSA 2010, Part II. LNCS, vol. 6017, pp. 206–214. Springer, Heidelberg (2010)
8. Carlini, M., Castellucci, S.: Modelling and simulation for energy production parametric dependence in greenhouses. Mathematical Problems in Engineering (2010)
9. Carlini, M., Castellucci, S., Guerrieri, M., Honorati, T.: Stability and control for energy production parametric dependence. Mathematical Problems in Engineering (2010)
10. Carlini, M., Honorati, T., Castellucci, S.: Photovoltaic greenhouses: comparison of optical and thermal behaviour for energy savings. Mathematical Problems in Engineering (2012)
11. Carlini, M., Castellucci, S.: Efficient energy supply from ground coupled heat transfer source. In: Taniar, D., Gervasi, O., Murgante, B., Pardede, E., Apduhan, B.O. (eds.) ICCSA 2010, Part II. LNCS, vol. 6017, pp. 177–190. Springer, Heidelberg (2010)
12. Carlini, M., Castellucci, S., Allegrini, E., Tucci, A.: Down-hole heat exchangers: modelling of a low-enthalpy geothermal system for district heating. Mathematical Problems in Engineering (2012)
13. Drescher, U., Brüggemann, D.: Fluid selection for the Organic Rankine Cycle (ORC) in biomass power and heat plants. Applied Thermal Engineering 27, 223–228 (2007)
14. Carlini, M., Castellucci, S.: Modelling the vertical heat exchanger in thermal basin. In: Murgante, B., Gervasi, O., Iglesias, A., Taniar, D., Apduhan, B.O. (eds.) ICCSA 2011, Part IV. LNCS, vol. 6785, pp. 277–286. Springer, Heidelberg (2011)
15. Boretti, A.: Recovery of exhaust and coolant heat with R245fa organic Rankine cycles in a hybrid passenger car with a naturally aspirated gasoline engine. Applied Thermal Engineering 36, 73–77 (2012)
16. Carlini, M., Cattani, C., Tucci, A.O.M.: Optical modelling of square solar concentrator. In: Murgante, B., Gervasi, O., Iglesias, A., Taniar, D., Apduhan, B.O. (eds.) ICCSA 2011, Part IV. LNCS, vol. 6785, pp. 287–295. Springer, Heidelberg (2011)
17. Winter, C.-J., Sizmann, R.L., Vant-Hull, L.L.: Solar power plants. Springer, New York (1991)
18. Saleh, B., Koglbauer, G., Wendland, M., Fischer, J.: Working fluids for low-temperature organic Rankine cycles. Energy 32, 1210–1221 (2007)
19. Delgado-Torres, A.M., García-Rodríguez, L.: Analysis and optimization of the low-temperature solar organic Rankine cycle (ORC). Energy Conversion and Management 51, 2846–2856 (2010)
20. US Department of Commerce, N.: Standard Reference Data, http://webbook.nist.gov/chemistry/fluid/

21. Roy, J.P., Mishra, M.K., Misra, A.: Performance analysis of an Organic Rankine Cycle with superheating under different heat source temperature conditions. Applied Energy 88, 2995–3004 (2011)
22. Bocci, E., Di Carlo, A., Marcelo, D.: Power plant perspectives for sugarcane mills. Energy 34, 689–698 (2009)
23. Villarini, M., Limiti, M., Abenavoli, R.I.: Overview and comparison of global concentrating solar power incentives schemes by means of computational models. In: Murgante, B., Gervasi, O., Iglesias, A., Taniar, D., Apduhan, B.O. (eds.) ICCSA 2011, Part IV. LNCS, vol. 6785, pp. 258–269. Springer, Heidelberg (2011)
24. Orecchini, F., Bocci, E., Di Carlo, A.: Process simulation of a neutral emission plant using chestnut's coppice gasification and molten carbonate fuel cells. Journal of Fuel Cell Science and Technology 5 (2008)
25. Bocci, E., Zuccari, F., Dell'Era, A.: Renewable and hydrogen energy integrated house. International Journal of Hydrogen Energy 36, 7963–7968 (2011)
26. Di Carlo, A., Bocci, E., Naso, V.: Process simulation of a SOFC and double bubbling fluidized bed gasifier power plant. International Journal of Hydrogen Energy 38, 5857–5874 (2013)
27. Tucci, A.: Modeling of an Air Conditioning System with Geothermal Heat Pump for a Residential Building. Mathematical Problems in Engineering (2013)
28. Poullikkas, A.: Economic analysis of power generation from parabolic trough solar thermal plants for the Mediterranean region—A case study for the island of Cyprus. Renewable and Sustainable Energy Reviews 13, 2474–2484 (2009)
29. Brenna, M., Falvo, M.C., Foiadelli, F., Martirano, L., Massaro, F., Poli, D., Vaccaro, A.: Challenges in energy systems for the smart-cities of the future. In: 2012 IEEE International Energy Conference and Exhibition (ENERGYCON), pp. 755–762 (2012)
30. Brenna, M., Falvo, M.C., Foiadelli, F., Martirano, L., Poli, D.: Sustainable Energy Microsystem (SEM): preliminary energy analysis. In: 2012 IEEE PES Innovative Smart Grid Technologies (ISGT), pp. 1–6 (2012)
31. Prudenzi, A., Silvestri, A., Lamedica, R., Falvo, M.C., Regoli, M.: A Domestic Electric Load Simulator Including Psychological Aspects of Demand. In: Proc. of IEEE-PES 2010 General Meeting, Minneapolis, Minnesota (2010)

Mathematical Analysis of Gasification Process Using Boubaker Polynomials Expansion Scheme

Andrea Colantoni[1], Elena Allegrini, Fabio Recanatesi[1], Manuela Romagnoli[1],
Paolo Biondi[1], and Karemt Boubaker[2]

[1] Department of Agriculture, Forest, Nature and Energy (DAFNE), University of Tuscia,
Via S. Camillo de Lellis snc, 01100 Viterbo, Italy
{colantoni,allegrini.e,fabio.rec,mroma,biondi}@unitus.it
[2] Unité de physique des dispositifs à semi-conducteurs, Tunis EL MANAR University,
2092 Tunis, Tunisia
mmbb11112000@yahoo.fr

Abstract. In this work, a mathematical model of gasification technology has been developed. The aim of the present paper is to analyze the total Gibbs free energy G(U) of the system during the process which strongly depends on the type of biomass CH_MO_N. In order to reach this goal, the Boubaker Polynomial Expansion Scheme (BPES) is exploited as an efficient optimization protocol. The results lead to find out the optimal values for biomass composition and clearly show that the latter and its substrate ratio influence the free energy trend.

Keywords: gasification process, free energy, biomass, mathematical analysis.

1 Introduction

In the recent years, research, development and industry have been addressed to the use of Renewable Energy Sources (RES)and have moved towards their exploitation. Actually they have been recognized as a promising and an extremely attractive alternative to fossil fuels. This becomes even more important if climate change and global warming have to be successfully faced. Moreover, in the current energy scenario it is fundamental to rely on environmental-compatible and sustainable technologies which may ensure the independence on fossil fuels with fewer impacts if compared with exhaustible sources. Thus, special attention has been paid to the use of biomass which can be efficiently converted via biological or thermo-chemical processes. Gasification belongs to the second category of conversion methods and consists in transforming a solid fuel into a gaseous energy vector which is called syngas. Gasification can be further considered as an interesting chance since small and medium size plants can be successfully integrated in a distributed energy system. Nowadays it is seen as one of the most efficient ways in which the energy embedded by the biomass itself can be transformed into useful and marketable products [1, 2, 3, 4].

Gasification process consists in a partial thermal oxidation which leads to the production of gases (such as CO_2, water, carbon monoxide, hydrogen and gaseous

B. Murgante et al. (Eds.): ICCSA 2013, Part II, LNCS 7972, pp. 288–298, 2013.
© Springer-Verlag Berlin Heidelberg 2013

hydrocarbons), small quantities of char, ash and condensable compounds (tars and oils). The reaction is developed by supplying an oxidizing agent, e.g. oxygen, steam or air. The syngas produced by the gasification is more versatile than the original biomass and can be successfully used to power gas engines and gas turbines [1, 2]. Biomass gasification is characterized by extremely complex chemical reactions. Broadly speaking, the below-listed stages can be identified during the process [1, 2]:

1. drying, which occurs at 100-200°C and leads to moisture content reduction in the biomass. Typical ranges of moisture content are from 5% to 35% and decrease to <5% thank to the present stage;
2. pyrolisis or devolatilisation, consisting in biomass thermal decomposition in absence of oxygen or air. Volatiles are removed and light hydrocarbon gases are released from the biomass so that the latter is reduced to solid charcoal. Moreover, if these gases condense (this occurs at very low temperatures), liquid tars are generated;
3. oxidation, which is a reaction between oxygen and solid carbonized biomass and results in formation of carbon dioxide. Water is generated due to the oxidation of hydrogen. The reactions of carbon and hydrogen oxidation release a large amount of heat;
4. reduction, which occurs in absence or in small quantities of oxygen and requires temperatures between 800 and 1000 °C.

The following reactions take place during the gasification process:

Table 1. Main reactions during the gasification process

Gasification stage	Chemical reaction	Reaction code
Partial-combustion reaction	$C + \frac{1}{2}O_2 \rightarrow CO$	a
Combustion reaction	$C + O_2 \rightarrow CO_2$	b
Bouduard reaction	$C + CO_2 \rightarrow 2CO$	c
Water-gas reaction	$C + H_2O \rightarrow CO + H_2$	d
Methanation reaction	$C + 2H_2 \rightarrow CH_4$	e
Water-gas shift reaction	$CO + H_2O \rightarrow H_2 + CO_2$	f
Reforming reaction	$CH_4 + H_2O \rightarrow CO + 3H_2$	g
Cracking reaction	Char $\rightarrow CH_4 + H_2O + C_MH_N + H_2$	h

Partial combustion and combustion are exothermic reactions so that thermal energy is released during their development. Then, several endothermic reactions occur and a specific amount of heat is absorbed.

Gasification process can be successfully developed using three different kinds of gas-solid contacts [1, 5, 6]:

a. fixed bed: the gasifier zone has a fixed position and the gas passes through the material. The bed consists of solid fuel particles. In the updraft gasifier, air and biomass are fed up at the bottom and the top of the reactor respectively. Air passes through a grate and goes upward. In the downdraft gasifier, both air and solid biomass move in the same direction. In the cross-flow gasifier, air is introduced from one side while biomass moves downward. The gas is withdrawn from the opposite side, at the same level of the air inlet;

b. fluidized-bed: biomass particles are kept in a state of suspension using a gasifying agent which is blown through the bed. Fuel particles enter the reactor from the bottom and are quickly mixed with the bed material. Depending on the fluidization pattern and combination character, three different solutions can be adopted: bubbling, circulating or dual fluidized bed;

c. entrained flow bed which are commonly used for coal since they can be slurry fed.

Considering the growing interest in biomass gasification, several models have been proposed and aimed at better explaining this process. This paper presents a mathematical modelling of the thermodynamic equilibrium in a gasifier. The aim of the-work is to identify the optimal conditions for the gasifier performance with specific regard to the total Gibbs free energy G(U) of the system. A downdraft fixed-bed reactor is considered in the simulation protocol since this solution is easy to construct and has robust structure. Moreover, it can operate at partial load and is characterized by low pressure drop. Carbon conversion and thermal efficiency can be considered high so that the amount of tar and phenols in product gas is limited. Product gas is suitable for direct firing but an extensive cleanup is needed [1, 2].

2   Material and Methods

The efficiency of a biomass gasifier strongly depends on several complex chemical reactions. In order to better understand these complicated processes –which are related to the rate of heating and to the residence time in the reactor- mathematical models are needed and have been developed. Several attempts have been carried out from time to time to explain the nature of gasification. Three different categories emerge from this effort [1, 2]:

- kinetic rate models, which provide basic and essential information on kinetic mechanisms since they play a fundamental role in designing gasifiers and evaluating and improving their performances. Moreover, they are accurate, although an intensive computation is required;
- thermodynamic equilibrium models, which are independent on the gasifier design. The chemical equilibrium is achieved when a reacting system is at its most stable composition, i.e. when the entropy of the system is maximised while the Gibbs free energy is at its minimum value. Although the thermodynamic equilibrium may not be reached, several models have been developed and used widely;
- neural network models, describing some characteristics of the process which involve a Multilayer Feed forward Neural Network (MFNN).

It has to be noted that some models combine and exploit the first two models [1, 2]. The attention will be focused on thermodynamic equilibrium models which are based on two general approaches: stoichiometric and non-stoichiometric. The first method requires a set of selected independent reactions and the definition of the species involved in the process. On the opposite, in the non-stoichiometric approach, reaction mechanisms are not directly involved in the numerical simulation so that a set of selected species are assumed to be present in the syngas. Past studies proved that both methods are equivalent [1, 2, 7, 8]. It is extremely important to highlight that thermodynamic equilibrium calculations do not depend on gasifier design and, as a consequence, are suitable to study the influence of fuel process parameters. However, it has to be considered that equilibrium might not be reached at low operation temperatures [7,8].

Equilibrium models are based on the following assumptions [7, 8]:

- Biomass is represented by the general formula $C\ H_M O_N$;
- Heat losses are neglected: the gasifier can be regarded as a perfectly insulated apparatus. Thus, the process can be considered adiabatic although in practice gasifiers have heat losses to the external environment;
- No chars living with the exit the gasifier products;
- Gasification reactions are fast enough and residence time is long enough to reach the thermodynamic equilibrium state (at P= 1 bar);
- Ashes are considered to be meaningless;
- Tars are not modelled.

Considering the above-listed items, it has to be considered that equilibrium models may lead to great disagreements if some circumstances occur [1].

In the considered process, pyrolisis or devolatilisation, gases are supposed not to condense. In this case, tar generation can be neglected. Moreover, it is known that the volatile yield of dry wood is subdivided intolight hydrocarbons tar carbon monoxide, carbondioxide, hydrogen and moisture yield. According to Ragland & Aerts (1991) tar fraction is negligible.

For normalization purposes and taking into account that the reaction considered in Table 1 are mainly c, e, f and h, the following reaction takes place in the case of wood-like materials:

$$
CH_M O_N + aH_2 O + bO_2 + cN_2 \rightarrow
$$
$$
x_1 C + x_2 H_2 + x_3 CO + x_4 H_2 O + x_5 CO_2 + x_6 CH_4 + x_7 N_2 \tag{1}
$$

As a first step, coefficients a, b and c are calculated using stoichiometric balance conditions. The partial pressure $p_{j=2...7}$ of each gas species j is defined as:

$$
p_j \Big|_{j=2..7} = \frac{x_j}{P_X} = p\,\frac{x_j}{\displaystyle\sum_{j=2}^{7} x_j} \tag{2}
$$

where P is the pressure within the gasifier and P_X is given by:

$$P_X = \frac{\displaystyle\sum_{j=2}^{7} x_j}{p} \qquad (3)$$

Hence, if adsorption constant, rate constant and equilibrium constant, for each reaction $R_{m=1\ldots4}$ are denoted as $K_{m=1\ldots4}$, $k_{m=1\ldots4}$ and $\tilde{K}_m\big|_{m=1..4}$ respectively, the net equation rate $\xi_m(U)\big|_{m=1..4}$ for each reaction $R_{m=1\ldots4}$, where $U=(x_1, x_2, x_3, x_4, x_5, x_6, x_7)$, can be formulated based on the Langmuir-Hinshelwood mechanism:

$$
\begin{cases}
\xi_1(U) = -A_1 \dfrac{x_5 - \dfrac{x_3^2}{P_X \tilde{K}_1}}{\displaystyle\sum_{j=2}^{7}(K_j + \dfrac{1}{p})x_i} \sqrt[3]{x_1^2} \\[4em]
\xi_2(U) = -A_2 \dfrac{x_4 - \dfrac{x_3 x_2}{P_X \tilde{K}_2}}{\displaystyle\sum_{j=2}^{7}(K_j + \dfrac{1}{p})x_i} \sqrt[3]{x_1^2} \\[4em]
\xi_3(U) = -A_3 \dfrac{x_2^2 - \dfrac{x_6 P_X}{\tilde{K}_3}}{P_X \displaystyle\sum_{j=2}^{7}(K_j + \dfrac{1}{p})x_i} \sqrt[3]{x_1^2} \\[4em]
\xi_4(U) = -A_4 \dfrac{x_4 x_6 - \dfrac{x_3 x_2^2}{P_X^2 \tilde{K}_4}}{P_X \displaystyle\sum_{j=2}^{7}(K_j + \dfrac{1}{p})x_i} \sqrt[3]{x_1^2}
\end{cases}
\qquad (4)
$$

with $A_m\big|_{m=1..4}$ constants.

These items are to linked the adsorption model of Irving Langmuir (1916). This model assumes for a complex process, a rate law can be written for each step of process, and when the rates become equal an equilibrium state will occur.

The rate of adsorption is proportional to the concentration of gas/liquids wich react at a given stage. The proportionality factor (between rate of adsorption and gas/liquids concentration) is defined as rate constant. The equilibrium constant for a given reaction is universally defined as: products to reactants adjusted ratio.

The first components of $U=(x_1, x_2, x_3, x_4, x_5, x_6, x_7)$ are solutions to the differential equations system:

$$\begin{cases} \dot{x}_1 = \xi_1(U) + \xi_2(U) + \xi_3(U) \\ \dot{x}_2 = -\xi_2(U) + 2\xi_3(U) - 3\xi_4(U) \\ \dot{x}_3 = -2\xi_2(U) - \xi_3(U) - \xi_4(U) \\ \dot{x}_4 = \xi_2(U) + \xi_4(U) \\ \dot{x}_5 = \xi_1(U) \\ \dot{x}_6 = -\xi_3(U) + \xi_4(U) \end{cases}$$

(5)

For given values of parameters, the resolution of equations 5 is carried out using the Boubaker Polynomials Expansion Scheme BPES. The BPES is applied by setting the expression:

$$x_i(t)\Big|_{i=1..6} = \frac{1}{2N_0} \sum_{k=1}^{N_0} \lambda_{k,i} \times B_{4k}(r_k t)$$

(6)

where:

- B_{4k} are the 4k-order Boubaker polynomials;
- r_k are B_{4k} minimal positive roots;
- N_0 is a prefixed integer;
- $\lambda_{k,i}$ are unknown pondering real coefficients;
- t has been introduced to define the substrate characteristic ratio as follows:

$$t = \frac{M}{N}$$

(7)

where M and N are hydrogen and oxygen coefficients in expression (1) respectively. The main advantage of the formulation expressed by (6) is the evidence of verifying the boundary conditions, in advance to problem resolution thanks to the properties of the Boubaker polynomials [9-16], besides proposing differentiable and piecewise continuous solutions [16-21]. The BPES protocol ensures the validity of the related boundary conditions expressed through biological conditions, regardless main equation features. In fact, thanks to Boubaker polynomials first derivatives properties:

$$\begin{cases} \sum_{q=1}^{N} B_{4q}(x)\Big|_{x=0} = -2N \neq 0; \\ \sum_{q=1}^{N} B_{4q}(x)\Big|_{x=r_q} = 0; \end{cases}$$

(8)

and:

$$\begin{cases} \displaystyle\sum_{q=1}^{N} \frac{dB_{4q}(x)}{dx}\bigg|_{x=0} = 0 \\ \displaystyle\sum_{q=1}^{N} \frac{dB_{4q}(x)}{dx}\bigg|_{x=r_q} = \sum_{q=1}^{N} H_q \end{cases}$$

(9)

where:

$$H_n = B'_{4n}(r_n) = \left(\frac{4r_n[2 - r_n^2] \times \displaystyle\sum_{q=1}^{n} B_{4q}^2(r_n)}{B_{4(n+1)}(r_n)} + 4r_n^3 \right)$$

(10)

the boundary conditions are inherently verified.

By introducing expression (6) in Eq. (5), boundary conditions become redundant since they are already verified by the proposed expansion. As a consequence, by majoring the terms $\xi_m(U)_{m=1\ldots4}$, the problem is transformed in a linear system with unknown real variables $\lambda_{k,i}$. Considering the Equations (5) and (6), the problem can be written as shown below:

$$\begin{cases} \begin{pmatrix} \theta_{1;1} & .. & .. & \theta_{1;N_0} \\ .. & \theta_{2;2} & .. & .. \\ .. & .. & .. & .. \\ \theta_{N_0;1} & .. & .. & \theta_{N_0;N_0} \end{pmatrix} \begin{pmatrix} \lambda_{1,i} \\ \lambda_{2,i} \\ .. \\ \lambda_{N_0,i} \end{pmatrix} = \begin{pmatrix} c_1 \\ c_2 \\ .. \\ c_{N_0} \end{pmatrix} \end{cases}$$

(11)

and using the matrix standard form:

$$\begin{cases} [\Theta] \times [\lambda]_i = [C] \\ [\Theta] = \begin{pmatrix} \theta_{1;1} & .. & .. & \theta_{1;N_0} \\ .. & \theta_{2;2} & .. & .. \\ .. & .. & .. & .. \\ \theta_{N_0;1} & .. & .. & \theta_{N_0;N_0} \end{pmatrix}; \quad [\lambda]_i = \begin{pmatrix} \lambda_{1,i} \\ \lambda_{2,i} \\ .. \\ \lambda_{N_0,i} \end{pmatrix}; \quad [C] = \begin{pmatrix} c_1 \\ c_2 \\ .. \\ c_{N_0} \end{pmatrix} \end{cases}$$

(12)

The system (12) is solved using Householder algorithm by testing the convergence coefficients represented by (13) and gradually incrementing of N_0.

$$\lambda_{k,i}^{(Sol.)}\Big|_{\substack{k=1..N_0 \\ i=1..6}} \qquad (13)$$

The final result is hence (for $N_0 = 157$):

$$x_i^{(Sol.)}\Big|_{i=1..6} = \frac{1}{2N_0}\sum_{k=1}^{N_0}\lambda_{k,i}^{(Sol.)}B_{4k}(r_k t) \qquad (14)$$

For a given substrate, the total dimension-less free energy G(U) is defined as:

$$G(U) = \sum_{j=1}^{7} x_j\left(\frac{g_j^0}{RT} + \ln\varpi_j\right) \qquad (15)$$

where g_i^0 and ω_i are chemical standard potential and activity of a species i respectively. Since ω_i is identified to x_j for gaseous species and equal to unity for condensed substances, and by taking into account the solution of Eq. (5), it gives:

$$G(U) = x_1\left(\frac{g_1^0}{RT}\right) + \sum_{j=2}^{7} x_j\left(g_j^0 + RT\ln x_j\right) =$$

$$\frac{g_1^0}{2RTN_0}\sum_{k=1}^{N_0}\lambda_{k,1}^{(Sol.)}B_{4k}(r_k t) \qquad (16)$$

$$+ \sum_{j=2}^{7}\frac{1}{2N_0}\sum_{k=1}^{N_0}\lambda_{k,j}^{(Sol.)}B_{4k}(r_k t)(g_j^0) + RT\ln\frac{1}{2N_0}\sum_{k=1}^{N_0}\lambda_{k,j}^{(Sol.)}B_{4k}(r_k t))$$

G is a sophisticated function of T since x_j are T-dependent too. Quick calculation using average parameters gives an optimal temperature T equal to ≈ 1044 K, which is in agreement with the values estimated by Ozgur Colpan et al. and ZA Zainal et al. (i. e. 850°C and 750°C, respectively).

3   Results and Discussion

Solution plots are presented in this section. In order to verify the analytical results given by the optimization protocol, some hypothetical sets of parameters are used, considering the abundant literature on this topic. Moreover, different values have been considered in the numerical simulation to observe biologically plausible scenarios dynamic model [23, 24].

Figure 1 shows a 3D graph representing the dimensionless free energy G(U) depending on M and N parameters. In the present study, it is confirmed that optimality is not obtained for wood-like biomass, where the substrate ratio is equal to 2.18. The optimal condition is efficiently reached for $M\approx33$ and $N\approx18$ and corresponds to a particular class of biomass. This record is in good agreement with the results of Zainal et al., Lorente et al., McKendry and Ozgur Colpan et al..

It is important to highlight the main advantages of the numerical simulation developed in the work –i.e. the presentation of analytical piecewise continuous solutions and the respect of asymptotic stability- if compared with previous studies, e.g. Yang *et al.*, Van den Driessche and Rogers, which focused the attention only on globally stable numerical solutions.

Since the monitored solution is the total dimension-less free energy G(U), it is noted thatminimal values are recorded for both ranges (N<5) and M> 30. These two ranges correspond either to hydrogen excess or oxygen default. In reference to the considered Langmuir-Hinshelwood scheme, low oxygen content in the substrate gives unfavourable conditions to Boudouard reaction while hydrogen excess disables reforming reaction. Moreover, the shape of the solution gives evidence to the existence of differently efficient combination of substrates. In fact, non-integer solutions correspond to intermediary chemical formulae which correspond to compound mixes rather than unique substrate. This feature attributes a realistic aspect to the whole model, since pyrolisis, devolatilisation and oxidation, which are not instantaneous processes, yield unavoidably heterogeneous mixtures.

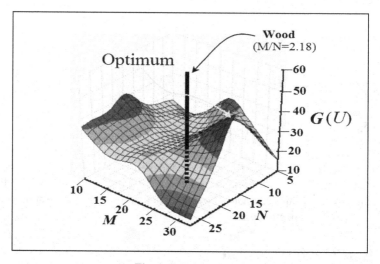

Fig. 1. Solution plot

4   Conclusion

Biomass can be further considered as a potentially reliable and renewable energy source. Considering that the lifetime of fossil fuels is decreasing and if emissions of greenhouse gases have to be successfully diminished, an appropriate and efficient use of biomass has to be taken into account and further investigated. Gasification technology becomes even more important if compared with ecological and environmental impacts of fossil fuels.

The present paper is aimed at preliminarily studying the relation between the gasification process and the type of biomass, with specific regard to Gibbs free energy

G(U). In order to reach this goal, an optimization protocol has been developed and is based on the use of the Boubaker Polynomials Expansion Scheme (BPES). The results given by the mathematical modelling show G(U) variations depending on the type of biomass. Thus, they might represent a useful tool for decision making and understanding energy conversion patterns.

References

1. Puig-Arnavat, M., Bruno, J.C., Coronas, A.: Review and analysis of biomass gasification models. Renewable and Sustainable Energy Reviews, 2841–2851 (2010)
2. Baggio, P., Baratieri, M., Fiori, L., Grigiante, M., Avi, D., Tosi, P.: Experimental and modeling analysis of a batch gasification/pyrolysis reactor. Energy Conversion and Management, 1426–1435 (2009)
3. Bocci, E., Villarini, M., Bove, L., Esposto, S., Gasperini, V.: Modeling Small Scale Solar Powered ORC Unit for Standalone Application. Mathematical Problems in Engineering, art. no. 124280 (2012)
4. Di Carlo, A., Bocci, E., Naso, V.: Process simulation of a SOFC and double bubbling fluidized bed gasifier power plant. International Journal of Hydrogen Energy 38, 532–542 (2013)
5. Monarca, D., Colantoni, A., Cecchini, M., Longo, L., Vecchione, L., Carlini, M., Manzo, A.: Energy Characterization and Gasification of Biomass Derived by Hazelnut Cultivation: Analysis of Produced Syngas by Gas Chromatography. Mathematical Problemsin Engineering, art. no. 102914 (2012)
6. Warnecke, R.: Gasification of biomass: comparison of fixed bed and fluidized bed gasifier. Biomass and Bioenergy, 489–497 (2000)
7. Ravikiran, A., Renganathan, T., Pushpavanam, S., Voolapalli, R.K., Cho, Y.S.: Generalized Analysis of Gasifier Performance using Equilibrium Modeling. Industrial and Engineering Chemistry Research, 1601–1611 (2012)
8. Zhang, K., Chang, J., Guan, Y., Chen, H., Yang, Y., Jiang, J.: Lignocellulosic biomass gasification technology in China. Renewable Energy, 175–184 (2013)
9. Awojoyogbe, O.B., Boubaker, K.: A solution to Bloch NMR flow equations for the analysis of homodynamic functions of blood flow system using m-Boubaker polynomials. Curr. App. Phys. 9, 278–288 (2009)
10. Barry, P., Hennessy, A.: Meixner-type results for riordan arrays and associated integer sequences. J. Integer Sequences 13, 1–34 (2010)
11. Belhadj, A., Bessrour, J., Bouhafs, V., Barrallier, L.: Experimental and theoretical cooling velocity profile inside laser welded metals using keyhole approximation and Boubaker polynomials expansion. J. Thermal Analysis Calorimetry 97, 911–920 (2009)
12. Belhadj, A., Onyango, O., Rozibaeva, V.: Boubaker polynomials expansion scheme-related heat transfer investigation inside keyhole model. J. Thermophys Heat Transf. 23, 639–642 (2009)
13. Fridjine, S., Amlouk, M.: A new parameter: An ABACUS for optimizing functional materials using the Boubaker polynomials expansion scheme. Modern Phys. Lett. B 23, 2179–2182 (2009)
14. Ghanouchi, J., Labiadh, H., Boubaker, K.: An attempt to solve the heat transfert equation in a model of pyrolysis spray using 4q-order m-Boubaker polynomials. Int. J. Heat Technol. 26, 49–53 (2008)

15. Kumar, A.S.: An analytical solution to applied mathematics-related Love's equation using the Boubaker polynomials expansion scheme. J. Franklin Institute 347, 1755–1761 (2010)
16. Labiadh, H., Boubaker, K.: A Sturm-Liouville shaped characteristic differential equation as a guide to establish a quasi-polynomial expression to the Boubaker polynomials. Diff. Eq. and Cont. Proc. 2, 117–133 (2007)
17. Milgram, A.: The stability of the Boubaker polynomials expansion scheme (BPES)-based solution to Lotka-Volterra problem. J. Theoretical Biolog. 271, 157–158 (2011)
18. Slama, S., Bessrour, J., Bouhafs, M., Ben, K.B., Num, M.: Heat Transf. Part A. 55, 401–404 (2009)
19. Slama, S., Bouhafs, M., Ben, K.B., Mahmouda, A.: Boubaker polynomials solution to heat equation for monitoring a3 point evolution during resistance spot welding. Int. J. Technol. 26, 141–146 (2008)
20. Tabatabaei, S., Zhao, T., Awojoyogbe, O., Moses, F.: Cut-off cooling velocity profiling inside a keyhole model using the Boubaker polynomials expansion scheme. Int. J. Heat Mass Transfer. 45, 1247–1255 (2009)
21. Yildirim, A., Mohyud-Di, S.T., Zhang, D.H.: Analytical solutions to the pulsed klein-gordon equation using Modified Variational Iteration Method (MVIM) and Boubaker Polynomials Expansion Scheme (BPES). Computers Math. Appl. 12, 2473–2477 (2010)
22. Lorente, E., Millan, M., Brandon, N.P.: Use of gassification syngas in SOFC: impact of real tar anode materials. International Journal of Hydrogen Energy 37(8), 7271–7278 (2012)
23. Zainal, Z.A., Ali, R., Lean, C.H., Seetharamu, K.N.: Prediction of performance of a downdraft gasifier using equilibrium modeling different biomass materials. Energy Conversion and Management 42(12), 1499–1515 (2001)
24. Ozgur Colpan, C., Yoo, Y., Dincer, I., Hamdullahpur, F.: Thermal modeling and simulation of an integrated solid oxide fuel cell and charcoal gasification system. Environmental Progress and Sustainable Energy 42, 380–385 (2009)
25. Ragland, K.W., Aerts, D.J.: Properties of Wood for Combustion Analysis. Bioresource Technology 37, 161–168 (1991)

Energy-Aware Control of Home Networks

Vincenzo Suraci[1], Alvaro Marucci[2], Roberto Bedini[2],
Letterio Zuccaro[3], and Andi Palo[3]

[1] Università degli studi e-Campus, Via Isimbardi 10, Novedrate, 22060, Italy
vincenzo.suraci@uniecampus.it
[2] Department of science and technology for Agriculture, Forests, Nature and Energy,
University of Tuscia, Via S. Camillo De Lellis, s.n.c., 01100 Viterbo, Italy
[3] Università degli Studi di Roma La Sapienza, Dip. di Ingegneria Informatica,
Automatica e Gestionale (DIAG), Via Ariosto 25, Rome, 00185, Italy

Abstract. Home networks have become heterogeneous environment hosting a variety of wireless and wired telecommunication technologies. Currently there no exist any intelligent energy saving mechanism to control the home networks. In this paper we propose an energy-aware strategy that integrate Wireless Sensor Network (WSN) with a convergent digital home network. We aim to demonstrate that a WSN can act as a dependable control plane to manage the high speed home network. While the home network nodes can be deactivated, the WSN is always on and, due to the low data rate required to properly work, it consumes a very limited quantity of energy. This mutual interaction leads to a substantial reduction of the energy consumptions. Simulation results show that this strategy is effective in different scenarios and provides a tangible economic benefit.

Keywords: Green networks, Sensor Networks, Home networks, Future Internet.

1 Motivation

The European countries are facing a hard period since the economic crisis coming from the US has passed the Ocean and has spread into the whole European economic and productive system. To avoid an endless crisis the EU has urgently approved a four years (2010-2013) European Economic Recovery Plan (EERP). The thrust of the EERP is to restore confidence of consumers and business. It is suggested to provide a demand stimulus of € 200 billion [1]. The major instrument identified to apply the EERP strategy is the Public Private Partnership (PPP): an industry-driven RTDI initiatives tackling the major socio-economic challenges that look for maximizing EU industrial capabilities in order to allow their best exploitation for the EU industry of tomorrow, enhancing its competitiveness through high impact actions on research and innovation. In particular the PPP on the Future Internet initiative [2] identifies as the most important needs for the research and innovation Green technologies and Gigabit ICT for all the European citizens.

B. Murgante et al. (Eds.): ICCSA 2013, Part II, LNCS 7972, pp. 299–311, 2013.
© Springer-Verlag Berlin Heidelberg 2013

Recent publications [3], [4],[5],[6], [7], [22] and [28] show that, despite the expectations, the ICT industry energy need cannot be neglected when compared to the other energy sectors. Indeed it is increasing with an exponential growth and it is about to exceed than the aviation industry. Thus the Green ICT technologies are an emergent research field that needs pragmatic solutions [8]. While Green ICT have been investigated in the context of core and access networks [9] [10], consistent improvements should be still done in the context of residential and office networks [26] and [27].

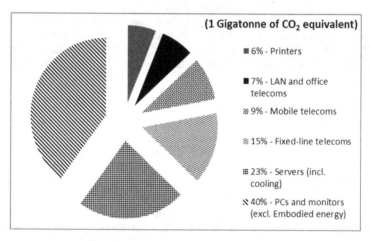

Fig. 1. Green IT power consumption [21]

A study carried out by Gartner in 2007 (Fig. 1) shows that the 7% of Green ICT power consumption is due to home and office networks. It is a small portion of the whole ICT power consumption. So far there not exist systematic and integrated solutions to green this market sector. Some isolated initiatives have been taken by the manufacturers to improve the network elements efficiency and to make them more green. However the Green ICT problem in home network have been never faced with an integrated approach.

Today's homes are equipped with a multitude of devices using several wired or wireless communication technologies forming a heterogeneous network environment. This environment may include distinct technologies such as Ethernet, Wi-fi, Power Line Communications. To improve the network resource exploitation the IEEE 1905.1 working group on Convergent Digital Home Network is designing a technology-independent abstraction layer called Inter-MAC able to work in conjunction with any present or near future technology dependent access technique ([24] and [24]). This layer is in charge of control the hOME Gigabit Access (OMEGA) network and provide services as well as connectivity to a multitude of devices. In [11] and [12] the heterogeneous technologies could converge below the network protocol layer, by means of the creation of an intermediate layer: the Inter-MAC. While the combined use of heterogeneous access technologies allows the home network to speed up to 1Gps without new wires, the energy consumption in high speed home networks remains an almost unexplored field.

In this paper we show how the combined use of wireless sensor network and OMEGA network can be applied for a mutual advantage. The wireless sensor network can rely on the gigabit network facilities when it is powered up. The gigabit network can be activated and deactivated dynamically relying on an always on and energy efficient wireless sensor network.

The document is organized as follows: in section 2 the overall research methodology is presented, in terms of architectural design of the proposed solution. In section 3 we describe the simulation environment used to deploy the designed solution. In section 4 the simulation environment and the related testbeds are the described in detail. In section 5 the simulation results are analyzed critically to verify the proof of concept and some business benefit in terms of Return Of Investment are highlighted. In section 6 some conclusions on the obtained results and needed future work are presented.

2   Gigabit Home Networks

The Inter-MAC architecture is presented in [13] and [14]. It is divided into data plane, control plane, and management plane. Data plane is responsible for transferring the user/application data packets. It manages the packets arriving at a device, both from the upper layer (network) and the lower MAC/LLC layer. Control plane performs short-term actions in order to manage the data plane behaviour. It is responsible for managing the correspondence between the higher layer application protocol requests and the establishment of new connections or paths to the desired destination with the appropriate QoS requirements. Management plane is concerned with long-term actions which describe the behaviour of the device itself.

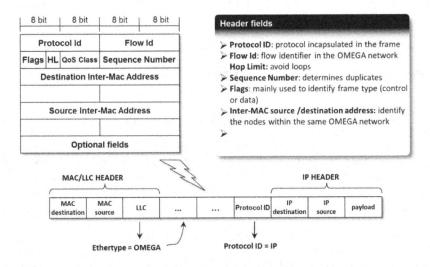

Fig. 2. Inter-MAC frame format and header

To be properly managed, Inter-MAC layer uses a layer 2.5 Inter-MAC protocol. Each OMEGA node, equipped with the Inter-MAC layer functionalities sends and receives proper data and control plane frames. The frame format is shown in Fig.2.

To let OMEGA network interoperate with a wireless sensor network for a synergic cooperation aiming to save energy, a twofold integration is needed. First of all each energy-consuming OMEGA node must be coupled to a wireless sensor node. Consequently the topology of an integrated OMEGA-WSN network is depicted in Fig. 3.

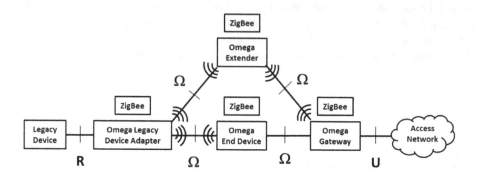

Fig. 3. Integrated OMEGA/WSN network: the OMEGA links are represented with letter "Ω", the last-mile connection is represented with letter "U", the connection between legacy devices and the OMEGA network is represented with letter "R".

The integration must be applied also in the protocol stack as shown in Fig. 4, where a ZigBee protocol stack has been used. Thus the Inter-MAC control and management planes must communicate to perform an intelligent energy save and a rational use of the OMEGA network facilities.

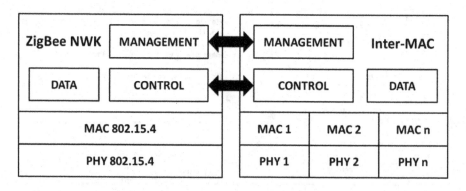

Fig. 4. Integrated protocol stack

More specifically each OMEGA node can communicate to the sensor node its operational status. The following mean operative status have been defined: (i) Transmission/Reception, (ii) Bridging and (iii) Idle. In the first case the node is actively participating to provide QoS-aware applications to an end user acting as a sender or as a receiver of the flow. When the node is acting as a network bridge, there is the possibility to deactivate the unused network interfaces. When the node is in idle, there is the possibility to switch off the whole node in a deep sleep mode. In the last two cases it is possible to save energy, depowering parts of the OMEGA communication systems or the whole node.

As in [15] whenever a new flow must be setup from a source OMEGA node A and a destination OMEGA node B, a broadcast (or unicast) path request control frame is sent at Inter-MAC level, to start up the reactive (or proactive) path selection. The path request control frame flows through the OMEGA network from the source node to the destination node to activate the most appropriate path selection algorithm (driven by QoS, energy-saving or fault-tolerant considerations [25]).

In case of an integrated OMEGA+WSN network, the above described message sequence must be adapted to cope with the necessity to intelligently activate the idled or partially operating OMEGA nodes. Whenever a source OMEGA node A needs to setup a new flow towards the destination OMEGA node B, the path request control frame is sent to the Wireless Sensor (WS) node A. The WS node A is in charge to send the broadcast path request over the sensor network that will be intercepted by the WSN node B associated to the destination OMEGA node B. Thus the WSN acts as a dependable control plane, where low data rate control packets can flow even when the OMEGA network is shut down. While the OMEGA nodes or the OMEGA node's network interfaces can be deactivate, the WSN is always on and, due to the low data rate required to properly work, consume a very limited quantity of energy.

When an intermediate WSN node receives a broadcast path request, it activates the OMEGA node and all its interfaces if deactivated, retransmits the broadcast packet (to reach even nodes that are out of range from the source node) and sends an acknowledge packet back to the source WSN node. This awakening phase is needed by the source OMEGA node A to ensure that all the OMEGA nodes are activated when it performs the QoS-aware path selection algorithm to discover the best available path from A to B. Once the path selection solution is known, the OMEGA nodes not involved in the new flow can turn back to their previous idled status. The WSN represents a backup control plane for the OMEGA network able to provide both, a reliable low data rate communication channel and an energy aware protocol to activate dynamically only the OMEGA nodes needed to provide QoS to the running flows.

Not all the WSN are feasible for a Green ICT scope. As shown in Fig. 5, there exist three main WSN topologies: star, tree and mesh. These topologies require an increasing amount of energy.

While in a star topology the always on node is the PAN coordinator, in a tree topology the full function device must perform relatively complex functionalities thus consuming more energy than a simple connected WSN device. In a mesh topology all the nodes are potentially full function devices, thus in the worst case the mesh topology is the less energy efficient. Unfortunately in a indoor environment as a home, or a small office, home office, the presence of walls reduces the communication range between the nodes composing the WSN, thus to ensure the maximum level of reliability and availability of the WSN overlay, a mesh topology (e.g. multicluster ZigBee) is highly recommended.

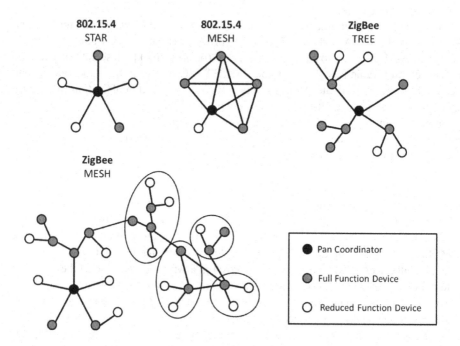

Fig. 5. WSN topology examples

3   Technology Description

To demonstrate the validity of the proposed approach, a simulation environment has been setup. The Modeler OPNET 14.5 Educational Version has been used as network simulator for all the tests. It has been preferred against other solutions (NS3, OMNET) for two main reasons. First of all OPNET natively support the technologies needed to perform intensive tests: Ethernet, Wi-Fi and ZigBee. The PLC channel has been approximated using a statistical approach described in [15]. On the other hand a particularly accurate OMEGA OpNET model was already available ([12] and [15]). As shown in Fig. 6, the Inter-MAC layer has been introduced in the OpNET simulator protocol stack as a new component called imac and located between the layer 3 protocol (IP) and the layer 2 protocols (arp and mac). As clearly shown in Fig. 6, the Inter-MAC is in charge to interface the IP layer with the heterogeneous underlying technologies: wireless LAN (wlan), powerline (bus) and Ethernet (hub).

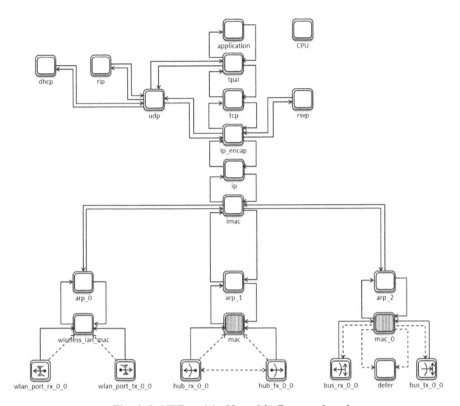

Fig. 6. OpNET model of Inter-MAC protocol stack

The OMEGA protocol stack has been modified to cope with the necessity of interfacing with the WSN protocol stack and to manage the operational status of the connected network interfaces and of the whole OMEGA node. This modification has affected mainly the imac finite state machine definition. As shown in Fig. 7 a new status has been introduced in the model: inactive. The inactive status means that the OMEGA node is temporary sleeping and needs to be awaken in order to be operative. This status can be modified only by the associated WSN node that, on the contrary, is always running. The developed WSN node model is depicted in Fig. 8. The control_interface process is in charge to manage the communication with the OMEGA node and to simulate the management and control plane interaction with the Inter-MAC layer. As shown in Fig. 9 the following main functionalities are provided by the control_interface process: APP_CALL – this interrupt is triggered by the OMEGA node control plane. When a new path setup procedure is needed, the OMEGA node invoke the broadcast functionality. The WSN node sends a broadcast wake-up packet over the sensor network. Consequently all the WSN nodes activate the relative OMEGA nodes. Each OMEGA node switch from the inactive status to an idle status. Once the whole OMEGA network is active, the Inter-MAC layer of the source OMEGA node sends over the OMEGA network a Path request packet to trigger the distributed path selection algorithms. BROADCAST_RX – this interrupt is triggered when a WSN node intercept a broadcast path request packet. If the relative OMEGA node is inactive an interrupt to the imac process is sent to wake up the OMEGA node or to activate all its network interfaces. IMAC_STATE_UPDATE – it is sent by the imac process to update the WSN node with the actual status of the OMEGA node.

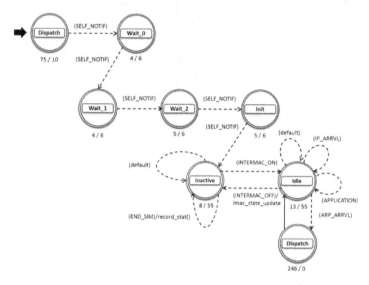

Fig. 7. Energy-aware imac process

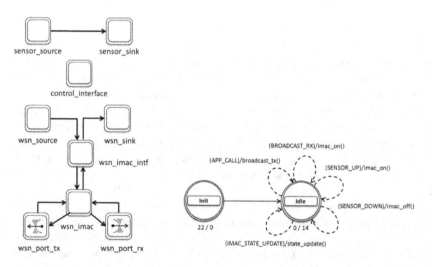

Fig. 8. WSN node

Fig. 9. Control-interface process

4   Developments

The simulation framework has been used to implement different usage scenarios: home network, Small Office Home Office (SOHO) network and Small Enterprise (SE) network. A typical home network is supposed to have less than 10 connected devices. A typical SOHO network is supposed to have up to 50 connected devices, while a SE network may have up to 100 connected devices. Several topologies have been used in the OpNET simulator to generate the most reasonable OMEGA network configurations.

In order to assess the energy saving due to the combined use of OMEGA and WSN networks, the following assumption have been done on the basis of literature and state of the art search (see [17], [18], [19] and [20] for more details):

- each OMEGA interface (Wi-Fi, Ethernet, Homeplug) has an average energy consumption of 1.5 W;
- each WSN interface has an energy consumption of 0.25 W;
- an average KWh cost of € 0.25;
- an average WSN node cost of € 4;
- an average OMEGA node energy consumption of 8 W.

Intensive tests have been done for the different scenarios. For each scenario multiple tests have been performed to obtain average results. In each test the overall energy consumption has been evaluated applying three different energy saving strategies:

1. No energy saving - the OMEGA network is operative 100% of the time and no WSN is acting to save energy;
2. Partial energy saving – the WSN can solely deactivate the idled interfaces, but not the whole OMEGA node, which is always operative;
3. Maximum energy saving – the WSN can deactivate the whole OMEGA node when it is in an idle status.

In particular the results have been analyzed in terms of differential benefit obtained applying the strategy 2 and 3 against the strategy 1. Thus the results represent the real energy and cost saving associated to the proposed solution.

5   Results

As depicted in Fig. 10, the more the OMEGA network is used during a 24 hours duty day, the less the energy saving strategies are effective. Nevertheless a more aggressive energy saving strategy can guarantee a result three times better than a less aggressive strategy.

The adoption of a maximum energy saving strategy has its own drawbacks. While an OMEGA node needs less than 1 second to power up and activate a single network interface, it takes from 20 up to 30 seconds to turn on the whole OMEGA node and to become fully operative. Thus the best energy saving strategy must be leveraged with the expected reactive time. It doesn't exist a unique applicable solution, thus a proper policy must be decided dynamically or statically to adopt the most appropriate energy saving strategy.

The efficiency of the proposed strategies can be estimated in terms of relative energy saving. As shown in Fig. 10, having as reference energy consumption the values associated to the "No energy saving" strategy, and considering an average network daily use ranging from 12 to 4 hours, the "Partial energy saving" shows an average efficiency ranging from 13% up to 27%, while the "Maximum energy saving" strategy ranges from 37% up to 75%.

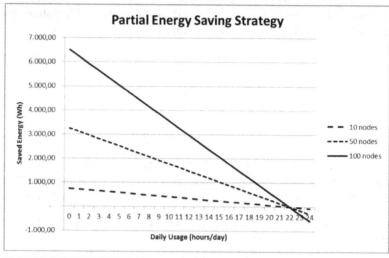

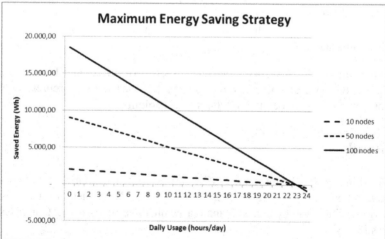

Fig. 10. left graph represents the partial energy saving strategy assuming that each OMEGA node is equipped with 2 interfaces; the right graph represents the maximum energy saving strategy

Integrating a wireless sensor network in the OMEGA network incurs an additional cost in terms of acquiring the sensor network and of energy spent to supply the sensor network. But this initial additional cost can be paid off from the profit gained by the energy saved in using the OMEGA network. Considering the above described scenarios, assuming an average use of 4 hours per day of the OMEGA network and an average of two interfaces for each node, some business benefits can be clearly quantified.

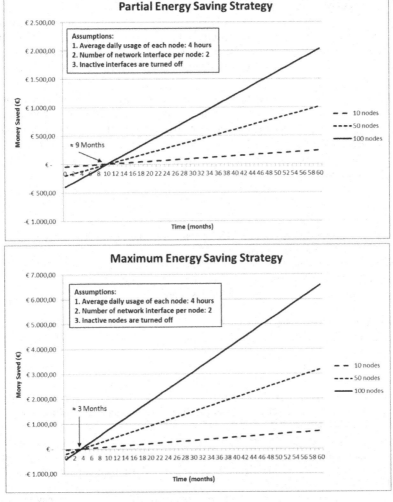

Fig. 11. Cost saving results

If the strategy adopted is to switch off only the interfaces of a device in idle state than the additional cost is paid off in a period of 9 months for all cases. The average life of the wireless sensor network is estimated to be approximately 5 years. So in this period the benefits are more effective for the scenario with 100 nodes. This fact is depicted in the left part of Fig. 11. Instead if the energy saving strategy is to totally switch off the inactive nodes the pay off period is reduced significantly to 3 months. In this case even the benefits for the smallest scenario with 10 nodes are substantial and should not be underestimated.

6 Conclusions

In this paper we considered the problem of adopting Green ICT technologies in future home networks. We proposed and evaluated an energy saving mechanism by integrating a wireless sensor network with the OMEGA home network. Our strategy dynamically activates and deactivates the gigabit OMEGA network, by means of an always on and energy efficient wireless sensor network. Simulation results show that this mechanism brings substantial long term benefits due to the energy saved in the overall network, in different usage scenarios.

As future works a better integration of the two networks can be achieved not only in terms of energy saving but also in terms of data plane in order to obtain a mutual benefit. In one hand, the wireless sensor network can transmit the control messages by using the OMEGA data plane when the nodes are active, and in the other hand the OMEGA nodes can use the wireless sensor network as signaling plane when their interfaces are powered off.

References

1. International Labour Organization, G20 Statistical update - EU: heterogeneous shocks and responses across countries, p. 2 (April 2009)
2. A Public-Private Partnership (PPP) for the Future Internet, p. 11 (April 2009)
3. Postnote "ICT and CO2 emissions", Number 319 (December 2008)
4. Forge, S.: Powering down: remedies for unsustainable ICT. Foresight, Journal 9(4), 3–21 (2007)
5. Lechner, R.: An Inefficient Truth. Global Action Plan, IBM (August 28, 2008)
6. Mckenna, P.: Can we stop the internet destroying our planet? New Scientist (2637) (January 03, 2008)
7. Santad, Ijcsns, Visual Basic Software Tool for FTTH Network Management System. International Journal of Computer Science and Network Security 9(2) (February 2009)
8. Riaz, M.T., Gutierrez, J.M., Pedersen, J.M.: Strategies for the next generation green ICT infrastructure. Applied Sciences in Biomedical and Communication Technologies (2009) ISBN: 978-1-4244-4640-7
9. Vereecken, W., Van Heddeghem, W., Colle, D., Pickavet, M., Demeester, P.: Overall ICT footprint and green communication technologies. In: 2010 4th International Symposium on Communications, Control and Signal Processing (ISCCSP), Belgium (2010) ISBN: 978-1-4244-6285-8
10. Zhang, Y., Chowdhury, P., Tornatore, M., Mukherjee, B.: Energy Efficiency in Telecom Optical Networks. IEEE Communications Surveys & Tutorials 12(4) (Fourth Quarter 2010) ISSN: 1553-877X
11. Suraci, V., Castrucci, M., Oddi, G., Cimmino, A., Colella, R.: Convergence in Home Gigabit Networks: Implementation of the Inter-MAC Layer as a Pluggable Kernel Module. In: IEEE PIMRC 2010 Services, Applications, and Business - 21st Annual IEEE International Symposium on Personal, Indoor and Mobile Radio Communication (2010)

12. Castrucci, M., Oddi, G., Tamea, G., Suraci, V.: Application QoS Management and Session Control in a Heterogeneous Home Net-work using Inter-MAC layer support. In: Future Network & MobileSummit 2010 Conference Proceedings (2010) ISBN: 978-1-905824-16-8
13. Javaudin, J.P., Bellec, M., Varoutas, D., Suraci, V.: OMEGA ICT project: Towards convergent Gigabit Home Networks. In: IEEE PIMRC 2008 - Gigabit Home Access Special Session, Proceedings PIMRC08-1191 (2008)
14. Meyer, T., Suraci, V., Langendörfer, P., Nowak, S., Bahr, M., Jennen, R.: An inter-MAC architecture for heterogeneous Gigabit home networks. In: IEEE International Symposium on Personal, Indoor and Mobile Radio Communications, PIMRC 2009 (2009)
15. Suraci, V., Oddi, G., Mattiacci, N., Angelucci, A.: Admission control and drop strategies in a UPnP-QoS controlled home network. In: IEEE International Symposium on Personal, Indoor and Mobile Radio Communications, PIMRC 2010 (2010)
16. Javaudin, J.-P., Bellec, M., Varoutas, D., Suraci, V.: OMEGA ICT project: Towards convergent gigabit home networks. In: IEEE International Symposium on Personal, Indoor and Mobile Radio Communications, PIMRC 2008 (2008)
17. Lin, T.-M.: Power Consumption Issues for WLAN Systems, http://mnet.cs.nthu.edu.tw/paper/tmlin/051209.pdf
18. Balani, R.: Energy Consumption Analysis for Bluetooth, WiFi and Cellular Networks, http://nesl.ee.ucla.edu/fw/documents/reports/2007/PowerAnalysis.pdf
19. Cisco, Ethernet Power Study of Cisco and Competitive Products, white paper
20. Homeplug powerline alliance, HomePlug Green PHY Specification, white paper
21. Kumar, R., Mieritz, L.: Conceptualizing Green IT and data center power and cooling issues, Gartner Research Paper (2007)
22. Colantoni, A., Giuseppina, M., Buccarella, M., Cividino, S., Vello, M.: Economical analysis of SOFC system for power production. In: Murgante, B., Gervasi, O., Iglesias, A., Taniar, D., Apduhan, B.O. (eds.) ICCSA 2011, Part IV. LNCS, vol. 6785, pp. 270–276. Springer, Heidelberg (2011)
23. Delli Priscoli, F.: Design and Implementation of a Simple and Efficient Medium Access Control for High Speed Wireless Local Area Networks. IEEE Journal on Selected Areas in Communications (JSAC) 17(11), 2052–2064 (1999)
24. Cusani, R., Delli Priscoli, F., Ferrari, G., Torregiani, M.: A Novel MAC and Scheduling Strategy to Guarantee QoS for the New-Generation Wireless LAN. Special Issue on "Mobile and Wireless Internet: Architecture and Protocols" of IEEE Wireless Communications (IEEE Personal Communications), IEEE's Computer and Vehicular Technology Societies (U.S.A.) (3), 46–56 (June 2002)
25. Pietrabissa, A., Castrucci, M., Palo, A.: A MDP approach to fault-tolerant routing. European Journal of Control 18(4), 334–347 (2012)
26. Di Giorgio, A., Pimpinella, L., Quaresima, A., Curti, S.: An event driven smart home controller enabling cost effective use of electric energy and automated demand side management. In: 19th Mediterranean Conference on Control and Automation MED 2011, Corfu, pp. 358–364 (June 2011)
27. Di Giorgio, A., Pimpinella, L., Liberati, F.: A model predictive control approach to the load shifting problem in a household equipped with an energy storage unit. In: 20th Mediterranean Conference on Control and Automation MED 2012, Barcelona, pp. 1491–1498 (July 2012)
28. Marucci, A., Monarca, D., Cecchini, M., Colantoni, A., Manzo, A., Cappuccini, A.: The Semitransparent Photovoltaic Films for Mediterranean Greenhouse: A New Sustainable Technology. Mathematical Problems in Engineering 2012, 1–16 (2012)

Development of an Energy System Model in Jiangsu Region with MARKAL: An Analysis of the Supply Side

Vincenzo Naso and Flavio Rottenberg

CIRPS Sapienza University of Rome
Piazza San Pietro in Vincoli 10, 00184 Rome, Italy
vincenzo.naso@uniroma1.it, flavio.rottenberg@gmail.com

Abstract. The Research, conducted in collaboration with Department of Mechanical Engineer of Southeast University of Nanjing situated in the Jangsu Region, has been developed over a period of three years in the ambit of Scientific Cooperation Project *New Interuniversity Network For Energy Management*. The main objective is to have a complete picture of the energy production and consumption in Jiangsu region of China, forecast the production and consumption level in the next 25 years in a "business as usual" hypothesis. Having achieved this result, it then will be possible to formulate hypothesis on future energy policies and valuate its results against the base case. The aforesaid research has been conducted after an evaluation of the modelling software available on the market within which Markal has been chosen as the more flexible and adequate for the research's purposes. At the moment a forecast of the Jiangsu final energy consumption from 2012 to 2027 (time-horizon for the model) by energy commodity and by economic sector has been developed.

1 Introduction

China has a population exceeding 1.3 billion and its economy has been growing in the range 6-10% per year for many years, reaching the maximum of 10.3 % in last quarter of 2010. China will need huge amounts of energy to feed this growth, but the mentioned trend is driving unavoidable constrains. On one hand, the country is concerned about security of supply, so it does not want to import too large portion of energy needs; on the other hand, its most available fuel is coal and heavy use of coal leads to severe pollution across the country. The question facing China is how to meet an ever-increasing demand for energy without continued degradation of air quality and damage to the environment in general. Jiangsu Province is situated in the center of the east coastal areas of China and it is adjacent to China's largest city of Shanghai. It covers an area of 102.600 km^2 (corresponding to 1.06% of China's territory). Jiangsu orography is carachterized by smooth terrain, extensive plains and interconnecting watercourses. The population around to 74.75 million makes the Jiangsu province one of the most densely populated province in China. Jiangsu governs 106 countries, cities and districts in total, with 13 cities directly under the provincial jurisdiction. After the reform and opening-up about thirty years ago, Jiangsu has attained sustainable rapid economic development and has implemented active strategies in economic

B. Murgante et al. (Eds.): ICCSA 2013, Part II, LNCS 7972, pp. 312–327, 2013.
© Springer-Verlag Berlin Heidelberg 2013

internationalization. Based on the outstanding regional privileges of the Yangtze River Delta, Jiangsu has made painstaking efforts to create fine investment environment and enhance open-up economy, by making full use of her complete infrastructure, advanced sci & tech and education as well as her solid industrial bases. At present, Jiangsu has established economic and trade relationship with 218 countries and regions, its export volume amounting to 227.9 billion US dollar in 2008. Accordingly, the open-up Jiangsu has become a homeland of investment for overseas investors with a total foreign capital of over 100 billion US dollar. Jiangsu is today one of the most developed regions of China and it usually boasts a GDP (Gross Domestic Product) figure 2% higher than the already high national GDP figure. Its energy balance sheet is highly unsustainable because the region is poor in energy sources and they have to be imported from other regions of China or from other Countries. The primary energy consumption distribution is the same of the other Chinese regions: coal is the most used followed by oil, gas and a small presence of hydropower. Nuclear is still in the early stage of their development in the region.Electric power generation is fed mainly by coal, with three main power plants with a unit size of: 300, 600 and 1000 MW.

The model developed in this research analyses the status of the Chinese energy system and gives an evaluation of the energy production and consumption trend in Jiangsu Region of China in the next 25 years in a "business as usual" hypothesis. The scenarios have been evaluated using the Markal software, that has been chosen because of its flexibility and compatibility to the requirements of the research. The collected data for the modeling of energy scenarios is organized in the so-called Reference Energy System (RES), that can be revised as the main structure of the model.

2   Materials and Methods

Jiangsu is an economic developed province in China, at the same time it is a big energy-consuming province. The lack of primary energy sources and the fast development of economy make the Jiangsu have an unbalanced energy sheet where the consumptions exceeds the production. Following the hierarchical organization of the Markal models, it has been chosen to organize the data relevant to the region through a Reference Energy System. Markal models are bottom-up models and are built up by descriptions of the components of the energy system and their interactions, rather than equations describing behaviour at an aggregated level. The components fall into two categories; commodities and processes. These make up the building blocks for the description of the energy system. Commodities are the flows through the system, while processes are the technologies that transform commodities from one form to another. The network they form thus closely resemble a flowchart, and is generally referred to as a Reference Energy System (RES).

Commodities represent energy carriers, materials, emissions, useful energy demands, money, etc. flowing through an energy system. A process transforms commodities from one form into another, and only in exceptional cases (e.g., exchange and storage processes) a process consumes the same commodity that it produces. The process groups are Combined Heat and Power, Demand devices, Electricity generation, Energy, Heat generation, Material and Miscellaneous. For instance, processes may represent power plants transforming fuels into electricity, or they may represent washing machines and aeroplanes producing washed laundry and travelled distance.

As mentioned, the structure of the model is organized as a Reference Energy System (RES), which is a network depicting flows of commodities through various processes. Commodities such as primary energy carriers are initially either imported to the model region or obtained by production of raw material in mining processes, for example. A RES describes the process of these primary resources through different stages consisting of transformation, transportation, storage and distribution of commodities, until they are either consumed by end-use demand or exported out of the region.

The flow of commodities through the RES is governed by three basic equations. The first is the transformation equation which relates the input flows to the output flows of a process. Second is the market allocation equation which limits the inflow or outflow share of a particular commodity. Finally there is the balance equation which ensures that there is a balance between production and consumption of all commodities. Individual plants are not modelled separately, but similar plants are aggregated into processes that represent a whole technology type. In addition to existing technologies, a RES usually contains a variety of alternatives and options of technologies that are not available today (or only at a very high cost) but that can be envisaged to be utilised in the future. In the following page the Reference Energy System of Jiangsu regions is shown, where the current technologies are coloured blue and the future technologies are coloured yellow. As it is possible to see, the main fossil resources are represented there (Hard coal, Crude Oil, Natural Gas). Also their path from mine to burner is shown, with all the various conversion and production technologies that are used to serve the needs of the final demand. In this study the demand is modeled as an aggregate, for instance Industry demand of Heat is served by both Hard Coal and Natural Gas.

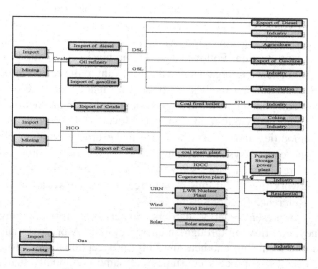

Fig. 1. Reference Energy System of Jiangsu

2.1   Description of Res Data

In the following section all the RES parameters are mentioned and described, some concepts about the actual situation are also presented and discussed.

Coal. As shown in the 'Energy Balance Sheet' of 2011, 90% of primary energy production in Jiangsu province is raw coal which is also the main energy resource of China. (Figure 2).

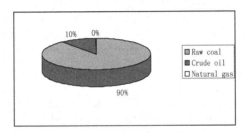

Fig. 2. Structure of primary energy production in Jiangsu province in 2011

All the production of primary energy cannot satisfy the demand with its fast development. most of the coal used is imported from other provinces in China and even abroad. In 2011, 20.13Mt standard coal was produced in Jiangsu province while the total consumption is 112.96 Mt standard coal. 95.99Mt standard coal is imported from other provinces and 1.83Mt is from abroad, while 82.58Mt standard coal are consumed for processing and conversion while the rest is for direct end-use consumption. According to the statistic of Nanjing customs, more than 0.7Mt coal was imported from abroad during the first 4 months in 2010 with a total cost 67.71 million US dollars, with growths of 72.2% and 100% respectively, while the average importation price is 96 US dollars/ton, with a growth of 18.3%.

Table 1. Overall production and consumption of primary energy in Jiangsu province

UNIT: MT STANDARD COAL	RAW COAL	CRUDE OIL	NATURAL GAS
PRODUCTION	20.13	2.35	0.09
IMPORT FROM OTHER PROVINCES	95.99	10.02	1.73
IMPORT ABROAD	1.83	20.15	0.00
FOR PROCESSING AND CONVERSION	82.58	31.80	0.45
END-USE CONSUMPTION	30.38	0.40	1.34

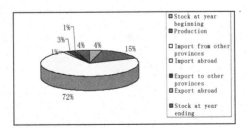

Fig. 3. Structure of raw coal available in Jiangsu province

As figure 4 shows, 86% of raw coal goes to steam electric power generation, and 9% is used for heat supply. As Jiangsu lies in the middle of China, and district heating is not so common in this area, almost all the heat is used for industry.

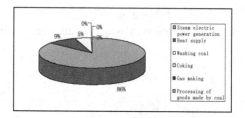

Fig. 4. Structure of raw coal processing & conversion input

As mentioned before, Jiangsu is an industrial province. As shown in figure 5, in 2010 94% of raw coal is used in secondary industry, which mainly includes building construction and industry.

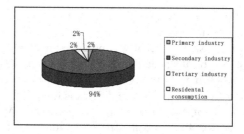

Fig. 5. Raw coal end-use comsumption

Crude Oil. Jiangsu has a lack of crude oil production. As figure 6 shows, 58% of crude oil available is imported from other Countries, and 29% from other Chinese provinces.

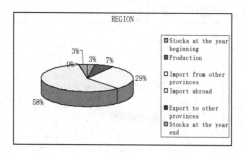

Fig. 6. - Structure of crude oil available in Jiangsu province in 2010

The end-use consumption of crude oil is small comparing with that for processing and conversion (Figure 7). In 2011, about 0.4 Mt standard coal of crude oil was used for end-use consumption, especially for industry (Figure 8).

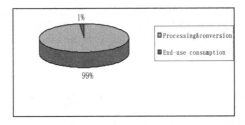

Fig. 7. Consumption of crude oil in 2011

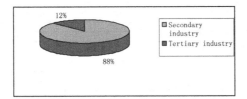

Fig. 8. Crude oil for end-use consumption in 2011

Natural Gas. As it is shown in figure 3.11 that in Jiangsu 95% of natural gas available was imported from other provinces during 2011.

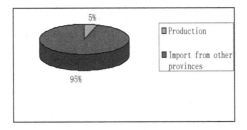

Fig. 9. Source of natural gas available in Jiangsu province

The situation may change somewhat, because 73.8 million tons reserves of rock oil and natural gas in Jiangsu have been found according to the conference of China geology academy 2007. A little amount of the natural gas consumed is imported from abroad, that is because LNG facilities are not yet developed in the region.

Electricity Generation. In 2011 the Jiangsu Province power installed capacity was 42,130 MW and the power production has been 210 TWh.

Electricity Consumption. Considering the high rate of growth of last years, Jiangsu consumes large quantity of electricity per year, more than 220 TWh per year.

Heat Consumption. Most of heat supply in Jiangsu province is used for industry, not for residential use partly because of the geographical location. Jiangsu is situated in the center of the east coastal areas of China and the climate is temperate, so according to the standards of building engineering, most of the areas in Jiangsu don't have the residential heating network.

Table 2. Main data of heat and electric power in Jiangsu in 2011

Unit: Mt standard coal	heating power	electric power
Quantity of energy available	0	94.3
1.Production		4.02
2.Import from other provinces		393.02
3.Export to other province		-302.74
Input–output+ofprocessing&conversion	615.16	2601.45
1.Steam electric power generation	-33.81	2601.45
2.Heat supply	648.97	
Loss	32.45	211.72
Loss of transportation and distribution		211.72
Consumption of end-using	582.71	2484.03
Primary industry		35.79
1.Agriculture,forestry,stock raising,fishery		35.79
Secondary industry	570.58	1992.29
1.industry	570.58	1965.18
2.architecture		27.11
Tertiary industry	0	209.26
1.traffic,storage,posts&telecommunication		20.22
2.wholesale,retailing,catering		68.23
3.others		120.81
Residental consumption	12.13	246.69
1.town	12.13	125.58
2,village		121.11

Tianwan Nuclear power Plant. Tianwan Nuclear Power Plant is the only nuclear power plant in Jiangsu province.(NPP, 2 × 1000 MW). The plant has four generating units and space for four more. The two units combined generate 14 TWh of electricity a year.

2.2   Definition of Energy Resources and Technologies in Jiangsu Region

After the definition of the Reference Energy Scheme (RES), it is possible to go into the details of the Resources and Conversion Technologies used in the region in order to define and describe their parameters, which are the data that will be inserted in the model. This is an exogenous data supplied as input into the model.

Definition of the Parameters in Resource Technology. The parameters reported in table 3 have been used in the description of "Import Technology" in the Markal environment.

Table 3. parameters used to describe an "Import Technology"

PARAMETER	DEFINITION	DOMAIN OF DEFINITION	UNITS
COST	Annual resource cost	Resource Technologies	Million yuan/Petajoules
ENV_SEP	Emissions coefficient/resource activity	SRCENCP Resource Technologies	Thousand tonnes/petajoules
OUT(ENT)r	Energy carrier output: resource technology (not export)	IMP_SEP-Import Resource Technologies MIN_SEP- Extraction Resource Technologies RNW_SEP-Renewable Resource Technologies STK_SEP - Stockpile Resource Technologies	petajoules/petajoules
CUM	Total resource availability	EXP_SEP - Export Resource Technologies IMP_SEP - Import Resource Technologies MIN_SEP - Extraction Resource Technologies RNW_SEP -Renewable Resource Technologies	petajoules
START	Start year	All Technologies (TCH+SRCENCP)	calendar year

Import of Crude Oil. The parameters relevant to the Import of Crude Oil have been defined as follows:

Table 4. parameters for "Import of crude oil" technology

	2007	2012	2017	2022	2027
env_sep	0.075	0.075	0.075	0.075	0.075
Cost	87.14	67.55	66.35	69.80	75.27
out(ent)r	1	1	1	1	1
Cum	-	-	-	-	-
Start	2007	-	-	-	-

Since the material and quality comprised by "Import of Crude Oil" it has not changed much, is has been maintained the value of 'env_sep' used in a previous model of China developed by University of Rome. The official forecast for the price of crude oil in the considered period of time is shown in the following table 5. The first column in this table is the year considered when building the model. The second column is the price of crude oil expressed in dollars 2010/barrel. In the third column, it is possible to find the price expressed in the model-accepted measure unit 'million yuan/PJ'.

Table 5. official forecast for price of crude oil

year	Price (2010dollars/barrel)	Price (million yuan/PJ)
2010	57.47	75.08
2013	49.99	65.30
2016	49.75	64.99
2019	51.95	67.86
2022	53.43	69.80
2025	56.37	73.64
2028	58.12	75.92
2030	59.12	77.23

Import of Diesel. As reported by press agency XinHua, the newest statistic realized by the "State Development and Reform Administration" shown that in the middle of 2007 the price of diesel reached 5237 yuan/ton. In order to forecast a pattern for the diesel's cost in the future years, the reference case published in the IEA Report

"World Oil Prices in Three World Oil Price Cases, 1980-2030" has been used,. This is obviously an approximation, although the price of crude oil is the most significant factor influencing the price of diesel, the influence of oil refinery capacity constraint in the neighboring regions of China is not taken in consideration because it is not yet known and beyond the scope of the present study. To take into consideration such an influence could be a legacy for the model development and fine-tuning work. Calculating from the price of crude the increasing rate, it is possible to derive the cost of diesel in the following years. For the parameter OUT(ENT)r, which means 'Energy carrier output: resource technology (not export)' the model's default value is one.

Table 6. parameters for "Import of diesel" technology

Scenario	Parameter	Technology	Commodity	2007	2012	2017	2022	2027
BASE	COST (millio-nyuan/PJ)	IMPDSL1	-	130.82	101.40	99.63	104.73	112.97
BASE	ENV_SEP	IMPDSL1	CO2	0.075	0.075	0.075	0.075	0.075
BASE	OUT(ENT)r	IMPDSL1	DSL	1	1	1	1	1
Scenario	Parameter	Technology	Value					
BASE	CUM	IMPDSL1	-	-	-	-	-	-
BASE	START	IMPDSL1	2007	-	-	-	-	-

Import of Gasoline. The process of getting the data in each parameter in this scenario is more or less the same as which in the scenario 'Import of diesel'. The only difference is the actual price of gasoline. Thus, for the parameter 'COST', the data used is what reported by the "State Development and Reform Bureau".

Table 7. forecast of gasoline price (millionyuan/PJ)

Year	increasing rate	Price of Gasoline
2007	1	128.26
2012	0.77	99.41
2017	0.76	97.68
2022	0.80	102.68
2027	0.86	110.7

Table 8. parameters for "Import of gasoline" technology

Scenario	Parameter	Technology	Commodity	2007	2012	2017	2022	2027
BASE	COST (millionyuan/PJ)	IMPDSL1	-	128.26	99.41	97.68	102.68	110.762
BASE	ENV_SEP	IMPDSL1	CO2	0.075	0.075	0.075	0.075	0.075
BASE	OUT(ENT)r	IMPDSL1	DSL	1	1	1	1	1
Scenario	Parameter	Technology	Value					
BASE	CUM	IMPDSL1	-	-	-	-	-	-
BASE	START	IMPDSL1	2007	-	-	-	-	-

Conversion Technology. The Conversion Technologies considered in this study are: Oil Refinery, Pumped Storage Power Plant, Coal Steam Power Plant (differentiated into plants with maximum power of 300 MW, 600 MW and 1000MW), Nuclear Power Plant, Wind Power and also some new technology which may be taken in the future , that are IGCC, Co-generation power plant , Photovoltaic technology. For each technologies has been considered the parameters shown in table 9.

Table 9. Definition of the parameters in Conversion Technology

PARAMETER	DEFINITION	DOMAIN OF DEFINITION	UNITS
AF	Annual availability	CON - Conversion Technologies PRC - Process Technologies	decimal fraction
FIXOM	Annual fixed O&M cost	CON - Conversion Technologies DMD - Demand Technologies PRC - Process Technologies	Millionyuan/gigawatts
INP(ENT)c	Energy carrier input: y	CON - Conversion Technologies *All Energy Carriers (ENT)	Petajoules/petajoules
INVCOST	Total cost of invest-ment in new capacity	CON - Conversion Technologies DMD - Demand Technologies PRC - Process Technologies	millionyuan/gigawatts
PEAK(CON)	Fraction of capacity in peak equations	CON - Conversion Technologies	decimal fraction
RESID	Residual installed capacity	CON - Conversion Technologies DMD - Demand Technologies PRC - Process Technologies	Gigawatts
VAROM	Annual variable O&M cost	CON - Conversion Technologies DMD - Demand Technologies PRC - Process Technologies	millionyuan/petajoules
AF_TID	Fraction of unavailabil-ity which is forced outage	CON - Conversion Technologies	decimal fraction
CAPUNIT	Units of activity/unit of capacity	CON - Conversion Technologies DMD - Demand Technologies PRC - Process Technologies	petajoules/gigawatts
LIFE	Lifetime of new capacity	CON - Conversion Technologies DMD - Demand Technologies PRC - Process Technologies	number of years
OUT(ELC)_TID	Electricity output	ELE - Electric (not District Heat) Conversion Technologies CPD - Electric and District Heat Conversion Technologies STG - Storage Conversion Technologies ELC - Electric Energy Carriers	decimal fraction
START	Start year	*All Technologies (TCH+SRCENCP)	calendar year
BOUND(BD)	Bound on capacity LO : Lower UP : Upper	CON - Conversion Technologies DMD - Demand Technologies PRC - Process Technologies Bounds	Gigawatts
CF(Z)(Y) Utilisa-tion for season, time of day	I-D : Intermediate Day I-N : Intermediate Night S-D : Summer Day S-N : Summer Night W-D: Winter Day W-N: Winter Night	XLM - Fixed Capacity Utilisation Conversion Technologies Time Slice: Season-Time of Day	decimal fraction

A brief description of main hypothesis and data sources used for the definition of parameters for each technologies has given in the next paragraphs.

Oil Refinery. From the statistic from "Jiangsu Statistic Bureau", up to 2008, the quantity of oil refined in Jiangsu region is equal to 20.6 million tons, while the total quantity of oil refined nationwide in China is 285 million tons. Thus, it is possible to infer that the oil refinery capacity installed in Jiangsu is roughly 10% of the overall Chinese refinery capacity. There are some new technologies which can bring up the efficiency of producing gasoline and diesel to 75% and 85% respectively, but due to the high investment cost they haven't been commonly used in China. So to evaluate the actual efficiency for each product the average value of the efficiency of the com-monly used technology has been used. Later in the following years, the efficiency may be improved , with the 75% and 85% for gasoline and diesel as its upper boun-dary. For the parameter 'invcost', has been taken as an example the 'Yangkou Ha-bour' program in Nantong, Jiangsu Province. The annually ability of oil refinery is 12.000.000 ton, the total investment cost has been of around 20 billion yuan RMB.

Pumped Storage Power Plant. The peak time for using electricity in Jiangsu region is from 9:00 to 11:00 am , and from 18:00 to 21:00 or 19:00 to 22:00 pm. So totally the peak hours can be considered to last 7 hours a day, thus the availability factor will be fixed AF = 0.3. The efficiency for this plant is known to be 0.76, so the parameter INP(ENT) here is 1/0.76 = 1.32. Normally in Jiang Su region pumped storage power plant are built using six power unit of 250MW each, the total investment cost is around 6 billion yuan for such a plant. So taking the plant in LiYang as an example, one of the most significant projects in Jiangsu's '11th five year plan', the total investment in the pumped storage power plant is 6.8 billion yuan with an installed capacity of 1.5GW. For the aforesaid plant building plans the Parameter Bound for the installed capacity is thus:

Table 10. Bound on capacity for Pumped storage plants

	2007	2012	2017	2022	2027
Bound(L) (MW)	250+100	100+1000	1100+1500	1100 +1500	2600
Bound(U) (MW)	350	1100+1500	1100+1500+1500	4100	4100+1500

For a plant with an installed capacity of 1000 MW, annually its production of electricity is 1.49 billion kWh. Then, the CAPUNIT factor can be calculated as follows CAPUNIT = 14.9 *100000000*1000w*3600s /1gw = 5.36PJ/GW

Coal Steam Power Plant. There are three different kinds of coal steam power plants being currently used in the region, with the a maximum power capacity of 300 MW, 600 MW and 1000 MW respectively. Since there is a significant difference between their efficiency and fundamental parameters it is possible to define 3 different technologies. For coal steam power plant, the differentiating factors are the cost and length of annual maintenance, the operating cost of the plant and their efficiency. Besides that, it is also possible to name the various plant present in Jiangsu region, their expected life and how many of each kind can be expected into service in the next years.

Nuclear Power Plant. The largest nuclear power station is located in Lianyungang, Jiangsu province whose AF is at least 0.8 as reported by the "Commission of science technology and industry for national defense" from its institutional website. It was reported by ChinaDaily.com.cn that Chinese government planned to start up the second stage of 'Tianwan nuclear project'. The program planned to start at the end of 2009 with a growing capacity of 2000MW. Regarding FIXOM and VAROM, was not possible to find detailed and mature data indicating the situation in the region subject of study.

Photovoltaic Power Plant. The general utilizing efficiency is 76%-80% for a photovoltaic power plant while normally the investment for this kind of technology is 5000yuan/kw and the operation and maintenance cost plays 0.43 to 0.47% of the investment cost annually. It was reported by the "Jiangsu Weather forecast Bureau".

Wind Power. As "Jiangsu Developing and reform bureau" stated, the average utilizing time is between 3000 hours to 4000 every year, so the parameter 'AF' is between

3000/(24*365) to 4000/(24*365). The wind power plant can be divided into two categories: inshore and offshore. The average efficiency for wind to transform into electricity is 30% to 50%.

IGCC- Integrated Gasification Combined Cycle. The only project of IGCC actually planned is the one in Xuzhou, Jiangsu province. The total investment is 1.37 billion yuan and the capacity is 300MW.

3   Result and Discussion

In the following paragraph the results of the proposed model are presented and described. The goal of this step of the work is to forecast the final energy consumption from year 2007 to 2027 (time-span of the model). The forecast should be made for each energy commodity and for each economic sector in order to input in the model the data for the sectorial demand, since in the current phase the demand data are not analyzed. The available data in order to perform this forecast are: GDP and inflation rate during 1997-2011; Energy intensity during 2003-2011. Using the regional energy consumption, the 'adjusted GDP-base 2010' of years 2003-2011 and the forecasted declining rate of energy intensity after 2011, the total energy consumption until 2027 can be derived. Therefore, the share of energy consumption for each energy commodity in the future scenario can be calculated.

3.1   Forecast of GDP

According to the "11th five-year planning", the forecasted GDP growth rate of China is presented below (table 11):

Table 11. China's GDP Growth rate

Period	China's GDP Growth rate
2006-2010	8%;
2011-2015	7%;
2016-2020	6%;
2021-2025	6%;
2026-2030	5%;

The GDP growth rate of Jiangsu region is usually 3% higher than the national value, so the first has been derived from the latter adding 3% for every period . This lead to the following table:

Table 12. Jiangsu's GDP Growth rate

Period	Jiang su's GDP Growth rate
2011-2015	10%;
2016-2020	9%;
2021-2025	9%;
2026-2030	8%;

Since China is developing so fast now, it is possible to predict that GDP's growing rate in Jiangsu will be higher than 10% for several years. With these data it is possible to calculate the GDP value for the next 25 years in China, these values should be adjusted to inflation, taking 2007 as base year. The inflation rate of 1997-2011 in China is got from State Statistical Bureau; inflation rate of 2007 in China is got by IMF. It is predicted by Asian Development Bank (ADB) that yearly average inflation rate in China during 2007-2013 will increase slightly but overall is lower than 3%. After 2012, the inflation rate is predicted to 2.5% for each year according to the government policy principle stipulating that inflation rate of China should be kept between 2-3% to maintain that RMB against the US dollar cannot be devaluated and holds a rising space. The measure unit of money has been chosen as Billion yuan.

3.2   Forecast of Energy Consumption

In order to forecast the future energy consumption per sector, the first step is to forecast the energy intensities per unit of GDP for sector. This will be later multiplied for the GDP value of the year, in order to get the desired result. Total energy consumption data in Jiangsu region during 2003-2011 have been collected, that of 2011 is the latest data that could be found. According to the 11th "five-year plan", at the end of 2012, the energy consumption per unit of GDP should decline by 20%. For the following years , according to the overall policy of China, has been assumed that the declining pattern in energy consumption may be stable in the period 2011-2015, and assume a lower value in the remaining model forecast years. As a result, the energy intensity declining rates of the following years showed in table 13.

Table 13. China Energy intensity's Growth rate

Period	Declining rate in energy intensities
2006-2010:	4,4%
2011-2015:	4,4%
2016-2020:	4%
2021-2025:	4%
2026-2030:	3%

All the Energy Intensities during 2006-2030 can be calculated as follows.

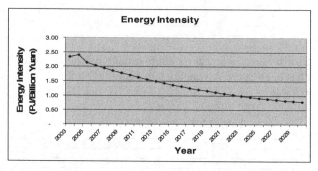

Fig. 10. China's forecasted Energy intensity growth rate

3.3   Energy Consumption for Each Commodity

To get the consumption of commodities, the proportion each commodity covers in the total energy consumption is forecasted. With the hypothesis that, the renewable energy consumption will be 16% of total energy consumption by 2030, it can be derived that the fossil energy would be 84%. Using these data and a constant pattern of approach to the goals, the share of each fuel can be forecasted for each year.

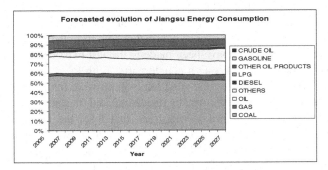

Fig. 11. Jiangsu's forecasted Energy consumption per fuel

3.4   Energy Consumption for Each Sector

According to the hypothesis that the industry sector share in energy consumption in Jiangsu region will fall, but will remain the biggest one, while the other sectors' share will grow steadily, it has been derived the sector's share in total consumption in the period 2005-2030.

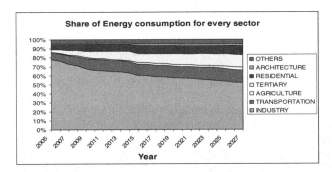

Fig. 12. Jiangsu's forecasted Energy consumption per sector

4   Conclusion

Energy sources of Jangsu Region are in the same respective proportions as in the overall nation: coal is the most used followed by oil, gas and a small presence of hydropower and nuclear, still in the early stage of their development in the region.

Electricity is mainly generated from coal, with three main stream types of plant, 300, 600 and 1000 MW. Clean coal technologies are beginning to be implemented as the construction of a IGCC plant is planned for 2008.

The climate in the region is not so rigid, so residential heating is not used at all. The demand for heat is thus limited to industrial uses, where cogeneration is not yet commonly used. As a mean of developing the model has been draw for the region a Reference Energy Scheme that shows the flow of energy commodities from well to final uses through processing and conversion technologies.

The present and future energy commodities and technologies in the supply side have been fully described by a series of parameters that form their Markal definition. Even though enormous effort have been made in the last two decades, China remains less energy-efficient than its Western counterparts. With the development of market economy and private enterprises, the situation is improving. Industrial sector remains the biggest energy consumer, followed by residential and transport sector.

The energy demand from heavy industries, that form the backbone of Chinese economy, and from the transport sector will continue to grow at double digit rates in the foreseable future. This poses enormous pressure on the supply side and call for the development of energy-efficient technologies in all end uses.

References

1. Bahree, B.: Economy improves, but oil prices carry risk: Extra supply won't come in time to tame impact of $50 crude on recovery. Wall Street Journal (Eastern edition), CCXLV(238), A3 (2004)
2. EIA (Energy Information Agency), International Energy Outlook, US Department of Energy, EIA, Washington, DC (2003)
3. Gates, D.F., Yin, J.Z.: Energy development strategy in China's urbanization. In: Chen, J., Chen, A. (eds.) An Analysis of Urbanization in China. Xiamen University Press, Xiamen (2002) (in Chinese)
4. IEA (International Energy Agency), China – an in-depth study. World Energy Outlook 2008 (2003)
5. Ehara, N.: IEA collaboration with India and China on oil security. IEA/ASEAN/ASCOPE (2004)
6. Great growth of imported coal in Jiangsu Province, Ministry of Commerce of the People's Republic of China Special Commissioner's Office in Nanjing (May 13, 2005)
7. Chen, W., Wu, Z., He, J., Gao, P., Xu, S.: Carbon emission control strategies for China: A comparative study with partial and general equilibrium versions of the China MARKAL model. Energy, Environment and Economy Research Institute/Institute of Nuclear and New Energy Technology, Tsinghua University (2005)
8. Wen, Zhou, H. (eds.): The Globalization of the Chinese Economy. Edward Elgar, Northampton
9. NBS (National Bureau of Statistics), China Statistical Yearbook. China Statistics, Beijing (2010)
10. Lu, J.-G., Huang, Y., Li, F., Wang, L., Li, S., Hsia, Y.: The investigation of ^{137}Cs and ^{90}Sr background radiation levels in soil and plant around Tianwan NPP, China. Journal of Environmental Radioactivity 90 (2006)

11. Carlini, M., Castellucci, S., Allegrini, E., Tucci, A.: Down-hole heat exchangers: modelling of a low-enthalpy geothermal system for district heating. Mathematical Problems in Engineering 2012, Article ID 845192, doi:10.1155/2012/845192, ISSN 1024-123X
12. Nancy, Y., David, F., Ke, X.: Improving Transport Fuel Quality in China: Implications for the Refining Sector. Trans-Energy Research Associates, Seattle, WA, Lawrence Berkeley National Laboratory, Berkeley, CA, China Petrochemical Consulting Corporation, Beijing (2002)
13. Carlini, M., Castellucci, S.: Efficient energy supply from ground coupled heat transfer source. In: Taniar, D., Gervasi, O., Murgante, B., Pardede, E., Apduhan, B.O. (eds.) ICCSA 2010, Part II. LNCS, vol. 6017, pp. 177–190. Springer, Heidelberg (2010)
14. Lin, J., Zhou, N., Levine, M.D., Fridley, D.: Achieving China's Target for Energy Intensity Reduction in 2010: An exploration of recent trends and possible future scenarios. Lawrence Berkeley National Laboratory (2006)
15. Sinton, J.E., Fridley, D.G.: Comments on Recent Energy Statistics from China. Lawrence Berkeley National Laboratory (2003)
16. Colantoni, A., Giuseppina, M., Buccarella, M., Cividino, S., Vello, M.: Economical analysis of SOFC system for power production. In: Murgante, B., Gervasi, O., Iglesias, A., Taniar, D., Apduhan, B.O. (eds.) ICCSA 2011, Part IV. LNCS, vol. 6785, pp. 270–276. Springer, Heidelberg (2011)
17. French, H.W.: China promotes another boom: Nuclear power. The New York Times (January 15, 2005)
18. Wang, X., Feng, Z.: Energy consumption with sustainable development in developing country: a case in Jiangsu, China. Energy Policy 31(15), 1679–1684 (2003)

Photovoltaics in Italy, Mechanisms of Promotion: A Cost-Benefit Analysis of the Italian "Conto Energia" and Evaluation of Externalities

Marco Lucentini and Diego Di Palma

CIRPS, Sapienza University of Rome
Piazza San Pietro in Vincoli, 10, 00184 Rome – Italy
{marco.lucentini,diego.dipalma}@uniroma1.it

Abstract. Italy is currently one of the fastest growing photovoltaic markets worldwide. The introduction of the feed-in-tariff mechanism called "Conto Energia" has raised and developed the market, contributing to the reduction of GHGs emissions and to increase the diffusion of the smart grid model.

This study aims develop a cost-benefit analysis of the "Conto Energia" feed-in-tariff mechanism, by measuring all costs and benefits of electric energy produced by photovoltaics in Italy, from 2006 up to 2020. Particular attention has been given to the analysis of the fast increase of photovoltaic power installed capacity in Italy by the end of 2010.

The analysis measures costs and benefits associated to "Conto Energia" system (the principal Italian photovoltaic policy support) and assesses the support profitability for the entire society by giving a monetary value to the socio-economic benefits of PV electricity promotion.

Keywords: Photovoltaics, Feed-in-tariff mechanism, "Conto Energia".

1 Introduction

In 2005, a specific policy instrument has been introduced in Italy to support PV energy, the so-called "Conto Energia". From a policy perspective is possible to affirm that a revolution took place with its set up, because it filled the gap of insufficient and unstable support provided until that time [1, 2]. Furthermore, it aligned Italy with the rest of European countries regarding photovoltaic promotion policies, widely relying on Feed-in Tariffs (FIT) [3, 4]. After the introduction of the first phase in 2005 (the so called "Old Conto Energia"), a second phase of "Conto Energia" (the so called "New Conto Energia") has been issued. In the "Old Conto Energia", the overall power expected to be supported was 500 MWp (360 and 140 MWp for smaller and bigger plants respectively), with an annual limit (cap) of 85 MW and a final target of 1.000 MWp by the year 2015 [5, 6]. The tariffs varied with the nominal power of the plant ranging from 0,445 €/kWh to 0,490 €/kWh. The duration of the support was 20 years and the tariffs were updated on a yearly basis, taking into account the consumer price index variation.

B. Murgante et al. (Eds.): ICCSA 2013, Part II, LNCS 7972, pp. 328–343, 2013.
© Springer-Verlag Berlin Heidelberg 2013

The first phase of the "Conto Energia" boosted the installation about 5˙500 plants, corresponding to a cumulative power around 160 MW, out of 380 MW admitted to the incentive. Reasons for that were due to administrative barriers, to bureaucratic problems linked to authorizations and grid connection permits and to a bubble effect cause by a "license trade" effect [7].

The first experience of FIP mechanism in Italy definitely contributed to increase PV market penetration, as well as revealed the necessity of reviewing the mechanism because of some barriers that hampered market growth and prevented the attainment of the target. Difficulties experienced with the "Old Conto Energia" have been partially solved with a new phase of the "Conto Energia" Programme (issued in February 2007) that established simplified administrative procedures, revamping the Italian photovoltaic sector by overcoming most barriers: elimination of the maximum eligible size of 1MW, promoting smaller and architectural integrated plants; increase o tariff granted for specific kind of plants (5%) and for plants with a capacity up to 200 kW, operating in net-metering system. The objectives were a total capacity to be supported of 1˙200 MW, a goal of cumulative power installed of 3˙000 MW by 2016.

As concerns the valorization of the electricity, alternatively to net-metering, PV owners can commercialize the electricity fed into the grid either through a direct sale on IPEX (Italian Power Exchange) or through bilateral contracts and otherwise indirectly through the "dedicated electricity intake" (Ritiro Dedicato) managed by GSE. The latter consists of a simplified way of selling the electricity fed into the grid through the establishment of a convention between the system operator and the GSE.

An alternative option is to benefit from the net-metering ("on-the-spot trading") scheme. The mechanism allows to deliver electricity produced by plants up to 200 kWp into the grid and to withdraw the same amount of electricity when it is needed but the plant is not producing (or production is not enough), using the grid as a sort of "electricity tank".

From the "Fourth Conto Energia" some photovoltaic systems have be assigned to an omni-comprehensive tariff mechanism (including the sale tariff and the incentive).

2   The "Conto Energia" Programme: 2006-2012 Results

Looking at preliminary results, the second phase of the "Conto Energia" boosted the installation of about 480˙000 plants up to January 2013, corresponding to a cumulative installed capacity slightly above 17 GWp.

Although the good growth of year 2009, the year 2010 registered an impressive increase of installations due to the imminent deadline of the "New Conto Energia" mechanism (31st December 2010). To insure the 2010-tariff for plants realized (the definition of "closed installation" has been published at the end of November 2010: the plants had to be completely finished except for the connection to the grid by the local distributor), a online procedure to communicate with the GSE was established.

As a consequence, a great number of plants have been realized, waiting the connection to the grid to become operative. The data published by GSE, in the early 2011, have shown more than 3,5GW already realized and to be connected to the grid before June 2011. As the matter of the fact, despite of the goal of cumulative power installed by 2016 of 3GW, the results of "New Conto Energia" were very discrepant from predictions, with more than 10,5GW operative by the end of 2011 (Fig. 1).

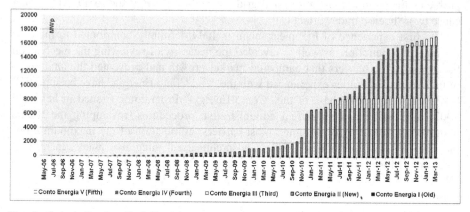

Fig. 1. Cumulative Installed Capacity under the three editions of "Conto Energia". Source: GSE (December 2011).

2.1   Market Development and Outlook

Thanks to the introduction of the "Conto Energia" the photovoltaic installed power increased significantly in most recent years, reaching 70% in the period 2005 – 2009, more than 300% in the period 2009 – 2010 and about 150% from 2010 to 2011. Similarly the electricity production passed from around 30 GWh in 2005 to 800 GWh in 2009, to 2,4 TWh in 2010 and to more than 20 TWh in 2012, gaining a greater but still limited portion of the domestic electricity demand (near to 6,5% of total electric energy yearly consumed in Italy). As concerns the technology breakdown, most of the modules installed are wafer-based crystalline silicon (c-SI).

For lower power ranges (1-3kWp) Mono Poly c-SI technologies cover roughly 98% of the market, while Thin-film and novel concept technologies reach higher penetration (around 8%) in larger applications.

The Long Term-Market. The incentive mechanism introduced by "Fifth Conto Energia" signed a remarkable stop to the previous market behavior. However Italy still remains the world's second largest PV market. Regarding the longer term, considering the decreased number of requests currently under the examination of GSE, the number of plants waiting for the connection to the grid, the effect of the fifth phase of "Conto Energia", the growing maturity of the technology (i.e. system prices cost reductions and consequent higher competitiveness) and adopting a conservative approach towards EPIA figures until 2020 it seems rationale to forecast an annual increase lower than 300 MW of newly installed capacity per year and a cumulative installed capacity of 22.000 MW up to 2020 (Fig. 2) [8, 9].

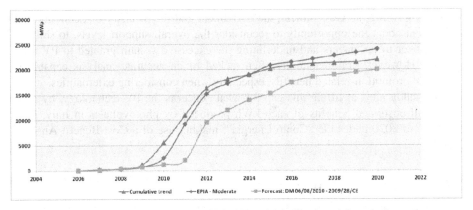

Fig. 2. Italian long-term PV Market outlook. (Author elaboration)

PV Modules and System Prices Evolution. In Table 1 the typical system prices (excluding VAT) and the best prices achieved reported by IEA are shown [10, 11, 12]. More specifically, typical prices derive from an average of about 10˙000 plants (both small and large) while the best price are the lowest ones and regard import products (for Crystalline silicon modules from China and for thin film modules). Note that for 2008, the standard price is referred to c-Si modules while the best to Thin-Film ones.

Table 1. National trends in system prices for small roof-mounted system (1-3 kW) *Author estimation

Year	2006	2007	2008	2009	2010	2011*	2012*	2013*	2014*	2015*	2016*	2017*	2018*	2019*	2020*
System price (€)	6400	6500	6000	5500	4900	3700	2450	1925	1750	1620	1390	1255	1106	901	853

The Administrative Process. According to the article 12 of 387/2003 Decree, the authorization for the construction and the execution of a PV plant is subjected to a unique (simplified) administrative license called "Autorizzazione Unica" (AU) which has to be issued by the Region (or the Province when the latter receives a delegation) within 180 days after the system operator request (maximum theoretical legal limit).

3   Cost-Benefit Analysis of the "Conto Energia" Programme

3.1   The Objective

The introduction of the "Conto Energia" has significantly changed the future of the photovoltaic technology in Italy. As already mentioned, the "Conto Energia" is rolled over to final electricity consumers through an additional charge (of the A3 component) on the electricity bills. There are growing concerns about costs (and above all of their distributional income effects) bored by the electricity consumers. The AEEG emphasized in several occasions the increasing charges deriving from RES promotion policies, warning the Government, the Parliament and the Ministry of Economic Development about the opportunity to reconsider the overall support levels, to shift the burden over to tax payers and underlining the excessive sustain granted to PV technology. However, argumentations often lacked of an accurate analysis capable to reveal all promotion-related benefits, especially when considering externalities.

This study aims at filling this gap for what concerns the PV technology, by measuring all costs and benefits of each kWh produced by photovoltaics in Italy, from 2006 up to 2020 under the "Conto Energia", making use of a Cost-Benefit Analysis (CBA).

3.2   The Cost-Benefit Analysis

Cost-Benefit Analysis (CBA) has long been used as a powerful appraisal technique for investments in the developing world where markets are seriously distorted and prices do not provide a satisfactory basis for assessing their overall impact on society. However, during the last thirty years, CBA is also widely applied in developed countries for the assessment of public projects and policies, because of the growing concern for distorted market prices and the existence of externalities, both indicating a divergence between private and social costs and an inefficient allocation of resources. In particular, due to an increasing environmental awareness, CBA is now routinely used in environmental policy assessment in order to secure that policy choices are as cost-effective as possible.

Central in the theoretical foundation of CBA is that costs and benefits reflect the preferences of individuals which are measured as Willingness To Pay (WTP) for benefits and Willingness To Accept compensation (WTA) for costs.

It is furthermore assumed that individual preferences can be aggregated so that social benefit is simply the sum of all individuals' benefits and social cost is the sum of all individuals' costs. CBA aims at aggregating all costs and benefit associated with each policy option, the former defined as reductions and the latter as increases in human wellbeing. For non-traded or intangible goods, for which measures do not exist in the marketplace, a range of approaches are in use today for deriving such estimates to be integrated in CBA. However, in many cases significant environmental impacts are not easily amenable to defensible monetary valuation and are therefore omitted from CBA. Given that policies affect societies over a long time horizon, costs and benefits should be also aggregated over time. Here also, there are no hard rules of how far into the future one has to go. Summing costs and benefits over time requires that 'time preferences' of individuals have also to be accounted for in a way that current choices do not disregard the interests of future generations [13, 14, 15]. This practically consists in selecting an appropriate discount rate.

For a project or policy to be selected benefits must exceed costs, while alternative options are ranked according to their net benefits. The most general formulation of this decision rule is the following:

$$\left\{ \sum_{i,t} B_{i,t} \cdot (1+r)^{-t} - \sum_{i,t} C_{i,t} \cdot (1+r)^{-t} \right\} > 0 \qquad (1)$$

B= Benefit i occurring in time t
C= Cost i occurring in time t
r =Discount rate to calculate the present value (PV) of all costs and benefits (social discount rate ranges 3-4%).

It is worth mentioning that due to the complexity of the electricity market functioning and the uncertainty about externalities monetary values, is highly difficult to reach a perfect measurement. Moreover, we underline that the CBA has been conducted only for the "Conto Energia" programme and therefore the PV related support.

Furthermore, due to insufficient and reliable data, the analysis does not consider benefits and costs related the valorization of the electricity, currently available for PV system operators. Consequently, for an adequate reconsideration of the RES support levels a more comprehensive CBA should be carried out and sustained by other robust public policy assessments (such as Cost-Effectiveness Analysis or Multi Criteria Analysis). Additionally, for a more complete CBA of PV promotion policy levels, also the support coming from the valorization of the electricity should be included.

Despite that, being the "Conto Energia" the most consistent and effective part of the incentive, the study can be considered robust enough to make known the real value of PV electricity promotion. The analysis measures as much costs and benefits associated to the Italian photovoltaic ("Conto Energia") policy support as possible, determines the value of each kWh of photovoltaic electricity production and robustly reveals the support profitability for the entire society.

To properly attain the target, the study takes into account the most recent publicly available externalities estimations (coming from the Cost Assessments for Sustainable Energy Systems Project – CASES [2]) and the most consistent proxy values for some RES promotion benefits related to the Italian context. Moreover, it provides with monetary values estimations about socio-economic costs and benefits linked to PV promotion and to higher PV penetration. This approach is particularly important considering that Italy, as other Member States, has released and recently submitted to the European Commission its Renewable Energy National Action Plan (RENAP) [16] and that, as widely accepted [17], will have to rationalise its support mechanism.

3.3   The Structure

In order to correctly assess the Net Present Value (NPV) of all costs and benefits linked to the promotion of photovoltaic electricity under the framework of the "Conto Energia" Programme from 2006 up to 2020, we divided the society in five main segments (Fig. 3) and calculated the flows both in € and in €/kWh. The proposed structure allows to compute 'sector CBAs', therefore to indicate who bears the most of the costs and gets the most of the benefits as well as the net benefit (or cost) for the entire society.

Although the fifth edition of the "Conto Energia" fixed the end of the incentive mechanism in 2016, the time horizon of the study was still set to 2020 to assess the effects of "Conto Energia" Programme, in relation to the objectives set by the European directive 2009/28/EC.

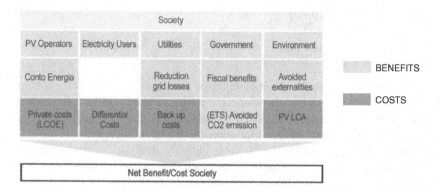

Fig. 3. "Conto Energia" CBA Structure. Source: author elaboration.

The PV system operators and the electricity users are considered the direct participants of the "Conto Energia" Programme, as represent who benefits and who bears the costs of the incentive respectively. Other sectors are indirectly involved and subjected to the consequences (positive and negative externalities) of the policy.

For the period 2006-2012, when possible, the values have been observed and for the future some estimations have been made. The CBA is indeed performed as the program started in late 2005 and will continue to produce its effect in the future.

To discount all future benefit and costs we used a social discount rate equal to 3,5% as suggested by the European Commission CBA Guidelines. The most important assumption refers to the possibility of incentives for new plants at least until 2013, when will be phase out due to the proximity to the almost full competitiveness with conventional energy sources (as called "grid parity").

The main assumptions regarding the future trends of the "Conto Energia" Programme are presented in Table 2. The value of "Conto Energia" tariffs for each year are calculated as weighted average for the different types of system (full integration on buildings, retrofit installations, on-ground installations, etc.) and for each power range.

Table 2. "Conto Energia" future trends estimations. Source: author elaborations.

Year	Installed	Production	"Conto Energia" tariffs
2006	9	2	0,500
2007	80	43	0,470
2008	418	200	0,420
2009	1140	796	0,415
2010	5500	5046	0,410
2011	11000	12508	0,313
2012	16300	18729	0,228
2013	18100	21013	0,148
2014	19000	22285	0,065
2015	20200	23933	
2016	20800	24892	
2017	21200	25624	
2018	21350	26060	
2019	21500	26500	
2020	22000	27379	

3.4   Methodology and Assumption

PV System Operators CBA. PV system operators benefit from the incentive which grants them a monetary support for each kWh of electricity produced. As a consequence, the incentive is considered as a positive flow in the PV system operators CBA. An average value of the tariffs for all kind of plants is assumed although the tariffs are normally differentiated by system size and level of integration. On the other hand, PV system operators have to sustain the costs of installing, operating, maintaining and dismantling the plant (i.e. private costs). The investment costs are considered as a negative flow in the PV system operators CBA.

In order to assess private costs we used the Average Levelised Cost of Electricity (LCOE) methodology (The notion of levelised costs of electricity (LCOE) is a handy tool for comparing the unit costs of different technologies over their economic life. It would correspond to the cost of an investor assuming the certainty of production costs and the stability of electricity prices. In other words, the discount rate used in LCOE calculations reflects the return on capital for an investor in the absence of specific market or technology risks. Given that such specific market and technology risks frequently exist, a gap between the LCOE and true financial costs of an investor operating in real electricity markets with their specific uncertainties is usually verified).

The methodology calculates the generation costs in €/kWh on the basis of the power supplied to the station bus bar where electricity is fed into the grid. This cost estimation methodology discounts the time series of expenditures to their present values by applying a discount rate.

According to the methodology used in the IEA study [11], the levelised lifetime cost per kWh of electricity generated is the ratio of total lifetime expenses versus total expected outputs, expressed in terms of present value equivalent. The total lifetime expenses include the value of capital, fuel expenses and operation and maintenance expenses. We propose here the formula we used, which does not include fuel expenditures and adapted to the specific characteristics of PV production:

$$LCOE = \frac{CAPEX + PV(O\&M)}{PRV(EP)} \qquad \text{with} \qquad (2)$$

$$PRV(O\&M) = \sum_{K=1}^{25} \frac{O\&M}{(1 + WACC)^k} \quad PRV(EP) = \sum_{K=1}^{25} \frac{EP \cdot (1 - DEGR \cdot k)}{(1 + WACC)^k}$$

Where CAPEX is capital expenditures (€), O&M is Operating and maintenance cost (€), EP Electricity Production (kWh), WACC is the Weighted Average Cost of Capital [%], k is the year, DEGR is annual degradation (0.8%) and PRV stays for Present Value. The LCOEs have been calculated for four different market segments from 2006 up to 2020 according to the following parameters. However, we used average LCOE values per each year considering them as a weighted reference for entire market segmentation (Table 3).

Table 3. LCOE System Parameters

	Residential 3kWp	Commercial 50kWp	Industrial 200kWp	Field >1000kWp
Cost of Debt	5%	5%	5%	5%
Cost of Capital	5%	12%	12%	12%
Equity share	20%	20%	20%	20%
WACC	5%	6,4%	6,4%	6,4%
System economic lifetime	25%			
O&M	1,5%	1,5%	1,5%	1,5%
Performance ratio	80%	80%	85%	85%

Assuming a learning ratio at system level of 0,80 and a worldwide cumulative in-stalled capacity of about 270 GWp by 2020 we estimated future PV system prices that we used as input for the CAPEX values (Table 4).

Table 4. PV system prices estimation (complete system: modules, structures, BOS and installation). Source: author.

Year	Residential 3kWp (€/kWp)	Commercial - 50kWp (€/kWp)	Industrial 200kWp (€/kWp)	Field (€/kWp)
2011	3425	3231	3134	3048
2012	2450	2665	2585	2515
2013	2156	2345	2275	2213
2014	1897	2064	2002	1947
2015	1670	1816	1762	1714
2016	1469	1598	1550	1508
2017	1293	1407	1364	1327
2018	1138	1238	1201	1168
2019	1001	1089	1057	1028
2020	881	959	930	904

In Fig. 4 is presented the trend of the assumed LCOE and tariff levels for the entire period of the analysis.

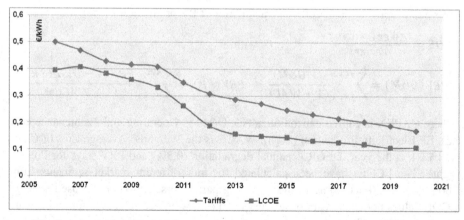

Fig. 4. Assumed LCOE and "Conto Energia" tariff levels. Source: author elaboration.

Electricity Consumers CBA. According to the methodology used by the German Federal Ministry for the Environment, Nature Conservation and the Nuclear Safety we used the 'differential costs' notion in order to properly estimate the net cost of the "Conto Energia" Programme [18], which constitutes a negative flow in the electricity consumers CBA.

The rationale is the following: the incentive is paid by electricity consumers who, however, would have paid and would pay the electricity in any case. Therefore, to obtain the net cost, the price of the electricity (the Prezzo Unico Nazionale – PUN) registered on the IPEX during the years 2006-2020 (estimating a 2% annual increase from 2010 onwards) has been subtracted to the costs bored by the electricity consumers.

Utilities CBA. The TSO (Transmission System Operator, meaning Terna Spa) and local grid operators are indirectly involved in the "Conto Energia" Programme. As a matter of fact, the additional back-up (or balancing) costs of uncertain generating power have been included as a negative flow, which causes fluctuation by producing energy from an intermittent energy source (i.e. solar). As in other electricity markets electricity production is planned one day ahead in the spot market (for Italy the "Mercato del Giorno Prima" - MGP) and eventual deviations from the planned operation may generate additional balancing costs of balancing. These additional costs are bored by the TSO in the so called "Mercato del Bilanciamento" (MB).

An estimation of this value is provided by the CASES Project which indentifies a value of 0,009 €/kWh when considering minimum back-up costs (a hard coal condensing power plant) and 0,015 €/kWh for maximum back-up costs (a gas-fired CCGT plant) [2]. We chose an average value in between equal to 0,012 €/kWh for the Italian case. On the other hand, we included as a positive flow the benefits due the distributed nature of PV electricity that contributes to reduce grid losses. To give a monetary value to this benefit we referred to the CTR component, a fee that is normally paid by the local grid system operators to power plant owners connected to medium and low voltages (i.e. the case of photovoltaics). The component takes into account the reduced costs linked to the fed of the electricity into the grid at lower voltages. In other words this value represents a proxy for the avoided transmission costs. The AEEG fixed such value at 0,0038 €/kWh for 2011 [17].

Government CBA. The Government indirectly benefits from the support mechanism as thanks to the fiscal regime applied to the PV business. Moreover, as PV does not emit CO_2 while producing electricity, it contributes to reduce compliance costs with the European Emission Trading Scheme (ETS) target. These two elements have been considered as a positive flow in the government CBA.

A value of the 'fiscal benefits' has been calculated by the Energy Strategy Group [19] for 2009 equal to 300 Million euro according to the following assumptions:

- Direct taxes paid by Italian (and foreign Italian-based) firms operating in the different steps of the PV value chain;

- The same direct taxes paid by the owners of PV systems whose operation generate revenues (i.e. the fraction of the electricity which is sold and the tax applied to the "Conto Energia" revenues);
- The tax paid by ground-mounted system owners, subjected to the municipal cadastral tax (ICI);
- The VAT tax at 10% applied to the electricity sold produced by plants above 20kWp;
- The VAT (always at 10%) of Italian and foreign Italian-based firms operating in the different stages of the PV value chain.

According to the information regarding the Italian PV industry and the hypothesis made about the "Conto Energia" future trends, the value for past and future years has been made for making use of a proportion.

On the other hand, for the estimation of the avoided purchase costs of CO_2 permits on the ETS, an average value of 15€/ton for the entire period of the analysis has been assumed.

Environmental Externalities CBA. Externalities arise, when the social or economic activities of a participant in the economy have negative or positive impacts on another participant and these impacts are not fully accounted for or compensated by the first participant. When producing electricity (energy), system operators incur not only in private costs but also generate external costs. The sum of the two constitutes the so called "social cost" or also full cost of generating energy even if the external costs are not bore by system operators.

In the environmental externalities CBA the environmental externalities associated to the PV electricity generation have been included as a negative flow and the environmental externalities associated to the CCGT electricity generation as a positive flow. The latter have been considered as a benefit considering the positive effect of avoided emissions obtained thanks to the production of the same amount of electricity with PV instead of natural gas. Due to the high reliance on gas of the Italian power mix (more than 50%) the comparison is a good conservative proxy values.

The values refer to the estimations provided by the CASES project and we specifically refer to monetary values of release of substances or energy (noise, radiation) into environmental media (air, soil, water) causing - after transport and transformation - considerable (not negligible) harm to ecosystems, humans, crops or materials. Externalities are calculated for all stages of the production process according to the LCA approach and on the basis of the ExternE methodology [2].

3.5   Results

According to our calculations the "Conto Energia" Programme offers a net present benefit of 6,95 € per each kWh of PV electricity produced (Fig. 5).

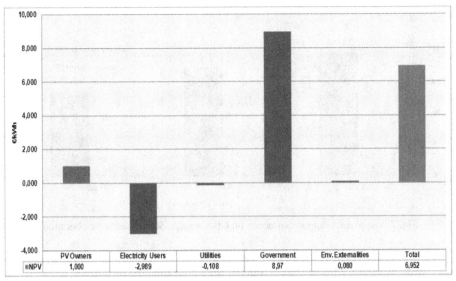

Fig. 5. CBA results in unitary terms (€/kWh). Source: author elaboration.

The cumulative net present benefits amount to more than 43 billion euro (Fig. 6).

The calculations show that the introduction of the policy has to be considered positive, as the benefits considerably outweigh the costs when considering the perspective of the entire society.

However we registered an important problem of distributional effects, regarding the costs of policy promotion. The electricity consumers indeed bear a noteworthy part of the overall costs without any income weighting.

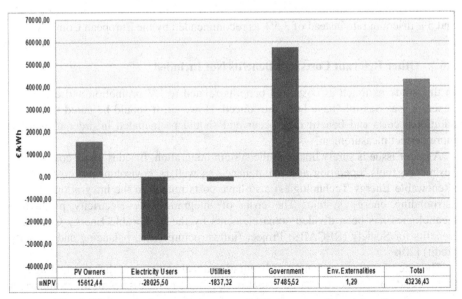

Fig. 6. CBA results in cumulative terms (M€). Source: author elaboration.

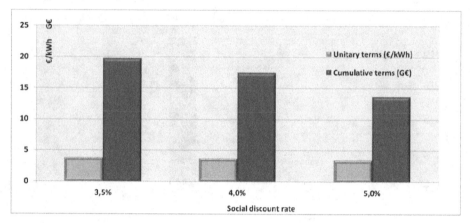

Fig. 7. Social discount rate influence on CBA results. Source: author elaboration.

This problem is indeed behind the decision of the Government to repeatedly act on the current regulatory frameworks reducing time after time the incentives.

As concerns PV system operators, they receive a considerable support in order to be incentivized to install their PV system. The support level is sensibly high and even if in the private costs taxes are not included, the calculations indicate that there may be margin to reducing the incentive levels and to preserving the boost effect (allowing a reduction of the burden on electricity consumers). The government clearly benefits from the introduction of the policy, receiving a net benefit of about 8,97 € per each kWh produced. The environmental externalities avoided almost double the externalities generated during the PV plant lifecycle.

In order to provide with more robust evaluation, we present in Fig. 7 the influence of the discount rate, showing the observed differences in the results when using a 4% and 5% discount rate instead of 3,5% as recommended by the European Commission.

3.6   Other Relevant Costs and Benefits Not Included

In this study as much as costs and benefits related to PV promotion and electricity generation as possible have been introduced. However, it should be noted that some additional costs and benefit may occur and should be included in order to reach a more perfect measurement.

A major issue is surely linked to the system integration. Besides direct costs for investments, fuels, operation and maintenance as well as environmental costs, RETs (Renewable Energy Technologies) also have costs related to the integration into the surrounding energy system. The costs of integrating the electricity production technologies has been divided soma elements by the Renewable Energy Costs and Benefits for Society (RECABS) Project (infrastructure costs, balancing and capacity credit) [20].

The size of the cost or benefit of integrating a technology into the surrounding system depends of course to a large extent on the design of the system. Most important factors are:

- The share of intermittent power in the electricity system
- The characteristics of the electricity system.
- The operational procedures and market design.

Furthermore, the costs will depend on the nature of the technology to be integrated. For ex-ample, solar PV typically has a production profile that has a better fit with the consumption profile than wind power. In the present study only the increasing balancing costs and the benefits from reduced grid losses due to a higher PV penetration have been included. However, it's important to underline that PV can potentially influence the grid infrastructure in two other ways:

- PV can contribute to defer networks upgrade needed to anticipate load growth. The benefit are very sit-depending but an estimation is provided by the RECAB Project equal to 14 €/MWh [20];
- In a scenario where the impact of higher RES penetration is considered, PV definitely contributes to higher infrastructure costs as it requires reinforcement at TSO and DSO level, improvements in communication infrastructure (smarter meters) plus distributed storage.

Moreover, it should be important to take into consideration the time dependency of the value of PV power.

Retail electricity prices vary considerable during the day and year due to variations, primarily in power demand. Peak load power is more costly than base load power, because the peak load power plants only run for a limited number of hours each year, thereby raising their capital costs.

For more sunny regions, such as Italy, where peak demand to a great extent is driven by electricity load from air-conditioning, the peak load value from solar power can be expected to be higher. The RECAB Project estimate of 10 €/MWh assuming that PV power is on average 20% more worth than average power prices of 40 €/MWh.

Another important added value of PV electricity not included in the CBA is the benefits occurred by the increase security of supply derived from a reduced fuel dependency.

Furthermore, there may be room for the verification of the so called merit-order effect which consist of a reduction of the wholesale electricity prices due to a higher penetration of RES.

Lastly, there are some local benefits and costs associated to the PV electricity production that is worth to mention and that we didn't include in the CBA. The benefit are related to the employment effect (meaning job creation) linked to the development of a new business and industries. Of course the number should be considered as a net one, without displace jobs. Similarly, the increase of PV electricity production displace profits of conventional energy producers.

4   Conclusions

This study analyses cost and benefits of the principal support scheme currently applied in Italy to promote photovoltaic electricity generation in Italy for the period

2006-2020. The analysis has been carried out making use of a Cost-Benefit Analysis methodology and taking into consideration positive and negative externalities originated from the policy support.

The calculations show that the "Conto Energia" Programme offers a net present benefit of 6,6 € per each kWh of electricity produced with photvoltaics and a cumulative net present benefit of 30,35 billion euro when considering the perspective of the entire society, assuming a 3.5% social discount rate as suggested by the European Commission.

Moreover, it has been registered an important problem of distributional effects regarding the costs of policy promotion, due to the lack of an income weighting of such cots; a relatively high benefit surplus for PV system operators, which may suggest a reduction of the support level in order to reduce the burden on electricity consumers while preserving the soundness of the incentive; a very positive result for the government fiscal budget which benefits of a considerable public revenues increase; a marginal negative result for the Utility sector; lastly a positive outcome concerning the reduction of environmental externalities.

In addition, the study revealed the importance of removing some key barriers that are hampering the development of the photovoltaic technology, namely the legal and especially the grid connection barriers.

A central assumption of the study is the reliability of the external costs estimations taken from the CASES Project and the optimality of the choices made on the different assumptions regarding the selection of cost and benefit flows for the different sectors of the society, the system prices parameters used to calculate private costs, the "Conto Energia" future trends and the monetary proxy values used to assess other externalities linked to the incentive programme [2].

The results do not pretend to be a perfect measurement and the limits of the analysis can be deduced by the assumptions made for each sector CBA and by the list of other relevant factors that could not be included. Besides that the study is supported by a comprehensive analysis on the photovoltaic investment framework conditions and by a sound investigation on the main existing barriers hosting the full deployment of the technology.

In conclusion the "Conto Energia" introduction has to be considered positive for the entire society but its revision, that is taking place during these days, should definitely take into consideration the above results in order to better shape the promotion policy.

References

1. Cost Assessment of Sustainable Energy Systems (CASES Project). Private costs of electricity and heat generation (deliverable no D.4.1) (2008a)
2. CASES. Full cost estimates of the use of different energy sources (deliverable D.6.1) (2008b)
3. European Commission (EC), Directive 2009/28/ec of the European Parliament and of the Council of 23 April 2009 on the Promotion of the use of energy from renewable sources and amending and subsequently repealing Directives 2001/77/EC and 2003/30/EC (2009)

4. EC, Directorate-General Regional Policy. Guide to cost-benefit analysis of investment projects Structural Funds, Cohesion Fund and Instrument for Pre-Accession (2008)
5. International Energy Agency (IEA), Energy Technology Perspective 2010. Scenarios and Strategies to 2050 (2010a)
6. Commissione Nazionale per l'Energia Solare (CNES). Rapporto preliminare sullo stato attuale del solare fotovoltaico nazionale (2008)
7. IEA Photovoltaic Power Systems Programme (PVPS), Annual Report (2009)
8. GSE, An interview to Italian PV market for detecting barriers and their overcoming to a full deployment (2009)
9. European Photovoltaic Industry Association (EPIA), Global Market Outlook for Photovoltaics until 2014. March 2010 update (2010)
10. IEA, National Survey Report of PV Power Applications in Italy (2008)
11. IEA, Projected Costs of Generating Electricity 2010 (2010c)
12. IEA, World Energy Outlook 2009 (2009)
13. Fraunhofer Institute Systems and Innovation Research (ISI), The Merit-order effect: A detailed analysis of the price effect of renewable electricity generation on spot market prices in Germany (2007)
14. Cocchi, S., Castellucci, S., Tucci, A.: Modeling of an Air Conditioning System with Geothermal Heat Pump for a Residential Building. Mathematical Problem in Engineering 2013, Article ID 781231 (2013), doi.org/10.1155/2013/781231, ISSN 1024-123X
15. Lucentini, M., Rottenberg, F., Di Palma, D.: A model for the evaluation of energy indicators in end users audit for hospital sector. In: ASME 2011 International Mechanical Engineering Congress and Exposition, IMECE 2011 4 (Parts A and B), pp. 305–311 (2011)
16. European Commission, Renewable Energy National Action Plan (RENAP) (2010)
17. IEA, Energy Policy of IEA Countries: Italy review 2009 (2010b)
18. German Federal Ministry for the Environment, Nature Conservation and the Nuclear Safety (BMU) (2007)
19. Energy Strategy Group (ESG), Solar Energy Report (2009)
20. Renewable Energy Deployment Implementing Agreement (IEA-REDT). Renewable Energy Costs and Benefits for Society (RECABS) (2008)

Application of Adaptive Models for the Determination of the Thermal Behaviour of a Photovoltaic Panel

Valerio Lo Brano, Giuseppina Ciulla, and Marco Beccali

DEIM, Dipartimento di Energia, Ingegneria dell'informazione e Modelli matematici,
Università di Palermo, viale delle scienze, Ed.9 90128, Palermo, Italy
{lobrano,ciullaina,marco.beccali}@dream.unipa.it

Abstract. The use of reliable forecasting models for the PV temperature is necessary for a more correct evaluation of energy and economic performances. Climatic conditions certainly have a remarkable influence on thermo-electric behaviour of the PV panel but the physical system is too complex for an analytical representation. A neural-network-based approach for solar panel temperature modelling is here presented. The models were trained using a set of data collected from a test facility. Simulation results of the trained neural networks are presented and compared with those obtained with an empirical correlation.

Keywords: Artificial Neural Network, photovoltaic, cell temperature.

1   Introduction

Renewable Energy Sources (RES) are important for promoting the competitiveness of the economy of countries, the security of energy supply systems and to improve the environmental protection [1, 2]. Generally, RES are easily accessible, inexhaustible and compatible with the environment. Among RES, solar energy has the greatest energy potential and photovoltaic (PV) arrays permit to produce electric power directly from sunlight with no fossil-fuel consumption, no noise, and posing no health and environmental hazards during the operational phase of life. This fact, together with the slow but ongoing decline of conventional energy sources, implies a promising role for PV power-generation systems in the near future. Despite the technological and environmental benefits granted by this technology, the development of PV panels is hindered by economic factors. The high cost of production and installation makes the PV technology feasible only when public funding is available [2, 3].

Furthermore, it is clear that the availability of reliable predictive tools is very important for the dissemination of all renewable energy technologies [4, 5]. In details, from the point of view of the designer and end-users of PV systems, the availability of reliable software tools is essential   to optimize the performance of PV systems in the planning phase and finally to correctly assess the economic gain [6]. In order to evaluate the real performance of PV panels the correct prediction of operating

B. Murgante et al. (Eds.): ICCSA 2013, Part II, LNCS 7972, pp. 344–358, 2013.

temperature is very important [7]; an increase of few degrees can considerably reduce the conversion efficiency of the system thus reducing the power output.

A reliable tool for predicting the temperature of PV systems is also particularly important in those hybrid systems called PV-Thermal (PVT), which allow the recovery of thermal energy that otherwise, would be wasted into the surrounding environment [8-10].

The aim of this work is to explore the possibility to offer an alternative method, respect to empirical correlations, which allows modelling the operating temperature of PV devices by using techniques based upon adaptive systems. Adaptive systems, such as Artificial Neural Networks (ANN) should allow to predict, with a fast and reliable procedure, the temperature of the PV module as weather conditions change.

2   Introduction to Artificial Neural Networks and Classification Problems

ANNs emulate some of the functions and capabilities of the human brain and their use is now widespread in the scientific literature, in particular in those physical models where, although the interconnections among some variables are widely demonstrated, the corresponding mathematical functions are not known or they are extremely complex [11-13]. One of the most important area of ANNs is the pattern recognition and the main task of this approach is the classification [14]. What is a classifier? A classifier is an expert system that builds a relation between a variable space A and a vector of labels B. A classifier is able to assign a label to a sample extracted from the space A. Nowadays, many expert systems are used as classifiers, e.g. in order to assign or not the label of "SPAM" to an email sent to our mailbox [15]. The classifier examines some features of the email such as the sender, the object, the text and other variables and then it assigns or not the label of SPAM. Another example of a classifier is an expert system able to recognize people from face images [16] or from the sound of voices [17].

Classification is different from clustering: in a cluster analysis, data are automatically separated in groups characterized by some similarities, called clusters. So, clustering is an unsupervised process that groups the data autonomously. Classification is a supervised process where a user decides the set of labels.

A Neural Network Classifier (NNC) simply judges the distance between a pattern of input variables and some given labels. There are two ways to measure this distance: by numerical methods and non-numerical ones. The numerical techniques measure the above-mentioned distances in deterministic or statistical way. The non-numerical techniques are linked to symbolic processes like fuzzy sets. Making these activities automatic permits to reach the target in a faster and more reliable way. Furthermore, the use of NNCs often makes it possible to identify correlations between data which are so complex that they would be hardly recognized even by an expert human operator.

In time series forecasting problems one of the approaches often used by researchers is represented by ANN based techniques that can be used as an alternative method in the analysis of complex and/or ill-defined engineering problems. ANNs do not require the formulation of a mathematical relation describing a complex natural and/or physical system and have the capability of detecting its hidden structure. Accurate forecasting of time series can be useful in many practical situations and the knowledge of variation in the operating temperature of PV can play a remarkable role in the assessment of power output.

3   The Operative Temperature of a PV Panel

To design and assess the performances of a PV system, an accurate PV model should predict reliable Current-Voltage (*I-V*) and Power-Voltage (*P-V*) curves under real operating conditions [18].

The "five-parameters model" represents the most common equivalent circuit that better describes the electrical behavior of a PV system. The equivalent circuit is composed of a photocurrent source I_L, a diode in parallel with a shunt resistance R_{sh}, and a series resistance R_s as shown in Figure 1.

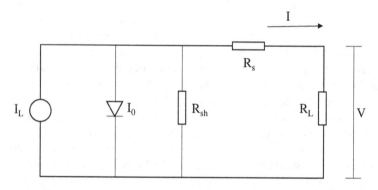

Fig. 1. Schema of one diode equivalent simplified circuit closed on a resistive load R_L

Based on this simplified circuit, the mathematical model of a photovoltaic cell can be defined in accordance with the following expression that permits to retrieve the *I-V* curve:

$$I = I_L - I_0 \left(e^{\frac{V + I \cdot R_s}{n T_c}} - 1 \right) - \frac{V + I \cdot R_s}{R_{sh}} \tag{1}$$

in which I_L depends on the solar irradiance, I_0 is the diode reverse saturation current and is affected by the silicon temperature, n is the ideality factor and T_c is the cell temperature [K].

As it is well known, the performance of a photovoltaic panel is defined according to the "peak power", which identifies the maximum electric power supplied by the panel when it receives an insolation of 1 kW/m^2 at a cell temperature of 25°C.

In actual conditions, it is essential to evaluate the operating conditions under all possible circumstances of irradiance G, cell temperature T_c, wind speed W, air temperature T_{air} and electric load R_L.

The T_c temperature thus is a key parameter that affects the energy conversion efficiency of a PV panel: increasing the temperature, the delivered power decreases. In literature, there are several available empirical correlations to obtain the PV panel operating temperature and these correlations have been developed for common geometries and weather conditions. From a mathematical point of view, the correlations for the PV operating temperature are either in explicit or implicit form; in the latter case, an iteration procedure is necessary for the calculation. Most of the correlations typically include the reference conditions and the corresponding values of the pertinent variables [19].

3.1   Impact of Solar Irradiance and Temperature on the *I-V* and *P-V* Curves

For given values of G, T_c and R_L, the operating point can be identified by drawing lines of the different loads R_L on the *I-V* characteristics.

In Figure 2 and Figure 3, it is possible to observe how the intersection between the load line and PV characteristics corresponds to the working point; with the same graphical method, it is possible to identify the working point in terms of electric power. The red circles indicate the locus of maximum power output points.

The solar energy conversion into electrical energy is obviously influenced by the operation point of the panel [20].

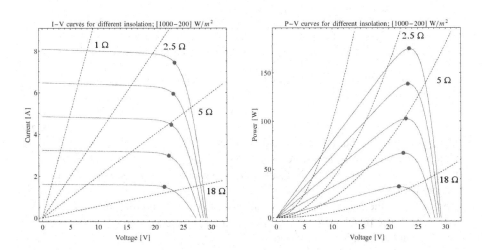

Fig. 2. Working point of a generic PV panel at constant temperature (25 °C) varying insolation and electric load

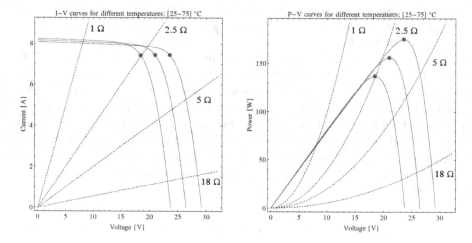

Fig. 3. Working point of a generic PV panel at constant irradiance (1000 W/m²) varying temperature and electric load

4 Artificial Neural Network Application

As shown by the previous considerations, the thermo-electrical behavior of a PV panel is a complex function of the actual climatic conditions and of the cell characteristics.

The complexity of this physical phenomenon does not allow a complete and accurate analytical representation of the thermo-electric balance of the PV panel and the performance assessment generally follows two paths:

1. a simplification of the thermo-electric balance by using empirical correlations;
2. an application of adaptive systems that learn from a large amount of monitored data.

ANNs are distributed, adaptive, generally nonlinear learning machines built from many different processing elements similar to biological neurons. Each artificial neuron (AN) receives connections from other ANs and/or itself. The structure of these connections defines the ANN topology. The signals flowing on the connections are scaled by adjustable parameters called weights and the values of these weights are updated during the training phase. The ANs sum all contributions and produce an output. The outputs of each AN can be either system outputs or inputs sent to the same or other ANs.

In this work, the authors have tested the use of ANNs to predict the operating temperature of a PV panel using the data monitored in a test facility. Different ANNs topologies and typologies were tested.

4.1   Data for Training and Testing

A large database of specific data that represent the analyzed physical system is required to construct an adaptive system. In order to forecast the temperature of a PV panel, an experimental system (Figure 4) was installed on the roof of the Energy Department of the University of Palermo. The test facility and the monitoring system consist of the following equipment:

- a photovoltaic panel (Kyocera KC175-GH-2),
- a precision resistance set,
- a multimeter Fluke189/FVF2,
- a Delta Ohm pyranometer mod. LP PYRA 02 AV linked to
- an Advantech ADAM 6024 module,
- a Davis Vantage PRO2 Plus Weather station; more details concerning the test facility are explained in [21].

The PV panels and the pyranometer were tilted at an angle that is equal to the latitude of the location (38° South). The electrical load R_L was obtained by precision resistances, and the current was calculated on the basis of the measured voltage, accepting the error due to the resistances value. The silicon temperature was measured using thermocouples (type T, copper-constantan) [22] installed at the rear film of the panel [23].

Fig. 4. Experimental system

All data were collected every 30 minutes and stored for further calculations and comparisons. The physical data used for the training of the ANN were:

- Air temperature [°C];
- Solar irradiance [W/m^2];
- Wind speed [m/s];
- Voltage [V];
- Power output [W];
- Electrical Load [Ω].

The process by which the knowledge of the physical system is transferred to the adaptive system is carried out in the training phase. During the training phase of an ANN, a set of known input–output vectors are presented to the network updating some mathematical entities. The subsequent testing phase will assess the quality of the neural network model comparing the output with the real data belonging to a dataset not used in the training phase.

4.2   Preliminary Analysis of the Collected Data

It is possible to find a wide range of ANNs characterized by different topologies and typologies. Before choosing the neural topology, all data are subject to a pre-processing step that consists in a preliminary analysis that permits to identify possible outliers, to remove unreliable values, to carry out a statistical analysis, and to perform a correlation analysis. After the pre-processing step, the database is validated and the correlation analysis permits a first evaluation of the mutual relationships among the considered variables.

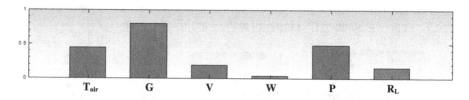

Fig. 5. Correlation analysis between operating temperature and all input data

Figure 5 shows the linear correlation between T_c and all the other features. The higher the bar goes, the more the features are correlated. Blue bars indicate a positive correlation while red bars indicate a negative correlation. The preliminary correlation analysis identified a strong positive correlation between T_c and the solar irradiance; on the contrary, a weak negative correlation with the wind speed was detected. Furthermore, a moderate positive correlation with electrical power, air temperature and voltage was found.

A statistical analysis permitted to assess the maximum (Max), mean (Mean) and minimum (Min) values and the standard deviation (StDev) of all considered features (Table 1).

In the following Training Step, the authors decided not to consider the Voltage and Electric load input because their values are already computed in the Electrical power value; the tested ANNs will consider as input only a vector with four components (T_{air}, G, P and W) and as output the T_c. The training dataset is composed by 2827 vectors and the testing dataset consists in 605 vectors to be used in the validation phase. After the pre-processing phase, it was possible to choose the topology of neural network; different simulations relating to several topologies of ANNs have been tested but in this work only the best ANN solutions will be described: the Gamma two Layer, the Recurrent one Layer and the MLP two Layer. For each topology the design and the algorithm are analyzed, each neural networks was trained and was validated with a post processing phase.

Table 1. Statistical Evaluation

	T_{air} [°C]	T_{cell} [°C]	G [W/m²]	V [V]	W [m/s]	P [W]	R_L [Ω]
Max	32.60	64.97	1221.00	28.01	8.00	167.09	18.00
Min	9.80	14.43	44.93	3.90	0.00	15.20	1.00
Mean	19.98	36.52	675.01	19.31	2.30	70.75	7.18
StDev	3.49	7.643	262.86	6.72	1.41	40.54	5.59
sample	3432	3432	3432	3432	3432	3432	3432

4.3   Gamma Two Layer

A Gamma ANN is characterized by special memory ANs. The memory AN receives several inputs and produces multiple outputs which are delayed versions of the combined input. This feature has a biological interpretation because, when the biological neuron receives multiple connections, the signal propagation is delayed [24].

As it is possible to see in Figure 6, the proposed Gamma topology is composed of two data sources block (input and output), three gamma memory blocks, three function layer blocks, three weight layer blocks and one delta terminator block. The function layer can be seen as non-linear thresholds for the propagation of the signals. They give the adaptive system its non-linear computing capabilities. The weight layer represents the long-term memory of the system and is adjusted during the learning phase. Finally, the delta terminator is an error criterion block that takes two signals and compares them according to a specific criterion. The appellation "Terminator" means that the signals terminate to flow across the system.

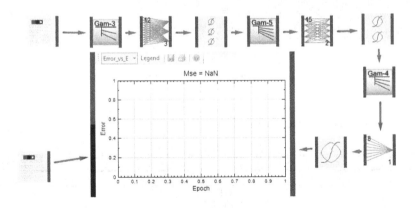

Fig. 6. Gamma two Layer layout

After the training, the post-processing phase evaluates the error (Figure 7) and the absolute error (Figure 8) between the measured operating temperature data and the calculated output employing the remaining 605 vectors.

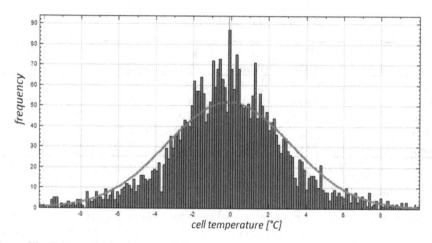

Fig. 7. Error distribution over 605 vectors of T_c with Gamma two Layer topology

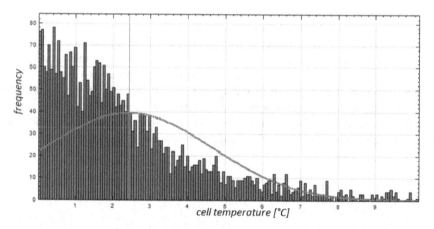

Fig. 8. Absolute Error distribution over 605 vectors of T_c with Gamma two Layer topology

The values of the Mean Error (ME) and Mean Absolute Error (MAE) of the Gamma two Layer topology are reported in Table 2.

Table 2. Mean Error and Mean Absolute Error of the Gamma two Layer topology

	Gamma two Layer	
	ME	**MAE**
[°C]	-0.119	2.428
StDv	3.286	2.217

As shown in Figure 9 a Confidence plot of ± 4.408 °C contains 95% of the outputs.

The black line, which represents the calculated output, well follows the trend of the purple line that represents the experimental output. The 95% confidence is delimited by the red (high) and blue (low) lines.

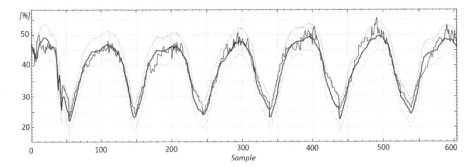

Fig. 9. Confidence Plot of calculated output *versus* measured data of Gamma two Layer topology

4.4   Multi-Layer Perceptron (MLP) Two Layer

A (MLP) is a kind of ANN consisting of multiple layers of ANs in a directed graph, with each layer fully connected to the next one. Except for the input ANs, each node is a neuron with a non-linear activation function. A MLP utilizes a common supervised learning technique called back-propagation for training the network. This topology is one of the simplest available for ANNs. In our work, the MPL two Layer is composed by: two data sources blocks (input and output), two function layer blocks, two weight layer blocks and one delta terminator block as shown in Figure 10.

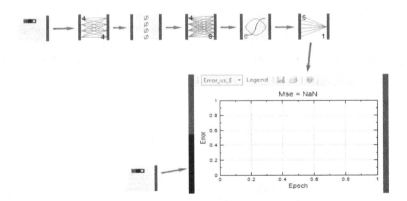

Fig. 10. MLP two Layer topology

After the training, the post-processing phase evaluates the error and the absolute error between the measured and the calculated operating temperature data. The cumulative results are reported in Table 3.

Table 3. Mean Error and Mean Absolute Error of the MLP two Layer

	MLP two Layer	
	ME	MAE
[°C]	0.207	2.476
StDv	3.407	2.349

For the MLP two Layer topology the confidence plot that contains the 95% of the outputs is of ± 3.936°C (Figure 11).

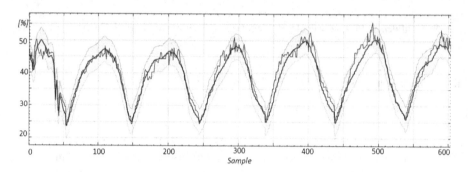

Fig. 11. Confidence Plot of calculated output *versus* measured data of MLP two Layer topology

4.5   Recurrent MLP One Layer

The Recurrent MLP one Layer is a simple ANN topology that employs a recursive flow of the signal to preserve and to use the temporal sequence of events as useful information. This topology (Figure 12) is composed of two data sources blocks (input and output), two weight layer blocks, two function layer blocks and one delta terminator block.

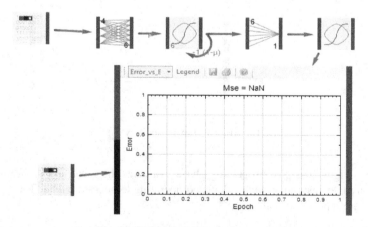

Fig. 12. Recurrent MLP one Layer topology

Figure 12 iconizes a feedback connection where μ is the weight of the feedback used to scale the input. Of course, there are different values of μ for each signal flowing into the first function block layer. After the opportune training phase, the following results are observed:

Table 4. Error Distribution of the Recurrent MLP one Layer topology

	Recurrent MLP one Layer	
	ME	**MAE**
[°C]	0.229	2.489
StDv	3.436	2.386

For the Recurrent MLP one Layer topology the confidence plot that contains the 95% of the outputs is ± 4.517° C, as shown in Figure 13.

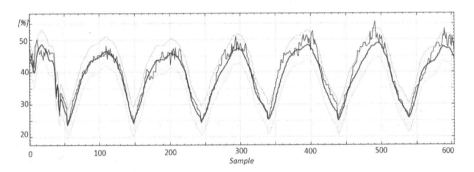

Fig. 13. Confidence Plot of Recurrent one Layer topology

5   Evaluation of the Results

Each neural network was characterized by the same Input Data (Air Temperature, Wind Speed, Solar Irradiance and Electric Power) and was trained with a dataset of 2827 vectors. The Gamma two Layer and the Recurrent MLP two Layer are two ANN typologies that, with different approaches, have the capability to preserve the temporal sequence of data (memory), while the MLP two Layer is a static ANN. The results show that all the considered ANNs provide a reliable model that is able to fit well with the experimental trends. Generally, the ME is about ± 0.2 °C and the MAE is close to 2.4°C. The training phase of Gamma two Layer ANN requested 1 minute, while the other two typologies requested a shorter time.

In literature [19, 25-29], there are different models that allow calculating the operating temperature. In order to validate the neural network approach, the authors made a comparison between the values of T_c obtained by the three previously illustrated networks and the operating temperature calculated by one of the most cited empirical correlation:

$$T_c = T_a + \alpha G \cdot (1 + \beta T_a) \cdot (1 - \gamma W) \cdot (1 - 1.053\eta_c) \tag{2}$$

where η_c is the efficiency of a PV panel and α, β and γ are three constants (α=38.0385, β=3.15126, γ=2.64173) [24].

The comparison, as represented in Table 5, shown that the empirical correlation achieves a MAE that is about twice compared to the ANN results.

Table 5. Comparison between the ANN and empirical correlation results

ANN	Training time	MAE
Gamma two Layer	≈ 1 min	2.428
Recurrent MLP one Layer	≈ 10 s	2.489
MLP two Layer	≈ 20 s	2.476
Empirical	-	**4.719**

6   Conclusions

In this paper, an artificial neural network approach has been proposed to determine the operative temperature of a PV panel.

The application of the artificial neural network model represents a simple and fast solution to correctly evaluate the operative regimen of a PV system. To this purpose, different network architectures have been tested and trained with experimental data consisting in: air temperature, wind speed, solar irradiance, power output and cell temperature. The three best solutions of ANNs are reported: Gamma two Layer, MLP two Layer and Recurrent MPL one Layer.

The results obtained by the ANNs demonstrates that this approach can be considered a reliable tool to forecast the cell temperature of the PV panel. Comparing the performances of these networks with a very often cited empirical model, used for determining the operating temperature of the panel, the ANN approach presents a significant lower MAE. Furthermore, the very short time required by MLP two Layer and Recurrent MPL one Layer for the training phase, suggests that ANNs could be integrated in a software for run-time evaluation.

References

1. IEA: International Energy Agency; World Energy Outlook (2012) ISBN: 978-92-64-18084-0
2. Pernick, R., Wilder, C.: Clean Tech Nation: How the US Can Lead in the New Global Economy. Harper Business (2012)
3. Cellura, M., Campanella, L., Ciulla, G., Guarino, F., Lo Brano, V., Cesarini, D.N., Orioli, A.: A net zero energy building in Italy: design studies to reach the net zero energy target. In: Building Simulation, an IBPSA–AIRAH Conference, Sydney, Australia (2011)

4. Beccali, G., Cellura, M., Lo Brano, V., Orioli, A.: Single thermal zone balance solved by Transfer Function Method. Energy and Buildings 37(12), 1268–1277 (2005)
5. Ciulla, G., Franzitta, V., Lo Brano, V., Viola, A., Trapanese, M.: Mini Wind Plant to Power Telecommunication Systems: A Case Study in Sicily. Advanced Materials Research 622, 1078–1083 (2013)
6. Di Dio, V., Favuzza, S., La Cascia, D., Miceli, R.: Economical Incentives and Systems of Certification for the Production of Electrical Energy from Renewable Energy Resources. In: Clean Electrical Power, ICCEP 2007, pp. 277–282. IEEE (2007)
7. Cellura, M., Ciulla, G., Lo Brano, V., Marvuglia, A., Orioli, A.: A Photovoltaic panel coupled with a phase changing material heat storage system in hot climates. In: 25th Conference on Passive and Low Energy Architecture, PLEA 2008, Dublin (2008)
8. Tiwari, A., Sodha, M.S.: Performance evaluation of hybrid PV/thermal water/air heating system: A parametric study. Renewable Energy 31, 2460–2474 (2006)
9. Zakharchenko, R., Licea-Jimenez, L., Perez-Garc, S.A., Vorobiev, P., Dehesa-Carrasco, U., Perez-Robles, J.F., Gonzalez-Hern, J., Vorobiev, Y.: Photovoltaic solar panel for a hybrid PV/thermal system. Solar Energy Materials & Solar Cells 82, 253–261 (2004)
10. Eicker, U.: Solar Energy for Buildings,1st edn. Wiley, ISBN:0-471-48637-X
11. Cellura, M., Lo Brano, V., Mistretta, M., Orioli, A.: To Assess the Validity of the Transfer Function Method: A Neural Model for the Optimal Choice of Conduction Transfer Functions. ASHRAE Transactions 116(2), 585 (2010)
12. Ciulla, G., Lo Brano, V., Orioli, A.: A criterion for the assessment of the reliability of ASHRAE conduction transfer function coefficients. Energy and Buildings 42(9), 1426–1436 (2010)
13. Haykin, S.: Neural Networks: A Comprehensive Foundation, 2nd edn. Prentice Hall (1999)
14. Bishop, C.: Pattern Recognition and Machine Learning. Springer (2006)
15. Yang, Y., Elfayoumy, S.: Anti-Spam Filtering Using Neural Networks and Baysian Classifiers. In: Proceedings of the 2007 IEEE International Symposium on Computational Intelligence in Robotics and Automation, Jacksonville, FL, USA (2007)
16. Rowley, H.A., Baluja, S., Kanade, T.: Neural Network-Based Face Detection. IEEE Transactions on Pattern Analysis and Machine Intelligence 20, 23–38 (1998)
17. Venayagamoorthy, G.K., Moonasar, V., Sandrasegaran, K.: Voice recognition using neural networks. In: Proceedings of the 1998 South African Symposium on Communications and Signal Processing. Cape Town (1998)
18. Tsai, H.-F., Tsai, H.-L.: Implementation and verification of integrated thermal and electrical models for commercial PV modules. Solar Energy 86, 654–665 (2012)
19. Skoplaki, E., Palyvos, J.A.: On the temperature dependence of photovoltaic module electrical performance: A review of efficiency/power correlations. Solar Energy 83, 614–624 (2009)
20. Lo Brano, V., Ciulla, G., Franzitta, V., Viola, A.: A novel implicit correlation for the operative temperature of a PV panel. In: 2012 AASRI Conference on Power and Energy Systems (2012)
21. Lo Brano, V., Orioli, A., Ciulla, G., Culotta, S.: Quality of wind speed fitting distributions for the urban area of Palermo, Italy. Renewable Energy 36, 1026–1039 (2011)
22. Cardona, E., Piacentino, A.: A measurement methodology for monitoring a CHCP pilot plant for an office building. Energy and Buildings 35(9), 919–925 (2003)
23. Lo Brano, V., Orioli, A., Ciulla, G.: On the experimental validation of an improved five-parameter model for silicon photovoltaic modules. Solar Energy Materials and Solar Cells 105, 27–39 (2012)

24. Principe, J.C., Euliano, N.R., Lefebvre, W.C.: Neural and Adaptive Systems, Fundamentals though Simulations. John Wiley & Sons, Inc. ISBN 0-471-35167-9
25. Skoplaki, E., Palyvos, J.A.: Operating temperature of photovoltaic modules: A survey of pertinent correlations. Renewable Energy 34(1), 23–29 (2009)
26. Servant, J.M.: Calculation of the cell temperature for photovoltaic modules from climatic data. In: 9th Biennal Congress of ISES-Intersol 1985, Montreal, Canada, p. 370 (1985)
27. Duffie, J.A., Beckman, E.A.: Solar energy thermal processes, 3rd edn. Wiley, Hoboken (1991)
28. Hove, T.: A method for predicting long-term average performance of photovoltaic systems. Renewable Energy 21, 207–229 (2000)
29. Tiwai, G.N.: Solar Energy-Fundamentals, design, modeling and applications, p. 450. Alpha Science, Pang Bourne (2002)

The Economic Evaluation
of Investments in the Energy Sector: A Model
for the Optimization of the Scenario Analyses

Gianluigi De Mare, Benedetto Manganelli, and Antonio Nesticò*

Faculty of Engineering, University of Salerno, Italy
{gdemare,anestico}@unisa.it,
benedetto.manganelli@unibas.it

Abstract. The well-known evaluation indices do not allow to take into account
the interaction between current investment alternatives and future decisions.
The real options theory provides answers to the limits that the evaluators discovered in the traditional techniques of *capital budgeting* and allows to give a
value to managerial flexibility, i.e. the ability of management to review its
decisions on the basis of changes in the economic context. Implementing the
traditional cash-flow analysis with the tool of real options, the study defines a
logical-operational model capable of verifying the financial viability of investments in the energy sector. The model is applied to the economic study of a
project to produce energy from renewable sources, specifically the construction
of a new wind farm. The different operational phases of the model used for the
optimization of the scenario analyses, return the value of the positive potential
that can result from management flexibility and innovation.

Keywords: Wind Farm Valuation, Real Options, Monte Carlo Simulation, Renewable Energy.

1   Introduction and Aim of Work

The economic evaluation of investment projects is crucial in the allocation of resources process, both public and private [1-2-3-4]. In order to express an opinion on
the economic implementation of the project initiative, *discounted cash flow* (DCF)
analysis can be used, with it comparing the monetary income and expenditure of an
investment for a predetermined period of time [5-6]. The result of the analysis is expressed through the well-known evaluation indices, the *Net Present Value* (NPV),
Internal Rate of Return (IRR) and *payback period*. These indices, however, do not
allow to take into account the interaction between current investment alternatives and
future decisions [7]. They are criteria that ignore the added value that could result
from the flexibility and innovation management, capable of changing the course
of the investments. According to Dixit A.K. and Pindyck R.S. [8]: «As a practical

---

* This paper is to be attributed in equal parts to the three authors.

B. Murgante et al. (Eds.): ICCSA 2013, Part II, LNCS 7972, pp. 359–374, 2013.
© Springer-Verlag Berlin Heidelberg 2013

matter, many managers seem to understand already that there is something wrong with the simple NPV rule as it is taught – that there is a value to waiting for more information and that this value is not reflected in the standard calculation. In fact, managers often require that an NPV be more than merely positive ... It may be that managers understand a company's options are valuable and that it is often desirable to keep those options open».

The real options theory provides answers to the limits that the evaluators discovered in the traditional techniques of *capital budgeting*[1] and is applied in uncertain management environments characterized by high dynamism, due to it allowing to give a value to managerial flexibility, i.e. the ability of management to defer a project, extend it, or leave it, as well as to review its decisions on the basis of changes in the economic context [11-12-13-14-15-18-52-53].

This paper first examines the benefits resulting from the application of real options to the economic evaluation of projects. Implementing the traditional cash-flow analysis with the tool of real options, the study will then define a logical-operational model capable of verifying the financial viability of investments in the energy sector.

The model is applied to the economic study of a project to produce energy from renewable sources, specifically the construction of a new wind farm. The description of the planned works is followed by the macroeconomic analysis of the area of interest, which shows that, in a difficult economic context such as the one currently being experienced by the world economy, the expansion of renewable energy sources is an opportunity not only for energy diversification and environmental protection, but also for research and employment. The DCF analysis shows the cost-effectiveness of the initiative. The evaluation is then implemented with the theory of real options. The real options of abandonment, postponement and expansion investment are considered. The Monte Carlo simulation is adopted for the calculations, using the Fairmat Academic software in two different scenarios: one in which the real options are considered and another in which all the decisions are made at the beginning. Comparing the results, it is clear that the average value of the extended NPV is greater than what is obtained in the base case. This means that the real options valorise the project. The highest value obtained is the monetary representation of the options. Fundamentally, the managerial flexibility that is taken into account with real options lengthens the probability distribution of the NPV toward positive values, since it offers the

---

[1] The limitations of the traditional techniques of *capital budgeting* in defining and evaluating some characteristic aspects of investments in real assets, can be described by three characteristics: uncertainty (technical and market), irreversibility of the investment, flexibility to dynamically adapt to changing choices environmental conditions [9]. In particular, the standard techniques are effective in short-term evaluations, but are not able to model the uncertainty of wider horizons. Barnett M.L. [10] emphasizes that these techniques do not consider the ability of management to control the risk and deal with situations of loss, so as to lead to discard those investments that are riskier but bearers of great strategic potential. See also Kogut B. and Kulatilaka N. [18].

For more details on the limits of the Net Present Value and the differences between the assumptions of the NPV and economic reality, see Mun J. [19] as well as Luehrman TA [20].

opportunity to reap the benefits of positive scenario developments and, simultaneously, allows to intervene in order to contain the negative impact of unfavorable changes in the variables that affect the investment.

2   The Investment as a Source of Real Options

In traditional DCF analysis, uncertainty is considered by adopting a probability distribution on the different possible scenarios or by adjusting the discount rate [16-17]. Consequently, a higher level of uncertainty reduces the value of the project. On the contrary, in the real options approach[2], greater uncertainty may result in a higher value of the investment, if managers use the options to respond flexibly to events. This is expressed by the criterion of $NPV_{extended}$, given from the sum of NPV_{base}, which expresses the NPV in the absence of strategic opportunities, and the value of real options OP, which represents the value of the project for management adaptability. Thus, management must look at the markets in terms of the evolution of uncertainty, determining the degree of exposure of investments and then placing them in a manner to derive maximum benefit from the uncertainty itself.

The literature on option pricing was first written in the 1970s with the works of Black F. and Scholes M. [21], di Merton R.C. [22] e di Cox J.C., Ross S.A. and Rubinstein M. [23]. The first references to real options appeared the first half of the 1980s. However, the term real options was founded in 1977 through the intuition of Myers S.C.[24]: «The value of the firm as a going concern depends on its future investment strategy. Thus it is useful for expositional purposes to think of the firm as composed of two distinct asset types: 1) *real assets*, which have market values independent of the firm's investment strategy, and 2) *real options*, which are opportunities to purchase real assets on possibly favorable terms».

Real options apply the theory of financial options to the analysis of real assets. The elements that distinguish an option are: The *stock price* (S), i.e. the value of the underlying asset; The *time to maturity*, time period of validity of the right; The *strike price* (X), i.e.the price to pay to exercise the right; The *volatility* (σ^2), given by the standard deviation of returns of the underlying asset; The *risk free rate* (r), provided by the performance of non-risky investments, typically government bonds.

In the application of real options to the economic evaluation of investments, the underlying asset S can be the NPV of the project. Obtained with traditional DCF analysis, the NPV is then the starting point to implement the real options. The time to maturity is the period within which the option is deemed exercised. Volatility is the standard deviation of implied returns or percentage changes in values that characterize the underlying project or stochastic variable.

Adopting a process perspective, the evaluation of investments aimed at enhancing strategic opportunities is divided into three main phases [7]:

---

[2] Copeland, T. and Antikarov, V. [25] define the real option «as the right, but not the obligation, to take an action (e.g., deferring, expanding, contracting or abandoning) at a predetermined cost, called exercise price, for a predetermined period of time, the life of the option». Similarly, Kogut, B. and Kulatilaka, N. [26].

1. analysis of the risk profile;
2. strategic analysis of the profile;
3. quantitative analysis.

The three phases are not strictly sequential. There are some connections between different times, but the logic is either circular or iterative.

Risk profile analysis is the basis for the quantitative analysis and mainly consists of two essential moments. The first is the estimate of the NPV basis, i.e. the present value of cash flows resulting from the deterministic component of the investment. The second is to identify the critical variables and the consequent modelling of the stochastic component of the project, which is the process that describes the evolution of the value of the initiative over time. The variation in time t of the value of an uncertain variable, which may coincide with the value of the underlying S, is describable by means of a stochastic diffusion process (geometric Brownian motion) and represented by the equation:

$$\frac{dS}{S} = \mu \cdot dt + \sigma \cdot dz \tag{1}$$

The two terms that describe the uncertain variable is the instantaneous expected return μ and instantaneous volatility σ of the value of the underlying S. The analysis of the instantaneous yield is based on the determination of a deterministic parameter μ that represents the trend value around which it is believed that the total value of the uncertain variable can evolve. From a geometric point of view, this value may coincide with the slope of the straight line interpolating the historical values of the variable. The volatility should then be estimated, which can be interpreted as the standard deviation of returns of the underlying asset.

The second step is to analyse the strategic profile of the project. It should recognise the possible areas of flexibility of the project in order to verify the predetermined scenarios. This leads to the identification of real options and their parameters (strike price, the value of the underlying asset, volatility, expiration date).

The third step is to analyse the quantitative profile. The aim is to arrive at the value of strategic opportunities, i.e. the estimate of the extended value of the project. To this end, the real options, once identified and defined, must be evaluated. Amram, M. and Kulatikala, N. [27] recognize three general evaluation methods: 1) the pde method solves a partial differential equation (pde) that equals the change of value in the option to the change of value in the equivalent portfolio, 2) the dynamic programming method outlines possible future outcomes, from which it then infers the future value of the optimal strategy, and 3) the simulation method establishes, for thousands of possible outcomes, the average value of the optimal strategy based on the decision made [28].

In the frequently used Monte Carlo simulation method [29-30], the optimal investment strategy at the end of each path is determined and the payoff calculated. The current value of the option is found by calculating the average of the payoffs and then discounting the average value thus obtained. This method can manage many aspects of the applications to the real world.

The real options theory is applied to evaluate investments in various sectors. According to Triantis A. and Borison A. [31], the main sectors are: intensive capital industries with uncertain cash flows (e.g., investments in natural resources and engineering), those that have suffered substantial structural changes and thus make the most traditional techniques evaluation unreliable (e.g., the production of electricity), as well as those in which the strategic prospects are the main determinants of value creation (e.g., high-technology sectors, innovation, research and development).

In recent years, not only in Italy, the energy sector has undergone regulatory changes as well as changes in its market logic, passing from being a regulated and monopolistic sector to a liberalized, uncertain and highly competitive one. These changes have pioneered the application of real options theory [9], making it possible to consider the intrinsic value of investment flexibility. For energy production from renewable sources projects, this ability has become particularly important [32], as demonstrated by the various applications in current literature [33-34-35].

Applications specific to the wind energy sector are, among others, those of Munõz-Hernández J.I., Contreras J., Caamaño J. and Correia P.F. [36], as well as Venetsanos K., Angelopouloua P. and Tsoutsos T. [37]. For the latter, the authors identify uncertainties and attributes of the resources that are used for the production of energy. Subsequently, the following real options have been identified: option to defer development, time to build option (staged investment), option to alter the operating scale, option to abandon, growth options.

3   DCF Analysis for the Economic Evaluation of a Wind Farm

Characteristics of the project. Through the installation of nine wind turbines, 7 with a power of 2.5 MW and 2 with a power of 1.5 MW, the project involves the construction and operation of a wind farm with a total capacity of 20.5 MW. The wind turbines are arranged along underground power lines that carry power to a transformer station (20/150 kV) and then to a new power station (380/150 kV), owned by Terna SpA, to enter the National Transmission Grid (NTG).

The area covered by the plant falls within the Municipality of Vallata (AV), in the eastern part of the Campania Region (Italy). The total area of the wind farm is about 4 km^2. The area covers an altitude of between 600 and 750 metres above sea level. The land has been obtained by signing preliminary agreements with the owners for the acquisition of surface rights and the right of way and conduits.

The sector. Renewable energy sources are going through a period of great development in the world. The sector has not been affected by the current financial crisis and is projected to achieve the objectives set by the EU for 2020 with the climate and energy package of January 24, 2012: 20% reduction of greenhouse gas emissions, increase up to 20% savings energy, increase to 20% of energy consumption from renewable sources. In the European Union, over the next 40 years further investment will be required [38]. The REN21 [39] report shows the significant growth of investments in renewable energy, indicating China with 48.9 billion dollars primarily used in the wind industry as well as research of new wind farm technologies. Wind energy can significantly contribute to solving the world's energy problems, as highlighted by the various reports published by the International Energy Association [40], the European Commission [41] and the European Wind Energy Association (EWEA).

In Italy, wind energy production is growing. Especially in the South, due to the topography and wind of its territory, as demonstrated by 84% of the number of national facilities and 98% of installed capacity. The untapped potential is also remarkable, as reported by SVIMEZ and SRM [42].

The study of the demand for electricity is carried out by processing data provided by Terna Rete Italy [43-44]. If the production of electricity is compared with its application in the Campania region, there is a deficit for 2011 of 9,136 GWh, or 47.7%. This deficit is now offset by imports. Starting form an annual consumption of 3,014 kWh per capita, the demand for electricity in the town of Vallata and surrounding municipalities in the first and second crown is estimated. From the results, there is domestic demand equal to 133.8 GWh/year.

Estimates of income and residual value of the works. The energy produced by the wind farm will be sold to the grid operator, which represents the direct user of the system. With the liberalization of the electricity production introduced in Italy by Legislative Decree no. N. 79/1999, selling electricity produced from renewable energy sources has become easier, with the guarantee of sale to the network. Producers can sell energy in two different ways: 1) through dedicated withdrawal, 2) on the free market. It is also possible to consume the energy produced.

Withdrawal involves supplying electricity to the Energy Services Operator (ESO)[3], which pays for it. This service from the ESO therefore avoids that producers have to manage the selling of energy, namely placement on the Power Exchange or drawing up of bilateral contracts. Once on the grid, the energy is sold. In particular, the Power Exchange, administered by the Electricity Market Operator (GME) has been operating in Italy since April 2004.

Currently, the energy produced by wind turbines connected to the grid, can be stimulated with two alternative support mechanisms: Green Certificates (GC) or the all-inclusive rates (TO), the latter only in the case of power plants with less than 200 kW. GCs are negotiable securities, issued on the basis of the amount of electricity produced by the plants. The commercial value of a CV is derived from the Bersani Decree (Legislative Decree no. 79/1999) and subsequent amendments. This framework has imposed on those operators who enter the network or import more than 50 GWh/year, an obligation for a percentage to come from renewable sources. This requirement may be satisfied through the purchase of GCs relative to energy from renewable sources produced by other operators. The unit price of the GC is updated monthly.

As an alternative to GCs, the TO, reserved for wind turbines with less than 200 kW of power, are fixed rates of remuneration of electricity fed into the grid. The TO

---

[3] The ESO is a holding company that supports the development of renewable energy sources through the management and delivery of the relevant incentives.

remunerate only the electricity fed into the grid, while the GC remunerate all the net energy produced, thus also rewarding any portion that is self-consumed. The GC and TO are recognized for a period of 15 years, with both mechanisms being managed by the ESO. In the case of GC, in addition to the incentive, manufacturers can count on an additional income: the value of electricity produced. Placing the energy produced into the network (for sale on the electricity market, "dedicated withdrawal" or "net metering") or auto-consumption. On the other hand, the TO are only a source of revenue.

In light of the above, for the wind farm assessed, the revenues are derived from: the sale on the energy market; the sale of Green Certificates.

To estimate the revenue from the sale of energy, the average price of electricity and GC in the year preceding the assessment were referred to. In 2012, these prices are respectively 75.48 €/MWh and 74.12 €/MWh [45].

The overall manufacturability of annual plants must also be known. To this end, maps of the producibility of electricity supplied taken from the Atlante Eolico Italiano were consulted. The project area has specific values of productivity between 3,000 and 3,500 MWh/MW. For the evaluation of the system being studied, the average value of 3,250 MWh/MW was used. From the calculations, this results in a net annual production for the entire system of 58,630 MWh.

Applying the productivity data, the unit value of the two items of revenue provided, it follows that the system can generate the following annual revenue:

$$\text{Annual revenue (from 1 to 15 years)} = (75.48 + 74.12) \text{ €/MWh} \times 58,630 \text{ MWh} =$$
$$\text{€8,771,048.00 ;}$$

$$\text{Annual turnover (from 16 to 20 years)} = 75.48 \text{ €/MWh} \times 58,630 \text{ MWh} =$$
$$\text{€4,425,392.40 .}$$

It is also worth considering the residual value of the works at the end of 20 years referred to in the evaluation. This residual value is equal to 0-5% of the initial investment [46]. In this case study, it is assumed 2.5% of the initial investment, i.e. €891,750.

Cost estimate. This is obtained as the sum of the investment and operating costs. The interest on the debt should also be considered.

Investment costs. The total cost of a wind turbine per kW installed varies significantly depending on the country of reference, ranging from 1,000 €/kW to 1,350 €/kW [41]. In Italy, the investment costs are higher than the average of other countries, since most of the plants are installed in hilly or mountainous areas, with sometimes being difficult to access them. For a typical configuration of wind turbine installed on the ground with a total average output of 20 MW at a site of medium complexity, the investment cost ranges from 1,550 €/kW for large systems installed in areas with low complexity, up to a maximum of 2,000 €/kW for small systems installed at sites with a complex orography [40]. Considering an average cost of 1,740 €/kW, the construction of the wind turbine under study requires an investment of € 35,670,000.

The useful life of a wind turbine system is usually over 20 years [37-47-36-38]. At the end of its useful life, there are two possibilities: 1) dismantling of the site, 2) replacement of the machines installed with new wind turbines. For the second solution, known as "repowering", the removal of the old machines should be considered. The removal costs of a single wind turbine are between € 20,000 and € 40,000. In case of withdrawal of the investment, the removal costs of the machines must be added the remediation costs of the site.

Operating costs. These include personnel costs, maintenance costs and royalties, as detailed in the financial plan. In particular, the maintenance costs are derived from Amicarelli V. and Tresca F.A. [48]. Royalties are paid to the local part of the private company that operates the plant, to the extent of 3.6% of the total revenue per annum.

Repayment schedule of the loan. The intention is to take out a loan for the amount of € 21,402,000, equal to 60% of the total investment costs. The remaining 40% is given by private capital. A ten-year amortization schedule at a constant rate is drawn up, with an annual interest rate of 7% currently applied in Italy for funding in the wind sector. The annual interest is included in the financial plan.

The financial plan. It is prepared over a 20 year period. The investment costs are assumed to be all concentrated in the 1st year. The evaluation is carried out at constant prices. The cash flows are discounted at the discount rate of 5%, already adjusted for inflation [49]

The pre and post-tax results are estimated, taking into account: - the IRES (corporate income tax), calculated ats the rate of 27.5% on profit before taxes - IRAP (Regional Tax on Productive Activities), which in the Campania region is equal to 4.97% of pre-tax gross staff costs and financial charges. The results of the financial plan are in Table 1. The processing of cash flows allows for the evaluation indices, both pre-tax and post-tax (Table 2).

Table 1. Financial Plan. All the amounts are in €uro

year	1	2	3	4	5	6	7
INCOME							
Sale of energy		4,425,392	4,425,392	4,425,392	4,425,392	4,425,392	4,425,392
Green Certificates		4,345,656	4,345,656	4,345,656	4,345,656	4,345,656	4,345,656
COSTS							
Investment	-35,670,000						
Staff		-40,000	-40,000	-40,000	-40,000	-40,000	-40,000
Maintenance	-255,041	-255,041	-510,081	-510,081	-510,081	-510,081	-510,081
Royalty		-315,758	-315,758	-315,758	-315,758	-315,758	-315,758
Interest on debt		-1,498,140	-1,389,708	-1,273,687	-1,149,543	-1,016,710	-874,578
GROSS PROFIT	-35,925,041	6,662,110	6,515,501	6,631,523	6,755,666	6,888,500	7,030,631
IRES		1,832,080	1,791,763	1,823,669	1,857,808	1,894,337	1,933,424
IRAP		325,808	318,595	324,303	330,411	336,946	343,939

Table 1. (*continued*)

NET PROFIT	-35,925,041	4,504,222	4,405,144	4,483,551	4,567,447	4,657,216	4,753,269

Year	8	9	10	11	12	13	14
INCOME							
Sale of energy	4,425,392	4,425,392	4,425,392	4,425,392	4,425,392	4,425,392	4,425,392
Green Certificates	4,345,656	4,345,656	4,345,656	4,345,656	4,345,656	4,345,656	4,345,656
COSTS							
Staff	-40,000	-40,000	-40,000	-40,000	-40,000	-40,000	-40,000
Maintenance	-510,081	-510,081	-510,081	-1,020,162	-1,020,162	-1,020,162	-1,020,162
Royalty	-315,758	-315,758	-315,758	-315,758	-315,758	-315,758	-315,758
Interest on debt	-722,497	-559,770	-385,653	-199,347			
GROSS PROFIT	7,182,712	7,345,439	7,519,556	7,195,781	7,395,128	7,395,128	7,395,128
IRES	1,975,246	2,019,996	2,067,878	1,978,840	2,033,660	2,033,660	2,033,660
IRAP	351,421	359,428	367,994	352,064	361,872	361,872	361,872
NET PROFIT	4,856,045	4,966,016	5,083,684	4,864,877	4,999,596	4,999,596	4,999,596

Year	15	16	17	18	19	20
INCOME						
Sale of energy	4,425,392	4,425,392	4,425,392	4,425,392	4,425,392	4,425,392
Green Certificates	4,345,656	4,345,656				891,750
COSTS						
Staff	-40,000	-40,000	-40,000	-40,000	-40,000	-40,000
Maintenance	-1,020,162	-1,020,162	-1,020,162	-1,020,162	-1,020,162	-1,020,162
Royalty	-315,758	-315,758	-159,314	-159,314	-159,314	-159,314
Interest on debt						
GROSS PROFIT	7,395,128	7,395,128	3,205,916	3,205,916	3,205,916	3,205,916
IRES	2,033,660	2,033,660	881,627	881,627	881,627	881,627
IRAP	361,872	361,872	155,763	155,763	155,763	155,763
NET PROFIT	4,999,596	4,999,596	2,168,526	2,168,526	2,168,526	2,168,526

Table 2. Evaluation Indices

	pre-tax	post-tax
NPV	40,690,170 €	16,428,785 €
IRR	17.97%	10.73%
Payback period	7 years	10 years

4   Real Options and Monte Carlo Simulation for the Case Study

The following real options have been identified:

1. *Abandonment Option*. Before the execution of the works, this option may be exercised for incorrect predictions about the characteristics of windiness or financial difficulties for the investor. If electricity prices become no longer acceptable, the project is deferred.
2. *Deferment Option*. If the investor fears a drop in prices of electricity and/or Green Certificates, may defer the investment over time until he gets more information or reassurance. The option can only be implemented prior to the execution of works and often depends on the perspective of regulatory changes.
3. *Expansion Option*. It expands the investment if the macroeconomic conditions for the wind energy sector improve, or if the legislation introduces more incentives. The option is exercisable after the completion of the works and until the end of the evaluation period (American option). It involves the installation of additional 5 wind turbines of 2.5 MW, with an increase in capacity of 61% and an additional investment cost amounting to € 21,750,000.

Having identified the possible real options, the project NPV is assumed as an underlying assets [25-20]. The value of the asset without options has already been estimated with the traditional DCF analysis (Market Asset Disclaimer).

4.1   Uncertainty Estimate

The uncertainty of the market is represented by the volatility of the underlying asset. In this regard, Méndez M., Goyanes A. and Lamothe P. [47] wrote that for the design of a wind farm: «Various alternative solutions are often used to assess volatility: 1) Taking the market return volatility of a similar company; an approximation would thereby be made that might lead to error, since finding a company whose characteristics exactly match those of the project would be no easy task. 2) Using the volatility of those elements that generate the project's cash flows, such as, for instance, the volatility of electric power prices; however, these elements only partially reflect the project's uncertainty. 3) We thus turn the market into a complete market, for we assume the project's market value to be its present value and assess volatility by simulating its expected returns from year 0 to year 1. This allows us to combine all market uncertainties into a single one: the volatility of the project».

For the case study, the volatility is estimated based on the NPV.

Estimate the volatility of the NPV with risk analysis. Risk analysis is designed to identify adverse events that may affect the feasibility of the project, in order to assess the extent to which uncertainties can affect the NPV of the project. The analysis starts from the identification and characterization of the uncertainties of the project. For the wind farm under investigation, the following uncertainties have been identified.

1. *Uncertainty about the overall annual manufacturability of the plant.* In the previous section, a specific average producibility of 3,250 MWh/MW was assumed, which corresponds to the value of 58,630 MWh used in the evaluation. For the Municipality of Vallata, the Atlante Eolico Italiano gives a range of 3,000-3,500 MWh/MW, upon which a triangular probability distribution for the overall manufacturability of the system is based, with minimum and maximum values respectively of 54,120 MWh and 63,140 MWh.

2. *Uncertainty in the selling prices of energy.* To construct the distribution of the probability of the selling price of electricity, the average monthly real prices in the period 2005-2012 were analysed. There is a minimum price of 54.94 €/MWh and a maximum price of 112.63 €/MWh. It assigns a normal type probability distribution, characterized by:

— an average of $\bar{p} = \frac{1}{n}\sum_{i=1}^{n} p_i = 80.53$ €/MWh ;

— a standard deviation $\sigma = \sqrt{\frac{1}{1-n}\sum_{i=1}^{n}(p_i - \bar{p})^2} = 12.59$ €/MWh, defined in the range from 54.94 €/MWh to 112.63 €/MWh.

3. Uncertainty on the price of Green Certificates. The real average annual prices of GCs in the period 2005-2012 were analysed. The trend line of prices has the equation $y = -6.5378\,x + 13,228$. Temporally extending this trend, estimates of 67.77 €/MWh for 2013 and 61.23 €/MWh for 2014 are obtained. Considering the trend in the figure, a triangular type probability distribution is assumed, where:

— the mean value is of   74.12 €/MWh used in estimating deterministic revenues;

— the minimum value is 61.23 €/MWh (forecast for 2014);

— the maximum value is of 81.80 €/MWh at 2011 (the possibility that in the future, there will be higher prices of 74.12 €/MWh, is confirmed by the data of January 2013 which sees a price of 79.52 €/MWh).

It is worth noting the probability distributions of the uncertain variables, the risk analysis is implemented with the Monte Carlo simulation method. This method involves extracting a large number of possible scenarios from a multivariate distribution determined by the probable hypotheses on the parameters of the analysis, that then allows for the calculation of statistics on the sample extractions. The Oracle Crystal Ball software is used. For the post-tax NPV, the results are shown in Figure 1. The standard deviation is 5.694.136 €. The volatility σ of the NPV is 31.24%.

4.2   The Value of Real Options

The flexibility of the wind farm project is a function of the real options of deferral, abandonment and expansion. To estimate the value of the options and their interactions, the Fairmat Academic software was used, which implements the Monte Carlo technique for the European option [29] and the least-squares approach for the American approach [50].

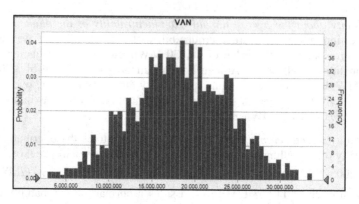

Fig. 1. Probability distribution of the post-tax NPV

The model requires the measurement of the risk-free discount rate, due to it being the rate of return on government bonds in the medium to long term. From the auction of ten-year BTP 30/01/2013, the rate is 4.17%. Table 3 shows the input data for the calculations, which were carried out in relation to two cases:

1. all the real options are exercisable;
2. all the decisions are already taken at time 0 (base case).

Table 3. Input data of the analysis model

Parameter	Value	Description
NPV	16,428,785 €	NPV without considering the value of the real options
σ	31.24 %	Volatility of the NPV
r_f	4.17 %	Risk-free discount rate (BTP 10 year)
I_1	1,783,500 €	Capital expenditure: for the study of the wind speed profile of the site, the design of systems for the administrative process (building permits, environmental impact assessment, etc.). Between 2 and 5% of the investment costs [38]
S	0	Recovery value in the abandonment option
e	61%	Rate of expansion capacity
I_e	21,750,000 €	Additional investment costs

The stochastic process that evolves the NPV of the project is characterized by a geometric Brownian motion [34-36-51].

In Fairmat, the interactions between the decisions are described in a window called Option Map. For the mapping of the options, the software allows to select different types of options, which are represented with different coloured diamonds: Pink diamond, decision already made; Green diamond, European option; Blue diamond, American option. For the case study, every kind of option is associated with the corresponding payoffs and the expiration time.

The study and design phases are identified with a pink diamond. It is therefore considered a *committed option*. The expiration time is one year and the payoff is equal to the cost of I_1. An operator *Or* can choose the best among the following options (investment or abandonment option).

The investment can be started immediately or deferred for up to two years: it is therefore considered an American style option (blue diamond), that is exercisable at any time until the expiration date. At the same time, it is considered an abandonment option (American put option, blue diamond), exercisable as an alternative investment. In this second case, the payoff coincides with the recovery value, i.e. is null.

Finally, it is possible to exercise an expansion option, after the investment. This is an American call option, with a payoff equal to $\max(eV_1-I_e,0)$.

At this point, the Monte Carlo simulation is implemented and the extended NPV calculated as an average over a sample of 5000 paths. It is a $NPV_{extended}$ of € 20,554,931, with a standard error of 342,558 and a standard deviation of 24,222,536.

The project is also evaluated in the base case, i.e. without considering the flexibility of the real options. In this case, the operator Or is disabled, without the possibility of abandoning the project, the options become already taken American decisions (*committed option*). There is an obligation to defer the investment at year 3 and expand the production capacity at year 20.

In this case, the NPV is € 15,344,549, lower than the value obtained with the deterministic analysis (€ 16,428,785) since the investment is deferred at year 3. The standard error is 380,166 and the standard deviation of 26,881,803.

Comparing the results of the two scenarios, it is clear that the average value of the NPV in the case of the real options is larger than what is obtained in the base case. The difference between the two values is of € 5,210,382. This difference shows that the real options approach allows to exploit the strategic opportunities and operational flexibility of the investment. The higher value is the monetary representation of the options identified through the examination of the dynamic aspects of the proposed wind farm.

In addition, comparing the probability distributions of the NPV, it is worth noting that in the case of the real options, the values are always positive, while in the base case without options, it is possible to have negative values among the simulated paths. In fact, management flexibility moves the probability distribution of the NPV towards positive values.

5   Conclusions

Traditional DCF analysis shows the limits in all the cases in which the possibility of management to vary the course of an investment as a result of changes occurred in the economic context is marked. This happens in a sector characterized by strong dynamism, and thus uncertain financial viewpoints, such as that of renewable energy. In such situations, it is appropriate to carry out the economic evaluation of projects by defining a logical-operational   process in order to understand the different effects of possible scenario evolutions. Improvement of the results is due to the implementation of analysis tools such as those described in this work. The reference is to the theory of real options, applicable today in a rather expeditious way with the software available, and the Monte Carlo simulation method, which makes it possible to trace a large number of development paths of the project.

The different operational phases of the model used for the optimization of the scenario analyses require a careful selection of the data sources. Only the rigorous processing of numerical information allows for the correct characterization and quantification of the parameters that determine the extent of the extended NPV of the project. This returns the value of the positive potential that can result from management flexibility and innovation.

In implementing a wind farm project, the economic evaluation shows great validity, allowing to express a value investment with real options that 34% higher than the value of the same project in the base case , i.e. in the case in which all decisions are taken at time zero.

References

1. Pennisi, G.: Tecniche di valutazione degli investimenti pubblici, Istituto Poligrafico e Zecca dello Stato, Roma (1991)
2. Saaty, T.L.: Fundamentals of Decision Making and Priority Theory with the Analytic Hierarchy Process. RWS Publications, Pittsburgh (1994)
3. Morano, P., Nesticò, A.: Un'applicazione della programmazione lineare discreta alla definizione dei programmi di investimento. Aestimum (50). University Press, Firenze (2007)
4. De Mare, G., Nesticò, A., Caprino, R.M.: La valutazione finanziaria di progetti per il rilancio del territorio. Applicazioni a casi reali, FrancoAngeli, Milano (a cura di, 2012)
5. De Mare, G., Lenza, T.L., Conte, R.: Economic evaluations using genetic algorithms to determine the territorial impact caused by high speed railways. World Academy of Science, Engineering and Technology, ser. ICUPRD 2012 (71) (2012)
6. De Mare, G., Morano, P., Nesticò, A.: Multi-criteria spatial analysis for the localization of production structures. Analytic Hierarchy Process and Geographical Information Systems in the case of expanding an industrial area. World Academy of Science, Engineering and Technology, ser. ICUPRD (71) (2012)
7. Micalizzi, A.: Opzioni Reali. Logiche e casi di valutazione degli investimenti in contesti di incertezza. EGEA, Milano (1999)
8. Dixit, A.K., Pindyck, R.S.: The Options Approach to Capital Investment. Harward Business Review (1995)
9. Fernandes, B., Cunha, J., Ferreira, P.: The use of real options approach in energy sector investments. Renewable and Sustainable Energy Reviews (15) (2011)
10. Barnett, M.L.: Paying attention to real options. R&D Management 35(1) (2005)
11. Kensinger, J.W.: Adding the value of active management into the capital budgeting equation. Midland Corporate Finance Journal 5(1) (1987)
12. Dixit, A.K., Pindyck, R.S.: Investment Under Uncertainty. Princeton University Press, Princeton (1994)
13. Trigeorgis, L.: Real Options: Managerial Flexibility and Strategy in Resource Allocation. MIT Press, Cambridge (1996)

14. de Neufville, R.: Real Options: Dealing With Uncertainty in Systems Planning and Design. Integrated Assessment 4(1) (2003)
15. Yeo, K.T., Qiu, F.: The value of management flexibility – a real option approach to investment evaluation. International Journal of Project Management 21 (2003)
16. Florio, M.: La valutazione degli investimenti pubblici, vol. I, Franco Angeli, Milano (2001)
17. De Mare, G., Nesticò, A., Tajani, F.: The Rational Quantification of Social Housing. An Operative Research Model. In: Murgante, B., Gervasi, O., Misra, S., Nedjah, N., Rocha, A.M.A.C., Taniar, D., Apduhan, B.O. (eds.) ICCSA 2012, Part II. LNCS, vol. 7334, pp. 27–43. Springer, Heidelberg (2012)
18. Kogut, B., Kulatilaka, N.: Options Thinking and Platform Investment: Investing in Opportunity. California Management Review 36(2) (1994)
19. Mun, J.: Real Options Analysis. Tools and Techniques for Valuing Strategic Investments and Decisions. John Wiley & Sons, Haboken (2002)
20. Luehrman, T.A.: Investment Opportunities as Real Options: Getting Started on the Numbers. Harvard Business Review (1998)
21. Black, F., Scholes, M.: The Pricing of Options and Corporate Liabilities. Journal of Political Economy 81 (1973)
22. Merton, R.C.: Theory of Rational Option Pricing. Bell Journal of Economics and Management Science 4 (1973)
23. Cox, J.C., Ross, S.A., Rubinstein, M.: Option Pricing: A Simplified Approach. Journal of Financial Economics 7 (1979)
24. Myers, S.C.: Determinants of corporate borrowing. Journal of Financial Economics 5 (1977)
25. Copeland, T., Antikarov, V.: Real Options. Texere LLC, New York (2001)
26. Kogut, B., Kulatilaka, N.: Capabilities as Real Options. Organization Science 12 (2001)
27. Amram, M., Kulatikala, N.: Real Options. Strategie d'investimento in un mondo dominato dall'incertezza. ETAS, Milano (2000)
28. Triantis, A.J.: Real Options. In: Logue, D., Seward, J. (eds.) Handbook of Modern Finance. Research Institute of America, New York (2003)
29. Boyle, P.P.: Options: A Monte Carlo Approach. Journal of Financial Economics 4 (1977)
30. Areal, N., Rodrigues, A., Armada, M.J.R.: Improvements to the Least Squares Monte Carlo Option Valuation Method. Derivatives Research 11 (2008)
31. Triantis, A., Borison, A.: Real Options: State of the Practice. Journal of Applied Corporate Finance 14(2) (2001)
32. Kumbaroglu, G., Madlener, R., Demirel, M.: A Real Options Evaluation Model for the Diffusion Prospects of New Renewable Power Generation Technologies. In: Proceedings of the 6th IAEE European Conference "Modelling in Energy Economics and Policy", Zurich, Switzerland (2004)
33. Ramirez, N.: Valuing Flexibility in Infrastructure Developments: The Bogota Water Supply Expansion Plan. MIT, Cambridge (2002)
34. Bøckman, T., Fleten, S.-E., Juliussen, E., Langhammer, H.J., Revdal, I.: Investment Timing and Optimal Capacity Choice for Small Hydropower Projects. European Journal of Operational Research 190 (2007)
35. Fleten, S.-E., Maribu, K.M., Wangensteen, I.: Optimal Investment Strategies in Decentralized Renewable Power Generation under Uncertainty. Energy 32 (2007)
36. Munõz-Hernández, J.I., Contreras, J., Caamaño, J., Correia, P.F.: Risk assessment of wind power generation project investments based on real options. In: Proceedings from the 13th International Congress on Project Engineering, Badajoz (2009)

37. Venetsanos, K., Angelopouloua, P., Tsoutsos, T.: Renewable energy sources project appraisal under uncertainty: the case of wind energy exploitation within a changing energy market environment. Energy Policy 30 (2002)
38. ENEA, Unità Centrale Studi e Strategie. Domanda e offerta di energia in Italia e nel mondo: situazione attuale e scenari futuri, Energia Ambiente e Innovazione (3) (2012)
39. REN21. Renewables 2012 Global Status Report (2012), http://new.ren21.net/
40. International Energy Association. IEA WIND 2011 Annual Report (2011)
41. European Commission. Wind Energy. The facts. Costs & Prices, vol. II (2010)
42. SVIMEZ and SRM. Energie rinnovabili e territorio. Scenari economici, analisi del territorio e finanza per lo sviluppo, Giannini Editore, Napoli (2011)
43. Terna Rete Italia. Dati Statistici sull'energia elettrica in Italia. L'elettricità nelle Regioni, Terna Rete Italia, Roma (2012a)
44. Terna Rete Italia. Previsioni della domanda elettrica in Italia e del fabbisogno di potenza necessario. Anni 2012-2022, Terna Rete Italia, Roma (2012b)
45. Gestore del Mercato Elettrico. Notiziario Borsa Italiana dell'energia (56) (2013)
46. Hebei Electric Power Design & Research Institute. Statement of maintenance cost & residual value (2009)
47. Méndez, M., Goyanes, A., Lamothe, P.: Real Options Valuation of a Wind Farm. Universia Business Review (2009)
48. Amicarelli, V., Tresca, F.A.: Considerazioni economiche sulla produzione di energia eolica. Energia Ambiente e Innovazione (6) (2011)
49. Commissione Europea, programmazione 2007-2013, Orientamenti metodologici per la realizzazione delle analisi costi-benefici
50. Longstaff, F.A., Schwartz, E.S.: Valuing American Options by Simulation: a Simple Least-Squares Approach. The Review of Financial Studies 14 (2001)
51. Blanco, C., Choi, S., Soronow, D.: Energy Price Processes Used for Derivatives Pricing & Risk Management. Financial Engineering Associates (2001)
52. De Mare, G., Nesticó, A., Tajani, F.: Building investments for the revitalization of the territory. A multisectoral model of economic analysis. In: Murgante, B., Misra, S., Carlini, M., Torre, C.M., Quang, N.H., Taniar, D., Apduhan, B.O., Gervasi, O. (eds.) ICCSA 2013, Part III. LNCS, vol. 7973, pp. 493–508. Springer, Heidelberg (2013)
53. De Mare, G., Manganelli, B., Nesticó, A.: Dynamic analysis of the property market in the city of avellino (Italy):The wheaton-di pasquale model applied to the residential segment. In: Murgante, B., Misra, S., Carlini, M., Torre, C.M., Quang, N.H., Taniar, D., Apduhan, B.O., Gervasi, O. (eds.) ICCSA 2013, Part III. LNCS, vol. 7973, pp. 509–523. Springer, Heidelberg (2013)

A Qualitative and Quantitative Analysis on Metadata-Based Frameworks Usage

Eduardo Guerra[1] and Clovis Fernandes[2]

[1] National Institute for Space Research (INPE) - Laboratory of Computing and Applied Mathematics (LAC) - P.O. Box 515 – 12227-010 - São José dos Campos, SP, Brazil
[2] Aeronautical Institute of Technology (ITA) - Praça Marechal Eduardo Gomes, 50 CEP 12.228-900 - São José dos Campos – SP, Brazil
guerraem@gmail.com, clovistf@uol.com.br

Abstract. The usage of metadata-based frameworks is becoming popular for some kinds of software, such as web and enterprise applications. They use domain-specific metadata, usually defined as annotations or in XML documents, to adapt its behavior to each application class. Despite of their increasingly usage, there are not a study that evaluated the consequences of their usage to the application. The present work presents the result of an experiment that aimed to compare the development of similar applications created: (a) without frameworks; (b) with a traditional framework; (c) with a metadata-based framework. As a result, it uses metrics and a qualitative evaluation to assess the benefits and drawbacks in the use of this kind of framework.

Keywords: framework, metadata, metric, experiment, software design, software architecture.

1 Introduction

A framework is a set of classes that supports reuses at larger granularity. It defines an object-oriented abstract design for a particular kind of application that does not enable only source code reuse, but also design reuse [1]. Frameworks can enable functionality extension by providing abstract methods in its classes that should be implemented with application-specific behavior. Other alternative is to provide methods to configure instances for which part of the functionality is delegated. This instances can be application-specific or from framework's built-in classes [2]. In the present work, the frameworks that use those approaches based on inheritance or composition to enable its extension are called Traditional Frameworks.

The framework structures have evolved and recent ones make use of introspection [3] [4] to access at runtime the application classes' metadata, like their superclasses, methods and attributes. As a result, it eliminates the need for the application classes to be coupled with the framework abstract classes and interfaces. The framework can, for instance, search in the class structure for the right method to invoke. The use of

B. Murgante et al. (Eds.): ICCSA 2013, Part II, LNCS 7972, pp. 375–390, 2013.

this technique provides more flexibility to the application, since the framework reads dynamically the classes structure allowing them to evolve more easily [5].

For some frameworks, however, once they need a domain-specific or application-specific metadata to customize their behavior, the information found in the class definition is not enough [6]. This kind of information can be represented and defined in code annotations [7], external sources, like XML files and databases, or implicitly by using naming conventions [8] [9]. In the present work this kind of framework is named Metadata-based Framework, which can be defined as the one that process their logic based on the metadata from the classes whose instances they are working with [10].

Before this study, not much information about the benefits of developing and using metadata-based frameworks were found in the literature, but some development communities are increasingly adopting them as standards. Consistent with that, there are many recent frameworks developed and APIs defined using this approach, such as Hibernate [11] , EJB 3 [12], Struts 2 [13] and JAXB [14].

The main goal of this study is to evaluate the benefits and drawbacks in the usage of metadata-based framework. In order to do that, an experiment was conducted aiming to compare the uses of traditional and metadata-based frameworks to create the same functionality. The experiment carried out during an undergraduate course of advanced topics in object-orientation. The students using three different approaches developed the same application: (a) without frameworks; (b) with a framework that do not use metadata; (c) with a metadata-based framework. Students also answered a questionnaire to register their impressions on the experience.

Metrics and visualization techniques were applied to the source code of the three applications in order to evaluate the design of each one. Issues like coupling, amount of code and complexity were considered in the analysis. Other more subjective issues like the facility to use the framework, easiness to evolve the application and the development time were addressed in the questionnaire and in observations during the implementation. The evaluation resulted in a set of consequences, both positives and negatives, concerning the use of metadata-based frameworks.

2   Metadata-Based Frameworks

Metadata is an overloaded term in computer science and can be interpreted differently according to the context. In the context of object-oriented programming, metadata is information about the program structure itself such as classes, methods and attributes. A class, for example, has intrinsic metadata like its name, its superclass, its interfaces, its methods and its attributes. In metadata-based frameworks, the developer also must define some additional application-specific or domain-specific metadata.

The metadata consumed by the framework can be defined in different ways [9]. One alternative is to define them in external sources, like XML files and databases. Another possibility that is becoming popular in the software community is the use of code annotations, which is supported by some programming languages like Java [7] and C# [15]. Using this technique the developer can add custom metadata elements directly into the class source code. The use of code annotations is also called attribute-oriented programing [6] [16].

Metadata-based frameworks can be defined as frameworks that process their logic based on the metadata of the classes whose instances they are working with [10]. The use of metadata changes the way frameworks are build and how software developers use them. In metadata-based frameworks there are some variable points in the framework processing which are determined by class metadata. Reflective algorithms in some cases cannot be applied due to more specific variations for some classes. In this context, metadata can be used to configure specific behaviors when the framework is working with that class.

From developer's perspective in the use of this kind of framework, there is a stronger interaction with metadata configuration than with method invocation or class specialization. That makes the number of method invocations in framework classes smaller and localized.

The following are examples of how metadata-based frameworks and APIs can be used in different contexts: Hibernate [11] is a framework for object-relational mapping; SwingBean [19] is a framework that generates forms and tables in Java Swing based on class structure and metadata; EJB 3 [12] is an standard Java EE API for enterprise development that uses metadata to configure concerns such as access control and transaction management; and JColtrane [20] is a XML parsing framework based on SAX which uses annotations for conditions to define when handler's methods should be invoked.

3   Experiment Description

One of the great difficulties to evaluate the benefits and drawbacks of the use of a metadata-based framework is the nonexistence of comparison basis. In other words, it is hard to find two frameworks with the same purpose, one build using traditional methods and other based on metadata, that can both be used for comparison. Four different scenarios abstracted from existent frameworks were used as reference for the case studies in this experiment.

The experiment main goal can be defined as: *"To create traditional and metadata-based frameworks for the same purpose and applications with the same behavior using them, aiming to generate a comparison basis and identify the benefits and drawbacks of the metadata-based approach."*

According to [21] classification, a Controlled Experimentation Method is used in the experiment, that can also be classified as a Synthetic Environment Experiment, since it is performed on an academic setting and simulates the creation of a piece of functionality in an application. Based on the taxonomy presented by [22], the experiment is designed to present cause-effect results, to be performed by novices and on an in-vitro environment. A similar approach to evaluate implementation approaches can be found in [23] and [24].

The following are the requirements that were considered in the elaboration of the experiment to reach its objectives: (a) two frameworks for the same purpose must be created using the traditional and the metadata-based approach; (b) solutions with the same external specified behavior must be developed using both frameworks and also without their use; (c) solutions with the same specified behavior to be compared must not be developed by the same persons; (d) neither the frameworks nor the solutions that implements the specified behavior must be developed by the present work's authors; (e) the development time of the solutions must be measured; (f) the design of the solutions must be assessed; and (g) the participants development experience to create the solutions must be assessed.

The experiment took place in *Advanced Topics in Object Orientation* discipline, which is an optional class in the fifth year of the *Computer Engineering* graduation coarse in the *Aeronautical Institute of Technology*. It was executed in the second semester of 2009, when twelve students attended the course. They were divided in four groups of three students, one for each scenario.

3.1   Experiment Stages

The development of the frameworks and the implementations using them, were divided in five distinct stages. The class was divided in four groups of three students, each responsible for the development of the first solution and both versions of the framework for one scenario. Other distinct groups developed the other solutions using the frameworks. Fig. 1 illustrates graphically the experiment stages and the software products generated in each one.

In **Stage 1**, students received a specification with the solution requirements that must be implemented by them. More than one class could compose this solution and the specification also defined how an external class should interact with them to use its functionality. Based on this defined protocol, the students also had to create an automated test suite to verify if the solution implements the specified requirements. They must not use frameworks and they should measure the development time for the solution's and test suite's implementation. This stage was executed at the student's home, and was carried on the beginning of the coarse when only testing techniques and basic concepts had been taught.

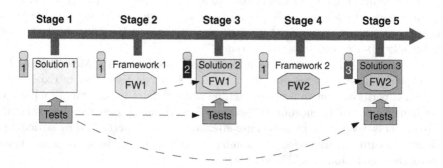

Fig. 1. Experiment stages representation

In **Stage 2**, the same group should develop a framework using traditional and reflection techniques in order to make easier the creation of similar solutions of the one developed in Stage 1. The framework scope and functionalities were specified in a document and used by the students as a reference. They also should provide documentation for the framework usage. The only restriction was that the framework must not use annotations or information defined externally. Nothing was said about reflection and code conventions in the specification, but their use were allowed. This stage was executed at their home as well, and happened at the middle of the coarse when reflection and object-oriented design techniques had already been taught.

In **Stage 3**, the group that worked in a different scenario should implement a solution similar to the one developed on Stage 1, but using the framework developed in Stage 2. The students received the same requirements specification used in Stage 1 and the framework documentation developed in Stage 2 to be read just before the implementation beginning. Students also received a project configured with: the tests; the framework in the classpath; and empty classes needed for the test to compile. The solution was considered implemented when the test suite executed successfully. This stage was executed in the class lab and the present work's author measured the implementation time.

In **Stage 4**, the same group that implemented stages 1 and 2 should develop a metadata-based framework for the same purpose of the one developed in Stage 2. The framework scope, functionalities and the role of metadata were specified in a document and used by the students as a reference. They also should provide documentation for the framework usage as they did in Stage 2. This stage was executed at their home, and was accomplish at the end of the coarse when annotations and techniques to develop frameworks with metadata had already been taught.

In **Stage 5**, a group that has not been worked already in the scenario should implement a solution similar to the ones developed in stages 1 and 3, but using the framework developed in Stage 4. The conditions were similar to Stage 3. This stage was executed in the class lab and the present work's author measured the implementation time.

The solutions developed by the students are not complete applications, but pieces of code that could potentially be a part of an architectural layer. They focused on a single concern, which is the domain aimed by the frameworks to be developed. Therefore, the specifications define simple problems to be implemented nevertheless with a lot of constraints to simulate the requirements of a real application.

The groups were free to use any strategy learned in the classes for the first framework implementation, since it fulfills the objective to make easier the development of that kind of solution. For the second framework, the specification defined more clearly for which purpose the metadata would be used and students did not have much freedom on their choices.

Each case study aimed to use respectively the following scenarios for the metadata usage: (a) mapping between command-line parameters and a class that represents them; (b) validation of method parameters and constraints; (c) stock market event handling; and (d) automatic generation of an HTML form. Each scenario focuses on a different architectural pattern documented for this kind of framework [25].

This scenario diversity is important to enable assessment of, not only the general characteristics, but also the specific ones from each distinct metadata usage.

After the implementation, the design of each solution was measured and evaluated using the metrics and the visualization techniques, such as polymetric views [26] and class blueprints [27]. The development experience was assessed through time measurements, the present work's author's observations and the student's answers to a questionnaire whose questions are presented in the next subsection.

3.2 Questionnaire

The students answered a questionnaire at the end of Stage 5 to evaluate their experience and impressions on the development of each solution.

The students filled a table answering for each solution development the following three questions: (1) how easy was the development of the application's source code; (2) how easy was the use of the framework; and (3) how easy would be to change the code to add new features. Each question could be answered as one of the following alternatives: (a) very easy; (b) easy; (c) average; (d) hard; and (e) very hard. The students also wrote a free text about their experience to justify his answers, which was also considered in the analysis.

To compare quantitatively the characteristics of each solution, the answers were turned into numbers using numeric scale from one to five respectively from very easy to very hard. This quantitative analysis was complemented with a qualitative one, using the author's observations and the answers to the open question.

3.3 Limitations

Despite the fact that the experiment achieved all the requirements, it still has some inherent limitations that can influence the implementations, which are used for the measurements and the conclusions. The following are the identified limitations that can have influence in the implementations: (a) the students learned about object-oriented design and frameworks from the beginning to the end of the Advanced Topics in Object Orientation discipline, which might have some influence in the source code quality; (b) the students did not have a wide experience in framework development and the difficulties in its use could have been from problems in the framework; (c) the solutions developed are not entire applications and the creation of only a functionality piece might not simulate the usage of a framework in a real system; and (d) the requirements in the specifications were not taken from real applications and were created to match the metadata usage scenarios, which might not be a precise simulation of a real development.

To deal with the two first limitations, the present work's author, who also observed the implementations, examined student's source code carefully. Whenever mistakes that can compromise the analysis were found, they were considered and referenced in the analysis in order to not invalidate the conclusions.

The two last limitations are related to the specifications and requirements used for each scenario. The requirements are based on concerns that might appear in real applications. The clear specification of how a class should interact with the solution simulates the framework usage encapsulation in order to minimize the effect of the implementation to cover only a piece of functionality.

The time measurements and the questionnaires can also suffer variations due to the following experiment characteristics: (a) the solution developed is a small piece of software and any unexpected fact, such as a bug, can increase greatly the relative development time; (b) students might have unconsciously evaluated the difficulty to develop each solution in comparison to the solutions of the other case studies developed; and (c) student could have more difficulty in software programming then others and this could interfere in the comparison between their answers.

To avoid the influence of those factors in the conclusions, the analysis of solutions developed was not strictly quantitative, but also qualitative. The present work's author, who took notes about students' difficulties and other events that could interfere with the results, observed the development of stages 3 and 5. The students also had an opportunity in the questionnaire to write their impressions about the development and justify their answers. This information was considered to the conclusions.

The solution was developed in class by the group, which helped to eliminate the influence of personal difficulties in programming, since the students helped each other to finalize the implementation. It was also important in the equality of each group's development capacity, to make the comparison of development time measurements more reliable.

4   Experiment Experience

The objective of this section is to present the questionnaire answers and the development time measurements. These data are used in the analysis performed in the next sections. Table 1 presents a summary of the the questionnaire answers.

Table 1. Questionnaire answers and development time

Scenario	Questions	Without Frameworks	Traditional Framework	Metadata-based Framework
A	Difficulty to Develop	5	11	10
	Difficulty to Use	-	8	11
	Difficulty to Modify	9	10	6
	Development Time	180 min	97 min	120 min
B	Difficulty to Develop	12	9	6
	Difficulty to Use	-	12	6
	Difficulty to Modify	12	11	5
	Development Time	300 min	144 min	43 min
C	Difficulty to Develop	8	9	7
	Difficulty to Use	-	9	12
	Difficulty to Modify	7	12	9
	Development Time	150 min	71 min	83 min
D	Difficulty to Develop	10	7	6
	Difficulty to Use	-	9	6
	Difficulty to Modify	14	8	9
	Development Time	360 min	128 min	68 min

The first column presents the number of the case study group with the development time for each case study in each phase. The questions are presented in a simplified way, but they represent the three questions described in the section 3.2. The three students answers in each experiment stage were summed and are presented at the table. It is important to highlight that for the implementation without frameworks, the time was measured by the students, but in the other phases, were measured by the present work's author.

5   Case Studies Metrics and Analysis

This section presents the metrics taken from the solutions and an analysis from the results of each case study. The metrics were based on the one from the overview pyramid [28], which is a metrics-based mean that both describe and characterize the structure of an object-oriented system by quantifying its complexity, coupling and usage of inheritance. The measured values for each version of each scenario are presented on Table 2.

Table 2.   Metrics values for the three versions of each scenario. The metrics are Number of Classes (NOC), Number of Methods (NOM), Cyclomatic Number (CYCLO), Lines of Code (LOC), Number of Operation Calls (CALLS) and Number of Called Classes (FANOUT).

	Scenario a			Scenario b			Scenario c			Scenario d		
	1	2	3	1	2	3	1	2	3	1	2	3
Simple Metrics												
NOC	4	4	4	5	5	5	14	11	14	2	2	2
NOM	26	29	23	18	16	16	58	48	71	35	21	21
LOC	163	172	119	82	61	28	271	204	249	287	108	76
CYCLO	41	48	30	26	9	9	80	65	83	74	21	21
CALLS	27	29	17	23	16	2	43	20	14	44	5	1
FANOUT	16	24	13	21	5	1	30	17	11	15	2	1
Computed Proportions												
NOM/NOC	6.5	7.25	5.75	3.6	3.2	3.2	4.14	4.36	5.07	17.5	10.5	10.5
LOC/NOM	6.26	5.93	5.17	4.55	3.81	1.75	4.67	4.25	3.50	8.2	5.14	3.61
CYCLO/LOC	0.25	0.27	0.25	0.31	0.14	0.32	0.29	0.31	0.33	0.25	0.19	0.27
CALLS/NOM	1.03	1.0	0.73	1.27	1.	0.12	0.74	0.41	0.19	1.25	0.23	0.04
FANOUT/CALLS	0.59	0.82	0.76	0.91	0.31	0.5	0.69	0.85	0.78	0.34	0.4	1.0

The analysis takes in consideration the metrics, a qualitative code analysis, questionnaire answers, students' observations, development time and the author's observations during the development. The following subsections present a detailed analysis of each case study.

5.1   Scenario 1 - Command-Line Parameters Mapping

Analyzing the absolute number of lines of code on Table 2, it is possible to verify that the lines of code increased a little comparing the solution without frameworks with the solution using the traditional framework. The solution using the metadata-based framework has the lower number of lines of code, even if seven additional ones used for annotations were considered.

The Intrinsic Operation Complexity (CYCLO/LOC) do not change much among the implementations, but considering the reduction in the lines of code, it is not the best metric to evaluate the solution's complexity. Since the quantity of methods remains more stable, the complexity per method is probably a better indicator. Calculating the Cyclomatic Complexity per Method of each solution, it is possible to observe that the solution with metadata has less value.

According to the development times presented in Table 1, the solution using the traditional framework was the fastest to implement followed by the one using the metadata-based framework. From the students notes and from author's observations, the following factors slowed down the development in Stage 5: (a) the framework was hard to understand and did not support the mapping of most of the situations; (b) the exceptions did not point out where were the problems; and (c) some framework exceptions had the same name of application exceptions in the test, which was a fact that took some time for the students to perceive.

According to Table 1, the solution without framework was considered easier to develop but the students recognized that it demanded a lot of manual work. Despite the second solution had been the fastest to implement, the students had the feeling that the framework did not help and increased the development complexity. The metadata-based framework was considered even more complex to understand and with a development difficulty similar to the second solution, but it was considered easier to maintain.

From the first to the second solution, the development time was reduced despite the framework being considered hard to understand and the solution having more lines of code and cyclomatic complexity. This fact can be assigned to the guidance that the framework usage provided for the developers to design the solution's structure.

In the metadata-based framework, the lean and less complex source code did not offset difficulty to understand and use the framework. The framework made difficult the development since the mapping functionalities did not support the application needs. In the case study, for seven properties to be mapped, only three could be mapped using only the metadata.

Observing the implementation, it is possible to notice that using the metadata-based framework, the implementation of some mapping methods was not necessary since the metadata was enough for the framework to execute the translation. This reduction of effort can be perceived in the metrics by the reduction of complexity and lines of code, but it was not enough to reduce the development time. Following this logic, if the framework supported more mapping functionalities only through metadata configuration, those benefits probably would be higher, consequently reducing the development time.

Another difficulty highlighted in the student's comments was the unclear messages in the exceptions threw by the framework. Those messages did not pointed out what was wrong in the metadata configuration, which hindered in the debug, taking a considerable piece of the development time away.

5.2   Scenario 2 - Method Invocation Constraints

Following the implementation's lines of code evolution on Table 2, it is possible to notice that using the frameworks they got reduced. Even considering the lines of code with annotations, that sums 21, the metadata-based framework is the shortest solution. The complexity was reduced using both frameworks, since they work with configurations and eliminate the implementation of rules of the application code.

In all the solutions, a proxy was used to implement the validation. In the first solution the proxy was implemented manually and used a lot of operation calls to implement the required validations in each method, which explains the coupling metrics. The number of calls in the solution that used the traditional framework were concentrated in the method that configured the proxy, which invoke a great number of operations on the framework classes. In the solution that uses annotations, only one call to a framework class was needed since it configured the validations based on annotations on the interface. Due to those annotations, this interface had with the framework a semantic coupling, which is not addressed by the metrics.

A large difference in the development time between the implementations can be found in Table 1. According to the students, the code creation was hard-working in the Stage 1 due to a lot of specifications for each method validation that did not allow code reuse. The creation of the method context validation was pointed by the students as specifically hard.

In the second solution, developed with the traditional framework, the students pointed the framework out as one great difficulty. This fact can be verified in Table 1. The lack of documentation for some features made necessary the consultation of the framework authors during the development. The framework also did not implement correctly the functionalities for method context validation, and consequently five unit tests were unable to execute successfully. The solution was not flexible to allow an extension to workaround this problem.

The students considered the implementation using the metadata-based framework easier and indeed the development time was significantly smaller. According to the group, the annotation names made them intuitive to use. They also felt that the code became a little polluted with the annotations, but they recognize that it was worth for the other benefits.

The development strategies used in Stage 1 and in Stage 3 were completely different. In the first solution the proxy implemented the validation rules in each method, using conditional rules to identify the invalid invocations, which explains its higher cyclomatic complexity. In the second solution, the proxy was created by the framework and configured by the application invoking methods to set the constraints in the proxy class. This configuration did not demand conditional logic, which reduced the solution complexity.

The solution implementation using the traditional framework had some problems that impacted in the development time, such as the framework lack of documentation and missing functionalities. The present work's author, which followed the implementation, judges that without those problems, the team probably would not had reached closer to the last development time.

The coupling had a remarkable difference between the second solution and the one that uses metadata. The source code that created the proxy using the traditional framework were completely dependent on that application class. Contrasting to this, in the third solution this source code was independent from the application class. For instance, in an application whose method invocations should be validated, using the metadata-based framework it would be possible to reuse the code for proxy creation for all classes. Notwithstanding, that would not be true using the traditional framework.

5.3   Scenario 3 - Stock Exchange Events

The size and complexity metrics do not change much among the implementations according to Table 2. The solution with the traditional framework has the lowest lines of code number, but amounts of unused code were found in the third solution. It was a student's attempt to implement the event representation that was not cleaned when another alternative was chosen.

The coupling is a characteristic that observing the metrics clearly changes among the implementations. The use of interfaces provided by the framework reduced the coupling between the event generator and event handlers. The metadata usage reduced even more this coupling, making easier changes in both sides. Fig. 2 presents the blueprint complexity [27] for the three developed solutions. The dark blue edges, that represent method invocations among classes, are clearly reduced following the implementations.

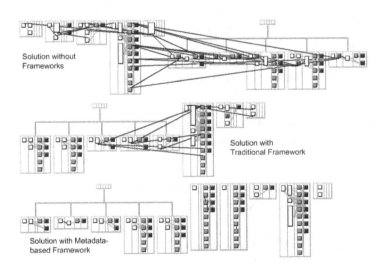

Fig. 2. Class Blueprint from the three implementations

It was expected that the complexity be reduced using the metadata-based framework, due to rules configuration using annotations. The students that implemented that solution misuse the framework and did not use the annotations to receive only the events with the desired property values.

Observing Table 1, the solution with the traditional framework was the one with the littlest development time. According to the student's observations, the framework guided the implementation using its interfaces in the application classes and the solution programming was simple.

In opposition to this, the implementation using the metadata-based framework was interfered by many problems that occurred. One of the problems was related to the environment configuration. The metadata-based framework version used an external library to create the dynamic proxies and that library was not included in the path of the project template used by the students. The exception thrown by the framework was not clear about this and it took some time for the students to perceive that a library was missing. The time used to copy the files and configure the project was not considered in the development time.

According to the framework documentation, for attributes in event classes, wrapper classes must be used instead of primitive types. For instance, Integer should be used instead of int. The students that implemented the solution did not attend to this and used primitive types. The framework did not throw an error and simply did not populate that attributes in the event. The group expended a long time finding out what was wrong.

The consequences of this fact can be observed in Table 1. For the metadata-based framework, the framework usage was considered more difficult than the application development itself. For the second implementation no exceptional fact was observed or reported by students, and even so they felt some difficulty in it. The implementation itself was not complicated, so the solution without frameworks, despite taking more time, was considered easy to create.

In this case study, the use of metadata did not reduce the amount of source code which should be developed. The reduction in complexity could not be evaluated since the framework functionalities that could impact on this were not used on the solution. Contrasting this, the reduction on the coupling between the event generator and its handler could be clearly observed by the metrics and the views.

The problems that occurred showed that using a metadata-based framework the developers lose even more control over the processing flow. Unexpected situations that happen inside the framework classes, even due an application class misconfiguration, are hard to be identified and understood by the developers. It highlights the importance of a good error handling strategy implemented by the framework, to validate the class structure and metadata.

5.4   Scenario 4 – HTML Form Generator

According to Table 2, the size metrics reduced through the implementations, especially from the solution without framework to the solution using a traditional framework. If the 24 lines of code with annotations were considered in the last solution, the difference comparing to the second implementation is not so significant.

The solution without frameworks used many methods defined in the same class to generate the HTML form. The inFusion tool, used to generate the overview pyramid, pointed this class out as a God Class and found two Feature Envies inside it [28].

Using the traditional framework, the functionality implementation was reduced to one method that invokes the framework many times to configure the form specific characteristics. The last solution was similar, but those configurations are in annotations on the target class, reducing even more the method size. This also explains the reduction in the coupling metrics.

For this case study, the development time using metadata-based frameworks was almost the half of the time to create the solution using the traditional framework, as presented in Fig. 2. Without using the frameworks, the development time was really longer, which confirms that developing a graphical interface can be a time consuming task.

The only exceptional situation observed was in the use of the traditional framework, when the students did not observe in the documentation that the class attributes should be public and then they took some time to find out what was wrong. By the observations during the development, without that setback the development time would not be largely reduced.

According to the students' evaluation, presented in Table II, the solution without framework was hard to create and could be considered even harder to modify. A student noted that the solution was not flexible and impossible to be reused to generate another form. Comparing the answers in the table, it is possible to affirm that the implementations had a technical draw in difficulty to develop and to change. The major difference was in the framework understanding, which was reinforced by some student's observations. According to them, the traditional framework use configurations in imperative code that is not much intuitive. The opposite was stated for the metadata-based framework.

This case study illustrates how the solutions with similar size metrics, considering the annotation's lines of code, could have a great difference in the development time. This was the only case study that did not have great issues that could interfere in the development time in the implementation with both frameworks. The reason can be found in the following student's observations: it is more intuitive to define metadata declaratively close to the code element which it is referring to, than using imperative code and referencing the code elements using strings.

The coupling in this case study also reduced through the implementations. Using the metadata-based framework, the same method could be used to generate the HTML forms for different classes since the difference between them can be found in their defined metadata. Contrarily, using the traditional framework different methods must be used for different classes, since the configurations should be made inside the methods.

6   General Analysis

A first conclusion that can be draw based on the metrics and development time is that the frameworks, traditional or metadata-based, bring benefits in the application design and can increase the productivity in those scenarios. However, its usage is inadvisable when it does not fulfill the application's needs and it is not flexible enough to be adapted to them. The frameworks provide an easy way to reuse functionality among features of the same application and even among different ones. Besides, they guide the development providing a ready-to-use design structure to the application, which can reduce the development time even when the lines of code are almost the same comparing to a solution without their use.

A metadata-based framework can potentially provide a solution in which the developer can add metadata to the existent classes intuitively increasing productivity, as it happened in the groups 2 and 4. In contrast to this, as evidenced by groups 1 and 3, the use of a framework based on metadata do not guaranties a high productivity. Consistent with this, in group 1 the solution that used the metadata-based framework took more time even having less lines of code.

According to [29], the lack of an explicit control flow in applications which uses frameworks can difficult the developer's understanding of it. In frameworks that use the metadata-based approach, where the adaptations are based on the class metadata, this problem is even worst since the flow of control is even more implicit. Because of that, it is difficult to find errors related to their usage in applications. For instance, metadata configuration errors, such as a missing property or a misspelled string, are pretty hard to detect. This difficulty to find errors can be a bottleneck in the team productivity. Those facts can be observed in the implementations with the metadata-based framework in groups 1 and 3.

This evidence makes the error handling and metadata validation important features for a metadata-based framework. The error or warning messages should be designed to help the developer to find a misconfiguration. Those frameworks were not automatically good just for using metadata. Best practices valid for every piece of software, such as good naming and clear documentation, are also important in this context. Specific best practices, such as those presented in [10], are also important to make the framework more flexible enabling it to be adapted to the application needs.

An interesting fact that happened in the traditional framework's implementations was that three of them used a programmatic approach to set additional information about the application classes into the framework, in other words, metadata. In Group 1, the application class had to implement an interface that had methods to return additional metadata about the class. In groups 2 and 4, the framework main class provides methods to set information referencing the application class elements directly in the framework. If inexperienced students in framework development had chosen a solution based on metadata definition even without its knowledge, which might evidence that defining metadata in those scenarios is an intuitive approach.

Despite all other facts, a constant characteristic of the solutions that used the metadata-based frameworks is the coupling reduction, which can be confirmed in all case studies. The use of this kind of framework decouples the application classes from framework since the need for them to implement interfaces or extend a superclass from the framework is eliminated. Its use also decouples the client class that invokes the framework functionalities from the application class that is processed from the framework. However, it is important to notice that it still exist an indirectly or semantic coupling between the framework metadata definition and the application class, which was not addressed by the metrics [30]. The use of an external metadata strategy or domain annotations mapped to framework annotations [31] can help to reduce this semantic coupling.

Other benefits also can be achieved by the use of metadata-based frameworks, which depends on the framework's functionality and domain. When the framework manages to encapsulate features that must be implemented by the application using a traditional approach, it probably would reduce the complexity and the lines of code number in the application where it is applied.

7   Conclusion

This paper presents an evaluation of metadata-based frameworks usage based on an experiment. The experiment created a comparison basis for applications without frameworks, using traditional frameworks and using metadata-based frameworks for distinct scenarios. As a result, it was possible to assess benefits and drawbacks in the use of this approach. The analysis used object-oriented metrics, questionnaire answers, observations, source code analysis, and development time measurements to reach the conclusions.

Further studies can explore the use of metadata-based frameworks with more features for more complete applications. In these scenarios, it would be possible to explore other issues, such as the reuse provided among different functionalities. Other future works could aim on solutions to common needs of this kind of framework, such as exception handling on metadata reading.

We thank for the essential support of FAPESP (Fundação de Amparo à Pesquisa do Estado de São Paulo) to this research.

References

1. Johnson, R., Foote, B.: Designing reusable classes. Journal of Object-Oriented Programming 1(2), 22–35 (1988)
2. Pree, W.: Hot-spot-driven development. In: Building Application Frameworks: Object-Oriented Foundations of Frameworks Design, ch. 16, pp. 379–393. Wiley, New York (1999)
3. Doucet, F., Shukla, S., Gupta, R.: Introspection in system-level language frameworks: meta-level vs. Integrated. Source Design, Automation, and Test in Europe, 382–387 (2003)
4. Forman, I., Forman, N.: Java reflection in action. Greenwich, Manning Publications (2005)
5. Foote, B., Yoder, J.: Evolution, architecture, and metamorphosis. In: Pattern Languages of Program Design 2, ch. 13, pp. 295–314. Addison-Wesley Longman, Boston (1996)
6. Schwarz, D.: Peeking inside the box: attribute-oriented programming with Java 1.5 (2004), http://missingmanuals.com/pub/a/onjava/2004/06/30/insidebox1.html
7. JSR 175: a metadata facility for the java programming language (2003), http://www.jcp.org/en/jsr/detail?id=175
8. Chen, N.: Convention over configuration (2006), http://softwareengineering.vazexqi.com/files/pattern.html
9. Fernandes, C., Ribeiro, D., Guerra, E., Nakao, E.: XML, Annotations and Database: a Comparative Study of Metadata Definition Strategies for Frameworks. In: XML: Aplicações e Tecnologias Associadas, Vila do Conde, Portugal (2010)

10. Guerra, E., Souza, J., Fernandes, C.: A pattern language for metadata-based frameworks. In: Conference on Pattern Languages of Programs, 16, Chicago (2009)
11. Bauer, C., King, G.: Java persistence with hibernate. Greenwich, Manning Publ. (2006)
12. JSR 220: Enterprise JavaBeans 3.0 (2006),
    http://www.jcp.org/en/jsr/detail?id=220
13. Brown, D., Davis, C., Stanlick, S.: Struts 2 in action. Greenwich, Manning Publ. (2008)
14. JSR 222: Java Architecture for XML Binding (JAXB) 2.0
    (2006), http://jcp.org/en/jsr/detail?id=222
15. Miller, J.: Common language infrastructure annotated standard. Addison-Wesley, Boston (2003)
16. Rouvoy, R., Pessemier, N., Pawlak, R., Merle, P.: Using attribute-oriented programming to leverage fractal-based developments. In: International ECOOP Workshop on Fractal Component Model, 5, Nantes (2006)
17. Gamma, E., Helm, R., Johnson, R., Vlissides, J.: Design Patterns: Elements of Reusable Object-Oriented Software. Addison-Wesley (1994)
18. O'Brien, L.: Design patterns 15 years later: an interview with Erich Gamma, Richard Helm and Ralph Johnson (2009),
    http://www.informit.com/articles/article.aspx?p=1404056
19. Swingbean: aplicações Swing a Jato!, http://swingbean.sourceforge.net
20. Nucitelli, R., Guerra, E., Fernandes, C.: Parsing XML Documents in Java Using Annotations. In: XML: Aplicações e Tecnologias Associadas, Vila do Conde, Portugal (2010)
21. Zelkowitz, M., Wallace, D.: Experimental validation in software engineering. Information and Software Technology 39, 735–743 (1997)
22. Basili, V.: The role of experimentation in software engineering: past, current and future. In: Proceedings of the 18th International Conference on Software Engineering, pp. 442–449. IEEE Computer Society Press, Washington, DC (1996)
23. Soares, S., Borba, P.: Towards progressive and non-progressive implementation approaches evaluation. In: Proceedings of Experimental Software Engineering Latin American Workshop (2004)
24. Noël, R.: Evaluating Design Approaches in Extreme Programming. In: Proceedings of Experimental Software Engineering Latin American Workshop (2005)
25. Guerra, E., Fernandes, C., Silveira, F.: Architectural Patterns for Metadata-based Frameworks Usage. In: Proceedings of Conference on Pattern Languages of Programs, 17, Reno (2010)
26. Lanza, M., Ducasse, S.: Polymetric views: a lightweight visual approach to reverse engineering. IEEE Transactions on Software Engineering 29(9), 782–795 (2003)
27. Ducasse, S., Lanza, M.: The Class Blueprint: visually supporting the understanding of classes export. IEEE Transactions on Software Engineering 31(1), 75–90 (2005)
28. Lanza, M., Marinesco, R.: Object-Oriented Metrics in Practice - Using Software Metrics to Characterize, Evaluate, and Improve the Design of Object-Oriented Systems. Springer (2006)
29. Fayad, M., Schmidt, D., Johnson, R.: Application frameworks. In: Building Application Frameworks: Object-oriented Foundations of Frameworks Design, ch. 1, pp. 3–27. Wiley, New York (1999)
30. Guerra, E., Silveira, F., Fernandes, C.: Questioning traditional metrics for applications which uses metadata-based frameworks. In: Workshop on Assessment of Contemporary Modularization Techniques, Orlando, vol. 3 (2009)
31. Perillo, J., Guerra, E., Silva, J., Silveira, F., Fernandes, C.: Metadata Modularization Using Domain Annotations. In: Workshop on Assessment of Contemporary Modularization Techniques, Orlando, vol. 3 (2009)

A Flexible Model for Crosscutting Metadata-Based Frameworks

Eduardo Guerra[1], Eduardo Buarque[2], Clovis Fernandes[2], and Fábio Silveira[3]

[1] National Institute for Space Research (INPE) - Laboratory of Computing and Applied Mathematics (LAC) - P.O. Box 515 – 12227-010 - São José dos Campos, SP, Brazil
[2] Aeronautical Institute of Technology (ITA) - Praça Marechal Eduardo Gomes, 50 CEP 12.228-900 - São José dos Campos – SP, Brazil
[3] Federal University of São Paulo (UNIFESP)
Rua Talim, 330 - CEP 12231-280 - São José dos Campos – SP, Brazil
guerraem@gmail.com, skyedu_b@yahoo.com, clovistf@uol.com.br,
fsilveira@unifesp.br

Abstract. Frameworks aims to provide a reusable functionality and structure to be used in distinct applications. Aspect-oriented frameworks address crosscutting concerns and provide ways to attach itself in the application in a transparent way. However, using aspects the variations in the behavior can only be customized by aspect inheritance, which can increase exponentially the number of aspects and difficult the pointcut management. This paper proposes a flexible model which combines techniques for the insertion of crosscutting functionality with the structure of a metadata-based framework. This model allows (a) the maintenance of the class obliviousness, (b) the independence of the crosscutting technology and (c) the framework customization by composition. Additionally, the paper presents SystemGlue, which is a crosscutting framework that implements the proposed concepts. A modularity analysis was performed in an application that uses this framework to evaluate if the objectives were achieved.

Keywords: framework, aspect orientation, metadata, software design, software architecture.

1 Introduction

A framework is a set of classes that supports reuses at larger granularity. It defines an object-oriented abstract design for a particular kind of application which does not enable only source code reuse, but also design reuse [1]. The framework's abstract structure can be filled with its own classes or application-specific ones, providing flexibility for the developer to adapt its behavior to each application. Besides flexibility, a good framework also increases the team productivity and makes application maintenance easier [2] [3].

A framework can contain points, called hot spots, where applications can customize their behavior [4]. They represent domain pieces that can change among applications. Points that cannot be changed are called frozen spots, which usually

B. Murgante et al. (Eds.): ICCSA 2013, Part II, LNCS 7972, pp. 391–407, 2013.
© Springer-Verlag Berlin Heidelberg 2013

define the framework's general architecture, which consists in its basic components and the relationships between them. There are basically two different types of hot spots, that respectively uses inheritance and composition to enable the application to add behavior [5]. The use of composition allows the creation of black box frameworks [6], which scales better and provides a more flexible structure than the ones that use inheritance.

An aspect-oriented framework [7], like object-oriented frameworks, can be considered a incomplete reusable application that must be instantiated to create a concrete software. They can be classified as cross-cutting framework, which implement non-functional requirements, and application frameworks, that implement business rules. To specialize the framework behavior to a target application, an abstract aspect should be specialized, implementing the abstract methods and configuring the desired pointcuts. This structure based on inheritance is not suitable for frameworks with a large number of possible behavior variations [8]. For instance, the number of necessary aspects can grow exponentially based on the number of possible variabilities.

This paper introduces a flexible model that can be used to create aspect-oriented frameworks which uses a metadata-based processing to eliminate the drawbacks of the approach based on aspect inheritance. The present work proposes the usage of metadata to configure framework variabilities, an internal structure to enable composition and a metadata definition technique to maintain obliviousness. To evaluate this properties a complex framework for system integration, ready to be used in production environments, was implemented and used in a case study. Based on that, a modularity analysis was performed to verify the proposed model properties.

2   Frameworks

This section aims to describe different kinds of frameworks highlighting their main characteristics and the way that they provide behavior adaptation. In subsection 2.1, the mechanisms based on inheritance and composition in object-oriented framework are described. Next, subsection 2.2 presents the aspect-oriented frameworks and the drawbacks of using only inheritance to implement the behavior variabilities. Further, subsection 2.3 introduces the metadata-based frameworks, how they work and how they are internally structured.

2.1   Object-Oriented Frameworks

A framework can be considered an incomplete software with some points that can be specialized to add application-specific behavior, consisting in a set of classes that represents an abstract design for a family of related problems. It provides a set of abstract classes that must be extended and composed with others to create a concrete and executable application. The specialized classes can be application-specific or taken from a class library, usually provided along with the framework [1].

Another important characteristic of a framework is the inversion of control [3, 9]. A framework's runtime architecture enables the definition of processing steps that can call applications handlers. This allows the framework to determine which set of application methods should be called in response to an external event.

An abstract class can define abstract methods that are invoked from a more general method in the same class. Those general methods are called template methods [10] and they define the skeleton of an algorithm in an operation, deferring to subclasses the redefinition of certain steps without changing the algorithm's structure. Those abstract methods are called hook methods [5] and must be implemented in the subclasses for framework adaptation [3].

The main framework class can also have some instance variables and delegate part of the execution to them. Those instances must obey a known protocol, extending an abstract class or implementing an interface, for the framework to be able to invoke methods on them. In this case, the hook methods invoked by the template methods are located in other classes, which are called hook classes. Thus, for framework adaptation it is not necessary to extend the framework main class and the developer must only change the instance that composes it. That instance can be taken from the framework's own class library or can be application-specific.

A framework is neither pure blackbox nor pure whitebox. The whitebox strategy is more difficult to use, because the developer must know details about the framework's internal structure. It is also more flexible, because it gives more freedom for choosing what should be overridden by the subclass. The blackbox strategy hides the implementation details and composes the application functionality with hook classes. It is also less flexible, since the application can interfere only in certain points. In whitebox, the implementation must be chosen when the class is instantiated, and in blackbox it can be changed later. A pattern language for framework evolution [6] suggests that a framework should start being whitebox, which is more flexible, and when the extension points became more clear, it should evolve to a blackbox strategy.

2.2   Aspect-Oriented Frameworks

Aspect-oriented programming [12] is a programing paradigm, whose main goal is to modularize cross-cutting concerns. The adoption of this paradigm by the software development community is still happening and it is usually used encapsulated inside tools and frameworks, such as JBoss Application Server [13, 14] and Spring [15,16].

The modularization capabilities of aspect-oriented programing can be used to improve object-oriented frameworks. Using aspects, it is possible to add features in an existent object-oriented framework without the modification of the original source-code [17]. This modularization of framework's features brings other benefits such as functionalities that can be easily disabled and potentially used in other contexts.

Other possibility is the creation of an aspect-oriented framework [7] that can be classified as cross-cutting framework, which implement non-functional requirements, and application frameworks, that implement business rules. Like object-oriented frameworks, those can be considered a incomplete reusable application that must be instantiated to create a concrete software. A framework's abstract aspect must be specialized to be weaved in the desired pointcuts and to add implementation in the hook methods, like represented in Fig. 1.

An abstract aspect cannot use composition in extension points, because its invocation is transparent for the application, which do not have direct access to the aspect to set the hook classes. The composition can be used in this context only if the hook classes are instantiated using a Factory Method [10], which is a type of hook method.

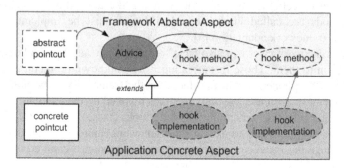

Fig. 1. The structure of an aspect-oriented framework.

A study about existent aspect-oriented frameworks [7] analyzed 13 frameworks and all of them contains a small number of functional variabilities. That can indicate that the existent model does not scale for a large number of possible behavior variations.

Indeed, based on this structure, every variability in those frameworks should be modeled as hook methods in the main abstract aspect. For variabilities whose behaviors can be combined the number of possible advices grows exponentially with the number of variabilities [8]. The concrete pointcuts also became granular and hard to manage.

2.3   Metadata-Based Frameworks

The framework structures has evolved and recent ones make use of introspection [18][19] to access at runtime the application classes metadata, like their superclasses, methods and attributes. As a result, it eliminates the need for the application classes to be coupled with the framework abstract classes and interfaces. The framework can, for instance, search in the class structure for the right method to invoke. The use of this technique provides more flexibility to the application, since the framework reads dynamically the classes structure allowing them to evolve more easily [20].

When a framework uses reflection [20][21] to access the class elements and execute its responsibilities, sometimes the class intrinsic information is not enough. If framework behavior should differ for different classes, methods or attributes, it is necessary to add a more specific meta-information to enable differentiation. For some domains, it is possible to use marking interfaces, like Serializable in Java Platform, or naming conventions [22], like in Ruby on Rails [23]. But those strategies can be used only for a limited information amount and are not suitable for situations that need more data.

Metadata-based frameworks can be defined as frameworks that process their logic based on the metadata of the classes whose instances they are working with [24]. In those, the developer must define into application classes additional domain-specific or application-specific metadata to be consumed and processed by the framework. The use of metadata changes the way frameworks are build and how they are used by software developers [25].

The developer's perspective in the use of those frameworks has a stronger interaction with metadata configuration than in method invocation or class specialization. In traditional frameworks, the developer must extend its classes, implement its interfaces and create hook classes for the behavior adaptation. He also have to create instances of those classes, setting information and hook class instances. Using metadata-based frameworks, programming focus is on declarative metadata configuration and the method invocation in framework classes is smaller and localized.

The basic processing in a metadata-based framework consists in the metadata reading from the target object, followed by its processing. In this process, the metadata read is used to adapt framework behavior and to apply introspection to access and modify the application object.

In [24], a pattern language for metadata-based frameworks was described, addressing the main issues about how to structure internally metadata-based frameworks. The patterns Delegate Metadata Reader and Metadata Processor combined enable the extension of the metadata schema, providing a solution that allow the insertion of new application-specific hook classes in the framework execution. This solution is used in APIs like Bean Validation [26] and frameworks like JColtrane [27].

The metadata consumed by the framework can be defined in different ways. Naming conventions [22] uses patterns in the name of classes and methods that has a special meaning for the framework. To exemplify this there are the Java Beans specification [28], which use method names beginning with 'get' and 'set', and the JUnit 3 [29], which interprets methods beginning with 'test' as test cases implementation. Ruby on Rails [23] is an example of a framework known by the naming conventions usage.

Conventions usage can save a lot of configurations but it has a limited expressiveness. For some scenarios the metadata needed are more complex and naming conventions are not enough. An alternative can be setting the information programmatically in the framework, but it is not used in practice in the majority of the frameworks. Another option is metadata definition in external sources, like XML files and databases. The possibility to modify the metadata at deploy-time or even at runtime without recompile the code is an advantage of this type of definition. However, the definition is more verbose because it has to reference and identify program elements. Furthermore, the distance that configuration keeps from the source code is not intuitive for some developers.

Another alternative that is becoming popular in the software community is the use of code annotations, that is supported by some programming languages like Java [30] and C# [31]. Using this technique the developer can add custom metadata elements directly into the class source code, keeping this definition less verbose and closer to the source code. The use of code annotations is called attribute-oriented programing [32].

Prior studies report a successful use of attribute-oriented programming in different contexts [33], like serialization, web service endpoints and interface to databases. It is also used in a fractal component model implementation [34] and in conjunction with Model-driven Development [35]. A recent experiment about the usage of metadata revels that the use of these frameworks reduces the application coupling and can increase the team productivity [36].

3   Proposed Model

This section presents the proposed model for metadata-based crosscutting frameworks. The word "crosscutting" was used instead of "aspect-oriented" since the model can also be applied to other implementation strategies like the use of dynamic proxies [19] and composition filters [37]. For simplification, in the model description the strategies are referenced as aspects, unless the differentiation is relevant in the context.

This model's goal is to provide a flexible structure for a crosscutting framework to be able to deal with a large number of variabilities. Other characteristics considered were the preservation of the class obliviousness and an easy framework adaptation for distinct architectures. The following subsections present the proposed practices to achieve these goals.

3.1   Metadata for Behavior Adaptation

Since an aspect can intercept the execution of different classes without their knowledge, it is hard to differentiate the execution for each one. The main strategy of the existent aspect-oriented frameworks for behavior differentiation is to provide different aspects for each possibility [7]. These aspects inherit from a framework abstract aspect specializing its behavior. As presented in the previous section, this model has serious drawbacks for a large number of variabilities, specially when they differ in a granular way among the classes and methods.

The foundation of the proposed model is to use class metadata to differentiate framework behavior. In aspects, the pointcuts are already defined based on class metadata, like class package, class name, method name, method return, parameter types and others. It can even use domain-specific or application-specific metadata defined in code annotations. Despite metadata defined can also be used for pointcut definition, this model proposes that this metadata should be consumed by the framework to enable differentiation of the execution logic among the classes.

It should define which variations are possible in the framework execution and provide a metadata schema to enable this differentiation. The metadata can be defined using code annotations, XML files, databases, code conventions or using a combination of this strategies. When a method execution is intercepted, the framework should read its intrinsic and domain-specific metadata and use it to parameterize its execution.

Among the benefits of this approach, it is possible to highlight that the use of metadata enable the existence of a single framework aspect. That aspect should be specialized only to define a more specific pointcut where it should be applied in the target application.

3.2   Intercepting Technology Independence

One of the requirements that should considered in the construction of a framework is the adaptability for different architectures and applications. The actual aspect implementations in Java language are not a standard adopted by all applications. Examples of aspect implementations in Java are AspectJ [15], Spring AOP [16] and JBoss AOP [38]. Additionally, other solutions provide functionality that allow the insertion of components that can intercept the execution of a component method, such as dynamic proxies [19], EJB 3 interceptors [39] and CDI interceptors [40].

To enable framework independence about how execution should be intercepted in the architecture, this model proposes the encapsulation of the framework main functionality in a component, like illustrated in Fig. 2. This component can be invoked by different kinds of software components which can intercept the application execution, such as aspects, filters and proxies.

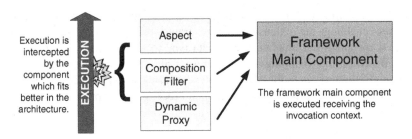

Fig. 2. Independence of the framework and intercepting component

This practice allow the application to choose how the crosscutting framework should be attached to it. It makes the framework invocations more flexible and enable it to adapt easily to distinct environments. The authors consider this practice is advisable, not only to frameworks based on metadata, but for every crosscutting framework.

3.3   Metadata Extension

As presented in the previous sections, one of the weaknesses of the current model adopted for aspect-oriented frameworks relies in the usage of aspect inheritance for behavior specialization. By using metadata for framework adaptation (subsection 3.1) and decoupling the main component from the execution interception (subsection 3.2), it is possible to use a model based on composition.

Fig. 3 illustrates the process proposed in this model. When the framework main component receives an invocation by one of the intercepting components, it invokes a class responsible for metadata reading that returns a representation of that information. This representation, called Metadata Container [24], can contain only the metadata retrieved, or moreover classes for which part of the execution can be delegated. These classes, created based on the class metadata, are called Metadata Processors [24].

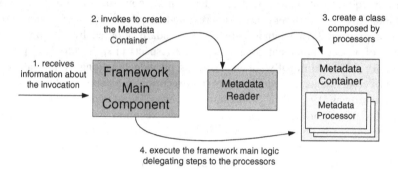

Fig. 3. Creation and execution of metadata processors

Using this structure, it is possible to create application-specific metadata processors, enabling the extension of the framework functionality. To make it possible, a mapping that links each metadata type to a class that reads it should be created. Based on that mapping, the class responsible for reading metadata delegate the reading of these types to the Metadata Reader Delegate classes [24]. These classes are responsible for the creation of the Metadata Processors, like presented on Fig. 4.

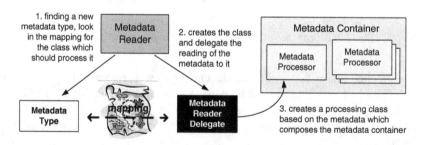

Fig. 4. Delegating the metadata reading

If an application needs to extend the framework functionality, the first step is to create a new metadata type, which can be for instance an annotation or an XML element. The next step is to create the reader delegate class and map it to the created metadata type. Further, the metadata processor with the desired behavior should also be implemented and returned as the result of the reader delegate execution. Since the processor would compose the metadata container, the execution of framework should be delegated to it.

This approach provides a solution that enables the extension of the framework behavior using composition. It allows processors to be combined in a more natural way without an explosion on the number of classes to support the combination of variabilities.

3.4   Domain Annotations

Especially when the metadata is defined using code annotations, the application class receives directly information about the framework concern. This creates a semantic coupling between the class and the framework, which reduces the application modularity.

To enable the usage of attribute-oriented programming without compromising the obliviousness, the present model proposes the use of domain annotations. The domain annotation concept was introduced by [41] in an attempt to introduce annotations in the context of Domain-driven Design [42]. The main idea is to represent domain concepts using annotations and not others related to non-functional and crosscutting concerns.

This model proposes the mapping of domain annotations to framework annotations, providing a decoupling of the application classes with the framework metadata. This mapping represents a translation of how the framework should deal with a class or a method which represents a given domain concept. This mapping also brings other benefits like a better modularization [43] and a reduction in the duplication of configurations [44].

Fig. 5 illustrates this mapping. The framework annotation should annotate the domain annotation instead of the class directly. The mapping can be called dynamic when the framework is prepared to search at runtime for its annotations inside other annotations. The mapping is static if a tool change the domain annotations to the framework annotations at compile time. For instance, Daileon is a tool which provides a function library that facilitate the implementation of a dynamic mapping and a tool for the static mapping [45].

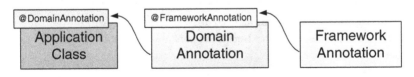

Fig. 5. Domain annotations mapping

4   Implemented Framework – Esfinge SystemGlue

The software developed to demonstrate the proposed model in the present work is Esfinge SystemGlue [46], which is an open-source framework which aims to enable the creation of distinct integration profiles for a given application. It was developed to solve a problem in a real application in which different clients needs to invoke distinct methods to integrate with their systems. The framework had already been functionally tested using ClassMock [47] and can be considered ready to be used in a production environment.

The next subsections describes the framework functionalities, its strategy for metadata definition and its internal structure.

4.1 General View

SystemGlue aims to provide an structure that allows the application to configure distinct integration profiles, enabling the invocation of different functionality according to the context. It uses metadata to define what should be executed after or before an application method execution. It supports method invocation, scheduling and message sending that can be executed based on conditions and asynchronously. The framework also uses metadata and code conventions to map the parameters and the returns among the invocations. The metadata definition can be defined in a flexible way using a combination of annotations and XML documents.

The following example exemplify the usage of annotations to configure the execution of functionality before and after a method execution. While the methods are invoked, their parameters and returns can be mapped and used among subsequent executions based on their names, which can be defined respectively by the parameter annotation @Param and the method annotation @ReturnName.

SystemGlue metadata configuration with annotations.

```
@Executions ({
    @Execute(clazz=InteligenceIntegration.class,
        method="getTargetInfo",
        when=ExecutionMoment.BEFORE,
        rule="order.targets.size==0") ,
    @Execute(clazz= UnitsIntegration.class,
        method="sendOrder",
        when = ExecutionMoment.AFTER,
async = true)
})
public void saveOrder(@Param("order") Order order){
    //core functionality implementation
}
```

4.2 Flexible Metadata Definition

The use of framework annotations direct in the application methods can be useful for executing functionality which should always be invoked. Since to change the code annotations the code should be re-compiled, it is not a good solution to allow the configurations to be changed for distinct integration profiles.

SystemGlue also supports the metadata definition using XML files. This approach allow a more decoupled definition, which is more suitable for define metadata in situations where more than one metadata set is necessary for one class [48], which is the case for integration profiles. The next code presents an example of the same metadata defined in the previus example represented in an XML file.

SystemGlue metadata configuration using XML.

```
<systemglue>
 <class name="expl.OrderService">
  <method name="saveOrder" params="expl.Order">
   <execute class="expl.InteligenceIntegration"
       method="getTargetInfo" when="BEFORE"
       rule="order.targets.size == 0"/>
   <execute class="expl.UnitsIntegration"
     when="AFTER" method="sendOrder" async="true"/>
   <execute/>
  </method>
 </class>
</systemglue>
```

For the framework to load an XML file it is necessary to invoke the method loadXMLFile() in the class MetadataRepository. This file can define metadata for more than one class and a class can have metadata defined in more than one file.

A drawback of the presented approaches is that if different methods needs the same metadata configuration, the code to define it should be duplicated. It reduces maintainability making difficult general modifications in the metadata definition.

To avoid this problem, an alternative for metadata definition is the usage of domain annotations [41], which represents concepts related to the application domain and are defined by the application. These annotations can be mapped to the SystemGlue metadata using annotations or in the XML file, providing an indirect configuration. Considering that the framework functionality is crosscutting, the domain annotations preserve the classes obliviousness [43], since they would not contain information about a crosscutting concern.

Next code listing presents an example of the domain annotation mapping using annotations. The SystemGlue annotations are used in the domain annotation @OrderModification instead of directly on the class method. The framework recognize this indirect configuration and add this metadata to all methods configured with it. This practice facilitate changes, since the modification of the framework annotations would affect all methods annotated with the domain annotation. A domain annotation can annotate other domain annotation providing an specialization mechanism.

SystemGlue configuration of domain annotations.

```
//annotation definition
@Executions({
  @Execute(clazz = InteligenceIntegration.class,
    when=ExecutionMoment.AFTER, method="getTargetInfo",
    rule="order.targets.size==0"),
  @Execute (clazz= UnitsIntegration.class,
    when = ExecutionMoment.AFTER, method="sendOrder",
    async = true)
})
```

```
public @interface OrderModification{}

//method definition
@OrderModification
public void saveOrder(@Param("order") Order o){}
```

The use of domain annotations can also be combined with XML definition. The metadata configuration can refer to an annotation instead of the method directly. Using this approach, the annotation can be simply defined without framework annotations. Next code listing presents an instance of the domain annotation metadata definition in the XML.

Referencing the domain annotation in the XML file.

```
<systemglue>
  <annotation name="expl.OrderModification">
    <execute class="expl.InteligenceIntegration"
        method="getTargetInfo" when="BEFORE"
        rule="order.targets.size == 0"/>
    <execute class="expl.UnitsIntegration"
        when="AFTER" method="sendOrder" async="true"/>
  </annotation>
<systemglue>
```

It is important to highlight that any combination of these techniques can be used together in the same method to define the invocation of distinct functionality. Despite the advantages and drawbacks, each one is more applicable to a different scenario.

4.3   SystemGlue Internal Structure

One of the requirements considered in the construction of SystemGlue is that it should to be adaptable for different architectures. To enable SystemGlue functionality to be inserted in the most natural way to the application architecture, the main functionality is encapsulated in a component, named SystemGlueExecutor, which does not crosscut the application functionality.

Other components, such as dynamic proxies or aspects, are responsible to intercept the application methods invocation and delegate the execution to the main component. This flexibility is important to allow the execution to be inserted in a way which fits better in the application architecture. SystemGlue provides implementation of reflection dynamic proxies [19], which creates proxies based only on interfaces, and CGLib proxies [49], which supports proxies based on classes. The framework was also tested using Spring AOP [15] and EJB3 Interceptors [39], however these implementations are not provided with the framework to avoid more dependencies.

The framework follows the basic structure proposed in the section 3, as presented in Fig. 6. The SystemGlue main component retrieves the metadata container from a

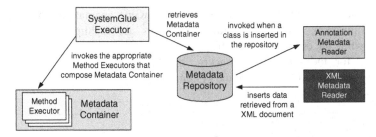

Fig. 6. SystemGlue internal organization

metadata repository when it receives an invocation. The repository is populated with information retrieved from XML files and from the class annotations. The metadata container is composed by instances of the type MethodExecutor, which represents the executions that should be made after and before the application method.

5   Modularity Analysis

This section presents an evaluation of the model modularity, by analyzing a case study that used Esfinge SystemGlue framework and verifying if it was able to achieve the proposed characteristics. As a tool to this analysis, this work used a Dependency Structure Matrix [50], which is a matrix that basically shows the dependence between all the elements in a given software.

The interpretation of a DSM is made by noticing that both rows and columns have the same information: they represent a complete list of system entities whose dependence should be mapped. Each cell of the matrix represent the number of dependences between the entity represented by the line to the entity represented in the column.

To evaluate if the model allows the fulfillment of the modularity requirements, a fictitious case study was prepared with an application that plays the role of a Hospital ERP and three other applications representing softwares that integrate with it. It uses Esfinge SystemGlue to integrate the applications by using the domain annotations functionality. Figure 7 shows the DSM created based on the developed software.

The domain annotations are in the package *br.com.lab.integration* (C, D, E, G, H, I and J), classes responsible to activate the main features of the application are in the package *br.com.lab.controller* (B and F), SystemGlue's annotations are in the package *net.sf.systemglue.annotations* (K, L, M, N, O and P), and the remaining packages represent the classes responsible for the integration functionality (Q, R and S).

Based on the DSM extracted from the case study, it is possible to draw some conclusions about the system modularity. The main application classes only depends on the domain annotations. This dependence is highlighted by the yellow rectangles. Since the domain annotations express domain information, the application classes does not contain even a semantic dependence with integration concerns.

The domain annotations depend on the SystemGlue annotations to define each one's configuration. The SystemGlue annotations are highlighted by the green rectangle, while the dependences are highlighted by the orange rectangles. The classes responsible for the integration concern, highlighted in the blue rectangle, are completely decoupled of the rest of the system.

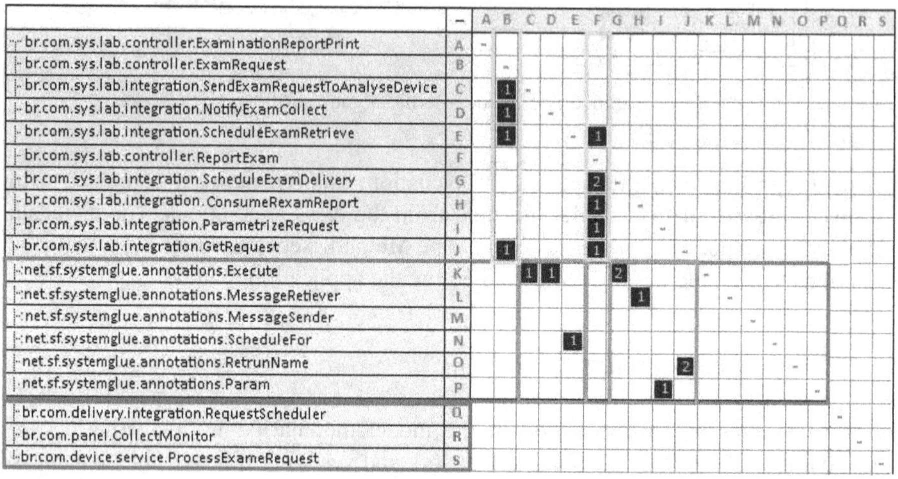

		-	A	B	C	D	E	F	G	H	I	J	K	L	M	N	O	P	Q	R	S
br.com.sys.lab.controller.ExaminationReportPrint	A	-																			
br.com.sys.lab.controller.ExamRequest	B		-																		
br.com.sys.lab.integration.SendExamRequestToAnalyseDevice	C		1	-																	
br.com.sys.lab.integration.NotifyExamCollect	D		1		-																
br.com.sys.lab.integration.ScheduleExamRetrieve	E		1			-		1													
br.com.sys.lab.controller.ReportExam	F						-														
br.com.sys.lab.integration.ScheduleExamDelivery	G						2	-													
br.com.sys.lab.integration.ConsumeRexamReport	H						1		-												
br.com.sys.lab.integration.ParametrizeRequest	I						1			-											
br.com.sys.lab.integration.GetRequest	J		1				1				-										
net.sf.systemglue.annotations.Execute	K			1	1			2				-									
net.sf.systemglue.annotations.MessageRetiever	L									1			-								
net.sf.systemglue.annotations.MessageSender	M													-							
net.sf.systemglue.annotations.ScheduleFor	N								1						-						
net.sf.systemglue.annotations.RetrunName	O													2		-					
net.sf.systemglue.annotations.Param	P												1				-				
br.com.delivery.integration.RequestScheduler	Q																	-			
br.com.panel.CollectMonitor	R																		-		
br.com.device.service.ProcessExameRequest	S																			-	

Fig. 7. Modularity analysis using a DSM

Hence, the framework enables configuration profiles on metadata integrations with domain annotations, since the application classes deal with the main features and have no syntactic or semantic dependencies of classes that perform the integrations. Then SystemGlue has the responsibility to activate the points of integration. Based on that it is possible to conclude that the proposed model allows the fulfillment of this modularity requirements.

6   Conclusions

This work proposes a new model for crosscutting frameworks which enables it to deal with a high number of behavior variations. It is probably not suitable for domains with a small number of behavior variations. It proposes the use of metadata to enable the framework to use composition as the strategy for behavior extension. The model also proposes techniques for decoupling the component responsible for the method interception and the use of domain annotations to enable the usage of attribute-oriented programming without compromising the obliviousness.

This model was used to build a framework named SystemGlue which aims to provide a flexible structure to enable the creation of distinct integration profiles for one application. It naturally deals with a high number of variations, including the possibilities of parameters and return mapping and the combinations of functions to be invoked before and after the application method execution. The integration functions can also be invoked conditionally and be executed asynchronously according to the configurations. The framework also provide flexible alternatives for metadata configuration and for attaching it in an architecture. A modularity analysis was performed in a case study that instantiated SystemGlue and the decoupling model requirements were evaluated.

We thank for the essential support of FAPESP (Fundação de Amparo à Pesquisa do Estado de São Paulo) to this research.

References

1. Johnson, R., Foote, B.: Designing reusable classes. Journal of Object-Oriented Programming 1(2), 22–35 (1988)
2. Wirfs-Brock, R., Johnson, R.: Surveying current research in object-oriented design. Communications of the ACM 33(9), 104–124 (1990)
3. Fayad, M., Schmidt, D., Johnson, R.: Application frameworks. In: Building Application Frameworks: Object-oriented Foundations of Frameworks Design, ch. 1, pp. 3–27. Wiley, New York (1999)
4. Pree, W.: Design Patterns for Object-Oriented Software Development. Addison Wesley/ACM Press (1995)
5. Pree, W.: Hot-spot-driven development. In: Building Application Frameworks: Object-Oriented Foundations of Frameworks Design, ch. 16, pp. 379–393. Wiley, New York (1999)
6. Don, R., Ralph, J.: Evolving Frameworks: A Pattern Language for Developing Object-Oriented Frameworks. In: Proceedings of the Third Conference on Pattern Languages of Programming (1996)
7. Camargo, V., Masiero, P.: Frameworks Orientados a Aspectos. In: 2005 Proceedings of the XIX Simpósio Brasileiro de Engenharia de Software – SBES 2005, Uberlândia, pp. 200–216 (2005)
8. Guerra, E., Silva, J., Silveira, F., Fernandes, C.: Using Metadata in Aspect-Oriented Frameworks. In: 2nd Workshop on Assessment of Contemporary Modularization Techniques (ACoM 2008) at OOPSLA. EUA, Nashville (2008)
9. Bosch, J., Molin, P., Mattsson, M., Bengtsson, P., Fayad, M.: Framework Problems and Experiences. In: Building Application Frameworks – Object-oriented Foundations of Frameworks Design. ch. 3, pp. 55–83. Wiley (1999)
10. Gamma, E., Helm, R., Johnson, R., Vlissides, J.: Design Patterns: Elements of Reusable Object-Oriented Software. Addison-Wesley (1994)
11. Jacobsen, E., Nowack, P.: Frameworks and Patterns: Architectural Abstractions. In: Building Application Frameworks – Object-oriented Foundations of Frameworks Design. ch. 2, pp. 29–54. Wiley (1999)
12. Kiczales, G., Lamping, J., Menhdhekar, A., Maeda, C., Lopes, C., Loingtier, J., Irwin, J.: Aspect-oriented programming. In: Akşit, M., Matsuoka, S. (eds.) ECOOP 1997. LNCS, vol. 1241, pp. 220–242. Springer, Heidelberg (1997)
13. Fleury, M., Reverbel, F.: The JBoss Extensible Server. In: Endler, M., Schmidt, D.C. (eds.) Middleware 2003. LNCS, vol. 2672, pp. 344–373. Springer, Heidelberg (2003)
14. Jamae, J., Johnson, P.: JBoss in Action: Configuring the JBoss Application Server. Manning Publications (2009)
15. Laddad, R.: AspectJ in Action: Enterprise AOP with Spring Applications, 2nd edn. Manning Publications (2009)

16. Walls, C., Breidenbach, R.: Spring in Action, 2nd edn. Manning Publications (2007)
17. Silva, M.T., Braga, R., Masiero, P.C.: Evolução Orientada a Aspectos de um Framework OO. In: 1º Workshop de Manutenção de Software Moderna, Brasilia, DF (2004)
18. Doucet, F., Shukla, S., Gupta, R.: Introspection in system-level language frameworks: meta-level vs. Integrated. In: 2003 Proceedings of the Source Design, Automation, and Test in Europe, [S.l.: s.n], pp. 382–387 (2003)
19. Forman, I., Forman, N.: Java reflection in action. Manning Publ., Greenwich (2005)
20. Foote, B., Yoder, J.: Evolution, architecture, and metamorphosis. In: Pattern Languages of Program Design 2. ch. 13, pp. 295–314. Addison-Wesley Longman, Boston (1996)
21. Maes, P.: Concepts and Experiments in Computational Reflection. In: Proceedings of the International Conference on Object-oriented Programming, Systems, Languages and Applications – OOPSLA 1987, pp. 147–169 (1987)
22. Chen, N.: Convention over configuration (2006), http://softwareengineering.vazexqi.com/files/pattern.html (accessed on dez. 17, 2009)
23. Ruby, S., Thomas, D., Hansson, D.: Agile Web Development with Rails, 3rd edn. Pragmatic Bookshelf (2009)
24. Guerra, E., Souza, J., Fernandes, C.: A pattern language for metadata-based frameworks. In: Proceedings of the Conference on Pattern Languages of Programs, 16, Chicago (2009)
25. O'Brien, L.: Design patterns 15 years later: an interview with Erich Gamma. In: Helm, R., Johnson, R. (eds.) InformIT (October 22, 2009), http://www.informit.com/articles/article.aspx?p=1404056 (accessed on dez. 26, 2009)
26. JSR 303: Bean Validation, http://www.jcp.org/en/jsr/detail?id=303 (accessed in dez. 17, 2009)
27. Nucitelli, R., Guerra, E., Fernandes, C.: Parsing XML Documents in Java Using Annotations. In: XML: Aplicações e Tecnologias Associadas, Vila do Conde, Portugal (2010)
28. JavaBeans(TM) Specification 1.01 Final Release (1997), http://java.sun.com/javase/technologies/desktop/javabeans/docs/spec.html (acessed in dez. 27, 2009)
29. Massol, V., Husted, T.: JUnit in Action. Manning Publications (2003)
30. JSR 175: a metadata facility for the java programming language (2003), http://www.jcp.org/en/jsr/detail?id=175 (accessed on dez. 17, 2009)
31. Miller, J., Ragsdale, S.: Common language infrastructure annotated standard. Addison-Wesley, Boston (2003)
32. Schwarz, D.: Peeking inside the box: attribute-oriented programming with Java 1.5. [S.n.t.] (2004), http://missingmanuals.com/pub/a/onjava/2004/06/30/insidebox1.html (accessed on dez. 17, 2009)
33. Cisternino, A., Cazzola, W., Colombo, D.: Metadata-driven library design. In: Proceedings of the Library-centric Software Design Workshop (2005)
34. Rouvoy, R., Pessemier, N., Pawlak, R., Merle, P.: Using attribute-oriented programming to leverage fractal-based developments. In: 2006 Proceedings of the International ECOOP Workshop on Fractal Component Model, 5, Nantes (2006)
35. Wada, H., Suzuki, J.: Modeling Turnpike Frontend System: a Model-Driven Development Framework Leveraging UML Metamodeling and Attribute-Oriented Programming. In: Proc. of the 8th ACM/IEEE International Conference on Model Driven Engineering Languages and Sytems, MoDELS/UML 2005 (2005)

36. Guerra, E., Fernandes, C.: An Experimental Evaluation on Metadata-based Frameworks Usage (unpublished)
37. Bergmans, L., Aksit, M.: Composing crosscutting concerns using composition filters. Commun. ACM 44(10), 51–57 (2001)
38. JBoss, AOP: Framework for Organizing Cross Cutting Concerns, http://www.jboss.org/jbossaop (acessed in June 01, 2010)
39. JSR 220: Enterprise JavaBeans 3.0 (2006), http://www.jcp.org/en/jsr/detail?id=220 (accessed on dez. 17, 2009)
40. JSR 299: Contexts and Dependency Injection for the JavaTM EE platform (2009), http://www.jcp.org/en/jsr/detail?id=299 (accessed on dez. 17, 2009)
41. Doernenburg, E.: Domain Annotations. In: The Thoughtworks Anthology: Essays on Software Technology and Innovation. ch. 10. Pragmatic Bookshelf, Raleigh (2008)
42. Evans, E.: Domain-Driven Design: Tackling Complexity in the Heart of Software. Addison-Wesley Professional (2003)
43. Perillo, J., Guerra, E., Silva, J., Silveira, F., Fernandes, C.: Metadata Modularization Using Domain Annotations. In: Proceedings of the Workshop on Assessment of Contemporary Modularization Techniques 3, [S.l.: s.n], Orlando (2009)
44. Perillo, J.: Daileon: Uma Ferramenta Para Habilitar o Uso de Anotações de Domínio. In: Trabalho de Curso (Engenharia de Software) - Curso de Especialização em Tecnologia da Informação, Instituto Tecnológico de Aeronáutica, São José dos Campos (2010)
45. Perillo, R., Guerra, E., Fernandes, C.: Daileon: A Tool for Enabling Domain Annotations. In: 6th ECOOP 2009 Workshop on Reflection, AOP and Meta-Data for Software Evolution, Genova (2009)
46. SystemGlue, http://systemglue.sf.net/ (accessed in mai 14, 2010)
47. Guerra, E., Silveira, F., Fernandes, C.: ClassMock: A Testing Tool for Reflective Classes Which Consume Code Annotations. In: Workshop Brasileiro de Métodos Ágeis (WBMA 2010), Porto Alegre (2010)
48. Fernandes, C., Ribeiro, D., Guerra, E., Nakao, E.: XML, Annotations and Database: a Comparative Study of Metadata Definition Strategies for Frameworks. In: XML: Aplicações e Tecnologias Associadas, Vila do Conde, Portugal (2010)
49. Code Generation Library - CGLIB, http://cglib.sourceforge.net/ (accessed in January 31, 2010)
50. Yassine, A.: An Introduction to Modeling and Analyzing Complex Product Development Processes Using the Design Structure Matrix (DSM) Method. Quaderni di Management (Italian Management Review) (9) (2004)

Improving the Quality of Software by Quantifying the Code Change Metric and Predicting the Bugs

V.B. Singh[1] and K.K. Chaturvedi[2]

[1] Delhi College of Arts & Commerce University of Delhi, Delhi, India
[2] Indian Agricultural Statistics Research Institute, New Delhi, India
{vbsinghdcacdu,kkcchaturvedi}@gmail.com

Abstract. "When you can measure what you are speaking about and express it in numbers, you know something about it; but when you cannot measure, when you cannot express it in numbers, your knowledge is of a meagre and unsatisfactory kind; it may be the beginning of knowledge, but you have scarcely, in your thoughts, advanced to the stage of science." LORD WILLIAM KELVIN (1824 – 1907). During the last decade, the quantification of software engineering process has got a pace due to availability of a huge amount of software repositories. These repositories include source code, bug, communication among developers/users, changes in code, etc. Researchers are trying to find out useful information from these repositories for improving the quality of software.

The absence of bugs in the software is a major factor that decides the quality of software. In the available literature, researchers have proposed and implemented a plethora of bug prediction approaches varying in terms of accuracy, complexity and input data. The code change metric based bug prediction is proven to be very useful. In the literature, decay functions have been proposed that decay the complexity of code changes over a period of time in either exponential or linear fashion but they do not fit in open source software development paradigm because in open source software development paradigm, the development team is geographical dispersed and there is an irregular fluctuation in the code changes and bug detection/fixing process. The complexity of code changes reduces over a period of time that may be less than exponential or more than linear. This paper presents the method that quantifies the code change metric and also proposed decay functions that capture the variability in the decay curves represented the complexity of code changes. The proposed decay functions model the complexity of code changes which reduces over a period of time and follows different types of decay curves. We have collected the source code change data of Mozilla components and applied simple linear regression (SLR) and support vector regression (SVR) techniques to validate the proposed method and predict the bugs yet to come in future based on the current year complexity of code changes (entropy). The performance of proposed models has been compared using different performance criteria namely R^2, Adjusted R^2, Variation and Root Mean Squared Prediction Error (RMSPE).

Keywords: Bug Prediction, Entropy, Software Quality, Software Repository, Complexity of Code Change.

B. Murgante et al. (Eds.): ICCSA 2013, Part II, LNCS 7972, pp. 408–426, 2013.
© Springer-Verlag Berlin Heidelberg 2013

1   Introduction

The changes in files of the software system are frequent over a period of time to meet the customer expectation. These changes are occurring due to bug repairs (BR), addition of new feature (NF) and enhancement/modification (EM) in the files. The maintenance task becomes quite difficult if these changes are not being recorded properly. These frequent changes are being maintained using software configuration management repository which makes maintenance task easy. The bugs occurring in the systems are recorded/reported in bug reporting systems configured at the development location. BugZilla [28] is the mostly used bug reporting system in the open source software community. This system also monitors the progress of the bug fix. The continual changes due to bug repair and feature enhancement further make the code complex due to increasing amount of changes in the files.

Bugs are generated due to mis-communication or no communication among active users, increasing software complexity, occurrence of programming errors, frequently changing requirements, early release pressures, addition of new feature, feature enhancement/modification, and bugs present in software development tools. During bug fixing, some new bugs are also generated (a case of imperfect debugging). These bugs are detected and reported on the configured bug tracking system and later on fixed. The bug detection/removal process has been modeled by different software reliability growth model in the literature [5, 6, 14, 15, 19, 20, 24, 27, 31, and 33].

The effect and contribution of earlier modifications/changes in a file get reduced over a period of time [2, 3, and 9]. The contributions due to these changes/modifications may follow different decay curves. The complexity/ uncertainty /randomness of the code change have been quantified and modeled using information theory based measures called entropy [9, 10 and 11].

Earlier authors have proposed decay functions which reduce the complexity of code changes either in exponential or linear fashion. In this paper, we consider that when the bugs are fixed, entropy/change complexity gets reduced and there is a correspondence between decay in entropy/complexity of code changes and fixing of bugs. Therefore, the rate at which entropy gets decayed/ reduced may be considered as bug's detection/ removal rate. Following research questions (RQ) have been set to conduct the present study.

Research question 1: Does the complexity of code changes can be quantified?

Research question 2: Does the proposed models are better over the linear and exponential in modeling the decay in complexity of code changes over a period of time?

Research questions 3: How does the complexity of code changes can be used to predict the future bugs?

To answer these questions, an empirical study has been conducted using sub components of Mozilla, an open source project. Rest of the paper is organized as follows.

The paper is divided into eight sections. Section 2 describes the review of work available in the literature. Section 3 deals with the code change process. In section 4, we have discussed the procedure of entropy calculation and different methods namely simple linear regression and support vector regression for building prediction models. This section also discussed the existing and proposed decay models to compute the change complexity metric. Section 5 describes the procedure for data collection and data pre-processing of the selected components of Mozilla project. In section 6, we have analyzed and discussed the results obtained by applying simple linear regression and support vector regression techniques on existing and proposed decay models for predicting the post release bugs in the software. Section 7 mentions the threat to validity of the results. Finally, paper is concluded in section 8.

2   Review of Work

During last decades, plethora of approaches have been proposed in the literature for bug prediction such as code metrics (lines of code) [1, 8, 21, 22 and 23], process metrics (number of changes) [9, 21] and previous defects [12, 16, and 25]. Many researchers have proposed that a prior modification to a file is a good predictor to determine the fault in the software [2, 7, 17 and 18]. It also signifies that if changes are frequent in a particular file, more likely it may become faulty for closed source software [7, 18] and open source software [13]. Authors concluded that code complexity metrics is highly correlated with LOC (Lines of Code) which is a very simpler metric. This metric has not been utilized in the literature as it supposed to be. Further, attempt has been made in evaluating the capability of SVM in predicting defect-prone software modules using binary classification [4]. In this study, each dataset contains 21 software metrics (independent variables) at the module-level and associated dependent binary variable (Defective) which determine whether or not the module has any defects. A study has been also conducted on predicting faults using complexity of code changes and proposed that these complexity metrics are better predictors of fault potential in comparison to other well-known historical predictors of faults i.e., prior modifications and prior faults [9]. History Complexity Metric (HCM) measure for an individual file/subsystem using exponential decay of complexity based model has been proposed and showed that HCM metric is better predictors of faults [9]. Further, a benchmark study has been conducted on defect prediction using publicly available dataset and provides a comparison of different bug-prediction approaches and proposed two more approaches for decay namely linear decay and logarithmic decay based models [2, 3]. The paper also evaluated the performance of bug prediction approaches at class level in the context of classification (defective/non-defective), ranking (most defective to least defective) and effort aware ranking (most defect dense to least defect dense) using the performance evaluation metrics. In the available literature for bug predictions, only Hassan [9] and D'Ambros et al. [3] have proposed the code change complexity based bug prediction. Recently, Singh and Chaturvedi [34] proposed to predict the future bugs of software based on current year entropy using support vector regression and compared with simple linear regression [3, 9].

3   Code Change Process

The changes are occurring in the code to meet the user demand and expectations. The changes are following a process by which changes are being occurred. It varies from assessing the increasing demand of user's requests to fixing of the reported bugs. There are three variants of the code change process as follows in [9, 10, and 11]. The basic code change level quantifies the patterns of changes instead of measuring the number of changes or measuring the effect of changes to the code structure. The changes are recorded based on the number of modifications in the file. Entropy is calculated based on the number of changes in a file for specific periods with respect to total number of changes in all files. This period can be taken as a day, week, month, year etc. based on the total duration of the project as well as the number of changes occurs in the system. Basic code change level is extended based on the variable length period and known as extended code change. The time period can be divided into three ways i.e., time based periods, modification limits based periods and burst based periods. Total length of the project is divided into equal length duration in time based periods. These partitions can be of any length. The periods are decided based on the equal number of modification in the modification limit based periods. Usually, the changes do not follow a specific pattern. It generally follows the burst based patterns i.e., the considered period is depend on the significant number of changes. The patterns of the changes occur in the project in contrast to the period based and modification based periods. It is our belief that files are modified during periods of high change complexity that also have the higher tendency to contain faults. This is called as file code change level.

4   Entropy Measurement

The information theory deals with assessing and defining the amount of information in the message [26]. The theory uses the information in measuring the amount of uncertainty or entropy of the distribution. The Shannon Entropy, H_n is defined as

$$H_n(P) = -\sum_{k=1}^{n} (P_k * \log_2 P_k) \tag{1}$$

Where $P_k \geq 0$ and $\sum_{k=1}^{n} P_k = 1$

For a distribution P where all elements have the same probability of occurrence ($P_k = 1/n$; $\forall k \in 1, 2, ..., n$), we achieve maximum entropy. On the other hand for a distribution P where an element i has a probability of occurrence $P_i = 1$ and $P_k = 0$ $\forall k \neq i$, we achieve minimal entropy.

Research questions 1.
Hassan [9] has used this concept to measure the complexity of code change, where the author has taken P_k as probability which is defined as ratio of the number of times k^{th} file changed during a period and the total number of changes for all files in that period.

For example, suppose that there are 15 changes occurred for four files and three periods. For a first period, there are five changes occurred across all four files. The probability of file F1, F2, F3, and F4 will be 1/5 (=0.2), 1/5 (=0.2), 2/5 (=0.4) and 1/5 (=0.2) respectively for the first period T1. These probabilities have been shown in fig. 1.

The entropy for this period can be calculated by substituting the values of these probabilities in the equation (1) which is useful in quantifying the code change using this entropy based metric. This quantification helps in answering the research question 1. From the definition, it is clear that the entropy will be maximum, if there are frequent changes in every file while it will be minimum if the changes are occurred in a single file [34]. Different complexity metric has been defined in order to predict the potential bugs in the software [9].

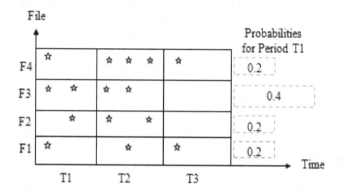

Fig. 1. Number of Changes in files with respect to a specific period of time

4.1   History Complexity Period Factor (HCPF)

Given a period i, with entropy Hi where a set of files, Fi are modified with a probability Pj for each file $j \in F_i$, we can define HCPFi for a file j during period i

$$HCPF_i = \begin{cases} C_{ij} * H_i & j \in F_i \\ 0 & \text{Otherwise} \end{cases} \qquad (2)$$

Where Cij is the contribution of entropy for the period i (Hi) assigned to file j. We have considered three existing and one proposed variants of HCPF using different weighting factors Cij which are given as follows

Case 1 (SimpleHCM): $C_{ij}=1$,
It means that equal weights are given to all files for the ith period.
Case 2 (WeightedHCM1): $C_{ij}=P_j$,
P_j is the probability of changes occurred in a particular file j with respect to all changes in the ith period.

Case 3 (WeightedHCM2): $C_{ij} = 1/F_i$,
F_i is the number of changed files in the ith period.
Case 4 (WeightedHCM3): $C_{ij} = P_j /F_i$,

The weight is the ratio of probability of changes occurred in a particular file j with respect to all changes in the ith period by number of changes files in i^{th} period.

4.2   History Complexity Metric (HCM)

HCM for a file over a set of evolution period {a,...,b} is defined as

$$HCM_{\{a,...,b\}}(j) = \sum_{i \in \{a,...,b\}} HCPF_i(j) \tag{3}$$

HCM definition indicates that complexity associated with a file keeps on increasing over a time as a file is modified. HCM for a subsystem (S) over a set of evolution period {a,...,b} defined as the sum of HCMs of all the files of that subsystem

$$HCM_{\{a,...,b\}}(S) = \sum_{j \in S} HCPF_{\{a,...,b\}}(j) \tag{4}$$

Earlier modifications would have their contribution to the complexity of the file reduced in an exponential fashion over time [9]. Similarly linear and logarithmic decay have their contribution reduced over time in linear and logarithmic way respectively [4]. The HCMs for these exponential, linear and logarithmic decay based models have been defined in equation (5), (6) and (7) respectively.

$$HCM_{\{a,...,b\}}(j) = \sum_{i \in \{a,...,b\}} e^{-\Phi*(CurrentTime-T_i)} HCPF_i(j) \tag{5}$$

$$HCM_{\{a,...,b\}}(j) = \sum_{i \in \{a,...,b\}} \frac{1}{(1+\Phi*(CurrentTime-T_i+1))} HCPF_i(j) \tag{6}$$

$$HCM_{\{a,...,b\}}(j) = \sum_{i \in \{a,...,b\}} \frac{1}{(1+\Phi*\ln(CurrentTime-T_i+1.01))} HCPF_i(j) \tag{7}$$

Where T_i is the end time of period i, ϕ is the decay factor and j is the jth file of the system/ subsystem.

In the above model, the complexity code changes have been reduced either in exponential or linear or logarithm form but in open source software development paradigm where the development team is geographical dispersed and there is an irregular fluctuation in code changes and bug detection/fixing process. The decay curve of the complexity of code changes over a period of time can not be fit to a fixed model but it can digress significantly.

The models we proposed in equation (8) and (9) capture the variability in decay curve depending upon the value of k. We have also proposed a polynomial decay model in equation (10) that shows decay less than exponential.

$$HCM_{\{a,...,b\}}(j) = \sum_{i \in \{a,...,b\}} e^{-\Phi*(CurrentTime-T_i)^k} HCPF_i(j) \qquad (8)$$

$$HCM_{\{a,...,b\}}(j) = \sum_{i \in \{a,...,b\}} \frac{1}{(1 + \Phi*(CurrentTime-T_i+1))^k} HCPF_i(j) \qquad (9)$$

$$HCM_{\{a,...,b\}}(j) = \sum_{i \in \{a,...,b\}} e^{-\Phi*\log((CurrentTime-T_i+1))} HCPF_i(j) \qquad (10)$$

Here k>=0, above generalized models provide more variations in decay in complexity of code changes. In this study, for calculating HCM using equation (5-10), we have used history complexity period factor as defined in equation (2) for contribution C_{ij}=1. In above equation (8) and (9) if we put k=1, model reduces to exponential and linear as given in equation (5) and (6).

4.3   Simple Linear Regression

The simple linear regression [30] have been extensively used to regress the dependent variable y using independent variable x with following equation

$$y = \beta_0 + \beta_1 * x \qquad (11)$$

Where β_0 and β_1 are regression coefficients

Once the parameters of this regression equation (11) has been obtained, the value of the dependent variable can be predicted based on the value of x. In this study, we have considered that x is HCM value of different decay models as independent variable and y is cumulated bugs likely to occur in the next year as dependent variable. The regression will be helpful in predicting the bugs based on the complexity of code change values of using the proposed as well existing decay based models.

4.4   Support Vector Regression

The empirical data modeling has been used to build up a model to deduce responses of the system that are yet to be observed in the system. Traditional approaches have faced generalization difficulties and produced models which can over-fit the data. This is a consequence of the optimization algorithms requires for parameter selection and helpful in selecting the 'best' model using statistical significance. The foundations of Support Vector Machines (SVM) have been laid down by Vapnik [32] and gaining popularity due to its attractive features. The formulation embodies the Structural Risk Minimization (SRM) principle over to traditional Empirical Risk Minimization (ERM) principle. SRM minimizes an upper bound on the Vapnik–Chervonenkis (VC) dimension (generalization error), as opposed to ERM that minimizes the error on the training data. SVM have been developed to solve the classification problem, but they have also shown the promising performance in solving the regression problems. The term SVM is typically used to describe classification with support vector methods and support vector regression is used to describe regression.

5   Data Collection and Pre-processing

Four subsystems "mozilla/layout/svg/", "mozilla/layout/base/", "mozilla/layout/tables/" and "mozilla/layout/xul/" of Mozilla project have been selected for the study [29]. The data collection, extraction and model building process are shown in fig. 2. These subsystems are selected as test cases and historical changes are extracted for different files of these subsystems. After extracting data from the repository, year wise number of changes has been recorded for all files of these subsystems from the CVS logs.

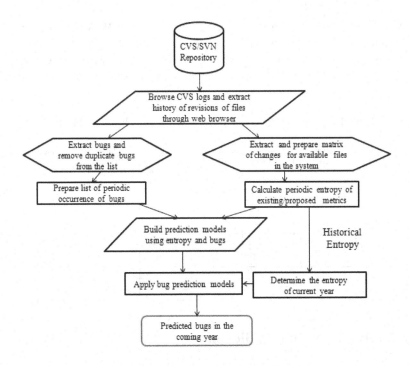

Fig. 2. Process for Building the Entropy Based Prediction Models

The identified bugs are extracted using pattern based searching and arranged using its first appearance of fix. The remaining duplicate entries of the bugs are discarded. In this study, we have taken the period as calendar year. There are 52 files in the subsystem "mozilla/layout/svg/", 60 files in the subsystem "mozilla/layout/base/", 29 files in the subsystem "mozilla/layout/tables/", and 106 files in the subsystem "mozilla/layout/xul/". This process has been given as follows

Step 1: Choose the project

Step2: Select the sub-systems

Step3: Browse the CVS logs to find the historical changes in various files for the selected subsystems

Step4: Arrange the files as per their revision number or commit date

Step5: Calculate year wise frequency of changes with respect to files in the chosen sub-system

Step6: Calculate entropy of the file

Step7: Compute the weights and recalculate the HCPF/HCMs for simple/weighted models

Step8: Build the Prediction Models for all HCMs and decay models

Step9: Calculate the entropy of the current year

Step10: Use the prediction Models to predict the bugs in the coming year based on the entropy of current year.

Year wise number of file changes as well as base value of HCM has been given in Table 1. These detected /removed bugs are extracted from the CVS log repository and confirms with the bug reporting system. We have extracted only those bugs in which the log message contains the prefix pattern as "Bug(s) fix(es) for" or "Bug(s) "or "bug #".

Table 1. Historical Changes in the Files and entropy based HCM values

Year	mozilla/layout/svg/		mozilla/layout/base/		mozilla/layout/tables/		mozilla/layout/xul/	
	Changes	HCM	Changes	HCM	Changes	HCM	Changes	HCM
1998	-	-	-	-	1261	1.170	-	-
1999	-	-	1149	1.035	835	2.383	731	1.373
2000	-	-	671	2.020	359	3.579	864	2.980
2001	-	-	687	3.145	428	4.778	676	4.732
2002	51	0.896	631	4.266	207	5.884	396	6.330
2003	15	1.474	474	5.470	191	7.107	351	7.983
2004	135	2.678	764	6.767	314	8.335	583	9.774
2005	258	4.084	515	7.924	154	9.577	396	11.536
2006	731	5.634	550	9.241	269	10.801	568	13.302
2007	529	7.166	689	10.430	227	12.025	621	15.002
2008	136	8.595	286	11.613	105	13.144	148	16.541
2009	-	-	37	12.558	-	-	-	-

6   Analysis and Bug Prediction

The effect of the complexity of code changes can be reduced over a period of time due to removal of bugs. Therefore, it shows a correspondence between decay in change complexity and detection/fixing of bugs. The rate at which change complexity get reduced may be considered as bug detection/fixing rate. This bug detection/removal rate has been estimated by using different growth models namely Goel-Okumoto (exponential)[6], Yamada S-Shaped[31], Ohba Inflected S-Shaped [24] and Kapur-Garg Model (Flexible model)[15] for different sub-systems of Mozilla. We have taken the value of model which gives the best fit in terms of different goodness of fit criteria [34]. The flexible model gives the best fit for the selected data sets under study. The values of parameter ϕ (bug detection rate or decay rate) are 0.880, 0.287, 0.347, and 0.284 for the sub-systems "mozilla/layout/svg/", "mozilla/layout/base/", "mozilla/layout/tables/" and "mozilla/layout/xul/" respectively. Entropy is calculated on yearly basis with respect to changes in the files.

The simple linear regression [30] has been fitted for the independent variable x and dependent variable y using equation (11). For each model, y represents the number of bugs in a subsystem. The variable x represents specific decay based metrics for each subsystem in the current year. The regression coefficients β_0 and β_1 are calculated using different values of the decay based models as independent variable and number of bugs as dependent variable.

We have varied the value of k from 0.1 to 1.0 for decay model in equation (8) and 0.1 to 2.0 for decay model in equation (9) with a step of 0.1. These decay models become exponential and linear decay when value of k is set to 1.0 respectively. The values of R-squared (R^2), Adjusted R Squared (Adj. R^2), Variation and Root Mean Squared Prediction Error (RMSPE) and regression coefficients have been calculated for various decay based as well as simple models as shown in table 2 to table 5 for the subsystems "mozilla/layout/svg/", "mozilla/layout/base/", "mozilla/layout/tables/" and "mozilla/layout/xul/" respectively.

Research question 2.
In support of answer for research question 2, we applied simple linear regression and support vector regression (SVR) using radial basis kernel function on decay based complexity of code change metrics of four datasets of Mozilla components. We varied the value of k from 0.1 to 1.0 for proposed model given in equation (8) and 0.1 to 2.0 for proposed model given in equation (9) with a step of 0.1 for linear regression.SVR has been applied for those cases where variation of k shows the significant improvement in terms of R^2 for linear regression.

We have observed from table 2 that the variation of k in equation (8) as 0.3 and 0.4 gives the significant improvement in terms of R^2 i.e., 0.982 and 0.982 in comparison with its counterpart model given in equation (5) whose value of R^2 is 0.971. For model in equation (8) if we increase the value of k from 0.4 to 1.0, performance degrades. In case of proposed model given in equation (9), no improvement has been observed for value of k ranging from 1.0 to 1.5. Proposed model given in equation (10) shows better performance over the existing model given in equation (7) i.e. the value of R^2 is 0.986 and 0.983. The decay in the complexity of code changes has

Table 2. Regression coefficients and other statistics for mozilla/layout/svg/ for different decay models

Models	k	R^2	Adj. R^2	Variation	RMSPE	Reg. Coefficients	
						β_0	β_1
SimpleHCM (without decay)	-	0.971	0.964	15.070	35.817	-41.486	76.364
Equation (8)	0.1	0.980	0.974	12.738	30.273	-112.757	162.333
	0.2	0.981	0.976	12.326	29.295	-123.454	171.795
	0.3	0.982	0.977	12.074	28.696	-134.989	182.120
	0.4	0.982	0.977	12.042	28.620	-147.184	193.195
	0.5	0.981	0.976	12.261	29.141	-159.782	204.831
	0.6	0.980	0.975	12.712	30.213	-172.468	216.769
	0.7	0.978	0.972	13.328	31.676	-184.901	228.699
	0.8	0.975	0.969	14.014	33.307	-196.753	240.301
	0.9	0.973	0.966	14.677	34.882	-207.760	251.286
	1.0	0.971	0.963	15.249	36.241	-217.740	261.431
Equation (9)	0.1	0.974	0.967	14.447	34.335	-47.804	86.937
	0.2	0.976	0.970	13.853	32.925	-54.220	98.817
	0.3	0.978	0.972	13.294	31.596	-60.727	112.140
	0.4	0.979	0.974	12.774	30.359	-67.314	127.054
	0.5	0.981	0.976	12.296	29.224	-73.972	143.717
	0.6	0.982	0.978	11.866	28.202	-80.688	162.302
	0.7	0.983	0.979	11.488	27.304	-87.452	182.990
	0.8	0.984	0.980	11.167	26.540	-94.250	205.978
	0.9	0.985	0.981	10.906	25.919	-101.070	231.476
	1.0	0.986	0.982	10.708	25.450	-107.900	259.708
	1.1	0.986	0.982	10.577	25.137	-114.724	290.910
	1.2	0.986	0.983	10.512	24.983	-121.530	325.336
	1.3	0.986	0.983	10.514	24.988	-128.303	363.254
	1.4	0.986	0.982	10.581	25.148	-135.029	404.946
	1.5	0.986	0.982	10.711	25.456	-141.695	450.714
	1.6	0.985	0.981	10.900	25.905	-148.287	500.876
	1.7	0.984	0.980	11.143	26.482	-154.791	555.765
	1.8	0.984	0.979	11.435	27.177	-161.195	615.736
	1.9	0.983	0.978	11.772	27.977	-167.485	681.163
	2.0	0.981	0.977	12.147	28.869	-173.651	752.438
Equation (10)	-	0.986	0.982	10.711	25.456	-141.695	450.714
Equation (7)	-	0.983	0.979	11.498	27.327	-98.962	136.192

been observed very low in case of model proposed in equation (9) with the increase or decrease in value of k from 1.0 with a step 0.1 because most of the changes are occurring in 7 files namely nsSVGClipPathFrame.cpp, nsSVGForeignObjectFrame.cpp, nsSVGGlyphFrame.cpp, nsSVGGradientFrame.cpp, nsSVGOuterSVGFrame.cpp, nsSVGPathGeometryFrame.cpp, nsSVGUtils.cpp out of 52 files.

The proposed model given in equation (8) with variation in k gives better performance in comparison with existing exponential decay model. The proposed model in equation (9) does not show the improvement over existing one because decay in complexity of code changes is very low for this data set. The proposed model in equation (10) gives similar performance with model in equation (6) which makes slow the decay in the complexity of code change.

Table 3 shows that the variation of k for proposed model given in equation (8) as 0.6 and 0.7 gives the significant improvement in terms of R^2 i.e., 0.994 and 0.994 in

Table 3. Regression coefficients and other statistics for mozilla/layout/base/ for different decay models

Models	k	R^2	Adj. R^2	Variation	RMSPE	Reg. Coefficients	
						β_0	β_1
SimpleHCM (without decay)	-	0.978	0.975	6.908	128.960	472.847	225.142
Equation (8)	0.1	0.982	0.979	6.346	118.465	366.231	311.654
	0.2	0.984	0.982	5.881	109.784	331.109	328.107
	0.3	0.987	0.985	5.320	99.314	288.010	348.273
	0.4	0.990	0.989	4.679	87.351	235.407	373.007
	0.5	0.993	0.992	4.033	75.277	171.753	403.295
	0.6	0.994	0.993	3.593	67.074	95.724	440.184
	0.7	0.994	0.993	3.756	70.105	6.631	484.638
	0.8	0.990	0.988	4.775	89.126	-95.024	537.289
	0.9	0.981	0.978	6.493	121.203	-207.015	598.095
	1.0	0.966	0.962	8.633	161.146	-325.279	666.050
Equation (9)	0.1	0.981	0.979	6.397	119.408	437.573	247.471
	0.2	0.984	0.982	5.901	110.148	401.453	271.728
	0.3	0.987	0.985	5.426	101.290	364.498	298.045
	0.4	0.989	0.987	4.981	92.976	326.718	326.559
	0.5	0.990	0.989	4.574	85.389	288.128	357.412
	0.6	0.992	0.991	4.219	78.756	248.744	390.752
	0.7	0.993	0.992	3.929	73.350	208.584	426.732
	0.8	0.994	0.993	3.721	69.464	167.667	465.508
	0.9	0.994	0.993	3.608	67.357	126.013	507.244
	1.0	0.994	0.993	3.599	67.184	83.647	552.107
	1.1	0.994	0.993	3.693	68.938	40.592	600.269
	1.2	0.993	0.992	3.881	72.454	-3.129	651.906
	1.3	0.992	0.991	4.149	77.457	-47.487	707.200
	1.4	0.991	0.990	4.481	83.645	-92.458	766.335
	1.5	0.989	0.988	4.861	90.735	-138.015	829.502
	1.6	0.987	0.986	5.276	98.490	-184.129	896.895
	1.7	0.985	0.983	5.717	106.724	-230.776	968.712
	1.8	0.983	0.980	6.176	115.289	-277.927	1045.158
	1.9	0.980	0.977	6.647	124.075	-325.557	1126.439
	2.0	0.977	0.974	7.125	132.995	-373.640	1212.768
Equation (10)	-	0.985	0.983	5.791	108.103	377.308	276.311
Equation (7)	-	0.988	0.986	5.133	95.814	296.441	328.781

comparison with its counterpart model given in equation (5) whose value of R^2 is 0.966. For model in equation (8) if we increase the value of k from 0.7 or decrease from 0.6, performance degrades. In case of proposed model given in equation (9), the value of performance measure is same as of the existing one.

It is observed from table 4 that the variation of k for proposed model given in equation (8) as 0.5 gives the significant improvement in terms of R^2 i.e., 0.998 in comparison with its counterpart model given in equation (5) whose value of R^2 is 0.928 but if we increase/decrease the value of k from 0.5, performance degrades. In case of proposed model given in equation (9), for value of k ranging from 0.6 to 0.8, it gives significant improvement in terms of R^2 i.e., 0.998 and 0.998 in comparison with its counterpart model given in equation (6) whose value of R^2 is 0.996. For model in equation (9) if we increase the value of k from 0.8 or decrease the value of k from 0.6, performance degrades.

Table 4. Regression coefficients and other statistics for mozilla/layout/tables/ for different decay models

Models	k	R^2	Adj. R^2	Variation	RMSPE	Reg. Coefficients	
						β_0	β_1
SimpleHCM (without decay)	-	0.987	0.986	6.021	24.808	38.145	56.956
Equation (8)	0.1	0.990	0.988	5.499	22.657	0.159	85.052
	0.2	0.992	0.991	4.783	19.708	-12.874	90.818
	0.3	0.995	0.994	3.922	16.157	-29.060	97.993
	0.4	0.997	0.997	3.002	12.368	-49.048	106.932
	0.5	0.998	0.998	2.459	10.130	-73.492	118.044
	0.6	0.997	0.996	3.183	13.113	-102.906	131.751
	0.7	0.991	0.990	5.147	21.206	-137.435	148.403
	0.8	0.979	0.976	7.844	32.315	-176.527	168.121
	0.9	0.958	0.953	11.007	45.349	-218.681	190.623
	1.0	0.928	0.919	14.425	59.433	-261.496	215.111
Equation (9)	0.1	0.990	0.989	5.326	21.945	26.935	63.548
	0.2	0.993	0.992	4.648	19.150	15.344	70.823
	0.3	0.995	0.994	3.999	16.478	3.371	78.839
	0.4	0.996	0.996	3.402	14.017	-8.990	87.658
	0.5	0.997	0.997	2.893	11.920	-21.741	97.349
	0.6	0.998	0.998	2.531	10.427	-34.883	107.981
	0.7	0.998	0.998	2.387	9.833	-48.418	119.629
	0.8	0.998	0.998	2.500	10.302	-62.346	132.370
	0.9	0.997	0.997	2.843	11.713	-76.666	146.289
	1.0	0.996	0.996	3.344	13.778	-91.378	161.471
	1.1	0.995	0.994	3.943	16.247	-106.480	178.007
	1.2	0.993	0.992	4.601	18.957	-121.970	195.993
	1.3	0.990	0.989	5.294	21.811	-137.844	215.529
	1.4	0.988	0.986	6.008	24.752	-154.099	236.718
	1.5	0.984	0.982	6.733	27.741	-170.731	259.670
	1.6	0.981	0.978	7.465	30.756	-187.737	284.497
	1.7	0.977	0.974	8.199	33.780	-205.111	311.318
	1.8	0.972	0.969	8.932	36.799	-222.848	340.256
	1.9	0.968	0.964	9.661	39.805	-240.945	371.439
	2.0	0.963	0.958	10.386	42.789	-259.396	405.001
Equation (10)	-	0.993	0.993	4.374	18.023	4.203	73.583
Equation (7)	-	0.995	0.995	3.639	14.991	-22.518	89.873

Table 5 shows that the variation of k in equation (8) as 0.2, 03, 0.4 and 0.5 give significant improvement in terms of R^2 i.e., 0.996, 0.996, 0.996, and 0.996 in comparison with its counterpart model given in equation (5) whose value of R^2 is 0.946 and if we increase the value of k from 0.8 or decrease the value of k from 0.2, performance degrades. In case of proposed model given in equation (9), for value of k ranging from 0.3 to 0.6 , it gives significant improvement in terms of R^2 i.e., 0. 0.996, 0996, 0.996 and 0.996 in comparison with its counterpart model given in equation (6) whole value of R^2 is 0.991 and if we increase the value of k from 0.6 or decrease the value of k from 0.3, performance degrades.

We have further plotted the values of R^2 for the studied data for all the decay based models in fig.3. It is observed that the proposed model given in equation (8) with value of k as 0.4 to 0.6 shows the maximum value of R^2 except in SVG dataset where it is for k as 0.2 to 0.4.

Table 5. Regression coefficients and other statistics for mozilla/layout/xul/ for different decay models

Models	k	R^2	Adj. R^2	Variation	RMSPE	Reg. Coefficients	
						β_0	β_1
SimpleHCM (without decay)	-	0.993	0.993	4.269	28.397	100.219	69.647
Equation (8)	0.1	0.995	0.994	3.759	25.009	54.914	95.717
	0.2	0.996	0.995	3.449	22.944	41.833	100.238
	0.3	0.996	0.996	3.205	21.320	26.066	105.689
	0.4	0.996	0.996	3.169	21.080	7.169	112.260
	0.5	0.996	0.995	3.527	23.462	-15.293	120.161
	0.6	0.993	0.992	4.393	29.225	-41.658	129.608
	0.7	0.988	0.986	5.758	38.304	-72.052	140.789
	0.8	0.979	0.976	7.560	50.289	-106.222	153.811
	0.9	0.966	0.961	9.726	64.702	-143.380	168.630
	1.0	0.946	0.939	12.168	80.941	-182.144	184.989
Equation (9)	0.1	0.995	0.994	3.874	25.772	86.993	76.093
	0.2	0.995	0.995	3.556	23.653	73.477	83.059
	0.3	0.996	0.995	3.337	22.200	59.676	90.578
	0.4	0.996	0.996	3.241	21.561	45.596	98.684
	0.5	0.996	0.996	3.280	21.819	31.242	107.412
	0.6	0.996	0.995	3.450	22.950	16.621	116.799
	0.7	0.995	0.994	3.734	24.839	1.740	126.884
	0.8	0.994	0.993	4.108	27.329	-13.391	137.703
	0.9	0.993	0.991	4.550	30.271	-28.765	149.299
	1.0	0.991	0.990	5.042	33.542	-44.372	161.711
	1.1	0.989	0.987	5.569	37.049	-60.203	174.983
	1.2	0.986	0.985	6.122	40.726	-76.246	189.155
	1.3	0.984	0.982	6.693	44.522	-92.491	204.274
	1.4	0.981	0.978	7.276	48.401	-108.926	220.382
	1.5	0.978	0.974	7.867	52.335	-125.541	237.527
	1.6	0.974	0.970	8.464	56.303	-142.325	255.754
	1.7	0.970	0.966	9.063	60.289	-159.264	275.110
	1.8	0.966	0.961	9.663	64.279	-176.347	295.645
	1.9	0.962	0.956	10.261	68.261	-193.563	317.406
	2.0	0.957	0.951	10.858	72.227	-210.900	340.443
Equation (10)	-	0.996	0.995	3.439	22.874	63.183	84.370
Equation (7)	-	0.996	0.996	3.140	20.890	30.687	99.679

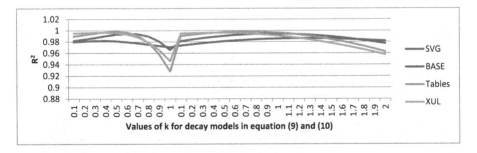

Fig. 3. The performance of decay models with different values of k for datasets under study using linear regression

All the values of k show the improved performance over exponential decay model having k as 1.0 in equation (9).

We have also applied Support Vector Regression (SVR) for those cases where the value of R^2 shows the improved performance in case of linear regression for the value of k. The performance parameters namely error mean, Standard Deviation ratio (S.D. ratio)

Table 6. Model Performance using SVR for mozilla/layout/svg/ for different decay models

Models	k	Error Mean	S.D. Ratio	R^2
Equation (8)	0.3	8.122	0.136	0.995
	0.4	7.340	0.132	0.996
	1.0	0.479	0.111	0.996
Equation (9)	1.0	5.722	0.142	0.997
	1.1	5.635	0.140	0.997
	1.2	5.520	0.137	0.997
	1.3	5.407	0.135	0.997
	1.4	5.278	0.132	0.997
	1.5	5.125	0.130	0.997
Equation (10)		12.138	0.158	0.996
Equation (7)		10.611	0.148	0.996

Table 7. Model performance using SVR for mozilla/layout/base/ for different decay models

Models	k	Error Mean	S.D. Ratio	R^2
Equation (8)	0.5	10.218	0.137	0.992
	0.6	8.413	0.133	0.992
	0.7	10.131	0.132	0.992
	1.0	64.094	0.195	0.986
Equation (9)	0.8	6.519	0.130	0.992
	0.9	6.436	0.129	0.992
	1.0	7.442	0.132	0.992
	1.1	6.868	0.131	0.992
Equation (10)		11.961	0.147	0.991
Equation (7)		13.161	0.142	0.991

Table 8. Model performance using SVR for mozilla/layout/tables/ for different decay models

Models	k	Error Mean	S.D. Ratio	R^2
Equation (8)	0.4	24.874	0.183	0.991
	0.5	19.917	0.195	0.991
	0.6	14.057	0.214	0.988
	1.0	26.158	0.292	0.979
Equation (9)	0.6	22.267	0.189	0.991
	0.7	19.834	0.195	0.990
	0.8	17.421	0.202	0.990
	1.0	12.662	0.218	0.988
Equation (10)		25.565	0.200	0.989
Equation (7)		67.227	0.390	0.996

Table 9. Model performance using SVR for mozilla/layout/xul/ for different decay models

Models	k	Error Mean	S.D. Ratio	R^2
Equation (8)	0.2	47.851	0.279	0.977
	0.3	46.990	0.269	0.979
	0.4	45.937	0.259	0.981
	0.5	42.052	0.263	0.981
	1.0	49.416	0.345	0.969
Equation (9)	0.3	46.513	0.279	0.977
	0.4	45.732	0.270	0.978
	0.5	45.138	0.263	0.980
	0.6	43.891	0.260	0.981
	1.0	33.185	0.282	0.977
Equation (10)		47.190	0.282	0.976
Equation (7)		46.305	0.265	0.980

and R^2 are shown in table 6 to 9 for "mozilla/layout/svg/", "mozilla/layout/base/", "mozilla/layout/tables/" and "mozilla/layout/xul/" data sets respectively. We observed that the performance variation is non-significant due to less number of data points but we also found that the value of R^2 is more than 0.99 in all the cases.

The above analysis confirms the research question 2 that the all proposed models given in equation (8), (9) and (10) has shown improved performance over the existing one and able to capture the variability in the complexity of code changes with greater accuracy.

Research question 3.
To answer the research question 3 i.e., how does the complexity of code changes can be used to predict the future bugs? The regression coefficients are calculated to predict the future bugs in the system due to changes in the files of the considered subsystem. These regression coefficients are used in predicting the future bugs occurring in the system using equation (11). The regression coefficients are shown in table 2 to table 5 for various decay models for different datasets. We have predicted the bugs for all components of Mozilla using different decay model based complexity of code changes. The predicted bugs of the mentioned subsystem using respective values of complexity of code change for different decay models are shown in table 6. For managerial applications, we can choose the result for which model gives the best fit.

Once we have the entropy/ complexity of code changes up to the current year which is denoted in equation (11) as independent variable then by using the value of regression coefficients, the value of next year cumulative bugs can be predicted.

For example, we have taken proposed decay model with k=0.3 as given in equation (8) for XUL components, we get the value of regression coefficients as 26.066 and 105.689.

The current year entropy of the xul component is 1.539623 and complexity of code change of the considered model is 11.23341. The cumulated predicted bugs will be 1213. The bugs so far detected/fixes for this component was 1112. The bugs predicted for the coming year will be 101 (=1213-1112) for current year entropy 1.539623.

The bug prediction will be helpful to the project manager in determining the resource required in fixing the bugs to maintain the quality of software.

7   Threats to Validity

In our study, we have considered only those bugs which are affecting the changes in the current subsystems. The data points are very less in number which can affect the accuracy of parameter estimation including the performance measure in SLR and SVR. The value of decay parameters have chosen as rate of bug detection which may also be considered as threats to validity for this paper because no proper statistical investigation has been done.

8   Conclusions

If we can predict the bugs which will occur in future then we can control the quality of software. In this paper, we firstly discussed how the complexity of code change can be quantified in terms of entropy and after that the existing as well as proposed decay models by considering practical bug occurrence and fixing environment has been discussed. We have applied bug vs complexity of code changes (entropy) linear regression and support vector regression for bug prediction using existing and proposed decay models. We have compared the performance of our proposed models on the basis of different comparison criteria namely, R^2, Adjusted R^2, Variation and Root Mean Squared Prediction Error (RMSPE). The experimental results show that our proposed models capture the variability in the complexity of code changes occur due to irregular fluctuations in the code change and bug detection/fixing process. Finally, the number of post release bugs has been predicted in the software by applying linear regression technique. The proposed models give better performance over the existing models depending upon the value of k, models capture the variability in the complexity of code changes. The value of R^2 is more than 0.99 in most of the proposed models which shows how accurately our proposed models can predict the future bugs based on current year entropy. This study can be further extended to capture all changes for the entire project at a more fine grain level of interval which will be helpful in predicting the actual number of future bugs. We will validate our theory on more data sets to get more confidence in complexity of code change based bug prediction.

Acknowledgements. The research work in this paper is partially supported by grants to the first Author from Department of Science and Technology (DST), Govt. of India with Grant No. SR/S4/MS: 600/09.

References

1. Arisholm, E., Briand, L.C.: Predicting fault prone components in a java legacy system. In: Proceedings of the 2006 ACM/IEEE International Symposium on Empirical Software Engineering, pp. 8–17. ACM (2006)
2. D'Ambros, M., Lanza, M., Robbes, R.: An extensive comparison of bug prediction approaches. In: MSR 2010: Proceedings of the 7th International Working Conference on Mining Software Repositories, pp. 31–41 (2010)
3. D'Ambros, M., Lanza, M., Robbes, R.: Evaluating defect prediction approaches: A benchmark and an extensive comparison. Empirical Software Engineering 17(4-5), 537–577 (2012)
4. Elish, K.O., Elish, M.O.: Predicting defect-prone software modules using support vector machines. The Journal of Systems and Software 81, 649–660 (2008)
5. Fenton, N.E., Ohlsson, N.: Quantitative analysis of faults and failures in a complex software system. IEEE Transactions on Software Engineering 26(8), 797–814 (2000)
6. Goel, A.L., Okumoto, K.: Time dependent error detection rate model for software reliability and other performance measures. IEEE Transactions on Reliability R-28(3), 206–211 (1979)
7. Graves, T.L., Karr, A.F., Marron, J.S., Siy, H.P.: Predicting fault incidence using software change history. IEEE Transactions on Software Engineering 26(7), 653–661 (2000)
8. Gyimóthy, T., Ferenc, R., Siket, I.: Empirical validation of object-oriented metrics on open source software for fault prediction. IEEE Transactions on Software Engineering 31(10), 897–910 (2005)
9. Hassan, A.E.: Predicting Faults based on complexity of code change. In: The Proceedings of 31st Intl. Conf. on Software Engineering, pp. 78–88 (2009)
10. Hassan, A.E., Holt, R.C.: Studying the chaos in code development. In: Proceedings of 10th Working Conference on Reverse Engineering (November 2003)
11. Hassan, A.E., Holt, R.C.: The chaos of software development. In: Proceedings of the 6th IEEE International Workshop on Principles of Software Evolution (September 2003)
12. Hassan, A.E., Holt, R.C.: The top ten list: Dynamic fault prediction. In: Proceedings of ICSM 2005, pp. 263–272 (2005)
13. Herraiz, I., Gonzalez-Barahona, J.M., Robles, G.: Towards a theoretical model for software growth. In: Proceedings of the 4th International Workshop on Mining Software Repositories, Minnesotta, USA (2007)
14. Kapur, P.K., Garg, R.B., Kumar, S.: Contributions to Hardware and Software Reliability. World Scientific Publishing Co. Ltd., Singapore (1999)
15. Kapur, P.K., Garg, R.B.: A software reliability growth model for an error removal phenomenon. Software Engineering Journal 7, 291–294 (1992)
16. Kim, S., Zimmermann, T., Whitehead, J., Zeller, A.: Predicting faults from cached history. In: Proceedings of ICSE 2007, pp. 489–498. IEEE CS (2007)
17. Khoshgoftaar, T.M., Allen, E.B., Jones, W.D., Hudepohl, J.P.: Data mining for predictors of software quality. International Journal of Software Engineering and Knowledge Engineering 9(5), 547–563 (1999)
18. Leszak, M., Perry, D.E., Stoll, D.: Classification and evaluation of defects in a project retrospective. The Journal of Systems and Software 61(3), 173–187 (2002)
19. Lyu, M.R.: Handbook of Software Reliability Engineering. McGraw-Hill (1996)
20. Musa, J.D., Iannino, A., Okumoto, K.: Software Reliability, Measurement, Prediction and Application. McGraw-Hill (1987)

21. Nagappan, N., Ball, T.: Use of relative code churn measures to predict system defect density. In: Proceedings of the 27th International Conference on Software Engineering, pp. 284–292 (2005)
22. Nagappan, N., Ball, T.: Static analysis tools as early indicators of pre-release defect density. In: Proceedings of ICSE 2005, pp. 580–586. ACM (2005)
23. Nagappan, N., Ball, T., Zeller, A.: Mining metrics to predict component failures. In: Proceedings of ICSE 2006, pp. 452–461 ACM (2006)
24. Ohba, M.: Inflection S-shaped software reliability growth model. In: Osaki, S., Hotoyama, Y. (eds.) Stochastic Models in Reliability Theory. LNEMS, vol. 235, pp. 144–162. Springer, Heidelberg (1984)
25. Ostrand, T.J., Weyuker, E.J., Bell, R.M.: Predicting the location and number of faults in large complex systems. IEEE Transactions on Software Engineering 31(4), 340–355 (2005)
26. Shannon, C.E.: A Mathematical Theory of Communication. The Bell System Technical Journal 27, 379–423, 623–656 (1948)
27. Singh, V.B.: A Study on Software Reliability Growth Modeling using Change Point and Fault Dependency. University of Delhi (2008)
28. The bugZilla project (2012), http://www.bugzilla.org
29. The Mozilla project (2012), http://www.mozilla.org
30. Weisberg, S.: Applied Linear Regression. John Wiley and Sons (1980)
31. Yamada, S., Ohba, M., Osaki, S.: S-shaped software reliability growth modelling for software error detection. IEEE Trans. on Reliability R-32 (5), 475–484 (1983)
32. Vapnik, V.: The Nature of Statistical Learning Theory. Springer, New York (1995)
33. Xing, F., Guo, P.: Support vector regression for software reliability growth modeling and prediction. In: Wang, J., Liao, X.-F., Yi, Z. (eds.) ISNN 2005. LNCS, vol. 3496, pp. 925–930. Springer, Heidelberg (2005)
34. Singh, V.B., Chaturvedi, K.K.: Entropy based bug prediction using support vector regression. In: Proceedings ISDA 2012 - 12th International Conference on Intelligent System Design and Applications, Kochi, India, November 27-29, pp. 746–751. IEEE Xplore, USA (2012)

Apply Agile Method for Improving the Efficiency of Software Development Project at VNG Company

Quoc Trung Pham[1], Anh Vu Nguyen[1], and Sanjay Misra[2]

[1] School of Industrial Management, HCMC University of Technology, HCMC, Vietnam
pqtrung@hcmut.edu.vn, anhvu_05@yahoo.com
[2] Department of Computer and Information Sciences, Covenant University, Nigeria
ssopam@gmail.com

Abstract. Software engineering process (SEP) is more and more considered the key factor for any software company to create a better quality software with low costs and high productivity. However, there is a gap between theory and practice of applying modern SEP, such as Agile method, for improving the efficiency of software project management, especially for software companies in a developing country like Vietnam. VNG Corporation, a software company in Vietnam with a lot of small and medium web based software projects, currently meets many difficulties in ensuring the success of these projects. The current approach for software development is no longer consistent with the increasing requirements and flexibilities of these projects and VNG is going to find a new method for improving the efficiency of their software projects. In this research, Agile method is applied and tested to check whether it can be a suitable method for VNG to overcome their problems in software development. Besides, a testing project based on Agile method is also conducted at VNG company to evaluate the solution. Results showed that Agile method can help to increase customers' satisfaction and it also helps to improve efficiency of project management by most KPIs.

Keywords: Agile, Software engineering process, Software development model, Project management, VNG Company.

1 Introduction

According to annual report of VNG corporation, about 1/3 of its software projects failed during 2009 to 2011 (these projects cannot be finished on time). There are many reasons for this failure, but according to their managers, the most reason is from frequent changing requests of customers during the software development process (average 5-10 changing requests/ project/ month). This reason also increased the total cost of software development project about 20% compared with original budget and it delayed the time to introduce new products to end-users.

Similar to other software companies in Vietnam, software engineering process at VNG is not clearly applied and there is a lack of standardized method for software

B. Murgante et al. (Eds.): ICCSA 2013, Part II, LNCS 7972, pp. 427–442, 2013.
© Springer-Verlag Berlin Heidelberg 2013

project management. Therefore, it is necessary for Vietnamese software companies to have a clearly stated and standardized method for software development to overcome above difficulties.

Recently, Agile method, a flexible software development process, is more and more considered a suitable method for developing software in a short time and it is flexible enough to allow frequent changing requests during project time. Agile method attracted many researchers and businessmen in discussing and sharing their experience in applying this method. Currently, Agile community in Vietnam has more than 300 members.

From above reasons, this research tries to apply Agile method for improving efficiency of software development project at VNG company and to evaluate the Agile solution in practice. This research aims at (1) Identify problems of software project management at VNG company, (2) Suggest a plan for applying Agile method in VNG, and (3) Conduct testing project for evaluation.

The research plan is: first, Agile method is reviewed and compared with other software development methods; then, data is collected and analyzed for understanding current problems of software project management and possibility of applying Agile method for solving these problems at VNG company; then, a plan for applying Agile method in VNG for improving efficiency of project management is suggested; and finally, a testing project is conducted based on above suggestion for evaluating the solution. The structure of this paper is organized as follows: (2) Research method; (3) Literature review; (4) Problems of project management at VNG; (5) Approach for solving problems of VNG; (6) Experimentation and results; and (7) Conclusion.

2　Research Method

2.1　Data Collection

- Secondary data: theory reviews, scientific journals, papers, related materials from the internet, internal documents of VNG company...

- Primary data: expert interviews, questionnaires to understand current problems of project management at VNG, discussing possibility for applying Agile method in VNG and feasibility of suggested solution in practice.

2.2　Data Analysis and Result Evaluation

- Qualitative analysis: lesson learnt, projects' document analysis, group discussion with project members (20 projects from 2009-2011, 10-15 projects' members), depth interviews with project managers (3-5 people)

- Experimentation: applying suggested plan for a testing project (2 months, 5 members), calculating KPIs for testing project and comparing with those KPIs of past projects with the same size and duration.

3   Literature Review

3.1   Agile Software Development Method (Agile)

Agile is a group of methodologies for software development based on iterative and incremental rule. In this method, requirements and solutions evolved through cooperation between self-managed and inter-functional groups [8]. Besides, Agile is also considered a philosophy or an approach for software development projects. Typical software development methods based on Agile include Extreme Programming (XP) or Scrum. In this research, latest version of Agile method based on Agile-manifesto is considered to be applied and tested. According to Agile-manifesto [1], main characteristics of Agile method can be summarized as follows:

− Iterative

− Evolutionary

− Adaptive

− Group architecture

− Empirical Process Control

− Direct interaction

− Value-based development

Agile focuses on quick responsibility to requirement changes, interaction of multi-parties, short time-frame, evolutionary and continuous development. Agile suggests a collection of rules, standards and practices, such as: source code management, coding standards, or prototype presentation to customers [10].

According to Agile method, each project (product) will be divided into small functional parts. Each part is executed as a complete product in a short time (2-4 weeks). Each finished part will be transferred to customer to use, test and feedback. As the same time, other parts will be developed and tested. This process continues until all parts are completed ([1]; [3]).

3.2   Comparison of Software Development Methods

Based on theory of software development, previous researches ([13], [15], [17]), and interviews with experts in IT field, advantages and disadvantages of popular software development methods (such as: Waterfall, V-shape, Prototype, Evolution, RUP, CMMI and Agile model) are summarized and showed in following table.

3.3   Success of Software Project Management

According to James S. and Shane W. [10], traditional way of thinking about success of a software project is based on 3 constraints: on time, within budget and satisfy users' requirements. In fact, many projects satisfied all constraints above, but they cannot be considered successful projects because final system is not suitable with

Table 1. Comparison of software development models

Model	Advantages	Disadvantages	Application
Waterfall model [17]	Clear process, step by step activities, clear documents for input and output of each step.	User requirements must be defined clearly. Testing phase is executed too late to discover problem during developing process. Many risks occurred near the end of project. High cost for changing and low speed for responding.	Should be applied for projects with low risk and low changing rate. Developers understand well user-requirements and methods.
V-shape model [2]	Testing phase is executed in parallel with analysis, design and implementation.	Similar to waterfall model	Similar to waterfall model
Prototype model [5]	User can see & understand key functions and features of new system. Improve communication between developers and end-users.	Prototype may not include all requirements of users. Therefore, this model leads to misunderstanding about new system. Because of short time for developing prototype, there will be a difference between requirement and prototype.	Mostly based on GUI Must be applied in case the customers can not define their requirements clearly at the first time.
Evolution model [5]	Reusing of prototype. A part of system can be implemented during system analysis and design phase.	Lack of a strict and clear process. Longer time for requirement analysis.	Applicable for short-term projects. Developing staff is not familiar with project field.
RUP model [5]	Less risk. Key requirement will be developed and transferred to customer in a short time. Including many versions, that will help improving the quality of final system.	High developing cost. High technological risk.	Applied for big system, in a long time. A part of system can be run sooner than other parts. Developers are familiar with project field.
CMMI model [15]	Reduce risk through process improvement, clear requirement for each step. Developing plan is controlled carefully, so product can be transferred to customer on time and easily with all documents.	High cost for evaluation. High cost for executing. Long time for documenting. Inflexible in project conducting.	Applied for big system, in a long time. Suitable for outsourcing projects. Clear hierarchy of developing staff based on position and profession.
Agile model [13]	High adaptability with changing requirements. Low risk through well managed by Sprint. Low time for interaction. Customer can monitor developing steps by continuously transferring module through sprint.	Developing staff must have skills and experience. Time needed for documents during project process is fairly long.	Suitable for small and medium sized projects (10-12 members) with short time for developing.

organization's purpose and it is not used by end-users in practice. James S. and Shane W. [10] also showed 3 kinds of the success of a software project, which are: individual success, technological success, and organizational success. Through their analysis, Agile method is considered a suitable method for software project and it can help software project managers to get all kinds of success above.

In this research, efficiency of software project management is measured based on ability to get 3 kinds of above success: Individual success, Technological success and Organizational success.

In the next section, problems of software project management at VNG company will be analyzed to understand the current situation of project management at VNG and to explore ability to solve these problems using Agile method. Based on this analysis, a plan for applying Agile method for improving efficiency of software project management will be suggested.

4 Problems of Project Management at VNG

4.1 VNG Company and Web Based Application Department

VNG Corporation (VNG) – former name: Vinagame – is a Vietnamese software company established in 2004 and specialized in developing online games and doing e-commerce business in Vietnam. It is the first company providing licensed games in Vietnam and now becomes a leading company in Vietnam online game market. Beside popular online games, such as: Vo Lam Truyen Ky, Zing Dance, Gunny, Boom online…, products of VNG also include Social networking site (ZingMe), Music site (ZingMp3), and e-Commerce sites (123Mua, ZingDeal). Web-based application department takes responsibility for developing web-based applications for both internal and external customers.

According to internal statistics and interviews with project managers, group leaders of Web based application department (3 people), total number of projects of this department from 2009 to 2011 is more than 20 projects, in which, 7 projects were failed or didn't finish on time. High rate of failed projects leads to low competitiveness and indirectly reduces total revenues of VNG.

Although, currently, Web-based application department has a strong background staff (more than 90% graduated from university with IT major) and good experience in software development (4-5 years working in IT field), these employees are not well managed and their responsibilities are not clear. This causes low quality of final products (40% of products are unsatisfied), and loss the trust of internal customers. Besides, members' role in each project is not clear, this causes unsatisfactoriness of projects' members and leads to difficulties in progress controlling and quality assurance.

Most products developed by Web-based application department are rich-content applications, which are main factor for attracting customers. Each web-based application has a lot of requirement changes during and after implementation or publish phase (avg. 5-6 changes). So, most products have to be changed in design or content every 3 months. This change is very important for the success of these projects, but Web-based application department cannot serve this demand thoroughly. As a result, changing queue is longer and longer and project management is considered inefficiency.

From end-user viewpoint, 3 main factors for attracting users and getting their loyalty are: rich content editing, attractive user interface, and short response time. For the first factor, VNG realized its importance and established Web-content department to care about this factor. For the two later factors, they are main responsibility of Web-based application department and VNG cannot satisfy their customers mostly because of these 2 factors. Web-based application performance is directly affected by analysis, design and implementation phase. Therefore, without a suitable software development method, Web-based application department cannot provide attractive user interface in a short time as customers' request.

4.2   Problems of Software Development Process at VNG

Based on internal demand for applications to satisfy VNG own requirements, Web-based application department was established with a few experienced employees. At the first time, because the number of projects is not much, there is no need for a standard process of software development. By the time, because of developing demand, the number of employees and projects increased very fast. So, it is necessary for Web-based application department to have an effective software development process to reduce errors, reuse project works, and shorten software developing time.

From 2009, waterfall model was applied and used by developers for improving the efficiency of project management in this department. At the earlier time, waterfall process brings many benefits to Web-based application department and it helps managing projects better. After a while (about 2011), all projects' member are familiar with waterfall process and all changing requests must be analyzed and approved before executing. This habit causes many difficulties for project management because current projects require higher quality, short time for development and more changing requests from customers. This problem requires a new software engineering process that can help shortening software developing time and making it easy for integrating new requests to final products.

According to year-end report of 2011, number of severe bugs (causes system halt more than 1 hour) of 2011 increased 20% higher than of 2010. This high rate of severe bugs affected directly to website performance and displayed a bad image about VNG's product. There are many reasons for this high rate, such as: server management, internet connection, inexperience of employees in solving problems and uncontrollable changing requests from customers. The last reason is considered the most important one (average 5-6 changing request/ product/ month). These problems increased project cost (avg. 20% higher than original budget) and affected to general target of VNG company in introducing new products to customers.

In order to know root causes of difficulties at Web-based application department, an interview with 15 projects' members was conducted. Through this interview, seven key reasons for their difficulties in software engineering process are summarized as follows:

- Current process is not suitable for short time development with frequent changing request

- High cost for maintenance of previous products

- Time span for each phase of software development process is not appropriate

- Improper setting of priority and value of changing requests

- Can not apply lesson learnt from previous projects to latter projects

- High leading time in communication between support department and projects' members.

- Complex project architecture and ineffective information flow

In general, at the current time (2012), Web-based application department meets a lot of difficulties in managing software development projects using waterfall model. This requires VNG company to find a new suitable model for software development to overcome above problems and to improve the efficiency of software project management at Web-based application department.

5   Approach for Solving Problems of VNG

From above analysis, main reason for inefficiency of software project management at Web-based application department stays in current waterfall model, improper software engineering process for VNG products at this time. Based on Table 1, two good candidate models for VNG to improve its current process are: CMMI model and Agile model.

According to Dr. Nguyen Long, General Secretary of Vietnam Informatics Association, in an interview with Saigon Businessman magazine [7], CMMI model is suitable for:

- Outsourcing oriented company because CMMI is required by foreigner customers.

- Big company because of high cost for certification.

- Big projects because it requires many members for documenting.

- Clear users' requirements at the first time because changing requests must be re-negotiated.

According to Barry B. and Richard T. [4], five criteria for evaluating suitability of Agile model with software development project include:

- Low criticality or low risk level.

- Experienced projects' members.

- Frequent changing requests.

— Small sized projects located in the same location.

— Open culture for changing and innovation.

In order to select suitable software developing model, an interview with 3 project managers at Web-based application department was conducted to identify key characteristics of VNG projects, budget allowed, developers' skill, changing request from customer and company culture.

Through this interview, CMMI model is realized not suitable and Agile model is proven to be a suitable model with VNG projects because:

— Most projects of Web-based application department are conducted to satisfy demands of internal customers and to support developing strategy of VNG.

— These projects do not have clear requirements at the first time and need a lot of changes during developing time.

— Most projects are small projects and located at the same place.

— Time allowed for software development must be short (less than 3 months)

— Developers at Web-based application department have a strong IT background and many experiences in software development.

— Web-based application department has a fairly young staff (average age is 30) and an open culture that facilitates free ideas and innovation.

From above analysis, the approach for solving problems of Web-based application department of this research will be applying Agile method for software engineering process. In order to know how to apply Agile method and to check whether Agile method can improve the efficiency of software project management of VNG or not, a pilot project will be conducted for evaluation.

6   Experimentation and Results

6.1   Experimentation Design

• Time for experimentation: from January 2012 to March 2012

• Pilot project for testing Agile method is SGN project (Social Gaming Network – phase 2): this software provides some additional features to a previous product, such as: manage users' profile, search for product information, friend connection between applications, activity notification, manage community page…

• Project members: 5 members (4 software engineer + 1 project manager), in which 3 members have 5-6 years experience and 2 other members have 1 year experience.

• This project is considered suitable with Agile method because it has:

— Low criticality: this project belongs to phase 2 of a current product of VNG to add more features and to increase value for current product, so it has a low risk level.

- Experienced project' members: all members have a bachelor degree in IT major and have enough experience for conducting this project.

- Frequent changing requests: because of social networking characteristics, this project allows many changing requests during project time to satisfy the customer more.

- Small size: 5 members working at the same location.

- Free culture: this is the long-term strategy of VNG for supporting creativity of employees.

- Plan for apply Agile method in VNG through pilot project

  - Training: combination of training course and self learning based on documents/ regulations from project managers to provide project' members key concepts of Agile method.

  - Habit changing: direct interaction with end-users/customers is a requirement for all project members; changing from organizing a few long meetings to daily short meetings to be able to quickly response to any necessary problem; self-responsibility is encouraged by assigning independent project works to each member and receiving complete results.

  - Setting KPIs: changing criteria for evaluating final results based on following factors

    o Concurrent users (CCU): number of customers using product at the same time. This is important factor to know whether new product attracts end-users or not. This number is based on system statistics.

    o Number of new users (system statistics)

    o Total time using product of end-users (system statistics)

    o Percentage of operational works decreased (compared with previous projects)

    o Satisfaction of internal customers

    o Satisfaction of external customers through comments, feedbacks

- Result comparison plan

  - In order to evaluate the efficiency of project management using Agile method, pilot project will be compared with previous projects (using traditional method) by some criteria, such as: time to finish, number of updated content, number of changing requests, time to response...

  - Projects (about 15 projects in the past 2 years) will be used for evaluating have the same size and duration with pilot project.

  - For comparison of project success, above criteria will be arranged in 3 kinds of project success: Individual success, Technological success and Organizatinal success.

6.2   Experimentation Results

After applying Agile method in software development of Web-based application department for testing project (SGN), some positive results in improving efficiency of project management could be summarized as follows:

- Concurrent users (CCU) of Gunny product (a system supported by SGN project) increased 1.5 times compared with CCU of this product before.
- Number of new users increased quickly through new features added by SGN project ("friend invitation" through users' connection network).
- Total time using product of end-users also increased through promotion events of SGN project.
- Total time and cost for operational works of Web-based application department, such as: inputting and updating web contents, has been decreased. Currently, these activities are run automatically due to a new feature added by SGN project. Quantity of web content needed to input manually is only 1/5 compared with before.
- Satisfaction of internal customers increased because product of SGN project connected existing products/services of VNG together through social networking, which facilitates internal communication and collaboration.
- Overall satisfaction of external customers also increased (90% satisfied, through an online survey) because of higher quality of final product (faster response time, better information, higher stability and simpler process).

In general, testing project showed that Agile method helps to increase satisfaction of customers, and to improve efficiency of project management through above KPIs. As a result, Agile method could be a suitable software engineering process for VNG to overcome its current problems and it should be applied in an enterprise-wide scope to increase software quality of VNG.

However, during process of testing project, some difficulties are also realized as follows:

- Unfamiliar of some project members with Agile process made it difficult for them to deal with overloaded works.
- It is difficult for project members to make a final decision and to remember all details of project works without an effective collaborating system.

Above difficulties can be overcome by

- Training and applying Agile model for a while makes developers familiar with new method,
- Getting supports from an effective communication and collaboration platform.

In order to check whether Agile method can help improving efficiency of project management, results of SGN project (Agile group) will be compared with previous projects (No-Agile group) by some criteria, such as: time to finish, number of updated content, number of changing requests, time to response…, which will be arranged into 3 kinds of success: Individual success, Technological success and Organizational success. The comparison results are summarized in following tables:

Table 2. Comparison of Agile group and No-Agile group by Organizational Success

ID	Criteria	No Agile Group	Agile Group
1	Time to finish	Later than plan 1-2 days.	Finnish on time
2	Customers' satisfaction about product	Average, because final product covered all users' requirements	High, because any feedback from customers will be added into final product
3	Possibility of project structural change or cancellation of some features of project.	Low, because there is a baseline of agreements between project group and customers.	High, because all users' requirements could be solved during project process.
4	Project efficiency	Meet requirement	Higher than requirement

Table 3. Comparison of Agile group and No-Agile group by Technological Success

ID	Criteria	No Agile Group	Agile Group
1	Technology complex	Average	High
2	Specification of users' requirements	Clear	Unclear, final result is different from original requirement
3	Request changing rate	Frequently (avg. 6 requests/ project)	Frequently (avg. 8 changing requests/ project)
4	Responding time	Average 1 day/ request	Average ½ day/ request
5	Extensibility of final product	Difficult because of fixed architecture	Easy because flexible and updatable architecture
6	Transferrable of product to another developer/ project group	Easy to transfer because of clear and multi-referenced documents	Difficult to transfer because of not priority of documentation and direct interactions between developers and end-users.
7	Difference of final product with original design	Low, because requirements and designs must be agreed at the first time of project.	High, because of high changing request during project time. Original design is just a draft version and can be changed gradually in implementation.
8	Product maintenance	Easy to maintain because final product is 90% similar to original design & documents are completed.	Fairly difficult to maintain because lack of documents and difference between final product and original design.

Table 4. Comparison of Agile group and No-Agile group by Individual Success

ID	Criteria	No Agile Group	Agile Group
1	Developing experience and skills	Average	High
2	Experience accumulation of project members	Average because each member only works in separated phase of project	Very high, because project members must take part in all phases of project
3	Working pressure of project members	Average, 6 hours/day, according to project leader's plan	Very high, because each member takes responsibility for whole module. Members are highly focused & task-oriented.

In summary, comparison results showed that projects of Agile group got higher efficiency than projects of No-Agile group in all kinds of success: Organizational, Technological and Individual success. Although there are still few disadvantages of Agile group in some criteria, it is proven to be a suitable method because overall benefit of Agile method is higher than its limitation to the efficiency of project management at Web-based application department of VNG Company.

7   Conclusion

From above results, Agile method is found to be suitable software development process for Web-based application department at VNG company. This method can help improving efficiency of project management and contributing to the success of software development project in all aspects: Organizational success, Technological success and Individual success.

Firstly, this research found some difficulties of software development at Web-based application department of VNG Company through internal data analysis and interviews with projects' members. These problems include: Inappropriate software engineering process, High cost for maintenance, Improper project schedule, Unsuitable setting of priority of changing request, Lost of past experience, High leading time, Complex project architecture.

Then, through discussion with project managers at Web-based application department, possibility of applying Agile method for improving efficiency of software project management at VNG was analyzed and testing projects was conducted for evaluation. Experimentation results showed that Agile method can help improving efficiency of project management at Web-based application department and solving problems of software development at VNG Company. This pilot project proved that Agile could be applied in VNG at enterprise-wide level with minor modifications.

However, Agile method should be applied for those projects with following characteristics: Low criticality, Experienced personnel, Dynamism, Small sized project, Open culture. Therefore, in order to apply this method in practice, companies should change their environment to fit with above conditions through some activities, such as: training, changing regulations, setting suitable KPIs, changing business culture...

Beside above advantages, this research also found some limitations of applying Agile method in practice and they should be improved to ensure the success of software development project. For example, applying Agile method may lead to difficulties in transferring product to another developing group and maintaining final product because of insufficiency of consistent project documents. These difficulties could be overcome by focusing on documenting phase after each project to have a complete project document. As a result, well organized documents will facilitate transferring of projects' work to other groups and make it easy for maintenance activities.

7.1   Limitations

This research explored the ability to apply Agile method for a small sample (1 project/ 1 department) of VNG Company. It is difficult to generalize the result to the whole company. Especially, it is very difficult for establishing a standard approach for applying Agile method for similar departments or companies because it depends on project size and characteristics of developing group.

Moreover, qualitative method used in this research is somewhat subjective because many conclusions are based on literature review and interviews with project members. This reduces the applicability of research results in practice.

7.2   Implication for Future Researches

Some directions for future research include:

— Apply Agile method for a bigger sample, such as: other departments of VNG, other software companies in Vietnam, service businesses…

— Quantitative research for measuring influence of Agile on the success of software project

— Combination of Agile method and CMMI for a better software engineering process.

Acknowledgements. Many thanks to employees and managers of VNG company, who provided internal data for analysis or participated in several interviews of this research for discussing possibility of applying Agile method for solving problems of Web-based application department and improving efficiency of software project management at VNG company.

References

[1]   Agile Alliance: Agile manifesto (2011), http://www.agilealliance.org/ (retrieved)

[2]   Andrew, R.: SAS Software Development with the V-Model. Ratcliffe Technical Services Limited, United Kingdom (2011)

[3]   Barlow, J.B., Justin, S.G., Mark, J.K., David, W.W., Ryan, M.S., Paul, B.L., Anthony, V.: Overview and Guidance on Agile Development in Large Organizations. Communications of the Association for Information Systems 29(1), 25–44 (2011)

[4]   Boehm, B., Turner, R.: Rebalancing Your Organization's Agility and Discipline. In: Maurer, F., Wells, D. (eds.) XP/Agile Universe 2003. LNCS, vol. 2753, pp. 1–8. Springer, Heidelberg (2003)

[5]   Cao, D.A.: Software development models. Global CyberSoft Vietnam (2005), http://www.pcworld.com.vn (retrieved)

[6]   Dean, L.: Agile Software Requirements: Lean Requirements Practices for Teams, Programs, and the Enterprise, 1st edn. Addison-Wesley Professional, USA (2011)

[7]   Dieu, T.: CMMi – Challenges of software quality. Doanh Nhan Saigon Online (2011), http://m.doanhnhansaigon.vn/ (retrieved)

[8]   Duong, T.T.: Scrum Introduction. FPT Company (2011), http://hanoiscrum.net (retrieved)

[9]   Granville, G.M.: The Characteristics of Agile Software Processes. TogetherSoft (2001)

[10]   James, S., Shane, W.: The Art of Agile Development. O'Reilly Media, Inc., CA (2008)

[11]   Ken, S., Jeff, S.: Scrum Guide. Scrum organization (2010), `http://Scrum.org` (retrieved)
[12]   Mountain Goat Software Company: Scrum Introduction (2011), `http://www.mountaingoatsoftware.com` (retrieved)
[13]   Pekka, A., Outi, S., Jussi, R., Juhani, W.: Agile software development methods - Review and analysis. VTT Publication 478 (2002)
[14]   RADTAC Company: Briefing paper of Agile unified process (2011), `http://www.radtac.co.uk` (retrieved)
[15]   Software Engineering Institute: CMMI model introduction. (2012), `http://sei.cmu.edu` (retrieved)
[16]   VNG Corporation: Company introduction (2011), `http://www.vng.com.vn` (retrieved)
[17]   Waterfall Model: Water-fall model Introduction (2011), `http://waterfall-model.com` (retrieved)

Appendices

A. Questionnaire for Exploring Problems of Project Management at VNG

Purpose: to find facts and figures for understanding about current difficulties of software project management at Web-based application department of VNG company from 2009 to 2011

Main objective: project team leaders and project managers

Number of interviewees: 3 people

Questions:

1. Number of projects conducted in your department from 2009 to 2011?
2. Please provide percentage of success project in this period according to following criteria
   a. Finnish on time
   b. Meet technological requirements
   c. Make customer satisfied
   d. Meet customers' requirements
3. How many projects cannot be finished by deadline?
4. How many projects cannot meet technological requirements?
5. How many projects finished on time but cannot satisfy customers?
6. As a project manager, are you satisfied with your group works? Rate your satisfaction with your project members (percentage)?
7. For you, which skills of your project members should be improved?
8. For you, within your success projects, how many percent of them could be considered:
   a. Individual success
   b. Technological success
   c. Organizational success
9. Is budget the most important problem of your projects?

10. Please tell me about average size of your projects by following criteria
    a. Number of project members
    b. Time constraint of your projects
11. How many changing requests your projects received from customers during or after each project? Please provide average number of changing requests/project/month?
12. Do you think most problems of your project management come from current software engineering process?
13. Do you think frequent changing requests or personal problems of project members have an effect on the success of your projects (time and quality)? Please give an example.
14. Please arrange following items in an priority order for improving efficiency of your project management
    a. Applying a more suitable software engineering process
    b. Improving problem solving skill of project members
    c. Closely collaborating with customers for receiving and analyzing requirements
    d. Providing more budget for your projects
    e. Changing projects' duration and scope

B. Questionnaire for Lesson Learnt of Project Management at VNG

Purpose: Extract lessons learnt after each software development project at Web-based application department from 2009 to 2011

Main objective: Members of software development projects

Number of interviewees: 15 people.

Questions:

1. How many projects did you participate at Web-based application department from January 2009 to December 2011?
2. Did you attend the project closure meeting after each project?
3. For you, what are main causes of late projects?
4. As a developer, do you have any experience in unsuccessfully integrating individual modules of a project?
5. Did you care about project cost? If yes, which kind of cost did you care about?
6. For you, does personal experience contribute to the success of your projects?
7. Is current software engineering process suitable with your projects? If not, please show some unsuitable points of current process.
8. How do you think about number of changing requests from customers in your projects?
9. Currently, does indirect communication with customers through access point (group/ project leaders) delay information flows in your projects?
10. How frequent of changing requests occurred in your projects? Is responding to these requests difficult to you/ your group? If yes, please give some main difficulties.

11. For you, are your customers satisfied with your products? If not, please give some reasons why you are unsatisfied.
12. What do you expect from your project manager?
13. According to your experience, which factor is the most important one affecting on the success/ failure of a software project?
14. What should be improved for ensuring the success of your projects?

C. Questionnaire for Exploring Suitability of Agile with Projects at VNG

Purpose: Collecting internal data and statistics of testing project and past projects for evaluating suitability of Agile method for improving efficiency of project management at Web-based application department of VNG company.

Main objective: Team leaders and project managers.

Number of interviewees: 3 people.

Questions:

1. Do you know about Agile method?
2. What is the average size of your projects? Do you think your projects' size is suitable with Agile method?
3. What is your current company culture? Do you think your company culture is suitable with Agile method? If not, what should you do to have an open culture for enabling Agile?
4. Do you think software engineering process affects on the success of your projects? Please give an example to explain your answer?
5. Agile method requires high discipline. Do you think your projects' members have a good discipline? If not, what should you do to improve the situation?
6. Agile method requires quick response time to customers' changing requests. Do you think there is any difficulty for direct interaction with customers? Which communication methods are currently in used at your department?
7. How do you think about current relationship between projects' members and customers? If not good, what is the most reason?
8. Ranking following items in priority order for applying Agile method in your department:
   a. Independent working ability of project members
   b. Analysis and planning skill
   c. Time management ability
   d. Skill for direct interaction with customers
   e. Programming skill
9. Agile method requires frequent short meetings. Do you think this kind of meeting is suitable with your project group? Is there any difficulty in applying short meetings in your project?
10. In your projects, did you receive changing requests frequently? Is there any difficulty in responding to these changing requests? Please give an example. For you, compared with total project cost, how many percent does cost for changing account for?

A Framework for Modular
and Customizable Software Analysis

Pedro Martins[1,*], Nuno Carvalho[2], João Paulo Fernandes[1,3],
José João Almeida[2], and João Saraiva[1]

[1] High-Assurance Software Laboratory (HASLAB/INESC TEC),
Universidade do Minho, Portugal
[2] Computer Science and Technology Center (CCTC),
Universidade do Minho, Portugal
[3] Reliable and Secure Computation Group ((rel)ease),
Universidade da Beira Interior, Portugal
{prmartins,narcarvalho,jpaulo,jj,jas}@di.uminho.pt

Abstract. This paper presents a framework for the analysis of software
artifacts. We revise and propose techniques that aid in the manipulation
and combination of target-language specific tools, and in handling and
controlling the results of such tools. We also propose to integrate under our framework techniques that are capable of performing language
independent analyses.

The final result of our work is an analysis environment that is modular
and flexible and that allows easy and elegant implementations of complex
analysis suites.

We finally conduct a proof of concept for our framework by analyzing
a well-known, widely used open-source software package.

Keywords: Software Analysis, Software Certification, Combinator Languages.

1 Introduction

Building software has historically always been considered a challenging engineering task. And this is particularly true nowadays, where programming involves
not only reusing libraries that are provided by our programming language of
choice, but also trusting and reusing libraries that have been built by other
programmers and that are available on the Internet as open source software.

While software reuse has evident benefits such as rapid development, one often
needs to make sure that the reused libraries satisfy certain properties. In our
context, we refer to analyzing a property of piece of software as its certification

---

* This work is partly funded by ERDF - European Regional Development Fund
through the COMPETE Programme (operational programme for competitiveness)
and by National Funds through the FCT - Fundação para a Ciência e a Tecnologia
(Portuguese Foundation for Science and Technology) within projects FCOMP-01-
0124-FEDER-010049, and FCOMP-01-0124-FEDER-022701.

B. Murgante et al. (Eds.): ICCSA 2013, Part II, LNCS 7972, pp. 443–458, 2013.

and indeed choosing between many available libraries with the same purpose is often influenced by the properties that each one holds.

This paper consolidates our ongoing effort to provide a customizable framework for the certification of (reusable) software packages under the Certification and Re-engineering of Open Source Software (CROSS) project[1]. Indeed, when one tries to come up with a solution for dealing with all packages, under any programming language and using potentially any analysis tool, multiple issues arise:

i) how can we provide a setting for users to easily and elegantly combine different tools, each one providing a concrete and desirable analysis for the same package?

   A solution for this problem was achieved by us with a combinator language that allows programmers to describe at an abstract level how software tools can be combined into powerful software certification processes [1].

ii) with such a language at hand, we turned our focus into providing a globally accessible framework where users could define and store their customized certification processes and that could work as a repository for a wide range of analysis tools that would otherwise need to be installed locally.

   With this in mind, we have developed a web portal that relies on the domain-specific language of i) as its core [2], and that provides a common storage location for multi-purpose analysis tools.[2]

iii) finally, we have realized the need to customize the results produced by different certifications. Indeed, in our context certification results may assume different formats ranging from simple text to complex images or charts, and producing impactful results (or reports) again may require manual customization.

   In order to satisfy this need, we have also developed an embedded domain-specific language for combining reports [3].

In this paper we now propose to build on these results and to improve on them. In particular, we make the following contributions:

1. we propose a single and coherent framework that elegantly integrates the combinator languages described in i) and iii). This is a framework that is currently under integration in the portal provided in ii);

2. while the tools that have already been integrated in our web portal allow for language-specific analyses only (i.e., they target one specific language such as C or Java and they focus on one specific characteristic of it), we now propose the integration of a set of analyses that are language independent. This analyses include inspecting elements other than source code that must be available on any software ecosystem, such as README files or even comments within the source code itself. This further extends the potential practical interest of our certification environment;

---

[1] http://twiki.di.uminho.pt/twiki/bin/view/Research/CROSS/
[2] While the portal already stores a significant number of analysis tools, still we rely on further inputs from the community to enlarge this set.

3. we provide a detailed case study that fully illustrates, one by one, the steps to undergo in order to certify a realistic software package. By this, we also hope to demonstrate the expressive and practical power of the global certification scenario that we envision in this paper.

This paper is organized as follows. In Sections 2 and 3 we revise the combinator languages described previously in items *i*) and *iii*), respectively. In Section 4, we propose and describe the set of analyses for elements other than source code that we have considered extending our certification environment with, i.e., we describe contribution 2 of the paper. In Section 5 we refer to contribution 1 of the paper, i.e., we describe how we have integrated in a single and coherent framework all the independent pieces that are necessary to provide customizable software certifications. The case study that we have used to demonstrate the power of the framework that we finally obtain, in the line of contribution 3 is described in Section 6. Finally, in Section 7 we conclude the paper and point some possible directions for future research.

2   Combining Software Analysis Tools

We start by revising the combinator language that is used in our web portal to allow the creation and customization of analysis schemas, that we call Certifications. Such combinator language, which has been proposed and thoroughly explained in [1], allows an easy and modular implementation of flows of information through different analysis tools that are integrated in our web server. Furthermore, the results of these tools always end up being collected and transformed into a report, contained on a XML file to which the users also have access.

In order to briefly introduce the reader to the combinator language of our portal, we present next a simple yet illustrative example of all the combinators in the language, using them in the construction of a concrete Certification. The example shown is a snippet of Haskell [4] code, which is the programming language that we have used to implement our combinators.

Example of a Certification defined with combinators:

```
certification =
    Input >- (slicer,"-j","-jpg")
            >- (jpeg2Report,"-jpeg","-r")   >|
    Input >- (memoryCheck,"-j","-r")          >|>
    (aggregator,"-r","-r")
    +> "Certification"
```

This example certification is called certification, and has two main flows of information, both started by the primitive Input. After Input is used, the user can create a flow of information by intercalating tools with the primitive >-. As long as he/she continues to use this primitive, the initial information will be consumed by the first tool, whose result will be consumed by the second one, and so on. In this particular example, we have one sequence: Input >- (slicer,"-j","-jpg") >- (jpg2Report,"-jpeg","-r"), where the initial input is processed by the tool 'slicer', which fuels information to jpeg2Report.

The combinators >| and >|> are used for parallel computation: the first one appears between linear flows of information, and splits the certification into parts that are autonomous and can run in parallel. The combinator >|> appears only once per parallel computation, and marks a point where all the results of all the parallel processes are combined using one tool whose responsibility is only to aggregate all these results.

In this particular example, we have two flows of information both started with the primitive Input. These are split with >-, meaning they will constitute two sets of tests that will run in parallel, and terminated with >|>, meaning that their result will be aggregated with the tool aggregator.

The last combinator is +>. The only purpose of this combinator is to create a Certification, by giving it a name, which in this case is "certification", and making it available on our web portal.

Our Certifications web portal is a collection of bash tools created and maintained by any user, with the only limitations of being capable of running in an UNIX-based shell and using the the standard UNIX streams, STDIN and STD-OUT to process and return information. Our combinator language works on top of this standard ambient and provides, through the web portal, a set of primitives that channels information through tools.

This combinators language is powerful enough to allow parallel chains of analysis, for example, when a user wants to integrate into the report two different results from two different types of analysis which are in no way related, while isolating them (failures in one chain do not necessarily imply the failure of the whole analysis, as the system tries to isolate errors). This is achieved through a meta-script that is generated by these combinators and controls systems calls and the flow of information through the tools that integrate the analysis.

Another important feature of our analysis combinators language is type checking. Tools by themselves are not type-constrained - they are just bash tools consuming and returning information through standard streams, but on practice tools have specific limitations according to the analysis they perform: some might work on Java or C, others on Haskell or even on XML, so there is a need to constrain these specifications on our combinators language.

In the previous example, we can see that tools are called with parameters, such as in: (slicer,"-j","-jpeg"). This parameters are used to indicate the types this tool will deal with. In this case, the tool will read Java code and produce a JPEG.

To avoid potential type errors, we force tools to have, in our web portal database, a set on input and output types, which can be triggered by the arguments when

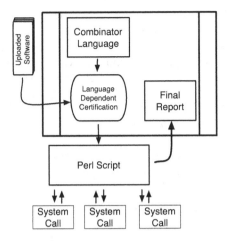

Fig. 1. The combinator language for tools

calling the tools. When creating a Certification, the user is forced to say explicitly the types of information one expects the tools to handle, and the machinery inherent to this language checks if the tools support the given types and if the types match between the flow on information throughout the tools.

Figure 1 shows an overview of the whole process: the user defines an analysis schema on the web portal using the combinators primitives, which is transformed into a Perl script, which does a set of systems calls to the tools that constitute the analysis and produces a report that the user can visualize on the web portal.

Through this combinator language one can easily create analysis suits on our web portal, which are self-contained but modular and can be further used to create more analysis. In [1] we explain in detail the creation, the usage and the inherent mechanisms that support this combinators, such as our Attribute Grammar-based type checker.

3   Report Combinators

In the previous section we have reviewed a combinator language that allows a user to easily create test suits using combinator primitives that specify an underlying script responsible for making controlled system calls and transform them into a report.

We have showed how powerful and simple a combinator language can be for controlling and creating test suits, but there is still one important part of that analysis - the report - where the user has no control whatsoever. This might not seem like a big problem for small analysis, but for situations when the user wants to perform a huge set of analysis, being able to organize their results is very important for being able to understand the analysis itself.

In this section we will review a combinator language that allows precisely this: through it, the user is capable of organizing and customizing the layout of the report and even personalizing it with custom titles and notes. This language, together with its implementation as well as the algorithms that support it is further analyzed in [3].

Next, we present an example of a report specified with our combinator language for reports. Similarly to the previous example, it is also written in Haskell.

Example of a report defined with combinators:

```
report = Init >| ("Memory Tests",
                  (beginSubsection $ cert1)
                 )
            >- ("Usability Tests",
                  ( beginSubsection $ cert2)
                    >-- ("Result of Cert3", cert3)
                    >-- ( cert4)
                 )
```

These combinators are used to specify reports, so through them the user can defined wether a certain certification should fit into a section, or a subsection or if should have a customized title. In the particular case of report, it is composed by 4 different certifications, here named cert1 to cert4 for simplicity, whose results are organized through sections and subsections.

The report starts with >|, whose only purpose is to start a report. This combinator is mandatory and its single usage represents the smallest report possible, composed by one section only.

In this particular example, the combinator is immediately followed by the string "Memory Tests", which represents the name of this section, and by another primitive, beginSubsection that, as the name clearly states, creates a subsection. This subsection is unnamed, because it is followed directly by a certification, and not by a string and a certification. The machinery responsibly for the implementation of these combinators gives standard names such as "Subsection 1" in cases like these.

The primitive >| can be followed by an infinite number of >-. Each of these create a new section, with the exact same rules we have seen: it's name can be customized by writing a string immediately after and it can be followed by the primitive beginSubsection to further structure the report. In this case, report is composed by two sections, named "Memory Tests" and "Usability Tests" respectively.

There is also the option to create an infinite number os subsections for each section in the report. In this case, the results is cert3 and cert4 are both integrated into subsections of "Usability tests", using the combinator >--. It is important to note that in the case of the result of cert3, a custom name is given to the subsection that integrates its result: "Result of Cert3". The programmer chose not to customize the title of the result of cert4.

Next, we present the XML file created by the combinators that implement report.

Example of a report generated with combinators:

```
<?xml version="1.0" encoding="ISO-8859-1"?>
<section title="Memory Tests">
    <subsection>
        c1_result
    </subsection>
</section>
<section title="Usability Tests">
    <subsection>
        c2_result
    </subsection>
    <subsection title="Result of Cert3">
        c3_result
    </subsection>
    <subsection>
        c4_result
    </subsection>
</section>
```

As stated earlier, reports in our web portal environment are represented by XML files. The user is presented with an HyperText Markup Language (HTML) report in our portal, but it is just an transformation using eXtensible Stylesheet Language Transformations (XSLT) of the XML file, to which the user always has access. This also holds for the tool combinators presented in the previous section.

4   Non Source Code Software Analysis

Besides source code, another fundamental source of information about open source software lies in documentation, and other non source code files, like README, INSTALL, or HowTo files, commonly available in the software ecosystem. These documents, written in natural language, provide valuable information during the software development stage, and also in future maintenance and evolution tasks.

The CROSS research project aims at developing software analysis techniques that can be combined to assess open source software projects. Although most of the effort is spent analyzing source code, non-source code content found in packages can have a direct impact on the overall quality of the software. Forward *et al* survey [5] about the general opinion of software professionals regarding the relevance of documentation and related tools, highlights the general consensus that documentation content is relevant and important. It also highlights that software documentation technologies should be more aware of professionals' requirements, opposed to blindly enforce documentation formats and tools.

Documentation analysis is also relevant in other research areas. Program Comprehension is an area of Software Engineering concerned with gathering information and providing knowledge about software, to help programmers understand how a program works in order to ease software evolution and maintenance tasks [6]. Many of the techniques and methods used rely on mappings between program elements and the real world concepts these elements are addressing [7]. Non-source code content included in software packages can provide clues and valuable information to enhance the creation of these mappings. Program maintainers often rely on documentation to understand some key aspects of the software [8].

DMOSS[3] is a toolkit designed to systematically assess the quality of non source code text found in software packages. The goal of the toolkit is to provide a systematic approach to gather metrics about this content and assess its quality. It starts by gathering content written in natural language found in the package, process this content to compute metrics, and finally reason about these metrics to draw conclusions about the overall software quality. The toolkit handles a software package as an attribute tree, and the major engines for processing a package are implemented using tree transversal techniques. The specific metric calculations are made using a specialized set of plugins, that are responsible for: (1) analyzing a specific chunk of text and produce a metric, (2) reduce and aggregate sets of metrics to produce intermediate and final results, and (3) use templates for creating report snippets. Adding features to the analysis workflow is just a matter of adding a new plugin. This approach has allowed the development of a modular and pluggable toolkit, easy to maintain and extend. The toolkit can process any package, regardless of programming language used, but the text extracting tool (from files) can require update for some specific archiving technologies or documentation formats.

Assessing software quality for any given definition of quality is not easy [9] mainly due to subjectivity. DMOSS evaluates the non-source code files included in a software package. This set of files can include README files, INSTALL files, HTML documentation pages, or even UNIX man(ual) pages. Instead of trying to come up with a definition for quality, we select three main traits that we are concerned about. We envisage that these characteristics have a direct impact in the overall documentation quality regardless of the degree of individual subjectivity.

- Readability: text readability can be subjective, but there are linguistic characteristics that generally make text harder to read. Some of them can even be measured, as for example, the number of syntax errors or the excessive use of abbreviations;
- Actuality: this is an important feature of documentation and other textual files, they should be up-to-date, and refer to the latest version of the software;

---

[3] Documentation Mining Open Source Software.

- Completeness: this trait tells us how much the documentation is complete, and if it addressees all the required topics.

DMOSS processes a software package to gather information about specific metrics that are related with these traits.

The dmoss-process tool provided by the toolkit is used to process a package. The result of processing a given package is a tree, decorated with attributes storing the calculated set of features. Another tool provided by the toolkit is dmoss-report, that uses the result of the previous tool to create a report in HTML format. An example report, created for the tree[4] software package (version 1.5.3) is illustrated in Figure 2.

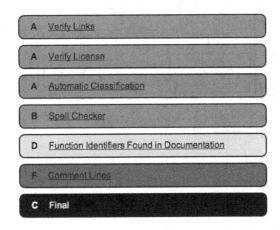

Fig. 2. Screenshot of a HTML report produced using DMOSS[5]

This report shows metrics that are used to grade key features about the package. For example, many documents in software packages contain links to official websites or discussion forums, one of the plugins included in the toolkit validates that these link are still working. If all links included in the documentation are working this feature is graded A. Another example is the number of comment lines in order to the total source code lines. In this specific case the percentage of comment lines per number of line codes is below 20%, which graded this feature of documentation with grade F. Some of these features are based on thresholds, that can be configured and adapted to specific contexts or packages. By clicking on each specific feature in the HTML report, more information is shown regarding each specific metric. A final grade is given to the package (C in this report), which is the features' grade average.

For more details about the DMOSS toolkit please refer to [10].

---

[4] Available from http://mama.indstate.edu/users/ice/tree/.
[5] Figure requires colored printing for optimal visualization.

5   Improving Software Analysis in CROSS

We have revised in Sections 2, 3 and 4 different technologies that aid in software analysis by implementing techniques that are applied in the analysis customization, in its resulting data and in the verification of important meta-information orthogonal to most software systems.

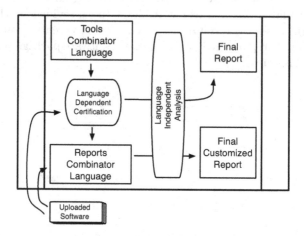

Fig. 3. A framework for software analysis

In this section we describe how the integration of different technologies creates an inter-dependent ecosystem that is capable of producing important artifacts whose information can help in understanding, improving and expanding a huge domain of programs written in various programming languages.

In Figure 3 we present the overall analysis framework that constitutes our analysis environment. We can clearly see three technologies in action and how they are interconnected:

– The tool combination language is being used to create language-dependent certifications, which can, alone, evaluate and analyze software artifacts. These certifications are built upon software tools existing in the web portal, and represent controlled flows of information through these tools, until a desired result is obtained.
– The report combination language is built upon the certifications created with the tool combination language. Working in a similar fashion, this time we are not controlling the analysis itself, but rather the data it provides. With the introduction of this technology, different certifications can be structured to create powerful and customized reports.
– The language-independent analysis works as a layer providing contextual information for uploaded software resources. It is independent of the type of analysis, being it a simple certification or a complex multi-certification set, and provides important results about the uploaded software.

The analysis framework suggested in this paper is the result of the integration of all these technologies into a setting that uses the main advantaged of which one of them to support powerful software analysis.

5.1   Integrating the Report Combinators

The technology to customize certifications by creating flows of information across heterogeneous tools in already integrated in our web portal [2], and is an important way in which software can be analyzed and studied in our environment. The structure of the reports that results from that analysis, on the other way, was predetermined and its structured was steady.

The machinery presented in Section 3 was explicitly created to solve this issued, with its integration designed to be natural to our environment and its usage intuitive and similar to the usage of other CROSS mechanisms, both for analysis of specifications.

Specifying a report in our web portal works in parallel with creating a certification with the tools combinators: all certifications can be used and combined into a custom report, but this task is done in two steps: first the user creates a set of certifications he wants to compose into an analysis suite (or uses the ones that already exist in the portal's database) and then he specifies how this suite of certifications is composed into a structured report.

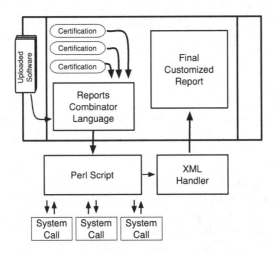

Fig. 4. The combinator language for reports customization

Figure 4 shows how this machinery integrates into our web portal. It starts with report customization based on previous existing certifications. Since certifications represent by themselves analysis suites, these are applied in order until all their results is obtained. The next step is to pass the results through an

XML Handler, whose responsibility is to arrange them according to the user specification.

Similarly to the architecture sketched in Figure 1, the final report is presented in HTML format, although it exists as raw XML to which the user has access. The analysis engines also try to isolate errors and apply the analysis even if there are problems with its constituting certifications.

With the introduction of this technology, we believe our analysis environment becomes more powerful while maintaining it easy to use, by facilitating the important step of analyzing the final data through which qualifiable results are obtained.

The access to the XML file that results from analyzing software is also an important feature since users might want to apply further automated analysis to the results produces by our framework, being XML an optimum medium for this analysis. This automation can even be performed by uploading the XML report into capable certifications specified in the framework itself.

5.2   Integrating Language-Independent Analysis

Since a language-independent analysis is an important technology to be integrated in any software analysis, the mechanics of our framework imply this techniques constitutes a layer which, by being orthogonal to any software artifact, is also orthogonal to any analysis.

The modular nature of our framework implies this analysis was easily integrated, and its results are as customizable as the results of any other certification, aiding in the quality and in the data of the final report.

6   Case Study

In this section, we apply our software quality framework to a software artifact of practical interest: we analyze the VLC media player (VLC)[6]. This is a free, open source and cross-platform multimedia player whose capability of playing various multimedia encoded files and various streaming protocols make it a well-known and widely used tool. We have used the source code from VLC version 2.0.5[7] that is available from SourceForge[8], a well-known, web-based source code repository that hosts more than 300000 projects. The VLC version that we have used is of size 18.4 MBytes and contains more than 3500 files.

To test VLC, we envisioned a test composed of:

1) Our default language-independent analysis to produce generic results regarding the software documentation and the overall quality of its source code.

---

[6] http://www.videolan.org

[7] http://sourceforge.net/projects/vlc/files/
2.0.5/vlc-2.0.5.tar.xz/download (accessed in 2013-2-14).

[8] http://sourceforge.net

2) A certification to compute the number of C source files (in particular, this certification searches for *.c files).

3) A certification that identifies C source-code files in which function strcat() is used. Unless care is taken, this function can cause memory overwriting, and in extreme cases allows hacking of the target machine through buffer overflow[9]. The use of this function in a program does not necessarily imply that it is unsafe, but may raise improvement concerns.

4) A certification that produces the number of C++ lines of code throughout the entire project to give a general overview of the amount of functionality that is implemented in this language.

Certifications 1) to 4) are composed using our combinator language for reports, with the results of certifications 1), 2) and 4) being shown in the first, second and third sections of the report, respectively. The result of certification 3) is intended to be shown as a subsection of the second section, providing the overall perspective illustrated as follows:

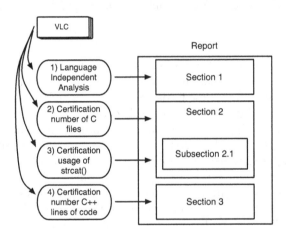

Fig. 5. An example of an analysis

The analysis that we have implemented is straightforward: VLC feeds a serious of certifications that analyze different aspects of it while producing a structured report. While certifications larger in number and in complexity and more structured reports are possible within our framework, we have opted to maintain our running example as simple and clear as possible.

In Figure 7 (page 16) we see the final resulting report that was produced after analyzing VLC. The first thing to notice is the structure of the report: it is easy

---

[9] http://en.wikibooks.org/wiki/C_Programming/C_Reference/string.h/strcat (accessed in 2013-2-14).

to read and to understand, and we see the subdivision specified in Figure 5. It is important to remember that the final report is always an XML file, which we do not show here due to size constrains, and the layout presented is the HTML file we choose to generate from the original XML report. The user is free to personalized this transformation as he finds better suits his/her needs, or even analyze the XML file directly.

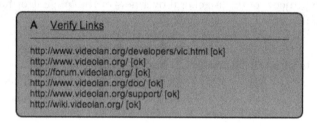

Fig. 6. One example of the results provided by the language-independent analysis[10]

The language-independent analysis is shown as a series of colored lists. This lists represent the various fields that are analyzed, and that go from licenses verifications to spell checking. These fields are colored depending on how critical the results are, and are fully expandable to show details about each specific analysis. Figure 6 shows an example of the expanded results of the field: "Verify Links". These are green, meaning the results are good, and show each link individually and its status, indicating if they are either online or offline. All the fields of the language-independent are expandable to show details about its analysis.

Another interesting result is the number of files our analysis found which use the function strcat(). We found thirteen files in the source code of VLC that use this function, indicating good points for possible optimizations of the software. It should be noticed that due to the high modularity of our framework, the user could easily change this certification to show simpler results, such as only indicating if it found any usage of strcat(), or more complex ones, where the exact line and column of each file is presented in the report. Making these modifications would imply only very simple transformations to the certification that searches for these parameters.

By looking at the results presented in Figure 5 we can obtain interesting information about a random software package, VLC, even though we choose a very simple analysis made by simple certifications. We believe such interesting results obtained by this simple analysis suite prove the potential of our framework to support complex examinations and, more important, to produce important information from software artifacts.

---

[10] Figure requires colored printing for optimal visualization.

This report was validated by our XML Schema.

Analysis report:

This report was generated on *2013-13-02*.
The file uploaded was: *vlc-2.0.5.tar.xz*

Language Independent Analysis

Description:

    Small descrition goes here. More information in:
    http://eremita.di.uminho.pt/~nrc/cross/dmoss.html.

Result:

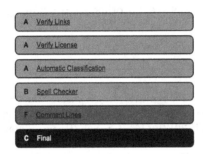

Number of C source files

Description:

    This certification gives the number of files whose extension is '*.c'.

Result:

    This program has 765 C files.

Files where the function strcat() is used

Description:

    This certification shows all the files where the function strcat() is used.
    A revision of these files is advised.

Result:

    ./doc/libvlc/vlc-thumb.c: 1
    ./modules/access/rtsp/real.c: 1
    ./modules/access/zip/zipaccess.c: 1
    ./modules/demux/mp4/drms.c: 1
    ./modules/demux/subtitle.c: 11
    ./modules/media_library/sql_monitor.c: 1
    ./modules/packetizer/vc1.c: 0
    ./modules/services_discovery/sap.c: 1
    ./modules/stream_filter/httplive.c: 1
    ./src/input/var.c: 1
    ./src/misc/update_crypto.c: 1
    ./src/playlist/loadsave.c: 1
    ./src/stream_output/sdp.c: 1

Number of C++ lines of code

Description:

    This certification gives us the number of lines of code
    in all the files whose extension is '*.cpp'.

Result:

    This program has 89116 lines of C++ source code.

End of the report!

Fig. 7. The report produced after analyzing VLC

7   Conclusion

In this paper we have revised different technologies that aid in software analysis, and which we have combined into an analysis framework that is modular and flexible, allowing an easy implementation of software analysis both by integrating new techniques and by re-organizing existing ones.

We have furthermore applied one example of such an analysis to a well known, medium sized but realistic software product from the open source community, whose results are promising and prove the potential of our framework.

As for future work, we will be focusing our attention on flexible and effective ways of spreading the analysis results other than simply showing them on a browser.

References

1. Martins, P., Fernandes, J.P., Saraiva, J.: A purely functional combinator language for software quality assessment. In: Symposium on Languages, Applications and Technologies (SLATE 2012). OASICS, vol. 21, pp. 51–69. Schloss Dagstuhl - Leibniz-Zentrum fuer Informatik (2012)
2. Martins, P., Fernandes, J.P., Saraiva, J.: A web portal for the certification of open source software. In: 6th International Workshop on Foundations and Techniques for Open Source Software Certification (OPENCERT 2012). LNCS (2012) (to appear)
3. Martins, P., Fernandes, J.P., Saraiva, J.: A combinator language for software quality reports. International Journal of Computer and Communication Engineering 2 (2013)
4. Simon, Hughes, J., Augustsson, L., Barton, D., Boutel, B., Burton, W., Fasel, J., Hammond, K., Hinze, R., Hudak, P., Johnsson, T., Jones, M., Launchbury, J., Meijer, E., Peterson, J., Reid, A., Runciman, C., Wadler, P.: The Haskell 98 Report (1999)
5. Forward, A., Lethbridge, T.: The relevance of software documentation, tools and technologies: a survey. In: Proceedings of the 2002 ACM Symposium on Document Engineering, pp. 26–33. ACM (2002)
6. Nelson, M.L.: A survey of reverse engineering and program comprehension. CoRR (2005)
7. Rajlich, V., Wilde, N.: The role of concepts in program comprehension. In: Proceesing of the 10th International Workshop on Program Comprehension, pp. 271–278. IEEE (2002)
8. Thomas, B., Tilley, S.: Documentation for software engineers: what is needed to aid system understanding? In: Proceedings of the 19th Annual International Conference on Computer Documentation, pp. 235–236. ACM (2001)
9. Kitchenham, B., Pfleeger, S.: Software quality: the elusive target [special issues section]. IEEE Software 13(1), 12–21 (1996)
10. Carvalho, N.R., Simões, A., Almeida, J.J.: Open source software documentation mining for quality assessment. In: WorldCIST 2013 - World Conference on Information Systems and Technologies Proceedings (to appear, 2013)

Complexity Metrics for ClassSheet Models[*]

Jácome Cunha[1,2], João Paulo Fernandes[1,3], Jorge Mendes[1], and João Saraiva[1]

[1] High-Assurance Software Laboratory (HASLab/INESC TEC) &
Universidade do Minho, Portugal
{jacome,jpaulo,jorgemendes,jas}@di.uminho.pt
[2] CIICESI, ESTGF, Instituto Politécnico do Porto, Portugal
jmc@estgf.ipp.pt
[3] Reliable and Secure Computation Group ((rel)ease),
Universidade da Beira Interior, Portugal
jpf@di.ubi.pt

Abstract. This paper proposes a set of metrics for the assessment of
the complexity of models defining the business logic of spreadsheets. This
set can be considered the first step in the direction of building a quality
standard for spreadsheet models, that is still to be defined.

The computation of concrete metric values has further been integrated
under a well-established model-driven spreadsheet development environment, providing a framework for the analysis of spreadsheet models under
spreadsheets themselves.

Keywords: Spreadsheets, Models, ClassSheets, Metrics, Quality.

1   Introduction

Spreadsheet systems are paradigmatic in terms of widespread use and success.
Indeed, spreadsheets are intensively used in industry in the development of business applications specially by non-professional programmers, often referred to as
end users.

The reasons for the tremendous commercial success that spreadsheets experience undergoes continuous debate, but it is almost unanimous that two key
aspects deserve to be recognized: i) spreadsheets are highly flexible, which inherently guarantees that they are intensively multi-purpose; ii) the initial learning
effort associated with the use of spreadsheets is objectively low.

It is also widely accepted that spreadsheets, in contrast with their success, tend
to be highly error-prone. In this line, several studies can be referenced [14,16,17],

---

[*] This work is part funded by ERDF - European Regional Development Fund
through the COMPETE Programme (operational programme for competitiveness) and by National Funds through the FCT - Fundação para a
Ciência e a Tecnologia (Portuguese Foundation for Science and Technology)
within projects FCOMP-01-0124-FEDER-010048, FCOMP-01-0124-FEDER-020484,
and FCOMP-01-0124-FEDER-022701. The three first authors were funded
by FCT grants SFRH/BPD/73358/2010, SFRH/BPD/46987/2008, and
BI2-2012_PTDC/EIA-CCO/108613/2008_UMINHO, respectively.

B. Murgante et al. (Eds.): ICCSA 2013, Part II, LNCS 7972, pp. 459–474, 2013.
© Springer-Verlag Berlin Heidelberg 2013

including one showing that up to 94% of all spreadsheets contain errors [13]. Also, there is a long and frequently-updated list of horror stories that directly involve spreadsheets, which is maintained by the *European Spreadsheet Risks Interest Group*[1]. It will only take a minute scrolling this list for the interested reader to be acquainted with how easy it is to cause great (mostly financial) damage using a simple spreadsheet.

In an attempt to address some of the issues that arise from the use of spreadsheets, Engels and Erwig [9] proposed the use of models, namely ClassSheets, to abstractly define the business logic of spreadsheet data. The key idea is that it is easier to understand, to maintain, and to develop such abstract an concise business logic models than the corresponding, possibly large and complex, spreadsheet data. Furthermore, they also proposed a first attempt to use Model-Driven Engineering (MDE) in the context of spreadsheets: from the ClassSheet model a first spreadsheet (i.e. instance) is produced. This a *standard* spreadsheet where some of the business logic, expressed in the model, is embedded as spreadsheet formulas and visual objects. Such a generated spreadsheet guides end users inputting correct data, thus avoiding errors. Recently, we have extended this work to provide a full MDE experience [5]: both ClassSheet models and spreadsheet data are defined in a widely used spreadsheet system where the same visual spreadsheet representation and user interaction is provided for both software artifacts. Most importantly, we developed techniques to allow end users to evolve either the model or its instance, having the correlated artifact automatically updated [4]. These techniques are implemented in the MDSheet framework [6].

The approach provided by the MDSheet framework highly resembles the way a civil engineer thinks, for example, of a house: first, a model is defined and thoroughly evolved, and only then a house is actually built. During this evolution process the engineer computes metrics to reason about the model/house: for example, determining its area, the number of stairs needed between floors, etc. Furthermore, he/she also needs to reason/understand the complexity and quality of the model so that the construction of the house does not become impossible or too expensive. When developing a ClassSheet model, or when evolving an existing one, we face the same problem: we need to reason about the model so that we evolve it in the right direction. That is to say that just like civil engineers, we need metrics for ClassSheet models so we can understand their complexity and quality. However, no such set of metrics has been proposed so far.

In this paper, we build on this limitation to make a first step towards the construction of the first quality standard for spreadsheet models. Indeed, we propose a comprehensive and representative set of metrics that can be used to provide complexity considerations of such models, being this the first major contribution of this paper. The second major contribution of the paper is the empirical analysis of the proposed metrics. Indeed, we apply our metrics to a significant set of spreadsheet models, and we suggest how the value of each metric may

---

[1] The horror stories are available at http://www.eusprig.org/horror-stories.htm.

positively or negatively influence the quality of the overall model. Finally, the third contribution of the paper is that we have implemented the computation of the metrics that we propose under the environment of [6]. This means that for any spreadsheet model under such environment, users can automatically obtain its corresponding value for each of the proposed metrics.

The remaining of this paper is organized as follows. In Section 2 we revise the spreadsheet modeling framework under which we propose to analyze the complexity of spreadsheet models. In Section 3 we introduce in detail the set of metrics that we propose to use in our analysis, and in Section 4 we describe its implementation. Finally, in Section 5 we compare our work with related works and in Section 6 we conclude the paper and point some directions for future research.

2 Modeling Spreadsheets with ClassSheets

In this paper, we propose a set of metrics for spreadsheet models. In particular, we focus our attention on a particular type of well-established and well-studied modeling framework for spreadsheets: the ClassSheet framework [9].

ClassSheets are a high-level, object-oriented formalism to specify the business logic of spreadsheets. ClassSheets allow users to express business object structures within a spreadsheet using concepts from the Unified Modeling Language (UML) such as *classes* and *attributes*, and in fact a mapping from ClassSheets to UML has already been proposed [8]. Using ClassSheet models, it is possible to define spreadsheet tables and to give them names, to define labels for the table's columns, to specify the types of the values such columns may contain and also the way the table expands (e.g. horizontally or vertically).

Using ClassSheets, spreadsheet development is triggered by the definition of a ClassSheet model, that abstractly defines the structure of a spreadsheet, from which a concrete spreadsheet instance, in which actual data is to be inputted, is derived. In order to achieve a practical spreadsheet development environment, we have in the past proposed to embed ClassSheet models in spreadsheets themselves [5]. This feature was further fully integrated in a widely used spreadsheet system [6], and our approach provided the first coherent and single environment for creating and evolving spreadsheet models while automatically obtaining conforming instances: the ClassSheet model is defined in one worksheet of the spreadsheet while the conforming data is updated in another worksheet of that same spreadsheet.

In Figures 1 and 2 we show an example that was built under the environment that we have developed: in Figure 1 a ClassSheet model for a personal budget spreadsheet was defined, and in Figure 2 we can already see that some concrete values for, e.g. incomes such as salary and expenses with housing were already inserted in the spreadsheet that one obtains from the model defined previously. Actually, incomes and housing expenses are ClassSheet **classes** that are **vertically expansible** (indicated in the model by rows 6 and 12 being filled with ellipsis): this means that in the data instances as many concrete incomes and housing expenses as necessary are possible. One may also note that the model foresees multiple years for a budget, with year being an **horizontally expansible class** (indicated in the model by column I being filled with ellipsis). Also, ClassSheet models are attribute-based: months, for example, are **attributes** of the year class. This is indicated with lower-case names such as jan or feb in the model, but derived attributes are also present, like profitsjan or balancejan. These derived attributes consist in spreadsheet formulas, where a standard formula is used as the attribute value, and references are attribute names instead of usual cell references. Moreover, ClassSheets can also have **relationships**: the attributes that intersect both a class that expands vertically and a class that expands horizontally form a relationship. This is the case of the cells B5 to H5 which are the attributes of the relationship between the class profits (that expands vertically) and the class year (that expands horizontally).

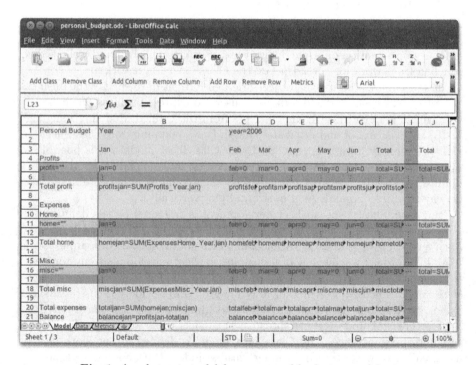

Fig. 1. An abstract model for a personal budget spreadsheet

In the remaining of this paper, we will use the budget spreadsheet model and its instance as a running example. Also, in our current work we follow the same philosophy that we have followed in the past: we have chosen to integrate the calculation of the metrics and the visualization of its results under a traditional spreadsheet environment. Indeed, one may already observe from Figures 1 and 2

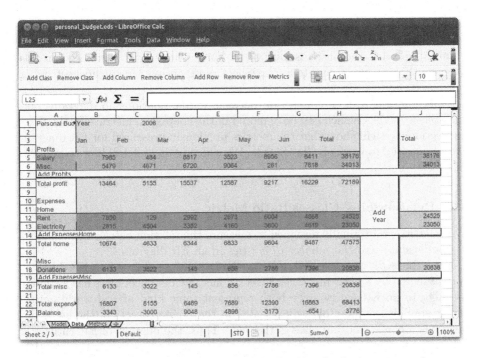

Fig. 2. A concrete spreadsheet for a person's budget for the first semester of a year

that a third worksheet named Metrics has been added to our running environment, and it is in this worksheet that the results of the metrics that we propose are presented (both numerically as well as graphically), in a way that we describe in detail in Section 4.

3   Metrics for Spreadsheet Models

In [10] the authors describe a set of metrics to analyze the complexity of Entity Relationship (ER) diagrams. These metrics are easy to understand and easy to be interpreted. Unfortunately, they were designed to work only with ER diagrams, and not in the context of spreadsheets.

In this section we will explain how the metrics presented in [10] can be adapted for spreadsheet models. More precisely, we will describe in detail how to adapt each metric to work with ClassSheets.

3.1   Why Entity Relationship Metrics Work for ClassSheets

As we described in Section 2, ClassSheets are a high level formalism with concepts like *classes* and *attributes*. This high level concepts allow to compare ClassSheet models with other paradigms, like, for example, entity relationship diagrams. In fact, in previous work we have shown that it is possible to infer

ClassSheet models from existing spreadsheets [3]. This inference technique has several steps, including the creation of an intermediate entity relationship diagram to represent the business logic of the spreadsheet data. It is based on this diagram that we can compute a ClassSheet representing the spreadsheet under consideration.

Thus, there is a close connection between entity relationship diagrams and ClassSheet models, which can be explored in the context of metrics. In the flowing sections we will show that the metrics previously presented for ER diagram can be adapted to work on ClassSheets. We will explain in detail how to perform this task.

3.2   Relationships–Classes Ratio Metric

This metric was originally designed to measure the relation between the *number of relationships* and the *number of entities* in an ER diagram.

From the work we presented in [3], we know that an entity in an ER diagram is represented by a *class* in a ClassSheet. In fact, a ClassSheet class has a similar meaning to an ER entity since both represent some kind of real world entity. In the spreadsheet example we presented in the previous section, the home expenses is a class. Another class is the year. Thus, the number of ER entities is calculated in ClassSheets as the *number of classes*. Next, we show the function that calculates this part of the metric. Note that we use Haskell [15] notation to express our functions.

```
nC = subtract 1 . length . filter (not . classExpandsB) . classes
```

This implementation is very simple: we get the classes from the model, filter out some special classes that do not represent entities, and subtract one unit (there is a class involving the model that should not be counted).

The second part of the metric, the number of relationships, is calculated in a similar way to the ER diagrams. In ClassSheets it is possible to have one attribute of one class referencing another attribute (similar to a foreign key in the database realm). Also, there are some special classes called *cell classes* which represent $M : N$ relationships. For our running example, the cells B5:H5 in the ClassSheet model (Figure 1) compose a cell class, that is, a relationship. Moreover, formulas with references also represent a kind of relationship. For instance, in the running example, the cell B20 references the two expenses classes, namely Home and Misc. Thus, representing a relationship from the class personal budget to the expenses classes.

So, to compute the *number of relationships*, we calculate the number of attributes that reference other classes, inside or outside formulas, plus the number of cell classes. If there is more than one attribute in one class referencing another class, this counts as several relationships, since each reference adds some complexity to the model.

Next, we show the function that calculates the number of relationships:

```
nR m = fromTabs + fromOthers
  where
    fromTabs = sum $ map ((+2) . length . nub . g m) cbs
    fromOthers = sum $ map (length . nub . f m) cs
    cbs = filter classExpandsB $ classes m
    cs = filter (\c -> not $ or $ map (classIntersects c) cbs)
      $ filter (uncurry (||) . (classExpandsH /\ classExpandsV))
      $ classes m
```

This implementation is more complex than the last one and we refer to the documentation of the tool for a better explanation of such function. In fact, we will not show the implementation when its complexity makes the comprehension of the work decrease, like in this case.

Finally, the metric is calculated using the following formula (all the formulas are the same as proposed in [10]):

$$RvsC = \left(\frac{nR}{nR + nC} \right)^2$$

The computed value for this metric, and for all the others we will present next, is always bounded by the interval $[0, 1]$. This helps to visualize and interpret the metric results [10].

Intuitively, the greater the number of relationships, the greater the complexity, specially if there are few classes.

For the example ClassSheet presented in Figure 1, this metric is $RvsC = (3/(3+4))^2 = 0.184$.

3.3   Class Attributes–Classes Ratio Metric

The second metric we adapt to ClassSheets was introduced in [10] to calculate the relation between the *number of entity attributes* and the *number of entities* in an ER diagram.

As we explained in the previous metric, we know that an ER entity is represented in ClassSheets by a class. Thus, the attributes of an entity are the attributes of a ClassSheet class. For instance, in the running example jan and fev are attributes of the class home expenses. Thus, the metric transposes well to the ClassSheet realm.

The attribute count is done using the following function:

```
nCA =
length . filter (cellIsFormula) . concat . grid_to_lists . grid
```

The function simply calculates the number of cells that are formulas, which, in this context, means the cells that are attributes.

The function to count of entities was already shown in the previous metric. Next, we show the formula to compute this metric:

$$CAvsC = \left(\frac{nCA}{nCA + nC}\right)^2$$

Intuitively, the greater the number of attributes in a class, the greater its complexity is.

For our running example, the metric evaluates as $CAvsC = (63/(63 + 4))^2 = 0.915$.

3.4   Relationship Attributes–Relationships Ratio Metric

This metric was originally designed to measure the relation existing between the *number of relationship attributes* and the *number of relationships* in an ER diagram.

As in entity relationship diagrams, ClassSheets can also have relationships with attributes. For instance, in the running ClassSheet example, cell B5 (with the content jan=0) is an attribute of the relationship between year and income. In fact, all the cells in the range B5:H5 are attributes of such relationship.

To compute the number of relationship attributes we use the following function:

```
nRA = length . filter (isRel) . concat . grid_to_lists . grid
  where isRel (CellFormula (Formula _ (ExpRef _ _ ))) = True
        isRel _ = False
```

The function gathers all the cell classes, that we previously identified as being relationships, and counts all the attributes composing them.

The function to compute the number of relationships was already identified. The formula to compute this metric is given next:

$$RAvsR = \left(\frac{nRA}{nRA + nR}\right)^2$$

The output of this metric means, intuitively, that the greater the number of attributes, the greater the relationship complexity.

For our running example, this metric evaluates to $RAvsR = (88/(88 + 3))^2 = 0.935$.

3.5   M:N Relationships–Relationships Ratio Metric

The metric we now explain was first introduced to measure the *number of M:N relationships* compared with the *total number of relationships* in an ER diagram.

In our realm, M:N relationships also exist. In the running example, the relationship between year and income is of this kind. This happens because for each year the spreadsheet can have several lines of income, and each kind of income can appear in several years. Thus, the metric can be adapted to work on ClassSheets.

To calculate the *number of M:N relationships*, we designed the function nMNR. Since it is a complex function, and since it would not help the reader to better understand our work, we do not show it here and refer to our implementation for more details.

The function to calculate the number of relationships was already described. To compute the complete metric, we use the following formula, again as in [10]:

$$MNRvsR = \frac{nMNR}{nR}$$

The higher the number of M:N relationships, the higher the metric will measure. In the case of ClassSheet models, and also in ER diagrams, M:N relationships are more complex to handle than other relationships.

In the running example, this evaluates to $MNRvsR = 3/3 = 1$.

3.6   1:N and 1:1 Relationships–Relationships Ratio Metric

Analogously to the previous metric, this one measures the relation between the *number of 1:N and 1:1 relationships* and the *total number of relationships*.

We have implemented in Haskell the function n1N&11, but as in the previous metric, given its complexity, we refer to the implementation for further details. Nevertheless, if a relationship is not of the kind M:N, then it must be of 1:N or 1:1 kind.

Finally, the ratio is calculated calculated using the formula:

$$1N\&11vsR = \frac{n1N\&11}{nR}$$

Intuitively, it is better to have more 1:N and 1:1 relationships than M:N. Thus, the higher this measure gets, the best.

Unfortunately, our running example does not have 1:N nor 1:1 relationships, and thus the metric measures 0.

3.7   N-Ary Relationships–Relationships Ratio Metric

It is common to have n-ary relationships in ER diagrams. Thus, this metric was introduced to measure the relation between the *number of n-ary relationships* and the *total number of relationships*.

As in the ER realm, in ClassSheets it is possible to have n-ary relationships. This can be computed counting the number of references outgoing from a binary relationship, which are given by cell classes, as explained in the next metric. To do this, it is necessary to add the number of classes that reference other classes and are referenced back by the same class.

To compute the number of n-ary relationships, we have implemented the function nNaryR:

```
nNaryR m = (length . filter ((>0). length . nub . g m) .
            filter classExpandsB . classes) m
```

Essentially, this function computes the number of cell classes, plus the number of classes that reference other classes and are referenced back by the same class.

The metric is computed calculating the ratio between the number of n-ary relationships and the total number of relationships:

$$NaryRvsR = \frac{nNaryR}{nR}$$

Similarly to M:N and 1:N relationships, n-ary relationships are more complex than binary relationships. Thus, intuitively, it is preferable to have this measure as low as possible.

Unfortunately, our running example does not have n-ary relationships, and thus the metric measures 0.

3.8 Binary Relationships–Relationships Ratio Metric

The last metric we adapt computes the relation between the *number of binary relationships* and the *total number of relationships*.

As ClassSheet models can have n-ary relationships, they can also have binary ones. In fact, if a relationship is not n-ary, it is binary. Thus, the function computes the total number of relationships minus the number of n-ary ones.

The metric is thus calculated through the formula:

$$BRvsR = \frac{nBR}{nR}$$

Once more, it is preferable to have this measure higher, meaning that most relationships in the model are binary and not n-ary.

For our running example, $BRvsR = 3/3 = 1$.

3.9 Formulas–Cells Ratio Metric

All the metrics we described until now are adapted from entity relationship diagrams and are quite interesting to give complexity related to entities/classes, relationships, and their attributes. But this is not enough for spreadsheet models. One of the main issues related to spreadsheets, and their models, is formulas and their complexity. Thus, we need to introduce new metrics that give some measures about them.

The first metric we introduce computes the relation between the *number of formulas* and the *number of cells*.

The number of formulas in a ClassSheet is given by the following function:

```
nF m = length $ catMaybes $ concat $ grid_map (aux) (grid m)
  where
    aux _ (CellFormula (Formula _ (ExpFun _ _))) = Just ()
    aux _ (CellFormula (Formula _ (ExpBinOp _ _ _))) = Just ()
    aux _ (CellFormula (Formula _ (ExpPar _))) = Just ()
    aux _ _ = Nothing
```

This functions traverses a ClassSheet model and returns the size of the list of formulas.

The function that calculates the number of cells is shown next, and counts all the cells in the model that are not empty:

```
nCe = length . filter (/= (CellValue $ VText "")) .
      concat . grid_to_lists . grid
```

The final metric if given by the formula:

$$FvsCe = \frac{nF}{nCe}$$

Intuitively, the greater the number of formulas in a model, the greater the complexity of such model.

For the running example, this metric computes the value $FvsCe = 41/130 = 0.315$.

3.10   Formula References–Formulas Ratio Metric

As the previous metric, this one reports to formulas, in particular, it gives the relation between the *number of references in formulas* and the *number of formulas*. Note that all the other references in the model are already used to calculate other metrics. In fact, the formula references were the only ones not being explored.

The function, **nRe**, that computes the references of a formula is complex, and thus we do not show it here.

The function that calculates the number of formulas in a ClassSheet model was described in the previous metric. The formula that computes this metric is now given:

$$RevsF = \left(\frac{nRe}{nRe + nF}\right)^2$$

Intuitively, the greater this measure is the more complex the model we have, that is, if each formula has many references, it will be more difficult to handle.

For our running example, the measure computed is $RevsF = (88/88 + 41)^2 = 0.465$.

3.11   Size Metrics

This last metric is also not adapted from [10]. In fact, it calculates a set of general measures about a model, namely (in parenthesis we show the metrics computed for our running example),

– the total number of non-empty cells (130);
– the width of the model, that is, the number of columns (10);
– the height, that is, the number of rows (21);
– the number of expandable classes (4);

– the number of input cells (22);
– the number of output cells (41).

These metrics are implemented traversing the ClassSheet model and counting the corresponding characteristic, which can be seen in the complete implementation.

In the next section we will show in detail the tool we have designed and implemented to make these metrics available to users in the environment they are used to, that is, in a spreadsheet system.

4   The MDSheet Framework

In the course of our work, we developed an OpenOffice/LibreOffice extension named MDSheet [6]. This tool was first created to implement the embedding of ClassSheet models, with the possibility to evolve them having the data automatically coevolved [5]. Then, we further improved it with bidirectional transformations allowing users to evolve the data without being concerned with the model since the tool synchronizes them automatically [4]. The metrics described in the previous section were also implemented in the MDSheet framework, being the implementation explained in this section.

In the user interface, a new button is available in the toolbar (see Figure 3) to evaluate the metrics for the current model. When such button is pressed, a new

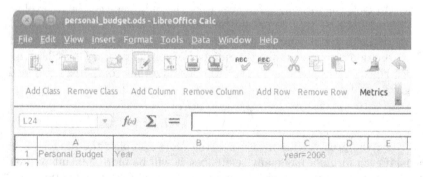

Fig. 3. MDSheet toolbar with button to evaluate the metrics of the current model

worksheet, named `Metrics`, is created with the metrics for the current model, but also with the average of the metrics for the models available in our public repository of ClassSheet models. An example of such worksheet is depicted in Figure 4, where column A contains the name of the metrics, column B contains the results obtained from running the metrics for the current model, and column C contains the average of the metrics gathered from the repository.

The goal of comparing the metrics of the current model with metrics from other models is to quickly provide a relative estimate of the quality of the current model. The repository is still a work in progress, and we aim to develop and collect a comprehensive set of models, such that new users can start working

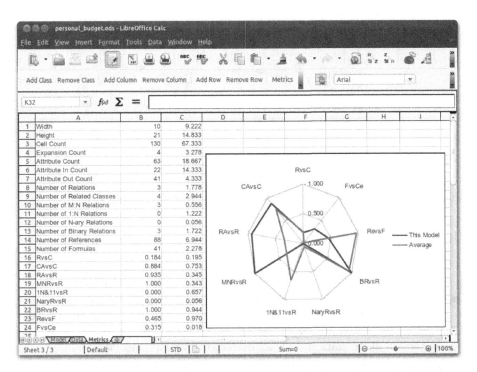

Fig. 4. Sheet with metrics generated by MDSheet for the running example

directly based on a model already available, but also that modelers can use as a starting or reference point for their work. When the repository becomes large enough, we will be able to group the models by domains (e.g., finances or database), and users will have the possibility to compare their model with others from the same domain, providing a better estimate of the model quality.

When the models repository is updated, so is the MDSheet framework, so that it contains the most recent metrics.

The MDSheet framework is developed in such a modular way that it can be very easily improved with the addition of new features, such as we did with the metrics evaluation. The framework is divided in three main parts, as can be seen in Figure 5:

Core — This part, developed entirely in Haskell, deals with the abstract representation of the models and data. This allowed to define a bidirectional transformation system at a high level, and also to specify and integrate the metrics for the models. Currently, this part is only composed by the already mentioned transformation system and the metrics evaluation code.

Interface — This part is the one that users interact with. It is mainly developed in OpenOffice Basic (an idiom of the Basic language). For the metrics evaluation, it sends a request to the core through the integration code to evaluate the metrics of the model and then creates the new worksheet with the result in it.

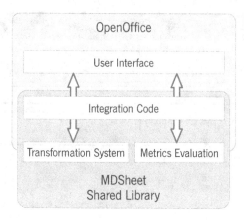

Fig. 5. Architecture of our OpenOffice/LibreOffice environment with the MDSheet extension

Integration Code — This part is used to connect the interface to the core of the MDSheet framework. This connection, developed in C/C++ and Haskell, is entrusted with the invocation of the core functions and also with the conversion of the data from the interface to the core and vice versa.

The MDSheet framework (both source code and compiled version) is available at:

http://ssaapp.di.uminho.pt

5   Related Work

Metrics for spreadsheets have been defined [1,7,12] by different authors. Unfortunately, such metrics do not work well for spreadsheet models. In this work, we adapted and extended existing metrics from the modeling realm to spreadsheet models. To the best of our knowledge, this is the first attempt of such a work.

In [2] the authors take a first step towards automated assessment of spreadsheet maintainability. They apply the selected metrics to the EUSES spreadsheet corpus in order to study their behavior. Their work aims to achieve a maintainability model for spreadsheets, whilst we defined a first set of metrics for spreadsheet models. These metrics can now be used to construct, for example, a maintainability model for spreadsheet models.

In [11] the authors sketch a new maintainability model that alleviates some problems reported by other techniques. Since this work was done in an industrial environment, the authors discuss their experiences with using such a model for IT management consultancy activities. Again, their model features maintainability of software. With our work we intend to make available the tools, i.e., the metrics, necessary to achieve a similar goal, but for ClassSheet models.

6   Conclusion

This paper presents a set of metrics for spreadsheet models, in this case, Class-Sheets. We adapted metrics from a different modeling paradigm, namely entity relationship diagrams, because there is an interesting similarity between both modeling fields.

Although all the metrics existing for ER diagrams were adapted, this was not enough. Formulas, which are a concerning issue in spreadsheets, and thus in spreadsheet models, were not considered in ER diagrams metrics, not even for computed values (considering an extended version of ER diagrams). Thus, we introduced some new metrics, in the same line of work of the other metrics, to compute some measures about ClassSheet formulas and their relationship with the underlying model. Moreover, we calculated a set of ClassSheet specific metrics such as the width of the model and the number of cells.

This work represents an interesting basis for constructing a quality standard for ClassSheets. In fact, it could be further developed and used to construct, for example, a maintainability standard or even a complete quality standard for ClassSheets.

References

1. Bregar, A.: Complexity metrics for spreadsheet models. In: Proceedings of the 2004 European Spreadsheet Risks Interest Group (EuSpRIG) CoRR abs/0802.3895, pp. 85–93 (2004)
2. Correia, J.P., Ferreira, M.A.: Measuring maintainability of spreadsheets in the wild. In: Proceedings of the 27th IEEE International Conference on Software Maintenance, ICSM 2011, pp. 516–519. IEEE (2011)
3. Cunha, J., Erwig, M., Saraiva, J.: Automatically inferring classsheet models from spreadsheets. In: Proceedings of the 2010 IEEE Symposium on Visual Languages and Human-Centric Computing, VLHCC 2010, pp. 93–100. IEEE Computer Society (2010)
4. Cunha, J., Fernandes, J.P., Mendes, J., Pacheco, H., Saraiva, J.: Bidirectional transformation of model-driven spreadsheets. In: Hu, Z., de Lara, J. (eds.) ICMT 2012. LNCS, vol. 7307, pp. 105–120. Springer, Heidelberg (2012)
5. Cunha, J., Fernandes, J.P., Mendes, J., Saraiva, J.: Embedding and evolution of spreadsheet models in spreadsheet systems. In: Proceedings of the 2011 IEEE Symposium on Visual Languages and Human-Centric Computing, VLHCC 2011, pp. 186–201. IEEE Computer Society (2011)
6. Cunha, J., Fernandes, J.P., Mendes, J., Saraiva, J.: MDSheet: A Framework for Model-driven Spreadsheet Engineering. In: Proceedings of the 34rd International Conference on Software Engineering, ICSE 2012, pp. 1412–1415. ACM (2012)
7. Cunha, J., Fernandes, J.P., Peixoto, C., Saraiva, J.: A quality model for spreadsheets. In: Proceedings of the 8th International Conference on the Quality of Information and Communications Technology, Quality in ICT Evolution Track, QUATIC 2012, pp. 231–236 (2012)
8. Cunha, J., Fernandes, J.P., Saraiva, J.: From Relational ClassSheets to UML+OCL. In: Proceedings of the Software Engineering Track at the 27th Annual ACM Symposium on Applied Computing, SAC 2012, pp. 1151–1158. ACM (2012)

9. Engels, G., Erwig, M.: ClassSheets: automatic generation of spreadsheet applications from object-oriented specifications. In: Proceedings of the 20th IEEE/ACM International Conference on Automated Software Engineering, pp. 124–133. ACM (2005)

10. Genero, M., Jiménez, L., Piattini, M.: Measuring the quality of entity relationship diagrams. In: Laender, A.H.F., Liddle, S.W., Storey, V.C. (eds.) ER 2000. LNCS, vol. 1920, pp. 513–526. Springer, Heidelberg (2000)

11. Heitlager, I., Kuipers, T., Visser, J.: A practical model for measuring maintainability. In: Proceedings of the International Conference on Quality of Information and Communications Technology, QUATIC 2007, pp. 30–39. IEEE Computer Society (2007)

12. Hodnigg, K., Mittermeir, R.T.: Metrics-based spreadsheet visualization: Support for focused maintenance. CoRR abs/0809.3009 (2008)

13. Panko, R.: Facing the problem of spreadsheet errors. Decision Line 37(5) (2006)

14. Panko, R.: Spreadsheet errors: What we know. what we think we can do. In: Proceedings of the 2000 European Spreadsheet Risks Interest Group (EuSpRIG) (2000)

15. Peyton Jones, S.: Haskell 98: Language and libraries. Journal of Functional Programming 13(1), 1–255 (2003)

16. Powell, S.G., Baker, K.R., Lawson, B.: A critical review of the literature on spreadsheet errors. Decision Support Systems 46(1), 128–138 (2008)

17. Rajalingham, K., Chadwick, D.R., Knight, B.: Classification of spreadsheet errors. In: Proceedings of the 2001 European Spreadsheet Risks Interest Group (EuSpRIG), Amsterdam (2001)

An Evaluation on Developer's Perception of XML Schema Complexity Metrics for Web Services

Marco Crasso[1,2,3], Cristian Mateos[1,2,3], José Luis Ordiales Coscia[2],
Alejandro Zunino[1,2,3], and Sanjay Misra[4]

[1] ISISTAN Research Institute
[2] UNICEN University
[3] Consejo Nacional de Investigaciones Científicas y Técnicas (CONICET)
[4] Department of Computer Engineering, Atilim University, Ankara, Turkey

Abstract. Undoubtedly, the Service-Oriented Computing (SOC) is not an incipient computing paradigm anymore, while Web Services technologies is now a very mature stack of technologies. Both have been steadily gaining maturity as their adoption in the software industry grew. Accordingly, several metric suites for assessing different quality attributes of Web Services have been recently proposed. In particular, researchers have focused on measuring services interfaces descriptions, which like any other software artifact, have a measurable size, complexity and quality. This paper presents a study that assesses human perception of some recent services interfaces complexity metrics (Basci and Misra's metrics suite). Empirical evidence suggests that a service interface that it is not complex for a software application, in terms of time and space required to analyze it, will not be necessarily well designed, in terms of best practices for designing Web Services. A Likert-based questionnaire was used to gather individuals opinions about this topic.

Keywords: Service-Oriented Computing, Web Services, Web Service Understandability, Web Service Complexity, Human Perception.

1 Introduction

Service-Oriented Computing (SOC) is a recent paradigm that allows developers to build new software by composing loosely coupled pieces of existing software, or *services* [1] . Services are usually provided by *providers*, who only expose services interfaces to the outer world, hiding technological details as much as possible. By means of these interfaces, potential *consumers* can determine what a service (functionally) offers and remotely invoke it from their applications. Consequently, SOC promotes not only code reuse but also process reuse, in the sense that the life cycle of invoked third parties services does not depend on consumers.

The high availability of broadband and ubiquitous connections nowadays allows users to reach the Internet from everywhere and at every time, enabling in turn the invocation of network-accessible services from within new software. In fact, the number of real published services has increased in the last few years. This has generated a global scale marketplace of services where providers offer their services

B. Murgante et al. (Eds.): ICCSA 2013, Part II, LNCS 7972, pp. 475–486, 2013.

interfaces, and consumers may invoke them regardless geographical aspects through the Internet [2] to ensure ubiquity. Therefore, services are implemented using standard Web-enabled languages and protocols, and thus are called *Web Services*. Web Services is on the other hand the most common technological materialization of SOC, and find their application in diverse contexts such as migrating legacy systems to modern platforms [3] or exposing remotely-accessible information to smartphones [4].

Regardless the context of usage, much of the success of an individual Web Service depends on the quality of its interface, because in practice this is the only information source consumers have available when reasoning about the functionality offered by the service [5] . Moreover, even when many approaches that aim at simplifying the task of finding appropriate services exists [6] , is the user who decides which service to select from a list of potential candidates. This decision unavoidably requires to manually inspect the description of the candidate services interfaces. Web Services interfaces are specified by using the WSDL, an XML-based format for describing a service as a set of operations, which can be invoked via message exchange.

Like any other software artifact, a WSDL document has many measurable attributes [7] . For example, [8]_ proposes a catalog of common bad practices found in WSDL documents, whereas in [7] and [9] two suites of metrics to assess the complexity of Web Services interfaces are proposed. There are almost two definitions of what it means for a WSDL document to be *complex*. One definition considers the execution time –time complexity– and space usage –spatial complexity– required by a software application to inspect and interpret a WSDL document. For instance, when automatically building a service *proxy* from a WSDL document, the complexity of a WSDL documents is directly related to the time and space needed to build the proxy. Furthermore, Kearney and his colleagues [10] alternatively state that complexity is defined *"by the difficulty of performing tasks such as coding, debugging, testing, or modifying the software"*. Under this definition, complexity relates to how complex is for humans to inspect WSDL documents.

Previous research works have emphasized on measuring some non-functional concerns associated with interfaces in WSDL [8 ,7 ,9] . These concerns, specially complexity, have been found to be related to interface understandability, i.e. under Kearney's complexity definition, and consequently some WSDL-level metrics have been proposed. Based on these catalogs, service developers can modify and improve their WSDL documents until desired metrics values and therefore certain understandability levels are met. Therefore, it is really important for services providers to consider these WSDL-level metrics in order to control WSDL documents attributes.

This paper presents an evaluation of developers' perception of the core complexity metric described in [9] , namely the Data Weight (DW) metric. The motivation of this study is that it is clear that DW reflects the time and space complexity of a data-type definition, i.e. DW is aligned with the first definition of complexity, whereas other subjective aspects of a WSDL document that impact on cognitive complexity, may not be so clearly reflected by DW. For instance, by definition of DW a definition having restrictions, attributes and extensions, which are typical constructors found in WSDL documents that allow developers to build well structured data-types, will be consider more complex than a primitive data-type, e.g. a string. In this paper we provide empirical evidence showing that the participants of our experiment perceive that WSDL documents having well structured data-types have higher DW values, which means that there is a trade-off between the "weightlessness" of data-type definitions and adopting well-design practices for defining WSDL documents data-types, as the one described in [11] .

The rest of the paper is organized as follows. Section 2 reviews related work focused on quality metrics for services interfaces and services interfaces improvement. The detailed experimental results are presented in Section 3. Section 4 concludes the paper and presents future work directions.

2   Background and Related Work

Web Services interfaces are described using WSDL, a language that allows providers to describe two main aspects of a service, namely what it does (i.e., its functionality) and how to invoke it (i.e., its binding-related information). Consumers use the former part to match external services against their needs, and the latter part to actually interact with the selected service. With WSDL, service functionality is described as a *port-type* $W = \{ O_0(I_0, R_0), .., O_N(I_N, R_N) \}$, which lists one or more operations O_i that exchange input and return messages I_i and R_i , respectively. Port-types, operations and messages are labeled with unique names, and optionally they might contain some comments.

Messages consist of *parts* that transport data between providers and consumers of services, and vice-versa. Exchanged data is represented by using data-type definitions expressed in XML Schema Definition (XSD), a language to define the structure of an XML construct. XSD offers constructors for defining simple types, restrictions, and mechanisms to define complex constructs. XSD code might be included in a WSDL document using the *types* element, but alternatively it might be put into a separate file and imported from the WSDL document or external WSDL documents so as to achieve type reuse. Therefore, it is commonly said that Web Services data-types are specified in XSD.

There have been different research efforts to measure quality in WSDL-described services interfaces, but also to improve it, though in a comparative smaller quantity. The next subsections summarize related works.

2.1   Quality Metrics for Services Interfaces Descriptions

Previous research has emphasized on the importance of services interfaces and more specifically their non-functional concerns. The work · of [8] identifies a suite of common bad practices or anti-patterns found in services interfaces, which impact on the understandability and discoverability of described services. Here, understandability is the ability of a service interface description of being self-explanatory, i.e., no extra software artifact apart from a WSDL is needed to understand the functional purpose of a service.

Discoverability refers to the ability of a service of being easily located when consumers inquiry the registry where the service is stored. The suite consists of eight bad practices that frequently occur in a corpus of public WSDL documents. To assess the understandability and discoverability of a WSDL document, one could account bad practices occurrences because the fewer the occurrences are, the better the resulting WSDL document is. The authors then offer a tool called Anti-patterns Detector [12] , which automatically computes the proposed metrics based on an input WSDL document.

[7] describes a metrics suite that consists of different kinds of metrics, ranging from common size measurements like lines of code and number of statements, to metrics for measuring the complexity and quality of services interfaces. All the involved metrics can be computed from a service interface in WSDL, since the metric suite is purely based on WSDL schema elements occurrences. The most relevant complexity metrics included in the suite are Interface Data Complexity, Interface Relation Complexity, Interface Format Complexity, Interface Structure Complexity, Data Flow Complexity (Elshof's Metric), and Language Complexity (Halstead's Metric). Moreover, the proposed quality metrics are Modularity, Adaptability, Reusability, Testability, Portability, and Conformity. Metrics results are expressed as a real coefficient in [0,1]. For complexity metrics, a value in [0-0.4) indicates low complexity, in [0.4-0.6) indicates medium complexity, in [0.6-0.8) indicates high complexity and in [0.8-1] indicates that the service is not well designed at all. Instead, for quality metrics [0-0.2) indicates no quality at all, [0.2-0.4) indicates low quality, [0.4-0.6) indicates medium quality, [0.6-0.8) indicates high quality, and [0.8-1.0] indicates very high quality.

Regarding services interfaces complexity, Baski and Misra [9] present a metric suite (BM metrics suite) whose cornerstone is that the effort required to understand data sent to and from the interfaces of a service can be characterized by the structures of the messages that are used for exchanging and conveying the data. Basing on this statement, Baski and Misra define five metrics: Data Weight (DW), Distinct Message Count (DMC), Distinct Message Ratio (DMR), Message Entropy (ME) and Message Repetition Scale (MRS), which can also be computed from a service interface in WSDL, since this metric suite is purely based on WSDL and XML schema elements occurrences. Below, we further explain these metrics since this paper is based on them.

Data Weight Metric. The definition of the Data Weight (DW) metric computes the complexity of the data-types conveyed in services messages. For the sake of brevity, we will refer to the complexity of a message $C(m)$ as an indicator of the effort required to understand, extend, adapt, and test a message m by basing on its structure. $C(m)$ counts how many elements, complex types, restrictions and simple types are exchanged by messages parts, as it is further explained in [9]. Formally:

$$DW(wsdl) = \sum_{i=1}^{n_m} C(m_i) \tag{1}$$

, where n_m is the number of messages that the WSDL document exchanges. The DW metric is a positive integer. The bigger the DW of a WSDL document, the more complex its operations messages are. For the purposes of this paper, we have assumed n_m to consider only those messages that are linked to an offered operation of the WSDL document, thus the DW metric does not take into account dangling messages.

Distinct Message Count Metric. Distinct Message Count (DMC) metric can be defined as the number of distinct-structured messages represented by $[C(m), n_{args}]$

pairs, i.e., the complexity value $C(m)$ and total number of arguments n_{args} that the message contains [9].

To better illustrate the DMC metric, in Figure 1 two WSDL documents defining an operation for returning the weather report in a city are shown. The WSDL document on the left defines two operations (**GetWeatherReportIn** y **GetWeatherReportOut**) pointing to wrapper data-types encapsulating one argument (**city** and **report**, respectively). It is easy to see that both messages have the same complexity, and thus computing DMC would output $[C(GetWeatherReportIn),1]$ and $[C(GetWeatherReportOut),1]$ as the pairs. As a result, DMC is zero since there are not distinct pairs. On the other hand, by looking at the WSDL document on the right, the GetWeatherReportRequest data-type has now two arguments. The associated pairs are thus $[C(GetWeatherReportIn),2]$ and $[C(GetWeatherReportOut),1]$, resulting in $DMC = 2$.

```
...                                              ...
<wsdl:types>                                     <wsdl:types>
  <xsd:element name="GetWeatherReportRequest">     <xsd:element name="GetWeatherReportRequest">
    <xsd:complexType>                                <xsd:complexType>
      <xsd:sequence>                                   <xsd:sequence>
        <xsd:element maxOcurrs="1" minOcurrs="1"         <xsd:element maxOcurrs="1" minOcurrs="1"
                name="city" type="xsd:string"/>                 name="latitude" type="xsd:float"/>
      <xsd:sequence>                                     <xsd:element maxOcurrs="1" minOcurrs="1"
    <xsd:complexType>                                            name="longitude" type="xsd:float"/>
  </xsd:element>                                        <xsd:sequence>
  <xsd:element name="GetWeatherReportResponse">      <xsd:complexType>
    <xsd:complexType>                                </xsd:element>
      <xsd:sequence>                                 <xsd:element name="GetWeatherReportResponse">
        <xsd:element maxOcurrs="1" minOcurrs="1"       <xsd:complexType>
                name="report" type="xsd:string"/>       <xsd:sequence>
      <xsd:sequence>                                     <xsd:element maxOcurrs="1" minOcurrs="1"
    <xsd:complexType>                                            name="report" type="xsd:string"/>
  </xsd:element>                                        <xsd:sequence>
<wsdl:types>                                          <xsd:complexType>
<wsdl:message name="GetWeatherReportIn">           </xsd:element>
  <wsdl:part element="tns:GetWeatherReportRequest" <wsdl:types>
          name="response"/>                        <wsdl:message name="GetWeatherReportIn">
</wsdl:message>                                       <wsdl:part element="tns:GetWeatherReportRequest"
<wsdl:message name="GetWeatherReportOut">                    name="response"/>
  <wsdl:part element="tns:GetWeatherReportResponse"</wsdl:message>
          rname="response"/>                       <wsdl:message name="GetWeatherReportOut">
</wsdl:message>                                       <wsdl:part element="tns:GetWeatherReportResponse"
...                                                          name="response"/>
                                                   </wsdl:message>
                                                   ...
```

Fig. 1. DMC: Examples

Distinct Message Ratio Metric. The Distinct Message Ratio (DMR) metric complements DW by attenuating the impact of having different messages within a WSDL document with the same structure. As the number of similarly-structured messages increases the complexity of a WSDL document decreases, since it is easier to understand similarly-structured messages than structurally different ones as a result of gaining familiarity with repetitive messages [9]. Formally:

$$DMR(wsdl) = \frac{DMC(wsdl)}{n_m} \quad (2)$$

The DMR metric provides a number in the range given by $[0,1]$, where 0 means that all defined messages are similarly-structured, and 1 means that all messages are dissimilar. Therefore, the WSDL documents of Figure 1 (left) and Figure 1 (right) have a DMR of $0/2 = 0$ and $2/2 = 1$, respectively.

Message Entropy Metric. The Message Entropy (ME) metric exploits the probability of similarly-structured messages to occur within a given WSDL document. Compared with the DMR metric, ME also bases on the fact that repetition of the same messages makes a developer more familiar with the WSDL document and results in ease of understandability, but ME provides better differentiation among WSDL documents in terms of complexity. Formally:

$$ME(wsdl) = \sum_{i=1}^{DMC(wsdl)} P(m_i) * (-log_2 P(m_i))$$

$$P(m_i) = \frac{nom_i}{n_m}$$

(3)

where nom_i is the number of occurrences of the ith message, and $P(m_i)$ represents the probability that such a message occurs within the given WSDL document. The ME metric has values in the range $[0, log_2(n_m)]$. A low ME value shows that the messages are consistent in structure, which means that data complexity of a WSDL document is lower than that of other WSDLs having equal DMR values.

Message Repetition Scale Metric. The Message Repetition Scale (MRS) metric analyzes variety in structures of WSDL documents. MRS measures the consistency of messages by considering frequencies of $[C(m), n_{args}]$ pairs, as follows:

$$MRS(wsdl) = \sum_{i=1}^{DMC(wsdl)} \frac{nom_i^2}{n_m}$$

(4)

The possible values for MRS are in the range $[1, n_m]$. When comparing two or more WSDL documents, a higher MRS and lower ME show that the developer needs less effort to understand the messages structures due to the repetition of similarly-structured messages.

2.2   Approaches to Improve Services Interfaces

As far as we know, two main approaches to improve services interfaces have been explored. One approach occurs at the deployment phase of services and WSDL documents. The other approach is called "early", since it deals with the improvement of interfaces during the implementation phase of the underlying services, i.e., prior to obtain their corresponding WSDL documents.

The work of [8] fits in the first approach mentioned, since it identifies WSDL bad practices and supplies guidelines to remedy these. A requirement inherent to applying these guidelines is following *contract-first*, a method that encourages designers to first build the WSDL document of a service and then supply an implementation for it. In this context, the term contract refers to *technical contract* and *interface*, indistinctly. However, the most used method to build WSDL documents in the industry is *code-first*, which means that one first implements a service and then generates the corresponding WSDL document by automatically extracting and deriving the interface from the implemented code. This means that WSDL documents are not directly created by humans but are instead automatically derived via language-dependent tools, which essentially map source code to WSDL code.

With regard to the early approach, the idea is to anticipate potential quality problems in services interfaces. Conceptually, the approach is to identify refactoring operations for services implementations that help to avoid problems in services interfaces. The main hypothesis of this approach is the existence of statistical relationships between two groups of metrics, one at the service implementation level and another at the service interface level. This implies that at least one metric in the former group could be somehow "controlled" by software engineers attempting to obtain better WSDL documents, with respect to what is measured by the metrics that belong to the latter group. This also means that if a software engineer modifies his/her service implementation and in turn this produces a variation on implementation level metrics, this change will be propagated to services interfaces metrics, assuming that both groups of metrics are correlated. Thus, key to this approach is understanding how implementation level metrics relate to interface level ones.

The work presented in [13] studies the relationships between service implementation metrics and the occurrences of the bad practices investigated in [8] , when such interfaces are built using the code-first method. This bad practices come with a number of metrics not to assess complexity but to measure service discoverability, i.e., how effective is the process of ranking or locating an individual service based on a meaningful user query that describes the desired functionality. Then, the authors gathered 6 classic OO metrics from several services implementations, namely Chidamber and Kemerer's [14] CBO (Coupling Between Objects), LCOM (Lack of Cohesion of Methods), WMC (Weighted Methods Per Class), RFC (Response for Class), plus the CAM (Cohesion Among Methods of Class) metric from the work of Bansiya and Davis [15] and the well-known lines of code (LOC) metric. Additionally, they gathered 5 ad-hoc metrics, namely TPC (Total Parameter Count), APC (Average Parameter Count), ATC (Abstract Type Count), VTC (Void Type Count), and EPM (Empty Parameters Methods). Then, the authors collected a data-set of publicly available code-first Web Services projects and analyzed the statistical relationship among metrics. Finally, the correlation analysis shows that some WSDL bad practices are strongly correlated with some implementation metrics.

As the reader can see, previous efforts present in the literature strongly suggest that employing correlation analysis among classical software engineering metrics and other metric suites is in the right direction towards predicting undesirable situations. In this sense, in [16] the authors analyze the correspondence between the metrics described in Section 2.1 and classical software engineering metrics. The achieved results show that there are statistically significant relationships among pairs of metrics. Accordingly, the authors conclude that by monitoring several metrics, which are present at services implementations, the complexity, in particular the Data Weight (DW), of the target WSDL documents can be reduced [16] .

Recent investigations show that the refactorings needed to reduce DW in WSDL documents affect other quality concerns, in particular the adoption of best design practices described in [11] . The experimental results showing these undesirable relationships are going to be explained in another paper, i.e., the explanation of the statistical correlations between DW and anti-patterns is out of the scope of this paper, but a few words about this phenomena are need to clearly explain the focus of this paper. The DW metric measures the complexity of a data-type in XSD, from the perspective of a software application that must interpret and process such data definition. For example, parses like Xerces, and xjc require more CPU processing for parsing bigger data-types definitions, in terms of lines of code. This is because DW increases as $C(m)$ counts how many elements, complex types, restrictions and simple types are exchanged by messages parts. However, not always a bigger data-type definition would be a bad design decision. In general, when using XSD for defining data-types, more lines of code means a more detailed definition, which in turn makes easier to represent a business object specifically. A contra-example of this situation is given when defining a data-type using general-purpose data-types, like the xsd:any, called "wild-card" in [17] , which allows defining almost any XSD content in just one line of code: <element name="theName" type="xsd:any"/>. Understanding business-object represented by this data-type is impossible, unless natural language documentation accompanying the data-type definition provides enough information. On the other hand, as explained in [18] when defining wild-cards developers are breaking best design practices of WSDL documents [11] .

Returning to the mentioned correlations between best design practices and the DW metric, this paper represents a step towards understanding why the refactorings needed to reduce DW may deteriorate the design quality of WSDL documents. In this sense, this paper analyzes human beings' opinions about the DW metric. This paper aims to answer whether there is a trade-off between achieving low DW values, i.e., a smaller complexity, and preserving the highest possible standards in terms of WSDL documents design quality.

3   Experimental Results

We performed a controlled experiment with humans to assess how they perceive Data Weight (DW) metric values. The experiment involved 27 participants that were asked to complete a survey to collect their opinions. The survey was designed as a Likert's based questionnaire for determining to whether there is a trade-off between services data-types design quality and how complex a service description is for a computer application, i.e., the DW metric values.

The survey was developed in the context of the "Service-Oriented Computing" [1] course of the Systems Engineering at the Faculty of Exact Sciences (Department of Computer Science) of the UNICEN during 2012. The course has been offered since in 2008, is optional, and its audience are last-year undergraduate students and postgraduate students (both master and doctoral programs) without knowledge on SOC concepts. The course requirements are excellent skills on programming and some experience with Java development. In the context of our experiment, this means that the participants could be regarded as software developers. After five lectures within one week of three hours each discussing the fundamentals of the SOC paradigm and enabling technologies the students were instructed to develop a code-first Web Service, then a contract-first Web Service, and finally an application that consumes the previous services. After that, the students were invited to complete the

---

[1] http://www.exa.unicen.edu.ar/~cmateos/cos

survey. The survey asks the participants to analyze different versions of a Web Service. Each version of the service has been chosen to present different levels of the DW metric along with data-types having different design quality. The design quality of each data-type was assessed in accordance to the catalog of best design practices of [11] . To do this, three Web Services specialist follow the conventions and recommendations of W3C XML Schema Definition 1.1 [2].

The questionnaire has been designed as a set of 18 statements. For each statement, the participant can choose among 6 alternatives, being *totally agree, agree, somewhat agree, somewhat disagree, disagree* and *completely disagree*. The reason to choose among six alternatives is that by being a pair number there is not chance for neutral opinions. We have included in the questionnaire 12 positive statements and 6 negative ones. A positive statement is a statement that textually confirms the existence of the hypothesis that the survey is testing, e.g., the existence of a trade-off between DW metric values and services interfaces design quality. On the other hand, a negative statement is one that denies the existence of the mentioned trade-off. This is important to note, because each statement is associated with a numerical score, which is known as Likert's score. The reason to have both positive and negative statements is to include contingent statements. A contingent statement is employed for re-formulating a statement to achieve more confident about a participant's response.

The numerical score of a statement depends on the polarity of the statement. For positive statements the numerical score ranges from 5 (totally agree) to 0 (totally disagree), whereas negative statements are inversely scored, i.e., ranging from 0 (totally agree) to 5 (totally disagree).

We computed the Likert score per participant. The numerical score for a participant's questionnaire is calculated by summing the individual score of all his/her statements. Then, the bigger the numerical score of a participant, the stronger is his/her confident about the existence of the trade-off between DW metric values and services interfaces design quality. A score equals to 0 means that the participant strongly disagrees with the trade-off existence, but a 90 score means that the participant strongly agrees with trade-off existence. For readability purposes, we translated the Likert's scores from a [0,90] scale to a [0,100] scale.

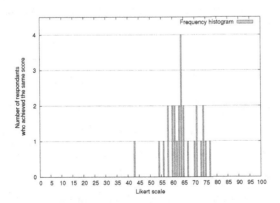

Fig. 2. Likert scale: Frequency of the scores

---

[2]   XML Schema Definition (XSD), http://www.w3.org/TR/xmlschema11-1/

Figure 2 shows the score histogram, where each bar contains the number of participants who had the same score. Figure 2 shows that nine participants did not achieve the same score as any other participant. As shown in the Figure, this situation corresponds to participants achieving extreme scores, i.e., the lowest or the highest scores. At the same time, four participants achieved the same score, as denoted by the highest bar of Figure 2 , and they were positioned in the middle of the score scale and not in the extremes.

Figure 3 shows that by smoothing the frequency results using Bézier curves, they tended to a normal distribution with an average $\mu=65$ and a standard deviation $\sigma=7.5$. Then, 95.4% of the students scored between $[\mu-2*\sigma, \mu+2*\sigma]$. In other words, 25 students scored in the range of $[50, 80]$, which manifests the existence of a trade-off between DW metric values and WSDL design criteria.

In other words, the previous results provide empirical evidence showing that the surveyed humans perceives that for a given WSDL document having high DW metric values will not necessarily mean that the service interface is complex for them. On the other hand, the surveyed humans confirmed that a well designed service interface could have a higher DW metric value in its associated WSDL document than a poorly designed one.

The questionnaire included a statement designed to assess humans' perception of two aspects of a service interface that impact on the service description complexity. One aspect is the use of descriptive names for denoting WSDL elements. This aspect is important for the purposes of the experiment as the DW metric only considers the syntactic information of a WSDL document. On the other hand, the other aspect relates to arranging error information using special WSDL fault messages. This aspect is important because for the DW metric another WSDL message would increase service complexity, whereas in [11] the authors identify *"using a separated (extra) message for exchanging fault information"* as a best design practice for WSDL documents. Figure 4 shows that the average score for the participants was 2. The y-axis of the Figure also shows the statement text. As the reader can see, the participants did not share a strongly defined opinion about the usage of descriptive names and fault messages as the only drivers for service interface quality. Accordingly, more work should be done to analyze the humans' perception of these concerns, in order to include them in the calculation of the DW metric.

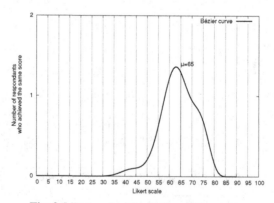

Fig. 3. Likert scale: Distribution of the scores

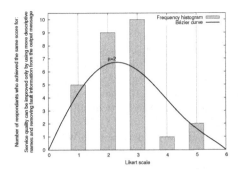

Fig. 4. Likert scale: Frequency of the scores

4   Conclusions and Future Work

This papers states that the core metric of the metric suite described in [9] is very useful for assessing the complexity of WSDL documents, when we refer to complexity as the computational effort that a software application (e.g., a parser) must put to interpret and "consume" the document, but the metric may be revised and improved in order to use it to assess the design quality of such a document. Historically, there has been controversy between software engineering metrics that are based on syntactical constructions and those based on subjective aspects. For example, when T. McCabe discusses the effectiveness of physical size driven metrics and use as an example a 50 line program consisting of 25 consecutive "IF THEN" constructs, he shows that those metrics based on LOC might not provide an accurate overview of the program complexity. In this case, we have identified a similar situation in which a WSDL document having few lines of code will have a smaller DW than another having more code and potentially a better design, in terms of data-type definitions.

A major limitation of the presented study is the number of participants that have been surveyed, although some tendencies could be observed from their responses. In this sense, the study shows that the participants perceive the existence of a trade-off between the DW metric and the design quality of services interfaces descriptions. All in all, more work should be done in order to determine and possibly adjust the DW metric definition, or define a complementary metric. Moreover, the study has shown that such a complementary metric may consider two important aspects ignored by DW, namely data-types names and fault-messages.

We are planning to extend the present paper by including a statistical analysis that shows the correlation between the DW metric and WSDL best practices. This could be done by quantifying the level of compliance of WSDL documents with the WSDL design practices proposed in [11] and performing appropriate statistical analyses.

Acknowledgments. We acknowledge the financial support provided by ANPCyT (PAE-PICT 2007-02311).

References

1. Bichler, M., Lin, K.-J.: Service-Oriented Computing. Computer 39(3), 99–101 (2006)
2. Wang, S., Sun, Q., Zou, H., Yang, F.: Reputation measure approach of Web Service for service selection. IET Software 5(5), 466–473 (2011)
3. Rodriguez, J.M., Crasso, M., Mateos, C., Zunino, A., Campo, M.: Bottom-Up and Top-Down Cobol System Migration to Web Services. IEEE Internet Computing 17(2), 44–51 (2013) ISSN 1089-7801
4. Ortiz, G., De Prado, A.G.: Improving device-aware Web Services and their mobile clients through an aspect-oriented, model-driven approach. Information and Software Technology 52(10), 1080–1093 (2010)
5. Crasso, M., Rodriguez, J.M., Zunino, A., Campo, M.: Revising WSDL documents: Why and how. Internet Computing 14(5), 30–38 (2010)
6. Crasso, M., Zunino, A., Campo, M.: A survey of approaches to Web Service discovery in Service-Oriented Architectures. Journal of Database Management 22(1), 103–134 (2011)
7. Sneed, H.M.: Measuring Web Service interfaces. In: 12th IEEE International Symposium on Web Systems Evolution (WSE), pp. 111–115 (September 2010)
8. Rodriguez, J.M., Crasso, M., Zunino, A., Campo, M.: Improving Web Service descriptions for effective service discovery. Science of Computer Programming 75(11), 1001–1021 (2010)
9. Baski, D., Misra, S.: Metrics suite for maintainability of extensible markup language Web Services. IET Software 5(3), 320–341 (2011)
10. Kearney, J.P., Sedlmeyer, R.L., Thompson, W.B., Gray, M.A., Adler, M.A.: Software complexity measurement. Communications of the ACM 29(11), 1044–1050 (1986)
11. Rodriguez, J.M., Crasso, M., Mateos, C., Zunino, A.: Best practices for describing, consuming, and discovering Web Services: a comprehensive toolset. Software: Practice and Experience (2012)
12. Rodriguez, J.M., Crasso, M., Zunino, A., Campo, M.: Automatically detecting opportunities for web service descriptions improvement. In: Cellary, W., Estevez, E. (eds.) Software Services for e-World. IFIP AICT, vol. 341, pp. 139–150. Springer, Heidelberg (2010)
13. Mateos, C., Crasso, M., Zunino, A., Coscia, J.L.O.: Detecting WSDL bad practices in code-first Web Services. International Journal of Web and Grid Services 7(4), 357–387 (2011)
14. Chidamber, S., Kemerer, C.: A metrics suite for Object Oriented design. IEEE Transactions on Software Engineering 20(6), 476–493 (1994)
15. Bansiya, J., Davis, C.G.: A hierarchical model for Object-Oriented design quality assessment. IEEE Transactions on Software Engineering 28, 4–17 (2002)
16. Coscia, J.L.O., Crasso, M., Mateos, C., Zunino, A., Misra, S.: Predicting Web Service maintainability via Object-Oriented metrics: A statistics-based approach. In: Murgante, B., Gervasi, O., Misra, S., Nedjah, N., Rocha, A.M.A.C., Taniar, D., Apduhan, B.O. (eds.) ICCSA 2012, Part IV. LNCS, vol. 7336, pp. 29–39. Springer, Heidelberg (2012)
17. Pasley, J.: Avoid XML schema wildcards for Web Service interfaces. IEEE Internet Computing 10, 72–79 (2006)
18. Crasso, M., Rodriguez, J.M., Zunino, A., Campo, M.: Revising WSDL documents: Why and How. IEEE Internet Computing 14(5), 48–56 (2010)

A New Approach
for Distributed Symbolic Software Testing

Nassima Aleb and Samir Kechid

USTHB-FEI
BP 32 Al ALLIA 16111 Bab Ezzouar Algiers Algeria
{naleb,skechid}@usthb.dz

Abstract. This paper presents a new parallel algorithm for backward symbolic execution. We use a program modeling allowing an easy distributed symbolic execution and a scalable program testing. A program is divided into several parts assigned to different nodes. A particular node: the Coordinator allocates tasks to workers and collects final results.

Keywords: Program Testing, Path coverage testing, Parallelization, Symbolic Execution; Backward Analysis.

1 Introduction

Software testing is an important part of the software engineering life cycle. However, conventional testing methods often fail to detect faults in programs. One reason for this is that a program can have an enormous number of different execution paths due to conditional and loop statements. Thus, it is practically infeasible for a test engineer to manually create test cases sufficient to detect subtle bugs in specific execution paths. In addition, it is a technically challenging task to generate test cases that cover different paths in an automated manner. To address such limitations, concolic (CONCrete + symbOLIC) testing combines concrete dynamic analysis and static symbolic analysis to automatically generate test cases to explore execution paths of a program, to achieve full path coverage. However, concolic testing may consume significant amount of time exploring execution paths, and this is an obstacle toward its practical application [4]. To address this limitation, a distributed concolic testing has been developed and studied by several research groups[17,18]. However the proposed solutions could still be ameliorated. In this paper, we present a contribution in this issue. First, we intervene on the symbolic execution process itself: we perform backwards symbolic execution rather than forwards ones as in concolic testing. The use of backward analysis is motivated by the following points:

1. It is a goal-directed analysis.
2. Backward analysis is more scalable than a forward symbolic execution because it doesn't explore unnecessary program paths (i.e., paths not leading to location of interest) and unnecessary regions of code (i.e., assignments that don't influence the truth value of the formula).

Second, we develop a distributed symbolic execution framework. Symbolic execution techniques usually represent executions as trees. Distributed process applied on trees

B. Murgante et al. (Eds.): ICCSA 2013, Part II, LNCS 7972, pp. 487–497, 2013.

has as weaknesses that it requires dynamic partitioning to be well-balanced. In this paper we use another program representation which allows an easier parallelization of the backward symbolic execution process, and which guarantee a static well-balanced portioning.

The rest of the paper is organized as follow: Section 2 describes related work. Section 3 introduces the program representation we adopt. The section 4 describes our backward symbolic execution methodology. Section 5 presents the approach of distributed generation of test cases. Finally the section 6 illustrates the method by a case study and the section 7 concludes the paper by highlighting contributions.

2   Related Work

Various testing methods and tools have been implemented to realize the core idea of symbolic execution [7]. The core idea behind Concolic testing is to obtain symbolic path formulas from concrete executions and solve them to generate test cases by using constraint solvers. Existing tools can be classified in terms of the approach they use to extract symbolic path formulas from concrete executions. The first approach for extracting symbolic path formulas is to use modified virtual machines on which target programs execute: PEX [2], KLEE [8] and jFuzz [9] are some tools using this approach. The second approach is to instrument the target source code to insert probes that extract formulas from concrete executions at run-time. Tools that use this approach include CUTE [1,14], DART [12], CREST [13], jCUTE [14,15] and SCORE [5,20]. There has been little research on employing distributed platforms to improve the scalability of concolic testing techniques. In [16] is proposed a static partitioning technique for parallelizing symbolic execution that uses pre-conditions/prefixes of symbolic executions to partition the symbolic execution tree. A limitation of this approach is that the resulting partitioned symbolic execution trees are not well-balanced. Thus, some nodes may finish exploring symbolic execution paths quickly and become idle while other nodes take long times to complete exploration, which degrades overall performance. In contrast, King [16] and ParSym [17] utilize dynamic partitioning of target program executions. King populates a queue of symbolic execution sub-trees dynamically, but the resulting speedup decreases as the number of nodes increases beyond six. ParSym uses a central server that collects test cases generated from nodes and distributes the test cases to the other nodes whose queues are empty. ParSym demonstrates speedup on grep 2.2 and a binary tree program on up to 512 nodes, but does not achieve linear speedup. Cloud9 is a testing service framework based on parallel symbolic execution techniques implemented on KLEE. Cloud9 uses dynamic partitioning to ensure that the job queue lengths of all nodes stay within a given range. SCORE distributes test cases among multiple nodes in a dynamic on-demand manner.

3   Modeling

Programs are represented by two models. The first, called: Data Model, records the different expressions computing program variables values. The second: Control Model describes the control structure of the program.

3.1   Data Model: DM

The Data Model is a table describing the program memory. Each row of DM is composed of: An integer representing the statement location in the program, the variable name; and the expression computing the corresponding variable value. An expression could be: An input, a constant or a function call. An input of a variable x is modeled by x=$x meaning that x has an unknown value: a symbolic constant of the adequate type.

3.2   Control Model: CM

The Control Model describes the program structure. It contains constraints that make possible the execution of each statement of DM.   CM models conditional and   loop statements.

Conditional Statements: There are two sorts of conditional statements: alternative statement (with the else branch) and the simple conditional (without the else branch).
An alternative statement is modeled by *(C,CT,CF,End)*:

- *C*: is the   condition of the statement.
- *CT*: The location of the first instruction to do if *C* is True.
- *CF*:The location of the first instruction to do if *C* is False.
- *End*:The location of the first instruction   after the conditional.   CM[i] is the element number i of CM, Then[i]= [CT,CF[and Else[i]=[CF,End[are its intervals. A simple conditional   statement is represented by *(C,CT,-1,End)* with *C,CT* and *End* having the same meaning as the alternative statement.

Loops: A loop statement is modeled   by *(C,CT,-2,End)*   with *C, CT* and *End* having the same meaning as in conditional   statements.
We note C[i] the condition of the block modeled by the row i of DM.

3.3   Program Modeling Example

```
1: scanf("%d%d",
          &z,&x)
       if (z==x)
3:     { t=x+1;
4:       y=x-1;
       }
       else
5:     { t=x-1;
6:       y=x+1 ; }
7:     y=t+y;
       if (z>=3)
8:       { t=z ; }
       else
9:   { t=-z;
       if (t=-1)
10:      {y=0;}
       else
11:      { y=1;}
       }
```

loc	Var	Expression
1	z	$z
2	x	$x
3	t	x+1
4	y	x-1
5	t	x-1
6	y	x+1
7	y	t+y
8	t	z
9	t	-z
10	y	0
11	y	1

	C	CT	CF	End
1	z==x	3	5	7
2	z>=3	8	9	12
3	t=-1	10	11	12

Fig. 1. Program Modeling Example

3.4   Path Modeling

Let Path[i] be a string expressing the required guard value of all the elements of CM coming before CM[i], to attain this element (CM[i]). Path [i] is called: string path. Let Path [i,k] be the character of the position k in Path[i], and let's note the expression "must be equal" by "$\equiv$". So we define Path [i,k] by:

$$
\text{Path } [i,k] = \begin{cases}
\text{`1'} & \text{if} \quad C[k] \equiv \text{True} \\
\text{`0'} & \text{if} \quad C[k] \equiv \text{False} \\
\text{`x'} & \text{if} \quad C[k] \equiv \text{True or False. \quad Both are possible} \\
\text{`*'} & \text{if} \quad C[k] \text{ value is not required}
\end{cases}
$$

Each character 'x' represents two possible paths. For example, in the figure 1: x00 represents the two paths: 100 and 000 leading to Else[3], we say that these two paths are derived from x00. So, if P is a string path containing nx characters 'x' then the number of paths represented by P is 2^{nx}.

The use of the character '*' is necessary in the situation when we follow a path in a 'else' branch, so all the conditions which are in the corresponding 'then' branch are not required and must not be evaluated.

Definition (Terminal Path): Let CM a control model having n elements. A path represented by a string path: Path[i] is said terminal if it does not exist j$\in$ [1,n] such that j$\neq$i and Path[i] is a prefix of Path[j].

The usual definition of string prefix is extended to take into account the special character 'x' in such a way that it could be matched with both '0' and '1'.

Terminal[i] is a predicate which truth value is true if and only if Path[i] is a terminal path. A terminal path represents a complete execution of the program (from the beginning of the program to its end).

4   Backwards Symbolic Execution (BSE) for Program Testing

Instead of performing as usual forward symbolic executions to compute gradually path conditions, we perform backward symbolic investigations. In fact, in forward analysis, the tree grows rapidly in width and depth. We perform backwards investigations to minimize the size of the execution tree and include only paths that are of interest to our analysis. Backward symbolic execution is based on the well-known concept of Weakest Precondition[6], so let's recall it. Let $v=e$ be an assignment, where v is a variable and e is an expression of the appropriate type. Let P be a predicate. WP($v=e,P$) is P with all occurrences of v replaced with e. For example: WP($y=x+2$, $y>8$) = $(x+2)>8$. We denote WP(l,P) the weakest precondition of the predicate P w.r.t. the statement having the location l in the table DM. In our analysis, we need to compute predicates weakest preconditions on intervals of program locations i.e.: a sequence of adjacent locations; on a whole element of CM and on a union of several intervals.

- **Sequence :** $WP([l_i,l_j[,P) = \begin{cases} P & \text{if no variable occurs in } P \\ WP(l_i,P) & \text{if } l_j=l_i \\ WP([l_i,l_{j-1}[,WP(l_{j-1},P)) & \text{Otherwise} \end{cases}$

- **Union**   : $WP([Si,Sj[\cup[Sk,Sl[,Cd)=WP([Si,Sj[,WP([Sk,Sl[,Cd))$.

- **If Then Else   statement :** E of the form (C,CT,CF,End), CF>0

  $WP(E,Cd)=C \wedge WP([CT,CF[,Cd) \vee \neg C \wedge WP([CF,End[,Cd)$

- **If Then Statement :   :**   E of the form (C,CT,CF,End), CF=-1

  $WP(E,Cd)=C \wedge WP([CT,CF[,Cd) \vee \neg\ C \wedge Cd$

- **Loops:** $(CF=-2)$: $C_0=C$; $P_0=P$; k=1;
  While(True)
  { $P_k=WP([CT,End[,P_{k-1})$; $C_k=WP([CT,End[,C_{k-1})$
    If $(C_k'=False) \vee (P_k=P_{k-1})$ { $WP(e,P)=P_k'$; exit}
    k=k+1; }
  C_k' and P_k' are obtained from C_k and P_k by replacing the variables modified in the loop by their initial values.

The algorithm using backward analysis to generate test cases performs the following steps:

1.   For each element of CM compute the string path: Path[i].
2.   For each Terminal path Path[i], concatenate 'x' at the end of Path[i] to represent the two possible alternatives in CM[i]; for each path $Path_r[i]$ derived from Path[i], compute the Weakest Precondition of $Path_r[i,l]$, where l is the length of $Path_r[i]$ along the path expressed by $Path_r[i]$, to deduce the required input values.

Example: Let's return to the program of the figure 1:

loc	Var	Expression
1	Z	$z
2	X	$x
3	T	x+1
4	Y	x-1
5	T	x-1
6	Y	x+1
7	Y	t+y
8	T	z
9	T	-z
10	Y	0
11	Y	1

I	C	CT	CF	End	Path[i]	Terminal (Y/N)
1	z==x	3	5	7	-	N
2	z>=3	8	9	12	X	N
3	t=-1	10	11	12	x0 : x0x	Y

Path[1] is designated by "-" representing the empty string in fact the block CM[1] does not required any constraint to be reached. Path[2]= "x" since Path [2,1] = C[1]= (z==x) ≡ True or False, both are possible. Path[3]="x0" i.e. Path[3,1]=C[1]=(z==x)≡True or False and Path[3,2]=C[2]=(z>=3)≡False : To access to the block CM[3] the condition (z>=3) must be false and the condition (z==x) could have any value. Terminal [3] is True; So we concatenate 'x' at the end to represent the two possible alternatives in CM[3]. The paths derived from Path[3] are $Path_1[3]$="000" ; $Path_2[3]$="001"; $Path_3[3]$="100" and $Path_4[3]$="101"

Let's generate a test case for $Path_1[3]$ by performing backward analysis:
WP([1,3[∪[5,8[, z<3)∧WP([1,3[,z≠x)∧WP([9,10[,t≠-1)=($z<3)∧($z≠$x)∧($z≠1)=True
➜$z=1∧ $x≠1. The same reasoning is performed for the other paths.

The number of program execution paths being generally great, it makes sense to parallelize the process of test cases generation and to distribute it over several computation units.

5   Parallel BSE (PBSE) for Program Testing

As we said before, path explosion problem is well-known in the software testing area. It constitutes a big obstacle to the application of symbolic execution to real world software. A suitable solution is the parallelization of path explorations. The automatic test generation is assigned to several workers. A particular worker called the Coordinator initiates and concludes the automatic test case generation by Parallel Backwards Symbolic Execution. The algorithm in the figure 2 describes the Coordinator; nbWorker is the number of workers. We aim to achieve the following objectives:

1. Perform a static well-balanced partitioning.
2. Reduce redundancy in computations.
3. Reduce dependencies between workers.
4. Reduce communications between the coordinator and the workers.

The role of the coordinator is to prepare the program by representing it with the pair (DM,CM); to compute the string path: Path[i] for each element of CM and to partition the tasks in a well-balanced way over the workers. Compute_NBP() computes for each element of CM the number of paths derived from it. The function Compute_NBC() computes the number of conditions in each path. Finally, the role of the procedure Divide () is to divide equitably the paths over the workers regarding the total numbers of paths and conditions. It regroups paths by sets each set will be assigned to one Worker.

The Procedure Assign-test() assigns a set of paths to each Worker. Receipt(k,Ck) is True if the Worker k has sent its results Ck to the Coordinator.

Coordinator

Input : Program : Prog, NbWorker;
Output: A set of test cases for Prog.
Initialization: NbPath=0;NBCond=0;TestCase=∅
(DM,CM)=Model(Prog)
For i=1 to CardDM
Do Begin   If Terminal(DM[i])
                    Then   Path[i]=Compute_Path(i);
                            NBP[i]=Compute_NBP(i);
                            NBC[i]=Compute_NBC(i);
                            NbPath=NbPath+NBP[i];
                            NBCond=NBCond+NBC[i];
        End;
Divide (NbPath, NBcond, nbWorker);
For k=1 to NbWorker
Do   Assign-test(k, Set);
IF   (Receipt(k, Ck)) **Then**   TestCase=TestCase∪Ck
End.

Fig. 2. Coordinator

The task of each worker consists to compute test cases of the paths assigned to it and to send the results to the coordinator.

For a given Worker to compute test cases of the element CM[i], it is not necessary to have the whole tables DM and CM but just the parts required to its computations, these parts could be easily determined statically.

6   Case Study

In this section we present an illustrative example as a case study elucidating our approach and highlighting its suitability. We suppose that we have two workers.

loc	var	exp
1	x	$x
2	y	$y
3	z	2*x
4	t	z
5	z	1+x
6	e	z+y
7	e	z+1
8	t	e+2
9	k	t+1
10	f	0
11	r	0
12	k	1
13	f	1
14	r	f+1
15	r	f-1
16	e	z+1
17	t	0
18	t	x
19	t	y
20	z	-x-1
21	t	-y-1
22	z	x+1
23	t	y+1
24	f	t+z

```
1      scanf("% d",&x)
2      scanf( « %d »,&y)
       if (x+y !=0)
       { if (x>y)
3      { z=2*x
          If (z>10)
4          { t=z
5            z=1+x
           }
6        Else{ e=z+y
            If (e>0)
7            { e=z+1
8              T=e+2
            If (t>0)
9            { k=t+1}
10           Else{f=0}
           }
11         Else {r=0}
12         K=1
           }
         }
13     Else { f=1
          If (f>x)
14        { r=f+1}
15        Else {r=f-1}
          If (f==r+1)
16        { e=z+1
17          T=0
          }
          Else If (t<x)
18             {t=x}
19             Else{t=y}
           }
         }
       Else  { If (x+y<-3)
20           { z=-x-1
21             t= -y-1
           }
22         Else { z =x+1
23               t= y+1
             }
24         f=t+z
         }
```

	C	CT	CF	End	PATH	Ter(y/n)	NBP	NBC
1	x+y!=0	3	20	25		N		
2	x>y	3	14	20		N		
3	z>10	4	6	13		N		
4	e>0	7	11	12		N		
5	t>0	9	10	11	1101x	Y	2	5
6	f>x	14	15	16		N		
7	f==r+1	16	18	20		N		
8	t<x	18	19	20	10***x0x	Y	4	5
9	x+y<-3	20	22	24	0*******x	Y	2	2

In the table, we have reported just Terminal paths. To obtain a well-balanced partitioning, a possible way is to assign CM[5] and CM[9] to a same worker (w1) and   CM[8] to another worker (w2).

Automated Test Cases Generation:

Worker W1:
Path[5]=1101x => $Path_1[5]$=11010   and $Path_2[5]$=11011.

$Path_1[5]$: $C[1] \equiv T$ and $C[2] \equiv T$ and $C[3] \equiv F$ and $C[4] \equiv T$ and $C[5] \equiv F$ (I)
WP([1,3[,x+y!=0)=($x+$y!=0)
WP([1,3[, x>y)=($x>$y)
WP([1,3[$\cup$[3,3[, z<=10)=(2*$x<=10)
WP([1,3[$\cup$[3,3[$\cup$[6,6[, e>0)=(2*$x+1>0)
WP([1,3[$\cup$[3,3[$\cup$[6,6[$\cup$[7,9[,t<=0)=(2*$x+3<=0)

So the formula expressing (I) is :

($x+$y!=0)$\wedge$($x>$y)$\wedge$(2*$x<=10)$\wedge$(2*$x+1>0)$\wedge$(2*$x+3<=0)

This formula is unsatisfiable so the path $path_1[5]$ is not feasible.

$Path_2[5]$: The formula is
($x+$y!=0)$\wedge$($x>$y)$\wedge$(2*$x<=10)$\wedge$(2*$x+1>0)$\wedge$(2*$x+3>0)
A possible solution is $x=4   and $y=2 which constitutes a test case for this path.
A similar reasoning is performed for path[9] (there is   2 conditions to verify)

Worker W2:

Path[8]=10***x0x=> $Path_1[8]$=10***000 ; $Path_2[8]$=10***001; $Path_3[8]$=10***100 ; $Path_4[8]$=10***101

The same reasoning is performed to compute the set of tests of the four paths.
   Finally, the set of test case of the entire program is the union of all   set test cases.

7   Conclusion

In this paper we have presented a new approach for symbolic distributed path testing. We have first defined the backward symbolic execution for test generation method performed to compute path formulas. Then we have proposed a distributed version of our solution allowing to take advantage of the adopted program modeling and to overcome the problem of path explosion. Our solution presents several advantages:

1. It is a backward method: So instead of executing the entire program as in the other methods, it just captures the impact of each statement on the considered predicates.
2. Our distributed solution is well-balanced without requiring a dynamic partitioning.
3. No dependencies between workers.
4. The amount of communication between the workers and the coordinator is minimal.
5. Each worker is required to have just the information concerning the paths it computes.

However, our solution presents some drawbacks, the most important is redundancy. This is due to the fact that we have chosen to keep the communication between workers minimal. In fact, we can reduce redundancy if we allow workers to share more information about possible common parts of paths they have to analyze.

References

[1]   Sen, K., Marinov, D., Agha, G.: CUTE: A concolic unit testing engine for C. In: European Software Engineering Conference/Foundations of Software Engineering (2005)
[2]   Tillmann, N., Schulte, W.: Parameterized unit tests. In: European Software Engineering Conference/Foundations of Software Engineering (2005)
[3]   Godefroid, P., Levin, M.Y., Molnar, D.: Automated whitebox fuzz testing. In: Network and Distributed Systems Security (2008)
[4]   Kim, M., Kim, Y.: Concolic testing of the multi-sector read operation for flash memory file system. In: Oliveira, M.V.M., Woodcock, J. (eds.) SBMF 2009. LNCS, vol. 5902, pp. 251–265. Springer, Heidelberg (2009)
[5]   Kim, Y., Kim, M.: SCORE: a scalable concolic testing tool for reliable embedded software. In: European Software Engineering Conference/Foundations of Software Engineering, Szeged, Hungary, September 5-9, pp. 420–423 (2011), tool demonstration track
[6]   Dijkstra, E.: A discipline of programming. Prentice Hall (1976)
[7]   Pasareanu, C., Visser, W.: A survey of new trends in symbolic execution for software testing and analysis. Software Tools for Technology Transfer 11(4), 339–353 (2009)
[8]   Lattner, C., Adve, V.: LLVM: A compilation framework for lifelong program analysis & transformation. In: Intl. Symp. on Code Generation and Optimization (2004)

[9]   Jayaraman, K., Harvison, D., Ganesh, V., Kiezun, A.: jFuzz: A concolic whitebox fuzzer for Java. In: NASA Formal Methods Symposium (2009)

[10] Visser, W., Havelund, K., Brat, G., Park, S.: Model checking programs. In: Automated Software Engineering (September 2000)

[11] Pasareanu, C., Mehlitz, P., Bushnell, D., Gundy-burlet, K., Lowry, M., Person, S., Pape, M.: Combining unit-level symbolic execution and system-level concrete execution for testing nasa software. In: International Symposium on Software Testing and Analysis (2008)

[12] Godefroid, P., Klarlund, N., Sen, K.: DART: Directed automated random testing. In: Programming Language Design and Implementation (2005)

[13] Burnim, J., Sen, K.: Heuristics for scalable dynamic test generation. EECS Department, University of California, Berkeley, Tech. Rep. UCB/EECS-2008-123 (September 2008)

[14] Sen, K., Agha, G.: CUTE and jCUTE: Concolic unit testing and explicit path model-checking tools. In: Ball, T., Jones, R.B. (eds.) CAV 2006. LNCS, vol. 4144, pp. 419–423. Springer, Heidelberg (2006)

[15] Staats, M., Pasareanu, C.: Parallel symbolic execution for structural test generation. In: International Symposium on Software Testing and Analysis (2010)

[16] King, A.: Distributed parallel symbolic execution. Kansas State University, Tech. Rep., MS thesis (2009)

[17] Siddiqui, J.H., Khurshid, S.: ParSym: Parallel Symbolic Execution. In: International Conference on Software Technology and Engineering (2010)

[18] Bucur, S., Ureche, V., Zamfir, C., Candea, G.: Parallel symbolic execution for automated real-world software testing. In: 6th ACM SIGOPS/EuroSys (2011)

[19] Deng, X., Lee, J., Robby: Bogor/kiasan: A k-bounded symbolic execution for checking strong heap properties of open systems. In: Automated Software Engineering (2006)

[20] Kim, M., Kim, Y., Rothermel, G.: Distributed concolic algorithm of the SCORE framework. KAIST, Tech. Rep. (2011), http://pswlab.kaist.ac.kr/publications/2012/Whitepaper-score.pdf

[21] de Moura, L., Bjørner, N.: Z3: An efficient SMT solver. In: Ramakrishnan, C.R., Rehof, J. (eds.) TACAS 2008. LNCS, vol. 4963, pp. 337–340. Springer, Heidelberg (2008)

[22] Hutchins, M., Foster, H., Goradia, T., Ostrand, T.: Experiments of the effectiveness of dataflow- and control flow based test adequacy criteria. In: International Conference on Software Engineering, pp. 191–200 (1994)

[23] National Security Agency (NSA), FIPS 180-3: Secure hash standard (SHS) (2008)

[24] Wilcoxon, F.: Individual comparisons by ranking methods. Biometrics Bulletin 1(6), 80–83 (1945)

Cross Project Validation for Refined Clusters Using Machine Learning Techniques

Veer Sain Dixit[1] and Shveta Kundra Bhatia[2,*]

[1] Computer Science Department, Atma Ram Sanatan Dharma College, University of Delhi,
New Delhi, India
veersaindixit@rediffmail.com
[2] Computer Science Department, Swami Shraddhanand College, University of Delhi,
New Delhi, India
shvetakundra@gmail.com

Abstract. Clustering is used for discovering groups and identifying interesting distributions and patterns in the underlying data whereas classification is a technique used to predict membership for data instances within a cluster. Correct classification of similar users in a cluster helps in better prediction of web pages. In the past lot of work has been done on original web log data whereas in this paper we intend to apply classification on refined clusters by implementing Modified Knockout Refinement Algorithm(MKRA). This approach leads to the improvement in cluster quality and prediction accuracy. After refining the clusters using MKRA we apply different learning techniques on refined clusters. Various performance measures of learning techniques are evaluated and compared. These days the machine learning community is trying to get better solutions for improving classification accuracy by applying ensembled classification. We further intend to apply ensembling on the classifiers used in our model to observe the betterment in the classification accuracy performance.

Keywords: Classification Techniques (CT), Refined Clusters (RC), Accuracy, Ensembling.

1   Introduction

Data Mining [22] is the most significant application of Machine Learning (ML). Machine Learning can be applied to establish relationships between attributes and improving the efficiency of systems. Learning techniques in which instances are given with known labels are supervised techniques otherwise unsupervised, where instances are unlabeled. By applying these unsupervised (clustering) algorithms, researchers hope to discover unknown, but useful, classes of items (Jain et al., 1999).The main issue in clustering is to generate groups of patterns allowing us to discover similarities, differences and deriving useful conclusions about them. There are no predefined

---

* Corresponding author.

B. Murgante et al. (Eds.): ICCSA 2013, Part II, LNCS 7972, pp. 498–512, 2013.

classes or examples that would show what kind of desirable relations should be valid among the data. On the other hand, classification is a procedure of assigning a data item to a predefined set of categories. The categories defined by using some clustering algorithm. Classification is a model based approach implemented on the training data where each instance has a class label. The trained model is tested for classification accuracy and performance which is then used to predict unlabelled data. Different classification algorithms result in variable accuracies making selection of a single best algorithm difficult. The solution is to combine multiple classifiers to observe the improvements in performance accuracy known as ensembling. Basic idea of ensemble [23 - 28] methodology is to combine a varied set of classifiers, each of which solves the same original task, thus obtaining a better composite global classifier model with more performance accuracy. An ensemble is used to make predictions to give better results by compensating for poor learning algorithms by compromising the computation time in the process of building strong learners.

In Web Usage Mining, a major problem is to obtain the similar group of URL's from web log data that can be used for prediction of URL's for similar session users. This kind of issue can be resolved with the aid of machine learning being applied on similar sessions in a group which can be used directly to obtain the end results. In this paper the aim of our work is to investigate the performance of different classification methods on refined clusters obtained by eliminating dissimilar session's from original clusters on the basis of access and time spent on common pages within a pair of sessions. Original clusters are obtained by using K-Means algorithm on the sessions from the web log file.

2   Related Work

A study named STATLOG given by King [1] et. al. was one of the best known studies. A brief review of what Machine Learning includes can be found in Dutton & Conroy (1996) [2]. De Mantaras and Armengol (1998) [3] also presented a historical survey of logic and instance based learning classifiers. Le Cun (1995) [4] compared several learning algorithms on a handwriting recognition problem. Lim et. al. (2000) [5] performed a comparison of decision trees and other classification methods. Perlich (2003) [6] conducted a comparison between decision trees and logistic regression. Kotsiantis et. al. (2003) [7] has compared six classification methods on an educational data. Phyu et. al. (2009) [8] gave a comprehensive survey of different classification techniques in Data Mining. Provost and Fawcett (1997) [15] discussed ROC other than accuracy for different learning algorithms. Caruana and Mizil (2006) [9] performed an empirical comparison of supervised learning algorithms. Kotsiantis (2007) [10] studied various machine learning techniques and compared them using various parameters. Minaei-Bidgoli et al. [16] have compared six classifiers (quadratic Bayesian classifier, 1-nearest neighbors, k-nearest neighbors, Parzen window, feedforward neural network and decision tree) to predict the course final results from learning system log data and found K-Nearest Neighbor giving best results. Othman and Yau (2007) [11] compared different classification techniques using WEKA for breast cancer concluding that Bayes network classifier has the potential to significantly improve the conventional classification methods for use in medical or in general, bioinformatics field. Thorbe (2012) [12] performed a comparison of four different learning techniques on four different data sets concluding that different algorithms perform better for different datasets. Hussain, Khan, Nazir and Iqbal [17] presented a comprehensive study of latest and most famous facial feature extraction techniques and as well as classification techniques concluding K-Nearest Neighbor giving the best classification. Most of the comparisons concluded that some methods perform better or worse than other methods on an average but with a dependency on the problems and metrics used. The best models give poor performance sometimes and the models with poor average performance perform occasionally well [9]. A combination of classifier models can be designed to increase the accuracy of a single classifier by training several different classifiers [21] and combining the output to generate an ensemble. Ensembling [18, 19] in machine learning has been used in a large number of applications with an aim to improve classification accuracy. Ensembles have been implemented with techniques such as bagging, boosting, random forests and cross validation on classifiers such as Support Vector Machines, Decision trees, Neural Networks etc.

All the above work has been performed on a single data file having the classes defined for the complete data set whereas we shall perform comparison of techniques on refined classes by evaluating various performance measures and finding out which technique helps in learning the refined web log sessions with maximum accuracy and minimum error. We further ensemble the classifiers implemented individually on refined clusters to design a new classifier to obtain better prediction accuracy.

3   Proposed Scheme

Machine Learning is the process of creating a classifier for identifying new instances into classes and learning a set of rules from the training set. Choosing a learning algorithm is a critical step. The classifier's evaluation is based on prediction accuracy, recall, precision, absolute error, relative error and root mean square error. The classification accuracy can be evaluated using three techniques [9]. One technique is to split the training set by using two-thirds for training and the rest for estimating performance. In cross-validation, the training set is divided into mutually exclusive (independent) and equal-sized subsets and for each subset the classifier is trained on the union of all the other subsets. Leave-one-out validation is a special case of cross validation. All test subsets consist of a single instance.

Our proposed scheme intends to use the cross validation technique for evaluating classification accuracy of the learning techniques. Most of the comparisons on classification techniques include data sets being divided into clusters and giving cluster labels for each entry. The clusters are then used for training on the basis of various learning algorithms. We propose to use refined clusters for learning to improve accuracy of the techniques. We use a web log file that contains entries listing the web pages accessed by different users. By performing preprocessing [13] steps web log file is converted into sessions.

Suppose for the given web site, there are n sessions S= {S_1, S_2, S_3... S_n} accessing the set of pages P= {p_1, p_2, p_3... p_m}. The pages are associated with the sessions as follows:

$$\text{Access } (p_i, S_j) = \begin{cases} 1 \text{ if } p_i \text{ is viewed by } S_j \\ 0 \text{ otherwise} \end{cases}$$

In this paper K-Means algorithm is used to define groups of similar sessions initially. For example we get a cluster having six sessions as follows:

	p_1	p_2	......	......	p_{366}
S_1	1	1	....	....	0
S_5	0	0	....	....	1
S_7	0	1	....	....	1
S_9	1	1	....	....	0
S_{13}	0	1	....	....	1
S_{15}	1	1	....	....	0

To improve the quality of a cluster we wish to eliminate sessions having a factor of dissimilarity within that cluster. Refinement is done using the Modified Knockout Refinement Algorithm (MKRA) [29] using a combination of dissimilarities based on access and time among session pairs within a cluster resulting in refined clusters. The dissimilarity is calculated as follows:

$$Ds(S_i, S_j) = ((1 + D_a(S_i, S_j)) * (1 + D_t(S_i, S_j))) - 1 \tag{1}$$

Where

$$D_a(S_i|S_j) = \frac{\Sigma p_k((p_{ka}|S_i)|(p_{kn}|S_j)) + \Sigma p_k((p_{kn}|S_i)|(p_{ka}|S_j))}{\Sigma p_k((p_{ka}|S_i)|(p_{ka}|S_j)) + \Sigma p_k((p_{ka}|S_i)|(p_{kn}|S_j)) + \Sigma p_k((p_{kn}|S_i)|(p_{ka}|S_j))} \tag{2}$$

Defined as the ratio of sum of the number of pages accessed in session S_i but not in S_j to the sum all the variables accessed in at least one session. The time dissimilarity is computed as given below:

$$D_t(S_i|S_j) = 1 - Sim_t(S_i|S_j) \tag{3}$$

Where

$$Sim_t(S_i|S_j) = \frac{\Sigma_k(t(p_k|S_i) * t(p_k|S_j))}{\sqrt{\Sigma_k(t(p_k|S_i))^2 * \Sigma_k(t(p_k|S_j))^2}} \tag{4}$$

Applying MKRA on the above example we get a new cluster by eliminating session S_5 as follows.

	p_1	p_2	......	......	p_{366}
S_1	1	1	....	....	0
S_7	0	1	....	....	1
S_9	1	1	....	......	0
S_{13}	0	1	....	....	1
S_{15}	1	1	....	....	0

Algorithm for Modified Knockout Refinement Algorithm

```
Input: Set of Original Clusters (OC).
Process: Step 1: Do
Pick up the first cluster (Ck)
Generate contingency table for every pair of sessions C (Si, Sj) in
the cluster.
Evaluate dissimilarity on the basis of access between all sessions
using the following:
```

$$d_k(s_i, s_j) = \sum_{\substack{k=1 \\ i \neq j}}^{n} r_{ki} + s_{ki} \Big/ q_{ki} + r_{ki} + s_{ki}$$

```
Evaluate dissimilarity on the basis of time between all sessions us-
ing the following:
```

$$Sim_t(S_i|S_j) = \frac{\sum_k (t(p_k|S_i) * t(p_k|S_j))}{\sqrt{\sum_k (t(p_k|S_i))^2 * \sum_k (t(p_k|S_j))^2}}$$

$$D_t(S_i|S_j) = 1 - Sim_t(S_i|S_j)$$

```
Combine the dissimilarities using the following formula:
    Ds(Si, Sj) = ((1 + Da(Si,Sj)) * (1+Dt(Si, Sj))) - 1

Generate a Symmetric Dissimilarity Matrix(SDM)
    If Threshold > average(SDM) for d(Si,Sj)
        Count++
            If(Count>0.8*max(count_SDM)
                Eliminate Si and Sj from the cluster Ck
            End If
        End If
        Generate Refined Clusters (RC)
    End Do
Repeat for all the clusters

Output: Refined Clusters (RC)
```

We then apply various learning algorithms on the refined clusters and evaluate our results. Used learning techniques in this paper are as follows:

3.1   Naive Bayes

Naive Bayesian networks (NB) [7, 9, 11] are networks that are simple and composed of acyclic graphs. The Bayes rule is applied to compute the probability of a class based on instances within the class. The class having highest posterior probability will be the predicted class for the instance. The probability model for a classifier is a conditional model. NB classifiers work better when there is a single underlying model of the dataset taking into account prior information about a given problem. NB has an advantage of short computation time for training along with little storage space and a disadvantage of computational difficulty of exploring a previously unknown network.

3.2   Naive Bayes (Kernel)

Naive Bayes (NB) can perform well with discrete and continuous variables comprising large number of variables and large data sets. An extension of the naive Bayes classifier uses kernel density estimation. This method is very similar to the naive Bayes, but the density of each continuous variable is estimated averaging over a large set of kernels.

3.3   K-Nearest Neighbor

The K-nearest neighbor [8] algorithm is amongst the simplest of all machine learning algorithms. K-nearest neighbor algorithm (K-NN) is a method for classifying objects based on closest training examples in the feature space. An object is classified by a majority vote of its neighbors, with the object being assigned to the class most common amongst its k nearest neighbors. K-NN computationally requires large storage with sensitivity to the similarity function. Varying the value of K affects the performance of the algorithm.

3.4   Decision Trees

Decision Trees [8, 7] are considered to be one of the most popular and powerful approaches for representing classifiers. They are recursive schematic tree-shaped diagrams used to determine a course of action or show a statistical probability. The algorithm for decision trees is defined as a greedy algorithm that constructs decision trees in a top down divide and conquer manner recursively. Decision Tree classification is a time consuming algorithm. Decision Trees work well with categorical/ discrete features but do not perform well with problems that require diagonal partitioning.

3.5   Decision Trees (Parallel)

Data is growing at very fast rates for which algorithms need to be developed that are capable of classifying large datasets and also being computationally efficient and scalable. One possible solution is to use parallelism to reduce the amount of time spent in building classifiers from very-large datasets and keeping the classification accuracy. A parallel decision tree learns a pruned decision tree which can handle both numerical and nominal attributes.

3.6   Support Vector Machines

Support Vector Machines (SVMs) [7] are one of the recent supervised machine learning techniques. SVMs focus on the notion of a separator on either side of a hyper plane that separates two data classes. Maximizing the separator and creating the largest possible distance between the separating hyper plane and the instances on either side of it has been proven to reduce an upper bound on the expected generalization error. When it is possible to linearly separate two classes, an optimum separating hyper plane can be found by minimizing the squared norm of the separating hyper plane. SVM technique reaches a global minimum and avoids ending in a local minimum which happens in other techniques like Back Propagation in Neural Networks. SVM methods are binary i.e. for a multiclass problem one must reduce it to binary classification problems. SVM does not perform well with discrete data [9].

3.7   Rule Induction

This operator works similar to the propositional rule learner named Repeated Incremental Pruning to Produce Error Reduction (RIPPER, Cohen 1995). Starting with the less prevalent classes, the algorithm iteratively grows and prunes rules until there are no positive examples left or the error rate is greater than 50%. In the growing phase, for each rule greedily conditions are added to the rule until the rule is perfect (i.e. 100% accurate).

3.8   Neural Net

This operator learns a model by means of a feed-forward neural network trained by a back propagation algorithm. The Back Propagation network learns by example. The activation function used is the usual sigmoid function where the Errors from the output neurons and running them back through the weights to get the hidden layer errors. Having obtained the error for the hidden layer neurons we modify the weights for the hidden layer. By repeating this method for any number of layers we can train a network for obtaining the desired output on the basis of some input.

3.9   Ensembling

The idea of ensembling is to combine multiple classifiers with a known fact of improvement in classification accuracy and prediction. The general architecture of ensembling includes combining the decisions by training different classifiers and to obtain a new classifier with better performance. The predictions with different classifiers are aggregated and implemented on new instances presented to the model.

4   Experiment and Results

The experiments are carried out using a log file containing information about all web requests to NASA's official website available in the archive from Jul 01 to Jul 31,

ASCII format, 20.7 MB. The raw web log file used for the experiment contained 131031 web requests. This file can be found at the following address: http://ita.ee.lbl.gov/html/contrib/NASA-HTTP.html. We create sessions from the log file that is used for clustering sessions with similar interests. We optimize the log file by including only those sessions that have an access count greater than equal to four, reducing the entries to 28213. To reduce the problem of sparsity in our session file we consider sessions with minimum page access as 10. Clustering has been performed by using the K-means algorithm. The value of K is taken to be 15. K-means is one of the simplest unsupervised learning algorithms that solve the well-known clustering problem. The above process results in clusters having sessions with similar access to web pages. These are termed as Original Clusters (OC). We now refine the OC using Modified Knockout Refinement Algorithm (MKRA) on the basis of a combination of access and time dissimilarity. We obtain Refined Clusters (RC) that has better cluster quality than original clusters. The refinement process reduces the number of sessions to 22301. For applying the various classification techniques we optimize by reducing sparsity in the session file and including only those session where the session count is greater than equal to ten. We used the above defined machine learning techniques on RC to make clusters learn the type of sessions in them. We used open source software Rapid Miner [14] to apply the learning techniques and to evaluate the various performance measures. Evaluation of models has been done by applying a 10-fold cross validation [20] with stratified sampling that divides the clusters in a 9:1 ratio for training and testing respectively that shall be repeated 10 times. Stratified sampling builds random subsets and ensures that the class distribution in the subsets is the same as in the whole example set. The various parameter settings for the learning techniques are described in subsequent lines. Naive Bayes uses Laplace correction to prevent high influence of zero probabilities. Naïve Bayes (Parallel) also uses Laplace correction along with estimation mode being Greedy and number of kernels as 10. The K-NN algorithm uses the Euclidean distance measure and also varies the value of K as 1, 3, 5, 7, and 10 to study its effect on the accuracy of the method. Decision Trees use the gain ratio criterion for making decisions with minimal size of split as 4, minimal leaf size 2 and a confidence value of 0.25. Parallel decision trees use the same parameters as decision trees along with the variation in number of threads to reduce the amount of time spent in building classifiers from the datasets. Support Vector Machine uses the C-SVC type SVM with kernel type as radial basis function. Rule induction uses a criterion of information gain, minimal prune benefit as 0.25 with a sample ratio of 0.9. Neural Net evaluates on the basis of variation in the number of training cycles as 100, 200, 300 and 400. The learning rate for the neural net is 0.3 with a momentum of 0.2. Selecting a good classifier that outperforms other classifiers is a difficult task. For solving this problem ensembled learning can be applied to obtain a new classifier which outperforms the base classifiers. We applied the cross validation technique on the refined log file where the file is divided for training and testing. The training dataset is given as the input to the voting module that further divides the training set into subsets based on the number of classifiers to be used. We used K-NN, Naïve Bayes (Kernel), Decision Trees and Rule Induction classifiers to design the ensemble as these classifiers had prediction accuracy greater than 75%.The predictions are based on the majority vote scheme where the classification of an unlabeled instance is performed according to the class that obtains the maximum number of votes. Also known as the plurality vote, this approach has been used as a combining method for comparing newly proposed methods. The above scheme is applied on the testing instances to evaluate accuracy and prediction of the model. The architecture used is as follows:

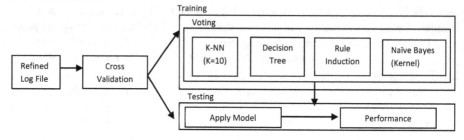

Fig. 1. Framework of the Experiment

The results for various learning techniques and their corresponding evaluation measures are as follows.

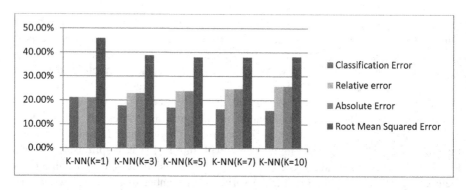

Fig. 2. Analysis of errors for different values of K in K-Nearest Neighbor

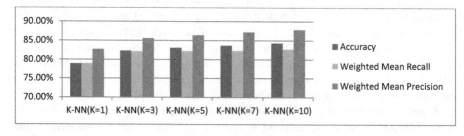

Fig. 3. Analysis of Accuracy, Recall and Precision for different values of K in K-Nearest Neighbor

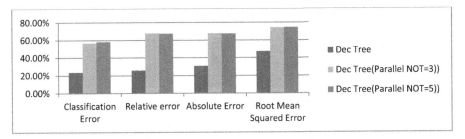

Fig. 4. Analysis of errors for different values of number of threads in Decision Trees

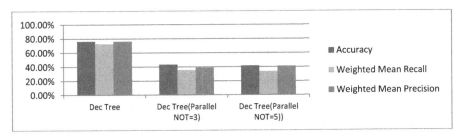

Fig. 5. Analysis of accuracy, Recall and Precision for different values of number of threads in Decision Trees

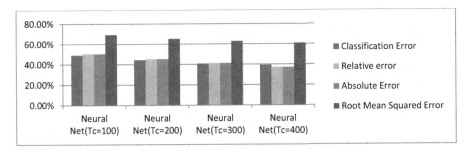

Fig. 6. Analysis of errors for different values of training cycles in a Neural Net

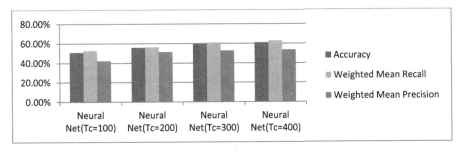

Fig. 7. Analysis of accuracy, Recall and Precision for different values of training cycles in a Neural Net

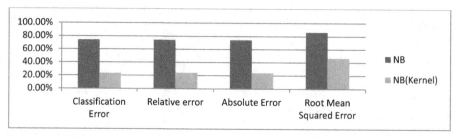

Fig. 8. Analysis of errors for Naïve Bayes and Naïve Bayes (Kernel)

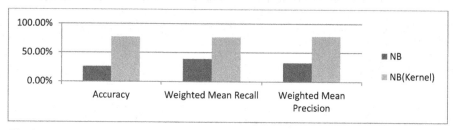

Fig. 9. Analysis of accuracy, Recall and Precision for Naïve Bayes and Naïve Bayes (Kernel)

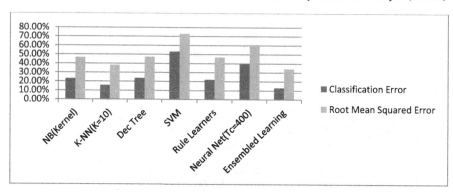

Fig. 10. Analysis of errors for various Learning Techniques

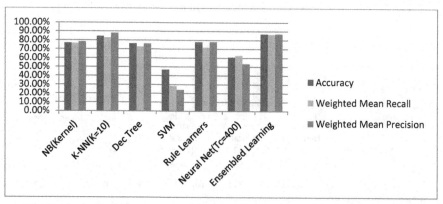

Fig. 11. Analysis of Accuracy, Recall, and Precision for various Learning Techniques

Table 1. Comparison of F-Measure scores for prediction into classes based on various learning techniques

Cluster Name	NB(Kernel)	K-NN (K=10)	Dec Tree	SVM	Rule Learners	Neural Net (Training Cycles=400)	Ensembled Learning
s2	84.61%	92.00%	65.21%	#	59.53%	10.71%	91.26%
s3	71.40%	81.61%	72.45%	#	82.50%	60.91%	77.61%
s4	78.41%	84.82%	82.27%	54.42%	82.54%	53.29%	89.97%
s5	83.63%	88.68%	73.33%	56.71%	74.94%	67.01%	87.57%
s6	73.01%	81.44%	76.86%	#	78.26%	79.82%	90.44%
s7	82.35%	85.19%	72.37%	#	72.51%	50.80%	85.94%
s8	77.74%	85.77%	82.86%	79.26%	80.36%	80.94%	89.68%
s9	74.79%	85.31%	78.81%	2.66%	76.98%	40.00%	87.67%
s10	69.93%	73.87%	62.86%	#	55.93%	25.66%	84.65%
s11	73.15%	79.42%	73.57%	54.86%	76.05%	35.85%	87.48%
s12	74.61%	91.63%	74.44%	#	70.94%	51.38%	84.06%
s13	85.20%	89.27%	76.21%	51.82%	82.07%	83.36%	80.50%
s14	68.45%	79.83%	65.17%	#	69.86%	69.63%	87.01%

The above graphs and table show that the outcome for K-NN classification (K=10) is the best classification technique in comparison to NB, Decision Trees, SVM, Rule Learners and Neural Net on used data set. K-NN classification shows maximum results for accuracy and minimum results for classification and root mean squared errors. The performance of NB, Decision Tree and Rule Learners is almost the same in the experiment. K-NN outperforms other learning techniques for the used web log data. The speed of learning for K-NN with respect to number of instances is as high as compared to other techniques with transparency as it appeals to the human users. The K-NN technique lacks a principled way to choose the value of K which can be solved using cross validation as in our experiments. The choice of K affects the performance of the algorithm and a large value of K helps in solving the problem of noisy instances. In our experiment the value of K being greater than equal to 10 tends to result in accuracy of classification above 83% which is higher than the rest of the techniques. The accuracy of Naïve Bayes, Rule Learners and Decision Trees is reduced by 7% to 8% as compared to K-NN. For SVM and Neural Net there is a remarkable reduction of 37.44% in accuracy as compared to K-NN. No single classification technique can uniformly outperform other classification techniques over all the data sets. So performance of classification techniques is dependent on the nature of the data set used. Since every classification technique is having their own merits and demerits, classifiers are combined for improvement of individual classifiers to obtain more certain, precise and accurate results. The ensembled classifier generates better prediction accuracy on refined clusters by 2.52%. The classification error and the root mean square errors for ensembled learning are reduced as compared to the errors generated when individual classifiers were applied on the refined clusters.

5   Conclusion

Classification methods for Web Usage are compared on refined clusters and K-Nearest Neighbor technique in particular was found to be best and relevant on web log data set than other classification techniques. The refined clusters were introduced for classification where the refining is performed on the basis of access and time dissimilarity between a pair of sessions within a cluster. The refined clusters are classified on the basis of various techniques such as NB, K-NN, Decision Trees, Rule Learners, Neural Net and SVM. The ratio of training and testing taken in the experiment is 90:10 where 90% of the data is used for training and the rest 10% for testing. We used a 10 fold cross validation technique with stratified sampling on all the classification algorithms. Variations in the K-Nearest Neighbor, Decision Trees, Naive Based and Neural Net were also evaluated and compared to find the best parameters for the techniques. K-Nearest Neighbor algorithm gave best results for K=10, simple decision trees performed better than parallel decision trees, naïve bayes kernel performed better than simple naïve bayes and neural networks performance improved with the increase in the number of training cycles. Comparing the above with SVM and Rule Learning techniques K-NN with K=10 gave the best results for prediction of instances in the classes. It is proven argument that some classification algorithms perform better or worse than other algorithms on an average but with a dependency on the problems, type of data set used and metrics used. Ensembling of Naïve Bayes (Kernel), K-Nearest Neighbors, Decision Trees and Rule Induction resulted in better classification of unlabeled instances with less classification error. In summary we have obtained the following facts from our analysis and outcome:

- Refined clusters result in better prediction accuracy for all individual classifiers than original clusters (Due to restriction on number of pages given in the proceedings instructions the results for original clusters is not provided).
- K-NN with K=10 provides best results among all the classifiers used.
- Ensembling of classifiers results in developing a composed classifier with improved prediction accuracy.

Acknowledgement. The authors would like to thank the reviewers for their valuable comments and suggestions that contributed to the improvement of this work.

References

1. King, R., Feng, C., Shutherland, A.: Statlog: comparison of classification algorithms on large real-world problems. Applied Artificial Intelligence (1995)
2. Dutton, D., Conroy, G.: A review of machine learning. Knowledge Engineering Review 12, 341–367 (1996)
3. Mantaras, D., Armengol, E.: Machine learning from examples: Inductive and Lazy methods. Data & Knowledge Engineering 25, 99–123 (1998)
4. LeCun, Y., Jackel, L.D., Bottou, L., Brunot, A., Cortes, C., Denker, J.S., Drucker, H., Guyon, I., Muller, U.A., Sackinger, E., Simard, P., Vapnik, V.: Comparison of learning algorithms for handwritten digit recognition. In: International Conference on Artificial Neural Networks, pp. 53–60. EC2 & Cie, Paris (1995)

5. Lim, T.S., Loh, W.Y., Shih, Y.S.: A comparison of prediction accuracy, complexity, and training time of thirty-three old and new classification algorithms. Machine Learning 40, 203–228 (2000)

6. Perlich, C., Provost, F., Simono, J.S.: Tree induction vs. logistic regression: a learning-curve analysis. J. Mach. Learn. Res. 4, 211–255 (2003)

7. Kotsiantis, S.B., Pierrakeas, C.J., Pintelas, P.E.: Preventing student dropout in distance learning using machine learning techniques. In: Palade, V., Howlett, R.J., Jain, L. (eds.) KES 2003. LNCS, vol. 2774, pp. 267–274. Springer, Heidelberg (2003)

8. Phyu, T.N.: Survey of classification techniques for Data Mining. In: Proceedings of the International Multi Conference of Engineers and Computer Scientists (2009)

9. Caruana, R., Mizel, A.: An empirical comparison of Supervised Learning Algorithms. In: Proceedings of the 23rd International Conference on Machine Learning (2006)

10. Kotsiantis, S.B.: Supervised Machine Learning: A Review of Classification Techniques. Informatica 31, 249–268 (2007)

11. Othman, M. F., Yau, T.: Comparison of Different Classification Techniques Using WEKA for Breast Cancer. In: Biomed 2006. IFMBE Proceedings, vol. 15, pp. 520–523. Springer, Heidelberg (2007), www.springerlink.com©

12. Thombre, A.: Comparing logistic regression, neural networks, C5.0 and m5' classification techniques. In: Perner, P. (ed.) MLDM 2012. LNCS, vol. 7376, pp. 132–140. Springer, Heidelberg (2012)

13. Bhatia, S.K., Dixit, V.S.: A Propound Method for the Improvement of Cluster Quality. IJCSI International Journal of Computer Science Issues 9(4(2)), 216–222 (2012)

14. http://www.rapid-i.com

15. Provost, F.J., Fawcett, T.: Analysis and visualization of classifier performance: Comparison under imprecise class and cost distributions. In: Knowledge Discovery and Data Mining, pp. 43–48 (1997)

16. Minaei, B., Kashy, D.A., Kortemeyer, G., Punch, W.: Predicting student performance: an application of data mining methods with an educational web-based system. In: Proceedings of 33rd Frontiers in Education Conference, pp. T2A13–T2A18 (2003)

17. Hussain, Khan, Nazir, Iqbal: Survey of various feature extraction and classification techniques for facial expression recognition. In: Proceedings of the 11th WSEAS International Conference on Electronics, Hardware, Wireless and Optical Communications, and Proceedings of the 11th WSEAS International Conference on Signal Processing, Robotics and Automation, and Proceedings of the 4th WSEAS International Conference on Nanotechnology, pp. 138–142

18. Sun, Y., Wong, A.C., Kamel, M.S.: Classification of imbalanced data: A review. Int. J. Pattern Recogn. 23(4), 687–719 (2009)

19. Kuncheva, L.: Combining pattern classifiers. Wiley Press, New York (2005)

20. Schaffer, C.: Selecting a classification method by cross-validation. Mach. Learn. 13(1), 135–143 (1993)

21. Woods, K., Kegelmeyer, W., Bowyer, K.: Combination of multiple classifiers using local accuracy estimates. IEEE Trans. Pattern Anal. Mach. Intell. 19, 405–410 (1997)

22. Yang, Q., Wu, X.: 10 challenging problems in data mining research. Int. J. Inf. Tech. Decis. 5(4), 597–604 (2006)

23. Rokach, L.: Ensemble-based classifiers. Artif. Intell. Rev. 33, 1–39 (2010)

24. Ho, T.K.: Multiple classifier combination: Lessons and next steps. In: Kandel, Bunke (eds.) Hybrid Methods in Pattern Recognition, pp. 171–198. World Scientific, Singapore (2002)

25. Demsar, J.: Statistical comparisons of classifiers over multiple data sets. J. Mach. Learn. Res. 7, 1–30 (2006)
26. Kittler, J., Hatef, M., Duin, R., Matas, J.: On combining classifiers. IEEE Trans. Pattern Anal. Mach. Intell. 20(3), 226–239 (1998)
27. Barandela, R., Valdovinos, M., Sanchez, J.S.: New applications of ensembles of classifiers. Pattern Anal. App. 6, 245–256 (2003)
28. Galar, M., Fernandez, A., Barrenechea, E., Bustince, H., Herrera, F.: A Review on Ensembles for the Class Imbalance Problem: Bagging-, Boosting-, and Hybrid-Based Approaches. IEEE Transactions on Systems, Man, and Cybernetics—Part c: Applications and Reviews 42(4) (2012)
29. Bhatia, S.K., Dixit, V.S., Singh, V.B.: Dissimilarity Measures: Web Session Cluster Refinement and Analysis. To be Published in the Proceedings of the 6th International Conference on Quality, Reliability, Infocom Technology and Industrial Technology Management Organized by Department of Operational Research, University of Delhi, November 26-28 (2012)

A Methodology and Framework for Automatic Layout Independent GUI Testing of Applications Developed in Magic xpa

Daniel Fritsi, Csaba Nagy, Rudolf Ferenc, and Tibor Gyimothy

Department of Software Engineering
University of Szeged, Hungary
fritsi@frontendart.com, {ncsaba,ferenc,gyimi}@inf.u-szeged.hu

Abstract. Testing an application via its Graphical User Interface (GUI) requires lots of manual work, even if some steps of GUI testing can be automated. Test automation tools are great help for testers, particularly for regression testing. However these tools still lack some important features and still require manual work to maintain the test cases. For instance, if the layout of a window is changed without affecting the main functionality of the application, all test cases testing the window must be re-recorded again. This hard maintenance work is one of the greatest problems with the regression tests of GUI applications.

In our paper we propose an approach to use the GUI information stored in the source code during automatic testing processes to create layout independent test scripts. The idea was motivated by testing an application developed in a fourth generation language, Magic. In this language the layout of the GUI elements (e.g. position and size of controls) are stored in the code and can be gathered via static code analysis. We implemented the presented approach for Magic xpa in a tool called Magic Test Automation, which is used by our industrial partner who has developed applications in Magic for more than a decade.

1 Introduction

Thoroughly testing an application via its user interface is not an easy task for large, complex applications with many different functionalities. Testers have to follow certain steps of thousands of test cases and need to evaluate the results manually. This hard work can be supported by automatic GUI testing tools, as these tools are able to follow and record user events (mouse, keyboard, etc.) generated by testers then play back these events to the application under test. This is a great help for regression tests, for example, where the aim is to re-test the application after a change. However, there remains still a lot of manual work to be done. Testers need to record the test case for the first time when they create it, and they need to maintain the recorded scripts as the application evolves.

Current tools support the most popular 3rd generation languages (e.g. C/C++, Java, C#), however higher level languages such as 4th generation languages

B. Murgante et al. (Eds.): ICCSA 2013, Part II, LNCS 7972, pp. 513–528, 2013.
© Springer-Verlag Berlin Heidelberg 2013

(Magic 4GL, ABAP, Informix) became also popular in software development. Developers programming in these languages do not write source code in the traditional way, but they develop at a higher level of abstraction, for instance, using an application development environment. In such languages the application code usually stores the description of the user interface too (e.g. structure of a window or a form and position, color or size of a control). In our paper we use this information to make the automatic testing process GUI layout independent. That is, a recorded test script does not depend on exact coordinates or the layout of the GUI, so the same test case can be reused later even when the developers make minor changes to the user interface of the application (e.g. they rearrange the buttons in a window).

One of the greatest problems with regression tests for GUI applications is that even a minor change in the GUI may result in rewriting all the test cases [4]. As a possible solution, our technique may significantly reduce the costs of maintaining regression tests to keep the quality of a GUI application assured.

The main contributions of this paper are:

- we propose a method to record and play back automatic GUI test scripts that are unaffected by minor changes of the GUI, hence they are layout independent;
- we present our approach in an "in vivo" industrial context, as our tool is used by our industrial partner for testing Magic xpa applications. The presented approach was implemented during a research project in co-operation with our industrial partner, SZEGED Software Inc. During the project the tool was experimentally used for automated GUI testing, and it was extended with additional features. For further details on the project please refer to its webpage[1].

2   Automated Software Testing

Sommerville introduces the main goal of software testing in [18] as follows: „*testing is intended to show that a program does what it is intended to do and to discover program defects before it is put into use*". In the field of software testing, automated software testing is a relevant software engineering topic nowadays, mostly motivated by the industry. As a result many papers and books have been published in this area [5], [6], [7], [12], [17]. Here we elaborate on the literature and on related tools focusing on those that are closely related to our work.

Testing automation frameworks are usually divided into 5 generations [10], [11]. 1st generation frameworks are so-called record/playback tools that are based on simple test scripts where one script relates to one test case. The 2nd generational tools have scripts that are better designed to use/reuse functions, for example. 3rd generation frameworks take data out of the test scripts so a test script may be re-executed several times on different data. This concept is called

---

[1] http://www.infopolus2009.hu/en/magic

data-driven testing [7], [19]. Another concept, usually referred to as 4th generation testing is called keyword-driven, where the test creation process is separated into a higher level planning stage and an implementation stage, thus keywords defined at higher level drive the executions [2], [3], [7]. New techniques sometimes bring test automation to an even higher, so-called scriptless level (5th generation), where automated test cases are designed by engineers instead of testers/developers [9].

Our approach can be considered as a 3rd generation approach, because with carefully designed test scripts, the data can be separated from the execution process. The idea of keyword driven testing is also similar, but our test script is still at lower level, close to the implementation.

The idea of supporting the recording and playback of test cases by using test scripts based on GUI information from source code is novel to our best knowledge in 4GL context. However, static analysis is a common tool to support GUI testing in other approaches, e.g. for generating test scripts [8], [13], [14], [16].

There are a number of automatic GUI testing tools available for software engineers. Just to mention some examples, GUITest[1] is a Java library for automated robustness testing, Selenium[2] is a GUI testing tool for Web applications. As an application testing a web page it also provides solutions to simplify test scripts by using the identifier of a control from the HTML code of the web page. This is a similar approach to ours for Web applications. TestComplete[3], HP Quality Center and Quick Test Professional (QTP)[4] tools are also a widely used for applications written in 4GLs. Microsoft also provides automated GUI testing for instance via GUI Automation of the .NET Framework[5].

3 Specialties of a Magic Application

In the early 80's Magic Software Enterprises (MSE) introduced a new fourth generation language, called Magic 4GL. The main concept was to program an application at a higher level meta language, and let an application generator engine create the final application. A Magic application could run on popular operating systems such as DOS and Unix, so applications were easily portable. Magic evolved and a new version of Magic has been released, uniPaaS and lately Magic xpa. The new version supports modern technologies such as RIA, SOA and mobile development too.

The unique meta model language of Magic contains instructions at a higher level of abstraction, closer to business logic. When one develops an application in Magic, she/he actually programs the Magic Runtime Application Environment

---

[2] http://seleniumhq.org/

[3] http://smartbear.com/products/qa-tools/automated-testing/

[4] HP Test Management (accessed 2013):
http://www8.hp.com/us/en/software-solutions/
software.html?compURI=1170256

[5] Microsoft UI Automation Overview (accessed 2013):
http://msdn.microsoft.com/en-us/library/ee684076%28v=vs.85%29.aspx

(MRE) using its meta model. This meta model is what really makes Magic a RADD (Rapid Application Development and Deployment) tool.

Magic comes with many GUI screens and report editors as it was invented to develop business applications for data manipulation and reporting. The most important elements of Magic are the various entity types of business logic, namely the data tables. A table has its columns which are manipulated by a number of programs (consisting of subtasks) linked to forms, menus and help screens. These items may also implement functional logic using logic statements, e.g. for selecting variables (virtual variables or table columns), updating variables, conditional statements.

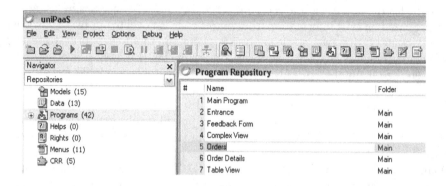

Fig. 1. A screen shot of the Magic xpa application development framework

Figure 1 is a screen shot of the Magic xpa development environment. Some major components of Magic xpa, as a 4th generation programming language are:

Data Objects. These are essentially the descriptions of the database tables. Just as the tables and their columns and primary or foreign keys are defined in a database, we can define these objects in Magic xpa too.

Programs. The logic of an application is implemented here. Programs are top-level tasks with several subtasks below them. A task always works on some Data Objects and performs some operations on them. We can define which database tables should the task use, and which operations should the task perform on them.

Menus. In the application, we can use different high-level menus and pop-up menus, which can be defined here.

Form Entries. Magic xpa has a form editor, where we can define the properties of a window (e.g. title, size and position) and we can place controls and menus on a form and customize them. A graphic window, a form is FormEntry in Magic xpa. In the Magic xpa development environment we can use many built-in controls or we can define our custom controls too. A form is always defined within a task. The form editor of Magic xpa is shown in Figure 2.

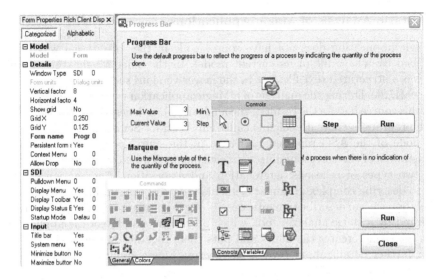

Fig. 2. A screen shot of the form editor of Magic xpa

4   Automatic GUI Testing of a Magic Application

We implemented a tool called *Magic Test Automation*, which enables the automatic GUI testing of applications implemented in Magic xpa. The automatic testing of a Magic application has three main steps (see Figure 3):

1. Analyzing the Magic application. Here we perform a static analysis of the application to gather all the required data of its GUI.
2. Recording GUI events. This is the step where we monitor the mouse and keyboard events and use them to create layout independent test scripts.
3. Playback recorded GUI events. We use the layout independent test scripts to simulate mouse and keyboard events on the application being tested.

In case of layout-independent testing, once the application gets changed in the future, it is enough to repeat the analyzing and the playback steps, and re-recording test cases is not necessary.

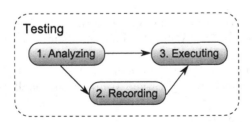

Fig. 3. Main steps of automatic GUI testing of a Magic application

4.1   Static Analysis of Magic Applications

A Magic application does not have source code in the "traditional way", it is described by a save file of its current model. In the older Magic versions this save was a structured text file, but in the newer versions such as Magic xpa, this is an XML file. During the analysis of a Magic application we extract information from this source file. As the result of the analysis, a graph describing the structure of the program is created, which is called an Abstract Semantic Graph (ASG).

A node of the ASG represents an item in the source code. All these nodes are instances of the corresponding source elements. Two nodes can be connected with two types of relations: aggregation and association. Aggregation can be used to describe complex grammar elements (edges of the syntax tree) and with association we can describe semantic details (e.g. identifier references). The graph is created by a static analyzer tool, which parses the save file of the application being analyzed, creates the nodes and puts them together in the ASG.

For further details about reverse engineering Magic applications please refer to our previous work [15].

4.2   Recording GUI Events

Recording GUI events is the process where we record the way the user interacts with the application under test into a certain script format. We catch the events generated by the user and we try to identify the related source element, then transform it to a command of a test script. Of course, user could write such a script manually, but for complex test cases it would be almost impossible.

Traditional, coordinate based automatic testing techniques record the event type and its position. In our layout-independent technique we record the event type and the identifier (in the source code) of the control on which the event occurred.

Hence, the most important task of recording is to identify the source element on which the actual user event happened. To be able to do this, we use dynamic traces of the executed application to identify the currently running tasks and form elements that are displayed on the screen. Once we catch a user event based on its position on the screen and the dynamic traces, we can identify the certain control of the source code, which is actually stored in the ASG. Figure 4 illustrates the process of the recording.

Recording is performed on Windows platform using Windows API. Catching a user event is based on Windows' hook mechanism (*SetWindowsHookEx*, *HookProc* functions).

In Figure 5 we illustrate the possible steps that a tester would perform testing a sample window of a Magic application. We recorded the illustrated steps with the Magic Test Automation tool and saved the script in Python format. Figure 6 shows the resulting Python script.

It can be seen that the Magic Test Automation tool connects the Magic code with the ASG and generates a script using the obtained identifiers. A traditional

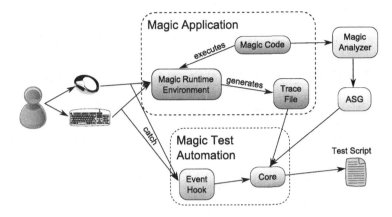

Fig. 4. Recording GUI Events

coordinate-based method would result in a script containing only coordinates, e.g. as it can be seen in Figure 7.

One can see that both scripts contain the same amount of instructions. When we execute the two scripts they will produce the same result, but what happens when we rearrange the window? (For illustration, see a rearranged window in Figure 8.) The application would work as before, but the controls would be in different positions. If we executed the layout independent Python script the result would be the same as before, because the Magic Test Automation tool recalculates the coordinates by the unique identifiers. In contrast, if we play the position based Python script then the result will be negative, because the controls are not in the positions as before.

4.3   Playback Recorded GUI Events

Once we have the test script, we need to be able to playback the recorded user events to the application, this is based on executing events of the script. However this is not enough, as the execution needs to be evaluated and we must make sure that the program under test behaves the same way as it did when we recorded the test script. This is done during the validation phase.

Executing Events In traditional, coordinate based techniques, executing a user event is simple, as the recorded event must be sent to the application with the recorded position. In our layout-independent technique we have no coordinates stored in the test script, but we store the identifier of the control.

Hence, during execution we calculate the coordinates of the control from the ASG, and transform these coordinates to positions on the screen.

If the application is modified, we can re-run the same test script, but with the ASG of the new version of the application (see Figure 9 for illustration).

To play back a recorded test script, first of all we need the script file, and the ASG to connect the unique identifiers of it with the corresponding Windows

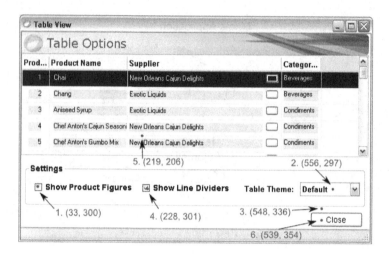

Fig. 5. An example window of a Magic xpa application with example steps for testing its GUI

```
from testrunner import *
from testrunner_ext import *

def runScript():
    assert waitFor(3014, findWindow(741, 379, "Table View")) == True
    mouseClickAtControl("Fe-Table View::Ct-Divider Box", MouseButtons.LEFT, 11, 23)
    mouseClickAtControl("Fe-Table View::Ct-Theme Combo", MouseButtons.LEFT, 44, 14)
    mouseClickAtControl("Fe-Table View::Ct-Theme Combo", MouseButtons.LEFT, 47, 33)
    mouseClickAtControl("Fe-Table View::Ct-Show Figures Box", MouseButtons.LEFT, 13, 17)
    mouseClickAtControl("Fe-Table View::Ct-Table", MouseButtons.LEFT, 218, 129)
    mouseClickAtControl("Fe-Table View::Ct-Close Button", MouseButtons.LEFT, 36, 22)
```

Fig. 6. A layout independent Python script for the steps in Figure 5

GUI elements. During the playback we must execute the Magic application in the Magic Runtime Environment and we must load the script file in the Magic Test Automation tool. The script file contains the recorded keyboard and mouse events which the Test Automation tool first interprets and then executes. (An illustration can be seen in Figure 10.)

During the interpretation we locate GUI elements in the ASG via their unique identifiers. After that, we identify the same Windows controls of the running application. This identification is sometimes quite complex as the lower level implementation of a control may be totally different than the simple Magic control. Suppose a complex tree control or a group box built from many smaller controls. In order to solve this identification problem we collect all information from the ASG that we need to identify a GUI element (position, size), but this is still not enough as the application can simultaneously display multiple windows and parent windows too. Therefore, we need all information from its parent elements too. This way we know that on which window the current element is located. Using the Windows API we can find windows and GUI elements by

```
from testrunner import *
from testrunner_ext import *

def runScript():
    assert waitFor(3014, findWindow(741, 379, "Table View")) == True
    mouseClickAXY(MouseButtons.LEFT, 33, 300)
    mouseClickAXY(MouseButtons.LEFT, 556, 297)
    mouseClickAXY(MouseButtons.LEFT, 548, 336)
    mouseClickAXY(MouseButtons.LEFT, 228, 301)
    mouseClickAXY(MouseButtons.LEFT, 219, 206)
    mouseClickAXY(MouseButtons.LEFT, 539, 354)
```

Fig. 7. A coordinate based Python script for the steps in Figure 5

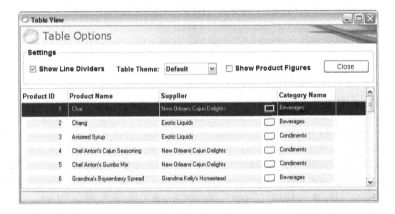

Fig. 8. A rearranged window of the example uniPaaS application (see Figure 5)

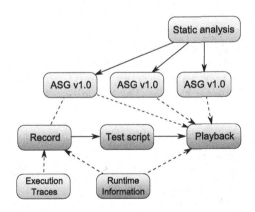

Fig. 9. After a new version, the same test script can be executed with the new ASG

header texts, positions and parent window identifiers. So, we get the handle of the window with the *FindWindow* and *FindWindowEx* functions by the header text and other attributes of it, which we read from the ASG. We can also calculate the relative coordinates to the window of the currently searched GUI element.

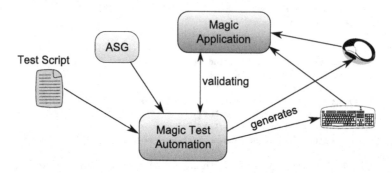

Fig. 10. Running Recorded GUI Events

As the GUI element can be within other GUI elements such as a group box, we start looking for it from the bottom of the Windows control tree and walk upwards to the top. We recalculate the relative coordinates until we get to the searched GUI element.

It is not always enough to know which Windows control matches a control with a unique ASG identifier because we must know the exact position where to click within the GUI element. In case of a button this is irrelevant, but in case of a tree view it is not. The Magic Test For complex controls, the automation tool generates script files where we store a position as the relative position to the identified Magic control. Based on these coordinates we can calculate the absolute position where we can generate the keyboard or mouse event using the Windows API.

Evaluating an Execution. Some steps of the evaluation can be done automatically after the test script was executed, however it is always necessary to tell the automation tool the validation steps manually after recording a test script. The tester can do it by inserting validation (e.g. assert) functions into the script file after the corresponding event handler. The Magic Test Automation tool supports the following validation possibilities:

- To check anywhere in the application's control tree, or in a particular window whether it contains a text or there is a window with a given title.
- Comparison of a specific GUI element's text with a given text.
- Verify that a GUI element is in focus or not.
- Verify that a GUI element is enabled or not.
- Verify that a check box or radio button is checked or not.

The Magic Test Automation tool will check these asserts and report the result of a test script accordingly.

Another advantage of these validation functions is that in addition to evaluate the results of an execution, one can use them in the previously mentioned delay functions too. For example, one can easily say that she/he wants to wait until a

```
from testrunner import *
from testrunner_ext import *

def runScript():
    assert waitFor(3014, findWindow(741, 379, "Table View")) == True
    mouseClickAtControl("Fe-Table View::Ct-Divider Box", MouseButtons.LEFT, 11, 23)
    assert validate(checkState("Fe-Table View::Ct-Divider Box", CheckStates.Checked)) == True
    mouseClickAtControl("Fe-Table View::Ct-Theme Combo", MouseButtons.LEFT, 44, 14)
    mouseClickAtControl("Fe-Table View::Ct-Theme Combo", MouseButtons.LEFT, 47, 33)
    assert validate(compareText("Fe-Table View::Ct-Theme Combo", "Rose")) == True
    mouseClickAtControl("Fe-Table View::Ct-Show Figures Box", MouseButtons.LEFT, 13, 17)
    assert validate(checkState("Fe-Table View::Ct-Show Figures Box", CheckStates.Checked)) == True
    mouseClickAtControl("Fe-Table View::Ct-Table", MouseButtons.LEFT, 218, 129)
    assert validate(checkFocus("Fe-Table View::Ct-Show Figures Box", False)) == True
    mouseClickAtControl("Fe-Table View::Ct-Close Button", MouseButtons.LEFT, 36, 22)
    assert validate(findWindow(741, 379, "Table View")) == False
```

Fig. 11. Examples for validations in a Python script

check box is checked or a specific text box contains a given text. Moreover, with Python scripts we can use them to control the execution of the test case. E.g. we can define complex test cases where we say that if a GUI element is activated then we want to do certain steps, otherwise we want to do a different chain of steps.

Figure 11 illustrates the Python script shown in Figure 6, extended with validation instructions. After clicking the check boxes there is a checkState function which checks that the check box is really checked or not. After selecting an item from the combo box there is a compareText function which checks that the combo box contains the correct text and after clicking in the table we check that the *"Fe~TableView::Ct~Show Figures Box"* has the focus or not. Finally, after clicking the *"Fe~Table View::Ct~Close Button"* button we check if the window closed successfully or not.

5   Comparison to Other Techniques

A comparison of some aspects of common techniques and our approach can be seen in Table 1. Here we elaborate on these techniques in details.

Keyword-Driven Testing. A keyword in its simplest form is an atomic test step or an aggregation of more atomic steps. It describes an action to be performed, hence keyword-driven testing is usually referred as action-word testing too. Most of the cases the keyword-driven testing is divided into two stages:

- planning stage,
- implementation stage.

In the planning stage test engineers determine the test steps for each test case (e.g. entering a text into a text field, clicking on a button, etc.). Later, in the implementation stage the engineers can use a framework to write the previously planned test scripts in a format which can be executed by the framework.

Table 1. Key features of different testing techniques that our tool can handle

	Keyword-driven	Data-driven	Modularity-driven	Coordinate-based	White-box based	Presented approach
no need of programming skills to design test scripts	X	X	X			
no hard-coded data in test scripts		X				X
combinable test scripts			X	X	X	X
no source code required to design test scripts	X	X	X	X		
test script execution handles rearrangements in windows					X	X

A special system under test may require unique actions and keywords which are important to be supported by the testing framework.

In some cases the planning stage and the implementation stage can be combined into one stage and engineers can write the scripts directly into the frameworks scripting format.

Our presented approach can be interpreted as keyword-driven testing because our implemented tool has its own scripting language which is able to understand specific keywords and translate them into mouse, keyboard or other input events. Similar to our tool, Selenium is also a record/replay tool. It is used for testing web applications. It has keywords like Goto WEBSITE or Enter "username", etc. TesComplete is also an automating testing tool which uses keywords to simulate input events. With TestComplete one can also record keyword-driven test scripts and edit them later manually. Another example for a keyword-driven test automation tool is TestArchitect[6] developed by LogiGear Inc.

Data-Driven Testing. Data-driven testing is based on the separation of testing data and execution logic, the tester specifies inputs and verifiable outputs for a test script so that the test script is executed several times on different inputs. Data-driven testing is usually used for testing a form of an application with specific data. So the tester has to specify the input data which the testing framework enters manually into the form under test and then compares the result to the expected output. The main difference between keyword-driven testing and data-driven testing is that in keywod-driven test scripts the data is hard-coded into the test script (e.g. enter "test text" to a textbox) and if one wants to test e.g. the same textbox with different data she/he has to create another test script.

---

[6] http://www.testarchitect.com/

Our approach relies on Python scripts resulting that it can be used for data-driven testing. With Python, the input data can be stored in variables, which can be initialized even in a separate script file, hence the input and the execution logic can be totally separated. Moreover by using arrays for storing input and expected output data, loops can be used to execute the same keyword several times with the input array. This way hard-coded data sets can be eliminated from our test scripts. Compared to other tools, TestComplete is also capable of specifying input data for test recorded test scripts so TestComplete can also be used for data-driven testing. Using extensions Selenium is also capable of executing test scripts on various input data.

Modularity-Driven Testing. Modularity-driven testing requires writing small, independent test scripts for each modules, packages and functions of the application under test. These small scripts are then used to create larger tests, realizing a particular test case. For example, if one wants to test one of the admin users' functions, she/he has to write a script for testing the login action and another separate script for testing the function itself. Then, in a larger test script, first the login script gets called and if it runs successfully the next script gets called which tests the admin's function.

One benefit of this technique is that one change in a module/function affects only its test cases and others might remain untouched during the maintenance of the test scripts.

With Python scripts, modularity-driven testing is also supported by our approach. One can write separate automated Python test scripts and combine them into a larger script by importing them.

Coordinate-Based and White-Box Testing. One common way for automated GUI testing is the coordinate-based testing, because the testing framework doesn't need to know anything about the tested application. Coordinate-based testing is a sort of keyword-driven testing. Usually a keyword contains a coordinate and a user action to be performed on the given coordinate. There are two kinds of coordinate based testing:

- Using absolute coordinates within the application window, where coordinates are relative to usually the upper left corner of the screen. This method does not appears to be very useful, but in many applications the position of the window is not important.
- Using coordinates that are relative to the upper left corner of the currently active window.

Coordinate-based test scripts are the solution if there is no available information about the application under test. However, coordinate-based test scripts are hard to maintain as it might easily change what is exactly on the same coordinate next time when we execute the application.

If we have access to the source code or some documentations of the application under test during its testing phases process it is called white box testing.

Basically our method is a white box testing because we use the layout description of the application to create test scripts.

5.1 Drawbacks of the Technique

Besides benefits, there are some important drawbacks which should be discussed here. First, we consider minor changes of the GUI those changes that simply rearrange the layout of the window and does not modify drastically the structure of it. Our method will recognize the control based on its unique identifier, which identifies the control based on its parents in the control tree. If the parent hierarchy changes, we will not be able to recognize the same control again.

Another important drawback is that the method works based on relative coordinates inside the identified controls. These coordinates may strongly depend on the internal layout of the control. For example, in tree controls if the order of the nodes varies between different executions, our tool may not follow the new structure. Similarly, our technique may fail in selecting an exact item from a listbox or a combobox if the list of elements changes.

Another way a developer can exploit our method is to change the size or position of a control at runtime. Since we read this information from the ASG, our method works as long as the size and the position of the control remains unchanged during execution.

6   Conclusions and Future Work

Our approach for layout independent automatic GUI testing is based on user interface descriptions stored in the source code. We use static code analysis to gather user interface descriptions and combine it with dynamic execution traces during the recording phase of a test case. The resulting test scripts contain only layout independent data which can be played back to the application later even if the user interface has been changed. This technique may dramatically lower the costs of regression tests where developers and testers have to maintain thousands of test cases.

We implemented our approach in a special 4GL environment called Magic xpa, and our implementation is currently used by our industrial partner where developers have been working with Magic for more than a decade. Our partner delivers wholesale products where high quality of the delivered product is top priority, which also requires thorough testing processes. We found our approach to be useful for our partner in their regression testing processes.

Using GUI information stored in the source code during automatic GUI testing is a novel approach for Magic 4GL. We note here that the idea can be easily generalized to other languages, where the GUI description can be extracted from the source code by static analysis (e.g. resource files of Delphi or C#) applications). However this might not stand for languages where the GUI is usually constructed dynamically, for instance in Java, where the dynamic nature of GUI generation makes our approach hardly applicable.

As future work we plan to improve our validation techniques and to support testing Magic applications with automatic test script and test input generations based also on the results of static analysis.

Acknowledgements. This research was supported by the Hungarian national grants GOP-1.2.1-08-2009-0005 and GOP-1.1.1-11-2011-0039.

References

1. Bauersfeld, S., Vos, T.E.J.: Guitest: a java library for fully automated gui robustness testing. In: Proceedings of the 27th IEEE/ACM International Conference on Automated Software Engineering, ASE 2012, pp. 330–333. ACM, New York (2012)
2. Buwalda, H.: Automated testing with action words, abandoning record and playback. In: Proceedings of the EuroStar Conference (1996)
3. Buwalda, H., Kasdorp, M.: Getting automated testing under control, software testing and quality engineering. STQE Magazine, Division of Software Quality Engineering (November/December 1999)
4. Dranidis, D., Masticola, S.P., Strooper, P.: Challenges in practice: 4th international workshop on the automation of software test report. SIGSOFT Softw. Eng. Notes 34(4), 32–34 (2009)
5. Dustin, E., Garrett, T., Gauf, B.: Implementing Automated Software Testing: How to Save Time and Lower Costs While Raising Quality, 1st edn. Addison-Wesley Professional (2009)
6. Dustin, E., Rashka, J., Paul, J.: Automated software testing: introduction, management, and performance. Addison-Wesley Longman Publishing Co., Inc., Boston (1999)
7. Fewster, M., Graham, D.: Software test automation: effective use of test execution tools. ACM Press/Addison-Wesley Publishing Co. (1999)
8. Ganov, S.R., Killmar, C., Khurshid, S., Perry, D.E.: Test generation for graphical user interfaces based on symbolic execution. In: Proceedings of the 3rd International Workshop on Automation of Software Test, AST 2008, pp. 33–40. ACM, New York (2008)
9. Hinz, J., Gijsen, M.: Fifth generation scriptless and advanced test automation technologies (2009)
10. Kaner, C.: Architectures of test automation (2000)
11. Kit, E.: Integrated effective test design and automation software development. Software Development Online (February 1999)
12. Li, K., Wu, M.: Effective GUI Test Automation. SYBEX Inc., Alameda (2005)
13. Lu, Y., Yan, D., Nie, S., Wang, C.: Development of an improved GUI automation test system based on event-flow graph. In: Proceedings of the 2008 International Conference on Computer Science and Software Engineering, vol. 02, pp. 712–715. IEEE Computer Society (2008)
14. Memon, A.M., Pollack, M.E., Soffa, M.L.: Hierarchical GUI test case generation using automated planning. IEEE Transactions on Software Engineering 27(2), 144–155 (2001)
15. Nagy, C., Vidács, L., Rudolf, F., Gyimóthy, T., Kocsis, F., Kovács, I.: Solutions for reverse engineering 4GL applications, recovering the design of a logistical wholesale system. In: 15th European Conference on Software Maintenance and Reengineering (CSMR), pp. 343–346 (March 2011)

16. Peleska, J., Löding, H., Kotas, T.: Test automation meets static analysis. In: GI Jahrestagung (2). LNI, vol. 110, pp. 280–290. GI (2007)
17. Posey, B.: Just Enough Software Test Automation. Prentice Hall PTR (2002)
18. Sommerville, I.: Software testing. In: Software Engineering, 9th edn. Addison-Wesley (2010)
19. Strang, R.: Data driven testing for client/server applications. In: Proceedings of the Fifth International Conference on Software Testing, Analysis and Reliability (STAR 1996), pp. 395–400 (1996)

A Semi-automatic Usability Evaluation Framework

Kornél Muhi[1], Gábor Szőke[1], Lajos Jenő Fülöp[1],
Rudolf Ferenc[1], and Ágoston Berger[2]

[1] University of Szeged
Department of Software Engineering
Árpád tér 2, 6720 Szeged, Hungary
{mkornel,kancsuki,flajos,ferenc}@inf.u-szeged.hu
[2] MONGUZ Ltd.
Nemestakács u. 12/A, 6722 Szeged, Hungary
aberger@monguz.hu

Abstract. Most of the software maintenance costs come from usability bugs reported after the release and deployment. A usability bug is really subjective, hence there is a large communication overhead between the end user and the developer. Moreover, the reputation of the software development company could be decreased as well. Therefore, proactively testing and maintaining software systems from a usability point of view is unambiguously beneficial.

In this paper we propose a research prototype, the Usability Evaluation Framework. The development of the framework is driven by well-defined requirements. It is built upon a usability model, it calculates usability metrics, it integrates questionnaires and it also ensures several meaningful reports. We have successfully applied the framework to evaluate and to improve the usability of two industrial software systems.

1 Introduction

A popular belief about software maintenance is that it primarily involves development of new features, testing, and fixing of programming bugs (e.g. a null pointer exception). Although, some researchers show that a significant part of software bugs are related to some kind of usability problems of the investigated applications. For example, Landauer [10] reports an interesting distribution of software bugs:

"About 80 percent of software life cycle costs occur not during development but during the maintenance period. In turn, 80 percent of these maintenance costs are a result of problems users have with what the system does, not programming bugs."

There are some prevalent methods to evaluate and test usability. Usability testing in labs is really useful because it can recover serious usability bugs. Albeit, it is expensive and time consuming because real users have to leave their current tasks

B. Murgante et al. (Eds.): ICCSA 2013, Part II, LNCS 7972, pp. 529–542, 2013.
© Springer-Verlag Berlin Heidelberg 2013

and they have to go the usability lab. It also requires experienced moderators and human factor experts, otherwise they probably face communication problems with the users. Namely, a usability lab is a simulation of a real production environment thus some usability bugs could be missed. Another really promising direction is remote usability testing but some of the usability bugs still could be missed.

Therefore, usability testing should be supported by tools that can be applied in production environments as well. Several usability testing tools are available, but these work only on one field, typically on web pages. To the best of our knowledge, there is no general solution available that supports the usability evaluation of arbitrary software applications, i.e. independently from programming language, operating system, and other factors (e.g. desktop, mobile and web applications).

In this paper, we present a research prototype, the Usability Evaluation Framework (UEF) that supports usability testing and usability maintenance in real production environments and can be applied on arbitrary software application as well. We have two goals with the framework. We want to develop a completely general framework that is widely applicable, and we want to test and improve the usability of two industrial software systems.

The paper is organized as follows. In the next two sections we describe the related work and elicit the requirements for the framework. In Section 4 we introduce the key component of the framework, the general usability model. Next, we show the details of the framework in Section 5. Section 6 shows our results about applying the framework in case of two industrial software systems. Finally, the last two sections present the evaluation of the framework and the conclusions.

2   Related Work

In our previous short paper [15] we described and introduced our general software quality model and framework with the underlying principles and methodology. UEF was just briefly introduced as one of the applications of this general softare quality framework, in less than a half page, so that paper did not introduced a lot of important details and results.

There are some very good and popular solutions available like Google Analytics [5] and TrackerBird [17]. These solutions collect data about user behaviour that can be used broadly, e.g. to derive usability problems [7]. Although, these solutions work on specific fields. For example, TrackerBird works on .NET applications while Google Analytics works on web pages.

Au et. al. [2] describes the aspects of usability testing, especially in case of mobile applications. They examined some systems and the results show that it is recommended to use an automated usability testing framework. This helps the developer to test the system more often, even in the early stages of the development, when the changes are cheaper. They also developed and presented an automated usability testing framework.

Ivory et. al. [9] compare and evaluate usability testing approaches. The authors suggest making a unified usability testing system, because the current approaches are giving very different results. The evaluation also shows that the current usability testing techniques use poor automation.

Harty et. al. [6] is about the automation of the usability testing of web-based applications. The authors describe why is it so hard to do usability and accessibility testing. They conclude that there are a lot of academic attempts for usability testing, but most of them are not usable in an industrial environment.

Several other papers [14] [8] [1] [4] [13] also deal with usability evaluation. To sum up, usability testing and evaluation is typically executed in a staging (testing) environment, and constrained to a special application field or domain. For example web log analyzers cannot be employed in case of desktop applications, .NET specific analyzers cannot be applied in case of Java applications, and so on.

3   Requirements

The following requirements are set up based on (i) the investigation of related work and (ii) after several discussions with professionals from the industry and our university.

Support Production Environments: Usability testing and evaluation are typically executed in a usability lab, where a moderator controls and observes the representative users. Several usability bugs can be detected with this technique but the laboratory circumstances also determine missed usability bugs. Therefore, the usability evaluation shall be performed in real-life *production environments*.

Detection of Patterns: The usability evaluation shall reveal typical and frequent sequences of user interactions, i.e. *patterns*. For example, such detected frequent sequences could be optimized and handled with more attention.

Detection of Usability Bugs: The usability evaluation shall provide tangible *usability bugs*. For example, it can be a frequent and complex user interaction sequence that could be re-engineered as a much usable and straightforward wizard; it can be a complex form that is really difficult to be filled out, therefore it is time-consuming for the users to work with, and so on.

Transparency: The usability evaluation shall be *transparent* and not affect the daily operational work of the users. If users suffer from any kind of interruption or disturbing factor (e.g. video analysis) then unacceptable extra costs are generated from a business point of view.

Automatic: The usability evaluation shall be *automatized* as much as possible, especially in users point of view. The rationale behind this requirement is to reduce extra costs generated by unnecessary manual work.

Wide Applicability: The usability evaluation shall be as *general* as possible. It should be domain independent, i.e. it should be applied arbitrary in different application domains, e.g. in case of financial applications, ERP applications, office applications and so on. It should be applicable in different kind of operating systems (e.g. Windows, Linux, etc.). Furthermore, it should be applicable in different environments (e.g. in case of desktop, web or mobile applications). Finally, the usability evaluation should be applicable in different kind of programming languages as well (e.g. Java, C++, C#, etc.).

4   Usability Model

A well defined data model (shown on Figure 1) is the key to our approach, and the basis of the other components of the UEF framework. The model is developed through several iterations. In each iteration, the model is evaluated and then improved based on the previously shown requirements. In this section, we introduce the final model in detail.

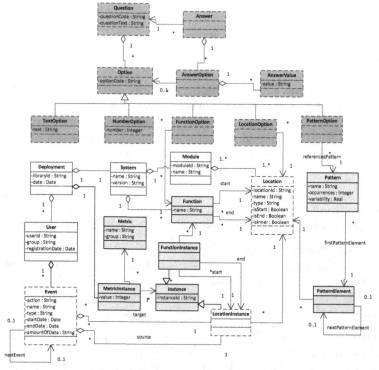

Fig. 1. The usability model

System nodes are denoted by non filled nodes with continuous borders in Figure 1. The software vendors usually deploy a customized build of the application to the customers to match their special needs. Based on this observation, the *Deployment* entity has a central role in the model, and it contains the *System* entity. Furthermore, it contains only those *Module* entities that the users have actually in the current deployment. Deployment also contains the registered *Users*.

Low level nodes are denoted by non filled nodes with dashed borders in Figure 1. Whenever a user clicks on a button in the application or presses a key on the keyboard, an event will be generated. Such activities are represented by the *Event* entity. It stores the event's creation and completion date, the name, the type, the executed action, and (optionally) the size of the processed data of the executed action. Besides storing such basic event data it is also required to store the source and target locations too. The source indicates where the event has been triggered (usually a menu item or window). The target represents the result of the event. Source and target information are represented through *Location* and *LocationInstance*. The Location entity represents the abstraction of a certain location with it's name, type and other information. LocationInstance represents a concrete instance of a Location. For example, in several applications it is possible to open multiple windows. Such windows have to be handled together in some point of view, and that is the reason behind Location. While in other aspects, we have to be able to distinguish them, and that is the reason behind LocationInstance. An event is generated by a user, hence User contains the corresponding Events in the model. The events are stored chronologically, which is represented by the nextEvent relation.

Derived nodes are typically based on low level nodes, and denoted by filled nodes with continuous borders in Figure 1. A functionality of the system can be described by a sequence of events between two locations. The model represents such functionalities with the *Function* entity. The *FunctionInstance* entity is a concrete instance of the referenced Function. It points to two LocationInstances according to the Location types defined by the Function. *Metric* represents the calculated properties of functions and locations. The *MetricInstance* entities assign concrete values to an instance (FunctionInstance or LocationInstance) based on the referenced metric. Because the metric entities are the same in case of every deployment, it is enough to store them just once.

Frequent user interaction sequences (i.e. patterns of user actions) are also represented in the model. A pattern consists of two parts: a main entity (*Pattern*), which represents the sequence's generated name, occurrence and variability; and *PatternElements* which links together the sequence's pieces. A pattern entity points to it's first element, and then a PatternElement refers the next element recursively.

Survey nodes are denoted by filled nodes with dashed borders in Figure 1. The model is also capable of storing data for generating online questionnaire, and for representing the results of the questionnaire. A survey usually contains a few *Questions* which have more *Options*. The user has to pick one or more of these options. An *Option* can be any kind of a field: a simple text (*TextOption*), a number (*NumberOption*) or a reference field to a function (*FunctionOption*) or location (*LocationOption*). The last two come in handy when we want to ask the users about the investigated system and about the automatically detected information (i.e. automatically calculated information can be related with subjective user opinions). Answer stores the options picked by the user (*AnswerOption*). The *AnswerValue* is introduced to store comments and justifications about the answers. Patterns can also be referenced in the survey (*PatternOption*).

5 Framework

In the followings, we shortly introduce the architecture of the framework that is also based on the requirements (see Figure 2). The arrows and numbers represent the working steps in chronological order.

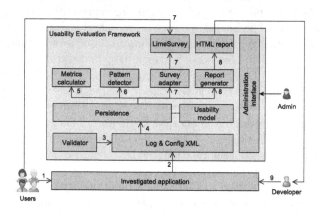

Fig. 2. Architecture of the UEF framework

The first step is to record user interactions that shall be performed by the investigated application. The second step is the generation of the framework input files by the investigated application. These two steps ensure that the framework and its services are completely independent from any kind of technological platform (e.g. operating system, programming language, etc.).

The framework has two kinds of input files. *Config XML* describes meta information about the system, e.g. registered users, available modules and available locations (windows, menu panels, etc.), see Figure 3. Based on these metadata, *Log XML* provides the concrete usage data (see Figure 4). The development of these XML formats is heavily influenced by the usability model.

The validator component checks the logfiles according to predefined syntactic and semantic rules. Then, the logged data are uploaded into the persistence framework for further processing. The schema of the persistence layer is configured with the *usability data model*. The model represents several information related to usability: events performed by the users, locations touched by the users, and so on (see Section 4).

The metrics calculator module can calculate several numeric or textual properties. This module can be extended, calculation of new metrics requires only a new metric calculator plugin to be developed.

```
 1<deployment libraryId="library1" date="2011-11-30"
 2   xmlns:xsi="http://www.w3.org/2001/XMLSchema-instance"
 3   xsi:noNamespaceSchemaLocation="extformat.xsd">
 4   <system name="Big Library" version="1.0.0"
 5     deploymentDate="2005-01-01">
 6     <module id="module01" name="reader"/>
 7     <module id="module02" name="operator"/>
 8   </system>
 9   <locations>
10     <location locationId="renting.books" name="Select books to
11       rent" type="window" module="operator"/>
12     <location locationId="renting.reader" name="Data of the
13       reader" type="window" module="operator"/>
14     <location locationId="renting.summary" name="Summary about
15       the rent" type="window" module="operator"/>
16     <location locationId="renting.help" name="Renting books help"
17       type="help" module="operator"/>
18     <location locationId="renting.fault" name="Renting books
19       error" type="error" module="operator"/>
20   </locations>
21   <functions>
22     <function name="Rent a book" startLocationId="renting.books"
23       endLocationId="renting.summary"/>
24   </functions>
25   <users>
26     <user id="1" group="administrator"
27       registrationDate="2011-11-30"/>
28     <user id="2" group="operator" registrationDate="2011-11-30"/>
29   </users>
30</deployment>
```

Fig. 3. Example for the configuration descriptor

To process the user events stored in the persistence framework, a pattern detection module has been developed based on the very efficient suffix array method [12]. This method has been widely applied and adopted in several fields of computer science, for example it is employed by software clone detection algorithms as well [3]. In case of UEF, all user interaction is stored and ordered in an array, and then, based on the suffix array method, the patterns of user interactions are detected.

Survey adapter module is capable of generating configuration files which can be imported into the LimeSurvey open source survey engine [11]. These surveys can be filled out by the users of the subject systems, and the results can be uploaded back into the persistence system. Questions can be easily added, modified and removed in the framework. These surveys collect subjetive opinions and can contain questions regarding the results of the current data analysis.

```
 1<log xmlns:xsi="http://www.w3.org/2001/XMLSchema-instance"
 2   xsi:noNamespaceSchemaLocation="logformat.xsd"
 3   libraryId="library1" date="2011-11-30">
 4   <events>
 5     <event action="rent" name="" type="button"
 6         startDate="2011-11-27T08:03:47.103"
 7         endDate="2011-11-27T08:03:47.103">
 8       <source locationId="main" instanceId="1"/>
 9       <target locationId="renting.books" instanceId="2"/>
10       <user id="1"/>
11     </event>
12     <event action="rent.reader" name="" type="button"
13         startDate="2011-11-27T08:03:52.163"
14         endDate="2011-11-27T08:03:52.163">
15       <source locationId="renting.books" instanceId="2"/>
16       <target locationId="renting.reader" instanceId="3"/>
17       <user id="1"/>
18     </event>
19     <event action="rent.summary" name="" type="button"
20         startDate="2011-11-27T08:04:00.271"
21         endDate="2011-11-27T08:04:00.271">
22       <source locationId="renting.reader" instanceId="3"/>
23       <target locationId="renting.summary" instanceId="4"/>
24       <user id="1"/>
25     </event>
26   </events>
27</log>
```

Fig. 4. Example for the events log descriptor

Based on the collected information (metrics, patterns and surveys) the UEF framework generates an HTML report (see UEF homepage [18]). Then the report can be employed by the developers to improve the usability of the investigated application.

We have implemented a prototype of the UEF architecture in Java.

6   Proof of Concept

Integrated collection management systems (ICMS) manage the business workflow of libraries and cultural institutes. Monguz Ltd. is one of the leading software development companies in Hungary in the field of ICMS. This company wanted to evaluate and to improve the usability of two ICMS systems. The usability evaluations were performed in the daily operative work of two selected Hungarian libraries with the help of the UEF. We are allowed to refer the investigated systems anonymously, so we refer them as System1 and System2. Both systems are implemented in Java, but System1 is a desktop while System2 is a web application. The study included all members of staff of the participant libraries who use the integrated system, without distinction to experience or position.

We have developed *specific loggers* inside the investigated applications to collect usage information. Finally, the collected information are exported into the input format of UEF (config and log XML files).

Based on the literature and discussions with professionals of Monguz Ltd. a few metrics have been implemented in UEF to help the evaluation of the two ICMS systems. Seffah and others [16] introduce more than one hundred metrics for usability in different aspects. We have employed and derived metrics from the productivity aspect. Some of the calculated metrics are summarized in Table 1. A complete collection of the calculated metrics (and other materials) are available on the project's homepage [18].

Table 1. Some of the calculated metrics

Metric	Metric name	Short description
NI	Number of Instances	Number of instances for a location type
RNI	Recursive Number of Instances	Number of instances for a location type plus instances of other windows opened from this location type
NT	Net Time	Net time spent on a location instance
ET	Existence Time	Time between opening and closing (location instance metric)
ATR	Active Time Ratio	The ratio of having focus to the time of existence (location instance metric)
PTR	Passive Time Ratio	The ratio of not having focus to the time of existence (location instance metric)
NE	Number of Errors	Number of error messages appeared from a given location instance
NHU	Number of Help Uses	Number of help uses initiated from a given location instance
RNE	Relative Number of Errors	The ratio of error messages appeared to the number of steps while performing a functionality starting from the given location instance
RT	Recursive Time	The time spent over the full functionality accessible from the given location instance (in seconds)

Based on several discussions with professionals of Monguz Ltd. a questionnaire has been set up as well. The survey has been filled out by 20 users in the case of System1, and by 50 users in the case of System2.

There is a question enquiring about the overall satisfaction of the users. Users of System2 gave an average 7.4 grade out of 10, while users of System1 marked their system with an average 6.89 score. Correlation models cannot be built between the questionnaires and calculated metrics because we evaluated only two software systems. Still, some relation can be observed when comparing these subjective results to the calculated objective metrics. Except of two cases (NT-Avg and NT-Med), the metric values of System2 are lower than the ones of System1 (see Table 2). Based on the other values, we can hypothesize that lower values for NI, RNI and ET indicate better overall satisfaction of the users.

Table 2. Most important metrics for the two examined systems. Lower or equal values of the comparison of System1 and System2 are denoted with gray background.

	System1			System2		
	Max	Avg	Med	Max	Avg	Med
NI	461	28	8	115	22	8
RNI	1753	292	18	291	34	8
NT	6628	169	10	4348	296	19.5
ET	95446	9269	33	37445	3755	7.9

There are some textual responses given by the users concerning concrete locations in the software. We compared these to the calculated metric values and it seems that in certain cases NI, RNI and ET could indicate the general usability of a software system, while NT could mark tangible usability bugs.

We have also examined the top 30 patterns (frequent user interaction sequences) found in both system (see Table 3). The average pattern length is significantly bigger in System1 (22 locations) than in System2 (14 locations). Based on this, we hypothesize that pattern length could be an indicator for usability problems as well (besides the metrics).

Table 3. Summary of the patterns found in the examined systems

	System1			System2		
	Min	Max	Avg	Min	Max	Avg
Pattern freq.	11	59	22.03	3	57	14.17
Pattern's length	10	16	10.7	5	16	6.3

6.1   Improvement of the Systems

We have made improvements to the ICMS systems based on the results of the UEF report. In the case of System1, some of the results point out exact usability bugs in the system, which were fixed in the following release. Other responses point out problems with the automatic notification system, which is also fixed. The data also pointed out possible usability improvements in certain windows of the circulation module, some of which are improved upon, other improvements are being planned in the next major release of the software.

7   Evaluation and Application Guidelines

In this section we evaluate UEF. First, we have to emphasize that the framework does not deal with application specific logging of user interactions. In some point of view it is a drawback because application developers have to extend their application to generate XML input files for the UEF framework. At the same time, it is a really big advantage in an other point of view. This feature guarantees a very important requirement of the framework, the wide applicability. By generating the necessary XML files, the framework provides several useful features: detection of frequent user interactions (i.e. patterns), calculation of metrics, detection of questionnaires based on predefined templates, and so on. In the followings, we show a detailed evaluation of the framework through the requirements introduced in Section 3.

- *Support production environments* is indirectly fulfilled by the XML-based configuration files. Nowadays, any kind of software application can produce simple XML text files that can be loaded into the framework. In fact, this requirement have to be fulfilled by the investigated application.
- *Detection of patterns* is fulfilled by the adaptation of the suffix array method and algorithm. In the proof of concept, several patterns have been detected and evaluated.
- *Detection of usability bugs* is fulfilled by the adaptation of a questionnaire engine and the automatically calculated metrics. However, manual investigation and evaluation of the questionnaire results and the calculated metrics is needed. Furthermore, we have calculated some basic metrics only, and validate them in a simple proof of concept experiment. Therefore, this requirement has been only fulfilled yet.
- *Transparency* is fulfilled because the data being recorded in the background, so the users are not affected. The framework ensures this requirement by its general XML input formats. The investigated application is responsible for silent logging and to provide the XML files.
- *Automatic* is mostly completed. UEF only requires one manual step from users: filling the questionnaires.
- *Wide applicability* is also fulfilled. It is domain- and application independent because its general usability model and interfaces do not contain any domain or application specific data, i.e. domain and application specific logging mechanisms are independent from the UEF. Platform-independency is guaranteed by that the framework is written in Java, and its interfaces are also platform-independent: the input can be given in XML format, while the HTML results can be displayed in any browser.

7.1 General Application Guidelines

Based on the experiencies collected during the proof of concept we have developed a general template that ensures the application of the UEF prototype in arbitrary domains as well.

1. **Generating UEF's XML inputs.** Either the XML files are directly created during the logging or if the investigated system logs elsewhere (e.g. to a database table) then the information have to be converted into UEF's XML files.
2. **Data collection (in production environment).** This step consists of the actual use of the observed systems, during which the logger components

collect the user interaction data into XML files. Depending on the amount of data you want to collect, this could take weeks or months.

3. **Surveying questionnaires.** A specified questionnaire will be generated as the metrics and the sequences have been calculated. This survey will be presented to the users of the observed systems. Depending on the number of the users, filling out the survey takes 1 or 2 days.

4. **Automatically generating reports.** The framework generates an HTML report based on the calculated information and the opinions given by the users. This step takes maximum 1 or 2 hours depending on the amount of data.

Apart from the fact that the survey has to be filled by the users, the UEF is automatic. It seems that specific loggers will have to be implemented for each new case study to produce specific XML files. In some degree it is true if a concrete software application should be investigated specifically with the logging of application specific data. Still, the extension of the corresponding software application would desire just a little extra work based on our current experiencies[1]. Moreover, specific libraries could be developed to collect technology specific data like TrackerBird [17] does it for .NET applications. This way, usability data could be collected easily from several software applications, which are in the same technology domain.

8   Conclusions

The contribution of the paper can be summarized as follows. Based on related works and several discussions with professionals from industry we have defined the requirements of a general usability evaluation framework. Next, we developed the Usability Evaluation Framework (UEF) conforming to the defined requirements. The provided framework architecture can be considered as a reference architecture for future (e.g. industrial) implementations as well.

We successfully applied UEF to evaluate and to improve the usability of two systems. They are evaluated in real-life production environments, at two libraries in Hungary. During the evaluation, the framework has been extended with newly implemented metrics and newly configured questionnaires based on the discussions with the industry partner. The developed metric calculator plugins and the configured questionnaires could be used by new potential partners and application developers in the future. Moreover, we have successfully demonstrated that the UEF framework can be extended easily with new metrics and questionnaires to satisfy the needs of an industrial partner.

The UEF framework is designed to be platform-, domain- and application-independent. Contrary, the evaluated two systems are a little bit homogeneous, i.e. they come from the same ICMS domain, both are written in Java while one run in a web browser and the other run as a standalone Windows desktop

---

[1] The discussed proof of concept has proved that it is easy to develop such XML generator components.

application. Therefore, we are currently working on the application of UEF in other contexts, i.e. the usability evaluation of an office software suite written in C++ and runs both on mobile devices and as a standalone Windows desktop application. We also plan to integrate further metrics and questionnaires into the framework, and to perform new case studies in the future.

Acknowledgement This research was supported by the Hungarian national grants GOP-1.1.2-07/1-2008-0007 and GOP-1.1.1-11-2011-0006.

References

1. Andreasen, M.S., Nielsen, H.V., Schrøder, S.O., Stage, J.: What happened to remote usability testing?: an empirical study of three methods. In: Proceedings of the SIGCHI Conference on Human Factors in Computing Systems, New York, NY, USA, pp. 1405–1414 (2007)
2. Au, F.T.W., Baker, S., Warren, I., Dobbie, G.: Automated usability testing framework. In: Proceedings of the Ninth Conference on Australasian User Interface, Darlinghurst, Australia, vol. 76, pp. 55–64 (2008)
3. Basit, H.A., Jarzabek, S.: Efficient token based clone detection with flexible tokenization. In: Proceedings of the 6th Joint Meeting of the European Software Engineering Conference and the ACM SIGSOFT Symposium on The Foundations of Software Engineering, ESEC-FSE 2007, pp. 513–516. ACM, New York (2007)
4. Chilana, P.K., Ko, A.J., Wobbrock, J.O., Grossman, T., Fitzmaurice, G.: Postdeployment usability: a survey of current practices. In: Proceedings of the 2011 Annual Conference on Human Factors in Computing Systems, New York, NY, USA, pp. 2243–2246 (2011)
5. Homepage of Google Analytics (May 7, 2013), http://www.google.com/analytics/
6. Harty, J.: Finding usability bugs with automated tests. Communications of the ACM 54, 44–49 (2011)
7. Hasan, L., Morris, A., Probets, S.: Using Google Analytics to Evaluate the Usability of E-Commerce Sites. In: Kurosu, M. (ed.) HCD 2009. LNCS, vol. 5619, pp. 697–706. Springer, Heidelberg (2009)
8. Hosseini-Khayat, A., Hellmann, T.D., Maurer, F.: Distributed and Automated Usability Testing of Low-Fidelity Prototypes. In: Proceedings of the AGILE Conference, pp. 59–66 (2010)
9. Ivory, M.Y., Hearst, M.A.: The state of the art in automating usability evaluation of user interfaces. ACM Computing Surveys 33, 470–516 (2001)
10. Landauer, T.K.: The Trouble with Computers: Usefulness, Usability, and Productivity. A Bradford Book (1995)
11. LimeSurvey - Open Source Survey Application (May 7, 2013), http://www.limesurvey.org
12. Manber, U., Myers, G.: Suffix arrays: a new method for on-line string searches. In: Proceedings of the First Annual ACM-SIAM Symposium on Discrete Algorithms, SODA 1990, pp. 319–327. Society for Industrial and Applied Mathematics, Philadelphia (1990)
13. Propp, S., Forbrig, P.: ViSE – A virtual smart environment for usability evaluation. In: Bernhaupt, R., Forbrig, P., Gulliksen, J., Lárusdóttir, M. (eds.) HCSE 2010. LNCS, vol. 6409, pp. 38–45. Springer, Heidelberg (2010)

14. Runge, M.: Simulation of Cognitive Processes for automated Usability Testing. Diploma, Deutche Telekom Laboratories, Berlin (2008)
15. Schrettner, L., Fülöp, L.J., Beszédes, A., Kiss, A., Gyimóthy, T.: Software Quality Model and Framework with Applications in Industrial Context. In: Proceedings of 16th European Conference on Software Maintenance and Reengineering (CSMR 2012) (2012)
16. Seffah, A., Donyaee, M., Kline, R.B., Padda, H.K.: Usability measurement and metrics: A consolidated model. Software Quality Control 14(2), 159–178 (2006)
17. Homepage of TrackerBird (May 7, 2013), http://www.trackerbird.com/
18. Homepage of UEF (May 7, 2013),
    http://www.inf.u-szeged.hu/~flajos/usability

Answers That Have Quality

Hendrik Decker[*]

Instituto Tecnológico de Informática, Universidad Politécnica de Valencia, Spain

Abstract. The lack of quality of stored data is reflected by violations of integrity constraints. Answers to queries in databases containing bad quality information usually cannot be trusted. Nevertheless, many answers given by such databases may still be useful, as long as they are derived from data the quality of which is sufficiently high. We formalize our intuition of answers that have quality on the basis of 'causes'. A cause of an answer is a minimal excerpt of the database that explains why the answer has been given. Thus, an answer has quality if the overlap of its causes with the causes of integrity violation is empty. Even if that overlap is not empty, but is sufficiently low, an answer may have sufficient quality. The amount of causes in the overlaps of causes of answers and integrity violations can be sized by quality metrics.

1 Introduction

In [20], we have shown how integrity checking in inconsistent databases can be achieved by metrics that compare the amount of constraint violation before and after an update. If the update would increase that amount, integrity checking methods that use those metrics do not sanction it. The metrics in [20] are based on 'causes'. Causes are the stored data that are responsible for the violations. Thus, the number or size of such causes provide reliable measures for quantifying the lack of integrity.

In [18], we have developed a concept for characterizing *answers that have integrity* (AHI) in databases that may violate their integrity constraints. The idea of AHI is also based on causes, since queries and constraints have the same syntax. Hence, causes may serve not only to construe why constraints are violated, but also to concisely explain why an answer is given to a query. Thus, an answer to a query has been defined to have integrity if it has a cause that does not overlap with any cause of integrity violation.

In [27,21,20,24], we have argued that, apart from stating necessary conditions for the semantic consistency and integrity of stored data, constraints are expressive enough to also serve for modeling, monitoring and maintaining quality attributes that are more general than consistency and integrity. In particular, constraints can express conditions of security, certainty, accuracy, reputation, trustworthiness and the like, in short, of quality.

Essentially, the difference of integrity constraints, in their original sense, and quality constraints is: the former require to be always satisfied in each state of the

---

[*] Supported by FEDER and the Spanish grants TIN2009-14460-C03, TIN2010-17139.

B. Murgante et al. (Eds.): ICCSA 2013, Part II, LNCS 7972, pp. 543–558, 2013.

database, while the latter may be violated to a certain extent at any time. Quality is a matter of more-or-less, unlike the original narrow-minded binary concept of integrity, to be either satisfied or violated. Thus, in general, quality constraints (as well as, in particular, a more open-minded vision of integrity constraints) call for violation tolerance. Apart from providing a generic paradigm for virtually all approaches to constraint maintenance, as argued in [25], the violation tolerance of metric-based methods for database integrity management is their main merit.

In this paper, we bring the aforementioned lines of research results together. However, we do not apply metric-based inconsistency-tolerant integrity mainte-nance to database quality management, since that has been addressed already in [26]. Rather, we generalize the approach to query answering in databases with violated constraints in [18] (which was focused on integrity in the traditional sense), in order to provide answers that have quality, rather than just integrity.

We go beyond what has already been achieved in [18], in two ways. Firstly, we simplify the concept of causes to a significant extent: instead of considering all clauses in the database including deduction rules and view definitions, we focus on basic causes, i.e., positive and negative facts, the presence or, resp., absence of which is sufficient for concisely explaining answers and constraint violations. Secondly, answers that have integrity are, by definition, devoid of any involve-ment with violated constraints. As opposed to that, it can be justified to concede sufficient quality to answers as long as their causes do not overlap too much with the causes of constraint violation. Hence, another technical contribution of the paper is a formal approach to what "(not) too much" means, in this context.

Thus, in short, we intend to provide a quantification of what it means that an answer has quality. In analogy to AHI, we abbreviate our approach to answers that have quality by *AHQ*.

In Section 2, we recapitulate some background of database logic. In Section 3, we first recall the original definition of causes and then introduce the new, simpler definition of basic causes. In Section 4, we first define AHQ, then relax this notion to ASQ, i.e., answers with sufficient quality, and then show how to compute AHQ and ASQ. In Section 5, we compare our work to related efforts. In Section 6, we conclude with an outlook to further work.

2   Preliminaries

In this section, we briefly address some foundational issues. Throughout, we use notations and terminology that are common for relational and deductive databases and their logical foundations [12] [1].

Since the syntax of integrity constraints and quality constraints is the same, also the latter can always be represented by *denial* clauses, i.e., universal sen-tences of the form $\leftarrow B$, where B is a conjunction of literals that asserts what *should not* hold in any database state. In other words, B expresses some lack of quality. If a constraint I expresses what *should* hold, then a denial form of I can be obtained by an equivalence-preserving transformation of $\leftarrow \sim I$, as pro-posed, e.g., in [14]. The clauses that define each fresh predicate that is needed for such transformations are simply added to the database. An illustration of such a transformation is given in Example 3*d*). The *quality theory* of a database D is a finite set QT of constraints imposed on D.

We are going to use symbols such as D, QC, QT, U to always stand for a database, a quality constraint, a quality theory and, resp., an update. For each sentence F, and in particular for each quality constraint, we may write $D(F) = true$ (resp., $D(F) = false$) if F evaluates to $true$ (resp., $false$) in D. Similarly, we write $D(QT) = true$ (resp., $D(QT) = false$) if each constraint in QT is satisfied in D (resp., at least one constraint in QT is violated in D).

Let $\mathcal{H}$ be a universal Herbrand base and $\mathcal{N}$ a universal set of constants in the language of each database, represented w.l.o.g. by natural numbers. As usual, we assume that the sets of intensional and extensional predicates in each database are disjunct, and that each database clause, each query and each constraint is range-restricted. We may use ';' to delimit elements of sets, since ',' also denotes conjunction in the body of clauses.

3   Causes

Causes have been introduced in [18]. The definition of causes had been specialized for definite databases, queries and integrity constraints, and for positive answers, in [20]. In [17], also causes for queries and constraints with negation and for negative answers in flat relational databases have been defined and studied. For the general case, the definition of causes has been simplified in [23]. In [16], basic causes for positive answers in definite databases with definite integrity constraints have been introduced. Basic causes are much less complex than causes in general, since they do not involve deductive rules or view definitions. Instead, they focus exclusively on extensional databases facts as the basic reasons for answers to queries or violations of integrity constraints.

However, reducing general causes to basic ones turns out to be much less straightforward for non-definite databases and queries. Among others, these difficulties are due to the non-monotonicity of database negation. Or to the circumstance that answers are negative, not necessarily due to the presence or absence of extensional facts, but also because queries containing literals with intensional predicates may fail, as shown by the following example.

Example 1. Let a two-place intensional predicate q be defined by the clause $q(x, b) \leftarrow r(x, x)$ in a database containing $r(a, a)$. Then, the answer *yes* to the query $\leftarrow q(a, b)$ can be explained by the presence of the fact $r(a, a)$. However, the answer *no* to the query $\leftarrow q(a, a)$ cannot be explained by recurring on any information about the extensional predicate r. Rather, the non-provability of each ground instance of $q(x, a)$ needs to be explained by its irreducibility to the presence or absence of any extensional fact in the database.

In 3.1, we recall the definition of general causes. In 3.2, we define basic causes, where also some more difficulties of using basic causes in non-definite databases are addressed.

3.1 General Causes

We assume some familiarity with the well-known *completion comp*(D) of a database D [10,39]. It essentially consists of the if-and-only-if completions of all predicates in the language, as defined in [10,39]. For a predicate p, let p_D denote the completion of p in D.

Definition 1. Let D be a database, p a predicate in $\mathcal{L}$, n the arity of p, $x_1, \ldots, x_n$ the $\forall$-quantified variables in p_D and θ a substitution of $x_1, \ldots, x_n$. For $A = p(x_1, \ldots, x_n)\theta$, the *completion* of A in D is obtained by applying θ to p_D and is denoted by A_D. Further, let $\underline{comp(D)} = \{A_D \mid A \in \mathcal{H}\}$, and *if*$(D)$ and *only-if*(D) obtained by replacing $\leftrightarrow$ in each $A_D \in \underline{comp(D)}$ by $\leftarrow$ and, resp., $\rightarrow$. Finally, let *iff*$(D) = $ *if*$(D) \cup$ *only-if*(D). The usual equality axioms of *comp*(D) be associated by default also to *iff*(D). The replacement of $\leftrightarrow$ by $\leftarrow$ and, resp., $\rightarrow$ in A_D is called the *if*-, resp., the *only-if-half* of A_D.

More detailed presentations of the items defined above can be found in [23,18]. Elements in *iff*(D) may be simplified by some obvious equivalence-preserving rewritings. For simplicity, we assume such rewritings throughout.

Example 2. Let $D = \{p(x,1) \leftarrow r(x);\ p(1,y) \leftarrow s(y,z);\ s(1,2);\ s(2,3)\}$. Then, *comp*$(D)$ contains the three completions below, each of which is a universally closed sentence with existentially quantified 'local' variables that do not occur in the head of any clause. We omit the universal quantifier prenex for all non-local variables in the completions below, and also the equality theory associated to *comp*(D) that interprets $=$ as identity.

$$p(x,y) \leftrightarrow (y = 1 \wedge r(x) \vee x = 1 \wedge \exists z(s(y,z))) \qquad\qquad r(x) \leftrightarrow \textit{false}$$
$$s(x,y) \leftrightarrow (x = 1 \wedge y = 2 \vee x = 2 \wedge y = 3)$$

The only-if half of p_D is

$$p(x,y) \rightarrow (y = 1 \wedge r(x) \vee x = 1 \wedge \exists z(s(y,z))).$$

Its instance $p(1,1) \rightarrow (1 = 1 \wedge r(1) \vee 1 = 1 \wedge \exists z(s(1,z)))$, obtained by the substitution $\{x = 1,\ y = 1\}$, is obviously equivalent to $p(1,1) \rightarrow (r(1) \vee \exists z(s(1,z)))$. Similarly, the instance $p(2,3) \rightarrow (3 = 1 \wedge r(2) \vee 2 = 1 \wedge \exists z(s(3,z)))$ of p_D simplifies to $p(2,3) \rightarrow \textit{false}$, which is equivalent to $\sim p(2,3)$. Similarly, the instance $s(2,3) \rightarrow (2 = 1 \wedge 3 = 2 \vee 2 = 2 \wedge 3 = 3)$ of the only-if half of s_D simplifies to $s(2,3) \rightarrow \textit{true}$, which is equivalent to *true* and thus can be omitted.

Definition 2. Let D be a database, $\leftarrow B$ a denial that either represents a query or a constraint, and θ a ground substitution of all variables in B such that $D(B\theta) = true$. Further, let C be a subset of *iff*(D) such that $C \models \exists B$, and for each proper subset C' of C, $C' \not\models \exists B$.

a) If $\leftarrow B$ is a query, then C is called a *cause* of the answer θ to $\leftarrow B$ in D.
b) If $\leftarrow B$ is a constraint, then C is called a *cause* of the violation of $\leftarrow B$ in D. C is also called a *cause* of $\exists B$ in D if C is a cause of the violation of $\leftarrow B$ in D.
c) For a quality theory QT, C is called a *cause* of the violation of QT in D if C is a cause of the violation of a denial form of the conjunction of all $QC \in QT$.
d) Let *vioCau*(D, QT) be the set of all causes of the violation of QT in D.

The preceding definition captures causes of positive answers and violations of constraints. The following one captures causes of negative answers.

Definition 3. Let D be a database, $\leftarrow B$ a query and C a cause of the answer *'yes'* (the identity substitution) to the query $\leftarrow \sim answer$ in $D \cup \{answer \leftarrow B\}$, where *answer* is a 0-ary predicate that does not occur in D. Then, E is also called a *cause* of the answer *no* to $\leftarrow B$ in D.

Several examples of causes of positive answers and constraint violations have already been presented in [20,23]. Some more, and in particular some examples of causes of negative answers, are displayed below.

Example 3.

a) Let $D = \{p \leftarrow r(x), s(x); r(1); s(2)\}$. Each cause of the answer *no* to $\leftarrow p$ in D contains the set $C_0 = \{p \to \exists x(r(x), s(x)); \sim s(1); \sim r(2)\}$. Moreover, each such cause contains, for each $i \in \mathcal{N}$, $i > 2$, either $\sim r(i)$ or $\sim s(i)$, and nothing else.

b) Let $D = \{p \leftarrow \sim q; q \leftarrow \sim r; q \leftarrow \sim s\}$. The two causes of the answer *no* to $\leftarrow p$ in D are $\{q \leftarrow \sim r; p \to \sim q; \sim r\}$ and $\{q \leftarrow \sim s; p \to \sim q; \sim s\}$.

c) Let $D = \{p \leftarrow q; p \leftarrow \sim q\}$ and $QC = \leftarrow p$. The two causes of the violation of QC in D are $\{p \leftarrow \sim q; \sim q\}$ and D.

d) Let $D = \{p(x) \leftarrow r(x); r(1)\}$ and $QC = \exists x(r(x) \wedge \sim p(x))$ a constraint. Clearly, $D(QC) = false$ (in fact, QC is violated in each database containing $p(x) \leftarrow r(x)$). A denial form of QC is $\leftarrow violated$, where *violated* is defined by $\{violated \leftarrow \sim s; s \leftarrow r(x), \sim p(x)\}$ (s is a fresh 0-ary predicate). Thus, the causes of *violated* in $D' = D \cup \{violated \leftarrow \sim s; s \leftarrow r(x), \sim p(x)\}$ explain the violation of QC in D. Thus, , for any $\mathcal{K} \subseteq \mathcal{N}$ such that $1 \in \mathcal{K}$, each cause of *violated* in D' is of the form $\{violated \leftarrow \sim s; s \to \exists x(r(x) \wedge \sim p(x))\} \cup \{p(i) \leftarrow r(i) \mid i \in \mathcal{K}\} \cup \{\sim r(i) \mid i \notin \mathcal{K}\}$.

e) Let $D = \{p \leftarrow q(1,x); q(2,y) \leftarrow r(y); r(1)\}$. The only cause of the answer *no* to the query $\leftarrow p$ in D is $\{p \to \exists x\, q(1,x)\} \cup \{\sim q(1,i) \mid i \in \mathcal{N}\}$.

f) Let $D = \{r(1), r(3), s(2), s(4)\}$ and $Q = \leftarrow r(x), s(x)$ a query. Depending on the extent of $\mathcal{N}$, there may be many causes of the answer *no* to Q in D. Each of them contains the set $\{answer \to \exists x\, (r(x), s(x)); \sim s(1); \sim s(3); \sim r(2); \sim r(4)\}$, and for each $i > 4$ in $\mathcal{N}$, either $\sim r(i)$ or $\sim s(i)$, and no other element.

3.2 Basic Causes

For constraint violations and positive answers to queries in definite databases and definite integrity theories, basic causes have been defined in [16]. For flat relational databases and denials that may contain negative literals, causes that are in fact basic (since there are no deduction rules) have been studied in [17,19]. In this section, we generalize basic causes for normal deductive databases and denials. In 3.2.1, we present the definition of basic causes. In 3.2.2, we illustrate that definition by several examples. In 3.2.3, we discuss the definition and its ramifications.

3.2.1 Defining Basic Causes

In Example 1, we have already seen that the predicates of negative basic causes cannot be exclusively extensional. The following definition characterizes the literals that are candidates for basic causes in general.

Definition 4. For a database D, a ground literal L is called *basic* in D if L either is a positive extensional fact in D or L is negative and the atom of L does not unify with the head of any clause in D.

In Example 1, $r(a, a)$, $\sim q(a, a)$, $\sim q(b, c)$ are basic, while $\sim r(a, a)$, $q(a, b)$, $\sim q(a, b)$ are not.

Definition 5. Let D be a database, $\leftarrow B$ a denial, and θ a ground substitution of all variables in B such that $D(B\theta) = true$. Further, let C be a set of basic literals in D such that $iff(D_R) \cup \{C\} \models \exists B$, where D_R is the set of deductive rules in D, and for each proper subset C' of C, $iff(D_R) \cup C' \not\models \exists B$.

a) If $\leftarrow B$ is a query, then C is called a *basic cause* of the answer θ to $\leftarrow B$ in D.

b) If $\leftarrow B$ is a constraint, then C is called a *basic cause* of the violation of $\leftarrow B$ in D. C is also called a *basic cause* of $\exists B$ in D if C is a basic cause of the violation of $\leftarrow B$ in D.

c) For a quality theory QT, C is called a *basic cause* of the violation of QT in D if C is a basic cause of the violation of a denial form of the conjunction of all $QC \in QT$.

d) Let $\mathsf{VioBas}(D, QT)$ be the set of all basic causes of the violation of QT in D.

We leave it to the reader to define the basic causes of negative answers in analogy to Definition 3.

3.2.2 Examples of Basic Causes

The databases, queries and constraints in Example 4, below, feature basic causes in correspondence to Example 3.

Example 4.

a) Let $D = \{p \leftarrow r(x), s(x); r(1); s(2)\}$. Each basic cause of the answer *no* to $\leftarrow p$ in D contains the set $C_0 = \{\sim s(1); \sim r(2)\}$. It also contains, for each $i \in \mathcal{N}$, $i > 2$, either $\sim r(i)$ or $\sim s(i)$, and nothing else.

b) Let $D = \{p \leftarrow \sim q; q \leftarrow \sim r; q \leftarrow \sim s\}$. The two basic causes of the answer *no* to $\leftarrow p$ in D are $\{\sim r\}$ and $\{\sim s\}$.

c) Let $D = \{p \leftarrow q; p \leftarrow \sim q\}$ and $QC = \leftarrow p$. The two basic causes of the violation of QC in D are $\{\sim q\}$ and $\{\ \}$.

d) Let $D = \{p(x) \leftarrow r(x); r(1)\}$ and $QC = \exists x(r(x) \wedge \sim p(x))$ a constraint. Analogously to Example 3d), the only basic cause of *violated* in the rewritten database $D' = D \cup \{violated \leftarrow \sim s; s \leftarrow r(x), \sim p(x)\}$ is $\{\ \}$.

e) Let $D = \{p \leftarrow q(1,x); \ q(2,y) \leftarrow r(y); \ r(1)\}$. The only basic cause of the answer *no* to the query $\leftarrow p$ in D is $\{\sim q(1,i) \mid i \in \mathcal{N}\}$.

f) Let $D = \{r(1), \ r(3), \ s(2), \ s(4)\}$. Each basic cause of the answer *no* to $\leftarrow r(x), s(x)$ contains $\{\sim s(1); \ \sim s(3); \ \sim r(2); \ \sim r(4)\}$ as well as, for each $i > 4$ in $\mathcal{N}$, either $\sim r(i)$ or $\sim s(i)$, and no other element.

3.2.3 Discussing Basic Causes

The following one-to-one correspondence is corroborated by comparing Examples 3 and 4.

Proposition 1. Let D be a database, $\leftarrow B$ a query and α an answer to $\leftarrow B$ in D, i.e., α is either some ground substitution or the answer *no*.

a) For each basic cause C_b of α, there is a cause C of α such that $C_b \subseteq C$.

b) For each cause C of α, there is a basic cause C_b of α such that $C_b \subseteq C$.

Note that Definition 5 improves previous definitions in [16,17,19] not only because it widens the classes of databases, queries and constraints for which basic causes are defined. It also remedies a deficiency of previous definitions, for which the logical consequences of the theory formed by the union of D_R and a basic cause C were not necessarily logical consequences of D. As opposed to that, the following corollary can be easily shown by applying Definition 5.

Corollary 1. For each database D and each basic cause C of some conjunction of literals B, and for each sentence F such that *iff* $(D_R) \cup \{C\} \models F$, we have that $comp(D) \models F$.

In fact, for the mentioned previous definitions, Corollary 1 holds, by definition, for $F = \exists B$, but not necessarily for arbitrary sentences F.

Apart from the circumstance that basic literals may not have extensional predicates, another difficulty for generalizing previous definitions of basic causes to negative answers and non-definite databases, queries and constraints had been the possible emptiness of basic causes, as, for instance, in Examples 4c, d. As opposed to that, general causes are never empty. Empty basic causes obviously cannot be used, as general causes are, for measuring inconsistency and the lack of quality in databases, by their number or size.

Empty basic causes are even more problematic since they never may overlap with any basic cause of constraint violation. Hence, the definition of AHI cannot be restricted to considering only basic causes in the overlap of causes of answers and of constraint violations, in general. However, for databases D and denials $\leftarrow B$ such that $D \cup \{\leftarrow B\}$ is *acyclic* [2] and *strict* [8], there is no such problem, as follows from Theorem 1c below.

Acyclicity and strictness are well-known properties in deductive databases. Acyclicity assures that the dependency graph of the set of all ground instances of clauses in D on which $\leftarrow B$ depends is loop-free. Strictness means that no predicate in B depends both positively and negatively on any predicate in D. Essentially, strictness bans the derivation of literals by rules without recurring on extensional facts, as, e.g., in $\{p \leftarrow q; \, p \leftarrow \sim q\}$ or $\{p \leftarrow q \sim q\}$. For instance, the databases and denials in Examples $4c\,d$ are acyclic but not strict, all others in Example 4 are acyclic and strict.

The following result sharpens Proposition 1.

Theorem 1. Let D be a database, $\leftarrow B$ a query such that $D \cup \{\leftarrow B\}$ is acyclic and strict, and α an answer to $\leftarrow B$ in D.

$a)$ For each basic cause C_b of α, there is a cause C of α such that C_b is the set of basic literals in C.

$b)$ For each cause C of α, the set of basic literals in C is a basic cause C_b of α.

$c)$ Each basic cause of α is non-empty.

A big advantage of basic over general causes is that the former are never larger and often much smaller in number and size than the corresponding latter ones. The importance of this difference will become obvious in 4.3.

4   AHQ

In 4.1, we re-define AHQ on the grounds of basic causes, i.e., we provide an improved formalization of what it means that an answer to a query in a database has quality. In 4.2, we relax AHQ to ASQ, which formalizes a concept of answers that have sufficient quality. In 4.3, we show how to compute AHQ and ASQ.

4.1   Re-defining AHQ

Consistent query answering (abbr. CQA) [3] provides answers that are correct in each minimal total repair of QT in D. CQA is based on semantic query optimization, using integrity constraints for speeding up query answering [31]. A similar approach is to compute consistent hypothetical answers together with a set of abduced hypotheses that can be interpreted as integrity-preserving repairs [37]

A new approach to provide answers that have integrity (abbr. AHI) and thus quality has been proposed in [18]. AHI determines two sets of data: the causes by which an answer is deduced, and the causes that lead to constraint violations. For databases D and queries without negation in the body of clauses, causes are minimal subsets of ground instances of clauses in D by which positive answers or violations are deduced. For negative answers and queries in databases that may contain clauses with negation, also minimal subsets of ground instances of the only-if halves of the completions of predicates in D form part of causes, as seen in 3.1. In general, causes are not unique.

Hence, an answer is defined to have integrity, and thus quality, if it has a cause that does not intersect with any of the causes of constraint violations, i.e., if it is deducible from data that are independent of those that violate constraints. Definition 6 below re-defines AHQ in terms of basic causes.

Definition 6. Let α be an answer to a query $\leftarrow B$ in (D, QT), i.e., α is either a substitution such that $D(B\alpha) = true$, or $D(\leftarrow B) = true$, i.e., $\alpha = no$.

a) Let B_α stand for $B\alpha$ if α is a substitution, or for $\leftarrow B$ if $\alpha = no$.

b) We say that α *has quality* in (D, QT) if there is a cause C of B_θ in D such that $C \cap \mathsf{VioBas}(D, QT) = \emptyset$.

4.2 Defining ASQ

Informally speaking, we are going to define that an answer has sufficient quality if one of its causes has a sufficiently low overlap with any cause of the violation of integrity in D.

It is fair to say that AHI is inconsistency-tolerant, since it provides correct results in the presence of constraint violations. However, each answer sanctioned by AHI is independent of any inconsistency in the database. Therefore, we are going to relax AHQ to ASQ: answers with sufficient quality. ASQ sanctions answers that are acceptable despite some lack of quality involved in their derivation.

To quantify that amount, some quality metric is needed. Unlike inconsistency measures which size the inconsistency in all of (D, QC) (cf. [23]), quality metrics only size the lack of quality involved in the derivation of given answers or constraint violations.

Definition 7. (*ASQ*)

a) For answers α to queries $\leftarrow B$ in (D, QT), a *quality metric* maps triples (D, QT, B_α) to $(\mathbb{M}, \preceq)$, where $\mathbb{M}$ is a metric space partially ordered by $\preceq$, and B_α be defined as in Definition 6a.

b) Let κ be a quality metric and *th* a threshold value in $\mathbb{M}$ up to which an impairment of quality is tolerable. Then, an answer α to some query $\leftarrow B$ in (D, QT) is said to *have sufficient quality* if $\kappa(D, QT, B_\alpha) \preceq th$.

A first, coarse quality metric κ could be to count the elements of the intersection $C_\alpha \cap \mathsf{VioBas}(D, QT)$, where C_α is the union of all causes of B_α. Or, the application designer may assign a specific weight to each element of each basic cause, similar to the tuple ranking in [5]. Then, κ can be defined by adding up the weights of elements in $C_\alpha \cap \mathsf{VioBas}(D, QT)$. Or, application-specific weights could be assigned to each ground instance QC' of each $QC \in QT$. Then, κ could add up the weights of those QC' that have a cause C' such that $C_\kappa \cap C' \neq \emptyset$.

For example, $\kappa(D, QT, B_\alpha) = |C_{alpha} \cap \mathsf{VioBas}(D, QT)|$ counts elements in $C_\alpha \cap \mathsf{VioBas}(D, QT)$, where $|\,.\,|$ is the cardinality operator. Or, $\kappa(D, QT, B_\alpha) = \sum \{ \omega(c) \mid c \in C_\alpha \cap \mathsf{VioBas}(D, QT) \}$, where ω is a weight function, i.e., κ sums up the weights of the lack of quality of elements in $C_\alpha \cap \mathsf{VioBas}(D, QT)$.

4.3 Computing Causes and ASQ

Perhaps the most important consequence of Theorem 1 is that, for databases and denials that are acyclic and strict, basic causes can be computed just like causes.

Computation of causes has been described in [18]. Thus, no new algorithm for computing basic causes has to be invented.

Hence, the main gain of basic causes comes to bear when applying the definitions of AHQ and ASQ, i.e., when checking sets of basic causes of answers and constraint violations for overlaps: these checks become simpler with basic causes, to the extend that their number and size are smaller than that of general ones.

In order to make the paper more self-contained, the approach in [18] to compute general causes, and thus basic causes, is recapitulated below.

As seen in [17], SLD resolution [36] provides an easy way to compute causes of positive answers and integrity violation of definite queries and denials in definite databases. Each cause of each answer corresponds to a refutation R that computes the answer: input clauses of R, instantiated with the substitution computed by R, are the elements of the cause. Hence, AHI can be computed by comparing causes drawn from refutations with causes of integrity violation. If the latter have been computed ahead of query time, then checking for overlaps can already be done while the answer is computed.

Similarly, for each SLDNF refutation R and each finitely failed SLDNF tree T of some query in a database, a cause for the answer computed by R or, resp., T can be obtained as described below in Definition 9. To prepare this definition, we first recall some basic SLDNF issues from [39] and ask the reader to agree on some denotations.

Let D be a database and $\leftarrow B$ a query. An *SLDNF computation* of $D \cup \{\leftarrow B\}$ is either an SLDNF refutation or a finitely failed tree of $D \cup \{\leftarrow B\}$. Each SLDNF computation involves one top-rank computation and possibly several subsidiary computations of lower rank, spawned by the selection of ground negative literals in goals of derivations.

It is easy to see that no finitely failed tree of rank n-1 that is subsidiary to some finitely failed tree T of rank n could contribute anything to explain the answer *no* computed by T. Thus, such subsidiary trees are ignored in Definition 8. It characterizes the set of computations involved in an SLDNF computation S that contribute to explaining the answer to the root of S by a cause.

Definition 8. Let S be an SLDNF computation of rank n $(n \geq 0)$.

a) The set S_r of *explanatory refutations* of S consists of each refutation R of rank k involved in S such that either $k = n$ and $R = S$, or $k < n$ and R is subsidiary to a tree in S_t of rank $k + 1$.

b) The set S_t of *explanatory trees* of S consists of each finitely failed tree T of rank k involved in S such that either $k = n$ and $T = S$, or $k < n$ and T is subsidiary to a refutation in S_r of rank $k + 1$.

Note that the mutual recursion in Definition 8a, b does not pose any problem if D is acyclic, i.e., the rank of each computation is bounded by the rank of the top-rank computation, and the rank of each subsidiary computation decreases iteratively until the lowest rank without subsidiary inferences is reached.

An SLDNF computation S is called *fair* if, in each tree $T \in S_t$ and each goal G in T, one of the most recently introduced literals is selected in G.

For each refutation R, let θ_R denote the substitution computed by R. The projection of θ_R to the variables in the root of R is the *computed answer* of R. For each database D, each query $\leftarrow B$ and each finitely failed tree T of $D \cup \{\leftarrow B\}$, the *computed answer* of T is *no*.

For each clause C, each only-if half H and each substitution θ, let $C\theta$, resp., $H\theta$ denote the formula obtained by applying θ to the $\forall$-quantified variables in C or, resp., H. For each only-if half H, let $h(H)$ denote the head of H.

Now, we are going to define computed explanations, which are supersets of causes.

Definition 9. For each SLDNF computation S, the *computed explanation* E_S of S consists of

$$E_S^+ = \{C\theta_R \mid C \in D,\ R \in S_r,\ C \text{ is input clause in } R\}$$

and

$$E_S^- = \{H\gamma \mid H\gamma \in D^-,\ \gamma \text{ is a ground substitution,}$$
$$h(H) \text{ is selected in some node of some tree in } S_t\}.$$

Thus, E_S^+ is obtained by instantiating the positive input clauses of each refutation $R \in S_r$ with θ_R. E_S^- is obtained by collecting the only-if-halves of all ground instances of each positive literal selected in any node of any tree in S_t.

Example 5. Let $D = \{p(x) \leftarrow q(x,x);\ q(1,2);\ q(2,3)\}$. The answer *no* to the query $\leftarrow p(x)$ is computed by a finitely failed tree consisting of a single branch rooted at $\leftarrow p(x)$, which is reduced to the goal $\leftarrow q(x,x)$, which fails. The only-if halves of the two selected positive literals in the tree are $p(x) \to q(x,x)$ and $q(x,x) \to (x{=}1 \land x{=}2 \lor x{=}2 \land x{=}3)$. The latter obviously is equivalent to $\sim q(x,x)$. Thus, $(\emptyset, \{p(i) \to q(i,i) \mid i \in \mathcal{L}^c\} \cup \{\sim q(i,i) \mid i \in \mathcal{L}^c\})$ is the computed explanation, which in fact is also a cause of $\sim p(x)$ in D.

Theorem 2 is easily inferred from Definition 9.

Theorem 2. For each database D, each query $\leftarrow B$ and each SLDNF computation S of $D \cup \{\leftarrow B\}$, the computed explanation of S is a superset of a cause of the answer computed by S. If $S_t = \emptyset$, then the computed explanation of S is a cause of the answer computed by S.

The following example illustrates that explanations computed by finitely failed SLDNF trees are not necessarily causes, since they may not be minimal, in the sense of Definition 5.

Example 6. Let $D = \{p \leftarrow q, r;\ r \leftarrow s\}$. Depending on the selection function, there are three SLDNF trees of $D \cup \{\leftarrow p\}$. In each, the goal $\leftarrow q, r$ is derived from the root. Then, if q is selected, the computation terminates with failure. If, instead, r is selected, the derived goal $\leftarrow q, s$ may fail in two ways, after selecting either q or s. Hence, depending on the selection, precisely one of the following three explanations $(\emptyset, \{p \to q \land r;\ \sim q\})$, $(\emptyset, \{p \to q \land r;\ r \to s;\ \sim q\})$, $(\emptyset, \{p \to q \land r;\ r \to s;\ \sim s\})$ can be drawn from the respective SLDNF tree. Only the first and the last of these explanations are causes of the answer no to the query $\leftarrow p$ in D, while the middle one is not because it properly contains the first one and thus is not minimal.

If, in Example 6, a fair selection policy is employed, the computation of a non-minimal explanation is avoided. However, fair selection alone is not enough, in general, as shown by the following example.

Example 7. Let $D = \{p \leftarrow q(x), r;\ q(1)\}$. Both left-to-right and right-to-left selection is fair for computing the answer no to $\leftarrow p$ in D. However, only the latter yields a cause, while the former computes the superset $(\emptyset, \{p \rightarrow \exists x(q(x) \wedge r)\} \cup \{\sim q(i) \mid i \in \mathcal{L}^c,\ i \neq 1\} \cup \{\sim r\})$, of a cause, in which each $\sim q(i)$ is superfluous.

Thus, each set C of basic literals drawn from an SLDNF computation must eventually be minimized, by checking if any proper subset of C satisfies Definition 7. Also, selection strategies such as those described in [15] can be used to obtain finitely failed trees with proper causes.

5 Related Work

In this section, we first address related work in the field of data quality, then other approaches to reasoning in the presence of inconsistency, and last we comment on work that is based on comcepts of causes that differ from ours.

5.1 Data Quality

For capturing the quality of data, an overwhelming number of attributes, characteristics, determinants, dimensions, factors, indicators and parameters have appeared in the literature, e.g., [44,35,41,43,7,28,4,6,42]. Although there is a puzzling diversity of definitions of data quality in the aforementioned papers, consistency is a key property in almost all attempts to characterize the quality of stored data. In databases, the standard way to specify consistency conditions is by expressing them as integrity constraints (a.k.a. assertions) in the database schema. Hence, consistency and integrity are usually synonymous in databases. Some authors even seem to regard database quality as synonymous to the compliance of the data with their constraints, e.g., [11,29].

However, we agree with [30] and others that, in databases, integrity is closely related to, but distinct from quality. Yet, many authors, including this one, advocate the point of view that integrity is just one of many aspects of quality. In fact, we have shown in [27,21,26] how the syntax of integrity constrains can also be used for modeling quality constraints in a wider sense. More precisely, constraints can not only be used to model, measure and monitor the integrity, but also many other desirable quality properties can be captured in terms of the compliance of data and constraints. That is also the baseline of this paper.

5.2   Reasoning in the Presence of Inconsistency

Apart from the use of causes in AHI and ASQ, causes also have proven useful for repairing databases that lack quality, as seen in [23]. In fact, basic causes are ideal candidates for reparing inconsistency, and thus damaged quality, since repairs that update basic literals correspond to solutions of the well-known view-update problem [13].

In [18], we have compared AHI with consistent query answering (CQA) [3], and also have related belief revision, knowledge assimilation and abduction to AHI. Due to Theorem 1, analogous relationships can be observed with regard to ASQ.

In [34], negative database consequences such as those that can be inferred from iff-completions are made explicit for inconsistency-tolerant reasoning, and, in particular, for query answering. The formalizations in [34] have to sacrifice several classical inference rules that are cornerstones of the logic framework assumed in this paper, such as modus ponens and reductio ad absurdum.

5.3   Other Concepts of Causes

A mathematical concept of causality was proposed in [33,9,32], which has been used to analyze the complexity of positive and negative answers to database queries in [40]. The underlying definition of 'actual causes' in that work, which is similar to but more complicated than Definition 2, is based on counterfactual reasoning. Roughly, actual causes correspond to literals in basic causes. Perhaps the most interesting feature of 'actual causes' is a quantification of the degree of 'responsability', which may serve to assign a weight to the elements of causes, for defining metrics for use in ASQ and quality maintenance.

There is a long-standing tradition of philosophical treatment of causality. More recently, it has also come to bear on informatics and AI, mostly in the form of probabilistic or counterfactual reasoning, as witnessed by a growing number of congresses about computing and philosophy. Those treatments of causality mostly go beyond the framework of causes as in this paper, which is confined by the comparatively simple theory of databases, the semantics of which is given by nothing but the completions of predicates.

6   Conclusion

We have generalized the concept of answers that have integrity (AHI) as presented in [18], to answers that have quality (AHQ) and answers that have suficient quality (ASQ). More precisely, we have generalized three features of AHI.

Firstly, quality is a more general concept than integrity. Secondly, AHI had originally been defined with regard to general causes. In this paper, we have used basic causes instead, on a larger scale than in [16], for defining AHQ. Thirdly, AHI and AHQ distinguish useful from doubtful answers, while ASQ distinguishes between more or less doubts in answers: an answer is doubtful, i.e., does not have integrity, iff its causes overlap with the causes of integrity violation in the given state; however, an answer that does not have total quality still has sufficient quality if the overlap of its causes with the causes that damage the quality of the database is sufficiently small. The size of the lack of quality can be measured by any quality metric, as characterized in [25,22].

In 4.3, we have only been concerned with sound ways to compute causes. In upcoming work, we intend to re-assess the issue of completeness of computing causes, which already had been addressed in [25]. In particular, we intend to reduce the possibly very high cost of computing complete basic causes by using answer set computing [38].

References

1. Abiteboul, S., Hull, R., Vianu, V.: Foundations of Databases. Addison-Wesley (1995)
2. Apt, K., Bezem, M.: Acyclic programs. New Generation Computing 9(3,4), 335–364 (1991)
3. Arenas, M., Bertossi, L.E., Chomicki, J.: Consistent query answers in inconsistent databases. In: Proceedings of PODS, pp. 68–79. ACM Press (1999)
4. Batini, C., Cappiello, C., Francalanci, C., Maurino, A.: Methodologies for Data Quality Assessment and Improvement. Comput. Surveys 41(3),16:1–16:52 (2009)
5. Berlin, J., Motro, A.: Tuplerank: Ranking discovered content in virtual databases. In: Etzion, O., Kuflik, T., Motro, A. (eds.) NGITS 2006. LNCS, vol. 4032, pp. 13–25. Springer, Heidelberg (2006)
6. Borek, A., Woodall, P., Oberhofer, M., Parlikad, A.K.: A Classification of Data Quality Asessment Methods. In: Proc. 16th ICIQ, pp. 189–203
7. Campbell, R., Zhang, L., Francis, L., Palenik, R., Popelyukhin, A., Scruton, G., Prevosto, V.: Survey of Data Management and Data Quality Texts. In: Proc. CAS Winter Forum, pp. 273–306. Casualty Actuarial Society (2007)
8. Cavedon, L., Lloyd, J.: A completeness theorem for SLDNF resolution. J. Log. Prog. 7(3), 177–191 (1989)
9. Chockler, H., Halpern, J.: Responsibility and blame: a structural-model approach. J. Artif. Intell. Res. 22, 93–115 (2004)
10. Clark, K.: Negation as failure. In: Gallaire, H., Minker, J. (eds.) Logic and Data Bases, pp. 293–322. Plenum Press (1978)
11. Cong, G., Fan, W., Geerts, F., Jia, X., Ma, S.: Improving Data Quality: Consistency and Accuracy. In: Proc. 33rd VLDB, pp. 315–326. ACM (2007)
12. Date, C.: The Relational Database Dictionary, Extended edn. Springer (2008)
13. Date, C.: View Updating and Relational Theory. O'Reilly (2012)
14. Decker, H.: The range form of databases and queries or: How to avoid floundering. In: Proc. 5th ÖGAI. Informatik-Fachberichte, vol. 208, pp. 114–123. Springer (1989)
15. Decker, H.: On explanations in deductive databases. In: Proc. 3rd Workshop on Foundations of Models and Languages for Data and Objects, Informatik-Bericht 91/3, pp. 173–185. Inst. f. Informatik, Tech. Univ. Clausthal (1991)
16. Decker, H.: Basic causes for the inconsistency tolerance of query answering and integrity checking. In: Proc. 21st DEXA Workshops, pp. 318–322. IEEE CSP (2010)

17. Decker, H.: Toward a uniform cause-based approach to inconsistency-tolerant database semantics. In: Meersman, R., Dillon, T., Herrero, P. (eds.) OTM 2010, Part II. LNCS, vol. 6427, pp. 983–998. Springer, Heidelberg (2010)

18. Decker, H.: Answers that have integrity. In: Schewe, K.-D., Thalheim, B. (eds.) SDKB 2010. LNCS, vol. 6834, pp. 54–72. Springer, Heidelberg (2011)

19. Decker, H.: Causes for inconsistency-tolerant schema update management. In: Proc. 27th ICDE Workshops, pp. 157–161. IEEE CSP (2011)

20. Decker, H.: Causes of the violation of integrity constraints for supporting the quality of databases. In: Murgante, B., Gervasi, O., Iglesias, A., Taniar, D., Apduhan, B.O. (eds.) ICCSA 2011, Part V. LNCS, vol. 6786, pp. 283–292. Springer, Heidelberg (2011)

21. Decker, H.: Data quality maintenance by integrity-preserving repairs that tolerate inconsistency. In: Proc. 11th QSIC, pp. 192–197. IEEE CSP (2011)

22. Decker, H.: Axiomatizing inconsistency metrics for integrity maintenance. In: Proc. 16th KES, pp. 1243–1252. IOS Press (2012)

23. Decker, H.: New measures for maintaining the quality of databases. In: Murgante, B., Gervasi, O., Misra, S., Nedjah, N., Rocha, A.M.A.C., Taniar, D., Apduhan, B.O. (eds.) ICCSA 2012, Part IV. LNCS, vol. 7336, pp. 170–185. Springer, Heidelberg (2012)

24. Decker, H.: Maintaining desirable properties of information by inconsistency-tolerant integrity management. In: Mayr, H.C., Kop, C., Liddle, S., Ginige, A. (eds.) UNISON 2012. LNBIP, vol. 137, pp. 13–24. Springer, Heidelberg (2013)

25. Decker, H.: Measure-based inconsistency-tolerant maintenance of database integrity. In: Schewe, K.-D., Thalheim, B. (eds.) SDKB 2013. LNCS, vol. 7693, pp. 149–173. Springer, Heidelberg (2013)

26. Decker, H.: Modeling, measuring and maintaining the quality of databases (to appear, 2013)

27. Decker, H., Martinenghi, D.: Modeling, measuring and monitoring the quality of information. In: Heuser, C.A., Pernul, G. (eds.) ER 2009. LNCS, vol. 5833, pp. 212–221. Springer, Heidelberg (2009)

28. Ehling, M., Körner, T.: Handbook on Data Quality Assessment Methods and Tools. European Commission, Eurostat (2007)

29. Fan, W.: Dependencies revisited for improving data quality. In: Proc. 27th PODS, pp. 159–170. ACM (2008)

30. Gertz, M.: Managing Data Quality and Integrity in Federated Databases. In: Integrity and Internal Control in Information Systems. IFIP Conference Proceedings, vol. 136, pp. 211–230. Kluwer (1998)

31. Godfrey, P., Grant, J., Gryz, J., Minker, J.: Integrity Constraints: Semantics and Applications. In: Logics for Databases and Information Systems. Engineering and Computer Science, vol. 436, pp. 265–306. Springer (1998)

32. Halpern, J.: Causality, responsibility, and blame: A structural-model approach. In: Proc. 3rd QEST, pp. 3–8. IEEE CSP (2006)

33. Halpern, J., Pearl, J.: Causes and explanations: a structural-model approach, part i: Causes. Brit. J. Phil. Sci. 56, 843–887 (2005)

34. Hinrichs, T., Kao, J., Genesereth, M.: Inconsistency-tolerant reasoning with classical logic and large databases. In: Proc. 8th SARA, pp. 105–112. AAAI Publications (2009)

35. Typology of Database Quality Factors. Software Quality Journal 7(3/4), 179–193 (1998)

36. Kowalski, R., Kuehner, D.: Linear Resolution with Selection Function. Artificial Intelligence 2(3-4), 227–260 (1971)

37. Kowalski, R.A., Sadri, F.: Teleo-Reactive Abductive Logic Programs. In: Artikis, A., Craven, R., Kesim Çiçekli, N., Sadighi, B., Stathis, K. (eds.) Sergot Festschrift 2012. LNCS, vol. 7360, pp. 12–32. Springer, Heidelberg (2012)

38. Lifschitz, V.: What is answer set computing? In: Proc. 23rd AAAI, pp. 1594–1597 (2008)

39. LLoyd, J.: Foundations of Logic Programming, 2nd edn. Springer (1987)

40. Meliou, A., Gatterbauer, W., Moore, K., Suciu, D.: The complexity of causality and responsibility for query answers and non-answers. In: Proc. 37th VLDB, pp. 34–45 (2011)

41. Pipino, L., Lee, Y., Wang, R.: Data quality assessment. CACM 45(4), 211–218 (2002)

42. Sidi, F., Panah, P., Affendey, L., Jabar, M., Ibrahim, H., Mustapha, A.: Data quality: A survey of data quality dimensions. In: Proc. CAMP 2012, pp. 300–304. IEEE CSP (2012)

43. Tejay, G., Dhillon, G., Chin, A.G.: Data Quality Dimensions for Information Systems Security: A Theoretical Exposition. In: Dowland, P., Furnell, S., Thuraisingham, B., Sean Wang, X. (eds.) Security Management, Integrity, and Internal Control in Information Systems. IFIP, vol. 193, pp. 21–39. Springer, Boston (2006)

44. Wang, R., Kon, H., Madnick, S.: Data quality requirements analysis and modeling. In: Proc. 9th ICDE, pp. 670–677. IEEE CSP (1993)

A Service-Oriented Software Development Methodology for Outsourced Working Force

Ricardo Puttini, Andre Toffanello, Armando Vidigal, Janaina Areal,
Gabriela Alves, R. Chaim, and Claynor Mazzarolo

University of Brasilia – Campus Universitário Darcy Ribeiro – Asa Norte
Brasília, DF – Brazil, 70910-000
{puttini,toffanello,avidigal,janalaguardia,
rchaim,mazzarolo}@unb.br

Abstract. This work presents a service-oriented software development methodology designed for use in enterprise with outsourced working force. The methodology is conceived as an evolution to traditional software factory methods and techniques, using manufacturing techniques and principles to establish a service factory unit that is responsible for delivering services organizes service orientation analysis, design, implementation, testing, and deployment processes and activities using a service factory approach.

Keywords: Service-oriented Architecture (SOA), Software Development Methodology, Service factory, Software factory, Outsourcing.

1 Introduction

Software engineering methodologies are derived from the principles, best practices and standards advocated by the software industry [1][2][3][4][5]. In this paper we present a software development methodology that aims at service-oriented software and leverage software factory [6][7][8] practices in order to establish clear outsourcing scopes and objectives.

Our approach in derived from research and field experience collected during the structuring of Service-Oriented Architecture (SOA) adoption initiatives in three major governmental organizations in Brazil (Ministry of Health, Ministry of Defense/Army, and Federal Sanitary Surveillance Agency) [9][10][11]. These organizations have different maturity level and leverage on outsourced working force for performing software development tasks.

2 Structure of the Methodology

2.1 Functional Units

Functional units are logical organizational units that provide *working force* with adequate professional skills required to perform related technical activities. Technical

B. Murgante et al. (Eds.): ICCSA 2013, Part II, LNCS 7972, pp. 559–573, 2013.

activities are grouped together in *work units*, which usually correspond to clearly defined tasks in the *software development lifecycle*. Each work unit produces formally defined *deliverables* (software artifacts). Professional skills are defined by *professional profiles*, which establish a type of professional that is capable of performing a selected set of technical activities. Functional units are classified as *corporate units* and *dedicate units*, according to the type of competence (jurisdiction) of the work being performed.

Corporate units performs corporate (or department) level activities and responsibilities, which are not exclusive to the service-oriented software development process. These units may be previously established in the organization.

Dedicate units performs activities and responsibilities that are exclusively related to the service-oriented software development processes. These units exist only for the purpose of SOA development in the organization.

Functional unit establish concrete management points for the working force and related work units. Therefore, they are used to manage and control outsourced resources and responsibilities in the software development process, in the sense that a complete functional unit can be provided by an outsourced working force. This notion is also useful from the outsource procurement, contracting and management viewpoint. For instance, contracts can be established with service providers for a whole functional unit service scope (i.e. a set of work units). These contracts are regulated by service-level agreements, which specify the professional profile required at each type of work unit, together with other service-specific quality-of-service attributes (e.g. time to complete; number of errors in software; etc.).

Outsourcing is used for clearly defined *work units* that require a technical background but do not represent strategic knowledge to the organization. In these sense, not all functional units are considered as objective for outsourcing, as all strategic and several management activities are usually assigned to professionals with a tighter commitment with the organization (i.e. employees). Although *functional units* can be

Table 1. Functional Units

Functional Unit	Type	Outsourcing[1]
Business Process Management Office	Corporate	No
Data Administration Office	Corporate	No
IT Infrastructure Office	Corporate	No
Enterprise Architecture Office	Corporate	No
Project Management Office	Corporate	Yes
SOA Governance Office	Dedicate	Yes
Service Factory	Dedicate	Yes
Software Factory	Dedicate	Yes
Software Test Factory	Dedicate	Yes

---

[1]   This is the outsourcing scope considered in this methodology. Applicability of the outsourcing model for functional unit marked as 'no' are not discussed in this paper.

compared to functional organization units, they do not actually exist as part of the formal organizational structure. Table 1 presents the *functional units* defined in this methodology, its classification, and eligibility for outsourcing.

2.2   Software Development Lifecycle

The service-oriented software development paradigm requires specific tasks. These are defined in a *software-oriented software development lifecycle*. In this methodology, the lifecycle is defined in order to favor splitting the technical work required in the process among different functional units that acts as service providers. The work is split in *stages*, which defines steps in the development process with clearly defined *deliverables*. Each stage is further classified according to the *software development discipline* involved in the work being performed, i.e. the body of software development knowledge required.

The work unit performed in each stage is defined in formal *processes*, which can be further decomposed in *sub-processes* and *tasks*. Table 2 provides the overview of the software development lifecycle stages, together with the respective software development discipline and sub-processes/tasks.

Table 2. Software Development Lifecycle

Stage (Process)	Discipline	Processes
Corporate Business Modeling	Business Modeling	Design Business Process Model Information Structure
Service Inventory Analysis	Analysis	Analyze Service Inventory Perform Service-Oriented Analysis Define Service Candidates Define Service Inventory Schema
Service-Oriented Analysis	Analysis	Analyze Service Perform Service-Oriented Analysis Model Service Candidate
Service-Oriented Design	Design	Design Service Contract
Service Logic Design	Design	Design Service Logic Design Composite Application
Service Development	Implementation	Implement Service
Service Test	Test	Test Service
Service Deployment and Publication	Deployment	Deploy Service Publish Service
Composite Application Implementation	Development	Implement Composite Application Discover Service Implement Service Orchestration
Application Test	Test	Test Composite Application Homologate Application
Application Deployment	Implementation	Deploy Composite Application

In order to assure the process visibility and correct interaction among service providers, each *process* in the lifecycle has a clearly defined set of *inputs* and *outputs*. These are further classified as *software artifacts* (SA), *development process artifacts* (DPA), *project management artifacts* (PMA) or *corporate artifacts* (CA). There are template documents for all artifacts presented in the methodology. However, presenting these templates does not fit the size of this technical publication. Appendix A presents a list of *artifacts* produced throughout the execution of this methodology.

2.3   Professional Profiles

Professional skills and capabilities are defined in *professional profiles*, which are used to describe the type of professional required for performing a defined work unit. Professional skills may be assessed by requiring industry accepted *professional certifications* and previous experience in SOA projects and roles. This methodology does not define specific *team members* (i.e. human resources) and *project team*. Human resources are allocated in a project-by-project basis. However, the methodology requires that team members fulfill the required professional profiles defined for each work unit. Profiles used in the methodology are shown in Table 3.

Table 3. Professional Profiles

Professional Profile	Professional Skills
Business Process Analyst	Business process management professional
SOA Analyst	SOA analyst professional
SOA Architect	SOA architect professional
SOA Infrastructure Specialist	Expert in SOA middleware and infrastructure
SOA Developer	Developer/Programmer (platform dependent) (SOA specialization desired)
SOA Tester	Software tester (SOA specialization desired)
SOA Auditor	Senior SOA expert/consultant
SOA Governance Specialist	Expert in SOA governance
Quality Assurance Specialist	Expert in software quality assurance (SOA specialization desired)
Project Manager	Project manager professional (e.g. PMP)
Data Administrator	Expert in data administration
Enterprise Architect	Expert in enterprise architecture

2.4   Overview of the Methodology

An overview of the methodology is presented in Figure 1. The elements presented in the figure will be explained in further details in the next sections.

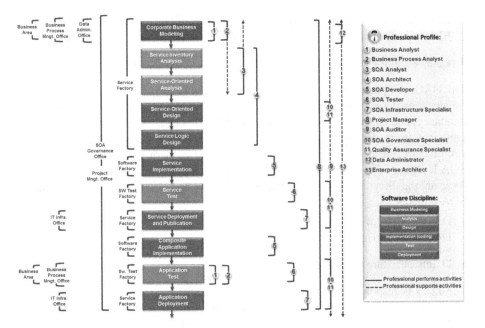

Fig. 1. Overview of the Service-Oriented Software Development Methodology

3 Functional Units

3.1 Corporate Units

SOA development requires several interactions with corporate-level structures, i.e. which perform tasks and have responsibilities that are not exclusively related to SOA projects. Corporate units are used to model these structures and interactions. In most organizations, there should be similar organizational structures that have equivalent responsibilities. Outsourcing strategies for services performed by corporate units are out of the scope of this paper (refer to Table 1 and footnote 1).

- Business Process Management Office (BPMO)

BPMO unit is responsible for the continuous business process improvement in the organization. This unit may exist inside or outside the IT enterprise. Besides of performing several activities related to business process modeling and design, BPMO usually has a normative role, establishing practices, policies and standards related to business process management (BPM). Note that there are several BPM technologies that have direct support in SOA-enabled platforms. Establishing a clear alignment between BPM and SOA practices and activities results in direct benefits for the development of service-oriented IT solutions [12]. Such alignment should be extended, with increased benefits, to the platform and technology level (e.g. compatible BPM and SOA middleware).

BPMO processes: *Design Business Process* and *Homologate Application.*
BPMO professional profile: *Business Process Analyst.*

• Data Administration Office (DAO)

The DAO unit plans, organizes, describes and controls corporate data resources. This organizational function is performed aiming at increasing consistency, avoiding redundancy, and ensuring the reuse of data models. This role can also exist within other units that are responsible for the corporate information architecture (e.g. Information Architecture Office), which are common units in modern organization structures. SOA leverage on corporate-level standardized data and information models [1][2] and these are the direct intersections between the jurisdiction of DAO and requirements of SOA projects.

DAO process: *Model Information Structure.*
DAO professional profile: *Data Administrator.*

• IT Infrastructure Office (ITIO)

The ITIO is responsible for the deployed IT infrastructure at the corporation. Most often, this unit also operates the running systems, including SOA-based information systems. ITIO has an important role in the deployment stages of the service-oriented development.

ITIO processes: *Deploy Service, Publish Service, Deploy Application.*
ITIO professional profile: *SOA Infrastructure Specialist*

• Enterprise Architecture Office (EAO)

Enterprise Architecture (EA) the is the process of translating business vision and strategy into effective enterprise change by creating, communicating and improving the key requirements, principles and operational models that describe the enterprise's current and future state, while enabling its evolution (Gartner definition). To date, many EA models exist. In the TOGAF framework, the EA encompasses business, information, application and technology architectures and highlights relationships and dependencies among them [13].

EAO unit is responsible for the design, development, establishment, evolution and management of the EA. This role has a direct relationship with more specific IT architectures (such as SOA), which shall be kept in sync with business goals and requirements. Therefore, the EA process oversees and constrains all other concrete architectures, including those related to software solutions (i.e. SOA). SOA models are usually used to describe software architectures and, therefore, SOA has been positioned within the application architecture quadrant. However, it has direct implications and dependencies with all other architectures in the EA framework.

Although there is not a process directly related to service-oriented development that is executed with direct participation of the EAO, enterprise architects oversee the development of specific business, data, information and technological architectures related to service-oriented solutions, which should be supported and constrained by the enterprise architecture [14].

EAO professional profile: *Enterprise Architect*

- Project Management Office (PMO)

PMO is common structure in IT enterprises, as most of development activities are organized as projects. PMO is responsible for the project management methodologies and practices [15]. It also manages the project portfolio in the organization and often hosts the project managers' pool. PMO oversees the *project management* processes, which is transversal to all other technical activities in the service-oriented development methodology. These processes and outsourcing approaches for PMO are not discussed in this paper.

PMO professional profile: *Project Manager*

3.2 Dedicate Units

Dedicate units are important structures in the methodology as they define the organization of a pool of human resources that are assigned through outsourcing contracts.

In the proposed methodology, each functional unit can be outsourced to a different service provider. Nevertheless, a common option is to outsource both Service Factory and Software Factory units to the same service. This approach usually reduces the governance burden related to hand overs between different service providers. Each service provider should have a *master contract* with the client organization, which specifies general terms and conditions, service specification and SLAs according to the scope of the outsourced services. These contracts are supplemented by project-specific subcontracts, which are specified in SDP artifacts called *Order of Service*, which establish actual project's terms, including project scope, scheduling, estimates and measurements, and defines the work units required at each specific service-oriented project.

Most of the technical work performed by the dedicate units is organized with a *factory* service model [6][7][8]. The *Factory* units are conceived as an evolution to traditional software factory methods and techniques, using manufacturing techniques and principles to deliver service-oriented software development services.

- SOA Governance Office (SGO)

The SGO is responsible for providing governance, compliance and quality assurance to SOA projects and products. It oversees all SOA projects and performs reviews, inspections, and audits for the lifecycle processes and technical work products. To protect the integrity of the software, the governance methodology also prescribes configuration controls over software, data, and technical documentation [16].

SOA governance processes should be established according to the organization maturity and are executed transversally to the SOA development lifecycle (similar to project management processes). A detailed discussion about processes and the outsourcing models for SGO is out of the scope of this paper.

SGO professional profiles: *SOA Governance Specialist, SOA Auditor, Quality Assurance Specialist*

- Service Factory

The *Service Factory* is a core functional unit in the methodology. It is responsible for performing all activities related to service-oriented analysis and design, and participates in most of the activities related to business modeling and deployment of services and solutions.

Service Factory processes: *Design Business Process, Model Information Structure, Analyze Service Inventory, Analyze Service, Design Service Contract, Design Service Logic, Design Composite Application, Deploy Service, Publish Service, Deploy Composite Application*

Service Factory professional profiles: *Business Process Analyst, SOA Analyst, SOA Architect, SOA Infrastructure Specialist*

- Software Factory

The *Software Factory* [6][7][8] is a functional unit responsible for software implementation activities. There are several organizations that use the software factory development model and conventional outsourcing *master contracts* are generally suited for allocation of programmers and developers. Note, however, that implementation may be related to a specific SOA middleware and experience in developing for those platforms are required. In such cases special outsourcing contracts may apply.

Software Factory processes: *Implement Service, Implement Composite Application*
Software Factory professional profile: *SOA Developer*

- Software Test Factory

The *Service Factory* unit performs activities related to software testing. In outsourced development models, test and measurements of software products should be executed by different service providers that are not involved in building the software product [6]. Therefore, the service provider outsourcing the Software Test Factory is usually distinct from the providers for the Service and Software factories.

Software Test Factory processes: *Test Service, Test Composite Application*
Software Test Factory professional profile: *SOA Tester*

4   Software Development Lifecycle

This section presents the processes systematizing the technical work of the proposed methodology. For each process, we present, in Table 4, the *process* name, main *input* and *output* artifacts, a short *process description* and related *professional profiles*. The methodology lifecycle is shown in Fig.1. Although this is represented as a waterfall software development lifecycle, the same processes can be used in a iterative development model.

Table 4. Processes Description

Design Business Process	
Input	Order of Service for Service Factory
Output	Business Process Design Specification
Description	Mapping and design of activities and information flow at business process-level. Processes must be modeled with sufficient details as to allow process automation. Process design may leverage BPMS tools.
Profiles	Business Process Analyst
Model Information Structure	
Input	Order of Service for Service Factory
	Business Process Design Specification
Output	Information Model Specification
Description	Modeling of information structure, according to corporate standards and information/data canonical models and vocabularies.
Profiles	Business Process Analyst, Data Administrator
Analyze Service Inventory	
Input	Business Process Design Specification
	Information Model Specification
	Service Catalog (corporate artifact)
Output	Service Candidate Document (partial)
Description	Services are identified through business process decomposition and collectively analyzed with service orientation paradigm. Current service inventory (represented in the service catalog) is considered in the analysis. Service normalization is usually performed.
Profiles	SOA Analyst, SOA Architect
Analyze Service	
Input	Business Process Design Specification
	Information Model Specification
	Service Candidate Document (partial)
Output	Service Candidate Document
	Service Profile (updated, one document for each service candidate)
Description	Services are individually analyzed with service orientation paradigm. This process includes information gathering steps and the service modeling sub-process that produces conceptual service definitions, called service candidates.
Profiles	SOA Analyst, SOA Architect
Design Service Contract	
Input	Business Process Design Specification
	Information Model Specification
	Service Candidate Document
	Service Profile
Output	Service Contract
	Service Profile (updated, one document for each service candidate)
Description	Formal design of the elements in the service contract (e.g. operations, messages, namespaces, usage policies/SLA, etc.). Specialized processes exist for each service model (e.g. task, entity and utility services).
Profiles	SOA Architect
Design Service Logic	
Input	Service Contract
	Service Profile
Output	Service Architecture Document
	Service Test Plan
	Service Profile (updated, one document for each service candidate)
Description	Design of the service logic architecture. Includes the specification of pre-processors, post-processors and wrapper service agents and design of the service composition logic. These may reside at SOA middleware (i.e. service bus).
Profiles	SOA Architect

Table 4. *(Continued)*

Design Composite Application	
Input	Business Process Design Specification Service Catalog (corporate artifact) Service Contract
Output	Application Architecture Document Application Test Plan Order of Service for Software Factory Order of Service for Test Factory
Description	Design of application logic architecture. Includes the specification of presentation layer and service composition (e.g. service orchestration) Components may reside at SOA middleware (i.e. orchestration engine, business rules engine, etc.).
Profiles	SOA Architect
Implement Service	
Input	Order of Service for Software Factory Service Contract Service Profile Service Architecture Document
Output	Service Profile (updated, one document for each service candidate) Service Source Code
Description	Implementation (coding) of service. May include development of components (service agents) that reside at SOA middleware (e.g. service bus).
Profiles	SOA Developer
Test Service	
Input	Order of Service for Software Test Factory Service Contract Service Source Code Service Test Plan
Output	Service Test Report
Description	Test of services. This is one type of unit test.
Profiles	SOA Tester
Deploy Service / Publish Service	
Input	Service Contract Service Profile Service Source Code Service Architecture Document
Output	Service Profile (updated, one document for each service candidate) Service Catalog (corporate artifact) (updated) Service Runtime Configuration
Description	Deployment of tested services into the runtime platform (SOA middleware) and publication of service contracts into the service catalog (and service registry).
Profiles	SOA Infrastructure Specialist, SOA Architect
Implement Composite Application	
Input	Business Process Design Specification Service Catalog (corporate artifact) Application Architecture Document
Output	Application Source Code
Description	Implementation (coding) of composite application. Includes implementation of presentation layer and service composition.
Profiles	SOA Developer

Table 4. *(Continued)*

Test Composite Application / Homologate Application	
Input	Order of Service for Software Test Factory Application Source Code Application Test Plan
Output	Application Test/Homologation Report
Description	Test and homologation of the application.
Profiles	SOA Tester, Business Process Analyst
Deploy Composite Application	
Input	Application Source Code Application Architecture Document
Output	Application Runtime Configuration
Description	Deployment of tested/homologated applications into the runtime platform (SOA middleware).
Profiles	SOA Infrastructure Specialist, SOA Architect

5   Professional Profiles

This section describes the professional profiles required to perform the technical work established in the methodology. These are shown in Table 5.

Table 5. Professional Profiles Description

Business Process Analyst	
Description	Is responsible for business process management activities. This profile exists both at BPMO and Service Factory. Typical activities include: business process mapping, modeling and design, continuous process improvement and definition of metrics and service level for processes
Skills	Business Process Management (BPM) practices and technologies Experience with BPMS platform (may include experience with BPMS platform used in the corporation)
SOA Analyst	
Description	Specializes in carrying out the analysis and definition of service inventory blueprints and the modeling and definition of service candidates, service capability candidates and service composition candidates
Skills	Proficiency in the service-oriented analysis of conceptual services, service blueprints, service modeling techniques and business service definition Experience with BPMS platform (desired)
SOA Architect	
Description	Specializes in the design of service-oriented technology architectures, service-oriented solutions, and related infrastructure.
Skills	Proficiency in mechanics of service-oriented computing through the mastery of patterns, principles, practices, and industry-standard technologies.
SOA Infrastructure Specialist	
Description	Is responsible for deploying and providing support throughout the software infrastructure of SOA (middleware SOA) and underlying resources.
Skills	Solid knowledge and experience in SOA software platform (may include expertise with SOA middleware used in the corporation) Proficiency in hardware and software infrastructure (desired)

Table 5. *(Continued)*

SOA Developer	
Description	Is responsible for developing code that encapsulates (Web) services and components defined in service and composite application architectures.
Skills	Proficiency in the programming framework adopted for SOA implementation
SOA Tester	
Description	Is responsible for testing services and composite applications
Skills	Knowledge and experience in software testing methodologies SOA specialization desired
Project Manager	
Description	Performs project management activities, including management of scope, schedule, resources, risks, cost, communication, acquisitions and stakeholders.
Skills	Proficiency in project management practices and tools
SOA Auditor	
Description	Is responsible for ensuring that the SOA development process occurs according to corporate standards and methodologies defined in the Ministry. Verifies if procedures established for each phase of the process are being strictly followed and if artifacts are generated accordingly. Applies the SOA governance principles defined.
Skills	Solid knowledge and experience in service orientation Knowledge of SOA middleware platform and SOA governance tools (may include experience with specific platforms used in the corporation)
SOA Governance Specialist	
Description	Is responsible for defining and maintaining the principles of service and composite application governance. Ensures compliance with the defined development process and the governance practices defined.
Skills	Proficiency in working with governance frameworks and processes in support of organizational and technological SOA governance requirements. Knowledge of SOA middleware platform and SOA governance tools (may include experience with specific platforms used in the corporation)
Quality Assurance Specialist	
Description	Performs quality-related verification of software products. Define and evolve best practices for quality assurance.
Skills	Expertise with testing techniques and practices specific to shared services and service-oriented solutions. Proficiency with required quality assurance processes and assessment criteria.
Data Administrator	
Description	Supports the creation of logical data models, data dictionaries in logical and physical process models, as well as data structure and related database objects.
Skills	Solid knowledge and experience in data modeling Knowledge of existing data models and database Good knowledge of the organization's business
Enterprise Architect	
Description	Is responsible for defining and evolving the enterprise architecture, as well as assuring compliance to it.
Skills	Proficiency in EA practices and methodologies Good knowledge of the business and operation of the organization

6   Methodology Implementation and Discussions

The proposed methodology was built from research and field experience related to SOA adoption initiatives in three large governmental organizations in Brazil (Ministry of Health, Ministry of Defense/Army, and Federal Sanitary Surveillance Agency) [9][10][11].

The main challenge related to these initiatives was related with the lack of qualified professionals available. As a result, existing human resources are becoming quite expensive and difficult to find. This is specially challenging in organizations with inflexible procurement and acquisition practices, which, in the studied case, results from the legal and regulatory framework for IT acquisition in Brazilian public sector.

Outsourcing is the only available choice, as to meet business requirements related to immediate instantiation of SOA processes in organizations that has low maturity in service orientation practices. However, this strategy must be followed with experienced consulting services, assuring expert knowledge at the SOA adoption initiative, including the definition of service oriented development methodology, reference architecture [14][17][18] and governance practices [16]. Methodologies and reference architectures should be clearly positioned as corporate design standards [19]. Formal definitions for these structuring elements in a SOA initiative and related compliance practices are key success factors for SOA adoption, especially in the case of outsourced working force [19].

In our methodology, these consultant experts are directly related to the SOA Governance Office unit, which has a distinct outsourcing model from the "factory" units. Our experience shows that when the SOA framework is defined and organization maturity is improved (at least to manageable SOA projects), SGO unit can be outsourced as a set of previously defined services.

Another important aspect learned from filed experience relates to the separation of roles in Service Factory and Software Factory units. There is important governance burden associated with this configuration. Therefore, in non-mature organizations, it may be better to merge these functional units and outsourcing these with the same service provider. Note, however, that if there is clear governance in place, the Software Factory and Software Test Factory units can be outsourced using existing contracts, which are possibly based on traditional software factory outsourcing models.

The outsourcing of the Service Factory and SOA Governance Office requires a clear definition of processes (work units), professional profiles and deliverables. These must be clearly specified in procurement documents (e.g. RFPs) and contracts. This methodology aims at providing a general definition for these, in relation to the Service Factory role. In a future paper we will present the outsourced SOA governance methodology, which is used for defining a viable outsourcing model for SGO unit.

7   Conclusions

In this paper we present a service-oriented software development methodology used as a general model for outsourcing SOA technical working force. The methodology arises as a synthesis of best practices learned from research and field experience establishing and experiencing development methodologies in three large organizations from the public sector in Brazil. Key findings are:

- A new outsourcing model is used for most of the technical work required by SOA developments. This is mainly represented by a "Service Factory" model, described in the paper.
- Existing outsourcing contracts for software factories and software testing can be used. However, there is a governance burden that requires maturity in SOA practices. If the maturity is not present, the service factory should encompass also the software factory role.
- Clearly defined processes (work units), professional profiles and deliverables, which must be present in procurement and contract documents, are essential for establishing the required governance of outsourced services.
- Positioning the development and governance methodologies, as well as the reference architecture as corporate design standards and establishing practices for assuring compliance with them are also key success factors for SOA projects with outsourced working force.

References

1. Erl, T.: SOA Principles of Service Design. Prentice Hall (2009)
2. Erl, T.: Service-Oriented Architecture – SOA Concepts, Technology, and Design. Prentice Hall (2008)
3. Gu, Q., Lago, P.: Guiding the selection of service-oriented software engineering Methodologies. Service Oriented Computing and Applications, SOCA 5, 203–223 (2011)
4. Ramollari, E., Dranidis, D., Simons, A.: A survey of service oriented development methodologies. In: Proceeding of Second European Young Researchers Workshop on Service Oriented Computing, pp. 75–80. University of Leicester, Leicester (2007)
5. Hotle, M., Landry, S.: Application Delivery and Support Organizational Archetypes: The Software Factory, Gartner Research Report G00167531 (May 2009)
6. Clements, P., Northrop, L.: Software Product Lines: Patterns and Practices, 3rd edn. Addison Wesley (2001)
7. Brown, A., Lopez, A., Reyes, L.: Practical Experiences with Software Factory Approaches in Enterprise Software Delivery. In: The Sixth International Conference on Software Engineering Advances, ICSEA 2011, Barcelona, Spain, October 23 (2011)
8. DATASUS/Ministry of Health, SOA Development Methodology (March 2012) (in Brazilian Portuguese)
9. CDS/Brazilian Army, SOA Development Methodology (preliminary) (December 2012) (in Brazilian Portuguese)
10. ANVISA, Service-oriented Methodology (April 2009) (in Brazilian Portuguese)
11. Jeston, J., Nelis, J.: Business Process Management, 2nd edn. Practical Guidelines to Successful Implementations. Butterworth-Heinemann (2008)
12. The Open Group - The Open Group Architecture Framework – TOGAF, version 9.1. The Open Group Standard (2011)
13. MacKenzie, C., Laskey, K., McCabe, F., Brown, P., Metz, R. (eds.): OASIS Reference Model for Service Oriented Architecture Version 1.0. OASIS Standard (October 12, 2006)
14. Project Management Institute, A Guide to the Project Management Body of Knowledge: PMBOK Guide, 4th edn. Project Management Institute (2008)

15. Erl, T.: SOA Governance. Prentice Hall (2011)
16. Estefan, J.A., Laskey, K., McCabe, F., Thornton, D. (eds.): OASIS Reference Architecture for Service Oriented Architecture Version 1.0. OASIS Draft 1 (2008)
17. The Open Group - SOA Reference Architecture. The Open Group Standard (2011)
18. David Norton, Case Study: O2 Uses Offshore Software Factory to Drive SOA Initiative (March 2010), Gartner (ID Number: G00170930)

Appendix A: Artifacts

Artifact	Type	Input of (process)	Output of (process)
Enterprise Architecture	CA	All processes	N/A
SOA Reference Architecture	CA	All processes	N/A
Enterprise Business Models	CA	All processes	N/A
Canonical Information Model	CA	All processes	N/A
Service Catalog	CA	All processes	(updated by) Publish Service
Order of Service for Service Factory	DPA	All processes	Project Planning
Business Process Design Specification	SA	All subsequent processes	Design Business Process
Information Model Specification	SA	All subsequent processes	Model Information Structure
Service Candidate Document	DPA	Design Service Contract Design Service Logic Design Composite Application	Analyze Service Inventory
Service Profile	SA	All subsequent processes	Analyze Service (updated by all subsequent processes)
Service Contract	SA	All subsequent processes	Design Service Contract
Service Architecture Document	SA	Implement Service Test Service Deploy Service	Design Service Logic
Service Test Plan	DPA	Test Service	Design Service Logic
Application Architecture Document	SA	Implement Composite Application Test Composite Application Deploy Composite Application	Design Composite Application
Application Test Plan	DPA	Test Application	Design Composite Application
Order of Service for Software Factory	DPA	Implement Service Test Service Implement Composite Application Test Composite Application	Design Composite Application
Order of Service for SW Test Factory	DPA	Test Service Test Composite Application	Design Composite Application
Service Source Code	SA	Test Service Deploy Service	Implement Service
Service Test Report	DPA	N/A	Test Service
Service Runtime Configuration	SA	N/A	Deploy Service
Application Source Code	SA	Test Composite Application Deploy Composite Application	Implement Composite Application
Application Test Report	DPA	N/A	Test Composite Application
Application Runtime Configuration	SA	N/A	Deploy Composite Application
SOA Governance artifacts	SA DPA	As per the SOA Governance Methodology	As per the SOA Governance Methodology
Project Management artifacts	PMA	As per the Project Management approach	As per the Project Management approach

Automatic Test Data Generation Using a Genetic Algorithm

Nassima Aleb and Samir Kechid

Computer Sciences Department University of Sciences and Technologies Houari Boumediene
BP 32 EL ALIA Bab Ezzouar, 16111, Algiers Algeria
{naleb,skechid}@usthb.dz

Abstract. The use of metaheuristic search techniques for the automatic generation of test data has been a burgeoning interest for many researchers in recent years. Previous attempts to automate the test generation process have been limited, having been constrained by the size and complexity of software, and the basic fact that in general, test data generation is an undecidable problem. Metaheuristic search techniques offer much promise in regard to these problems. Metaheuristic search techniques are high-level frameworks, which utilize heuristics to seek solutions for combinatorial problems at a reasonable computational cost. In this paper, we present a new evolutionary approach for automated test data generation for structural testing. Our method presents several noteworthy features: It uses a newly defined program modeling allowing an easy program manipulation. Furthermore, instead of affecting a unique value for each input variable, we assign to each input an interval. This representation has the advantage of delimiting first the input value and to refine the interval progressively. In this manner, the search space is explored more efficiently. We use an original fitness function, which expresses truthfully the individual quality. Furthermore, we define a crossover operator allowing to effectively improving individuals.

Keywords: Search-Based testing, Automated test data generation, Code Coverage Testing, Branch Coverage Testing, evolutionary testing.

1 Introduction

The need to automate software testing has provided rich set of challenging problems for the research community. One approach to software test automation that has achieved a great deal of recent attention is Search-Based Software Testing (SBST). SBST uses meta-heuristic algorithms to automate the generation of test inputs that meet test adequacy criterion. Despite the large body of work in SBST, the state of the art has changed little since the early seminal work on the Daimler Automated Software Testing System, which has been in use for more than a decade [7]. Though there have been many developments in SBST, these focus on changing the search algorithms and the way in which they are used, rather than the overall framework: program and problem modeling, program execution, and underlying fitness functions on which all metaheuristic search relies. This paper is focused on all these aspects. In particular, we use a form of partial backward symbolic execution to statically collect information available at compile time.

B. Murgante et al. (Eds.): ICCSA 2013, Part II, LNCS 7972, pp. 574–586, 2013.

We do not perform a complete symbolic execution, as this would be computationally expensive. Rather, we compute smaller amounts of symbolic information. This is rendered possible by using weakest preconditions. In this paper we combine backward symbolic execution and genetic algorithms to automatically generate test cases covering some adequacy criterion. We investigate two coverage criteria: Code coverage and branch coverage. The first aims at covering a given location in the source code of a program without specifying a particular path, while the second focused on a specific branch of the program. To explore more efficiently the large domains of input values, we use intervals instead of individual values. So, we use the same idea as in abstract interpretation [8], each input variable is assigned an interval. In this manner, an input variable is assigned an interval of values. If by using this interval we cannot achieve our purpose then we narrow it gradually. Using intervals is also advantageous for defining a powerful crossover operator. In fact, we have defined a recombination and mutation operators that guarantees that generated individuals are effectively better than their predecessors. The contributions of this paper are:

1. The Program modeling approach is different from the usual one in SBST. In fact, Usually, programs are modeled by control graphs flow, (CFG), we have adopted another modeling allowing an easier program manipulation.
2. Backward symbolic execution by weakest precondition allows to do not execute entirely the program but just the desired parts. Furthermore, contrary to true executions, symbolic one allow deducing some program's execution properties, which leads to
3. Assigning intervals to input variables instead of simple values is very valuable. It allows to explore more efficiently the search space, and to define efficient recombination and mutation operators.
4. The fitness function we define is very truthful, while being simple to compute.

The rest of the paper is organized as follow: Section 2 exposes some related work. Section 3 presents our modeling approach. In the section 4, we present our genetic algorithm for structural testing. Section 5 exposes some preliminary results, while section 6 concludes by highlighting contributions of our work and exposing some future directions.

2   Related Work

Recently, search– based approaches to test data generation have proved to be a popular application of Search–Based Software Engineering. Many test data generation scenarios can be attacked using a search–based approach, with examples in the literature including stress testing [6], finite state machine testing [10] and exception testing [27]. The present paper is concerned with the problem of generating test data for structural testing. Korel [16] was one of the first authors to apply search–based techniques to the problem of test data generation. Korel used a variation of hill climbing called the alternating variable method. Like other authors, for Korel the goal was to cover some 'difficult' branch not yet covered by a more lightweight random search. Xanthakis et al. [31] were the first authors to apply evolutionary computation algorithms to test data generation problems. McMinn [20] provides a comprehensive survey of search–based test data generation. In the past decade many authors have also addressed the problem of automated search for branch adequate test sets [5, 14,

16, 17, 19, 25, 28, 29, 31, 32]. For example, Baresel et al. and Bottaci consider the problems of fitness function definition [4, 5]. Jones et al. [14], McGraw et al. [19], Pargas and Harrold [25] and Wegener et al. [29] introduce approaches to evolutionary search for branch adequate test data. Other authors address closely related structural test adequacy criteria. For example, Xiao et al. [32] compare Evolutionary Testing with simulated annealing for the problem of condition-decision coverage, an alternative structural test data generation goal which can be reformulated as a branch adequacy problem using a testability transformation [11]. Mansour and Salam [17] consider the problem of path coverage, comparing evolutionary testing, hill climbing and simulated annealing. Many other non–structural test data generation goals have been considered in the literature [6, 10, 27], and these have also been formulated as search-based problems.

3   Program Modeling

Usually, programs are represented by control flow graphs, we adopt another representation which facilitates program manipulations. It separates a program into two models: The first describes all operations affecting variables values, we call it: Variable table; the second is called Control table, it summarizes the control structure of the program in a compact way.

3.1   Variable Table: VT

It models variable's declarations, assignments, and inputs. These instructions are numbered with integers representing locations.

1. **Variable declarations**: A declaration of the form: type idf is modeled by idf = type_idf0, meaning that idf has the type type and has not yet a known value. Global variables declarations are all designed by location 0. Local declarations are numbered by the location where they are performed.
2. **Assignments**: Are represented in the same way as in the source program.
3. **Simultaneous declarations and assignments**: A statement of the form: type idf=val is modeled by idf=type_val. For example: float x=2.5 : x=float_2.5
4. **Inputs**: An input assigns some value to a variable. So, the input of a variable v is modeled by v=$v, where $v is interpreted as an unknown constant.
5. **Predefined functions rand and malloc:** A call having the form: v=rand(..) or v=malloc(..)is modeled by v=£v where £v is an unknown constant.

3.2   Control Table: CT

It models conditional statements and loops. It describes the program structure. It contains constraints that make possible the execution of each statement of VT. CT models conditional and loop statements.

Conditional Statements: There are two sorts of conditional statements: alternative statement (with the else branch) and the simple conditional (without the else branch). An alternative statement is modeled by *(C,CT,CF,End):*

- *C*: is the condition of the statement.
- *CT*: The location of the first instruction to perform if *C* is True.

- *CF*: The location of the first instruction to perform if *C* is False.
- *End*:The location of the first instruction after the conditional.

CT[i] is the element number *i* of CT, *Then[i]= [CT,CF[* and *Else[i]=[CF,End[* are its intervals. A simple conditional statement is represented by *(C,CT,End)* with *C,CT* and *End* having the same meaning as the alternative statement. *CF* is set to (-1).

Loops: A loop statement is modeled in a similar way than conditional statements, by*(C,CT,CF,End)* where: *CF* is set to (-2). We note *C[i]* the condition of CT[i].

3.3   Modeling Example

(The CFG is given just for illustration, the left arrow of each condition stands for the condition 'True' and the right for 'False'.

Source Code

```
1,2,3,4: int x,y,z,t;
5:x=1 ;
6: scanf("%d",&y);
7:scanf("%d",&z);
    if (x>y+z)
8:{ x=z;
9:  t=y-1;
    if(y>0)
10:   {x=t ;
11:   t=0 ; }
    else
12:  t=t+x ;
    } else
13: { x=y;
14   t=z+x ;
    if (t>y)
    { While(x<t)
15: { x=x+2;
16:  t=t-1; }
    } if (x==t)
17:  { t=t+x;
18 :   x=x+1 ;
    } else
19:   t=x ;
    }
    if (t>0)
    if (t>10)
20: t=t-10;
```

VT

Loc	Variable	Expression
1	X	Int_x0
2	Y	Int_y0
3	Z	Int_z0
4	T	Int_t0
5	X	1
6	Y	$y
7	Z	$z
8	X	Z
9	T	y-1
10	X	T
11	T	0
12	T	T+x
13	X	Y
14	T	Z+x
15	X	X+2
16	T	t-1
17	T	T+x
18	X	X+1
19	T	X
20	T	t-10

CT

	C	CT	CF	End
1	x>y+z	8	13	20
2	y>0	10	12	13
3	t>y	15	-1	17
4	X<t	15	-2	17
5	X=t	17	19	20
6	t>0	20	-1	21
7	t>10	20	-1	21

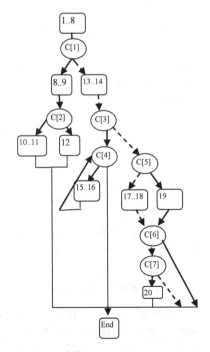

Fig. 1. Program Example Prog with its modeling

3.4   Modeling of the Adequacy Criterion

As we said before, we are investigating code coverage and branch coverage criterion. For the first criterion, we adopt a goal-oriented approach. So, having some location in the source code of the program, we aim to generate a test case that reaches this location, without defining a specific branch or path. Let L be the targeted location, let's call TP the "targeted path" leading to L. TP is a string constituted of characters '1','0', 'x' or '*'. Let's note the expression "must be equal" by "≡". So:

$$TP[i] = \begin{cases} \text{`1'} & \text{if } C[i] \equiv \text{True} \\ \text{`0'} & \text{if } C[i] \equiv \text{False} \\ \text{`x'} & \text{if both `1' or `0' are possible} \\ \text{`*'} & \text{if the value of } C[i] \text{ is not required} \end{cases}$$

If CT[i] represents an if-then-else statement then if L∈Then(i) then C[i] ≡True else if L∈Else(i) then C[i]≡False. A same reasoning is performed on the other CT subsets. **Examples**: In the program Prog: L =12: TP=10; L=14: TP=0*; L=20: TP= 0*0*x11

For a branch coverage testing, the criterion is introduced as some branch represented by a targeted path composed of the characters: '0', '1', or '*' : since we are interested in a precise sequence of blocks.

Examples: The branch represented in the CFG by dotted arrow has a TP=0*0110

4   A Genetic Algorithm for Test Data Generation

Having a program represented by its variable table VT and its control table CT, and a code coverage criterion expressed by a targeted path TP, we try to find a set of input variables values that allow covering TP. Our technique is presented by the algorithm of the figure 2. It starts with a population of individuals each one representing a possible initialization of input variables. Each input value is represented by an interval. This allows us to "correct" gradually some undesirable behavior instead of rejecting systematically each unwanted results. For each individual i, we compute its string path SPi recording the sequence of CT elements executed by the individual i. The fitness function computes the distance between TP and SPi. The objective is reached if we find an individual i* such that the distance between SPi* and TP is zero.   In the contrary case, the population must be improved. We define a combination operator and three mutation operators.

4.1   Individuals String Path Computing

To compute the execution path followed by some individual i, we must execute the program with the data of i. We use the concept of weakest precondition to perform "symbolic executions". This allows computing just the considered guard in the desired point of the program instead of running all the statements of the program.

4.1.1 Weakest Precondition

For a statement S and a predicate C, let WP(S,C) denotes the weakest precondition of C with respect to (w.r.t.) the statement S. WP(S,C) is defined as the weakest predicate whose truth before S entails the truth of C after S terminates. Let v=exp be an assignment, where v is a variable and exp is an expression. Let C be a predicate, by definition WP(v=exp,C) is C with all occurrences of v replaced with exp, denoted C[exp/v]. For example: WP(v=v+2, v>8) = (v+2)>8 = (v>6). In the subsequent, we denote WP(Si,C) the weakest precondition of the predicate C w.r.t. The statement having the location Si in the Variable table. We use the concept of weakest precondition to evaluate CT element's conditions which represent positions values of individual's string path. Hence, since we use intervals to design different parts of CT elements: Then(I), Else(I) and Body(I). We call the two first intervals: simple intervals and the last one iterative interval. So, in the subsequent, we extrapolate the definition of weakest precondition to be applied to simple locations intervals. Weakest precondition for iterative intervals will be exposed afterward. We define the weakest precondition of a predicate C w.r.t. an interval [Si,Sj[, denoted by WP([Si,Sj[,C), as the weakest predicate whose truth before Si entails the truth of C after Sj-1 terminates. The idea is to compute successively the weakest preconditions of C with respect to each location within [Si,Sj[starting by the end until we attain Si, or we obtain a constant meaning that there are no variables occurring in C. For each location Sk∈ [Si,Sj[, the result obtained from computing WP(Sk,Ck) is given as predicate to compute its weakest precondition w.r.t. Sk-1 and so on. So:

$$
WP([Si,Sj[,C) = \begin{cases} C & \text{If no variable occurs in C} \\ WP(Si,C) & \text{If Sj=Si} \\ WP([Si,Sj-1[,WP(Sj-1,C)) & \text{Otherwise} \end{cases}
$$

Example: WP ([1,5[,x>y+z)=WP([1,4[,WP(4,x>y+z))=WP([1,3[,WP(3,x>y+$z))
=WP([1,2[,WP(2,x>$y+$z))=WP([1,1[,1>$y+$z)=1>$y+$z
Which means that the condition x>y+z is satisfied just before the location 4 if and only if the input values of variables x and y verify the constraint 1>$y+$z.
We define also : WP([Si,Sj[∪[Sk,Sl[,C)=WP([Si,Sj[,WP([Sk,Sl[,C))

4.1.2 Computing of CT Guards for an Individual

First, let's notice the two following points:

1- The value of a CT guard is not at all times known. For example, for the individual i such that y=[-10,50] and z=[-200,100], the value of C[1]: x>y+z is not known.

2- It is not always required to know the truth values of all CT guards. This is due to the fact that CT elements are often opposite. So, for example, if an individual is such that C[1] is false, it is not necessary to evaluate C[2].

Consequently, the string path of an individual contains the characters: '1','0', 'u' or '*' meaning respectively: True, False, Unknown or not required. When we find the first 'u', we stop the computing by completing all the remainder positions by 'u'. Let i be an individual, and let $\omega i(C[k])$ be the valuation of C[k] for the individual i. The string path of i noted by SPi is the string : $a_1a_2a_3..a_n$ computed by the following algorithm:

If ($\omega i(C[k])$=True) **Then** a_k='1'
Else If ($\omega i(C[k])$ =False) **Then** a_k='0'
    Else If ($\omega i(C[k])$ is unknown Or $\omega i(C[k-1])$='u') **Then** a_k='u'
        Else a_k='*'

Let's call Pi_k the prefix having the length k of the path Pi, and let's introduce the formulas computing the truth value of C[k] with respect to a path Pi.

If (CT[k].CF• -2) **Then** **If** k=1 **Then** $\omega i(C[k])$ = WP([0,Begin(k)[)
                        Else $\omega i(C[k])$ = WP($Pi_{(k-1),}C[k]$)
        Else Let CT[k]=(C,Sb,-2,se), and let $\omega i(C[k])_j$ be the value of
                C[k] in the path P in the iteration j:
                If (j=1) **Then** $\omega i(C[k])_j = \omega i(C[k])$
                        Else **if** (j>1) and ($\omega i(C[k])_{j-1}$=True)
                            Then $\omega i(C[k])_j$ =WP($Pi_{(k-1)} \cup [Sb,Se[^{j-1} ,C[k])$

4.1.3   Evaluation of Program Expressions
We use the same definitions for arithmetic operations as in interval abstract interpretation [8]. Logical operations are our own definitions since we use a three valuated logic.

A.   Arithmetic Expressions Evaluation

We perform expression evaluations in the following manner:

---

1. Constant: Transform each constant c by the interval [c,c]
2. Somme: [a,b]+[c,d]=[a+b,c+d]
3. Opposite: -[a,b]=[-b,-a]
4. Substraction: [a,b]-[c,d]=[a-b,d-c]
5. Multiplication: [a,b]*[c,d]=[min(ac,ad,bc,bd), max(ac,ad,bc,bd)]

B. Logical Operations Evaluation

Logical expressions are then evaluated in the following manner:

If (a=b=c=d) Then ([a,b]=[c,d]) = True
Else if ([a,b]∩[c,d]=Ø) Then ([a,b]=[c,d]) =False
  Else ([a,b]=[c,d]) = U)
If (b<c) Then ([a,b]<[c,d]) = True
Else if (d<a) Then ([a,b]<[c,d]) =False
  Else ([a,b]<[c,d]) = U)
If (b<c) Then ([a,b]>[c,d]) = False
Else if (a>d) Then ([a,b]>[c,d]) =True
  Else ([a,b]>[c,d]) = U)

Intervals union, intersection, inclusion and appurtenance are defined as ordinary intervals operations.

4.1.4   Individual String Path Computing Example

Let's consider the program Prog, each individual is composed of two intervals having the type integer representing the variables y and z. A possible individual i is [-10,-5[; [-200,0[, let's compute its string path SPi to the location 6, let's note SPi =a_1a_2 . So, let's compute $\omega i(C[1])$ and $\omega i(C[2])$. We use the formulas defined in 4.1.2. $\omega i(C[1])$=WP([0,4[,x>y+z)=1>$y+$z=True=>a_1=1 (since $y+$z=[-210,-5]) $\omega i(C[2])$=WP((([0,4[∪[4,6[),y>0)=$y>0=False=>$a_2$=0. Consequently, SPi =10

4.2   Initial Population

To guarantee some desirable properties, initial population is generated in such a way that ensures the following points:

1- Diversity
2- Acceptable quality

Each created individual is first evaluated to verify the two precedent properties. To force diversity, we privilege individuals having different values in TP positions represented by 'x' since they are those positions which can provide different access paths. An individual has an acceptable quality if its string path has at least one correct position. It is also advantageous to use large size intervals in the initialization stage, and to narrow them progressively. Thus, let IPopI be the number of initial population individuals. So, as in [21], in the initialization phase, we generate IPopI+K (K>0) individuals but we preserve only the IPopI "best" ones.

4.3   Fitness Function

For an individual i, the fitness function, Fitness(i) measures the distance between TP and SPi. The computing of the fitness is performed in the following manner: Let Fit be a string such that:

If (TP[k]='x') OR (TP[k]=SPi[k]) then Fit[k]=0 else Fit[k]=1.

Fitness(i) is the decimal number obtained by converting the binary number represented by Fit. Example: Let TP=10xx1; and let i an individual such that: Chain i=1101U ; Fit=01001 ; Fitness(I)= 9. We remark that the Fitness function represents truthfully the distance between the desired behavior and the behavior of the considered individual. In fact, if we consider for example two individuals i1 and i2 such that Fit1=1000 and Fit2=0001 despite the fact that these two individuals have both one faulty position which does not match with TP, their Fitness must be different because the first individual has failed in the first guard so it has taken a path completely different from TP and it represents an execution that is completely deviating. While i2 has matched with TP until the last position so it is closer to TP. So, i2 is better than i1, which is effectively expressed by our fitness function since: Fitness(i1)=8 and Fitness(i2)=1. The goal is to find an individual i* such that Fitness(i*)=0.

4.4   Population Improvement

To ameliorate the population, we adopt a guided approach which increases the probability of obtained individuals to be effectively better than their parents. However, since recombination and mutation may be performed many times in a genetic algorithm, thus, they must be as simple as possible. So, we perform a gradual amelioration. It consists to 'correct' the first faulty position of each individual of P: We call faulty position a position whose value in the string path and in TP are different, and its value in TP is either '0' or '1'. A faulty position could be an unknown position or an erroneous one. We categorize individuals considering their fitness (fit or unfit) and their faulty positions values (wrong or unknown). Let's call FU, FE, UU, UE respectively the individual's categories: Fit with Unknown positions, Fit with Erroneous positions, Unfit with Unknown positions and finally, Unfit with Erroneous positions. We notice that by 'fit' we mean having an acceptable quality.

4.4.1   Recombination

Recombination operator is applied on the class FU to correct progressively individual's faulty positions. Let i1 and i2 be two individuals such that i1 has the position p as first unknown position and p is not a faulty position for i2. The idea is to use the data of the individual i2 to correct the unknown position p of i1. However, since in a program variables are strongly correlated to each other, modifying some data of an individual to correct some guard could in the same time alter negatively other guards. To avoid this situation, we modify data corresponding to some faulty position in a "conservative" way. So we perform an intersection between the data, occurring in the position p, of i2 and those of i1. Consequently, all the guards which had a known value conserve their value, those which had unknown values could have a known ones. Hence, the recombination operator is defined as follow: Let the individuals : i1,i2 and i3.such that p is a faulty position of i1, and let x1i,x2i,x3i the values intervals of the input variable xi respectively for i1,i2 and i3.

Input: i1,i2: individuals; p: integer representing a position.
Output: i3,i2: created individuals.
for all input variable xi :
If (xi ∈ C[p])
Then   x3i = x1i ∩x2i
Else   x3i=x1i
End.

Example: Let TP= 1x1101; let i1 an individual such that SPi1 = 011001, so the faulty positions of i1 are: 1 and 4. The recombination point will be the position 1. Let i2 be an individual such that SPi2= 101000, so, 1 is not a faulty position of i2. So, we use i2 to correct i1. Input variables occurring in C[1] are y and z, so : Recombination (i1,i2,1)=(i3,i2) such that : y3=y1∩y2 and z3=z1∩z2, where yi and zi are the intervals of variables y and z of the individual i.
I1=([-10,50],[15,100]); i2=([0,5],[80,200]);i3=([0,5],[80,100])

4.4.2   Mutation Operators

We define three mutation operators: Weak mutation, Strong mutation and Narrowing. Weak mutation makes a little perturbation on individuals of the class FE. It consists to modify the interval of a unique variable. The candidate variable to change must appear in the faulty position. To improve the category UE, we use Strong mutation. It consists to change randomly all variables values. The narrowing operator consists to reduce, in various ways, the input values intervals. It is used to eliminate faulty positions in individuals of category UU, since unknown values are due to intervals large sizes.

5   Experiments

We perform several experimentations to evaluate our approach. Genetic algorithm parameters like the maximal number of iterations, population size, and the proportion of "fit' individuals, are adjusted during experimentation. We report here the results obtained for 7 programs. We report the iterations number, the population size and if the tool finds a case test or not. Through several experiments, we have remarked that the quality of initial population is decisive for the tool performances. It is suitable to begin with large individual intervals and to narrow them progressively.

Program	Predicate Number	Population Size	Iterations Number	Covered
P1	6	50	3	Y
P2	9	50	3	Y
P3	12	100	4	Y
P4	15	100	4	Y
P5	20	100	200	Y
P6	23	100	200	N
P7	25	150	200	N

6   Conclusion and Future Work

We have presented an original approach of Search-Based program testing. The results obtained are encouraging. Our work presents several contributions:

1. Our program modeling approach is very powerful; it is as simple as advantageous. In fact, it permits to manipulate a program very easily. We can characterize, by a set of conditions,   each region of the program independently of the rest of the program and without being forced to cross the program entirely from the beginning.
2. Backward symbolic execution process is novel. It presents the advantage of computing only on the considered guards in the needed path, and not over the entire program.
3. The fitness function is novel and it quantifies truthfully the distance between a path and some required criterion.

Several future directions are possible to our work. The most direct one is to investigate, with the same modeling, other metaheuristics, like scatter search.

References

[1]   The Software-artifact Infrastructure Repository,
       http://sir.unl.edu/portal/index.html
[2]   Baker, J.E.: Reducing bias and inefficiency in the selection algorithm. In: Grefenstette, J.J. (ed.) Proceedings of the Second International Conference on Genetic Algorithms. Lawrence Erlbaum Associates Publishers (1987)
[3]   Baresel, A., Binkley, D., Harman, M., Korel, B.: Evolutionary testing in the presence of loop-assigned flags: a testability transformation approach. In: Avrunin, G.S., Rothermel, G. (eds.) ISSTA, pp. 108–118. ACM (2004)
[4]   Baresel, A., Sthamer, H., Schmidt, M.: Fitness function design to improve Evolutionary structural testing. In: GECCO 2002: Proceedings of the Genetic and Evolutionary Computation Conference, July 9-13, pp. 1329–1336. Morgan Kaufmann Publishers, San Francisco (2002)
[5]   Bottaci, L.: Instrumenting programs with flag variables for test data search by genetic algorithms. In: GECCO 2002: Proceedings of the Genetic and Evolutionary Computation Conference, July 9-13, pp. 1337–1342. Morgan Kaufmann Publishers, New York (2002)
[6]   Briand, L.C., Labiche, Y., Shousha, M.: Stress testing real-time systems with genetic algorithms. In: Beyer, H.-G., O'Reilly, U.-M. (eds.) Proceedings of the Genetic and Evolutionary Computation Conference, GECCO 2005, June 25-29, pp. 1021–1028. ACM, Washington DC (2005)
[7]   Baresel, A., Wegener, J., Sthamer, H.: Evolutionary test environment for automatic structural testing. Information & Software Technology 43(14), 841–854 (2001)
[8]   Cousot, P., Cousot, R.: Abstract interpretation: A Unified lattice model for static analysis of programs by construction or approximation of fixpoints. In: Principales of Programming Languages, POPL 1977, pp. 238–252 (1977)
[9]   Dijkstra, E.: A discipline of programming. Prentice Hall (1976)

[10] Derderian, K., Hierons, R., Harman, M., Guo, Q.: Automated Unique Input Output sequence generation for conformance testing of FSMs. The Computer Journal 49(3), 331–344 (2006)

[11] Harman, M., Hu, L., Hierons, R.M., Wegener, J., Sthamer, H., Baresel, A., Roper, M.: Testability transformation. IEEE Transactions on Software Engineering 30(1), 3–16 (2004)

[12] Hermadi, I., Ahmed, M.: Genetic algorithm based test data generator. In: Sarker, R., Reynolds, R., Abbass, H., Tan, K.C., McKay, B., Essam, D., Gedeon, T. (eds.) Proceedings of The 2003 Congress on Evolutionary Computation CEC 2003, Canberra, December 8-12, pp. 85–91 (2003)

[13] Aguirre, A.H., Rionda, S.B., Coello Coello, C.A., Lizárraga, G.L., Montes, E.M.: Handling Constraints using Multiobjective Optimization Concepts. International Journal for Numerical Methods in Engineering 59(15), 1989–2017 (2004)

[14] Jones, B., Sthamer, H., Eyres, D.: Automatic structural testing using genetic algorithms. Software Engineering Journal 11(5), 299–306 (1996)

[15] King, J.C.: Symbolic execution and program testing. Communications of the ACM 19(7), 385–394 (1976)

[16] Korel, B.: Automated software test data generation. IEEE Transactions on Software Engineering 16(8), 870–879 (1990)

[17] Mansour, N., Salame, M.: Data generation for path testing. Software Quality Journal 12(2), 121–134 (2004)

[18] Martello, S., Toth, P.: Knapsack Problems: Algorithms and Computer Implementations. Wiley, New York (1990)

[19] McGraw, G., Michael, C., Schatz, M.: Generating software test data by evolution. IEEE Transactions on Software Engineering 27(12), 1085–1110 (2001)

[20] McMinn, P.: Search-based software test data generation: A survey. Software Testing, Verification and Reliability 14(2), 105–156 (2004)

[21] McMinn, P.: IGUANA: Input generation using automated novel algorithms. A plug and play research tool. Technical Report, Department of Computer Science, University of Sheffield (2007)

[22] McMinn, P., Binkley, D., Harman, M.: Testability transformation for efficient automated test data search in the presence of nesting. In: Proceedings of the Third UK Software Testing Workshop, pp. 165–182 September (2005)

[23] Mühlenbein, H., Schlierkamp-Voosen, D.: Predictive models for the breeder genetic algorithm: I. continuous parameter optimization. Evolutionary Computation 1(1), 25–49 (1993)

[24] Jefferson Offutt, A.: An integrated system for automatically generating test data. In: Ng, P.A., Ramamoorthy, C.V., Seifert, L.C., Yeh, R.T. (eds.) Proceedings of the First International Conference on Systems Integration, pp. 694–701. IEEE Computer Society Press, Morristown (1990)

[25] Pargas, R., Harrold, M., Peck, R.: Test-data generation using genetic algorithms. Software Testing, Verification and Reliability 9(4), 263–282 (1999)

[26] Radio Technical Commission for Aeronautics. RTCA DO178-B Software considerations in airborne systems and equipment certification (1992)

[27] Tracey, N., Clark, J., Mander, K.: Automated program flaw finding using simulated annealing. In: International Symposium on Software Testing and Analysis (ISSTA 1998), pp. 73–81 (March 1998)

[28] Wang, H.-C., Jeng, B.: Structural testing using memetic algorithm. In: Proceedings of the Second Taiwan Conference on Software Engineering, Taipei, Taiwan (2006)

[29] Wegener, J., Baresel, A., Sthamer, H.: Evolutionary test environment for automatic structural testing. Information and Software Technology 43(14), 841–854 (2001)

[30] Whitley, D.: The GENITOR algorithm and selection pressure: Why rank-based allocation. In: Schaffer, J.D. (ed.) Proc. of the Third Int. Conf. on Genetic Algorithms, pp. 116–121. Morgan Kaufmann, San Mateo (1989)

[31] Xanthakis, S., Ellis, C., Skourlas, C., Le Gall, A., Katsikas, S., Karapoulios, K.: Application of genetic algorithms to software testing (Application des algorithmes génétiques au test des logiciels). In: 5th International Conference on Software Engineering and its Applications, Toulouse, France, pp. 625–636 (1992)

[32] Xiao, M., El-Attar, M., Reformat, M., Miller, J.: Empirical evaluation of optimization algorithms when used in goal-oriented automated test data generation techniques. Empirical Software Engineering 12(2), 183–239 (2007)

Genetic Algorithm for Oil Spill Automatic Detection from Envisat Satellite Data

Maged Marghany

Institute of Geospatial Science and Technology (INSteG)
Universiti Teknologi Malaysia
81310 UTM, Skudai, Johor Bahru, Malaysia
maged@utm.my, magedupm@hotmail.com

Abstract. The merchant ship collided with a Malaysian oil tanker on May 25, 2010, and spilled 2,500 tons of crude oil into the Singapore Straits. The main objective of this work is to design automatic detection procedures for oil spill in synthetic aperture radar (SAR) satellite data. In doing so the genetic algorithm tool was designed to investigate the occurrence of oil spill in Malaysian coastal waters using ENVISAT ASAR satellite data. The study shows that crossover process, and the fitness function generated accurate pattern of oil slick in SAR data. This shown by 85% for oil spill, 5% look–alike and 10% for sea roughness using the receiver –operational characteristics (ROC) curve. It can therefore be concluded crossover process, and the fitness function have the main role in genetic algorithm achievement for oil spill automatic detection in ENVISAT ASAR data.

Keywords: Oil Spill, ENVISAT ASAR data, Crossover Process, Fitness Function Genetic algorithm.

1 Introduction

Deepwater Horizon oil spill in 2010 is the most serious marine pollution disaster has occurred in the history of the petroleum industry. This disaster has dominated by three months of oil flows in coastal waters of the Gulf of Mexico. In this regard, the Deepwater Hoizon oil spill has serious effects on feeble maritime, wildlife habitats, Gulf's fishing, coastal ecologies and tourism industries. As a result, human health problems are caused because of the spill and its cleanup [25]. Consistent with Marghany and Hashim [7], Synthetic aperture radar (SAR) is a precious foundation of oil spill detection, surveying and monitoring that improves oil spill detection by various approaches. The different SAR tools to detect and observe oil spills are vessels, airplanes, and satellites [5]. Vessels can detect oil spills at sea, covering restricted areas, say for example, (2500 m x 2500 m), when they are equipped with the navigation radars [20]. On the other hand, airplanes and satellites are the main tools that are used to record sea-based oil pollution [22] [7].

Scientists which have agreed with that oil spill detection with SAR data is based on: (i) all dark patches present in the SAR images are isolated [3][16]; (ii) features

B. Murgante et al. (Eds.): ICCSA 2013, Part II, LNCS 7972, pp. 587–598, 2013.
© Springer-Verlag Berlin Heidelberg 2013

for each dark patch are then extracted [1][12]; (iii) these dark patches are tested against predefined values [22]; and (v) the probability for every dark patch is calculated to determine whether it is an oil spill or a look-alike phenomenon [1][14][20][22][25]. Therefore, Topouzelis et al., [21] and Topouzelis [22] reported that most studies use the low resolution SAR data such as quick-looks, with the nominal spatial resolution of 100 m x 100 m, to detect oil spills. In this regard, quick looks' data are sufficient for monitoring large scale area of 300 km x 300 km. On the contrary, they cannot efficiently detect small and fresh spills [22].

Further, SAR data have distinctive features as equated to optical satellite sensors which makes SAR extremely valuable for spill watching and detection [5] [7] [9] [13]. These features are involved with several parameters: operating frequency, band resolution, incidence angle and polarization [10] [12]. Marghany and Hashim [7] develop comparative automatic detection procedures for oil spill pixels in Multimode (Standard beam S2, Wide beam W1 and fine beam F1) RADARSAT-1 SAR satellite data post the supervised classification (Mahalanobis), and neural network (NN) for oil spill detection. They found that NN shows a higher performance in automatic detection of oil spill in RADARSAT-1 SAR data as compared to the Mahalanobis classification with a standard deviation of 0.12. In addition, they W1 beam mode is appropriate for oil spill and look-alikes discrimination and detection [10] [11] [12]. Recently, Skrunes et al., [9], nevertheless, reported that there are several disadvantages are associated with Current SAR based oil spill detection and monitoring. They stated that, SAR sensors are not able to detect thickness distribution, volume, the oil-water emulsion ratio and chemical properties of the SAR data. In this regard, they recommended to utilize multi-polarization acquisition data such as RADARSAT-2 and TerraSAR-X satellites. They concluded that the multi-polarization data show a prospective for prejudice between mineral oil slicks and biogenic slicks.

This work has hypothesized that the dark spot areas (oil slick or look-alike pixels) and its surrounding backscattered environmental signal complex looks in the ENVISAT ASAR data can detect using Cellular Genetic Algorithm. The contribution of this work concerns with designing Cellular Genetic Algorithm based on multi-objective optimization. However, previous work has implemented post classification techniques [4][9][16][18] or artificial neural network [19],[21],[24] which are considered as semi-automatic techniques. The objective of this work can divide into two sub-objectives: (i) To examine CGA [27] for oil spill automatic detection in ENVISAT ASAR data; and (ii) To design the multi-objective optimization algorithm for oil spill automatic detection in ENVISAT ASAR that are based on algorithm's accuracy.

2   Data Acquisition

The SAR data acquired in this study are from ENVISAT ASAR data on June 3, 2010, in single look complex format. On May 25, 2010 a merchant ship (Fig.1) collided with a Malaysian oil tanker on Tuesday morning, puncturing the tanker's hull and spilling 2,500 tons of crude oil into the Singapore Strait (Fig.2), maritime officials reported. The damage appeared to be limited to one compartment in the double-hulled tanker, the Bunga Kelana 3, with the spill amounting to about 18,000 barrels.

Fig. 1. The merchant ship collided with a Malaysian oil tanker

Fig. 2. Location of oil spill event

These data are C-band and had the lower signal-to-noise ratio owing to their HH polarization with a wavelength range of 3.7 to 7.5 cm and a frequency of 5.331 GHz. ASAR can achieve a spatial resolution generally around 30 m. The ASAR is intended for applications which require the spatial resolution of spatial resolution of 150 m. This means that it is not effective at imaging areas in depth, unlike stripmap SAR. The azimuth resolution is 4 m, and range resolution is 8 m (Table 1).

Table 1. ENVISAT ASAR Satellite Data Acquisitions

Mode Type	Resolution (m)		Incident Angle(°)	Swath width (km)	Date
	Range	Azimuth			
ASAR	4	8	19.2° - 26.7°	58-110	2010/6/3

3   Genetic Algorithm

On the word of Kahlouche et al., [27], Genetic algorithms (GA) differ from classification algorithm. In classification algorithm, a single point at every iteration is generated. Moreover, classification algorithms correspondingly choose the next point in the classification by a deterministic computation. In contrast, the genetic algorithm (GA) generates a population of cells at every iteration, where the superlative cell in the population approaches an optimal solution. Moreover, GA, implements probabilistic transition rules not deterministic rules as compared to classification algorithms [22] [28]. Consequently, Cellular Automata (CA) are mathematical algorithms that involve a large number of relatively simple individual units, or "cells," which is connected only locally, without the existence of a central control in the system. Each cell is a simple finite automaton that repeatedly updates its own state, where the new cell state depends on the cell's current state and those of its immediate (local) neighbors. The limited functionality of each cell, and the interactions, however, being restricted to local neighbors. Thus the system as a whole is capable of producing intricate patterns, and performing complicated computations [29].

A constrained multi-objective problem for oil spill discrimination in SAR data deals with more than one objective and constraint namely look-alikes, for instance, currents, eddies, upwelling or downwelling zones, fronts and rain cells). The general form of the problem is adapted from Sivanandam and Deepa [29] and described as

$$\text{Minimize } f(\beta) = [f_1(\beta), f_2(\beta), ..., f_k(\beta)]^T \tag{1}$$

Subject to the constraints:

$$g_i(\beta) \leq 0, \quad i = 1, 2, 3, ..I \tag{2}$$

$$h_j(\beta) \leq 0, \quad j = 1, 2, 3, ...J \tag{3}$$

$$\beta_s \leq \beta \leq \beta_U \tag{4}$$

where, $f_i(\beta)$ is the *i-th* pixel backscatters β in SAR data, $g_i(\beta)$ and $h_j(\beta)$ represents the *i-th* and *j-th* constraints of backscatter in raw direction and column direction, respectively. β_L and β_U are the lower and upper limit of values of the backscatter. The transition rules for the cellular automata oil spill detection is designed using the input of different backscatter values β to identify the slick conditions required in the neighbourhood pixels of keneral window size of 7x7 pixels and lines for a β pixel to become oil slick. These rules can be summarized as follows:

1. IF test pixel is sea surface, OR current boundary features THEN $\beta \geq 0$ not oil spill.
2. IF test pixel is dark patches (low wind zone, OR biogenic slicks OR shear zones) $\beta \leq 0$ THEN It becomes oil slick if its :

3.1   Data Organization

Let the entire backscatter of dark patches in ENVISAT ASAR are $[\beta_1, \beta_2, \beta_3, \ldots \ldots, \beta_K]$ where K is the total number backscatter of dark patches in the ENVISAT ASAR data. Therefore, K is made up from genes which is representing the backscatter β of dark patches and its surrounding environment and genetic algorithms is started with the population initializing step.

3.2   Population Initializing

Let P_i^j is a gene which corresponds to backscatter of dark pixels and its surrounding pixels. Consequently, P_i^j is randomly selected and representing both of backscatter variations of dark patches and its surrounding environmental pixels. Moreover, i varies from 1 to K and j varies from 1 to N where N is the population size.

3.3   The Fitness Function

Following Kahlouche et al., [27], a fitness function is selected to determine the similarity of each individual backscatter of dark patches in ENVISAT ASAR data. Then the backscatter of dark patches in ENVISAT data be symbolized by β_i where $i=1,2,3, \ldots, K$ and the initial population P_i^j where $j =1,2,3, \ldots, N$ and $i =1,2,3\ldots, K$. Formally, the fitness value $f(P^j)$ of each individual of the population is computed as follows [27]:

$$f(P^j) = [\sum_{i=1}^{K} |P_i^j - \beta_i|]^{-1} \qquad j=1, \ldots, N. \qquad (5)$$

where, N and K are the number of individuals of the population used in fitness process. Generally, Equation 5 used to determine the level of similarities of dark patches that are belong to oil spill in ENVISAT ASAR data.

3.4   Selection Step

The key parameter in the selection step of genetic algorithm which is chosen the fittest individuals $f(P^j)$ from the population P_i^j. The threshold value τ is determined by the maximum values of fitness of the population $Max\, f(P^j)$ and the minimum values of fitness of the population of $Min\, f(P^j)$ Indeed, in the next generations, this step serves the populations P. Therefore, the values of the fittest individuals dark patches in ENVISAT ASAR data are greater identifies threshold τ which is given by

$$\tau = 0.5\,[Max\, f(P^j) + Min\, f(P^j)] \tag{6}$$

Equation 6 used as selection step to determine the maximum and minimum values of fitness of the population, respectively. This is considered as a dark patches' population generation step in GA algorithms.

3.5   The Reproduction Step

According to Sivanandam and Deepa [29], Genetic algorithm is mainly a function of the reproducing step which involves the crossover and mutation processes on the backscatter population P_i^j in SAR data. In this regard, the crossover operator constructs the P_i^j to converge around solutions with high fitness. Thus, the closer the crossover probability is to 1 and the faster is the convergence [27]. In crossover step the chromosomes interchange genes. A local fitness value effects each gene as

$$f(P_i^j) = \left|\beta_i - P_i^j\right| \tag{7}$$

Then the crossfire between two individuals consists to keep all individual populations of the first parent which have a local fitness greater than the average local fitness $f(P_{av}^j)$ and substitutes the remained genes by the corresponding ones from the second parent. Hence, the average local fitness is defined by:

$$f(P_{av}^j) = \frac{1}{K}\sum_{i=1}^{K} f(P_i^j) \tag{8}$$

Therefore, the mutation operator denotes the phenomena of extraordinary chance in the evolution process. Truly, some useful genetic information regarding the selected population could be lost during reproducing step. As a result, mutation operator introduces a new genetic information to the gene pool [27].

3.5.1 Morphological Operations

Morphological operation on the selected individuals is performed prior to the cross-over and the mutation process. This is to exploit connectivity property of the ENVISAT ASAR data [28]. The morphological operators are implemented through reproduction step: (i) closing followed by (ii) opening. In this regard, the accuracy of dark patch segmentations are function of the size and the shape of the structuring element. Therefore, kerneal window size of a square of structuring of 7 x 7 is chosen to preserve the fine details of oil spill in ENVISAT ASAR data [29].

4   Results and Discussion

Fig.3 shows the ENVISAT ASAR was acquired on June 3rd 2010 after the Merchant ship collided with a Malaysian oil tanker near the Singapore and Malaysian coastal waters. Clearly, there are various of dark patches which are scattered over a large area of coastal waters. The lowest backscatter of -40 dB , -45dB, -50dB are noticed in areas A, B and C, respectively. The highest backscatter of -10 dB is represented ships in area D (Fig.4). In fact, oil spills change the roughness of the ocean surface to smoothness surface which appear as dark pixels as compared to the surrounding ocean [1-22]. Therefore, the speckle caused difficulties in dark patch identifications in SAR data [14][16].

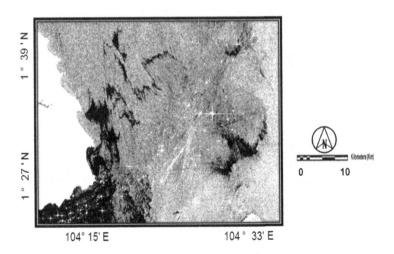

Fig. 3. ENVISAT ASAR data was acquired on June 3rd 2010

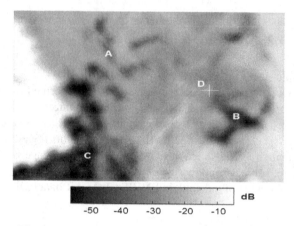

Fig. 4. Backscatter variations in ENVISAT ASAR data

However, the result of backscatter values is different as compared to previous studies of Marghany et al., [10] [11] and Marghany and Hashim [12]. This is because of previous studies used different radar sensor of the RADARSAT-1 SAR and these studies have done under different weather and ocean conditions compared to recent work. Fig.5 shows the crossover process with 10 individuals. In this 10 individuals, the positive dark patches are represented oil spill pixels while negative dark patches represent the surrounding pixels. Accordingly, every cellular is compared with the corresponding cell in the others to determine either to be positive or negative.

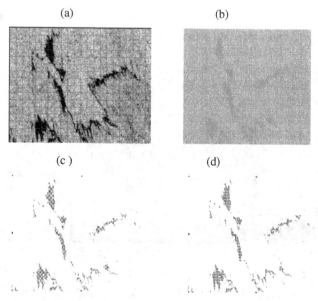

Fig. 5. Crossover procedures (a) original data, (b) first individual, (c) resulting from an individual prior cancellation, and (d) after cancellation

In these procedures, cell has a positive value and must be strengthened when cell in the intermediate prototype has a value larger than zero and greater than threshold 's value. In this regard, these cells are represented an oil spill event in ENVISAT ASAR data. On the contrary, the cell represents look-alikes when it has a negative value. As a result, the cell in the intermediate prototype is less than zero and threshold 's value. In this regard, this cell must be diminished. The variation cell value (positive or negative) is a function of dissimilarity of the comparable cells. This study confirms and extends the capabilities of GA introduced by Kahlouche et al., [27].

Clearly the genetic algorithm is able to isolate oil spill dark pixels from the surrounding environment. In other words, look-alike, low wind zone, sea surface roughness, and land are marked by white colour while oil spill pixels are marked all black (Fig.6). Further, Fig. 6 shows the results of the GA, where 100% of the oil spills in the test set were correctly classified. his study is not similar to previous work done by Marghany and Hashim [12]. The dissimilarity is because this work provides the automatic classifier based on GA but Marghany and Hashim work is considered as a semi-automatic tool for oil spill detection.

Fig. 6. Oil spill automatic detection by Genetic Algorithm (GA)

The receiver–operator characteristics (ROC) curve in Fig. 7 indicates a significant difference in the discriminated between oil spill, look-alikes and sea surface roughness pixels. In terms of ROC area, the oil spill has an area difference of 85% and 5% for look –alike and 10% for sea roughness and a ρ value less than 0.0005 which confirms the study of Marghany et al., [10] [11]. This suggests that genetic algorithm is an excellent classifier to discriminate region of oil slicks from surrounding water features.

It is found that ENVISAT ASAR with HV polarization allowed better discrimination of oil slick from surrounding sea features. In fact, ENVISAT ASAR has a steeper incident angle 20° with swath cover of 400 km and geometric resolution of 150 m. In general, genetic algorithm contains the crossover procedure. In this regard, a new population is generated in each crossover process. As a result, individual populations are examined by the fitness function and added to the population. Thus, new populations are continuously generated based on the dissimilarities between the two successive fitness values. In addition, the crossover procedure produced a more refined oil spill pattern by despeckle and maintenance the morphology of oil spill pattern features. This is because of the fitness function is used to sustenance the oil spill pixel classification. Indeed, fitness function selected oil spills morphological pattern which is close to the requested spill prototype.

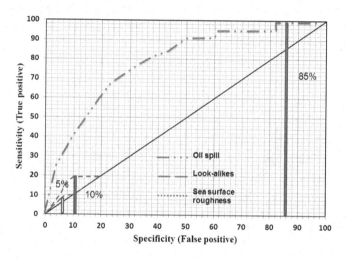

Fig. 7. ROC for Oil Spill Discrimination from Look-alikes and Sea Surface Roughness

In contrast to previous studies of Fiscella et al., [4] and Marghany and Mazlan [13], the Mahalanobis classifier provides a classification pattern of oil spill where the slight oil spill can distinguish from medium and heavy oil spill pixels. Nevertheless, this study is consistent with Topouzelis et al., [20-22]. In consequence, the genetic algorithm extracted oil spill pixels automatically from surrounding pixels without using different segmentation algorithm as stated at Solberg et al., [19]; Samad and Mansor [18];Marghany and Mazlan [12].

5   Conclusions

This study has demonstrated a design tool for oil spill detection in ENVISAT ASAR data using genetic algorithms. The genetic algorithm, is an essential function of the reproducing step which involves the crossover and mutation processes on the

ENVISAT ASAR selected backscatter populations. The study shows that the genetic algorithm provides accurate information about oil spill occurrences in ENVISAT. With ROC area, it could be inferred that oil spill, look-alikes and sea surface roughness were perfectly discriminated, as provided by the area difference of 85% for oil spill, 5% look–alikes and 10% for sea roughness. In conclusion, crossover process and the fitness function play as essential role in genetic algorithm to provide an excellent pattern classification of oil spills with despeckle reduction and morphological feature preservations. It can be said that a genetic algorithm can be suggested as an excellent algorithm for oil spill detection in such SAR data of ENVISAT ASAR.

References

1. Adam, J.A.: Specialties: Solar Wings, Oil Spill Avoidance, On-Line Patterns. IEEE Spect. 32, 87–95 (1995)
2. Aggoune, M.E., Atlas, L.E., Cohn, D.A., El-Sharkawi, M.A., Marks, R.J.: Artificial Neural Networks For Power System Static Security Assessment. IEEE Int. Sym. on Cir. and Syst. Portland, Oregon., 490–494 (1989)
3. Brekke, C., Solberg, A.: Oil Spill Detection by Satellite Remote Sensing. Rem. Sens. of Env. 95, 1–13 (2005)
4. Fiscella, B., Giancaspro, A., Nirchio, F., Pavese, P., Trivero, P.: Oil Spill Detection Using Marine SAR Images. Int. J. of Rem. Sens. 21, 3561–3566 (2000)
5. Frate, F.D., Petrocchi, A., Lichtenegger, J., Calabresi, G.: Neural Networks for Oil Spill Detection Using ERS-SAR Data. IEEE Tran. on Geos. and Rem. Sens. 38, 2282–2287 (2000)
6. Hect-Nielsen, R.: Theory of the Back Propagation Neural Network. In: Proc. of the Int. Joint Conf. on Neu. Net., vol. I, pp. 593–611. IEEE Press (1989)
7. Marghany, M., Hashim, M.: Comparative algorithms for oil spill detection from multi mode RADARSAT-1 SAR satellite data. In: Murgante, B., Gervasi, O., Iglesias, A., Taniar, D., Apduhan, B.O. (eds.) ICCSA 2011, Part II. LNCS, vol. 6783, pp. 318–329. Springer, Heidelberg (2011)
8. Marghany, M.: RADARSAT Automatic Algorithms for Detecting Coastal Oil Spill Pollution. Int. J. of App. Ear. Obs. and Geo. 3, 191–196 (2001)
9. Marghany, M.: RADARSAT for Oil spill Trajectory Model. Env. Mod. and Sof. 19, 473–483 (2004)
10. Marghany, M., Cracknell, A.P., Hashim, M.: Modification of Fractal Algorithm for Oil Spill Detection from RADARSAT-1 SAR Data. Int. J. of App. Ear. Obs. and Geo. 11, 96–102 (2009)
11. Marghany, M., Cracknell, A.P., Hashim, M.: Comparison between Radarsat-1 SAR Different Data Modes for Oil Spill Detection by a Fractal Box Counting Algorithm. Int. J. of Dig. Ear. 2, 237–256 (2009)
12. Marghany, M., Hashim, M., Cracknell, A.P.: Fractal Dimension Algorithm for Detecting Oil Spills Using RADARSAT-1 SAR. In: Gervasi, O., Gavrilova, M. (eds.) ICCSA 2007, Part I. LNCS, vol. 4705, pp. 1054–1062. Springer, Heidelberg (2007)
13. Marghany, M., Hashim, M.: Texture Entropy Algorithm for Automatic Detection of Oil Spill from RADARSAT-1 SAR data. Int. J. of the Phy. Sci. 5, 1475–1480 (2010)

14. Michael, N.: Artificial Intelligence: A guide to Intelligent Systems, 2nd edn. Addison Wesley, Harlow (2005)
15. Migliaccio, M., Gambardella, A., Tranfaglia, M.: SAR Polarimetry to Observe Oil Spills. IEEE Tran. on Geos. and Rem. Sen. 45, 506–511 (2007)
16. Mohamed, I.S., Salleh, A.M., Tze, L.C.: Detection of Oil Spills in Malaysian Waters from RADARSAT Synthetic Aperture Radar Data and Prediction of Oil Spill Movement. In: Proc. of 19th Asi. Conf. on Rem. Sen., Hong Kong, China, November 23-27, vol. 2, pp. 980–987. Asian Remote Sensing Society, Japan (1999)
17. Provost, F., Fawcett, T.: Robust classification for imprecise environments. Mach. Lear. 42, 203–231 (2001)
18. Samad, R., Mansor, S.B.: Detection of Oil Spill Pollution Using RADARSAT SAR Imagery. In: CD Proc. of 23rd Asi. Conf. on Rem. Sens. Birendra International Convention Centre in Kathmandu, Nepal, November 25-29. Asian Remote Sensing (2002)
19. Skrunes, S., Brekke, C., Eltoft, T.: An Experimental Study on Oil Spill Characterization by Multi-Polarization SAR. In: Proc. European Conference on Synthetic Aperture Radar, Nuremberg, Germany, pp. 139–142 (2012)
20. Topouzelis, K., Karathanassi, V., Pavlakis, P., Rokos, D.: Potentiality of Feed-Forward Neural Networks for Classifying Dark Formations to Oil Spills and Look-alikes. Geo. Int. 24, 179–219 (2009)
21. Topouzelis, K., Karathanassi, V., Pavlakis, P., Rokos, D.: Detection and Discrimination between Oil Spills and Look-alike Phenomena through Neural Networks. ISPRS J. Photo. Rem. Sens. 62, 264–270 (2007)
22. Topouzelis, K.N.: Oil Spill Detection by SAR Images: Dark Formation detection, Feature Extraction and Classification Algorithms. Sens. 8, 6642–6659 (2008)
23. Trivero, P., Fiscella, B., Pavese, P.: Sea Surface Slicks Measured by SAR. Nuo. Cim. 24C, 99–111 (2001)
24. Trivero, P., Fiscella, B., Gomez, F., Pavese, P.: SAR Detection and Characterization of Sea Surface Slicks. Int. J. Rem. Sen. 19, 543–548 (1998)
25. Velotto, D., Migliaccio, M., Nunziata, F., Lehner, S.: Dual-Polarized TerraSAR-X Data for Oil-Spill Observation. IEEE Trans. Geosci. Remote Sens. 49, 4751–4762 (2011)
26. Chaiyaratana, N., Zalzala, A.M.S.: Recent developments in evolutionary and genetic algorithms: theory and applications. In: Second International Conference On Genetic Algorithms in Engineering Systems: Innovations and Applications, GALESIA 1997, Glasgow, September 2-4, pp. 270–277 (1997)
27. Kahlouche, S., Achour, K., Benkhelif, M.: Proceedings of the 2002 WSEAS International Conferences, Cadiz, Spain, June 12-16, pp. 1–5 (2002), http://www.wseas.us/e-library/conferences/spain2002/papers/443-164.pdf
28. Gautam, G., Chaudhuri, B.B.: A distributed hierarchical genetic algorithm for efficient optimization and pattern matching. Pattern Recognition Journal 40, 212–228 (2007)
29. Sivanandam, S.N., Deepa, S.N.: Introduction to Genetic Algorithms. Springer, Heidelberg (2008)

Three Dimensional Coastline Deformation
from Insar Envisat Satellite Data

Maged Marghany

Institute of Geospatial Science and Technology (INSteG)
Universiti Teknologi Malaysia
81310 UTM, Skudai, Johor Bahru, Malaysia
maged@utm.my,
magedupm@hotmail.com

Abstract. The paper is focused on three-dimensional (3-D) coastline deformation from interferometry synthetic aperture radar (InSAR). In doing so, conventional InSAR procedures are implemented to three repeat passes of ENVISAT ASAR data. Further, three-dimensional sorting reliabilities algorithm (3D-SRA) is implemented with phase unwrapping technique. Consequently, the 3D-SRA is used to eliminate the phase decorrelation impact from the interferograms. The study shows the performance of InSAR method using the 3D-SRA is better than InSAR procedure which is validated by a lower range of error (0.06±0.32 m) with 90% confidence intervals. In conclusion, integration of the 3D-SRA with phase unwrapping produce accurate 3-D coastline deformation.

Keywords: InSAR, fringe, interferogram, three-dimensional sorting reliabilities algorithm (3D-SRA), Digital Elevation Model (DEM), coastline deformation, ENVISAT ASAR.

1   Introduction

In the last two decades, scientists have developed a powerful technique to measure the millimeter-scale of the Earth 's surface deformation by comparing complex synthetic aperture radar (SAR) data that were acquired a few days or a few years apart. This technique is known as interferometric synthetic aperture radar (InSAR). Accurate Earth 's surface deformation or digital elevation maps can be produced by implementing the single look complex synthetic aperture radar (SAR) images that are received by two or more separate antennas. The phase image is produced by multiplying the complex SAR image by the coregistered complex conjugate pixels of the other SAR data. Incidentally, the phase difference of the two SAR data is processed to acquire height and or deformation of the Earth's surface. Therefore, scientists have agreed that the accurate results of InSAR are required standard criteria for data acquisitions. According to Hanssen [3], short temporal baseline, appropriate spatial baseline, good weather conditions and ascending and descending SAR data are regular criteria to reduce decorrelation and noise and produce a reliable DEM.

B. Murgante et al. (Eds.): ICCSA 2013, Part II, LNCS 7972, pp. 599–610, 2013.
© Springer-Verlag Berlin Heidelberg 2013

Nonetheless, alternative SAR datasets must obtain at high latitudes or in zones of rundown coverage [6] [13]. The baseline decorrelation and temporal decorrelation, nevertheless, make InSAR measurements unrealistic [8] [9] [10] [11]. Incidentally, Gens [12] stated the length of the baseline designates the sensitivity to height changes and sum of baseline decorrelation. Additional, Gens [12] reported the time difference for two data acquisitions is a second source of decoration. Indeed, the time differences while compare data sets with a similar baseline length acquired one and 35 days apart suggests only the temporal component of the decorrelation. Therefore, the loss of coherence in the same repeat cycle in data acquisition is most likely because of baseline decorrelation. According to Roa et al. [7], uncertainties could arise in DEM because of limitation InSAR repeat passes. In addition, the interaction of the radar signal with troposphere can also induce decorrelation. This is explained in several studies [3] [8] [15].

Commonly, the propagation of the waves through the atmosphere can be a source of error exist in most interferogram productions. When the SAR signal propagated through a vacuum it should theoretically be subjected to some decent accuracy of timing and cause phase delay [3] [21]. A constant phase difference between the two images caused by the horizontally homogeneous atmosphere was over the length scale of an interferogram and vertically over that of the topography. The atmosphere, however, is laterally heterogeneous on length scales both larger and smaller than typical deformation signals [9] [23]. In other cases the atmospheric phase delay, however, is caused by vertical inhomogeneity at low altitudes and this may result in fringes appearing to correspond with the topography [24]. Under this circumstance, this spurious signal can appear entirely isolated from the surface features of the image, since the phase difference is measured other points in the interferogram, would not contribute to the signal [3][17]. This can reduce seriously the low signal-to-noise ratio (SNR) which restricted to perform phase unwrapping. Accordingly, the phases of weak signals are not reliable. According to Yang et al., [11], the correlation map can be used to measure the intensity of the noise in some sense. It may be overrated because of an inadequate number of samples allied with a small window [9]. Weights are initiated to the correlation coefficients according to the amplitudes of the complex signals to estimate accurate reliability [11][18].

In this paper, we address the question of utilization three-dimensional phase unwrapping algorithm to estimate rate changes of shoreline deformation. In fact, there are several factors could impact the accuracy of DEMs was derived from phase unwrapping [21]. These factors are involved radar shadow, layover, multi-path effects and image misregistration, and finally the signal-to –noise ratio (SNR) [11]. This demonstrated with ENVISAT ASAR data. The main contribution of this study is to implement three three-dimensional phase unwrapping algorithm with InSAR technique. Three hypotheses examined are: (i) three three-dimensional phase unwrapping algorithm can be used as filtering technique to reduce noise in phase unwrapping; (ii) 3-D shoreline reconstruction can be produced using satisfactory phase unwrapping by involving the three three-dimensional phase unwrapping algorithm; and (iii) high accuracy of deformation rate can be estimated by using the new technique.

2   InSAR Data Processing

Two methods are involved to perform InSAR from ENVISAT SAR SAR data (i) conventional InSAR procedures; and (ii) three-dimensional phase unwrapping algorithm i.e. three-dimensional sorting reliabilities algorithm (3D-SRA) [25]

2.1   Conventional InSAR Method

According, Zebker et al., [4], two complex SAR data are required to achieve InSAR procedures. These complex SAR data have a real (cosine) and an imaginary (sine) components. Then real and an imaginary components are combined as vectors to acquire information of phase and intensity of the SAR signal [10]. In Marghany [26], the surface displacement can estimate using the acquisition times of two SAR data S_1 and S_2. The component of surface displacement thus, in the radar-look direction (Fig.1), contributes to the further interferometric phase (φ) as [9]

$$\phi = \frac{4\pi(\Delta R)}{\lambda} = \frac{4\pi(B_h \sin\theta - B_v \cos\theta)}{\lambda} \qquad (1)$$

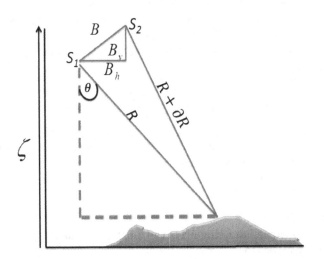

Fig. 1. InSAR Geometry

where ΔR is the slant range difference from satellite to target respectively at different time, θ is the look angle (19.2-26.7°), (Table 1) and λ is the ENVISAT ASAR wavelength Single Look Complex (SLC) which is about 5.6 cm for C- band. Therefore, B_h, B_v are horizontal and vertical baseline components[19].

Table 1. ENVISAT ASAR characteristics were used in this study [19]

Parameters	Values
Radar Wavelength (cm)	5.6
Orbit Repeat Time, days	14 11/35
Pulse Repetition (PRF), Hz	1316.000000
Ground Resolution	30
Swath Width (km)	105
Incident Angle (°)	19.2-26.7
Polarization	HH/VV

According to Lee [9], for the surface displacement measurement, the zero-baseline InSAR configuration is the ideal as $\Delta R = 0$, so that

$$\phi = \phi_d = \frac{4\pi}{\lambda}\zeta \tag{2}$$

In actual fact, zero-baseline, repeat-pass InSAR configuration is hardly achievable for either spaceborne or airborne SAR. Therefore, a method to remove the topographic phase as well as the system geometric phase in a non-zero baseline interferogram is needed. If the interferometric phase from the InSAR geometry and topography can strip off from the interferogram, the remnant phase would be the phase from block surface movement, providing the surface maintains high coherence [5]. Then, the phase difference $\Delta\phi$ between the two ENVISAT ASAR data positions and the pixel of target of terrain point is given by

$$\Delta\zeta = \frac{\lambda R \sin\theta}{4\pi B}\Delta\phi \tag{3}$$

Equation 3 is a function of normal base line B, and the range R. In addition, equation 3 can provide information about the heights and phase differences estimations. In fact, the estimated height of each pixel of ENVISAT ASAR data is an important task to generate a raster form of the DEM.

2.2 DEM Reconstruction Using a Three-Dimensional Sorting Reliabilities Algorithm (3D-SRA)

Hussein et al., [25] have proposed a new algorithm for three-dimensional phase unwrapping the algorithm that is called a three-dimensional sorting reliability algorithm (3D-SRA). The quality of each edge of phase unwrapping is a function of the connection of two voxels in 3-D Cartesian axis e.g. ,x ,y, z. Starting, to carry out the unwrapping path from high quality voxels to bad quality voxels [25]. In addition, following a discreet path, the 3D-SRA algorithm unwraps the phase volume which is significant to determine the 3-D volume change rate of shoreline. In this regard, the voxels connects the highest reliable edges that are unwrapped first with border surfaces. Consistent with Hussien et al., [26], the reliability value of an edge that connects a border voxel with another voxel in the phase volume is set to zero.

Let E_x, E_y, and N are the horizontal, vertical, and normal second differences, respectively which are given by

$$E_x(i,j,k) = \gamma[\phi(i-1,j,k) - \phi(i,j,k)] - \gamma[\phi(i,j,k)$$
$$-\phi(i+1,j,k)], \tag{4}$$

$$E_y(i,j,k) = \gamma[\phi(i,j-1,k) - \phi(i,j,k)] - \gamma[\phi(i,j,k)$$
$$-\phi(i,j+1,k)], \tag{5}$$

$$N(i,j,k) = \gamma[\phi(i,j,k-1) - \phi(i,j,k)] - \gamma[\phi(i,j,k)$$
$$-\phi(i,J,k+1)], \tag{6}$$

where i,j,k are the neighbors' indices of the voxel in 3 x 3 x 3 cube, and γ defines a wrapping operator that wraps all values of its argument in the range $[-\pi,\pi]$. This can be done by adding or subtracting an integer number of 2π rad to its argument [26]. γ can calculate from the wrapped-phase gradients in the x, y, and z directions as follows [25,26]

$$\gamma(i,j,k) = \frac{\partial \phi_{i,j,k}{}^x}{[\phi_{i+1,j,k} - \phi_{i,j,k}]}, \tag{7}$$

$$\gamma(i,j,k) = \frac{\partial \phi_{i,j,k}{}^y}{[\phi_{i,j+1,k} - \phi_{i,j,k}]}, \tag{8}$$

$$\gamma(i,j,k) = \frac{\partial \phi_{i,j,k}{}^z}{[\phi_{i,j,k+1} - \phi_{i,j,k}]}, \tag{9}$$

Equations 7 to 9 are represented 3-D array of the wrapped-phase gradients $\partial\phi^x, \partial\phi^y, \partial\phi^z$ and each has the same dimensions as the wrapped-phase volume. In addition, the maximum phase gradient measures the magnitude of the largest phase gradient that is, partial derivative or wrapped the phase difference in a $v*v*v$ volumes [25]. Using the sum of equations 4 to 6, the second difference quality map Q can be obtained [26]

$$Q_{i,j,k} = \sqrt{E_x^2(i,j,k) + E_y^2(i,j,k) + N^2(i,j,k}} \tag{10}$$

The unwrapping path is performed based on equation 10 where entirely the edges are stored in a 3D array and sorted with their edge quality values [26]. Further, unwrapping a voxel or a group of voxels concerning another group may require the addition or subtraction of multiples of 2π [25]. In addition, the badness of each voxel in a $v*v*v$ volume is determined by

$$b = \max \begin{Bmatrix} \max\left\{\left|\partial\phi_{i,j,k}{}^x\right|\right\} \\ \max\left\{\left|\partial\phi_{i,j,k}{}^y\right|\right\} \\ \max\left\{\left|\partial\phi_{i,j,k}{}^z\right|\right\} \end{Bmatrix} \tag{11}$$

b is the badness of voxel (m,n,l), which is given by

$$b = v^{-3} * [\sqrt{\sum (\partial\phi_{i,j,k}{}^x + \overline{\partial\phi_{i,j,k}{}^x})^2}$$
$$+ \sqrt{\sum (\partial\phi_{i,j,k}{}^y + \overline{\partial\phi_{i,j,k}{}^y})^2} \tag{11.1}$$
$$+ \sqrt{\sum (\partial\phi_{i,j,k}{}^z + \overline{\partial\phi_{i,j,k}{}^z})^2}]$$

where $\overline{\partial\phi_{i,j,k}}{}^{x,y,z}$ are the mean value of wrapped phase gradients in x,y,z directions, respectively through a $v*v*v$ volume that are centered at at that voxel [25].

2.3   Ground Survey

At the rear of Marghany [14], the GPS survey used to: (i) to record the exact geographical position of the shoreline; (ii) to determine the cross-sections of shoreline slopes; (iii) to corroborate the reliability of InSAR data co-registered; and finally, (iv) to create a reference network for future surveys. The geometric location of the GPS survey was obtained by using the new satellite geodetic network, IGM95. After a careful analysis of the places and to identify the reference vertexes, we thickened the network around such vertexes to perform the measurements for the cross sections (transact perpendicular to the coastline). The GPS data collected within 50 sample points scattered along 10000 m coastline. The interval distance of 2000 m between each sample location is considered. In every sample location, Rec-Alta (Recording Electronic Tachometer) was used to acquire the coastline elevation profile. The ground truth data were acquired on January 25   2011 during satellite passes.

3   Results and Discussion

In the present study, InSAR methods are implemented on ENVISAT ASAR data sets of 5 March 2003 (SLC-1), 28 April 2003 (SLC-2) and January 25, 2011, (SLC-3) of Wide Swath Mode (WSM) (Fig.2). They are acquired from ascending (Track: 226, orbit: 5290), descending (Track:490, Orbit: 6055) and descending (Track 420,Orbit 4655), respectively.

(a)   **(b)**   **(c)**

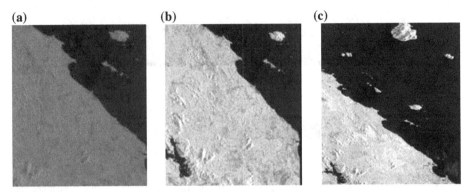

Fig. 2. ENVISAT ASAR data used in this study (a) 5 March 2003; (b) 28 April 2003; and (c) January 25 2011

These data are C-band and had the lower signal-to-noise ratio owing to their VV polarization with a wavelength range of 3.7 to 7.5 cm and a frequency of 5.331 GHz. ASAR can achieve a spatial resolution generally around 30 m. The ASAR is intended for applications which require the spatial resolution of spatial resolution of 150 m. This means that it is not effective at imaging areas in depth, unlike stripmap SAR. The azimuth resolution is 4 m, and range resolution ranges between 8 m.

Fig3a and F.g.3b present the reference DEM which has been generated from topographical 1:50000, and in situ measurements, respectively while Fig.3c shows the ENVISAT coherence data. It is clear that the maximum elevation of 50 m found in land. The maximum elevation of 10 m is shown along the coastline. The high coherence rates are existed in urban zone and along infrastructures with 0.9 while low coherence of 0.2 is found in a vegetation zone along the coastline. Since three ASAR data acquired in the wet northeast monsoon period, there has been an impact of wet sand on a radar signal penetration which causing weak penetration of radar signal because of dielectric. Indeed, the total topographic decorrelation effects along the radar-facing slopes are dominant and highlighted as lowest choherence value of 0.2. According to Marghany [27] the micro-scale movement of the sand particles driven by the coastal hydrodynamic, and wind speed of 12 m/s during northeast monsoon period [30] could change the distribution of scatters resulting in rapid temporal decorrelation which has contributed to lowest coherence along coastline. This result agreed with Marghany [27][31].

Cleary, there is huge differences between InSAR DEM and DEM generated from in situ measurements with 9 m differences while in situ measurements are concurred

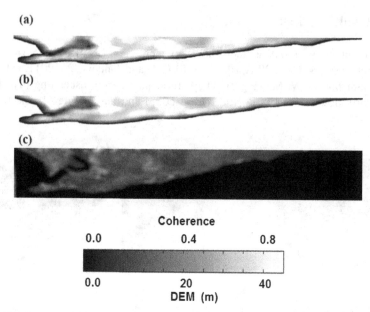

Fig. 3. DEM generated from (a) topographic map; (b) in situ measurements; and (c) coherence
ENVISAT ASAR

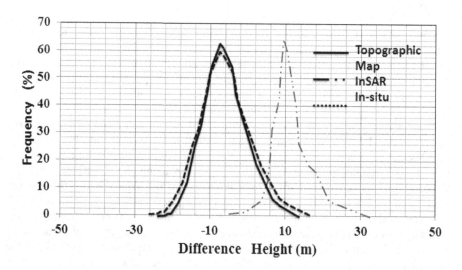

Fig. 4. Height differences between DEM generated from InSAR, topographic map and in situ
measurements

with DEM produced from topographic map within 1.3 m differences (Fig.4). This is
because the impact of   decorrelation. This result confirms the work done by
Marghany [27] [29]. The overall scene is highly incoherent (Fig.3) which extremely
effected the accuracy of InSAR DEM. This decorelation caused poor detection of

InSAR DEM which induce large ambiguities because of poor coherence and scattering phenomenology. Indeed, the signatures of the interferometric coherence and phase of vegetation are extremely impacted by temporal decorrelation in ASAR c-band data.

According to Marghany [27], the ground ambiguity and ideal assumption that volume-only coherence can be acquired in at least one polarization. This assumption may fail when vegetation is thick, dense, or the penetration of an electromagnetic wave is weak. This is agreed with the studies of Lee [9]; Marghany [14]; and Marghany [28][29]. This can be seen clearly in InSAR interfeogram has been made from ASAR data (Fig.5).

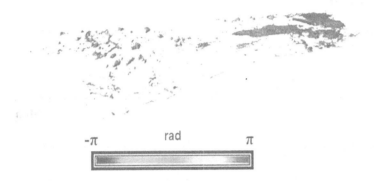

-π   rad   π

Fig. 5. Interferogram generated from InSAR data

Fig. 6 shows the interferogram created using three-dimensional sorting reliabilities algorithm 3-DSR. The full color cycle represents a phase cycle, covering range between –π to π. In this context, the phase difference given module 2 π; is color encoded in the fringes. Seemingly, the color bands change in the reverse order, indicating that the center has a critical coastline erosion of -3.5 m/year. This shift corresponds to 2 of coastal deformation over the distance of 10000 m.

-π   **rad**   π

-0.002   **m/year**   **4.0**

Fig. 6. Fringe interferometry generated by a three-dimensional sorting reliabilities algorithm

Table 2 shows the statistical comparison between the simulated DEM from the InSAR, real ground measurements and with using three-dimensional sorting reliabilities algorithm. This table represents the bias (averages mean the standard error, 90 and 95% confidence intervals, respectively. Evidently, the InSAR using three-dimensional sorting reliabilities algorithm has bias of -0.08 m, lower than ground measurements and the InSAR method. Therefore, three-dimensional sorting reliabilities algorithm has a standard error of mean of ± 0.05 m, lower than ground measurements and the InSAR method. Overall performances of InSAR method using three-dimensional sorting reliabilities algorithm is better than conventional InSAR technique which is validated by a lower range of error (0.06±0.32 m) with 90% confidence intervals.

Table 2. Statistical Comparison between InSAR and InSAR- three-dimensional sorting reliabilities algorithm

Statistical parameters	InSAR techniques			
	InSAR		Three-dimensional sorting reliabilities algorithm	
Bias	3.3		-0.08	
Standard error of the mean	2.2		0.05	
90% (90% confidence interval	**Lower**	**Upper**	**Lower**	**Upper**
	2.3	3.8	0.06	0.24
	1.3	3.2	0.04	0.32

This study confirms the work have been done by Abdul-Rahman et al., [25] [26]. The three-dimensional sorting reliabilities algorithm provides an excellent 3-D phase unwrapping which leads to high quality of 3-D coastline reconstruction. This could be contributed to quality map. Indeed, the 3-DSR algorithm is guided by including the maximum gradient quality maps. Therefore, quality maps guide the unwrapping path through a noisy region so that the interferogram patterns are in completing cycle as compared to InSAR interferogram. Moreover, as stated by Hussien et al., [26], changing the cube size has a great effect to reduce the effect of noise and improve the calculated quality. Additionally, the 3-DSR algorithm follows discrete unwrapping paths to ensure the processing of the highest quality regions even if they are separated from each other. In other words, within the 3-DSR algorithm, the edges are stored in an array which is based on the terms of the terms of their edge quality values. This means it relies on edge quality to guide the unwrapping path that produces accurate 3-D coastline reconstruction as compared to InSAR techniques. Generally, the 3-DSR algorithm can be excellent solution for decorrelation problems in such tropical area as Malaysia.

4   Conclusions

The paper has demonstrated InSAR phase unwrapping using the three-dimensional sorting reliabilities algorithm (3-DSR). The main purpose of this technique is to solve the decoration problem that is a critical issue with InSAR techniques. In addition, three-dimensional (3-D) coastline deformation from interferometry synthetic aperture radar (InSAR) is estimated. The study shows that InSAR produces discontinues interferogram pattern because of the high decorelation. On the contrary, the three-dimensional sorting reliabilities algorithm generated 3-D coastline deformation with bias of -0.08 m, lower than ground measurements and the InSAR method. Therefore, the three-dimensional sorting reliabilities algorithm has a standard error of mean of ± 0.05 m, lower than ground measurements and the InSAR method. In conclusion, the 3-DSR algorithm can be used to solve the problem of decorrelation and produced accurate 3-D coastline deformation using ENVISAT ASAR data.

References

1. Massonnet, D., Feigl, K.L.: Radar interferometry and its application to changes in the earth's surface. Rev. Geopyh. 36, 441–500 (1998)
2. Burgmann, R., Rosen, P.A., Fielding, E.J.: Synthetic aperture radar interferometry to measure Earth's surface topography and its deformation. Ann. Rev. of Earth and Plan. Sci. 28, 169–209 (2000)
3. Hanssen, R.F.: Radar Interferometry: Data Interpretation and Error Analysis. Kluwer Academic, Dordrecht (2001)
4. Zebker, H.A., Rosen, P.A., Hensley, S.: Atmospheric effects in inteferometric synthetic aperture radar surface deformation and topographic maps. J. Geophys. Res. 102, 7547–7563 (1997)
5. Askne, J., Santoro, M., Smith, G., Fransson, J.E.S.: Multitemporal repeat-pass SAR interferometry of boreal forests. IEEE Trans. Geosci. Remote Sens. 41, 1540–1550 (2003)
6. Nizalapur, V., Madugundu, R., Shekhar Jha, C.: Coherence-based land cover classification in forested areas of Chattisgarh, Central India, using environmental satellite-advanced synthetic aperture radar data. J. Appl. Rem. Sens. 5, 059501-1–059501-6 (2011)
7. Rao, K.S., Al Jassar, H.K., Phalke, S., Rao, Y.S., Muller, J.P., Li, Z.: A study on the applicability of repeat pass SAR interferometry for generating DEMs over several Indian test sites. Int. J. Remote Sens. 27, 595–616 (2006)
8. Rao, K.S., Al Jassar, H.K.: Error analysis in the digital elevation model of Kuwait desert derived from repeat pass synthetic aperture radar interferometry. J. Appl. Remote Sens. 4, 1–24 (2010)
9. Lee, H.: Interferometric Synthetic Aperture Radar Coherence Imagery for Land Surface Change Detection. Ph.D theses, University of London (2001)
10. Luo, X., Huang, F., Liu, G.: Extraction co-seismic Deformation of Bam earthquake with Differential SAR Interferometry. J. New Zea. Inst. of Surv. 296, 20–23 (2006)
11. Yang, J., Xiong, T., Peng, Y.: A fuzzy Approach to Filtering Interferometric SAR Data. Int. J. of Remote Sens. 28, 1375–1382 (2007)
12. Gens, R.: The influence of input parameters on SAR interferometric processing and its implication on the calibration of SAR interferometric data. Int. J. Remote Sens. 2, 11767–11771 (2000)

13. Anile, A.M., Falcidieno, B., Gallo, G., Spagnuolo, M., Spinello, S.: Modeling uncertain data with fuzzy B-splines. Fuzzy Sets and Syst. 113, 397–410 (2000)

14. Marghany, M.: Simulation of 3-D Coastal Spit Geomorphology Using Differential Synthetic Aperture Interferometry (DInSAR). In: Padron, I. (ed.) Recent Interferometry Applications in Topography and Astronomy, pp. 83–94. InTech - Open Access Publisher, Croatia (2012)

15. Spangnolini, U.: 2-D phase unwrapping and instantaneous frequency estimation. IEEE Trans. Geo. Remot. Sensing 33, 579–589 (1995)

16. Davidson, G.W., Bamler, R.: Multiresolution phase unwrapping for SAR interferometry. EEE Trans. Geosci. Remote Sensing 37, 163–174 (1999)

17. Marghany, M., Sabu, Z., Hashim, M.: Mapping coastal geomorphology changes using synthetic aperture radar data. Int. J. Phys. Sci. 5, 1890–1896 (2010)

18. Marghany, M.: Three-dimensional visualisation of coastal geomorphology using fuzzy B-spline of dinsar technique. Int. J. of the Phys. Sci. 6, 6967–6971 (2011)

19. ENVISAT, ENVISAT application, http://www.esa.int (accessed Febraury 2, 2013)

20. Zebker, H.A., Werner, C.L., Rosen, P.A., Hensley, S.: Accuracy of Topographic Maps Derived from ERS-1 Interferometric Radar. IEEE Geosci. Remote Sens. 2, 823–836 (1994)

21. Baselice, F., Ferraioli, G., Pascazio, V.: DEM reconstruction in layover areas from SAR and Auxiliary Input Data. IEEE Geosci. Rem. Sensing Letters 6, 253–257 (2009)

22. Ferraiuolo, G., Pascazio, V., Schirinzi, G.: Maximum a Posteriori Estimation of Height Profiles in InSAR Imaging. IEEE Geosci. Rem. Sensing Letters, 66–70 (2004)

23. Ferraiuolo, G., Meglio, F., Pascazio, V., Schirinzi, G.: DEM reconstruction accuracy in multichannel SAR interferometry. IEEE Trans. on Geosci. and Rem. 47, 191–201 (2009)

24. Ferretti, A., Prati, C., Rocca, F.: Multibaseline phase unwrapping for InSAR topography estimation. Il Nuov Cimento 24, 159–176 (2001)

25. Hussein, S.A., Gdeist, M., Burton, D., Lalor, M.: Fast three-dimensional phase unwrapping algorithm based on sorting by reliability following a non-continuous path. In: Osten, G., Novak (eds.) Optical Measurements System for Industrial Inspection IV Conference. Proc. SPIE, vol. 5856-Part 1, pp. 32–40 (2005)

26. Hussein, S.A., Gdeisat, M.A., Burton, D.R., Lalor, M.J., Lilley, F., Moore, C.: Fast and roubust three-diemsional best bath phase unwrapping algorithm. Appl. Opt. 46, 6623–6635 (2007)

27. Marghany, M.: DInSAR technique for three-dimensional coastal spit simulation from radarsat-1 fine mode data. Acta Geophy. 61, 478–493 (2013)

28. Marghany, M.: Three dimensional Coastal geomorphology deformation modelling using differential synthetic aperture interferometry. Verlag Der Zeitschrift fur Naturforschung. 67a, 419–420 (2012)

29. Marghany, M.: DEM reconstruction of coastal geomorphology from DINSAR. In: Murgante, B., Gervasi, O., Misra, S., Nedjah, N., Rocha, A.M.A.C., Taniar, D., Apduhan, B.O. (eds.) ICCSA 2012, Part III. LNCS, vol. 7335, pp. 435–446. Springer, Heidelberg (2012)

30. Marghany, M.: Intermonsoon water mass characteristics along coastal waters off Kuala Terengganu, Malaysia. Int. J. of Phys. Sci. 7, 1294–1299 (2012)

31. Marghany, M.: Modelling shoreline rate of changes using holographic interferometry. Int. J. of Phys. Sci. 6, 7694–7698 (2011)

Mangrove Changes Analysis by Remote Sensing and Evaluation of Ecosystem Service Value in Sungai Merbok's Mangrove Forest Reserve, Peninsular Malaysia

Zailani Khuzaimah[1], Mohd Hasmadi Ismail[2], and Shattri Mansor[1]

[1] GISRC, Faculty of Engineeering, Universiti Putra Malaysia, 43400 UPM, Serdang Selangor, Malaysia
[2] Faculty of Forestry, Universiti Putra Malaysia, 43400 UPM, Serdang Selangor, Malaysia
{Zailani,mhasmadi}@putra.upm.edu.my,
shattri@eng.upm.edu.my

Abstract. Mangrove forests are an important ecosystem which provides socioeconomic value to humankind. Despite their great value, mangroves have one of the highest rates of degradation of any global habitat, which is about 1% of the existing area per year. In fact, the socioeconomic value and ecosystem services of mangroves as a natural product are underestimated. The ecosystem services provided by mangroves are often ignored by the ongoing process of mangrove conversion. This is a major reason why conservation of this ecosystem is not a popular alternative. Thus, the main objective of this study is to evaluate the changes in mangrove forests and valuation of their ecosystem services. SPOT 5 imageries of years 2000 and 2010 have been used for change detection analysis. The vegetation index such as NDVI and AVI and unsupervised classification technique were employed in image processing. In order to obtain the value of socioeconomic impact from the mangrove changes and biodiversity disturbances, the ecosystem service valuation (ESV) model was applied. Results show that the total value of the existing mangrove forest ecosystem service was RM1,901,859.84. The value per unit area is about RM 1,650.92 /ha. The total values of others were RM161, 33.2 (crop land) and RM3,107,500 (water bodies), respectively. It is evident that Sungai Merbok's Mangrove Forest Reserve is very important for coastal ecology, where the orientation of mangrove ecosystem is huge and serves to provide essential services for the community. It also plays a crucial role in providing ecological balance to the coastal environment.

Keywords: Mangrove, ecosystem service valuation, changes detection, remote sensing.

1 Introduction

Ecosystems around the globe create and preserve an environment suitable for human survival. Today, ecosystem services are increasingly faced with threats globally. This trend is partially due to a lack of valuation, because resources are

B. Murgante et al. (Eds.): ICCSA 2013, Part II, LNCS 7972, pp. 611–622, 2013.
© Springer-Verlag Berlin Heidelberg 2013

not valued in the market and are hence ignored in management decisions. The world's tropical mangrove forests are disappearing at an alarming rate with the rapid growth in human population and conversion to other land uses. Mangrove ecosystems represent natural resources which are capable of producing a wide range of goods and services for coastal environments and communities. Ecosystem services provided by mangroves comprise all goods and services and modified ecosystems that benefit human well-being as well as support sustainable resource management. These include food production, provision of building materials and medicines, regulation of microclimate, disease prevention, generation of productive soils and clean water resources as well as landscape opportunities for recreational and spiritual benefits (Daily, 1997; Costanza and Folke, 1997; MEA, 2005; Boyd and Banzhaf, 2007 and Wallace, 2007). The usefulness of mangrove forests stems from the diversity of the forests as well as diversity of goods and services they provide.

Despite this need for understanding the components of mangrove ecosystem value, it is prohibitively expensive and unrealistic to conduct a detailed empirical non-market valuation. The need for ecosystem valuation information is especially great for public good services of ecosystems that are not popular in the market (Barbier et al., 1997; Carson, 1998). In particular, it is crucial that the value of mangrove ecosystems in developing countries be assessed. According to Katherine et al. (1998), understanding the importance and best use of different parts of a forest may help in formulating management policies that enable continuous supply of essential goods and services. The Economics of Ecosystems and Biodiversity (Sukhdev, 2008), increasingly recognise the critical role of ecosystem service valuation for sustainable development.

The key to sustainable resource development and management is regular access to updated information. Today, the interdisciplinary research approach has received much attention. This is attributed to complex problems in applied science that almost always span across various disciplines. The launch of remote sensing satellite with high resolution imagery provides resource and environmental planners with data and information to meet the sustainable development goal. Recognising the continuous degradation of mangrove (or wetland), goods and ecosystem services must be given a quantitative value if their conservation is to be appreciated. Therefore, geographical information system and remote sensing offer opportunities to facilitate continuous monitoring and assessment of ecosystem changes as well as effects taking place in mangrove areas.

Remote sensing data and GIS are used as tools for the following purposes: to assess biodiversity and land cover change as well as classification index at the mangrove landscape; to identify the relationship between the degree of disturbance and the nature of fragmentation processes in the study area; and to develop methodology that allows integration of land cover change processes and environmental changes into decision-making. All these are carried out together with strategies in the context of conservation biology and sustainable forest management at the landscape/within community level. Therefore, the objective of this study is to evaluate impact of mangrove changes on the socioeconomic value. This is achieved by conducting ecosystem service evaluation in Sungai Merbok's Mangrove Forest Reserve, Peninsular Malaysia.

2   Materials and Methods

2.1   Study Area

This study focuses on the Sungai Merbok Forest Reserve (SMFR) in the district of Sungai Petani, Kedah, Peninsular Malaysia (Figure 1). The location of the study area is within the following geographic coordinates: latitude of 50 25' N to 50 39' N and Longitude of 1000 19' E to 1000 32' E. The mangrove area covers an area of 4,037 hectares with 18 compartments. SMFR is the most extensive and most well-managed mangrove in the state of Kedah; it has been gazetted as a permanent forest reserve since 1951. The average daily temperature varies from 220C to 320C, humidity ranges from 80–90%, and annual rainfall ranges between 200 cm and 250 cm. Mangrove species that dominate the area are Rhizophora apiculata, Rhizophora mucronata, Brguiera parvifolia, Avecennia spp., Brugueria spp and Sonneratia spp. (Ong et al., 1991).

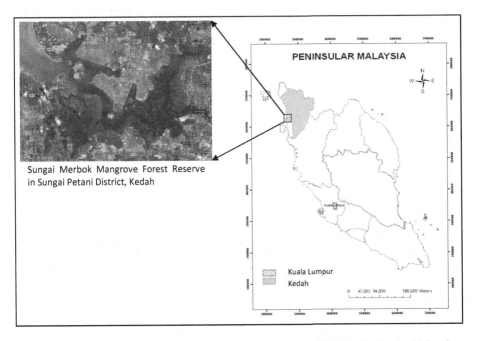

Fig. 1. The location of Sungai Merbok Forest Reserve, State of Kedah, Peninsular Malaysia

2.2   Methodology

Data Collection and Ecosystem Classification

The land use data of the study area were obtained from Malaysian Remote Sensing Agency (MARS) in Kuala Lumpur. The data were extracted from high resolution 2.5 m SPOT 5 satellite images and SPOT 5 multispectral (10 m) resolutions obtained in 2000 and 2010. Land use map of the area was obtained from Department of Agriculture Malaysia and forest map was from Kedah State's Forestry Department. Based on the characteristics of prevailing land cover and actual condition of SMFR, a total of four generic ecosystems were identified, namely Forest, Crop Land, Water Bodies and Barren Land.

Image Processing

SPOT 5 imageries were geo-rectified to 1:50,000 scale topographical map. After geo-correction, the next step is to extract information about the greenness of the study area. Multitemporal SPOT 5 imageries from the years 2000 and 2010 were analysed using vegetation index such as Normalised Difference Vegetation Index (NDVI) =B3-B2/B3+B2) and Advanced Vegetation Index (AVI) = [(B1+1) x (256-B2) x B5)] ^ (1/3) &B5=B1-B2)]). As a result, AVI was applied to the study area, because AVI offers more detailed information about all vegetation elements. AVI has proven to be more sensitive to forest density and hysiognomic vegetation classes in this area (Mohd Hasmadi et al., (2011). The NDVI and AVI results are shown in Figure 2 and Figure 3. The unsupervised classification technique (Figure 4) was applied to AVI image. Unsupervised classification is based on K-mean algorithms (four classes were developed based on ecosystem classification).

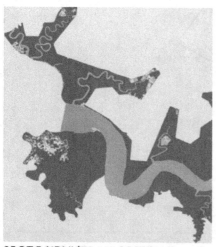 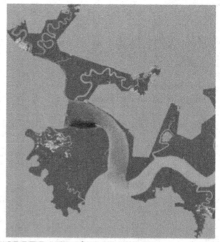

SPOT 5 NDVI image_2000   SPOT 5 NDVI image _2010

Fig. 2. NDVI maps index of SPOT 5 for year 2000(left) and year 2010 (right)

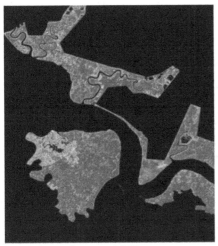 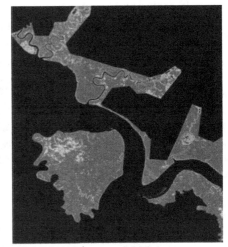

SPOT 5 AVI image_2000   SPOT 5 AVI image_2010

Fig. 3. AVI maps index of SPOT 5 for year 2000(left) and year 2010 (right)

 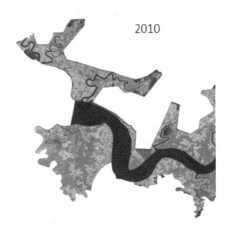

Fig. 4. Image maps of ecosystem classification from unsupervised classification

Mangrove Ecosystem Changes Estimation

In order to understand how a mangrove ecosystem changes, information is needed on what changes occur, where and when they occur as well as the rates at which they occur. Despite ongoing research efforts on ecosystem patterns, there is much room for research on development of basic land-cover datasets providing quantitative, spatial land-cover information (Xavier and Szejwach, 1998). The method applied to estimate the mangrove ecosystem changes was by conducting plot sampling on satellite image maps to detect changes in forest cover and land use during 2000 and 2010 for the study areas; field verification was also carried out to identify causes of the changes.

Then, GIS analysis was used to analyse the land use change and the transition matrix between land use types for the above-mentioned period of 10 years.

Assignment Model of Mangrove Ecosystem Service

In order to obtain ecosystem service values for each of the four land-cover categories, each category was compared with the 17 biomes identified in ecosystem service valuation model. The total value of ecosystem service represented by each land-cover category was obtained by multiplying the estimated size of each land-cover category by the coefficient value of the biome used as the proxy for that category. The principal method for assessing ecosystem service value is adopted from Costanza et al.(1997); it is shown below:

i.   Ecosystem service value

$$ESV = \sum (A_k * VC_k) \tag{1}$$

Where ESV is the estimated ecosystem service value, Ak the area and VCk the value coefficient (RM/ha/yr) for land use category 'k'. The change in ecosystem service value is estimated by calculating the difference between the estimated values for each land- cover category in 2000, 2010 and 2020 (predicted).

ii.   Ecosystem service function Value type f (value of services provided by individual ecosystem functions)

$$ESV_f = \sum (A_k * VC_{kf}) \tag{2}$$

Where ESVf is the estimated ecosystem service value of function, A_k = the area (ha), VC_k = the value of coefficient (RM/ha/yr), k = land-use category and VC_{kf} = the value coefficient (RM/ha/yr) for land use category k with ecosystem service function type f.

iii.   Value of combined ecosystem service

$$ESV_c = \sum w_f * A_k * VC_{kf} \tag{3}$$

Where, W_f = Equivalent weight factor ecosystem services per hectare, A_k = the area (ha), and VC_{kf} = the value coefficient (RM/ha/yr) for land use category k with ecosystem service function type f.

3   Results and Discussion

3.1   Land Use Changes

Land use change analysis in 2000 and 2010 were derived from digital image processing and classification on SPOT 5 imageries. The land use changes for each of

Table 1. Land use change from 2000–2010 (Note: (+) increased, (-) decreased)

Land use type	2000		2010		Changes (ha)	Area change since 2000 (%)
	Area (ha)	Area (%)	Area (ha)	Area (%)		
Crop land	435	13.3	548	16.8	113	3.5
Forest	1246	38.1	1152	35.3	-94	-2.9
Water bodies	1352	41.4	1243	38.0	-109	-3.3
Barren land	235	7.2	325	9.9	90	2.8
Total	3268	100.0	3268	100.0		0.0

Table 2. Prediction of land use change from 2010–2020 (Note: (+) increased, (-) decreased)

Land use type	2010		2020		Changes (ha)	Area change since 2010(%)
	Area (ha)	Area (%)	Area (ha)	Area (%)		
Crop land	548	16.8	661	20.2	113	3.5
Forest	1152	35.3	1058	32.4	-94	-2.9
Water bodies	1243	38.0	1134	34.7	-109	-3.3
Barren land	325	9.9	415	12.7	90	2.8
Total	3268	100.0	3268	100.0		0.0

the four generic classes are presented in Table 1. Meanwhile, the values of land use change for the next 10 years (2020) have been predicted based on the area change percentages in the past 10 years; this prediction assumes that there are no major activities or development in the study area (Table 2).

Based on the image analysis, ground verification and survey, the values of rate of change in land use were found to range from -2.9% to 3.5%. The most affected land use type was crop land and mangrove forest, which increased from 435 ha in 2000 to 548 ha in 2010, and decreased from 1246 ha in 2000 to 1152 ha in 2010, respectively. The other land use types also experienced changes during the 10-year period. Water bodies decreased from 1352 ha to 1243 ha, and barren land increased from 235 ha to 325 ha. By assuming that the rate of change does not vary significantly for the next 10 years, the prediction of land use in 2020 for each land use type is as follows: 661 ha for crop land, 1058 ha for forest area, 1134 ha for water bodies and 415 ha for barren land. Classified land use maps were verified by carrying out accuracy assessment exercise. The accuracy assessment for data in 2000 and 2010 recorded results of 91.55% and 91.56% respectively. Remote sensing products offer many advantages; the most important aspect is the provision of data of actual areas on the ground. The results obtained through vegetation indexes provide an accurate land use mapping of the real situation on the ground.

3.2   Estimation of Ecosystem Service Value

The ecosystem service value (ESV) for each land use type was estimated with the ecosystem coefficient value. Ecosystem service value coefficient for different land use types are as follows: crop land is RM294.00 per/ha, forest area RM3,100.80 per/ha and water bodies RM2,500.00 per/ha. Table 3 shows the ecosystem service value for each land use type.

Table 3. Ecosystem service value for each land use type (Note: (+) ESV increased, (-) ESV decreased)

Land use type	ESV (RM/yr)			Change (2000–2020)	
	2000	2010	2020	RM	%
Crop land	128,064.00	161,331.20	194,598.40	66,534.40	+51.9
Forest	3,863,596.80	3,572,121.60	2,049,628.80	1,813,968.00	-46.9
Water bodies	3,380,000.00	3,107,500.00	2,835,000.00	545,000.00	-16.1
Barren land	-	-	-	-	0.0
Total	7,371,660.80	6,840,952.80	5,079,227.20	2,292,433.60	31.0

Ecosystem service value for crop land increased from RM128,064.00 in 2000 to RM161,331.20 in 2010 and is predicted to increase to RM194,598.40 in 2020. The total change in crop land ecosystem service value was about RM66,534.40 (51.9%). The value for forest area decreased from RM3,863,596.80 in 2000 to RM3,572,121.60 in 2010; its value is predicted to be RM2,049,628.80 in 2020. Meanwhile, the value for water bodies also decreased from RM3,380,000.00 in 2000 to RM3,107,500.00 in 2010; its value is predicted to be RM2,835,000.00 in 2020. The decrease in total ESV for forest area and water bodies during 2000 to 2010 was 46% and 16.1% respectively. Barren land was not assigned any ESV since this land was unproductive; however, its actual ecological value was underestimated. The total ESV in the year 2000 was RM 7,371,660.80 and RM 6,840,952.80 in the year 2010; ESV is predicted to be RM 5,079,227.20 in 2020. On the whole, the change in ESV from 2000 to 2020 results in losses of about RM 22,292,433.60 (31%).

3.3   Ecosystem Service Function for Different Land Use Types

There are eight different ecosystem service function (ESF) parameters that have been assigned to the study area: gas regulation; climate regulation; water supply; soil formation and retention; water treatment; biodiversity protection; raw material; and recreation and culture. The value for each ecosystem service function for different land use type (RM/ha/year) was based on literature review and adapted to suit the Malaysian ecosystem for each ecosystem service function. The value of ecosystem service function for different land use type is presented in Table 4. Meanwhile, the change in the ecosystem service function from 2000 to 2020 for each land use type is presented in Table 5.

Table 4. Value for each ecosystem service function for different land use types

Ecosystem service function	Cropland/ha	Forest/ha	Water bodies/ha	Barren land/ha
Gas regulation	RM235.90	RM1,650.92	RM -	RM -
Climate regulation	RM419.80	RM1,273.57	RM216.98	RM -
Water supply	RM283.05	RM1,509.41	RM9,613.10	RM14.20
Soil formation and retention	RM688.67	RM1,839.60	RM4.72	RM9.43
Waste treatment	RM773.58	RM617.90	RM8,575.37	RM4.76
Total	RM3,259.54	RM10,306.44	RM21,683.75	RM198.29

Table 5. Change in the ecosystem service function (ESF) (Note: (+) ESF increased, (-) ESF decreased)

Land use type	2000 (RM)	2010 (RM)	2020 (RM)	Changes from 2000–2020	
				RM	%
Crop land	1,476,571.62	1,786,227.92	2,154,555.94	677,984.32	45.9
Forest	12,841,824.24	11,873,018.88	10,904,213.52	-1,937,610.72	-15.1
Water bodies	29,316,430.00	26,952,901.25	24,589,372.50	-4,727,057.50	-16.1
Barren land	46,598.15	64,444.25	82,290.35	35,692.20	76
Total	43,681,424.01	40,676,592.30	37,730,432.31	-5,950,991.70	-13.6

The value of the ecosystem service function for crop land in 2000 was RM1,476,571.62, and the value increased to RM1,786,227.92 in 2010; it is predicted to be about RM2,154,555.94 in 2020. The change in ESF is valued at RM677, 984.32 (45%) for the 20-year period covering the years 2000 to 2020.

For the forest area, the value of the ecosystem service function has decreased from RM12,841,824.24 in 2000 to RM11,873,018.88 in 2010; it is predicted to be about RM10,904,213.52 in 2020. This amounts to a loss of approximately RM1,937,610.72 (15%). ESF of water bodies also decreased from RM29, 316,430.00 in 2000 to RM26, 952,901.25 in 2010 and is predicted at RM24,589,372.50 in 2020. On the other hand, the ESF for barren land increased from RM46,598.15 in 2000 to RM64,444.25 in 2010 and is predicted at RM82, 290.35 in 2020. The total change in the ecosystem service function for all land use types decreased from RM43,681,424.01 in 2000 to RM40,676,592.30 in 2010 and is predicted to further decrease to RM37,730,432.31 in 2020. The decrease is expected to be about RM5,950,991.70 or 13% within 20 years.

3.4   Value of Combined Ecosystem Service Function

The combination of ecosystem service function in the same land use type with the equivalent weight factor is calculated based on the function for each parameter for each land use type; the top three or the higher rate of weight factor is then calculated

Table 6. Equivalent weight factor for each land use type for their ecosystem service function

Ecosystem service function	Cropland	Forest	Water bodies	Barren land
Gas regulation	0.07	0.16	0.00	0.00
Climate regulation	0.13	0.12	0.01	0.00
Water supply	0.09	0.15	0.12	0.07
Soil formation and retention	0.21	0.18	0.00	0.05
Waste treatment	0.24	0.06	0.07	0.02
Biodiversity protection	0.10	0.15	0.05	0.81
Food production	0.14	0.00	0.00	0.02
Raw material	0.01	0.12	0.00	0.00
Recreation and culture	0.00	0.06	0.09	0.02
Total	1.00	1.00	0.35	1.00

for each land use category. The contributions of combined ecosystem service function to overall value were ranked based on their estimated ESVf in 2000, 2010 and 2010. The equivalent weight factor ecosystem service per hectare is presented in Table 6.

Table 7 and Table 8 show the combined value of ecosystem service function and tótal change for all the ecosystem services.

Table 7. Combined ecosystem service function in 2000, 2010 and 2020

Year	Cropland (RM)	Forest (RM)	Water bodies (RM)	Barren land(RM)
2000	172,849.26	1,024,492.88	696,806.33	30,828.04
2010	217,750.33	947,203.69	502,799.77	42,634.53
2020	262,651.40	869,914.50	584,451.47	54,441.01

Table 8. Total change in the combined ecosystem service function from 2000–2020

Land Use Type	Cropland (RM)	Forest (RM)	Water bodies (RM)	Barren land(RM)	Total
Change	89,802.14	-154,578.38	-112,354.87	23,612.97	-153,518.13

The combined ESV for the study area in 2000 is RM172,849.26 for cropland, RM1,024,492.88 for forest area, RM696,806.33 for water bodies and RM30,828.04 for barren land. In terms of ecosystem service function, cropland and barren land types show increasing value while the values for forest and water bodies seem to be decreasing. Thus, the total change in the combined ESV for each land use type from 2000 to 2020 have the following values: RM89, 802.14 (increased) for cropland, RM154, 578.38(decreased) for forest, RM112, 354.87(decreased) for water bodies and RM23, 612.97(increased) for barren land.

4   Conclusion

There is a huge and growing interest as well as awareness and need in Malaysia for ecosystem service valuation and ecological economics studies. This study showed that satellite data are very useful and inexpensive for estimating changes in different ecosystem types and for valuing ecosystem services at the local level. In many cases, remotely sensed data may be the only economically feasible way to gather regular land cover and land use information with high spatial, spectral, and temporal resolution over large areas. Results revealed that the potential income from one hectare of mangroves over a 20-year period is expected to decreased at a rate of 15%, ranging between RM85,374.40 and RM72,492.87 per month. The decrease in the economic value of mangroves was largely influenced by the decrease of 2.9% in land use change from 2000 to 2010. This study has demonstrated the role of interdisciplinary practices in addressing natural resource valuation. In particular, data sources from socioeconomic as well as geospatial data were integrated to estimate the economic value of the mangroves in the Sungai Merbok's forest reserve. It is recommended that for future studies, ESV should be applied to a large number of sites around the country in support of ESV systems, carbon trading and national accounting in collaboration with local agencies. This strategy will make ESV a widely used, trusted, and evolutionary system for ecosystem service modelling and evaluation. Integrated knowledge base and policy formulation towards ecosystem service valuation for local condition should be developed in order to obtain more meaningful coefficient values that affect land use.

References

1. Barbier, E.B., Acreman, M., Knowler, D.: Economic Valuation of Wetlands: A Guide for Policy Makers and Planners. Ramsar Convention Bureau, Department of Environmental Economics and Management, University of York, Cambridge, UK (1997)
2. Boyd, J., Banzhaf, S.: What are ecosystem services? The need for standardized environmental accounting units. Ecological Economics 63, 616–626 (2007)
3. Carson, R.T.: Valuation of tropical rainforests: Philosophical and practical issues in the use of contingent valuation. Ecological Economics 24, 15–29 (1998)
4. Costanza, R., d'Arge, R., de Groot, R., Farber, S., Grasso, M., Hannon, B., Limburg, K., Naeem, S., O'Neill, R.V., Parauedo, J., Raskin, R., Sutton, P., van den Belt, M.: The value of the world's ecosystem services and natural capital. Nature 15(387), 253–260 (1997)
5. Daily, G.: Nature's Services: Societal Dependence on Natural Ecosystems. Island Press, Washington, DC (1997)
6. Ewel, K.C., Twilley, R.R., Ong, J.E.: Different Kinds of Mangrove Forests Provide Different Goods and Services. Global Ecology and Biogeography Letters 7(1), 83–94 (1998)
7. MEA. Millennium Ecosystem Assessment: Ecosystems and Human Well-being: Synthesis. Island Press, Washington, DC (2005)
8. Mitchell, R.C., Carson, R.T.: Using Surveys toValue Public Goods: The Contingent Valuation Method. Resources for the Future, Washington, DC (1989)

9. Mohd Hasmadi, I., Pakhriazad, H.Z., Norlida, K.: Remote Sensing for Mapping RAMSAR Heritage Site at Sungai Pulai Mangrove Forest Reserve, Johore, Malaysia. Sains Malaysiana 40(2), 83–88 (2011)
10. Ong, J.E., Gong, W.K., Wong, C.H., Zubirh, J.D.: Characterization of a Malaysian Mangrove Estuary. Estuaries 14(1), 38–48 (1991)
11. Costanza, R., Folke, C.: Valuing Ecosystem Services with Efficiency, Fairness and Sustainability as Goals, pp. 49–70. Island Press, Washington, DC (1997)
12. Sukhdev, P.: The Economics of Ecosystems and Biodiversity: An Interim Report (2008)
13. Wallace, K.J.: Classification of ecosystem services: problems and solutions. Biological Conservation 139, 235–246 (2007)
14. Xavier, B., Szejwach, G.: LUCC Data Requirements Workshop: survey of needs, gaps and priorities on data for land-use/land-cover change research. Institut Cartogra. c de Catalunya, Barcelona (1998)

Feature Selection Parallel Technique
for Remotely Sensed Imagery Classification

NhienAn LeKhac[1], Bo Wu[2], ChongCheng Chen[2], and M-Tahar Kechadi[1]

[1] School of Computer Science & Informatics, University College Dublin
Belfield, Dublin 4, Ireland
{an.lekhac,tahar.kechadi}@ucd.ie
[2] Key Lab. of Spatial Data Mining and Information Sharing, Fuzhou University
Fuzhou, P.R. China
{b.wu,chencc}@fzu.edu.cn

Abstract. Remote sensing research focusing on feature selection has long attracted the attention of the remote sensing community because feature selection is a prerequisite for image processing and various applications. Different feature selection methods have been proposed to improve the classification accuracy. They vary from basic search techniques to clonal selections, and various optimal criteria have been investigated. Recently, methods using dependence-based measures have attracted much attention due to their ability to deal with very high dimensional datasets. However, these methods are based on Cramer's V test, which has performance issues with large datasets. In this paper, we propose a parallel approach to improve their performance. We evaluate our approach on hyper-spectral and high spatial resolution images and compare it to the proposed methods with a centralized version as preliminary results. The results are very promising.

Keywords: feature selection, parallel algorithm, Cramer's V Test, min-max association, image classification.

1   Introduction

Remote sensing research that focuses on feature extraction and selection has attracted much attention of for quiet long time because feature extraction and selection constitute core prerequisites for various applications (e.g., image processing). Tremendous efforts have been dedicated to developing various feature extraction and selection methods to improve image processing effectiveness and classification accuracy in the last few decades [1][2][3]. Previous research generally suggests that effective extraction and utilization of potential multiple features of remotely sensed data, such as spectral signatures, various induced indices, and textural or contextual information, can significantly improve classification accuracy [4][5]. However, not all extracted features are equivalent in their contribution to classification tasks; some of them perhaps are superfluous and useless because either they are trivial or present high

B. Murgante et al. (Eds.): ICCSA 2013, Part II, LNCS 7972, pp. 623–634, 2013.
© Springer-Verlag Berlin Heidelberg 2013

correlations between them. Accordingly, the use of all possible features in a classification procedure may add unnecessary information redundancy and significantly decrease the classification accuracy [6]. As a consequence, employing feature selection techniques to extract the most effective subset from the candidate features is critical to classification of remotely sensed data into a thematic map [5]. Given an input data with N samples and m features $X = \{x_1, ..., x_m\}$, and the target classification variable c, the problem is to find a subset S of n features from m ($n \leq m$) features that optimally characterizes c.

Conceptually, feature selection in general requires a search strategy and criterion functions [7]. The search algorithm generates and compares possible feature selection solutions by calculating their criterion function values as a measure of the effectiveness of each given subset. Sequential search techniques, classical feature selection methods, look for the best feature subset with the prefixed number of features by adding to (resp. removing from) the current feature subset one feature at a time. These include sequential forward-selection (SFS) and sequential backward selection (SBS) [8]. Recently stochastic search algorithms, such as genetic algorithms [9] and clonal selection algorithms [10] have also been attempted for feature selection. Besides the search strategies, an optimal subset always depends of the evaluation function. Various optimal criteria, such as distance-based [11], entropy-based [12] and dependence-based measures [13], etc., have been widely investigated. In this paper, we focus on dependence based techniques. These techniques have some key advantages: 1) they can easily handle very high dimensional datasets (scalability), 2) they are computationally simple and fast, and 3) they are independent of the classification algorithm used. However, it has been recognized that a direct combination of individually good features in terms of certain criteria do not necessary lead to the best overall performance. As an alternative, some researchers have studied indirect or direct means to reduce redundancy among features and select features with minimal redundancy. [12] proposed a heuristic mutual information based max-dependency criterion (mRMR) method to minimize redundancy, which uses a series of intuitive measures of relevance to select very good features. The mRMR method was proven to be an effective technique for feature selection for remote sensing image classification [14]. Recently, we proposed two feature selection indices based on maximal association and minimal redundancy derived from *Cramer's V* test. We also showed its performance in terms of classification accuracy [15]. However, the main drawback of this method is the computational time. It is more severe in case of very large number of features. To analysis big-data of hundreds of features, we propose a parallel approach for feature selection. In order to evaluate this approach on large scale systems, we firstly show its performance with multi-threading paradigm and MPI programming model.

The rest of the paper is organised as follows. In Section 2 we present the background of our research where we discuss feature selection, the max-min associated indices for feature selection, and parallel/distributed paradigm. Section 3 summarises briefly a centralisation-based approach for feature section, then we described in detail a parallel/distributed model for max-min associated indices for feature selection in

Section 4. In Section 5 we evaluated the results of our model on sensored datasets and compared it to our previous approach. In Section 6 we discuss future work and conclude.

2   Background

In this section, we start by briefly describing an approach of feature selection based on two associated indices, we then present models for parallel algorithms. Finally, we discuss some related work in the context of our approach

2.1   Indices for Feature Selection

Cramer's V Test. The Chi-square test is one of the widely used measures to define dependence of variables and was proven to be effective in feature selection [13]. However, it is known that the Chi-square test of dependence is very sensitive to the sample size [16]. *Cramer's V* is the most popular nominal association that is used to measure the strength of the relationship between variables regardless of table size [17]. It has the advantage of not being affected by sample size and therefore is very useful in situations where one suspects a statistically significant Chi-square was the result of a large sample size rather than any substantive relationship between the variables [17]. Therefore, *Cramer's V* test is employed to measure the association between target and variables. Given a *r-row* × *s-column* cross-tabulation, *Cramer's V* can be directly derived from the Chi-square statistic:

$$V = \sqrt{\frac{\chi^2}{N \min\{(r-1),(s-1)\}}} \tag{1}$$

The value of *Cramer's V* varies between 0 and 1. If its value is large, then there is a tendency for particular categories of the first variable to be associated with particular categories of the second variable. It has been suggested in practice that a *Cramer's V* of 0.1 provides a good minimum threshold for suggesting there is a substantive relationship between two variables [18].

Max-Min Associated Indices. In [15], we explored the possibility that a combination of the *Cramer's V* coefficients can be further exploited for optimal feature selection. Two max-min-associated indices derived from the *Cramer's V* test coefficient were developed. For convenience, we firstly presented some important notations employed in this paper in Table 1.

Intuitively, selected features must have maximal target class associated ability. Therefore, a max-associated criterion is used to search for features satisfying (2) with *Cramer's V* test measurement between individual features x_i and class c (A condition):

$$\max A(S,c), A = \frac{1}{|S|} \sum_{x_i \in S} V(x_i,c) \tag{2}$$

where S is the number of subset features. It is likely that features selected according to the max-associated condition (2) will result in rich redundancy, i.e. the dependency among these features could be larger.

Table 1. Notations

Notation	Description	Notation	Description
X	Feature set	C	Number of classes
N	Total samples	x_i	The ith feature
m	Number of features	N_k	Sample number of the kth cross validation
S	Selected subset	$V(x_i,c)$	Cramer's V test between x_i and target c
$|S|$	Number of element of set S	$V(x_i,x_j)$	Cramer's V test between x_i and x_j

When two features highly depend on each other, the respective class discriminated power would not change much if one of them was removed. Therefore, the following minimal associated condition (R condition) among selected features could be added to select mutually exclusive features:

$$\min R(S), R = \frac{1}{|S|^2} \sum_{x_i,x_j \in S} V(x_i,x_j) \tag{3}$$

The max-min-associated indices for feature selection are derived directly from the above two criteria. Two combined methods, referred to as MMAIQ and MMAIS, are designed. These combinations are expressed as equation (4) and (5):

$$max \phi(A,R), \phi = A/R \tag{4}$$

$$max \phi(A,R), \phi = A - \lambda R \tag{5}$$

It is apparent that both (4) and (5) simultaneously satisfy the constraints on A and R. That is, a good feature should be one with maximal target class associated ability, and at the same time with minimal association among the selected features. In equation (5), there is a regularization parameter λ, whose function is to balance the functions of the two constraints in (2) and (3).

2.2   Parallel Computing Models

Dividing the problem into smaller tasks and assigning them to different processors for parallel execution are the two key steps in the design of parallel algorithms. Normally, there are two parallel models: data-parallel and task-parallel model. The data-parallel model is one of the simplest. In this model, the tasks are statically or semi-statically mapped onto processors and each task performs similar operations on different data. This type of parallelism that is a result of identical operations being applied concurrently on different data items is called data parallelism. Since all processors perform similar computations, the decomposition of the problem into tasks is usually based on data partitioning because a uniform partitioning of data followed by a static mapping is sufficient to guarantee load balancing [19]. Data-parallel algorithms can be implemented in both shared-address and message-passing paradigms. A key characteristic of data-parallel problems is that for most problems, the degree of data parallelism increases with the size of the problem, making it possible to use more processes to effectively solve larger problems.

The computations in any parallel algorithm can be viewed as a task-dependency graph. The task-dependency graph may be either trivial, as in the case of matrix multiplication, or complicated. In the task graph model, the interrelationships among the tasks are utilized to promote locality or to reduce interaction costs. This model is typically employed to solve problems in which the amount of data associated with the tasks is large relative to the amount of computation associated with them. This type of parallelism that is naturally expressed by independent tasks in a task-dependency graph is called task parallelism [19].

3   Feature Selection Algorithm

In [15], we presented a feature selection approach that can be summarised as follows: to select the candidate feature set, an incremental method is used to find the suboptimal features defined by Eq. (4) or (5). Although this search strategy does not allow the features to be reselected once they have been selected, it can usually ensure that the selected features with relevance and redundancy constraints are the most prominent features not to be removed. In addition, the incremental search method is rather fast. Suppose we already have S_{p-1}, the set with p^{th} features, the task is to select the pth feature from set $\{X - S_{p-1}\}$, such that the feature maximizes Eq. (4) or Eq. (5). The incremental algorithm optimizes the following conditions:

$$\max_{x_j \in X-S_{p-1}} [V(x_j,c) - \frac{\lambda}{p-1} \sum_{x_i,x_j \in S} V(x_j,x_i)] \tag{6}$$

$$\max_{x_j \in X-S_{p-1}} [V(x_j,c) / \frac{\lambda}{p-1} \sum_{x_i,x_j \in S} V(x_j,x_i)] \tag{7}$$

These optimizations can be computed efficiently in $O(|S| \cdot m)$ complexity. As a result, we can obtain the ranked features rapidly even if the dimension of features is possibly very high.

In [14], two important issues were also solved before the classification process. One is how to obtain the cross-tabulation, such that *Cramer's V* can be calculated if the concerned features contain continuous variables. In this case, a pre-processing step of discretization is required to obtain cross-tabulation. Another critical problem is how to optimise the best number of feature subsets. The best number of features is usually estimated by the K folds cross-validation of the correct classification rate (CCR).). Let $CCR_{K,cross}$ be K cross-validation repetitions. The $CCR_{K,cross}$ is given by

$$CCR_{K,cross} = \frac{1}{K}\sum_{k=1}^{K} CCR_k = \frac{1}{K}\frac{1}{N_k}\sum_{k=1}^{K}\sum_{i=1}^{N_k} L(c_i,\hat{c}_i) \qquad (8)$$

where $L(ci,\hat{c}i)$ denotes the zero-one loss function of the ith sample between the ground truth label c_i and the predicted label $\hat{c}_i$, and $N_k = (K-1) \times N / K$ is the number of test sets during the kth iteration. It is apparent that $CCR_{K,cross}$ is actually the average over K CCRs of cross-validation repetitions. Because CCR is not an accurate indicator with limited samples [15], the lower CCR limit index with compensatory information is thereby adopted to improve CCR. The lower CCR limit index criterion measures the variation between Chi-square and beta to optimize the number of selected features associated with the Gaussian maximal likelihood classifier (GMLC) [15]:

$$CCR_{K,cross}^{Lower} = CCR_{K,cross} - Loss_{cross}(N_k,m) \times [CCR_{K,cross} - 1/C] \qquad (9)$$

where C is the number of classes, more details on the loss information $Loss_{cross}$ can be found in [15].

Let the input dataset be $DisX$ that is a matrix of $r \times c$ where r is the number of samples and c is the number of features. Moreover, let $DisXc_i$ be the column i of $DisX$, Y be the label set, kk be the number of repetitions, $genCT$ be the function that generates a cross tabulation table and $CVTest$ be the function that performs the Cramer's V test. The feature selection algorithm can be resumed as in the Algorithm 1.

As shown in [15], this algorithm have been compared in terms of overall accuracy and kappa coefficient with the SFS and mRMR methods, which are known for their general abilities and good performances. When compared with SFS and mRMR, MMAIQ performs the best feature selection, and offers better or comparable classification accuracy in two experiments with different types of image. MMAIS also achieves satisfactory results in the same experiments. These results testify that MMAIQ and MMAIS provide new and effective options for feature selection.

Despite this approach is efficient in terms of accuracy in the feature selection, its running time is still an issues with large datasets. The overall complexity of this approach is $O(c3 \times r2 \times rlog(r))$ where r is the number of samples and c is the number of features and $rlog(r)$ is the complexity of sorting algorithm used in $genCT$ function that creates a relevant cross-table. When the number of samples as well as the number of features is increased, the running time is exposed with exponential complexity. Therefore, we propose a parallel approach for this algorithm in this paper.

Algorithm 1 (*pseudo code*) Optimal Cramer's V
Input: DisX, r, c, Y, kk.
Output: List the feature selected fea

```
1:  crm ← {0};
2:  for i ← 1 to c do
3:      crm_i ← CV Test(getGT (Y, r,DisXc_i));
4:  end for
5:  last ← 1; fea1 ← crm_1; curidx ← crm_2;
6:  cmi ← {0}; tmi ← {0};
7:  for i ← 1 to kk do
8:      fcln_i ← DisXc_fealast ;
9:      left ← c - last;
10:     for j ← 0 to left do
11:         tmi_j ← crm_j;
12:         tbl ← genCT(fcln_i, r, DisXc_curidxj );
13:         cmi_curidxj ← cmi_curidxj + CV Test(tbl);
14:     endfor
15:     m ← i × tmi_1 / cmi_curidx1 ; midx ← 1;
16:     for k ← 2 to left do
17:         tmp ← i × tmi_k / cmi_curidxk;
18:         if (tmp > m) then
19:             m ← tmp;
20:             midx ← k;
21:         end if
22:     endfor
23:     fea_i ← curidx_midx;
24:     curidx_midx ← curidx;
25:     incr(last); incr(curidx);
26: endfor
27: for i ← 1 to kk do
28:     incr(fea_i);
29: end for
```

4 Parallel Approach

As mentioned in Section 2.2, a parallel algorithm can be implemented in either data-parallelism or task parallelism. In our approach, we apply the data-parallelism. The reason is that the generation of a cross table as well as the calculation of *Cramer's V Test* require the analysis of whole datasets. Therefore, the overhead of exchanging large size of datasets among different sites should be taken into account in a data-parallel approach. As a consequence, we base on the data-parallel paradigm to design our parallel algorithm. We also presume that two function *genCT* (creating relevant cross-tables) and *CVTest* (Cramer's V Test) are atomic i.e. a parallel versions of *genCT* and *CVTest* are not in the context of this paper.

In analysing the sequential version of optimal *Cramer's V* [15], we notice that the cross validation step is an impact on runtime performance. The complexity of this algorithm is also based on this step as shown in the previous section. We propose hence, a parallel approach for this step. Algorithm 4.1 describes our approach. Let the input dataset be *DisX* that is a matrix of $r \times c$ where r is the number of samples and c is the number of features. Moreover, let $DisX_{ci}$ be the column i of *DisX*, Y be the label set, *kk* be the number of repetitions, *genCT* be the function that generates a cross tabulation table and *CVTest* be the function that performs the *Cramer's V* test. The feature selection algorithm can be resumed as in the Algorithm 2 (pseudo code).

Moreover, there is a threshold $P_{THRESHOLD}$ in this algorithm. The purpose of this threshold is to handle the granularity of this algorithm. When the number of iterations left is under this threshold, a sequential computation will be applied. The function *incr(x)* in this algorithm is equivalent to $x \leftarrow x + 1$. In this algorithm, the function $Par_{Fold()}$ performs the parallel computation of cross-table different parts $DisX_k$ of input datasets on in p processors. It can be described as in the algorithm $Par_{Fold()}$.

Algorithm 2 *(pseudo code)* Parallel Optimal Cramer's V
Input: DisX, r, c, Y, kk.
Output: List the feature selected fea

```
1:   crm ← {0};
2:   for i ← 1 to c do
3:       crmᵢ ←  CV Test(getGT (Y, r,DisXcᵢ));
4:   end for
5:   last ← 1; fea1 ← crm₁; curidx ← crm₂;
6: cmi ← {0}; tmi ← {0};
7:   for i ← 2 to kk do
8:       fclnᵢ ← DisXc_fealast ;
9:       left ← c - last;
10:    if left > P_THRESHOLD then
11:       FORK in k process p₁,p₂,...,p_k
12:          Each process p_k do
13:              tbl_k ←ParFold(DisX_k, curidx,fcln_k);
14:          endo
15:       JOIN k process: {tbl} = ∪ tbl_k
16:       for j ← 0 to left do
17:          cmi_curidxj ←  cmi_curidxj + CV Test(tblᵢ);
18:       endfor
19:    else
20:       for j ← 1 to left do
21:          tmiⱼ ←  crmⱼ;
22:          tbl ←  genCT(fclnᵢ, r, DisXc_curidxj );
23:          cmi_curidxj ←  cmi_curidxj + CV Test(tbl);
```

```
24:        endfor
25:      end if
26:      m ← i × tmi₁ / cmi_curidx1 ; midx ← 1;
27:      for k ← 2 to left do
28:          tmp ← i × tmi_k / cmi_curidxk;
29:          if (tmp > m) then
30:              m ← tmp;
31:              midx ← k;
32:          end if
33:      endfor
34:      fea_i ← curidx_midx;
35:      curidx_midx ← curidx;
36:      incr(last); incr(curidx);
37:  endfor
38:  for i ← 1 to kk do
39:      incr(fea_i);
40:  end for
```

Par_Fold () function
Input: startPos, endPos, curidx.
Output: DisXk

```
1:  for j ← startPos to endPos do
2:      cln ← GetMatrixColumn(curidx+j);
3:      DisXk_j ← genCT(cln);
4:  end for
```

5   Performance Evaluation

5.1   Experiments

In order to exploit efficiently different parallel computing platforms from shared-memory architecture to distributed architecture such as cluster and grid, we design two versions of our parallel approach. The first version is based on the multithreading paradigm where we can deploy on the shared-memory systems to exploit the multi-core architecture. This version is portable as it is based on the PTHREAD library [19]. The second version is based on a message-exchange paradigm to exploit the distributed architecture such as cluster or grid platforms. Concretely, it uses MPI library as the communication library because of it is simple and widely used as a standard of communication by exchanging message [19].

Besides, we use different datasets to evaluate our approach. The first one is a hyperspectral remote sensing image (PHI) Xiaqiao PHI (*xq*). The data set used in this experiment was collected in September 1999 of the Xiaqiao test site, a mixed agricultural area in Changzhou city, Jiangsu province, China, and is airborne push broom

Fig. 1. The experimental hyperspectral image cube of Xiaqiao

hyperspectral imagery (PHI). A sub-scene (346 × 350 pixels) of the PHI image with 80 bands was tested, and their spectral ranges were from 417 to 854 nm.

Fig. 1 shows the experimental PHI image cube. The ground truth spectral data were collected in September, 1999 by field spectrometer SE590. The observed image was expected to classify into eight representative classes, i.e. corn, vegetables-sweet potato, vegetable-cabbage, soil, float grass, road, water and puddle/polluted water. This dataset has 80 features. The second dataset is *fcl1* that is a historically significant data set, is located in the southern part of Tippecanoe County, Indiana. It has more than a few spectral bands, contains a significant number of vegetative species or ground cover classes, includes many regions (e.g., fields) containing a large numbers of contiguous pixels from a given class. This dataset has 12 features. The last one is *India* that is a district boundaries dataset prepared for FAO by Dept. of Energy and natural resources, University of Illinois, including Coastlines, national-subnational boundaries, lakes, and Islands. This dataset has 185 features. Indeed, we test our approach on different platforms varied from multi-core (up to quad-core) to cluster computing architecture (up to four computational nodes).

Table 2. Performance of parallel optimal *Cramer's V*

Dataset	Feature Selection Time (s) of optimal Cramer's V				
	Sequential	*PThread*		*MPI*	
		2 Threads	*4 Threads*	*2 Nodes*	*4 Nodes*
xq	25.53	15.77	8.42	22.73	19.92
fcl1	0.37	0.28	0.15	0.39	0.41
India	293.18	168.73	98.61	243.19	139.81

5.2   Analysis

As described in [15], the sequential optimal Cramer's V has been shown that it is the most efficient algorithm on Xiaqiao PHI dataset (*xq*) comparing to other selected algorithms with an improvement of 10.1% of overall accuracy. However, its computational time is significant. Table 2 compares the running time of two algorithms: sequential optimal *Cramer's V* and parallel optimal *Cramer's V* on different platforms with different datasets. By observing this table, we notice that the speed-up for multithreading approach is around 60% i.e. we can improve the running time by 1.6 times. Moreover, the speed-up for cluster computing architecture depends on the testing dataset. It's 20% and 12% for the *India* and *xq* datasets respectively. However, we do not gain the speedup of *fcll* dataset. In this case, the communication time dominates the computational time. Analysis on the communication and computational time of parallel approaches can be found in [20]. This means our solution gets significant performance for big datasets (e.g. *India*) in terms of computational time.

Besides, experimental results also show that the running times of two functions *genCT* and *CVTest* take are insignificant compared to the overall computational time. It also proves our presumption at the beginning of Section 4.

6   Conclusion and Future Work

In this paper, we present a parallel approach for improving the performance of an optimal solution MMAIQ and MMAIS for the feature selection in the classification of remotely sensed imagery. We also evaluate our approach with different datasets and comparisons of the proposed methods with a centralisation version as preliminary results. These experimental results consistently show that the proposed methods can provide an effective solution for feature selection in improving significantly the computational time that is a drawback of most feature selection approaches.

A Hadoop/MapReduce version of this proposed method is implementing and testing to evaluate the robustness of our solution on large scale distributed systems of dozen to hundreds of computational nodes in the context of the feature selection for big data. On the other hand, a knowledge map [21][22] is being used to improve the proposed approach in terms of combining both the data-parallelism and task-parallelism..

Acknowledgment . The authors gratefully acknowledge the research support from the EU Framework Programme 7, Marie Curie Actions under grant No. PIRSES-GA-2009-247608.

References

1. Foody, G.M.: Approaches for the production and evaluation of fuzzy land cover classification from remotely sensed data. International Journal of Remote Sensing 17, 1317–1340 (1996)
2. Stuckens, J., et al.: Integrating contextual information with per-pixel classification for improved land cover classification. Remote Sensing of Environment 71, 282–296 (2000)

3. Zhang, L., et al.: Texture Feature Fusion with Neighborhood Oscillating Tabu Search for High Resolution Image. Photogrammetric Engineering and Remote Sensing 74, 323–332 (2008)

4. Platt, R.V., Goetz, A.F.H.: A comparison of AVIRIS and Landsat for land use classification at the urban fringe. Engineering and Remote Sensing 70, 813–881 (2004)

5. Lu, D., Weng, Q.: A survey of image classification methods and techniques for improving classification performance. Journal of Remote Sensing 28, 823–870 (2007)

6. Price, K.P., et al.: Optimal Landsat TM band combinations and vegetation indices for discrimination of six grassland types in eastern Kansas. International Journal of Remote Sensing 23, 5031–5042 (2002)

7. Tan, P.N., Steinbach, M., Kumar, V.: Introduction to data mining. Pearson Education, Inc. (2006)

8. Jain, A., Zongker, D.: Feature selection: evaluation, application, and small sample performance. IEEE Transaction on Pattern Analysis and Machine Intelligence 19, 153–158 (1997)

9. Raymer, M.L., Punch, W.F., Goodman, E.D.: Dimensionality reduction using genetic algorithms. IEEE Transactions on Evolutionary Computation 4, 164–171 (2000)

10. Zang, L., et al.: Dimensionality Reduction Based on Clonal Selection for Hyperspectral Imagery. Transaction on Geoscience and Remote Sensing 45, 4172–4186 (2007)

11. Jensen, J.R.: Introduction to Digital Image Processing: A remote sensing perspective, 2nd edn. Prentice Hall, Piscataway (1996)

12. Peng, H., Long, F., Ding, C.: Feature selection based on mutual information: criteria of max-dependency, max-relevance, and min-redundancy. IEEE Transaction on Pattern Analysis and Machine Intelligence 27, 1226–1238 (2005)

13. Liu, H., Weng, Q.: A survey of image classification methods and techniques for improving classification performance. International Journal of Remote Sensing 28, 823–870 (2007)

14. Wu, B., et al.: Classification of QuickBird Image with Maximal Mutual Information Feature Selection and Support Vector Machine. Procedia Earth and Planetary Science 1, 1165–1172 (2009)

15. Wu, B., et al.: Feature selection based on max-min-associated indices for classification of remotely sensed imagery. International Journal of Remote Sensing (to be published, 2013)

16. Agresti, A., Finlay, B.: Statistical methods for the social sciences, 3rd edn., ch. 8. Prentice Hall (1997)

17. Garson, G.D.: Nominal association: Phi, contingency coefficient, Tschuprow's T,Cramer's V, lambda, uncertainty coefficient. Statnotes: Topics in Multivariate Analysis, http://www2.chass.ncsu.edu/garson/pa765/assocnominal.htm

18. Martinez-Casasnovas, J.A., et al.: Comparison between land suitability and actual crop distribution in an irrigation district of the Ebro valley. Spanish Journal of Agricultural Research 6, 700–713 (2008)

19. Gramma, et al.: Introduction to Parallel Computing, 2nd edn. Addison-Wesley (January 2003)

20. Le-Khac, N.-A.: Studying the performance of overlapping communication and computation by active message: Inuktitut case. In: International Conference on Parallel and Distributed Computing and Network (PDCN 2006), Innsbruck, Austria, February 12-14 (2006)

21. Le-Khac, N.-A., Aouad, L.M., Kechadi, M.-T.: An efficient Knowledge Management Tool for Distributed Data Mining. International Journal of Computational Intelligence Research 5(1) (2009) ISSN 0974-1259

22. Le-Khac, N.-A., Aouad, L.M., Kechadi, M.-T.: Handling large volumes of mined knowledge with a self-reconfigurable topology on distributed systems. In: IEEE Seventh International Conference on Machine Learning and Applications (IEEE ICMLA 2008), CA, USA, December 11-13 (2008)

Data Usability Processor for Optical Remote Sensing Imagery: Design and Implementation into an Automated Processing Chain

Erik Borg[1], Bernd Fichtelmann[1], and Hartmut Asche[2]

[1] German Aerospace Center, German Remote Sensing Data Center,
Kalkhorstweg 53, 17235 Neustrelitz, Germany
[2] University of Potsdam, Geoinformation Group, Department of Geography
Karl-Liebknecht-Strasse 24/25, 14476 Potsdam, Germany
{erik.borg, bernd.fichtelmann}@dlr.de,
{hartmut.asche}@uni-potsdam.de
http://www.dlr.de/dlr/en/desktopdefault.aspx/tabid-10260/,
http://www.geographie.uni-potsdam.de

Abstract. A range of global environmental and social problems, such as climate change or social transformation processes, are aggravated by diverse anthropogenic impacts. To monitor, analyse and combating these processes, topical information on the status, development, spatial and temporal dynamics of them is an indispensable prerequisite. The growing, frequently rapid demand for global and regional data in relevant geographical, geometric, semantic and temporal resolution can only be met by remote sensing data the majority of which are available on an operational scale. Not only does the availability of data present a major obstacle for the above applications, but also rapid processing of the acquired remote sensing data is a severe bottleneck for the provision of the required data for, e.g. time-critical investigations. These problems can be addressed by developing an automated processing chain to derive value-added data producing from the remote sensing input data. Effective automated data processing necessitates a data quality assessment prior to actual processing. This paper deals with a processor for an automated data usability assessment that can be integrated into an automated processing chain for operative value adding.

Keywords: Cloud cover, data quality, data usability processor, processing chain, remote sensing, Landsat.

1 Introduction

Increasing scarcity of environmental resources, such as fresh water or fertile land, coupled with constant build-up of man-made pressure on land use, holds unforeseeable risks for sustainable development of the environment, forcing careful and sustainable use of the limited environmental resources. One major cause of environmental risks results from conflicting claims on land/soil utilisation. To be able

B. Murgante et al. (Eds.): ICCSA 2013, Part II, LNCS 7972, pp. 635–651, 2013.
© Springer-Verlag Berlin Heidelberg 2013

to address and eventually solve such problems it is vital to strike a fair balance between the diverging user requirements to limit, and, if possible, reduce the mounting strain on nature and agriculture. The availability of environmental and geographic information resources is an important prerequisite if progress is to be achieved on this issue. Starting from the 1970s, an ever-growing part of these information needs can be met by remote sensing data from space.

To respond to the global land-use conflict the European Union (EU) and the European Space Agency (ESA) have jointly initiated the Global Monitoring for Environment and Security (GMES) program. This initiative is aimed at the development and provision of substantial, accurate and stable geo-information services based on remote sensing data products and ancillary spatial data, e.g. in-situ-information [1]. For that purpose, a range of current remote sensing technologies, including radar and multi-satellite systems, have been employed to establish a geographical database on environmental information. Many of the environmental parameters required can, however, be detected by optical remote sensing systems only. Regardless of this asset, one relevant constraint of optical remote sensing imagery is the high probability of clouds and haze during data acquisition limiting data quality. If cloud-obscured optical remote sensing data are to be processed the above mentioned limitations require mandatory an interactive assessment of each suboptimal dataset by an operator. Such operator-based image evaluation and processing to extract geo- and biophysical parameters is time-consuming and requires considerable expertise and manpower. As a consequence, only cloud-free or marginally clouded optical remote sensing data only are processed into usable information.

When cloud-free data only are processed, the requirements of the GMES initiative for information cannot be met. Important reasons for this include:

- Interactive processing of suboptimal optical time series data will not generate usable value-added products conforming to defined quality standards, such as GMES. In addition, manpower and time requirements will significantly impact the production costs.
- Interactive processing of optimal data will, in contrast, ensure the generation of usable quality products of bio and geophysical information.
- Provision of full-coverage value-added products for a given time or period can only be ensured if optical data for a region are recorded free of cloud cover at a given acquisition time.

As a consequence, cloud-free data only can be used for generation of valued-added products. Such an approach does not use the complete available data pool. One solution to this problem is the development and application of automatic processors and processing chains for operation of sub-optimal data at acceptable time and costs which is presented here.

In automated as well as in interactive data processing data quality is the key parameter to decide on the processing strategy to be pursued. This quality parameter is, in fact, applied to determine the utilisation of a particular processing strategy of optical remote sending data, namely:

- Interactive data quality interpretation by an operator.
- Selection of available remote sensing data from a data pool by the user.

- Setup and control of an automatic processing chain by choosing appropriate processing modules.

Controlling automatic processing of remote sensing data requires decision criteria in order to facilitate event-driven processing of data by selecting the corresponding processing modules. Such criteria can be derived from metadata of the respective remote sensing datasets. Relevant control parameters may also include technical system parameters, such as gain and offset, as well as data acquisition parameters, such as acquisition time, geographical coordinates, sun azimuth angle or sun elevation angle. Data quality is a particularly important control parameter when it comes to event-driven processing. The data quality parameter can directly be assessed from a given remote sensing dataset. It can either be expressed in terms of the cloud cover index[1] or the data usability index[2].

2 Material and Methods

2.1 Data Base

Development of the processing chain presented here is based on a data pool of 2,957 data-reduced and JEPG-compressed quick-look-data[3] with the corresponding metadata[4] of Landsat-7 from the period of 2000 to 2003. A short description of the preparation of the quick-look-data is given in [2], [3]. First, all bands were converted to a common ground resolution of 180 m in a preprocessing step. Second, the quick-look data were JPEG-compressed to minimize the storage volume. This data compression method is, regrettably, accompanied by a loss of data quality and information. The data used in this study were compressed with a 10:1 ratio of [2]. Notwithstanding the fact that the level of compression depends on the image content of a remote sensing scene, this represents a JPEG quality metric Q-factor of 35 [4].

2.2 Processing Chain

The European ground segment of Landsat-7 is a network of the receiving stations at Maspalomas (Spain), Kiruna (Sweden), Matera (Italy), and Neustrelitz (Germany).

The station network guarantees perfect reception and storage of data on behalf of the European Space Agency. Processing and data delivery was carried out on behalf of EURIMAGE enterprise [6][5]. Fig.1 shows a block diagram of the German Landsat ground segment (DFD). Data processing starts during the receiving phase. In this step data are preprocessed, augmented by metadata generated on-the-fly and stored in a database [8]. The interactive processing steps in the automatic processing chain are

---

[1] Cloud cover degree: Ratio of cloud pixels to total pixels of an unit (e.g. scene or quadrant).
[2] Data usability: Combination of cloud cover and cloud distribution as well as data errors.
[3] Quick-look data: are preview images derived from original remote sensing data.
[4] Metadata: describes remote sensing data (e.g. satellite mission, orbit, track, frame).
[5] LANDSAT 7/ETM+ data receiving was stopped at the end of 2003 [7].

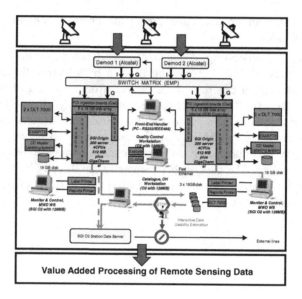

Value Added Processing of Remote Sensing Data

Fig. 1. Block diagram of the Landsat ground segment. Highlighted in red: interactive data usability assessment (adopted and changed from [5]).

highlighted in red. Here data quality has been assessed visually by (human) interpreters. Landsat quick-look data are displayed on the monitor.

The assessment framework for the imagery interpretation has been provided by ESA [9]. Key criteria to be addressed, are, i.a.:

- artifact (90 = data not usable),
- estimate haze, clouds, and cloud shade per quadrant,
- assessment of cloud distribution,
- differentiation of clouds covering land or water,
- estimation of data usability for land applications.

After completion of metadata with the assessment results these were transferred to ESA [8].

3   Data Usability Processor for a Processing Chain

Apart from data filtering measures, such as geographical or atmospheric data correction (cf. [10], [11]) the data usability processor for optical remote sensing data can be integrated in an automated value added processing chain.

Quick-look data of Landsat 7/ETM+ imagery with accompanying metadata are used to train the processor. Metadata are essential to control the processing and internal data transfer inside the processor. In a first processing step, the data are analysed for data errors, such as scan mirror errors, missing pixels, lines and areas [14]. In case errors are identified in the filtering process, processing of data is terminated. Data free from errors are subjected to further processing by the cloud cover assessment (CCA) module.

The CCA module comprises the calibration, referencing to satellite projection and classification sub-modules and can be characterised as the core of the processor. Metadata of the dataset are, in a second step, evaluated for data calibration using the calibration parameter of slope and gain as well as geographical coordinates of the scene to calculate the sun elevation angle required to transform the digital number of a pixel into a physical standard (reflectance). In a third step, a land-water mask is generated using the geographical scene coordinates to provide topographical information. The land-water mask is subsequently used to control the sensitivity of the ensuing classification modules. Inverse georeferencing to satellite projection is necessary to optimise the structure analysis.

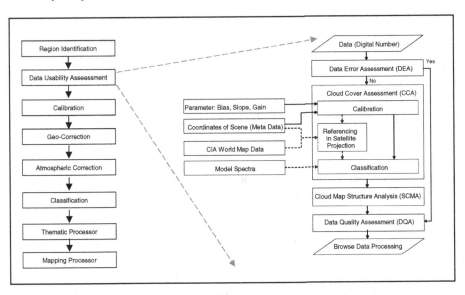

Fig. 2. Schematic diagram of a processing chain to derive value-added products (left) and of data usability processor (right) (adopted and modified from [12], [13])

The transformed land-water mask as well as the quick-look constitute the input data for cloud and haze classification to generate a cloud-haze-mask. A range of classification algorithms is available for Landsat data [15], [16], facilitating the selection of an optimal processing module according to target characteristics or geographic location. The cloud-haze mask facilitates the computation of cloud cover degree. This feature serves as a data input for the subsequent cloud map structure analysis (SCMA). This module produces parameters describing cloud distribution in an assessment unit (scene or quadrant). Both cloud cover and cloud distribution are important quality parameters of optical remote sensing data [17].

The different quality parameters discussed above are combined by the data quality assessment module (DQA) in order to provide a data use quality metric. It goes without saying that cloud cover degree is the principal factor in this measure. Cloud distribution is considered as an additional parameter to refine data usability in an assessment unit[6].

3.1   Preprocessing

Preprocessing of the data is an essential process step in the processor introduced here. Preprocessing includes subroutines, such as data error assessment, calibration, and transformation of topographic information into satellite projection. A brief description of the different sub-processors is given below.

3.1.1   Data Error Assessment

In accordance with the requirements of ESA [9] faulty data have to be designated unusable and excluded from further processing. This specification necessitates the development and application of a reliable method to detect data errors [13]. Key features to distinguish faulty from accurate data are:

- disturbed lines or data sections display smaller correlation to neighboring undisturbed lines than undisturbed lines to neighboring undisturbed lines,
- erroneous image data sections, lines or a number of pixel, are characterised by the digital number 0, (DN) in contrast to usable data.

A possible application of the correlation on selected erroneous data, such as Landsat-7/ETM+ JPEG-compressed quick-look data showed that:

- missing lines cannot be identified reliably at very low sun angle,
- artifacts caused by JPEG-compression trigger misinterpretations and incorrect conclusions,
- determination of erroneous datasets requires extensive computing time.

As a consequence, a simplified method has been developed based on the specification to characterise erroneous pixels by a digital number of DN = 0. Pixel values of a line are examined for this criterion and, at confirmation, are summed up. To minimize faulty indications, a threshold value has empirically been derived for the identification of erroneous scenes. To detect unusable data, an optimal threshold value of 10 could be determined empirically. Eventually, the second criterion (DN=0) has proved to be adequate to distinguish disturbed and undisturbed datasets. Following data analysis data usable are forwarded to data processing. Faulty data are excluded from further treatment.

3.1.2   Calibration

Calibration and transformation of Landsat-7/ ETM+ data to top-of-atmosphere reflectance ρ and to effective at-sensor brightness temperature T [K] is dealt with in substantial detail in the Landsat handbook [14]. Additional aspects to be taken into

---

[6] Assessment unit: scene, quadrant.

account for the calibration procedure are given by [17] to compute planetary top-of-atmosphere reflectance ρ_p based on the digital number of a pixel (DN) [18], [19]. The relation for approximated Earth-Sun-distance d is given by [20]. Transformation of digital number of a pixel (DN) into planetary top-of-atmosphere reflectance ρ_p in a specified band k can be calculated by using eq. 1 [18], [19]:

$$\rho_p = (\pi L(\lambda_k)\, d^2)/(E_S(\lambda_k)\, cos\, \Theta_S) \tag{1}$$

with $L(\lambda_k)$ spectral radiance at sensor's aperture [$W/(m^2 sr\ \mu m)$], $E_S(\lambda_k)$ denotes exo-atmospheric solar irradiance [$W/(m^2\mu m)$], d Earth-Sun distance [astronomical units], and Θ_S solar zenith angle [$degrees$].

Transformation (digital number) of thermal band 6 (wavelength: 10.4 μm to 12.5 μm) into surface temperature T can be calculated by using eq. 2.

$$T = K_2/ln((K_1/L(\lambda_k))+1) \tag{2}$$

with: $L(\lambda_k)$ spectral radiance at the sensor's aperture [$W/(m^2\, sr\ \mu m)$], K_1 calibration constant 1 [$666,09\ W/(m^2 sr\ \mu m)$] [7], K_2 calibration constant 2 [$1282,71\ K$] [8].

3.1.3   Calculation of Pixel Coordinates

Use of solar zenith angle in eq. (1) assumes that geographic coordinates of all pixels are available and usable. However, as a rule, only corner coordinates of the scene are given. In addition, coordinates can be used to select topographic information to support data analysis.

A satellite orbit in a static (non Earth rotation) satellite system Σ' can be characterised by a reference track in a system Σ, in which the track can be described by length λ only, while the latitude φ is generally 0. This satellite system is rotated by an angle $\delta = (180^o\text{-}\iota)$ (ι = inclination of the real orbit and reference satellite orbit in Σ') against the x axis, which cross exactly by ($\varphi=0,\ \lambda=0$), opposite to the geographical system.

The key advantage of an equatorial orbit in system Σ is that the distance between two points can be calculated with a simple relation (eq. 3) for a corresponding satellite image:

$$\varphi_1 = \varphi_0 + i\, \Delta\varphi$$
$$\lambda_1 = \lambda_0 + j\, \Delta\lambda \tag{3}$$

where $\Delta\varphi$ and $\Delta\lambda$ are pixel size in latitude and longitude, respectively, and i and j are the distance both pixels in columns and lines. For example, if an even number of lines is given it is of advantage to use for numerical solution of eq. 3 the relation $\varphi_0 = \Delta\varphi/2$ with ($-m/2 \leq i \leq m/2\text{-}1$). Transformation of the coordinates φ and λ of system Σ into coordinates φ' and λ' of geographical system Σ' is described by eq. (4), where $r = r'$ [21]:

---

[7] Landsat Handbook: chapter 9.2.4, Table 9.2 ETM+ Thermal Constants.
[8] Landsat Handbook: chapter 9.2.4, Table 9.2 ETM+ Thermal Constants.

$$cos \; \varphi' \; cos \; \lambda' = cos \; \varphi \; cos \; \lambda$$
$$cos \; \varphi' \; sin \; \lambda' = cos \; \delta \; cos \; \varphi \; sin \; \lambda - sin \; \delta \; sin \; \varphi$$
$$sin \; \varphi' = sin \; \delta \; cos \; \varphi \; sin \; \lambda + cos \; \delta \; sin \; \varphi \qquad (4)$$

In each case φ' can be considered as known since the latitude of reference track is identical to the geographical latitude of the real track in the geographical system. However, coordinates of the track in Σ, defined by $\varphi=0$, $sin \; \varphi=0$, and $cos \; \varphi=1$, are not given. In contrast, coordinates for UL, UR, LR, and LL in the system Σ' are known. Disregarding Earth rotation, UL and LL are on parallel small circle tracks to satellite track. Difference to longitude of LL is given by earth rotation and will be calculated in one of the next steps.

The corresponding 4 geographical latitudes define the reference track in geographical coordinates of which the corner coordinates on the left side of eq. (4) and the angle δ on the right sight are known. Coordinates on the right-hand side will have to be determined to use at least eq. (3) for determination of all pixel coordinates of the scene in system Σ. Subsequently all coordinates in Σ can be transferred into the coordinates of Σ'. Deviation from the real track in the rotating system will have to be determined at least. Knowledge of the respective longitudes λ' is not mandatory for eq. 3 in eq. (4). Transformation of a small circle is sufficient for segmentation of an image in constant longitudes and latitudes sections. Transformation of the 2 left corner coordinates within eq. (5) $\varphi=\varphi_G$, however, has to calculated as a prerequisite.

$$\lambda_k = arcsin \; [(sin \; \varphi_k - cos \; \delta \; sin \; \varphi_G) / (sin \; \delta \; cos \; \varphi_G)] \qquad with \; k=UL, LL \qquad (5)$$

φ_G as latitude in the satellite system Σ is the spherical distance between a small circle and a great circle which represents the reference track. This distance is identical in every system Σ'. It is precisely half the spherical distance of the two upper corner pixels located on the great circle. Pass the great circle (orthodrome) on a globe the points $A(\varphi_A, \lambda_A)$ and $B(\varphi_B, \lambda_B)$ the spherical distance α can be calculated to:

$$\alpha = arcos \; [sin \; \varphi_A - sin \; \varphi_B + cos \; \varphi_A \; cos \; \varphi_B \; cos(\lambda_B - \lambda_A)] \qquad A=UL, LL \; and \; B=UR, LR \qquad (6)$$

For both pairs A, B the result is same with: $\varphi_G = \alpha / 2$ \qquad (7)

Using the difference of both longitudes of eq. 5 and the corresponding line number n the line spacing $\Delta\lambda$ in eq. (3) can be computed for the scene by eq. (8).

$$\Delta\lambda = [(\lambda_{UL} - \lambda_{LL}) / n] \qquad with \; n = line \; number \qquad (8)$$

The distance $\Delta\varphi$ (eq. 3) between image elements of a line can be calculated using the pre-calculated spherical distance α of line limits and the number of m image elements (columns) using eq. (9):

$$\Delta\varphi = \alpha / m \qquad with \; m = number \; of \; column \qquad (9)$$

Use of m and n assumes that the coordinates of the UL corner possibly correspond to the upper left corner of the UL pixel. If the center of the UL pixel is given as upper left corner of scene $(n-1)$ and $(m-1)$ have to be used. A scanning geometry with constant distance between line elements is a prerequisite for applying eq. (9). In other cases, the respective distance between image elements will have to be determined. Subsequently it is possible to estimate the geometric assignment (φ, λ) for each image element of the appropriate reference track in the satellite system using eq. 3. Eq. (4) is used to transform the coordinates (φ, λ) into the satellite system Σ' with coordinates (φ', λ') the. The third eq. in (4) can be solved for φ'.

$$\varphi' = arcsin\ (sin\ \delta\ cos\ \varphi\ sin\ \lambda + cos\ \delta\ sin\ \varphi) \tag{10}$$

Working with geographical longitude requires that the definition interval is given with $-180° < \lambda \leq 180°$. Using eq. 11 a value of λ or λ' outside this interval can reset into it:

$$\lambda_{red} = 2arctan(tan(\lambda / 2)) \tag{11}$$

Derivation of λ' the first eq. necessitates to divide by the second eq. in (4).

$$((cos\ \varphi'\ sin\ \lambda')(cos\ \varphi'\ cos\ \lambda')) = ((cos\delta cos\varphi sin\lambda - sin\delta sin\varphi)/(cos\varphi cos\lambda)) \tag{12}$$

Following reduction and introduction of the tangency function

$$tan\ \lambda' = (cos\ \delta\ tan\ \lambda) - ((sin\ \delta\ tan\ \varphi)/cos\ \lambda) \tag{13}$$

eq. (12) can be solved for λ':

$$\lambda' = arctan(cos\ \delta\ tan\ \lambda - (sin\ \delta\ tan\ \varphi)/cos\ \lambda) \tag{14}$$

φ' and λ' are the coordinates of a scene for an orbit trough the geographic coordinates $[\varphi=0, \lambda=0]$. Subsequently, image coordinates for the reference track in the geographical system Σ' relating to real coordinates in the rotating geographical system Σ' can be calculated. Rotating around the z axis (north-south axis of the earth) with

$$\Delta\lambda_0 = ABS(\lambda^0_{UL} - \lambda'_{UL}) \tag{15}$$

the reference image can be positioned in a way that the first line covers the first line in the original image, when placed in the system Σ'. λ^0_{UL} is the longitude of the available *UL* corner coordinate and λ'_{UL} is the corresponding longitude of *UL* following transformation. All additional lines of the two images which correspond to each other are moved because of the earth rotation. The additional shift caused by the earth rotation $\Delta\lambda^0_j$ of each single image line can be computed with the help of the differences in the length of the two left real corner coordinates $(\lambda'_{UL}, \lambda'_{LL})$ to the respective corresponding length $(\lambda'_{UL}, \lambda^0_{LL})$ in the reference picture:

$$\Delta\lambda^0_j = j*ABS((\lambda^0_{UL}-\lambda'_{UL}) - (\lambda^0_{LL}-\lambda'_{LL}))/n \qquad n = number\ of\ lines \qquad (16)$$

The geographical latitude remains constant despite of earth rotation. The coordinates of the scene are completely available for all pixels after execution of the corrections of longitude λ':

$$\lambda"_{i,j} = \lambda'_{i,j} + \Delta\lambda_0 + \Delta\lambda^0_j \qquad for\ all\ i, j \qquad (17)$$

3.1.4 Calculation of Local Time for Pixels

The following task includes the calculation of local time for each pixel of the scene. For a satellite with a sun synchronous orbit it can be shown (e.g. [22]) that local crossing time t_{LC} for the respective nadir point of the satellite orbit in dependence of its geographical latitude φ_N and a constant equator crossing time t_{LEC} on basis of the previously known parameters.

$$t_{LC} = t_{LEC} - arcsin(tan(\varphi_N)\ cot(\iota))/15 \qquad \iota = Inclination\ of\ track \qquad (18)$$

This minus '-' in eq. (18) refers to the descending node of the track. A problem with equator crossing time is that it only can be restricted as assumed constantly [22], [23], [24]. However, a well-known difference exists between true and mean sun. This means that local noon, e.g., will change between -14 and +16 minutes in one year. This is obvious as part of diurnal drift in [25]. Disregarding orbit drift the time of equator crossing will change in the same order within one year. This is caused by the characteristics of an earth orbit around the sun. This relation is described by the eq. of time (t_E) can be found in [26].

The STK orbit determination software[9] [27] is used to calculate equator crossing of a satellite in UTC (Coordinated Universal Time). Seasonal variation of the equator crossing of the Landsat-7 platform has been simulated for this presentation with corresponding real measured orbit parameters at Neustrelitz for each 21st of all twelve months over year 2000. The values simulated for the descending node are marked with asterisks in fig. 3. Considering these 12 points only it will become obvious that the variation of LECT is very similar to that of Equation of Time (t_E). The nearest value to the 10:06 a.m. mean of these 12 values is given for the 21st of June. Using the t_{LEC} and t_E for this date as basis (index 2106) makes it possible to describe the seasonal variation of t_{LEC} with eq. (19):

$$t_{LEC} = t_{LEC\ 2106} + (t_E - t_{E\ 2106}) / 60 \qquad (19)$$

In addition, possible disturbances as orbital drift have to be included into eq. (19). Inclusion of t_{LEC} of eq. (19) into eq. (18) results in eq. (20).

---

[9] STK software: software of company AGI to model, analyse and visualise space systems (2004) http://www.agi.com/default.aspx

$$t_{LC} = t_{LEC\,2106} + (t_E - t_{E\,2106}) / 60 - arcsin(tan(\varphi_N)\,cot(\iota)) / 15 \qquad (20)$$

Eq. (15) applies to the nadir point of the satellite orbit only. As a consequence, local time for all pixels of an image line (eq. 20) will have to be corrected by using their respective length difference $\Delta\lambda'' = \lambda'' - \lambda''_N$ to the nadir point of this line by an additional term.

$$t_{LC} = t_{LEC\,2106} + (t_E - t_{E\,2106}) / 60 - arcsin(tan(\lambda_N)\,cot(\iota)) / 15 \; + \Delta\lambda''/15 \qquad (21)$$

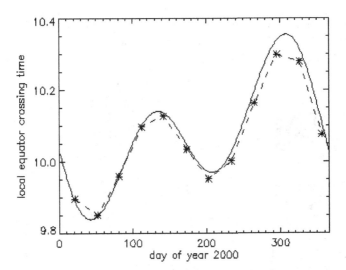

Fig. 3. Variation of t_{LEC} determined for each 21st day of a month by help of acquisition facility (dashed line) of descending node of Landsat 7 for year 2000 and connecting the 21st of March with the variation of Equation of Time by eq. (19).

Hence the geographical latitude and longitude as well as the corresponding local time for each pixel are available it is possible to calculate the solar zenith angle Θ_S. It depends on season and local time and can be calculated as

$$\Theta_S = arccos(sin(\delta_S)\,sin(\varphi') + cos(\delta_S)\,cos(\tau)\,cos(\varphi')) \qquad (22)$$

where φ' is the geographic latitude (eq. 10), δ_S is the solar declination [26], and τ is the local time t_{LC} (eq. 21) as angle.

3.1.5   Transformation of Topographic Information into Satellite Projection

Generally speaking, satellite data are corrected on a topographical basis to facilitate further processing of the data (e.g. atmospheric correction or thematic classification and value adding). To allow for fast processing, the preparatory "Transformation of topographic Information into Satellite Projection" procedure has been developed and used here. Advantages pertaining to the processing of remote sensing data are:

- An essential processor component is a structure analysis which analyses the connectivity of cloud pixels. This can be performed rapidly if missing pixels are not contained in the data to be analysed. By the geo-correction being used usually gabs of non-information pixels are computed within the data by transfer because of the difference of the satellite orbit orientation to the north- south direction.
- A land-water mask is a binary image unlike the remote sensing data. Therefore, a clear allocation can be performed on in the transformation of the data.

Under the consideration laid out above, topographic data can be transformed into a satellite projection as described in [28]. Fig. 4 schematically demonstrates operation steps required to deliver the topographic information in satellite projection.

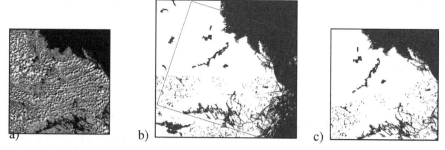

Fig. 4. Topographic information in satellite projection (blue: water; white: land) (adopted and modified by [28])a) Landsat-data (Stockholm-region 2002/06/01) (RGB: band 7, 5, 3) b) Map with inscribed region of quick-look-data as in a) c) Resulting land-water mask after transformation (b) on basis of coordinates

3.2   Value-Added Processing

3.2.1   Classification of Quick-Look Data

The main objective of a classification is the grouping of objects with identical feature properties and the differentiation against other objects which do not show these feature properties. In addition to this the classification labels the aggregated objects to corresponding pre-defined property scale.

Methods available for classification can be divided into supervised and unsupervised methods all of which can be applied interactively or automatically to a dataset.

Classifications procedures used in the data usability processor operate automatically on a pre-defined classification scheme. The Landsat-7 / ETM+ classification schemes employed are the NASA-ACCA (NASA - National Aeronautics and Space Administration) [15] and the ACRES-ACCA-procedure (National Earth Observation Group, previously known as ACRES) [16]. The NASA-procedure was developed as a component of the operative processing chain of Landsat ground station at EROS Data Center in Sioux Falls [15]. The ACRES procedure has been integrated in the operative Landsat processing chain of ACRES [16]. The potential to process JPEG-compressed quick-look data of both established automatic cloud cover assessment (ACCA) procedures has been analysed in [3]. Classification results have proved satisfactory for the application to JPEG-compressed quick-look data. Processing time was minimal.

3.2.2   Structure Analysis of Classified Data

The procedure employed for (a) analysing cloud deviation structure within an assessment unit and (b) for deriving indicators to estimate usability of the remote sensing data is described by [13]. Direction filters [29], [30] are used for the determination of the cloud distribution in the scene. This allows for estimation of the distance of undisturbed pixels of the image border or of cloud boundary in the predefined direction and for storing the result in a temporary direction matrix. Subsequent to filtering 8 result matrices for both the cloud mask and for the cloud-free comparison matrix exist. The minimal distance value is determined by means of a minimum operator and stored into a temporary result matrix for both masks (the cloud mask and the cloud-free mask). Following this the results of the structure analysis are provided to the sub-module for assessing the data usability value.

3.3   Adding Value to Landsat-7 / ETM+ Data and Browse Product Mapping

The process presented here yields information on the technical and scientific parameters selected of a scene in the form of quick-look data products in addition to the numerical values of data usability for individual quadrants. The final step in the usability evaluation for Landsat quick-look data is to summarise all information extracted and computed in a browse product. This is the final result of the processing chain.

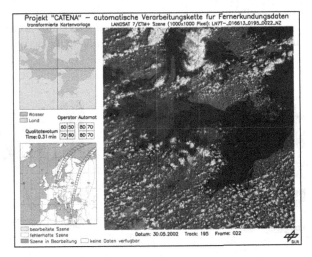

Fig. 5. Browse product to support interactive visual data quality assessment

In addition, other information can be made available to cater for the requirements of the various processing strategies of optical remote sensing data mentioned above. These results can be provided to the user as information on the remote sensing data quality as well as an orientation in extensive data archives. Furthermore, the data quality information can be used as matrix to control post-processing and processing of remote sensing data by an automatic process chain. An example of a browse product to support an interactive visual data usability assessment by interpreters and operators, respectively, is given in fig. 5. This product includes the transformed topography map in satellite perspective (left, top), the satellite orbit with track and frame, and the quick-look with clouds rendered in red (right).

4   Value-Added Information Products

Demand for full-coverage environmental information is increasing permanently. Based on earth observation data provision of this information can only be guaranteed by automatic and operative processing. Processing chains will have to include a post-processing component in addition to pre-processing and processing of the data. In this respect, the processor for the assessment of data usability introduced in this paper meets all the criteria of a process chain. The method developed for the fast, automated quality assessment of Landsat data can also be applied to other remote sensing data. For this purpose, the algorithms have been generalised by, e.g., introduction of a varying data size and adoption to remote sensing systems, such as ALOS, with a tilting sensor (fig. 6). In addition to data details (left site below), representation of the quick-look data (right site below), geographical information about the path received, to the scenes having been processed before and the scene immediately situated in the processing (left site below), information about the data usability is represented for the different assessment segments.

The cloud mask is shown separately above the corresponding RGB representation of the scene. The water mask has been derived according to the described procedure is represented on the left above.

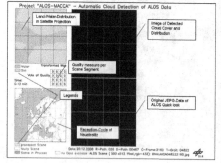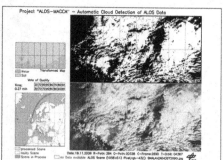

Fig. 6. Result of automated data usability assessment in ALOS data (adopted and modified by [28])

5   Conclusion

The current status of remote sensing and ICT offer a wealth of new opportunities to produce value adding remote sensing data for diverse applications. Basically this is the result of the following developments:

- permanently increasing number of remote sensing missions, sensors, and data
- considerably improved computing power and storage performance,
- permanent improvement of automated remote sensing data interpretation.

Compared against the requirements of ESA relevant to the functionality of the processor introduced here it can summarised:

1. Automation of remote sensing data interpretation for provision of real-time appropriation of data information products for time-critical applications is a service to be released urgently.
2. However, even if the respective interpretation processor serve only the generation of some pre-information it has to be made sure that a greatest measure in accuracy has to be guaranteed at the data processing.
3. The processor in this respect shows all essential features of a processing chain; these more highly are required for the generation for value added information products of reduced spatial resolution.

Acknowledgements. The authors wish to thank V. Beruti, G. Pitella and R. Biasutti (all of European Space Agency) for the provision of test data, constructive discussions, and interest shown in our investigations. The authors also wish to thanks Egbert Schwarz from the Neustrelitz remote sensing station for his efforts in determining the actual equator crossing time.

References

1. European Commission,
   http://ec.europa.eu/enterprise/policies/space/gmes/
   (last access: August 6, 2012)
2. Borg, E., Fichtelmann, B., Asche, H.: Assessment for Remote Sensing Data: Accuracy of Interactive Data Quality Interpretation. In: Murgante, B., Gervasi, O., Iglesias, A., Taniar, D., Apduhan, B.O. (eds.) ICCSA 2011, Part II. LNCS, vol. 6783, pp. 366–375. Springer, Heidelberg (2011)
3. Borg, E., Fichtelmann, B., Asche, H.: Cloud classification in JPEG-compressed remote sensing data (LANDSAT 7/ETM+). In: Murgante, B., Gervasi, O., Misra, S., Nedjah, N., Rocha, A.M.A.C., Taniar, D., Apduhan, B.O. (eds.) ICCSA 2012, Part II. LNCS, vol. 7334, pp. 347–357. Springer, Heidelberg (2012)
4. Lau, W.-L., Li, Z.-L., Lam, K.W.-K.: Effects of JPEG compression on image classification. Int. J. Remote Sensing 24(7), 1535–1544 (2003)
5. Beruti, V.: LANDSAT 7 ESA Stations Network Report - Focus on Product Generation, LTWG 11, LANDSAT Technical Working Group Meeting (USGS/NASA), Camberra, Australia, February 4-8, p. 21 (2002)

6. Bettac, H.-D., Reiniger, K., Brieß, K., Borg, E.: DLR/DFD presentation to the LGSOWG-30 Meeting, LGSOWG Meeting, Orlando, Florida, USA, November 12-15, p. 22 (2001)

7. Pollex, J.: Oral Communication (2001)

8. Schwarz, J., Bettac, H.D., Missling, K.-D.: Das DFD-Bodensegment für LANDSAT-7. In: Mehl, H., Dech, S. (eds.) DLR-Mitteilung 1999-03, Wessling, Germany, October 20-21, pp. 49–56 (2000)

9. Biasutti, R.: Cloud Cover Evaluation, LTWG 8, LANDSAT Technical Working Group Meeting (USGS/NASA), Ottawa, Canada, July 17-22, p. 10 (2000)

10. Richter, R., Schläpfer, D.: Atmospheric/topographic correction for satellite imagery. In: DLR Report DLR-IB 565-02/11, Wessling, Germany, p. 202 (2011)

11. Huang, L., Li, Z.: Feature-based image registration using the shape context. International Journal of Remote Sensing 31(8), 2169–2177 (2010)

12. Borg, E., Fichtelmann, B., Böttcher, J., Günther, A.: Processing of remote sensing data. EP 1 637 838 B1 (2009)

13. Borg, E., Fichtelmann, B.: Verfahren und Vorrichtung zum Feststellen einer Nutzbarkeit von Fernerkundungsdaten. Deutsches Patent Nr. 10 2004 024 595 B3 (2004)

14. NASA (2011), http://landsathandbook.gsfc.nasa.gov/ (last access: August 6, 2012)

15. Irish, R.: LANDSAT 7 Automatic Cloud Cover Assessment. In: Sylvia, S.S., Descour, M.R. (eds.) Algorithms for Multispectral, Hyperspectral, and Ultraspectral Imagery VI, Proceedings of SPIE, vol. 4049, pp. 348–355 (2000)

16. Xu, Q., Wu, W.: ACRES Automatic Cloud Cover Assessment of LANDSAT 7 Images. In: Spatial Sciences Conference 2003 – Spatial Knowledge Without Boundaries Canberra, September 23-26, p. 10 (2003)

17. Slater, P.N., Biggar, S.F., Holm, R.G., Jackson, R.D., Mao, Y., Moran, M.S., Palmer, J.M., Yuan, B.: Reflectance and Radiance-Based Methods for the In-Flight Absolute Calibration of Multispectral Sensors. Remote Sens. Environ. 22(1), 11–37 (1987)

18. Markham, B.L., Barker, J.L.: Spectral characterization of the LANDSAT Thematic Mapper Sensors. Int. J. Remote Sensing 6(5), 697–716 (1985)

19. Chander, G., Markham, B.L., Helder, D.L.: Summary of Current Radiometric Calibration Coefficients for Landsat MSS, TM, ETM+, and EO-1 ALI Sensors. Remote Sens. Environ. 113(5), 893–903 (2009), http://landportal.gsfc.nasa.gov/ Documents/Landsat_Calibration_Summary.pdf (last access: February 20, 2012)

20. Gurney, R.J., Hall, D.K.: Satellite-derived surface energy balance estimates in the Alaskan Sub-Arctic. J. Clim. Appl. Meteor. 22(1), 115–125 (1983)

21. Gellert, W., Küstner, H., Mellwich, M., Kästner, H.: Mathematik –, p. 837. Kleine Enzyklopädie, Verlag Enzy-klopädie, Leipzig (1970)

22. Wu, Z.-J., McAvaney, B.: Sampling methods for climate model calculated brightness temperatures. BMRC Research Report No. 118, Bureau of Meteorology Research Centre, Australia, p. 43 (2006), http://cawcr.gov.au/bmrc/pubs/researchreports/RR117.pdf (last access: March 1, 2013)

23. Ignatov, A., Lazlo, I., Harrod, E.D., Kidwell, K.B., Goodrum, G.P.: Equator crossing times for NOAA, ERS and EOS sun-synchronous satellites. International Journal of Remote Sensing 25(23), 5255–5266 (2004)

24. Johnson, D.B., Flament, P., Bernstein, R.L.: High-resolution satellite imagery for mesoscale meteorological studies. Bulletin of the American Meteorological Society 75, 5–34 (1994)

25. Mears, A.M., Schabel, M.C., Wentz, F.J., Santer, B.D., Govindasamy, B.: Correcting the MSU middle tropospheric temperature for diurnal drifts. In: International Geophysics and Remote Sensing Symposium 2002, vol. III, pp. 1839–1841 (2002)

26. Duffett-Smith, P.: Practical astronomy with your calculator, 3rd edn., p. 188. Cambridge University Press (1988)

27. AGI Company: Software (2004), http://www.agi.com/default.aspx

28. Fichtelmann, B., Borg, E., Kriegel, M.: Verfahren zur operationellen Bereitstellung von Zusatzdaten für die automatische Fernerkundungsdatenverarbeitung. In: Strobl, J., Blaschke, T., Griesebner, G. (eds.) 23. AGIT-Symposium, Salzburg, pp. 12–20. Wichmann, Berlin (2011)

29. Lehmann, T., Oberschelp, W., Pelikan, E., Repges, R.: Bildverarbeitung für die Medizin–Grundlagen, Modelle, Methoden, Anwendungen, p. 462. Springer, Heidelberg (1997)

30. Haberäcker, P.: Praxis der digitalen Bildverarbeitung und Mustererkennung, p. 350. Hanser, München (1995)

Satellite Time Series and in Situ Data Analysis for Assessing Landslide Susceptibility after Forest Fire: Preliminary Results Focusing the Case Study of Pisticci (Matera, Italy)

Antonio Lanorte, Claudia Belviso, Rosa Lasaponara, Francesco Cavalcante,
Fortunato De Santis, and Angelo Aromando

CNR-IMAA, C.da S. Loja 85050 Tito Scalo (PZ), Italy
rosa.lasaponara@imaa.cnr.it

Abstract. In this paper we present the preliminary results obtained from our investigations addressed to the estimation of landslide susceptibility after fire conducted using both satellite time series and in situ analysis.

The investigations were carried on in a test area in south of Italy, municipality of Pisticci, which is characterized by a high hydrogeological risk and affected by a severe forest fire between 26 and 28 August 2012 The total area burned amounted to 1100-1300 hectares, one of the largest areas burned in Basilicata in the last 15 years. Form the geological point the investigated site is quite fragile and the fire could affect and increase the land slide susceptibility of the area. Satellite data were analyzed in order to evaluate fire severity for the whole surface affected by fire using MODIS data available free of charge from the NASA web site. Satellite investigations were completed with in situ analysis and measurements.

Field measurements and laboratory mineralogical analysis, made on several sample sites selected in area characterized by different levels of fire severity, well fit together confirming the increase in landslide susceptibility after the fire event.

Keywords: satellite based analysis, MODIS, Fire Severity, clay minerals, XRD analyses.

1   Introduction

Many literature data describe the effects of fire on soil properties and how these effects depend upon fire severity [9]. Fire events generally modify physical, chemical and biological soil properties [9], [18] and alter soil minerals [11], [13]) sometimes irreversibly according to the heating process during the fire event. The combination of combustion and heat transfer produces steep temperature gradients in soil. In moist soil the latent heat of vaporization prevents temperature from exceeding 95 °C until water completely vapourises [2], [3] and the temperature then typically rises to

B. Murgante et al. (Eds.): ICCSA 2013, Part II, LNCS 7972, pp. 652–662, 2013.
© Springer-Verlag Berlin Heidelberg 2013

200-300 °C [9]. In presence of heavy fuel, the temperature of soil surface is 500-700 °C [6] and instantaneous values up to 850 °C occasionally [5]. Soil structure degradation and organic matter loss occur at 300 °C and from 100 °C until 450 °C, respectively whereas dehydroxylation and structural breakdown of clay minerals take place between 460 °C and 980 °C [6].

In this study, the mineralogical composition of samples involved by different level of fire severity was investigated analyzing the effects of the heating process on the changes in minerals.

Fire severity [15], [16] is a qualitative indicator of the effects of fire on ecosystems, since it affects soil, forest floor, canopy, etc. Assessing and mapping fire severity is important to monitor fire effects, to model and evaluate post-fire dynamics and to estimate the ability of vegetation to recover after fire (generally indicated as fire-resilience). In an operational context, fire severity estimation is critical for short-term mitigation and rehabilitation treatments.

After a fire, the spectral behaviour of vegetation changes due to consumption of fuel, presence of ash, reduced transpiration of vegetation and increased surface temperature. All these effects increase reflectance in mid-infrared and reduce surface reflectance in near-infrared. This is the reason because NBR index is computed on the basis of the two burn sensitive bands, infrared (NIR) and shortwave infrared (SWIR). For this reason, it may be one of the best indexes to detect a burn area.

2   Data set and Study Area

2.1   Study Area

The fire that covered the territory of the municipality of Pisticci between 26 and 28 August 2012 had its ignition point near the Basentana highway at a height of around 50 meters above sea level, where, aided by the strong wind and conditions of particular dryness of vegetation, spread quickly to the south direction going up the slope leading from the highway to about 350 meters of Pisticci town. The fire was later extended also on the south and east of Pisticci. The total area burned amounted to 1100-1300 hectares, one of the largest areas burned in Basilicata in the last 15 years (Fig 1).

The fire particularly struck the areas occupied by reforestation with conifers implanted a few decades ago to counter the instability phenomena which are very pronounced in this area (Fig. 2 and Fig. 4).

The vegetative state of these stands often with the presence of dead biomass in excess, combined with their high fire potential and climatic conditions of extreme dryness, favored a very vigorous fire spread.

In addition to reforested conifer plantations, the fire also hit areas occupied by xerophytic grasslands, mediterranean maquis and cultivated areas, in particular olive groves (Fig. 3).

The fire took also the form of wildland-urban interface fire, since several isolated houses were threatened by the flames that have also lapped the town of Pisticci.

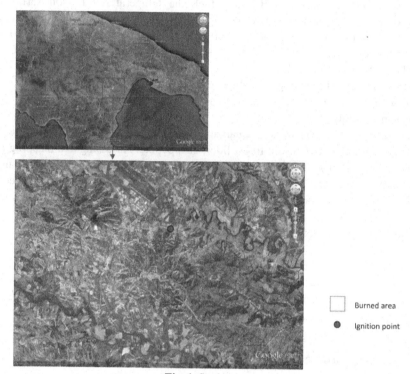

Burned area

● Ignition point

Fig. 1. Study area

Fig. 2. Burned areas in forest pine on slopes of clay

Fig. 3. Fire affected mediterranean shrubs, xerophytic grasslands and cultivated areas (here olive groves)

Fig. 4. Burnt area with surface instability phenomena

2.2   MODIS Data

MODIS is a key instrument aboard Terra (EOS AM) and Aqua (EOS PM) satellites.

These data will improve our understanding of global dynamics and processes occurring on land, in oceans, and in lower atmosphere. MODIS is playing a vital role in the development of validated, global, interactive Earth system models able to predict global change accurately enough, to assist policy makers in making sound decisions concerning the protection of our environment.

Terra's orbit around the Earth is timed so that it passes from north to south across the equator in the morning, while Aqua passes south to north over the equator in the afternoon. Terra MODIS, launched on December 18, 1999 and Aqua MODIS, launched on May 4, 2002, are viewing the entire Earth's surface, acquiring data in 36 spectral bands ranging in wavelength from 0.4 µm to 14.4 µm, with a high radiometric sensitivity (12 bit).

Two bands are imaged at a nominal resolution of 250 m at nadir, five bands at 500 m, and the remaining 29 bands at 1 km. A ±55-degree scanning pattern at the EOS orbit of 705 km achieves a 2,330-km swath and provides global coverage every one to two days.

MODIS bands used in this work are the first seven ones, corresponding to a spatial resolution of 250 and 500 m (Fig. 5). These spectral bands are suitable for the study of vegetation characteristics.

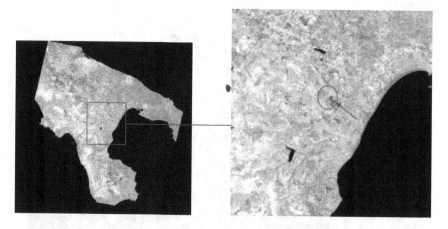

Fig. 5. MODIS image (band 2 – NIR) - August 28, 2012 (red arrow indicates burned area)

3   Method

3.1   Mineralogical Analysis

23 samples were collected in an area close to Pisticci village (Basilicata, Southern Italy) affected by a large forest fire in August 2012. In particular, soil layers with a

thickness of 2-4 cm were sampled in the pine forest designing area with different levels of fire severity: low (the plants have remained free), moderate (involving stem and leaves) and high (affecting the entire root system).

The mineralogy of the whole rock and fine fraction was determined by X-ray diffraction (XRD) using a Rigaku/Rint 2200 powder diffractometer with Cu-Kα radiation; graphite secondary monochromator; sample spinner and 40 kV x 30 mA.

The samples were analysed for bulk mineralogy on randomly oriented powders of whole rocks [22]. The <2 µm grain-size fraction was separated by gravimetric-settling in distilled water [1] and the mineralogy was investigated on oriented specimen after Mg++ saturation, by evaporation of clay-water suspension in order to obtain a concentration of 6 mg/cm2 [18]. XRD analysis on oriented amounts was carried out on air-dried specimens, glycolated at 60° for 8 hours and heated at 375 °C for 1 hour [20].

Quantitative mineralogical analysis of the bulk rock was carried out measuring peak area on random powder using WINFIT computer program [17]. Percentage of clay minerals in the < 2 µm fraction was estimated by measuring the peak areas on both glycolated and heated oriented amounts.

3.2 MODIS Data Processing

Remote sensing can be potentially a sound choice to map burn severity, since vegetati on removal, soil exposure, changes in soil and vegetation moisture content imply changes in reflectance [14]. Indeed, fire--related decreases in chlorophyll content and vegetation moisture lead to decreases in the visible and near-infrared (NIR) reflectance and increases in the mid-infrared (SWIR) reflectance [24].

Since the amount of green biomass destroyed by fires depends upon the burn sever ity, several authors have found good correlations between vegetation indices, compute d from post-fire remotely sensed data, and burn severity [7], [8], [10], [12], [21], [23].

The most effective NIR/SWIR index for burn severity available in the literature is t he Normalized Burn Ratio (NBR) [16].

MODIS data were used to compute the following index, by using MODIS spectral bands 2 (841-876 nm) and 7 (2105-2155 nm) with a spatial resolution of 500m.

$$NBRModis = (MODIS2 - MODIS7)/(MODIS2 + MODIS7) \qquad (1)$$

Using these two bands we obtained a NBR map which is particularly sensitive to changes occurring in vegetation cover affected by fire, such as amount of live green vegetation, moisture content, etc. NBR values generally range between 1 and −1 as well as NDVI. Strongly negative NBR values would indicate a larger reflectance in SWIR than NIR band, and this only occurs over not vegetated areas where fire cannot occur.

Multidate (pre and post fire) of the relative version of the delta Normalized Burn Ratio (dNBR) were investigated to measure and quantify the ecological effects of fire using pre-fire and post-fire MODIS satellite images.

4   Results and Discussion

4.1   Mineralogy

The mineralogical assemblages of bulk samples showed the presence of quartz, phyllosilicates, carbonate (calcite and dolomite, subordinately) and feldspar.

The < 2 μm grain-size fraction was mainly composed of illite and illite/smectite (I/S) mixed layers (about 50-90% in weight); kaolinite and chlorite were also present.

The relative amount of clay minerals showed a variation according to the severity of fire. In particular, the samples collected in areas less affected by the fire showed a progressive decrease of illite/smectite mixed layers from the deeper samples (less subjected to heat) to the surface samples (mostly subjected to higher temperatures) (Fig. 6).

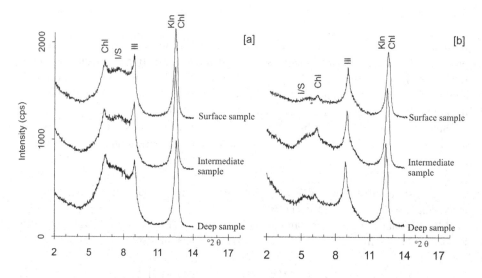

Fig. 6. Samples collected in the area characterized by low levels of fire severity. XRD patterns of the < 2 μm grain-size fraction [a] air dried; [b] glicolated

A progressive reduction of kaolinite and chlorite characterized instead the sediments of soil collected in the area where the fire was more persistent. These two mineralogical phases in fact, appeared to be more abundant in the upper levels, while their quantity decreased progressively from the top downwards until disappearing near the root completely burned (Fig. 7).

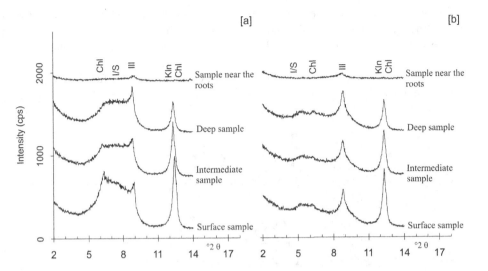

Fig. 7. Samples collected in the area characterized by high levels of fire severity. XRD patterns of the < 2 μm grain-size fraction [a] air dried; [b] glicolated

The mineralogical data clearly indicated the effect of fire on clay minerals with probably consequence on properties of soil and following increase in landslide susceptibility of the area after the fire event.

4.2   Fire Severity Assessment

In order to well characterize and identify burned area, we computed and used delta NBR (dNBR) expected to perform better than other methods in capturing the spatial complexity of severity within fire perimeters. Positive dNBR values represent a decrease in vegetation while negative values represent increased vegetation cover.

Maps obtained by the difference between pre- and post-fire indexes provide a measure of change which then can be used to estimate biomass loss, carbon release, aerosols production, etc.. Moreover, the difference in pre/post-burn NBR index could reflect surface change and characterize burn severity degree.

$$dNBR = NBRprefire - NBRpostfire \qquad (2)$$

In this study, to compute dNBR we used two MODIS images (Fig.8a): the first image is of August 15, 2012 (pre-fire) and the second image is of August 31, 2012 (post-fire).

Fire severity was classified into four levels as low, moderate-low, moderate-high, and high. The severity class of high indicates fire-damaged stands in which all the vegetation are totally burned and killed and even the ground is heavily modified. The moderate severity class means the area was damaged by fire but part of vegetation was still alive after fire and the effects on the soil are more limited . The low severity class is the area burned with little damage to vegetation and soil.

The fire severity map (Fig.8b) yields a gradient of detected change in relation to the characteristics of fire propagation: greater severity (red/marron) near and around the ignition point, moderate severity (cyan) in the central part of the fire, low severity (yellow) at the end.

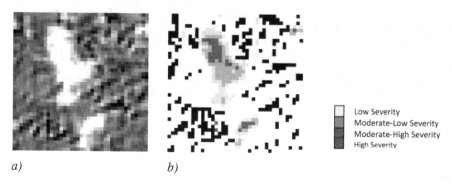

a)   *b)*

Low Severity
Moderate-Low Severity
Moderate-High Severity
High Severity

Fig. 8. .a) dNBR (August 15-August 31, 2012); *b)* Fire Severity Map

5   Final Remarks

Moderate Resolution Imaging Spectroradiometer (MODIS), time series and in situ data analysis have been conducted to assess landslide susceptibility after the large forest fire which affected the Pisticci Municipality (Basilicata, Southern Italy) in August 2012. Field measurements and laboratory mineralogical analysis, made on several sample sites selected in area characterized by different levels of fire severity, well fit together confirming the increase in landslide susceptibility after the fire event.

The preliminary results presented here show that a detailed study of changes in the mineral soil component after the passage of a fire can be a great tool to improve the interpretation of post-fire signals using spaceborne remote sensing data.

These results need to be confirmed after a careful analysis of correlated data and after using remote sensing data at higher spatial resolution.

References

1. Cavalcante, F., Belviso, C.: Trattamenti e metodi di preparazione di campioni di materiali per l'analisi diffrattometrica. In: Fiore, S. (ed.) Analisi di Materiali Argillosi per Diffrazione di Raggi X e Microscopia Elettronica a Scansione: Teoria ed Applicazioni, Argille e Minerali delle Argille, 5, VII Corso di Formazione AISA onlus, pp. 23–50 (2005)
2. Campbell, G.S., Jungbauer Jr., J.D., Bidlake, W.R., Hungerford, R.D.: Predicting the effect of temperature on soil thermal conductivity. Soil of Science 158, 307–313 (1994)

3. Campbell, G.S., Jungbauer Jr., J.D., Bristow, K.L., Hungerford, R.D.: Soil temperature and water content beneath a surface fire. Soil of Science 159, 363–374 (1995)

4. Certini, G.: Effects of fire on properties of forests soil: A review. Oecologia 143, 1–10 (2005)

5. DeBano, L.E.: The role of fire and soil heating on water repellence in wildland environments: a review. J. Hydrol. 231, 195–206 (2000)

6. DeBano, L.E., Neary, D.G., Folliot, P.F.: Fire's effects on ecosystems, 333 p. John Wiley & Son, New York (1998)

7. Díaz-Delgado, R., Lloret, F., Pons, X.: Influence of fire severity on plant regeneration by means of remote sensing imagery. Int. J. Remote Sensing 24(8), 1751–1763 (2003)

8. Doerr, S.H., Shakesby, R.A., Dekker, L.W., Ritsema, C.J.: Occurrence prediction and hydrological effects of water repellency amongst major soil and land-use types in a humid temperate climate. Eur. J. Soil Sci. 57, 741–754 (2006)

9. Franklin, S.B., Robertson, P.A., Fralish, J.S.: Small-scale fire temperature patterns in upland Quercus communities. J. Appl. Ecol. 34, 613–630 (1997)

10. Garcia–Haro, F., Gilabert, M.A., Melia, J.: Monitoring fire-affected areas using Thematic Mapper data. International Journal of Remote Sensing 22(4), 533–549 (2001)

11. Ghuman, B.S., Lal, R.: Soil temperature effects of biomass burning in windrows after clearing a tropical rainforest. Field Crops Research 22, 1–10 (1989)

12. Hammill, K.A., Bradstock, R.A.: Remote sensing of fire severity in the Blue Mountains: influence of vegetation type and inferring fire intensity. International Journal of Wildland Fire 15, 213–226 (2006)

13. Hatten, J., Zabowski, D., Schere, G., Dolan, E.A.: Comparison of soil properties after contemporary wildfireand fire suppression. Forest Ecology and Management 220, 227–241 (2005)

14. Jabukauskas, M., Lulla, K., Mausel, P.: Assessment of vegetation change in a fire- altered forest landscape. Photogrammetric Engineering and Remote Sensing 56, 371–377 (1990)

15. Key, C.H.: Ecological and sampling constraints on defining landscape fire severity. Fire Ecology 2(2), 178–203 (2006)

16. Key, C.H., Benson, N.C.: Landscape assessment: remote sensing of severity, the Normalized Burn Ratio. In: Lutes, D.C., et al. (eds.) FIREMON: Fire Effects Monitoring and Inventory System. USDA Forest Service, Rocky Mountain Research Station, Ogden, UT, General Technical Report, RMRS-GTR-164-CD:LA1-LA51 (2005)

17. Krumm, S.: WINFIT 1.2: version of November 1996 (The Erlangen geological and mineralogical software collection) of "WINFIT 1.0: a public domain program for interact ive profile-analysis under WINDOWS". In: XIII Conference on Clay Mineralogy and Petrology, Praha (1994); Acta Univers itatis Carolinae Geologica 38, 253–261 (1996)

18. Lezzerini, M., Sartori, F., Tamponi, M.: Effect of amount of material used on sedimentation slides in the control of illite "crystallinity" measurements. European Journal of Mineralogy 7, 819–823 (1995)

19. Lugassi, R., Ben-Dor, E., Eshel, G.: A spectral-based method for reconstructing spatial distribution of soil surface temperature during simulated fire events. Remote Sensing of Environment 114, 322–331 (2010)

20. Moore, D.M., Reynolds Jr., R.C.: X-ray Diffraction and the Identification and Analysis of Clay Minerals, 2nd edn., 378 p. Oxford University Press, Oxford (1997)

21. Ruiz-Gallardo, R., Castano, S., Calera, A.: Application of remote sensing and GIS to locate priority intervention areas after wildland fires in Mediterranean systems: A case study from south eastern Spain. International Journal of Wildland Fire 13(3), 241–252 (2004)

22. Srodon, J., Drits, V.A., McCarty, D.K., Hsieh, J.C.C., Eberl, D.: Quantitative X-ray diffraction analysis of clay-bearing rocks from random preparations. Clays and Clay Mineral 49, 514–528 (2001)
23. Sunar, F., Özkan, C.: Forest fire analysis with remote sensing data. International Journal of Remote Sensing 22(12), 2265–2278 (2001)
24. White, J.D., Ryan, K.C., Key, C., Running, S.W.: Remote sensing of forest fire severity and vegetation recovery. International Journal of Wildland Fire 6, 125–136 (1996)

Airborne Lidar in Archaeology: Overview and a Case Study

Nicola Masini[1] and Rosa Lasaponara[2]

[1] CNR-IBAM (Institute of Archaeological and Architectural Heritage), Potenza, Italy
n.masini@ibam.cnr.it
[2] CNR-IMAA (Institute of Methodologies for Environmental Analysis), Potenza, Italy
rosa.lasaponara@imaa.cnr.it

Abstract. Airborne laser scanning (ALS) is an optical measurement technique for obtaining high-precision information about the Earth's surface including basic terrain mapping (Digital terrain model, bathymetry, corridor mapping), vegetation cover (forest assessment and inventory), coastal and urban areas. Recent studies examined the possibility of using ALS in archaeological investigations to identify features of cultural interest, although the ability of this technology in this context has not yet been studied in detail. In this paper we provide an overview of past and present applications of ALS, as tool for detecting archaeological and palaeo-environmental features on bare surface as well as on vegetated and wooded areas. Moreover, a case study, related to an Etruscan site in Northern Latium, is showed. The LiDAR data were used for the first time, to our best knowledge, to identify looted tombs covered by vegetation.

Keywords: Overview, LiDAR, archaeology, looting, full-waveform, hill shading.

1 Introduction

The identification of archaeological features (from earthworks to surface structures), for both bare and densely vegetated areas, needs a DTM with a high accuracy. This can be obtained using ALS. It is an active remote sensing technique that provides direct measurements of earth's topography, mapped into 3D point clouds.

The laser scanner, mounted to an aeroplane or helicopter, emits in the near infrared range (for land applications), at a frequency rate of 30.000 to 100.000 pulses per second, into different directions along the flight path towards the terrain surface (see figure 1).

Each pulse could be reflected one or more times from objects (ground surface, vegetation, buildings, etc.), whose position is determined by computing the time delay between emission and each received echo, the angle of the emitted laser beam, the position of the scanner (determined using differential global positioning system and an inertial measurement unit).

B. Murgante et al. (Eds.): ICCSA 2013, Part II, LNCS 7972, pp. 663–676, 2013.
© Springer-Verlag Berlin Heidelberg 2013

There are two different types of ALS: (i) conventional scanners based on discrete echo and (ii) FW scanners. The first detect a representative trigger signal for each laser beam. The second digitize the complete waveform of each backscattered pulse; thus allowing us to improve the classification of terrain and off terrain objects, such as low vegetation, buildings, and other man-made structures lying on the terrain surface [1]. This enables to obtain DTMs with accuracy less than 0.1m and to detect archaeological structures and earthworks even under dense vegetation cover.

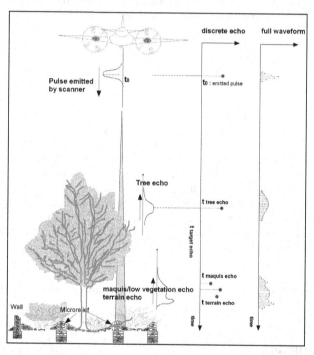

Fig. 1. Figurative examples of the main difference between the two types of ALS: (i) conventional scanners based on discrete echo and (ii) Full Wave scanners which digitize the complete waveform of each backscattered pulse.

For this aim, it is crucial to carry out accurate pre-processing namely classification of terrain and off terrain objects and post-processing to facilitate the detection of emerging archaeological remains.

This paper provides an overview on Lidar applications in archaeology (section 2) in Europe and in America since 2002. A state-of-the-Art of data processing methods follows in section 3. In section 4 a case study focused on the detection of looted tombs in an Etruscan site in Northern Latium will be showed and discussed; finally, conclusions follow.

2   The Use of Aerial LiDAR in Archaeology: An Overview on Applications

Since 2002, LiDAR dataset has been used for archaeological prospection. One of the first applications was done in UK for the identification and recording of earthwork traces of a Roman Fort in West Yorkshire [2]. The investigated site was characterized by earthworks, less than 1 m in height, and, for that, it had been missed by previous traditional aerial surveys.

In 2003 in Netherlands, LiDAR data were used for predictive modelling in the Holocene parts of the site of Eigenblok, to integrate conventional data source such as, soil, geomorphologic, geologic with palaeo-geographic maps [3].

In 2004 an accurate DEM for the entire Germany, in the framework of the project "Land and Survey bureau of the Baden-Wurttemberg State" was obtained by Sittler [4]. In Southern England, in the heart of Wiltshire Barnes, extensive remains and earthworks dating back from the Neolithic (ca. 4000–2400 BC) to the late Roman period (5th century AD) were investigated by using LiDAR [5]. Limits of conventional ALS in discriminating the low vegetation and underlying terrain have been dealt with by Pfeifer et al. [6].

One of the application on very famous sites was that performed on Stonehenge and its surrounding by Bewley et al. in 2005 who detected unknown sites hidden by vegetation and analyzed the "inter-visibility" of the monumental area as well as their spatial relationships [7]. In the same year Devereux et al. published the results of a study in a prehistoric hill-fort at Welshbury Hill, in the Forest of Dean, Gloucestershire [8] and, in Italy, Coren et al. used LiDAR data along with hyperspectral images to improve information on the archaeological area of Aquileia (North-East of Italy) [9]

In 2006 Humme et al. proposed a method based on a kriging interpolation to enhance the micro-relief like road beds, foot-paths and the earth walls surrounding the Celtic field system and filter the large scale topography component out. The application was performed on a Bronze Age village and 2500-year-old Celtic field system, near Doorwerth (East of the Netherlands) [10]. Harmon et al. [11] assessed the utility of 1 m resolution LiDAR for studying historic landscapes in two eighteenth-century plantation sites located near the Chesapeake Bay, in the state of Maryland. DSM LiDAR used in this study was obtained from the first return, whilst DTM from the last return. Relief detection was carried out by a visual analysis and also using enhancement of DEM based on hillshade surface models and contours maps.

In Ireland, in the framework of the Discovery Programme, LiDAR data were acquired in 2007 using FLI-MAP 400 system. Some studies were carried out on the basis of this dataset to map and identify archaeological features. LiDAR data were acquired for two different sites: (i) abandoned medieval settlements in Newtown Jerpoint (Kilkenny) and (ii) a prehistoric hillfort in Dún Ailinne (Kildare) [12]. In Greece, Rowland and Sarris in 2007 [13] used LiDAR combined with multi-sensor airborne remote sensing data from CASI and Airborne Thematic Mapper in order to locate the presence of exposed and known buried archaeological remains in Itanos (Eastern Crete).

In 2008 Doneus et al. [1] investigated the potential of full-waveform airborne laser scanning (RIEGL Airborne Laser Scanner LMS-Q560), to investigate an Iron Age hillfort located in a forested area called Purbach, in Austria. In the same year, Gallagher and Josephs [14] used LiDAR data to detect pre and post-European sites in the heavy woodland of Isle Royale National Park (Michigan, USA); whereas Challis et al. assessed the potential of LiDAR to enhance existing records of the historic environment of the River Dove valley [15]

In 2009, the State Office for Cultural Heritage Management of Baden-Wurttemberg, launched a three-year project aimed at obtaining the archaeological mapping of Baden-Wurttemberg using high resolution ALS data, covering an area of 35.751 km2. Within this project, Hesse developed and implemented a new tool for archaeological prospection: the Local Relief Model (LRM), based on the removal of large-scale landscape forms from the data [16]. In the same year, Danese et al. focused on the processing of DTM obtained from LiDAR using the Viewshed Analysis to obtain information about the extension of the area under the "visual control" of a medieval castle clinging on the top of hill [17].

In 2010 Chase et al. applied LiDAR-derived images in a tropical region, the jungle in Caracol, Belize, to study a very important ancient Maya site [18]. Stal et al. focused on investigating remains of trenches of the First World War around Mount Kemmel, in Flanders in Belgium [19].

In the last years Lasaponara and Masini focused on the potentiality of the latest generation of Airborne Laser Scanning (ALS), assessing its capability to detect and discriminate micro-topographic relief linked to archaeological remains from natural geomorphological features. The investigations were carried for medieval abandoned villages and an Etruscan site, exploiting also other remotely sensed date, including aerial infrared thermography and satellite multispectral imagery [20-22].

Finally, Chase et al [23] suggested that "the magnitude of change enabled by the use of LiDAR technology is particularly apparent in the archaeology of Mesoamerica" as also for other remote sensing technologies, as satellite (Lasaponara and Masini, 2012), which enabled investigations in remote areas, providing reliable data to optimize all efforts addressed from the discovery, to documentation and preservation of archaeological sites.

3   The Use of Aerial LiDAR in Archaeology: Brief Overview on Data Processing

3.1   Processing

The identification of archaeological features (from earthworks to surface structures), for both bare and densely vegetated areas, requires a very accurate DTM. To this aim, it is crucial to carry out the classification of terrain and off terrain objects by applying adequate filtering methods, which discriminate terrain and off-terrain points and eliminate outlier, such as, low points and aerial points.

Many different algorithms have been published for filtering ALS data. A list of the most commonly used filters is reported below: (i) Morphological filtering, (ii) Progressive densification, (iii) Surface based filtering, (iv) Segment based filtering, (v) Spline interpolation filtering.

(i) Morphological filtering: this group is based on the concept of mathematical morphology, a set of theoretical method of image analysis which provides a quantitative description of geometrical structures based on a set of operators Morphological filtering was published by Vosselmann, Sithole, Roggero, Lohmann et al. [24, 25, 26, 27]. Some of them are applied to point clouds [24, 25, 26], other are applied to raster data [27].

(ii) Progressive densification: this group is based the classification of the whole data set starting with a small given point cloud and increasing them iteratively. The most popular progressive densification method is the progressive triangular irregular network (TIN) densification created by Axelsson [28].

(iii) Surface based filtering: this group is based on the assumption that the whole point cloud belongs to terrain surface and then, iteratively, points are removed according a step-by-step refinement of the surface description. Surface based filtering methods were proposed by Kraus and Pfeifer and Elmqvist et al. [29, 30]

(iv) Segment based filtering: this group is based on the concept that classification is not based on single points but on segments, a set of neighbouring points with similar properties. In general, the point cloud segmentation is performed in object space or features space. In a step to step process, neighbouring points are merged to form a segment as long as their properties are similar with respect to some thresholds [31, 32].

(v) Spline interpolation filtering: to classify point cloud, first a bilinear spline interpolation and later a bicubic spline interpolation are performed, both of them with Tychonov regularization [31, 33]

3.2   Post-processing

In order to emphasize archaeological features with particular reference to micro-relief a shading procedures could be used. Several routines embedded in commercial softwares allow different solutions, such as the visualization of the elevations by using color graduations and the slope of the terrain, in order to identify the portions of the terrain that are relatively flat vs those that are relatively steep [33]. Shaded relief is obtained lighting the DTM by a hypothetical light source. The selection of the direction parameters (zenith angle z and azimuth angle) depends on the difference in height and orientation of the micro-relief of potential archaeological interest. Single shading is not the most effective method to visualize and detect micro-relief. If features and/or objects are parallel to the azimuth angle, they will not rise a shade. As a result, it would not be possible to distinguish them [33].

The problem could be solved by observing and comparing DTM scenes shaded by using different angles of lighting, as done for the case study showed in the following section.

In addition, because the different shaded DTM scenes are highly correlated Principal Components Analysis (PCA) could be used to reduce the redundancy of information [34]. PCA is a linear transformation which decorrelates multivariate data by translating and/ or rotating the axes of the original feature space, so that the data can be represented without correlation in a new component space [35]. For our application, the PCA transformed the input shaded DTMs in new components in order to make the identification of distinct features and surface types easier. The major portion of the variance is associated with homogeneous areas, whereas localized surface anomalies will be enhanced in later components, which contain less of the total dataset variance. This is the reason why they may represent information variance for a small area or essentially noise and, in this case, it must be disregarded.

4   LiDAR Application in an Etruscan Site in Blera

4.1   Study Area and State-of-the Art of Archaeological Investigations

The study area is located in the territory of Blera, province of Viterbo, about 60 km NW of Rome. It is characterized by two tufa plateaus (San Giovenale and Vignale), result of a pyroclastic flow of ignimbrite. They raise 210-230 m above sea level, and are delimited to the north by the brook Fosso del Pietrisco and to the south by the river Vesca (fig. 2).

The archaeological record certifies a long human occupation from 13th century BC to Medieval age. In particular, the excavations and surveys, started in 1870s and continued in the 1950s, revealed a centre of immense importance for the understanding of the Etruscan culture. Its development from later Bronze Age communities to Late Etruscan period is well attested [36-39].

In the San Giovenale plateau the archaeological remains show evidence of human occupation from the proto-Villanovan culture up to medieval period. In particular the main acropolis is characterized by habitation settlements, from 13th to 3rd centuries BC, in various parts of the site. San Giovenale is surrounded by several necropoleis such as Porzarago and Casale Vignale (Fig. 2), which indicate the periods represented in the habitation remains.

In the plateau of Vignale, surveys and excavations unearthed the foundations of Hellenistic houses and cellars and some archaic wells [36]. On the northern slope of the same plateau in 1959-60, the Swedish mission discovered some remains of an Etruscan bridge, which has been thought to have a connecting function between San Giovenale and Vignale. This hypothesis was recently confirmed by investigations conducted by using LiDAR [40]. These investigations allowed us to detect traces of a road on the northern slope which connected the plateau with bridge [40].

In the 70s-90s, the interest and the investigations in this area decreased. This encouraged the devastating activity of plunderers who pillaged tombs and tumulus to find artefacts, in spite of the efforts of the regional superintendence to prevent the looting.

A renewed interest in investigating this area started in 2007 thanks to the cooperation between Vignale Archaeological Project and two institutes (IMAA and IBAM) of the Italian CNR [41].

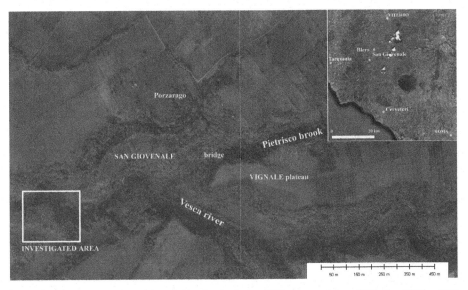

Fig. 2. Upper left: study area located in the territory of Blera, province of Viterbo (60 km NW of Rome). Main scene: the investigated area, denoted by a white rectangular box, nearby San Giovenale and Vignale plateau and the confluence between Vesca river and Pietrisco brook

The integrated use of different airborne and space remote sensing methods, along with GPS measurements, ultralight aerial photography and archaeological recognition allowed us to provide more information on the Vignale plateau and its surrounding.

In particular, as above-mentioned, the Full waveform laser scanning allowed us to detect and map a connecting road, covered by woodland, between the Vignale plateau and the main acropolis of San Giovenale. Moreover, satellite data processing, LiDAR and aerial infrared thermography provided new data on the presence of a habitation site on the westernmost part of the plateau and some necropoleis on the eastern part of the Vignale hill [40].

4.2   Investigation Aims, Data Processing and Discussion of Results

The Etruscan archaeological heritage is sorrowfully famous to have been and still be affected by illegal diggings that are strongly connected with the illicit trade of antiquities [42, 43].

Presently a rich remote sensing dataset is available for this area [40] and already used to analyze landscape as well as to detect and map bare soil and vegetated areas damaged by anthropic activities, including the plundering of tombs.

Therefore, the processing and the interpretation of LiDAR data has been herein oriented to the identification of areas affected by looting. The work is in progress, and, unfortunately for the archaeological heritage (!), the LiDAR is unearthing wide areas damaged by old and recent illegal diggings.

In this paper, we show and discuss the results obtained on a little plateau covered by dense woodland, named Montevangone, where, in the 50s, some tombs were excavated. It is located at south of Vesca river and at 0,5 km South-East of San Giovenale (see Figure 2 and 3).

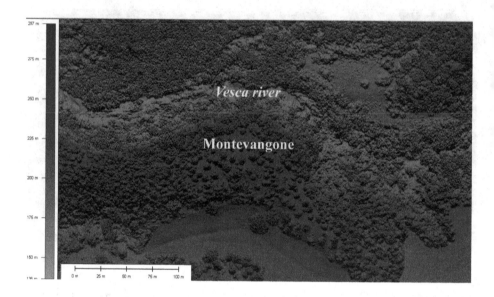

Fig. 3. The investigated area: Montevangone plateau. The DSM evidences the dense woodland cover.

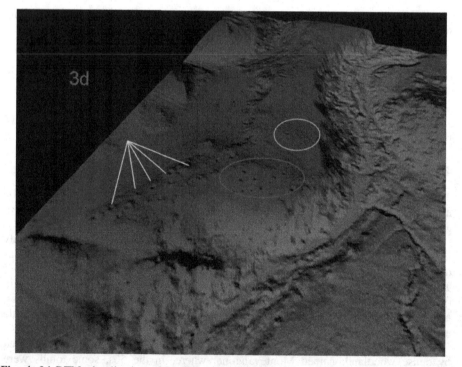

Fig. 4. 3d DTM visualization of Montevangone plateau. The removal of vegetation put in evidence some circular features referable to looted tombs.

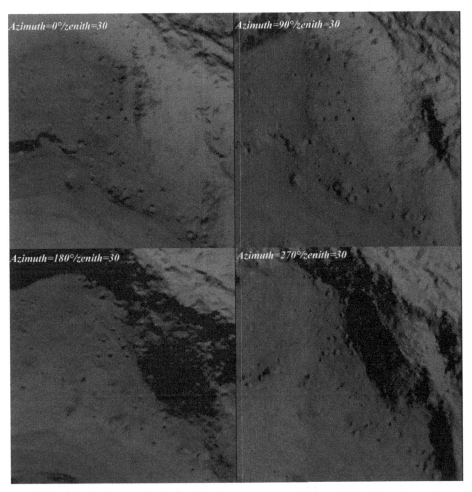

Fig. 5. Subset of Montevangone plateau DTM. Hill shading procedure has been performed in order to improve the visibility of the circular features.

The area has been surveyed by GEOCART on September 2010, by using a FW scanner, RIEGL LMS-Q560 on board a helicopter. The scanner acquired data in South-North and East-West directions, adopting a pulse repetition rate of 180.000Hz, with a divergence of the radius 0.5mrad. The average point density value has been about 20 points/m2 with an accuracy of 25 cm in xy and 10 cm in z (altitude).

The processing of point clouds aimed at classifying terrain and off terrain objects has been performed using a strategy based on a set of "filtrations of the filtrate". Appropriate criteria and threshold for the classification and filtering were set to gradually refine the intermediate results in order to discriminate and classify low, isolated and air points. Then, classification of ground points and points by class (building, different height vegetation) followed.

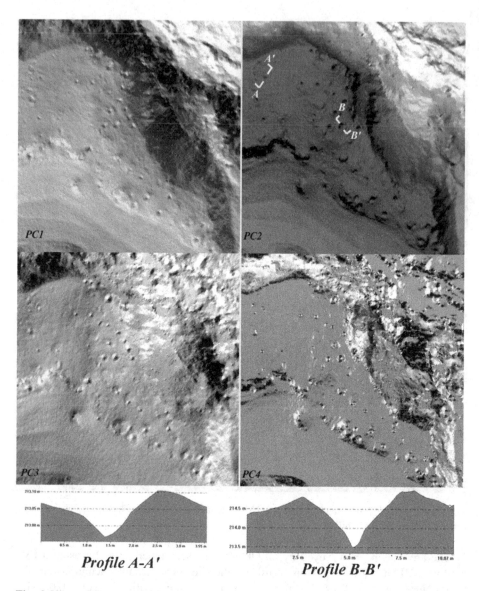

Fig. 6. Upper: PCA of shaded DTMs of Montevangone plateau; lower: profiles of two circular features indicated in the PC2 scene.

Finally the extraction of DTM has been obtained by using the progressive Triangulation Irregular Network (TIN) densification method (see section 3.1) by Axelsson [28] embedded in Terrasolid's Terrascan commercial software. The procedure starts from a coarse TIN surface. Then new points are added, though an iterative way, if they meet criteria based on distances to TIN facets and angles to the vertices of the triangle.

The accuracy of DTM made possible the identification of a large number of circular 'craters', clearly referable to looted tombs or attempts to find and plunder tombs.

In order to emphasize the visibility of these circular microrelief we used shading procedures (see section 3.2), employing routines embedded in ENVI and Global Mapper to visualize elevations and slopes by using colour graduations.

Hill shading has been performed by using different lighting azimuth (0-90-180-279°) and zenit angles (30°, 45°).

In addition, the different shaded DTM scenes have been processed by using the Principal Components Analysis (PCA) [33] to represent them in a new component space, emphasizing the less correlation and identifying better the borders of circular features. As a whole the results of PCA, in particular the second component improve the visibility of microrelief (compare Figure 6 upper, with Figure 5). This made possible to carefully identify and map the features linked to looting activity. In the area object of investigation (1.12 Ha) about 45 circular 'craters' have been identified and surveyed. Their measuring show diameters of the edges ranging 4 to 6.5 m and depth which vary from 0.20 to 1.40 mt. The measure of depths (Figure 6, lower) allowed us to distinguish two groups: 1) one with depths ranging 0.20 to 0.50 m, 2) another one with depths which vary from 0.70 to 1.40mt. The first is mostly located in the central part of the plateau, the second nearby the south-western slope. The different depth could be explained by different time of plundering, the deeper craters should be more recent, vivecersa the shallower craters should indicate an older looting.

5   Conclusions

In this paper, we provided on overlook on the use of Airborne Laser Scanning for archaeological purposes. The overview on the State of Art of ALS in Archaeology is also enriched with a focus on details and rational basis of LiDAR data processing (from data filtering to classification). Finally, we also showed the potential of using point clouds obtained by an aerial full-waveform laser scanner for an archaeological area in Middle of Italy. For the study area, the vegetation cover strongly limited the effectiveness of remotely sensed optical data, and, therefore, the use of LiDAR data enabled to overcome this limit.

From the methodological point of view, we set up a data-processing chain based on two main steps: 1) the classification of terrain and off terrain objects performed using local thresholds values to discriminate terrain and off terrain points; and 2) the post processing based on both illumination manipulations (DTM shaded by using different angles of lighting) and statistical analyses, such as, PCA and geospatial tools.

Such approach allowed us to to detect traces of looting under densely wooded area.

References

1. Doneus, M., Briese, C., Fera, M., Janner, M.: Archaeological prospection of forested areas using full-waveform airborne laser scanning. Journal of Archaeological Science 35(4), 882–893 (2008)

2. Holden, N., Horne, P., Bewley, R.: High-resolution digital airborne mapping and archaeology. In: Bewley, R.H., Raczkowski, W. (eds.) Aerial Archaeology: developing future practice, NATO Science Series, sub-series I: Life and Behavioural Sciences, pp. 173–180. IOS Press, Netherlands (2002)

3. Van Zijverden, W.K., Laan, W.N.H.: Landscape reconstructions and predictive modeling in archaeological research, using a LIDAR based DEM and digital boring databases. In: Archeologie und Computer, Workshop 7, Vienna, Austria (2003),
http://www.archeologie.leidenuniv.nl/content_docs/research/
vanzijverden_laan_2005_landscape__reconstructions.pdf

4. Sittler, B.: Revealing Historical Landscapes by Using Airborne Laser Scanning. A 3-D Modell of Ridge and Furrow in Forests near Rastatt (Germany). In: Thies, M., Koch, B., Spiecker, H., Weinacker, H. (eds.) Proceedings of Natscan, Laser-Scanners for Forest and Landscape Assessment - Instruments, Processing Methods and Applications. International Archives of Photogrammetry and Remote Sensing, vol. XXXVI, pp. 258–261, Part 8/W2 (2004)

5. Barnes, I.: Aerial remote-sensing techniques used in the management of archaeological monuments on the British Army's Salisbury Plain Training Area, Wiltshire, UK. Archaeological Prospection 10, 83–91 (2003)

6. Pfeifer, N., Gorte, B., Elberink, S.O.: Influences of vegetation on laser altimetry e analysis and correction approaches. In: Thies, M., Koch, B., Spiecker, H., Weinacker, H. (eds.) Proceedings of Natscan, Laser-Scanners for Forest and Landscape Assessment - Instruments, Processing Methods and Applications. International Archives of Photogrammetry and Remote Sensing, vol. XXXVI, pp. 283–287, Part 8/W2 (2004)

7. Bewley, R.H., Crutchley, S.P., Shell, C.: New light on an ancient landscape: lidar survey in the Stonehenge World Heritage Site. Antiquity 79, 636–647 (2005)

8. Devereux, B.J., Amable, G.S., Crow, P., Cliff, A.D.: The potential of airborne lidar for detection of archaeological features under woodland canopies. Antiquity 79, 648–660 (2005)

9. Coren, F., Visintini, D., Fales, F.M., Sterzai, P., Preparo, G., Rubinich, M.: Integrazione di dati laser scanning ed iperspettrali per applicazioni archeologiche. In: Atti 9a Conferenza Nazionale ASITA, Catania, Novembre 15-18 (2005)

10. Humme, A., Lindenbergh, R., Sueur, C.: Revealing Celtic fields from lidar data using kriging based filtering. In: Proceedings of the ISPRS Commission V Symposium, Dresden, September 25-27, vol. XXXVI, part 5, pp. 25–27 (2006)

11. Harmon, J.M., Leone, M.P., Prince, S.D., Snyder, M.: Lidar for archaeological landscape analysis: a case study of two eighteenth-century Maryland plantation sites. American Antiquity 71(4), 649–670 (2006)

12. Corns, A., Shaw, R.: High resolution LiDAR for the recording of archaeological monuments & landscapes. In: Lasaponara, R., Masini, N. (eds.) Advances in Remote Sensing for Archaeology and Cultural Heritage Management, Aracne, Roma, pp. 99–102 (2008)

13. Rowlands, A., Sarris, A.: Detection of Exposed and Subsurface Archaeological Structures Using Multi-sensor Remote Sensing. Journal of Archaeological Sciences 34(5), 795–803 (2007)

14. Gallagher, J.M., Josephs, R.L.: Using LiDAR to Detect Cultural Resources in a Forested Environment: an Example from Isle Royale National Park, Michigan, USA. Archaeological Prospection 15, 187–206 (2008)

15. Challis, K., Kokalj, Z., Kincey, M., Moscrop, D., Howard, A.J.: Airborne lidar and historic environment records. Antiquity 82(318), 1055–1064 (2008)

16. Hesse, R.: LiDAR-derived Local Relief Models (LRM) – a new tool for archaeological prospection. Archaeological Prospection 17, 67–72 (2010)

17. Danese, M., Biscione, M., Coluzzi, R., Lasaponara, R., Murgante, B., Masini, N.: An Integrated Methodology for Medieval Landscape Reconstruction: The Case Study of Monte Serico. In: Gervasi, O., Taniar, D., Murgante, B., Laganà, A., Mun, Y., Gavrilova, M.L. (eds.) ICCSA 2009, Part I. LNCS, vol. 5592, pp. 328–340. Springer, Heidelberg (2009)

18. Chase, A.F., Chase, D.Z., Weishampel, J.F., Drake, J.B., Shrestha, R.L., Slatton, K.C., Awe, J.J., Carter, W.E.: Airborne LiDAR, archaeology, and the ancient Maya landscape at Caracol, Belize. Journal of Archaeological Science 38, 387–398 (2010)

19. Stal, C., Bourgeois, J., De Maeyer, P., De Mulder, G., De Wulf, A., Goossens, R., Nuttens, T., Stichelbaut, B.: Kemmelberg (Belgium) case study: comparison of DTM analysis methods for the detection of relicts from the First World War. In: Proc. 30th EARSeL Symposium: Remote Sensing for Science, Education and Culture (2010)

20. Lasaponara, R., Masini, N.: Full-waveform Airborne Laser Scanning for the detection of medieval archaeological microtopographic relief. Journal of Cultural Heritage 10S, e78–e82 (2009)

21. Lasaponara, R., Coluzzi, R., Gizzi, F.T., Masini, N.: On the LiDAR contribution for the archaeological and geomorphological study of a deserted medieval village in Southern Italy. Journal Geophysics Engineering 7, 155–163 (2010)

22. Lasaponara, R., Masini, N., Holmgren, R., Backe Forsberg, Y.: Integration of aerial and satellite remote sensing for archaeological investigations: a case study of the Etruscan site San Giovenale. Journal of Geophysics and Engineering 9, S26–S39 (2012)

23. Chase, A.F., Chase, D.Z., Fisher, C.T., Leisz, S.J., Weishampel, J.F.: Geospatial revolution and remote sensing LiDAR in Mesoamerican archaeology. PNAS 109(32), 12916–12921 (2012), http://www.pnas.org/cgi/doi/10.1073/pnas.1205198109

24. Vosselman, G.: Slope based filtering of laser altimetry data. International Archives of the Photogrammetry. Remote Sensing and Spatial Information Sciences XXXIII (pt. B3), 935–942 (2000)

25. Sithole, G.: Filtering of laser altimetry data using a slope adaptive filter. International Archives of the Photogrammetry, Remote Sensing and Spatial Information Sciences XXXIV(pt. 3/W4), 203–210 (2001)

26. Roggero, M.: Airborne laser scanning: clustering in raw data. International Archives of Photogrammetry and Remote Sensing XXXIV(B3/W4), 227–232 (2001)

27. Lohmann, P.: Segmentation and filtering of laser scanner digital surface models. In: IAPRS, Xi'an, China, August 22-23, vol. 34(2), WG II/2, pp. 311–315 (2000)

28. Axelsson, P.: DEM generation from laser scanner data using adaptive TIN models. In: IAPRS, Amsterdam, Netherlands, vol. XXXIII, B4, pp. 111–118 (2000)

29. Kraus, K., Pfeifer, N.: Determination of terrain models in wooded areas with airborne laser scanner data. ISPRS JPRS 53, 193–203 (1998)

30. Elmqvist, M.: Ground estimation of laser radar data using active shape models. In: Proc. OEEPE Workshop on Airborne Laserscanning and Interferometric SAR for Detailed Digital Elevation Models, March 1-3, p. 8. OEEPE Publication no. 40 (on CD-ROM) (2001)

31. Sithole, G., Vosselman, G.: Experimental comparison of filtering algorithms for bare-earth extraction from airborne laser scanning point clouds. ISPRS Journal of Photogrammetry and Remote Sensing 59(1-2), 85–101 (2004)

32. Tovari, D., Pfeifer, N.: Segmentation based robust interpolation - a new approach to laser data filtering. In: ISPRS Workshop Laser Scanning 2005 (2005)

33. Brovelli, M.A., Reguzzoni, M., Sansò, F., Venuti, G.: Modelli matematici del terreno permezzo di interpolatori a spline. Bollettino SIFET, Supplemento Speciale 2, 55–80 (2001)

34. Masini, N., Coluzzi, R., Lasaponara, R.: On the Airborne Lidar Contribution in Archaeology: from Site Identification to Landscape Investigation. In: Wang, C.-C. (ed.) Laser Scanning, Theory and Applications, pp. 263–290. Intech (2011) ISBN 978-953-307-205-0

35. Kokalj, Z., Zaksek, K., Ostir, K.: Visualizations of lidar derived relief models. In: Opitz, R.S., Cowley, D.C. (eds.) Interpreting Archaeological Topography. 3d Data, Visualization and Observation, pp. 100–114. Oxbow Books, Oxford (2012)

36. Richards, J.A.: Remote Sensing Digital Image Analysis. Springer (1986)

37. Boethius, A., Fries, C., Gjerstad, E., Hanell, K., Östenberg, C.E., Thordeman, B., Welin, E., Wetter, E.: Etruscan Culture: Land and People: Archaeological Research and Studies Conducted in San Giovenale and its Environs by Members of the Swedish Institute. Columbia University Press, New York (1962)

38. Thomasson, B.: San Giovenale 1:1. General introduction (ActaRom-4o, 26:1:1), Stockholm (1972)

39. Pohl, I.: Nuovi contributi alla storia dell'abitato Etrusco di San Giovenale nel periodo fra il 500–200 a.C. Parola Passato 40, 43–63 (1985)

40. Karlsson, L.: San Giovenale, 4:1. Area F East. Huts and houses on the Acropolis (ActaRom-4o, 26:4:1), Stockholm (2006)

41. Lasaponara, R., Masini, N., Holmgren, R., Backe Forsberg, Y.: Integration of aerial and satellite remote sensing for archaeological investigations: a case study of the Etruscan site San Giovenale. Journal of Geophysics and Engineering 9, 26–39 (2012), doi:10.1088/1742-2132

42. Backe Forsberg, Y., Holmgren, L.A., Lasaponara, R., Masini, N.: Airborne and satellite multispectral imagery at the Etruscan site of San Giovenale, Blera (Lazio)—preliminary results Advances. In: Lasaponara, R., Masini, N. (eds.) Remote Sensing for Archaeology and Cultural Heritage Management, pp. 225–228. Aracne, Roma (2008)

43. van Velzen, D.T.: The world of Tuscan tomb robbers: Living with the local community and the ancestors. International Journal of Cultural Property 5, 111–126 (1996)

44. Elia, R.J.: Looting, Collecting, and the Destruction of Archaeological Resources. Non Renewable Resources 6(2) (1997)

A Model of Controlling Utilization
of Social Grants in South Africa

Qhayisa S. Cwayi[1] and Okuthe P. Kogeda[2]

Tshwane University of Technology, Computer Science Department,
Faculty of Information Communication Technology
Private Bag X680, Pretoria 0001, South Africa
{cwayiqs,kogedaPO}@tut.ac.za

Abstract. South Africa is one of the most unequal countries in the World. Social grant in South Africa is supported by the Social development macro policy framework since 1994. It was aimed at poverty alleviation that combines social and economic goals. Government over the years has been faced with a number of challenges in social grants and benefits administration such as fragmented institutional arrangements and a lack of uniformity, fraud and corruption. Despite the fact that the government took steps in minimizing fraud and corruption associated with the social grants administration, the challenges of utilization of funds disbursed still exist. This research seeks to design and implement a system that controls utilization of social grants using Near Field Communication technology. In this paper, we present the system design and architecture of the system to be implemented. We present the findings of a preliminary study of the utilization of social grants by recipients. We collected data by interviewing social grant recipients in South Africa. The results of this preliminary study shows that 62% of social grant recipients use their funds within a week, 82% within two weeks, and 92% within three weeks. Only 8% use the funds until the fourth week.

Keywords: Social Grant Utilization, Mobile Phone, Contactless card, Near Field Communication (NFC), Short-range Communication, interviews.

1 Introduction

Social grants are monies paid to the needy of the society. South Africa is one of the unequal societies in the World. Despite high economic growth witnessed since 2000, it has not trickled down to the large population of the country. An official unemployment rate of 26%, for poverty rate estimated at approximately 50% are the most severe measures of inequality in the world [1]. South Africa faces substantial challenges in addressing poverty, inequality and unemployment [5, 8]. While South Africa ranks as an upper-middle income country based on average income, some of the national social indicators are similar to those of the poorest countries of the world [2]. South African social security system is the government chief initiative in tackling these challenges. South African social security system expands coverage of grants and

B. Murgante et al. (Eds.): ICCSA 2013, Part II, LNCS 7972, pp. 677–692, 2013.
© Springer-Verlag Berlin Heidelberg 2013

limit increase in inequality. The South African Social Security Agency (SASSA) is responsible for the management, administration and payment of the social assistance grants. Social assistance is an income transfer in a form of grants provided by government [17]. Over the years, there has been endemic corruption in the department administering the funds. There were fraudulent claims, impersonation of recipients, dead claimants, etc. awash in the system [14, 17]. In 2012, the department developed and implemented a biometric-based system, which weeded a lot of fraudulent beneficiaries [14, 15, and 16]. The system has greatly minimized fraud and corruption related to the administration of the funds. However, there still remains a big challenge of utilization of the funds disbursed. Most of the recipients find themselves without cash to sustain themselves before the next payment.

There has been alleged misuse of funds by the recipients. Many social grant recipients purchase impulsively and nonessential items (i.e., alcohol, designer clothes, drugs, etc.). Some recipients lack vital skills of money management. Therefore, in this paper, we propose a model for controlling the utilization of social grants using Near Field Communication (NFC) technology that seeks to spread the spending of the monies received over the month. In this paper, we present the design, architecture and findings of a preliminary study of how social grant recipients spend their monies. Interviewing the recipients at the pay points (oral), at their places of work (web based and questionnaire); we collected data, compiled and analyzed it. The design, architecture and implementation of the proposed system have been based on the findings of the preliminary study, which provides solution to the utilization of social grant funds.

The rest of this paper is organized as follows: In Section 2, we provide the evolution of social grants in South Africa. In Section 3, we present different types of social grants in South Africa. In Section 4, we present the spatial distribution of social grants in South Africa. In Section 5, we present Near Field Communication and discuss reasons why we chose it. In Section 6, we present related work. The system design and architecture are presented in Section 7. In Section 8, we present the methodology. In Section 9, we present data analysis and results, and then conclude in Section 10.

2   Evolution of Social Grants in South Africa

The South African Constitution of 1996 confirms the country rejection of the cultural politics of exclusion, which characterized South Africa under apartheid [5]. The Constitution states that, within its available resources the state is appreciative to take reasonable legislative and other measures to progressively realize the right to social security to assist those who are unable to support themselves and their dependents [5]. With regards to children, the Constitution is even stronger as it does not include any limitation to this right in terms of the availability of resources [6].

To begin to give effect to these rights, a white paper for social welfare was drawn up in 1997 that sought to move away from healing services towards preventative programs [7]. In December 1995, the South African government established the Lund Committee [22] to evaluate the current social protection system and provide advice on ways to improve it. The committee recommended an Old Age Person grant (OAG), Disability grant, War Veterans grant, Grant-in-Aid grant, Care Dependency grant, and

Child Support Grant (CSG) that reaches a greater number of children and families. The aim was to target and provide the benefit for families in the poorer 30% of households [22]. Previously, the government provided a limited State Maintenance Grant (SMG) that did not target the poorest children and families equitably, as it went largely to poor colored and Indian women and their children at the cost of poor "black" women and their children [8]. In an effort to address this difference, the Lund Committee had to develop a grant that would reach significantly more poor families than the existing SMG.

However, it was instructed by the government to do so within the existing child grant budget and the other grants like Old Age pension grant, which ruled out simply expanding the reach of the SMG [7]. To meet these specific requirements, the committee proposed that the current SMG, which was then at a monthly rate of R127 per child (up to the age of 18) and R410 for a parent, should be phased out and replaced by a new Child Support Grant (CSG) at a monthly rate of R70 for children up to the age of nine. In March 1997 the South African Cabinet agreed to the phasing out of the SMG in favor of the CSG but changed the amount to R75 a month, and the suitability to children only up to the age of seven. The modification made the grant more accessible to caregivers and children when enabled the program to substantially increase the participation rate. Recipients of the grant were required to pass a mean test which was based on household income. Families living in rural areas earning a household income below R800 per month and families in urban areas earning a household income below R1100 were eligible to receive the grant. This was extended in 2004 and 2005 to include children up to the age of eleven and fourteen respectively. In 2008 children under the age of fifteen became eligible to receive the grant. Currently a child is eligible until their eighteenth birthday [9].

In 2012 the government implemented a Biometric-based registration system, which reduced fraud and corruption in social grants disbursement [14]. The Biometric-Based system is also able to identify the right beneficiaries. However, the spending period of social grant by beneficiaries is still a challenge and needs to be addressed so as to reduce crime and poverty in South Africa.

3   Types of Social Grants

South Africa classified social grants into eight. These include: Old Age Grant, War Veteran's Grant, Disability Grant, and Grant in Aid, Care Dependency Grant, Foster Child Grant, Child Support Grant and Social Relief of Distress [17, 20].

3.1   Old Age Grant

An old age grant is a monthly income provided by SASSA to older males and females who are 60 years old and above. The spouse must comply with the means test, must not be maintained or cared for in a state institution and must not be in receipt of another social grant for him or herself to qualify. Since April 2011, SASSA pays an amount of R1, 140.00 per month and R1, 160.00 per month for beneficiaries older than 75 years is granted. A total of 2, 862,570 people are receiving this type of grant by February 2013.

3.2   War Veteran's Grant

War Veteran's Grant is a monthly income provided by SASSA to former soldiers who fought in the First World War (1904-1918), Second World War (1939-1945), the Zulu uprising (1906) or the Korean War and who are unable to support themselves.

In addition to the above requirement, one has to be 60 years old and above or must be disabled, the spouse must meet the requirements of the means test, must not be maintained in a state institution, and must not be in receipt of another social grant in respect of himself or herself to qualify. As from April 2011 an amount of R1, 160.00 per month is granted. By February 2013 there were 589 recipients of this grant.

3.3   Disability Grant

Disability Grant is a monthly income provided by SASSA to people who have a physical or mental disability which makes them unfit to work, for a period of longer than six months. Permanent disability grant is granted to people whose disability continues for more than a year and a temporary disability grant if the disability continues for a continuous period of not less than six months and not more than twelve months. A permanent disability grant does not mean that a person will receive the grant for life, but just that it will continue for longer than 12 months.

An individual must be 18 to 59 years old of age, submit a medical report (not older than three months) confirming disability, spouse meets the requirements of the means test, must not be maintained in a state institution and must not be in receipt of another grant in respect of him or herself to qualify for this type of grant. An amount of R1, 140.00 is granted for this type of grant per month as from April 2011. A total of 1,168,464 recipients are on this grant by February 2013.

3.4   Grant in Aid

Grant in Aid is a monthly income provided by SASSA to people living on a social grant i.e., either disability grant, war veteran's grant or grant for older persons but cannot look after themselves owing to physical or mental disabilities, and therefore need full-time care from someone else. It is an additional grant to pay the person who takes full-time care of such a person.

The recipient must not be cared for in an institution that receives subsidy by state for housing of such beneficiary. An amount of R260.00 is granted per month as from April 2011. By February 2013, there were 72,767 recipients of this grant.

3.5   Foster Child Grant

Foster Child Grant is a monthly income provided by SASSA to a care giver of a fostered child. A foster child is a child who has been placed in the custody of a care giver by a court as a result of being either orphaned, abandoned, at risk, abused or neglected. An amount of R740.00 is granted per month as from April 2011. A total of 522,181 recipients were on this type of grant by February 2013.

3.6   Care Dependency Grant

Care Dependency Grant is a monthly income provided by SASSA to a care giver who takes care of a child who has a severe disability and is in need of full-time and special care. A child must be under the age of 18 years old, must submit a medical report confirming permanent or severe disability, applicant and spouse must meet the requirements of the means test (except for foster parents), the care dependant child or children must not be permanently cared for in a state institution to qualify. An amount of R1, 140.00 is granted per month as from April 2011. A total of 119,384 recipients were on this grant by February 2013.

3.7   Child Support Grant

Child Support Grant is a monthly income provided by SASSA to needy child's primary caregiver e.g., parent, grandparent or a child over 16 years heading a family. The child and applicant must reside in South Africa, must be the primary care giver of the child or children concerned, the applicant and spouse must meet the requirements of the means test, the applicant cannot apply for more than six non biological children and child cannot be cared for in state institution to qualify for this type of grant. An amount of R260.00 is granted per month as from April 2011. A total of 11,314,128 recipients were on this type of grant by February 2013.

3.8   Social Relief of Distress Grant

This is a monthly income provided by SASSA to persons in such dire need that they are unable to meet their and or families' most basic needs for a maximum period of 3 months. An extension for further 3 months may be granted in exceptional cases. The applicant must be awaiting payment of an approved social grant, applicant has been found medically unfit to undertake remunerative work for a period of less than 6 months, no maintenance is received form a spouse, parent or child obliged in law to pay maintenance and proof is furnished that efforts made to obtain maintenance have been unsuccessful, the bread winner is deceased, the bread winner of that person's family has been admitted to an institution funded by the state (i.e., prison, psychiatric hospital, state home for older persons, treatment centre for substance abuse or child and youth care centre, applicant has been affected by a disaster as defined in the Disaster Management Act or the Fund Raising Act, 1978 [23], and the person is not receiving assistance form any other organization as requirements to qualify for this type of grant.

4   Spatial Distribution of Social Grants

South Africa consists of 9 provinces governed by provincial governments headed by Premier. These provinces include: Eastern Cape (EC), Free State (FS), Gauteng Province (GP), KwaZulu Natal (KZN), Limpopo Province (LP), Mpumalanga Province (MP), North West (NW), Northern Cape (NC), and Western Cape (WC). The most common grant is Child Support Grant (CSG) followed by Old Age Grant (OAG), etc. This spatial distribution of social grants in South Africa is shown in Table 1 and Figure 1.

Table 1. Number of Grant Recipients by Grant Type and Region as at Feb. 2013 [17]

Region	Grant Type							
	OAG	WVG	DG	GIA	CDG	FCG	CSG	Total
EC	507,573	75	185,459	9,261	18,264	115,133	1,841,399	2,677,164
FS	171,320	8	86,522	1,185	5,825	40,118	633,776	938,754
GP	422,265	148	123,880	1,609	15,630	57,826	1,573,790	2,195,148
KZN	589,547	86	313,946	29,079	35,875	134,024	2,751,183	3,853,740
LP	394,150	47	88,784	11,044	11,782	56,909	1,581,874	2,144,590
MP	226,558	28	81,211	2,832	8,566	34,594	1,048,041	1,401,830
NW	216,524	19	86,296	4,043	8,278	41,382	748,365	1,104,907
NC	74,604	17	49,319	4,180	4,435	13,885	275,935	422,375
WC	260,029	161	153,047	9,534	10,729	28,310	859,765	1,321,575
Total	2,862,570	589	1,168,464	72,767	119,384	522,181	11,314,128	16,060,083

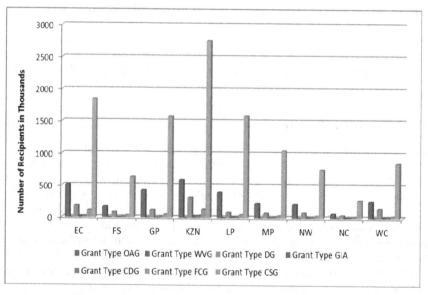

Fig. 1. Spatial Distribution of Social Grants in South Africa

5   Near Field Communication

This is a type of communication technology that wirelessly communicates with another at a maximum distance of around 20 centimeters [4]. NFC was invented by Nokia, Philips and Sony. It is defined by ISO/IEC 14443 as a standard that uses magnetic fields induction to enable communication between devices when they are touched together or closer [19]. NFC can be used in 3 ways:

- Taking pictures with mobile phone with a built in camera, and touch an enabled computer or television set to transmit the images for display.

- Downloading applications or games to a handheld device by touching the computer.
- Works in conjunction with another wireless technology, you could transfer large files between two devices, such as a laptop and a desktop, simply by touching the two together.

NFC operates at 13.56 MHz and rates ranging from 106 Kbit/s to 848 Kbit/s. NFC always involve an initiator (battery·powered) and a target (radio energy powered). The initiator actively generates a Radio Frequency (RF) field that can power a passive target. There are three main ways of using NFC:

- Card Emulator: the NFC device behaves exactly like a contactless card and can be used in transport fare payment systems based on Felica as well as open banking payment systems based on Visa, MasterCard or SASSA card.
- Reader Mode: the NFC device is active and reads a passive Radio Frequency Identification (RFID) tag; for example reading and storing a Web address or voucher from the poster from interactive advertisement.
- Person-to-Person (P2P) mode: two NFC devices communicate with each other exchanging information.

5.1   Why Near Field Communication?

The reasons why we decided on NFC technology over other available technologies include [4, 18, 19]:

- NFC is simple. It is a short-range, low power wireless link developed from radio-frequency identification (RFID) tech that can transfer small amounts of data between two devices held a few centimeters from each other.
- Ease of use – with NFC there is no need to carry multiple credit cards; reward cards from different stores, etc. instead load them on your phone.
- Security – it creates secure channel for communication and uses data encryption when sending sensitive messages between your phone and another device. If your wallet containing all your debit cards, credit cards, etc. is stolen that person will have instant access to them while you can password protect your Smartphone.
- Versatility – it covers a variety of uses including checking out a store, purchase and load concert tickets to their Smart phones, read information from smart poster, etc. from a single device. NFC strives to embrace both work and play.
- NFC technology does not need any pairing code, and because it is very low power, no battery in the device being read.
- It is affordable – it costs less to install and operate as most people already own smart phones.

6   Related Work

Achieving social grant utilization control is not a recent problem. There are many proposed solutions with very little practicability. Some of the related works include:

6.1   Cash Payment Programs

In traditional cash transfer programs, cash is physically delivered to a set of pay points like post offices or government offices [11]. Program recipients travel to these pay points to collect cash payments at a set time. Physically delivering cash incurs high transport costs and security risks for the program provider. In addition, these pay points are often infrequent, especially in rural or remote areas, so recipients often have to travel long distances to get to the pay points. This can also involve costs that eat into the value of the cash transfer if recipients have to pay for transport or spend hours travelling and queuing to collect the cash. This has proven to be a considerable barrier for the most vulnerable recipients, especially older people, people with disabilities, or those who are unable to travel due to ill-health. While Cash Payment System concerns was with the disbursement of cash to the social grant recipients, our system goes further to help the recipient in utilizing the funds available to him/her by restricting the amount of money he/she can use in a week and the type of items he/she can buy depending on the type of social grant.

6.2   SASSA Biometric-Based Payment Card System

South Africa began rolling out a new biometric based payment system for social grant beneficiaries that is aimed at cut down on fraud and corruption in social grant administration system [17]. The SASSA Biometric based payment card system can be used to access social grant anywhere in the country and at any time using multiple payment channels, such as ATMs, cash pay-points, and selected merchant stores. The card may also be used to purchase goods at a participating payment vendor (such as Boxer, Pick & Pay, Checkers, Shoprite, Spar) having a point of-sale (POS) device, purchase airtime, pay water and electricity accounts, or open accounts. You need not withdraw all your money at once and can save money for a rainy day. However, the Biometric-Based Payment card system is different to the proposed research, because we propose to control the utilization of the social grant money. The Biometric Based System does not monitor the utilization of the funds by the beneficiaries. Biometric Based System is helping in reducing fraud and corruption, but remaining challenge is how the beneficiaries spending their monies.

6.3   Mobile Device Based e-Payment System

Mobile Device Based e-payment System was proposed to use a mobile device for ticketing. The idea was to us Bluetooth for Personal Area Network (PAN) communications. Bluetooth was used, because it is a good solution for PAN applications with less time-critical transactions, transfer of large amount of data or transfer of the transition process to the mobile device. Ticketing and payment usually have different requirements: the amount of data is usually small, but the transaction needs to be performed fast (less than a second) [10]. Bluetooth is quite popular for PAN transaction as it is already integrated into lots of mobile devices. The Mobile Device Based e-payment System differs from the proposed system because it focuses on less time critical transactions; while the proposed system focuses on controlling utilization of grant monies. The proposed system uses NFC technology to secure transactions, whereas Mobile Devise Based e-payment System is using Bluetooth for PAN transactions.

7   System Design and Architecture

The system is designed to monitor the classical architectural design of social grant payments to control the utilization of the social grant money used by social grant beneficiaries. The details of all the social grant beneficiaries are stored and accessible from the SASSA database. Social grant beneficiaries, service providers (i.e., retail outlets) can request for information from SASSA database, which requires authorization for either completion of a transaction or confirmation of payment. The transaction can only be completed if there are enough funds; items to be purchased are in the list of the social grant beneficiary to purchase, among other conditions. The service can also be declined depending on the details (insufficient funds, type of product to purchase or beneficiary not registered) of the beneficiary. Beneficiaries have two options; it is either the beneficiary use NFC card or use mobile phone as shown in Figure 2.

The beneficiaries using NFC mobile phone, which fall under Peer-To-Peer (P2P) mode, uses a phone as a buying tool, sends the grant details to retailer. The retailer has an NFC adapter, and then gets beneficiary details (Account, grocery items) to make a request from SASSA. SASSA gets purchase request from the retailer for grant holder, and then checks if the beneficiary has money and buying allowed items. SASSA denies if the beneficiary do not qualifies or allow if qualifies. The Proposed Social Grant Architecture is not limited with other services like security and management is being addressed in every service.

The functions performed by beneficiary, SASSA and retailers are shown in Use Case form in Figure 3. The designs of the classes used in this system are shown in Figure 4.

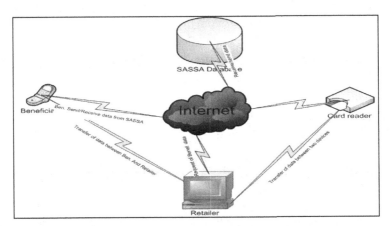

Fig. 2. Proposed Social Grant Architecture

The system operates following some business rules. These rules include:

- The system is requiring allowing the Beneficiaries to register account, to view their accounts, to make payments to different retail shops. Beneficiaries can also change their passwords, and make request for their account balance.

- It must also allow the Retailers to confirm if the beneficiary has got enough money to make payment, has to get approval from the SASSA database.

- The SASSA will manage all the accounts of the beneficiaries, register and delete beneficiaries, approve decline of payments. SASSA will also do the update of beneficiaries every year.

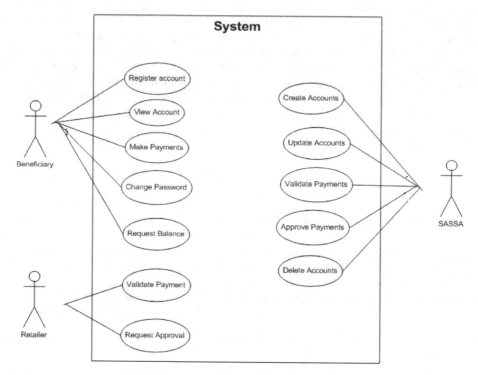

Fig. 3. Social Grant system USE CASE diagram

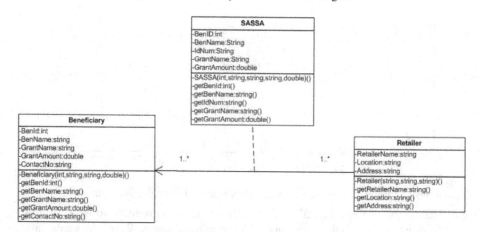

Fig. 4. Social Grant System Class Diagram

8   Methodology

In order to ascertain how social grant beneficiaries utilize their funds we have chosen interview as our research methodology. We created web based questionnaire, paper based questionnaire and interviewed some recipients orally. The sampling was based on three different types of interviews, which are: (1) Oral interview, (2) Paper-based interviews and (3) Web based interviews. This survey covered eight provinces of South Africa and used three-stage sampling procedure.

8.1   How Did We Design Our Research?

Research designs are the plans, structures and strategies of investigations which seek to obtain answers to various research questions. We followed three steps in designing the research which include:

- Selection of a sample of interest to study
- Collecting data
- Analyzing the Data

8.2   Sampling Strategy

We were unable to cover the whole of South Africa to collect data due to lack of resources (i.e., money, time, etc.). We collected data in the following provinces: Gauteng, Free State, North West, Kwa-Zulu Natal, Limpopo, Mpumalanga, Western Cape, and Eastern Cape.

8.2.1   Oral Interview

We conducted this type of interview using a tape recorder hidden from the social grant recipients for them to be free to give information. The reason we conducted this type of interview is because some of the beneficiaries were unable to read and write. The oral interview was taken from two different pay point areas but both from Gauteng province. We conducted the first oral interview in Soshanguve area called Zone M pay point, which is where most Soshanguve and the surrounding beneficiaries and caregivers collect their monies. The second oral interviews were conducted at SASSA in Pretoria where most beneficiaries and caregivers from around Pretoria and surroundings register their grants. We have experienced some difficulties during this interview because of the language and suspicion. Most people from Gauteng province are Sesotho, Tsonga and Venda, etc. speaking people. The language problem limited us in interviewing many social grant beneficiaries.

8.2.2   Written Interview

This interview was the most successful interview as we managed to conduct about 75% interviews. We designed some questioners and distributed to beneficiaries who were able to read and write. The questionnaires were designed using a Sogosurvey web-site

Table 2. Number of social security grant beneficiaries by province and grant spent period

Province	Grant spent in weeks				Total
	<1	<2	<3	<4	
Eastern Cape	102	41	18	15	176
Free State	6	4	2	1	13
Gauteng	83	19	12	9	123
KwaZulu-Natal	37	6	6	2	51
Limpopo	12	7	2	4	25
Mpumalanga	32	4	7	1	44
North West	1	2	3	1	7
Western Cape	59	15	10	10	94
Total	332	98	60	43	533

Eastern Cape Province had the highest social grant beneficiaries interviewed at 102, who spent their grant monies in less than a week, followed by Gauteng Province with 83 social grant recipients and then Western Cape Province with 59 social grant recipients, while North West Province reported the lower of one social grant recipient, as shown in Table 2. We interviewed a total of 533 social grant recipients across the country with the distribution on how long they took with the monies received before the next payment by SASSA.

8.2.3   Web-Based Interview
To organize the web-based interview, we created questionnaire using Sogosurvey website. We then used the email to send the questionnaires to the social grant beneficiaries that have access to internet. There was no much response from the social grant beneficiaries and caregivers. The reasons for this dismal response include: most beneficiaries and caregivers do not use internet especially from deep rural areas and old people like Old Age Pension (OAP) grant beneficiaries who are not familiar with internet; some social grant recipients were afraid of giving information over the internet even though we explained what the interview was all about; illiteracy and lack internet access, etc. The interview had about 18% of the total social grant recipients interviewed. In summary we interviewed 533, out of which 16 responded through the web, 420 through the written questionnaire and the rest being oral as shown in Figure 5.

9   Data Analysis and Results

This data analyses involve the use of some technical tools for weighing evidence and they also provide easily under stable and precise answers to questions of the study. To conclude our study properly it was necessary to analyze the data that we can correctly test our suggestions as well as to answer our research questions and we present the results of the study to our readers in an understandable and convincing form.

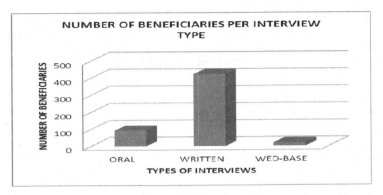

Fig. 5. Grant beneficiaries per Interview type

9.1   Characteristics of the Sample

Data on gender, age, provinces, salaries, educational qualifications, marital status, and exhausting period by spend period were collected. The data processed from the background question were represented graphically.

9.1.1   Age Group of the Beneficiaries

Table 1 provides a summary of social grant spent period by age group. We noted that there was a highest grant beneficiary of 332 who spent all their grants money before they finish seven days (within a week), followed by 98 grant beneficiaries who finished their monies before 14 days and 60 grant beneficiaries who spent all their monies before 21 days while 43 beneficiaries who spent their monies before 28 days were recorded as the lowest. We also observed that in age group 26-35, there were 92 social grant beneficiaries who spent their monies before seven days (within a week), followed by 72 social grant beneficiaries from age group 36-45, and 55 social grant beneficiaries from age group 16-25, while two social grant beneficiaries were the lowest from age group 0-15. According to this research analysis teenagers or say youth are the ones facing problems in terms of grant spending; the reason could be their level of education, demands and their maturity status, etc.

9.1.2   Marital Status

There were 117 single social grant beneficiaries who spent and finish theirs monies before seven days (within a week), followed by married social grant beneficiaries with 108, widowed social grant beneficiaries with 68 and lastly divorced social grant beneficiaries with 39 spending their monies within a week. Table 3 depicts social security grant beneficiaries by marital status and grant spent period.

Table 3. Number of social security grant beneficiaries by marital status and grant spent period

Marital status	Grant spent in weeks				Total
	<1	<2	<3	<4	
Divorced	39	12	8	3	62
Married	108	37	19	23	187
Single	117	28	22	11	178
Widow	68	21	11	6	106
Total	332	98	60	43	533

9.1.3   Ways of Surviving after Beneficiaries Exhausts Grant

This study has exposed the cracks in social security system in South Africa. Since most social grant beneficiaries exhaust their monies within a week, we sought their ways of survival thereafter. Figure 6 shows a total of 164 out of 533 of social grant beneficiaries interviewed borrow from friends and relatives, followed by 145 out of 533 of social grant beneficiaries who receive money from relatives and 121 out of 533 of social grant beneficiaries who use their salary to supplement the social grant monies. The remaining 133 social grant recipients may be engaging in other activities to supplement their income from department of social development. Some of these activities can be invested further and may include crime, etc.

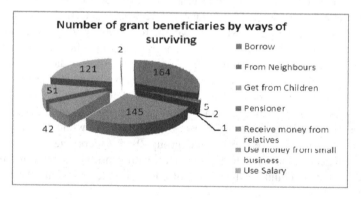

Fig. 6. Grant beneficiaries by way of surviving after exhausting the social security grant money

9.1.4   Gender and Level of Education versus Spent Period

Figure 7 and Figure 8, shows the number of social security grant beneficiaries by level of education and grant spend in weeks. It is noted that there were highest number of 143 social grant beneficiaries do not have matric level of education and spend all grant monies before seven days (within a week), followed by 124 social grant beneficiaries who have senior Certificate but spent all grant monies during the same period. And it is observed that there were 210 female social grant beneficiaries spend grant monies before seven days (within a week), while 36 social grant beneficiaries of the same gender spent money within 4 weeks. There were also 122 male social grant beneficiaries who exhausted their grant monies within a week, while in less than 4 weeks there were only seven male social grant beneficiaries.

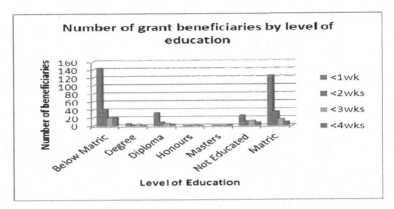

Fig. 7. Number of social security grant beneficiaries by level of education

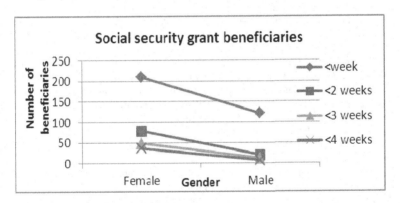

Fig. 8. Grant beneficiaries by gender

10   Conclusion and Future Work

In this research, we identify the positive developmental impact of the social grant in communities. We also have identified the problem of utilization of social grants by recipients and highlighted the need to provide applicable mechanisms. Social grant is an ongoing program that needs more and more monitoring not only to beneficiaries and caregivers but also to the staff tasked its administration. In this paper, based on gender, level of education and marital status significantly determined the number of weeks social grant recipients took with their monies. We propose a weekly limit as we identified that 62% of social grant beneficiaries are not able to keep their grant monies for more than a week and 92% of social grant recipients are not able to reach the next grant payment date. In future we will implement and test the system.

Acknowledgment. This work is undertaken with the department of Computer Science, Faculty of ICT at the Tshwane University of Technology (TUT). The authors would like to thank TUT for financial support that has enable this work.

References

[1] National Treasury, p 08 budget speech ch. 6 (2011),
     http://www.treasury.gov.za/documents/national%20budget/
     2011/speech/speech2011.pdf
[2] Economic Policy Research Institute working paper number 40 (November 6, 2007)
[3] Social Security and National Health Insurance by National Treasury (2011),
     http://www.treasury.gov.za/documents/national%20budget/
     2012/review/chapter6.pdf
[4] Dominikus, S., Aigner, M.: mCoupons: An Application for Near Field Communication
     (NFC). In: Proceedings of the 21st International Conference on Advanced Information
     Networking and Applications Workshops, Niagara Falls, Ontario, Canada, May 21-23,
     pp. 421–428 (2007) ISBN: 978-0-7695-2847-2
[5] Dutschke, M., Monson, J.: Children's Constitutional Rights to Social Services. In:
     Proudlock, P., Dutschke, M., Jamieson, L., Monson, J., Smith, C. (eds.) South African
     Child Gauge (2007/2008)
[6] South African Constitution, Section 27 (1 and 2) (1996), http://
     www.info.gov.za/documents/constitution/1996/96cons2.htm
[7] Hall, K., Woolard, I., Lake, L., Smith, C.: South African Child Gauge. Children's
     Institute, University of Cape Town (2012)
[8] Streak, J.C.: Child Poverty in South Africa and the Performance of the Child Support
     Grant Programme. PhD Thesis, University of Stallenbosch (2011)
[9] Budlender, D., Woolard, I.: The Impact of the South African Child Support and Old
     Age Grant on Children's Schooling and work. International labor Office, Geneva (2006)
[10] Barrientos, A., Wheeler, R.: Local economic effects of Social transfer. Institute of
     Development Studies, University of Sussex (2006)
[11] Schwarzer, H.: Impactos socioeconomicos do sistema de aposentadoris rurais no Brazil.
     Discussion Paper 729, IPEA (2000)
[12] Sadoulet, E., De Janvry, A., Davis: Cash transfer programmes with income multipliers.
     PROCOMPA in Mexico, World Development 29, 1043
[13] Anglucci, M., et al.: Extended family Networks in Rural Mexico. Harris School of
     Public Policy, University of Chicago (2006)
[14] http://www.corruptionwatch.org.za/content/
     biometric-cardsbeat-social-grant-fraud
[15] http://www.dsd.gov.za/index.php?option=com_content&task=vie
     w&id=379&Itemid=82
[16] http://www.southafrica.info/about/social/
     grants-200212.htm#.UVh8ARemh30
[17] http://www.sassa.gov.za/HOME-613.aspx
[18] Fischer, J.: NFC in Cell Phones: The New Paradigm for an Interactive World. IEEE
     Communications Magazine 47(6), 22–28 (2009)
[19] NFC forum, http://www.nfc-forum.org/resources/white_paper
[20] http://www.sassa.gov.za/ABOUT-SOCIAL-GRANTS/
     MEANS-TEST-650.aspx
[21] http://www.unicef.org/infobycountry/southafrica_62535.html
[22] Report on the Lund Committee on Child and Family Support (August 1996),
     http://www.info.gov.za/otherdocs/1996/lund.htm
[23] Fund Raising Act of 1978, http://
     www.greengazette.co.za/acts/fund-raising-act_1978-107

Testing Computational Methods to Identify Deformation Trends in RADARSAT Persistent Scatterers Time Series for Structural Assessment of Archaeological Heritage

Deodato Tapete[1] and Nicola Casagli[2]

[1] National Research Council (CNR) of Italy, Institute for the Conservation and Valorization of Cultural Heritage (ICVBC), Sesto Fiorentino, Florence, Italy
deodato.tapete@gmail.com
[2] Earth Science Department, University of Florence, Florence, Italy
nicola.casagli@unifi.it

Abstract. RADARSAT-1 data stacks processed by means of Persistent Scatterer Interferometry (PSI) over the central archaeological area of Rome, Italy, and radar-interpreted according to the procedure of radar mapping by Tapete and Cigna (2012), were re-analyzed by applying the Deviation Index DI_1 defined by Cigna et al. (2012). Our tests aimed to assess how an early computational identification of deformation trends within the displacement time series can support strategies of preventive conservation. The suitability of such semi-automated method for trend recognition is discussed with regard to a traditional approach of manual check of PSI time series, at the scale of single measurement point. Results from the case studies of Palatine and Oppian Hills are presented in this paper, examining both advantages and drawbacks offered by the implementation of such type of computational approach.

Keywords: Synthetic Aperture Radar, Persistent Scatterer Interferometry, Deformation Analysis, Deformation Trend, Archaeological Remote Sensing, Conservation.

1 Introduction

In the last years Persistent Scatterer Interferometry (PSI) became one of the new diagnostic techniques to assess the structural stability of archaeological monuments and historical buildings, thanks to its capabilities to couple information about past displacements back to 1992 with millimeter indication of ongoing movements [1-7].

The detected deformation is often evaluated based on the yearly rate of motion (expressed in mm yr^{-1}) of its time series. Nevertheless, this velocity value can be a misleading parameter whenever it is the average of different phases of displacement and does not mirror a constant deformation trend. So that it is not infrequent that an apparently stable point-wise ('Persistent Scatterer' – PS) or distributed ('Distributed

B. Murgante et al. (Eds.): ICCSA 2013, Part II, LNCS 7972, pp. 693–707, 2013.
© Springer-Verlag Berlin Heidelberg 2013

Scatterer' – DS) target on the ground (hereinafter called more generally 'Measurement Point' – MP) actually hides a non-linear displacement history throughout the monitoring period. If on one hand this aspect can be of minor concern in environmental applications (the analysis usually looks more at deformation patterns of larger MP clusters rather than single MPs), on the other side it is crucial for investigations which reach the scale of single building and manmade object, such as heritage conservation studies. Full characterization of the deformation behavior of each MP is required.

Furthermore, the increasing availability of High Resolution (HR) to Very High Resolution (VHR) Synthetic Aperture Radar (SAR) imagery (e.g., TerraSAR-X and COSMO-SkyMed scenes with spatial resolution of a few meters) providing unprecedented MP density than the Medium Resolution (MR) SAR one over the same area [8-9] requires the development of methods to ease the handling and management of such huge reservoirs of information, as well as their radar-interpretation. With this regard, the European Space Agency [10] has recently highlighted the usefulness of descriptors which might speed up the identification of trend variations within the MP time series (e.g., accelerations/decelerations and inversions of the trend).

Towards this direction, Cigna et al. [11] have recently proposed a computational method of semi-automated extraction of Deviation Indexes (DI) enabling the operators to describe the patterns recorded within the MP time series, thereby simulating the process of trend identification usually performed by visual inspection.

To verify the applicability of this method (and more generally of such computational approaches) also on cultural heritage exposed to structural and ground instability, by following the method of radar mapping defined by Tapete and Cigna [1] we performed some tests on RADARSAT-1 PSI time series from radar monitoring campaigns carried out over the central archaeological area of Rome, Italy, previously analyzed and discussed by Tapete et al. [3].

2   Input Data and Computation for Trend Detection

81 RADARSAT-1 images ($\lambda = 5.6$ cm), acquired along ascending orbits in Fine Beam Mode 3 (F3), with nominal ground resolution of 8 m, repeat cycle of 24 days, look angle θ of 41.3° and orbit inclination α of -8.1° at the latitude of central Rome, were processed by Tele-Rilevamento Europa (TRE) by means of PSInSAR (Permanent Scatterers InSAR) technique [12], to back-monitor the deformation field over the archaeological site of Palatine Hill in the period 07/03/2003 - 01/10/2009.

Whereas one RADARSAT-1 F3 ascending data stack ($\theta = 41.8°$) consisting of 87 images with last date of acquisition 22/02/2010 was processed by TRE with SqueeSAR algorithm [13], to image recent deformation over the Nero's Golden House and Oppian Hill. PSI processing also included 84 RADARSAT-1 descending scenes ($\theta = 35.2°$; $\alpha = +12.3°$) acquired in Standard Beam Mode 3 (S3) in the period 15/03/2003 - 22/08/2009, with nominal ground resolution of 30 m.

The Deviation Index (DI) was computed for the MPs by following the semi-automated method developed by Cigna et al. [11] (to which the reader should refer for full details).

In particular we tested the approach no.1 to calculate the Deviation Index DI_1:

$$DI_1 = \frac{\dfrac{1}{N_U} \displaystyle\sum_{i=N_H+1}^{N_H+N_U} |\Delta_i|}{s} \tag{1}$$

where:

- N_H: number of scenes constituting the historical interval (H), i.e. the sub-interval of the MP time series preceding a certain time of reference (the so-called 'breaking time', tb) which is selected according to the specific scope of the temporal analysis to detect a change of the deformation trend;
- N_U: the corresponding number of acquisitions of the updated interval (U) which follows tb;
- Δ_i: the difference between each displacement estimate i retrieved along the Line Of Sight (LOS) of the satellite and the corresponding value which would be obtained if the deformation trend which is recorded during H did not change during U;
- s: standard error of the regression as an estimate of the variability of the MP time series during H.

DI_1 is therefore used to quantify the deviation of the new deformation trend in U from the previous linear one in H, with regard to the selected tb.

While in environmental and hazard studies DI_1 is useful to identify precursor movements which can anticipate the occurrence of major natural hazards (e.g., landslides, earthquakes, volcanic phenomena), in heritage conservation applications calculation of DI_1 can support the preventive identification of deformation trends which may evolve into severer structural instability, up to the collapse of the monitored building or archaeological ruin (cf. the concept of 'early-stage warning'; [3-4]).

Concerning the estimation of the phase components (which are related to the detected deformation), a simple linear model of phase variation was applied to the PSInSAR processing of RADARSAT-1 scenes. Seasonal variations were instead added within the deformation model imposed in the SqueeSAR processing. Such difference in terms of deformation model applied during the PSI processing was purposely chosen to obtain a wider spectrum of deformation patterns on which DI_1 could be tested. Fig. 1 shows MP displacement trends commonly found during monitoring campaigns of historical buildings. While DI_1 is expected to assume low values (close to 0) for MPs characterized by constant linear deformation trends throughout the time series (Fig. 1A-D), sensible changes or inversions of the deformation trend (Fig. 1E-F) should be highlighted by higher DI_1 (only positive values, at least > 1).

The results from the implementation of this semi-automated method to the MPs identified over the archaeological area of Rome were compared with the corresponding evidence retrieved by manual check, undertaken time series by time series from a global view to the scale of single building according to the procedure of Tapete and Cigna [1].

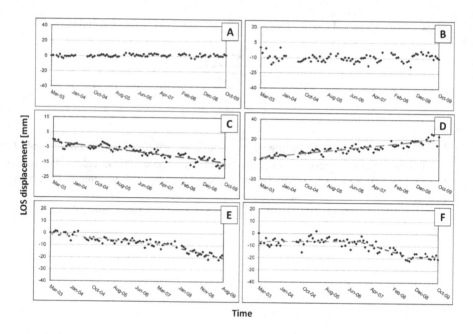

Fig. 1. Types of MP time series commonly retrieved from PSI processing over urban areas and historical buildings: A) stability and absence of LOS displacements; B) seasonal oscillations without displacement trend; C) seasonal oscillations defining a clear trend (away from the satellite); D) constant displacement trend (towards the satellite); E) change of the deformation trend; F) sequence of stability and acceleration phases

3   Case Studies in Rome, Italy

The central archaeological area of Rome is dominated by two hilltop sites called Palatine and Oppian Hills, located north-east and south-west of the Colosseum, respectively (Fig. 2). The geological context, as well as the anthropogenic factors which modified the landscape through centuries, nowadays create hazard conditions for the preservation of archaeological structures, in addition to the expectable deterioration processes which commonly affect heritage within urban environment. For this reason a site-specific PSI analysis was performed by Tapete et al. [3] to discover which sectors were actually critical for the conservation.

3.1   Palatine Hill

Palatine Hill still preserves several ruins of the ancient Roman palaces and temples from the monarchic period (7th century BC) to the imperial age (1st-4th century AD), such as the *Temple of Magna Mater* ('Great Mother'), together with additions in modern times (Fig. 2). Tuff strata, mainly competent tuffs and layers of *pozzolana*, produced after various eruptions from Sabatini and Laziale volcanic districts, represent the geological substrate of the hilltop ruins. They contain some horizons (e.g., the Lionato Tuff – VSN1), which tend to fracture severely when they outcrop and are exposed to weathering. Further factor of instability is due to historical anthropogenic factors, above all the superimposition of construction phases over natural caves, as well as galleries dug within the tuff strata for provision of building materials.

One of the main critical sectors was indeed recognized at the south-western corner of the hill, close to the *Temple of Magna Mater*, where the above mentioned hazard factors are clearly observable (Fig. 3A-C). The visual inspection of the time series of downwards moving PSs had showed an average value of yearly LOS displacement rate up to −3.3 mm yr^{-1} away from the satellite, with a LOS acceleration to −20.2 mm yr^{-1} recorded in the last year of monitoring, i.e. 2009 [3]. Such satellite evidence had confirmed the criticality hypothesized by the site conservators for this sector, also highlighting a worsening in recent times.

Based on these results, we calculated DI_1 of these PS time series, selecting two different *tb*, i.e. 02/01/2006 and 03/02/2009. These *tb* (hereinafter indicated as *tb$_{2006}$* and *tb$_{2009}$*) correspond with the dates of the first RADARSAT-1 acquisition for the years 2006 and 2009, i.e. the year when the technical report by archaeologists [14] warned about the critical condition of the *Temple of Magna Mater* and the year when the Italian Government declared the state of emergency for the Palatine Hill, respectively.

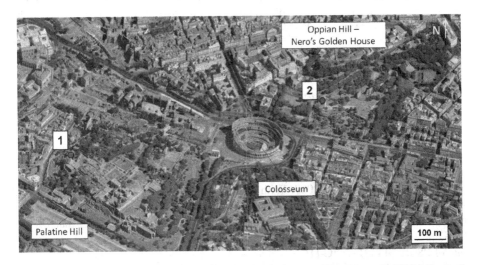

Fig. 2. Aerial view of the central archaeological area of Rome (Bing Maps © 2013 Microsoft Corporation) with indication of the Palatine and Oppian Hills and the archaeological ruins of 1) the *Temple of Magna Mater* and 2) the Neronian rooms, the latter sited at the north-western corner of the Nero's Golden House

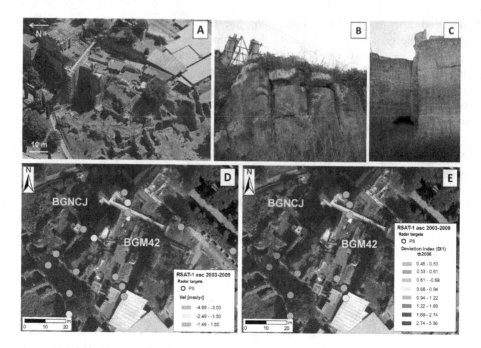

Fig. 3. Temple *of Magna Mater*: (A) aerial view (Bing Maps © 2013 Microsoft Corporation), (B) fractured outcrop of Lionato tuff, and (C) superimposition of archaeological ruins over caves and galleries. (D) LOS velocity of the identified RADARSAT-1 ascending (2003-2009) PSs and (E) the corresponding DI_1 distribution with *tb* fixed on 02/01/2006 (tb_{2006}), with indication of PS BGM42 and PS BGNCJ, the time series of which are reported in Fig. 4.

As already highlighted by the spatial distribution of LOS velocity values based on the yearly rate of motion (Fig. 3D), the distribution of DI_1 marks these two downwards moving PSs (BGM42 and BGNCJ) with regard to the surrounding MPs (Fig. 3E), and confirms the occurrence of superficial deformation over the archaeological structures located along the hill scarp due to instability of the local rock substrate. The high values of DI_1 retrieved with tb_{2006} (2.74 and 5.56, respectively) clearly suggest a trend deviation (Fig. 4A-B), which was not particularly appreciable at the visual inspection, which had instead allowed the detection of the above cited acceleration in 2009. With this regard, the calculation of DI_1 with tb_{2009} still distinguishes the two PSs from the surrounding ones. Nevertheless, it is worth noting that the higher value of DI_1 is found for the PS BGNCJ (1.04 rather than 0.78), the U pattern of which, at the visual inspection, seems to be less linear and clear than that of PS BGM42 (Fig. 4C-D).

As previously mentioned, the choice of breaking the time series in January 2006 and February 2009 was justified in light of the background information about the recent conservation history of the monitored monuments. If on one hand this approach answers to specific needs of analysis, on the other side it does not guarantee that major trend deviation occurred at those dates.

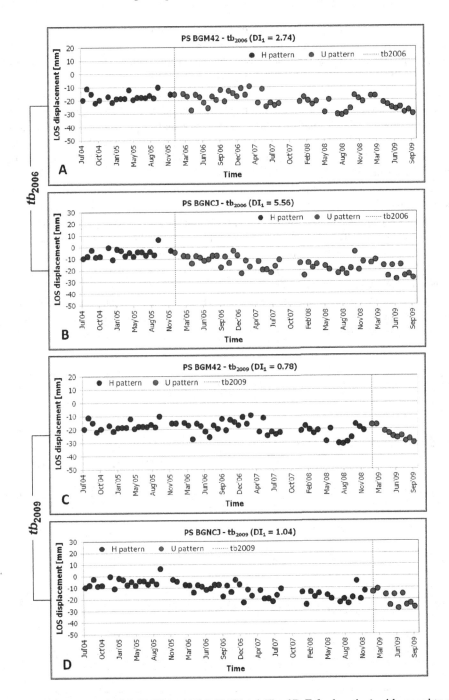

Fig. 4. Time series of PS BGM42 and PS BGNCJ (cf. Fig. 3D-E for location) with regard to *tb* on 02/01/2006 (A and B) and 03/02/2009 (C and D), respectively

To verify the appropriateness of this choice, DI_1 of the PS time series was calculated over the entire monitoring period and its values were plotted versus *tb*, thereby observing how DI_1 changes while *tb* shifts. Unexpectedly the curve of DI_1 vs. *tb* for the PS BGM42 (Fig. 5A) highlights two peak values at 3.16 on 15/03/2006 and 2.12 on 14/06/2007, which suggest splitting the time series into three intervals representing three different deformation trends (Fig. 5B). Conversely, no peak value is found for the LOS estimates in 2009, where the visual inspection and calculation of DI_1 with tb_{2009} highlighted a change of the deformation trend. Nevertheless, the slight increase of the DI_1 visible at the end of the curve might be correlated to an upcoming trend change in 2010.

As a general remark, the hump-shape of the DI_1 curve in June 2006, i.e. few acquisitions after the peak value in March 2006 (Fig. 5A), offers a good example of the effects on the calculated value of DI_1 due to sudden and high variations of LOS displacement estimates. A high difference between the values of LOS displacement estimates and those of the immediately previous records (Fig. 5B) can be perceived by the computational method as the beginning of a new deformation trend and consequently the corresponding DI_1 increases.

The DI_1 curve retrieved for the PS BGNCJ (Fig. 6A) seems instead to confirm that the major change of deformation trend within the PS time series occurred at the end of the year 2005 (Fig. 6B).

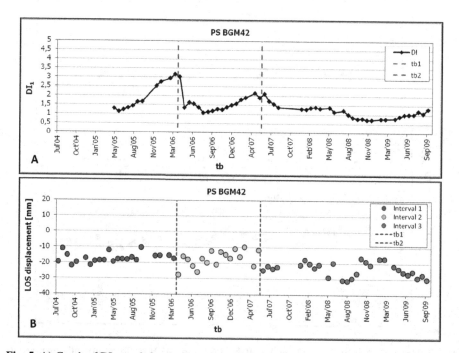

Fig. 5. A) Graph of DI_1 vs. *tb* for the time series of PS BGM42 (cf. Fig. 3D-E for location). B) Tri-partition of the PS time series based on the peak values of DI_1 retrieved from the computational method.

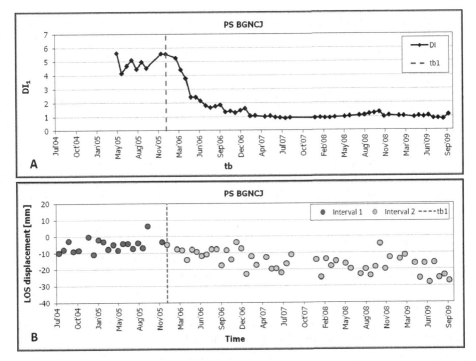

Fig. 6. A) Graph of DI_1 vs. *tb* for the time series of PS BGNCJ (cf. Fig. 3D). B) Bi-partition of the PS time series based on the peak values of DI_1 retrieved from the computational method

The peak values reach 5.58 and 5.56 on 09/12/2005 and 02/01/2006, respectively, which are not so different from the value 5.56 previously calculated by breaking the PS time series at tb_{2006}. In this case we can conclude that the two methods are equivalent and the arbitrary choice of breaking the time series at the beginning of the year 2006 was correct.

3.2   Oppian Hill and Nero's Golden House

The archaeological park of the Oppian Hill is renowned for the artificially buried remains of the *Domus Aurea* ('Golden House'), the luxurious suburban-type villa built by the Emperor Nero in 64 AD. After the landscape interventions carried out in 1930s which transformed the hill into a densely vegetated garden, the archaeological structures were definitely exposed to hazards related to water seepage and mechanic damage of tree roots. The impacts that this superimposition of buried structures and coverage terrain can have on heritage preservation, as well as on public safety, have been recently revealed by the collapse occurred on 30 March 2010. About 60-80 m² of the vault ceiling of a Trajan gallery neighboring the Nero's Golden House collapsed, opening a huge chasm in the overlying garden.

Referring to [3] for full details about the back-monitoring of the collapse event, we discuss the results of DI_1 calculation performed on two clusters of MPs identified over the Neronian rooms at the north-western corner of the Nero's Golden House (Fig. 2). These archaeological structures are typologically similar to the gallery collapsed in 2010. Hence, the early detection of trend deviations within MP time series can inform about ongoing instability and support a strategy of early-stage warning.

Over this sector of the Nero's Golden House, the manual check had highlighted a roughly common acceleration phase at the end of the monitoring period within the time series of some RADARSAT-1 descending MPs [3]. In particular, LOS yearly rates of about -5.1 mm yr^{-1} and -7.5 mm yr^{-1} away from the satellite were estimated for the DS A5X8K and DS A5X8X, respectively. The distribution of LOS displacement records throughout the year 2009 also suggested that ground motions and structural movements were still progressing at that time towards a significant deformation.

Based on these results of the time series analysis performed manually and taking into account the date of occurrence of the 2010 collapse, we can assume that the deformation recorded by RADARSAT-1 satellite throughout 2009 are crucial to assess the stability condition of the Nero's Golden House.

DI_1 calculation with tb_{2009} fixed on 18/01/2009, i.e. the date of the first RADARSAT-1 descending acquisition for the year 2009, clearly distinguishes the MPs identified over the Neronian rooms from PS A5X9L which is located over the sector of the archaeological park damaged in March 2010 (see yellow square and light blue star in Fig. 7A, respectively). As expected, this PS has a DI_1 value of 0.44, i.e. close to 0, since any deviation and/or acceleration/deceleration with regard to the previous trend can be found in its time series throughout the year 2009. Therefore DI_1 provides the same results of the manual check.

Conversely, the results obtained for the MPs over the Neronian rooms are only partially satisfactory. The DI_1 values span from 1.01 to 1.48 and highlight that most of these MPs were affected by trend deviation in 2009. Nevertheless, the spatial distribution of DI_1 over the entire terrain covering the Nero's Golden House and the relatively narrow field of DI_1 variability (from 0.24 to 2.90 as minimum and maximum values, respectively) do not sufficiently enhance the potential criticality of this sector (Fig. 7A-B). So that, from DI_1 we can only partially retrieve an immediate impression of potential instability for this sector.

Looking carefully at the corresponding time series broken at tb_{2009}, the highest values of DI_1 are actually found for those MPs which show an acceleration phase and/or a deviation trend highly perceivable even at the visual inspection. Hence, at the scale of single MP cluster located over a same monument or sector, the method is suitable to distinguish time series with trend deviation from those without.

Notwithstanding, similar values of DI_1 can be the result of different type of deformation trend, as exemplified by the time series of PS A5X8G and DS A5X8K (Fig. 7C). Although the two MPs show almost identical DI_1 (1.28 and 1.30, respectively), the deformation trend identified within the time series of DS A5X8K is more classifiable as an inversion of the trend, from a relatively stable condition to relevant LOS motions away from the satellite (LOS velocity changes from 0.4 mm yr^{-1} in H to -15.6 mm yr^{-1} in U).

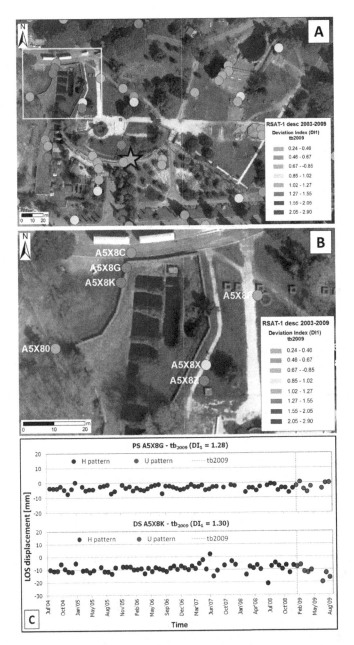

Fig. 7. Neronian *rooms*: A) Aerial view of the Nero's Golden House (Bing Maps © 2013 Microsoft Corporation) with DI_1 distribution of RADARSAT-1 descending (2003-2009) MPs with *tb* on 18/01/2009, and B) a zoomed view over the archaeological structures at the north-western corner (cf. yellow square). Light blue star indicates the location of the 2010 collapse and PS A5X9L. C) Time series of PS A5X8G and DS A5X8K (cf. labels in picture B for location).

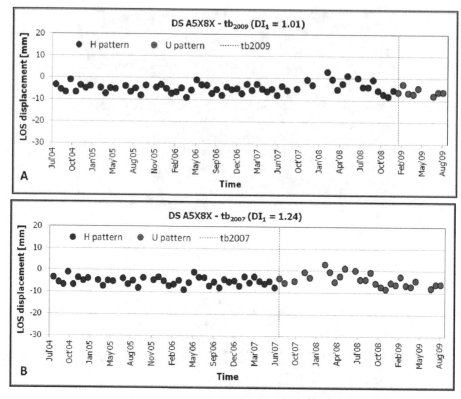

Fig. 8. Comparison of DI_1 calculation and trend identification for the time series of DS A5X8X (cf. Fig. 7B for location) with *tb* fixed on: A) 18/01/2009 and B) 16/07/2007

Conversely, the deformation trend in 2009 for PS A5X8G is more properly a phase of acceleration without change of LOS displacement direction (from +0.2 mm yr[-1] in H to +2.6 mm yr[-1] in U towards the satellite; Fig. 7C). Similarly to the case of the *Temple of Magna Mater*, DI_1 were also calculated with different *tb*, to verify when the major trend deviations occurred in the time series.

Fig. 8 shows the comparison between breaking the time series of DS A5XBX (see its location in Fig. 7B) at *tb₂₀₀₉* and on 16/07/2007 (i.e. *tb₂₀₀₇*) as an example of the effects of *tb* on the final value of DI_1. In this case, *tb₂₀₀₇* is a more appropriate breaking time with regard to which we can analyze the MP time series, since it distinguishes between a relatively stable period in 2004-2007 (average LOS rate of motion of about +0.3 mm yr[-1]) from a sequence of linear deformation trends starting from July 2007 (average LOS velocity of –2.3 mm yr[-1]; Fig. 8B).

4   Discussion and Conclusions

The tests performed on both exposed and partially/totally buried ruins of the central archaeological area of Rome demonstrate that the computational method of DI_1 (and more generally the DI approach) is highly suitable for site-specific analyses of PSI time series, where the main aim is not only the assessment of deformation macro-patterns, but also the identification of displacement trends related to localized deformation of single MPs. Such information is crucial to understand the structural condition of historical buildings and archaeological ruins.

The implementation of the DI approach can be event-driven, i.e. any deformation trend can be evaluated directly with regard to a specific event of instability or a certain period of time. This feature evidently makes the computational method strongly adherent to the background retrieved from the conservation history of the monument. It is therefore possible to confirm or modify a hypothesis formulated based on visual inspections and/or other-source data (e.g., archaeological prospection, surface monitoring surveys).

Nevertheless, DI_1 can be evaluated at various tb throughout the monitoring period. Consequently further deformation trends can be recognized, in addition to those expected or supposed to have occurred.

DI extraction for the PS identified over the *Temple of Magna Mater* provided interesting results with regard to the manual checks of the time series, and significantly improved the capability to identify deformation trends. The acceleration occurred since 2006 was indeed clearly recognized, especially for the PS BGNCJ, while a tri-partition for the time series of PS BGM42 was proved.

No less useful was the experimental implementation on the Nero's Golden House. With specific regard to the 2010 collapse, this case study of course confirms that neither manual nor computational methods of trend identification can provide a timely warning at an early stage of an instability process if any precursor occurs within the time series. In such contexts, we face phenomena which likely are not necessarily predictable using a PSI approach with such temporal sampling and spatial resolution.

On the other hand, the DI_1 calculations for the MPs over the Neronian rooms show the advantages and limits of this approach to give an immediate impression of the potentially critical sectors to which a particular attention should be paid, thereby assigning a higher level of priority. This is certainly dependent on the selected tb, the length of U, and the variability and noise of the LOS displacement records, both in H and U. Further factor is the type of PSI processing performed on the satellite radar data. In the case of the Nero's Golden House, the deformation model imposed also included a seasonal component of the deformation, and the DI_1 method is based on comparison between two linear trends of deformation, respectively before and after tb.

As a general remark, it is worth noting that the length of H has a relevant impact on the final value of DI_1. The possibility of assessing the occurrence of a deformation trend in U with regard to the previous trend persistent over a long H allows us to overcome one of the major limits of the visual inspection of the time series, i.e. the evaluation of a deformation trend in U by comparison only with the immediately preceding LOS displacement records. Such type of optical approach can therefore lead us to emphasize a trend which is not real. On the contrary, the visual inspection can sometimes better intercept a real deformation trend rather than a computational approach. This is due to the fact that a computational method compares with a longer

H, during which temporary variations of the displacement records can generate apparent trend changes. Therefore the reliability of the calculated DI_l can be affected.

With this regard, a standardized implementation of DI approaches would reduce the risk to evaluate deformation trends over too short periods of time. In such a case, the selection of a default length of H should always take into account the temporal coverage of the PSI dataset, its temporal sampling frequency and, not least, the specific conservation history of the studied cultural heritage. In the case of RADARSAT-1, a default length of H of 2 years would correspond to about 12 scenes, which are a sufficiently long period over which a deformation trend can be reliably defined and with regard to which a deviation trend or acceleration/deceleration can be assessed with a good degree of certainty.

The results obtained in these first experiments encourage further tests. Indeed the topic of an automatic examination and analysis of PSI time series is currently of high interest in the scientific community, as demonstrated for instance by the recent paper [15] which proposes an automatic classification to be used for studies of ground motions, based on a conditional sequence of statistical tests and using pre-defined deformation trends. Testing and adjusting these methods formerly set up for environmental applications can provide useful tools for archaeological remote sensing and conservation science. Nevertheless, ground truth and cross-validation with terrestrial surveys still remain an essential step during the MP interpretation, in reason of the high detail required by applications at the scale of single manmade object.

Acknowledgements. PSI RADARSAT-1 data were obtained in the framework of the research projects between the Italian Ministry of Cultural Heritage and Activities and the Earth Sciences Department, University of Florence. The authors thank Tele-Rilevamento Europa (TRE), Milan, Italy, for data processing with PSInSAR and SqueeSAR techniques.

References

1. Tapete D., Cigna F.: Rapid mapping and deformation analysis over cultural heritage and rural sites based on Persistent Scatterer Interferometry. International Journal of Geophysics 2012, Article ID 618609, 1–19 (2012)
2. Tapete, D., Cigna, F.: Site-specific analysis of deformation patterns on archaeological heritage by satellite radar interferometry. Materials Research Society Symposium Proceedings 1374, 283–295 (2012)
3. Tapete, D., Fanti, R., Cecchi, R., Petrangeli, P., Casagli, N.: Satellite radar interferometry for monitoring and early-stage warning of structural instability in archaeological sites. Journal of Geophysics and Engineering 9, S10–S25 (2012)
4. Tapete, D., Casagli, N., Fanti, R.: Radar interferometry for early stage warning on monuments at risk. In: Margottini, C., et al. (eds.) Landslide Science and Practice. Early Warning, Instrumentation and Monitoring, 2, pp. 619–625. Springer, Heidelberg (2013)
5. Ciampalini, A., Cigna, F., Del Ventisette, C., Moretti, S., Liguori, V., Casagli, N.: Integrated geomorphological mapping in the north-western sector of Agrigento (Italy). Journal of Maps 8, 136–145 (2012)

6. Cigna, F., Del Ventisette, C., Gigli, G., Menna, F., Agili, F., Liguori, V., Casagli, N.: Ground instability in the old town of Agrigento (Italy) depicted by on-site investigations and Persistent Scatterers data. Natural Hazards and Earth System Sciences 12, 3589–3603 (2012)

7. Cigna, F., Osmanoğlu, B., Cabral-Cano, E., Dixon, T.H., Ávila-Olivera, J.A., Garduño-Monroy, V.H., DeMets, C., Wdowinski, S.: Monitoring land subsidence and its induced geological hazard with Synthetic Aperture Radar Interferometry: A case study in Morelia, Mexico. Remote Sensing of Environment 117, 146–161 (2012)

8. Bovenga, F., Wasowski, J., Nitti, D.O., Nutricato, R., Chiaradia, M.T.: Using COSMO/SkyMed X-band and ENVISAT C-band SAR interferometry for landslides analysis. Remote Sensing of Environment 119, 272–285 (2012)

9. Wang, Y., Xiang Zhu, X., Bamler, R.: Retrieval of phase history parameters from distributed scatterers in urban areas using very high resolution SAR data. ISPRS Journal of Photogrammetry and Remote Sensing 73, 89–99 (2012)

10. European Space Agency (ESA): Session Summary: Terrain subsidence and Landslides. In: Proceedings of FRINGE 2011 Workshop, ESA-ESRIN, Frascati, Italy, September 19-23, 2011, ESA SP-697 (2012)

11. Cigna, F., Tapete, D., Casagli, N.: Semi-automated extraction of Deviation Indexes (DI) from satellite Persistent Scatterers time series: tests on sedimentary volcanism and tectonically-induced motions. Nonlinear Processes in Geophysics 19, 643–655 (2012)

12. Ferretti, A., Prati, C., Rocca, F.: Permanent scatterers in SAR interferometry. IEEE Transactions on Geoscience and Remote Sensing 39, 8–20 (2001)

13. Ferretti, A., Fumagalli, A., Novali, F., Prati, C., Rocca, F., Rucci, A.: A new algorithm for processing interferometric data-stacks: SqueeSAR. IEEE Transactions on Geoscience and Remote Sensing 49, 3460–3470 (2011)

14. Archaeological Superintendence of Rome: Valutazione e monitoraggio dello stato di conservazione del patrimonio archeologico: interventi per la riduzione del rischio e miglioramento sismico. Settore II: Tempio della Magna Mater. Internal technical report, 24 p. (2006)

15. Berti, M., Corsini, A., Franceschini, S., Iannacone, J.P.: Automated classification of Persistent Scatterers Interferometry time-series. Natural Hazards and Earth System Sciences Discussions 1, 207–246 (2013)

Author Index